Coal and Peat Fires: A Global Perspective

Captions for Front Cover Photos

Top Photo: View looking east at the burning Dębińsko Coal Mine waste pile in Czerwionka-Leszczyny, Rybnicki County, southwest Silesia, Poland. The underground Dębińsko mine (not visible here) is about 700 m to the west of this dump. The dump has ignited by spontaneous combustion for over 30 years in different locations that burn themselves out. Pennsylvanian bituminous coal recovered from unburnt and burned-out waste (reddish-white colored area) in the dump is excavated and used for highway construction, railroad ballast, and in the ceramics and building-construction industries. The black-colored rock covered with the white crust in the foreground is coal waste that is covered with bitumen and overlain by salammoniac. These nucleated on the coal waste during combustion. Photo by Ádám Nádudvari , 2016.

Bottom-Left Photo: Lower Permian bituminous (W-III grade) coal (504°C), possibly ignited by spontaneous combustion in an open pit mine at Bharat Coking Coal Limited, Bokapahari, Dhanbad district, Jharkhand State, Jharia coalfield, India. To extinguish the fire, the burning coal was removed with a hydraulic excavator and put into a nearby human-made pond. Photo by Varinder Saini, 2014.

Bottom-Second from Left Photo: A thrips collected in November of 2012 near a gas vent in St. Ignatius Cemetery at the Centralia Mine Fire, Pennsylvania, USA. This microarthropod and others found at Centralia and at the Healy Mine Fire, Alaska, are discussed in Chapter 2 of this book. Photo by Yelena White and Glenn B. Stracher, 2016.

Bottom-Third from Left Photo: A smoldering peat fire in a 40,000-ha peatland in the Wicklow Mountains, Ireland. This burn is on the Kippure House Estate, adjacent to the Wicklow Mountains National Park and near the Liffey Head bog complex along the floodplain of the River Liffey. It is part of a larger Special Area of Conservation designated for its mosaic of peatland habitats. Drainage cuts were made in the peat during World War II, and the peat was mined for use as a fuel. Ongoing drainage cuts were made over the years for commercial forestry of Sitka spruce (*Picea sitchensis*) and for plantings on the Kippure House Estate. Because of the drainage, the 50–100 cm thick top layer of peat dried out and periodically caught fire over the years when vegetation fires of questionable origin ignited the peat. The dead trees at the center of the photo remain from a peat fire that spread through this area for about a week in April of 2000. It was extinguished by the National Parks and Wildlife Staff, Coillte Forestry, the Fire Service, Civil Defense, and Irish Army. Drainage is ongoing, as can be seen by the 15-cm cut into the peat at the center of the photo. In 2015, a vegetation fire spread throughout this area and ignited the peat 24 h before this photo was taken. The living trees in the background are on the Coronation Plantation, in the Wicklow Mountains National Park. The River Liffey, visible in the photo, flows between the Kippure House Estate and Coronation Plantation. Photo by Nuria Pratt, 2015.

Bottom-Right Photo: Land occupied by the Datong Mining Group Co., LTD., Shanxi Province, northern China. At the Tashan Coal Mine, which belongs to the Datong Group, vertical shafts near the blue-colored gas-pumping station, visible in the distant left, are used to move explosive methane from underground mined-out gob areas. The gas is transferred through the metal pipes and preferentially put into storage tanks so it can be sold or else discharged into the atmosphere. The coal currently mined underground in this area is Permian coking coal (low ash, low-sulfur bituminous). Both underground coal-mine and coal-outcrop fires occur in this area. Removing the explosive methane helps reduce the risk of a mining catastrophe. Photo by Jianwei Cheng, 2017.

Coal and Peat Fires: A Global Perspective

Volume 5: Case Studies – Advances in Field and Laboratory Research

Edited by

Glenn B. Stracher
Division of Science and Mathematics, East Georgia State College, University System of Georgia, 131 College Circle, Swainsboro, Georgia 30401 USA

ELSEVIER

Elsevier is an imprint of Elsevier
Radarweg 29, PO Box 211, 1000 AE Amsterdam, Netherlands
The Boulevard, Langford Lane, Kidlington, Oxford OX5 1GB, United Kingdom
50 Hampshire Street, 5th Floor, Cambridge, MA 02139, United States

Library of Congress Cataloging-in-Publication Data
A catalog record for this book is available from the Library of Congress

British Library Cataloguing-in-Publication Data
A catalogue record for this book is available from the British Library

ISBN: 978-0-12-849885-9

For information on all Elsevier publications visit our website at https://www.elsevier.com/books-and-journals

Publisher: Candice Janco
Acquisition Editor: Amy Shapiro
Editorial Project Manager: Tasha Frank
Production Project Manager: Divya Krishna Kumar
Cover Designer: Victoria Pearson Esser

Typeset by TNQ Technologies

Dedication

This book is dedicated to my wife, Janet L. Stracher and my brother, Donald G. Stracher; to the memory of my parents, Bernard and Lillian Stracher and my sister, Iris Baum; and to my cousin, Gerald A. Klein. Their love and inspiration gave me the fortitude to carry on in the face of boundless adversity.

Preface to Volume 1

COAL AND PEAT FIRES: A GLOBAL PERSPECTIVE, Volumes 1–4, is a comprehensive collection of diverse and pioneering work in coal and peat fire research conducted by scientists and engineers around the world. It contains hundreds of magnificent color photographs, tables, charts, and multimedia presentations. Explanatory text is balanced by visually impressive graphics.

This work is devoted to all aspects of coal and peat fires. It contains a wealth of data for the research scientist, while remaining comprehensible to the general public interested in these catastrophic fires. Amateur and professional mineralogists, petrologists, coal geologists, geophysicists, engineers, environmental and remote sensing scientists, and anyone interested or involved in the technical aspects of coal and peat mining, coal and peat fires, and the effects of burning, from human health to combustion metamorphism, will find these four volumes useful. Although the technical level varies, the science-attentive audience will be able to understand and enjoy major portions of this work.

The four volumes are also a valuable source of information about the socioeconomic and geoenvironmental impacts of coal and peat fires. As an example, the mineral and select-gas analyses presented will be of great interest to environmental scientists, academicians, people employed in industry, and anyone interested in minerals and pollution.

The content of this work can be used to design and teach courses in environmental science and engineering, coal geology, mineralogy, metamorphic processes, remote sensing, mining engineering, fire science and engineering, among other subjects. A variety of case studies on a country-by-country basis, including prehistoric and historic fires, encompass a wide range of geoscience disciplines. These include mineralogy, petrology, geophysics, engineering, geochemical thermodynamics, medical geology, numerical modeling, and remote sensing—all making this work a cutting-edge publication in global coal and peat fire science.

Volume 1, before you, contains 19 chapters illustrated in full color. Chapter 1 discusses the origin of coal and coal fires. Chapter 2 discusses the techniques used for mining coal in addition to coal fires that occur in association with such mining. In Chapter 3, the connection between spontaneous combustion and coal petrology is discussed. Chapter 4 is about the utilization of coal by ancient man. Geotechnical and environmental problems associated with burning coal are discussed in Chapter 5. The general effects of coal fires that are burning around the world are discussed in Chapter 6, and Chapter 7 examines the environmental and human-health impacts of coal fires. Chapter 8 is devoted to explaining the laboratory procedure of gas chromatography, used to analyze samples of coal-fire gas collected in the field. Numerous complex processes associated with the nucleation of minerals from coal-fire gas and sampling techniques are presented in Chapter 9, and in Chapter 10 some analytical methods used to identify such minerals are discussed. Chapter 11 presents a synopsis of the analytical procedures used to identify the semivolatile hydrocarbons that nucleate from coal-fire gas. In Chapter 12, the magnetic signatures recorded by rocks and soils affected by the heat energy from burning coal are examined. Chapter 13 presents a synopsis of the historical utilization of airborne thermal infrared imaging for examining coal fires, and in Chapter 14, a more in-depth synopsis of the use of remote sensing technology for studying coal fires is presented. In Chapter 15, the historical and political implications for US government policy regarding coal fires are presented. The former US Bureau of Mines' role in controlling coal fires in abandoned mines and spoils piles is presented in Chapter 16. Chapters 17 and 18, respectively, present engineering fire-science studies of smoldering-coal combustion and the suppression of smoldering-coal fires. Volume 1 concludes with Chapter 19, in which the use of compressed-air-foam injection, for extinguishing coal fires, is discussed.

Volume 2 presents hundreds of color photos of coal and peat fires burning around the world as well as multimedia presentations that include movies, radio talk shows, and presentations given at professional meetings and elsewhere. Volume 3 presents case studies about fires on a country-by-country basis. Volume 4 is devoted to all aspects of peat and peat fires.

The editors of this four-volume book believe that scientists and engineers as well as the general public will find that the information presented herein reveals the complexity of coal and peat fire science, the effects of these fires, and useful methods for investigating them. We hope that the information presented will create global awareness about these fires and trigger new research ideas and methods for studying them, accelerate efforts to mitigate and extinguish them, and build a better living environment in mining areas around the world.

Glenn B. Stracher
Anupma Prakash
Ellina V. Sokol

Preface to Volume 2

COAL AND PEAT FIRES: A GLOBAL PERSPECTIVE, Volumes 1–4, is a comprehensive collection, both in hard cover and online editions, of diverse and pioneering work in coal and peat fire research conducted by scientists and engineers around the world. The four-volume set, illustrated in full color, contains the largest collection of coal and peat fire research available in any book ever published. Although the technical level varies, the science-attentive audience interested in these catastrophic fires will be able to comprehend and enjoy major portions of this work.

Each volume or the entire set can be used as a supplement for teaching courses in earth science, environmental engineering, mining engineering, fire science and engineering, among other subjects. In addition, the socioeconomic and geoenvironmental impacts of coal and peat fires are discussed and illustrated in color.

Volume 2 (Photographs and Multimedia Tours), before you, contains 24 chapters. Additional chapters, coal-fire gas and field data, and a wealth of multimedia materials are available on the companion Elsevier website for this book at http://booksite.elsevier.com/9780444594129. The multimedia materials include short movies, radio talk shows, entire conference proceedings, and presentations given at scientific meetings and elsewhere. Each chapter in Volume 2 presents a synopsis of select fire localities for the country discussed, a photo tour of those fires, a list of journal and book references, and a list of Internet addresses for additional reading. Chapters 1–24 discuss fires in the following respective countries: Australia, Azerbaijan, Canada, China, Colombia, the Czech Republic, England, France, Germany, India, Indonesia, Israel, Italy, Kazakhstan, Poland, Portugal, Romania, Russia (coal fires), Russia (peat fires), Scotland, South Africa, Spain, the United States, and Venezuela. Although combustion in Azerbaijan, England, and Israel are potentially related to hydrocarbon reservoirs other than coal or peat, the effects of combustion are analogous as illustrated in the remaining chapters, and so these three colorful and intriguing chapters were included for the greater reading audience.

Volume 1 discusses coal, coal combustion, and analytical techniques for studying coal fires and the by-products of combustion. Volume 3 presents case studies about fires on a country-by-country basis. Volume 4 is devoted to all aspects of peat and peat fires. In addition, an online interactive world map of coal and peat fires by Rudiger Gens, University of Alaska Fairbanks, is available on the Elsevier companion website mentioned earlier.

The editors of this four-volume book believe that scientists and engineers as well as the general public will find that the information presented herein reveals the complexity of coal and peat fire science, the effects of these fires, and useful methods for investigating them. We hope that the information presented will create global awareness about these fires and trigger new research ideas and methods for studying them, accelerate efforts to mitigate and extinguish them, and build a better living environment in mining areas around the world.

Glenn B. Stracher
Anupma Prakash
Ellina V. Sokol

Preface to Volume 3

COAL AND PEAT FIRES: A GLOBAL PERSPECTIVE, Volumes 1–4, is a comprehensive collection of diverse and pioneering work in coal and peat fire research conducted by scientists and engineers around the world. It contains hundreds of magnificent color photographs, tables, charts, and multimedia presentations. Explanatory text is balanced by visually impressive graphics.

This work is devoted to all aspects of coal and peat fires. It contains a wealth of data for the research scientist, while remaining comprehensible to the general public interested in these catastrophic fires. Amateur and professional mineralogists, petrologists, coal geologists, geophysicists, engineers, environmental and remote sensing scientists, and anyone interested or involved in the technical aspects of coal and peat mining, coal and peat fires, and the effects of burning, from human health to combustion metamorphism, will find these four volumes useful. Although the technical level varies, the science-attentive audience will be able to understand and enjoy major portions of this work.

The four volumes are also a valuable source of information about the socioeconomic and geoenvironmental impacts of coal and peat fires. As an example, the mineral and select-gas analyses presented will be of great interest to environmental scientists, academicians, people employed in industry, and anyone interested in minerals and pollution.

The content of this work can be used to design and teach courses in environmental science and engineering, coal geology, mineralogy, metamorphic processes, remote sensing, mining engineering, fire science and engineering, among other subjects. A variety of case studies on a country-by-country basis, including prehistoric and historic fires, encompass a wide range of geoscience disciplines. These include mineralogy, petrology, geophysics, engineering, geochemical thermodynamics, medical geology, numerical modeling, and remote sensing—all making this work a cutting-edge publication in global coal and peat-fires science.

Volume 3 contains 29 chapters illustrated in full color. Additional chapters, photos, and data will be available on the companion website for this book at http://booksite.elsevier.com/9780444594129. Chapter 1 discusses spontaneous combustion in association with open pit coal mining in Australia. Chapter 2 examines nanominerals and particulate matter from Brazilian coal fires. In Chapter 3, case studies utilizing remote sensing and in situ mapping for examining coal fires in China and India are presented. Chapter 4 discusses coal combustion and associated mineralization in the Helan Shan Mountains of northern China. Chapters 5 and 6 about the Czech Republic discuss, respectively, mineralization associated with burning coal in colliery-waste piles and combustion metamorphism in the Most Basin. The burning Anna I coal-mine dump in Alsdorf, Germany, and mineral nucleation mechanisms are discussed in Chapter 7. Chapter 8 explores the geothermal uses of smoldering coal-waste dumps. Mining impacts in the Jharia coalfield of India, the world's most complex coal-fire system, are discussed in Chapter 9. The physical properties of stone tools affected by hydrocarbon combustion and their use by ancient people in Israel are the subject of Chapter 10. Chapter 11 is devoted to a geophysical study of pyrometamorphic and hydrothermal rocks in Israel, along the Dead Sea transform fault. For the first time in any publication, in Chapter 12, the coal fires in the eastern African country of Malawi are assessed. Chapters 13 through 17 are devoted to case studies about coal fires in the upper Silesian coal basin of Poland, including fire prevention associated with the Rymer Cones (13); thermal transformations in the Starzykowiec coal dump (14); a thermal history of waste dumps (15); a general overview of coal mining and combustion in coal-waste dumps (16); and a study of mineral transformations, actinide mobility, and combustion metamorphism in the Wojkowice coal-waste dump (17). Chapter 18 presents a study about the mineralogical and magnetic effects due to coal mining and the use of coal from the Douro Coalfield in northwest Portugal. Case studies about Russia are presented in Chapters 19 through 21, and these examine ancient coal fires along the southwestern border of the Kuznetsk basin in Siberia (19); combustion metamorphism and the ellestadite-group minerals (20); and fayalite paralavas and combustion metamorphic complexes in the Kuznetsk coal basin (21). The Ravat coal fire in central Tajikistan and fayalite-sekaninaite paralavas are the subject of Chapter 22. Venezuela's coal-fire volcanoes are explored in Chapter 23. The hazards posed by coal fires in the interior of Alaska appear in Chapter 24. Anthracite-mine fires in northeastern Pennsylvania are covered in Chapter 25. In Chapter 26, a historical account of coal fires in the Richmond basin of Virginia is provided. The infrequently heard of coal fires in Oregon and Washington state are discussed in Chapter 27. Chapter 28 presents a study about the combustion, mineralogy, and petrology of oil-shale slags in Lapanouse-de-Severac, France, along with coal-fire analogies. The final chapter in Volume 3, Chapter 29, provides readers with a review of sampling techniques used to study coal fires.

Although combustion at Lapanouse-de-Severac and along the Dead Sea Transform is associated with hydrocarbon reservoirs other than coal or peat, the effects of combustion are analogous; so, these chapters were included for the reading audience.

Volume 1 discusses coal, coal combustion, and analytical techniques for studying coal fires and the by-products of combustion. Volume 2 presents hundreds of color photos of coal and peat fires burning around the world as well as multimedia presentations that include movies, radio talk shows, and presentations given at professional meetings and elsewhere. Volume 4 is devoted to all aspects of peat and peat fires. In addition, an online interactive world map of coal and peat fires by Rudiger Gens, University of Alaska Fairbanks, is available on the Elsevier companion website mentioned above.

The editors of this four-volume book believe that scientists and engineers as well as the general public will find that the information presented herein reveals the complexity of coal and peat fire science, the effects of these fires, and useful methods for investigating them. We hope that the information presented will create global awareness about these fires and trigger new research ideas and methods for studying them, accelerate efforts to mitigate and extinguish them, and build a better living environment in mining areas around the world.

Glenn B. Stracher
Anupma Prakash
Ellina V. Sokol

Preface to Volume 4

COAL AND PEAT FIRES: A GLOBAL PERSPECTIVE, Volumes 1–4, is the most comprehensive collection of global research ever published about coal and peat fires. The hardcover and electronic editions of each volume are illustrated in full color and contain an enormous amount of information useful to coal and peat geologists, mining and fire-science engineers, environmental scientists and engineers, mineralogists, petrologists, geophysicists, remote sensing scientists, and anyone interested or involved in coal and peat mining, coal and peat fires, and the effects of burning, from human health to combustion metamorphism. The volumes are also a valuable source of information about the socioeconomic, political, and environmental impacts of coal and peat fires.

The four volumes contain thousands of photographs, tables, charts, and illustrations linked to numerous case studies on a country-by-country basis, including prehistoric and historic fires. As such, this tome can be used to design and teach courses in environmental science and engineering, geophysical modeling, coal geology, mineralogy, metamorphic processes, remote sensing, mining engineering, fire science and engineering, among other subjects.

Volume 4, before you, contains six chapters, illustrated in full color, about the geology of peat and peat combustion. Additional material about peat and peat fires will be available on the companion website for this book at http://booksite.elsevier.com/9780444595102. Chapter 1 discusses the largest fires on Earth, smoldering peat fires. Chapter 2 presents a synopsis of the origin of peat, its characteristics, and its transformation with time. In Chapter 3, Italian peat and coal fires are examined and contrasted. Chapter 4 is devoted to a study of the geochemistry, chronology, and paleoreconstruction of peat fires in northern Mexico. Chapter 5 presents mathematical probabilistic models of peat-fire hazards. In Chapter 6, infrared analysis is used to study the horizontal propagation of laboratory peat fires.

Volume 1 discusses coal, coal combustion, and analytical techniques used to study coal fires and the by-products of combustion. Volume 2 presents hundreds of color photographs of coal and peat fires burning around the world as well as multimedia presentations that include movies, radio talk shows, entire conference proceedings, and presentations given at professional meetings and elsewhere. Volume 3 is devoted to case studies about coal fires burning the world over.

Periodically updated companion websites for each volume contain additional chapters; a wealth of multimedia material; gas and field data; and an interactive online world map of coal and peat fires by Rudiger Gens, University of Alaska Fairbanks.

The editors of this four-volume book believe that scientists and engineers as well as the general public will find that the information presented herein reveals the complexity of coal and peat fire science and the interdisciplinary approach required to investigate and model the nature and effects of these fires with useful methods for doing so. We hope that the information presented will create global awareness about these fires and trigger new research ideas and methods for studying them, accelerate efforts to mitigate and extinguish them, and build a better living environment in mining areas around the world.

Glenn B. Stracher
Anupma Prakash
Guillermo Rein

Preface to Volume 5

COAL AND PEAT FIRES: A GLOBAL PERSPECTIVE, Volumes 1–5, is the most comprehensive collection of global research ever published about coal and peat fires. The hardcover and electronic editions of each volume are illustrated in full color and contain an enormous amount of information useful to coal and peat geologists, mining and fire-science engineers, environmental scientists and engineers, mineralogists, petrologists, geophysicists, remote sensing scientists, and anyone interested or involved in coal and peat mining, coal and peat fires, and the effects of burning from human health to combustion metamorphism. The volumes are also a valuable source of information about the socioeconomic, political, and environmental impacts of coal and peat fires.

The five volumes contain thousands of photographs, tables, charts, and illustrations linked to numerous case studies on a country-by-country basis, including prehistoric and historic fires. As such, this tome can be used to design and teach courses in environmental science and engineering, geophysical modeling, coal geology, mineralogy, metamorphic processes, remote sensing, mining engineering, fire science and engineering, among other subjects.

Volume 5 contains 20 chapters devoted to advances in field and laboratory research about coal and peat fires. Additional material about these fires is available on the companion website for this book at https://www.elsevier.com/books-and-journals/book-companion/9780128498859. Chapter 1 discusses the earliest known uses of coal as a fuel. Chapter 2 presents a review of the arthropods and exciting images and data about microarthropods discovered at coal fires in Pennsylvania and Alaska. Chapters 3 and 4 examine coal fires in Pennsylvania. In Chapter 5, Modern and ancient coal fires in the Powder River basin of the United States are discussed. Chapter 6 discusses the occurrence and use of coal cinders in Washington state. Chapter 7 examines the effects of spontaneous combustion and coal-mine waste on streams in the El Bierzo coalfield of Spain. Chapter 8 analyzes thermal effects in coal-mine gob, and Chapter 9 examines the gases generated and the effects on methane-air flammability during the oxidation and pyrolysis of coal. In Chapter 10, the characteristics of coal-spontaneous combustion are analyzed. Chapter 11 quantifies the environmental impact of coal fires in Xinjiang, China, and Chapter 12 discusses colloid technology for preventing and extinguishing the spontaneous combustion of coal. The crystallochemical behavior of slag minerals and occurrence of potentially new mineral species from Lapanouse-de-Sévérac, France, are discussed in Chapter 13, and the burning coal heap of La Ricamarie in the Loire basin of France is discussed in Chapter 14. In Chapter 15, the locations of burning collieries in the Jharia coalfield of India are presented, and in Chapter 16, evidence is presented for the human health impacts of uncontrolled coal fires in Jharia. Chapter 17 discusses environmental monitoring of coal fires in Jharia using vegetation indices and surface temperature measurements. Chapter 18 presents remote sensing techniques for detecting self-heated hot spots on coal waste dumps in Upper Silesia, Poland. The geochemical behavior of trace elements in coal-gob piles in the Silesian basins of Poland is discussed in Chapter 19. Chapter 20 discusses peat fires in Ireland and their effects on peatlands.

Volume 1 discusses coal, coal combustion, and analytical techniques used to study coal fires and the by-products of combustion. Volume 2 presents hundreds of color photographs of coal and peat fires burning around the world as well as online multimedia presentations that include movies, radio talk shows, entire conference proceedings, and presentations given at professional meetings and elsewhere. Volume 3 is devoted to case studies about coal fires burning the world over. Volume 4 is devoted to the geology of peat and peat fires.

A periodically updated companion website for the volumes contains additional chapters; a wealth of multimedia material; gas and field data; and an interactive online world map of coal and peat fires by Rudiger Gens, University of Alaska Fairbanks.

The editors of Volumes 1–4 and the editor of Volume 5 believe that scientists and engineers as well as the general public will find that the information presented herein reveals the complexity of coal and peat fires and the interdisciplinary approach required to investigate and model the nature and effects of these fires with methods for doing so. We hope that the information presented will create global awareness about these fires and trigger new research ideas and methods for studying them, accelerate efforts to mitigate and extinguish them, and build a better living environment in mining areas around the world.

Glenn B. Stracher

Acknowledgments

I thank all contributors to ***Coal and Peat Fires: A Global Perspective***, Volumes 1–5, for the submission of their research for publication. I am also grateful for the permission granted by publishers to use data and reproduce figures from their journals, books, and websites. In addition, for Volume 5, I thank Elsevier's Senior Editorial Project Manager, Tasha Frank; Acquisitions Editor, Amy Shapiro; Project Manager, Divya Krishna Kumar; Outsourced Senior Project Manager, Shiva Meenakshi Sundaram; Copyrights Coordinator, Ashwathi Aravindakshan; Designer, Vicky Pearson; and Contracts Associate, Revathi Krishnamurthy for their assistance with this unique project. I also thank all the other people at Elsevier who assisted in the publication of this work and the development of its companion website, located at https://www.elsevier.com/books-and-journals/book-companion/9780128498859.

Glenn B. Stracher

Contents

List of Contributors

Wasim Akram Shaikh, Birla Institute of Technology, Mesra, Jharkhand, 835215, India.

Nicole M. Andel, Penn State Schuylkill, 200 University Drive, Schuylkill Haven, Pennsylvania, 17972, USA.

Conchi O. Ania, Conditions Extrêmes et Matériaux: Haute Température et Irradiation - Centre national de la recherche scientifique UPR 3079, Université d'Orléans, CS 90055, 1D Avenue de la Recherche Scientifique, Orléans Cedex 2, 45071, France.

Manoj K. Arora, PEC University of Technology, Department of Civil Engineering and Indian Institute of Technology Roorkee, Roorkee, Uttarakhand, 247667, India.

Harold Aurand Jr., Penn State Schuylkill, 200 University Drive, Schuylkill Haven, Pennsylvania, 17972, USA.

Claire M. Belcher, wildFIRE Lab, Hatherly Laboratories, University of Exeter, Exeter, EX4 4PS, United Kingdom.

Jean-François Carpentier, Université de Rennes, CNRS, ISCR, 35065 Rennes, France 2, rue du Thabor CS 46510, Rennes Cedex, 35065, France.

Sukalyan Chakraborty, Birla Institute of Technology, Mesra, Jharkhand, 835215, India.

Jianwei Cheng, College of Safety Engineering, China University of Mining and Technology, Xuzhou, Jiangsu, 221116, China.

Fangming Cheng, School of Safety Science and Engineering, Xi'an University of Science and Technology, No. 58 Yanta Road, Xi'an, 710054, China.

Justyna Ciesielczuk, Faculty of Earth Sciences, University of Silesia, Będzińska Street 60, 41-200 Sosnowiec, Poland.

Jun Deng, School of Safety Science and Technology, Xi'an University of Science and Technology, No. 58 Yanta Road, Xi'an, 710054, China.

Upasana Dhar, University of Texas at Arlington, Arlington, Texas, 76109, USA.

Robert B. Finkelman, University of Texas at Dallas, Richardson, Texas, 75080, USA.

Deolinda Flores, Departamento and Centro de Geologia, Faculdade de Ciências, Universidade do Porto, Rua do Campo Alegre, 687, Porto, 4169-007, Portugal.

Pierre Gatel, Centre de recherche de l'école des Mines Douai, Département génie civil et environmental and French Association of Micromineralogy (AFM) 37, rue Richer, Paris, F-75009, France.

Hartwig Gielisch, DMT GmbH & Co. KG, Geo Engineering & Exploration, GEE 5: Exploration- & Hydrogeology, Am Technologiepark 1, Essen, 45307, Germany.

Ravi P. Gupta, Retired, India Institute of Technology Roorkee, 2/201, Deepak Apartments Sahara Estate, Jankipuram, Lucknow, Uttar Pradesh, 226021, India.

Bernard Guy, Centre SPIN, département PEG École des Mines de Saint-Étienne, Institut Mines Télécom, UMR CNRS EVS (Environnement, Ville, Société), Université de Lyon, Saint-Étienne, F42000, France.

Aarinola Halimah Balogun, University of North Texas Health Science Center, Fort Worth, Texas, 76107, USA.

Donna Hawthorne, AOC Archaeology Edinburgh, Edgefield Road Industrial Estate, Loanhead, Midlothian, EH209SY, United Kingdom.

Edward L. Heffern, 813 Evergreen Street, Cheyenne, Wyoming, 82009, USA.

Nie Jing, Institute for Arid Ecology and Environment (IAEE), Xinjiang University, No. 666 Shengli Road, Urumqi, Xinjiang, 830046, China.

Li Jun, Department of Energy and Power Engineering, School of Mechanical Engineering, Tianjin University (Peiyang Campus), Jinnan District, Tianjin, 300350, China.

William Kombol, Palmer Coking Coal Company, LLP, P.O. Box 10, 31407 Highway 169, Black Diamond, Washington, 98010, USA.

Łukasz Kruszewski, Polish Academy of Sciences, Institute of Geological Sciences, Twarda 51/55 street, Warsaw, 00-818, Poland.

Danuta Kusy, Polish Academy of Sciences, Institute of Geological Sciences, Twarda 51/55 street, Warsaw, 00-818, Poland.

Nancy Lindsley-Griffin (late), 1315 Westmont Drive, Jacksonville, Oregon, 97530, USA.

Zhenmin Luo, School of Safety Science and Engineering, Xi'an University of Science and Technology, No. 58 Yanta Road, Xi'an, 710054, China.

J. Marion Wampler, 4053 Commodore Drive, Chamblee, Georgia, 30321, USA.

Fraser J.G. Mitchell, School of Natural Sciences, Botany Building, Trinity College, Dublin 2, Ireland.

Izabela Moszumańska, Polish Academy of Sciences, Institute of Geological Sciences, Twarda 51/55 street, Warsaw, 00-818, Poland.

Enda Mullen, National Parks and Wildlife Service, Kilafin, Laragh, Wicklow County, A98K286, Ireland.

Ádám Nádudvari, University of Silesia, Będzińska Street 60, Sosnowiec, 41-200, Poland.

Melissa A. Nolter, 1426 E. Center St., Mahanoy City, Pennsylvania 17948, USA; 101 Park Place Road, Mahanoy City, Pennsylvania, 17948, USA.

Ciaran Nugent, Forest Service, Reen Point, Blennerville, Tralee, Kerry County, V92X2TK, Ireland.

Olalekan Olanipekun, University of Texas at Dallas, Richardson, Texas, 75080, USA.

Anupma Prakash, Geophysical Institute, University of Alaska, 903 Koyukuk Dr., PO Box 99775-7320, Fairbanks, Alaska, 9977, USA.

Nuria Prat-Guitart, School of Biology and Environmental Science, Earth Institute, University College Dublin, Dublin, Dublin 4, Ireland; Pau Costa Foundation, Career Mossèn Cinto Verdaguer 42B-A, Taradell, Barcelona, 08552, Spain.

Zeng Qiang, Institute for Arid Ecology and Environment (IAEE), Xinjiang University, No. 666 Shengli Road, Urumqi, Xinjiang, 830046, China.

Peter W. Reiners, Department of Geosciences, University of Arizona, 1040 East 4th Street, Tucson, Arizona, 85721, USA.

Joana Ribeiro, Departamento and Centro de Geologia, Faculdade de Ciências, Universidade do Porto, Rua do Campo Alegre, 687, Porto, 4169-007, Portugal.

Catherine A. Riihimaki, Associate Director, Science Education, Council on Science and Technology, 234 Lewis Library, Princeton University, Princeton, New Jersey, 08544, USA.

Varinder Saini, Department of Earth Sciences, Indian Institute of Technology Roorkee, Roorkee, Uttarakhand, 247667, India.

John P. Shields, Managing Director, Georgia Electron Microscopy, 152 Barrow Hall, 115 D.W. Brooks Drive, University of Georgia, Athens, Georgia, 30602, USA.

Breana Simmons, School of Science and Mathematics, East Georgia State College, University System of Georgia, 131 College Circle, Swainsboro, Georgia, 30401, USA.

Glenn B. Stracher, Division of Science and Mathematics, East Georgia State College, University System of Georgia, 131 College Circle, Swainsboro, Georgia 30401, USA.

Isabel Suárez-Ruiz, Organic Petrography Laboratory, Instituto Nacional del Carbón (INCAR-CSIC), C/ Francisco Pintado Fe, 26, Oviedo, 33011, Spain.

Vincent Thiery, IMT Lille-Douai, Laboratoire de genie civil et de géo-environnement (LGCGE), Départment of Civil & Environmental Engineering, Douai, F59508, France.

Daniel H. Vice, Penn State Hazleton, 76 University Drive, Hazleton, Pennsylvania, 18202, USA.

Caiping Wang, School of Safety Science and Technology, Xi'an University of Science and Technology, No. 58 Yanta Road, Xi'an, 710054, China.

Tao Wang, School of Safety Science and Engineering, Xi'an University of Science and Technology, No. 58 Yanta Road, Xi'an, 710054, China.

Jimmy Wedincamp, School of Science and Mathematics, East Georgia State College, University System of Georgia, 131 College Circle, Swainsboro, Georgia, 30401, USA.

Yelena White, School of Science and Mathematics, East Georgia State College, University System of Georgia, 131 College Circle, Swainsboro, Georgia, 30401, USA.

Yang Xiao, School of Safety Science and Technology, Xi'an University of Science and Technology, No. 58 Yanta Road, Xi'an, 710054, China.

Zhai Xiaowei, School of Safety Science and Engineering, Xi'an University of Science and Technology, No. 58 Yanta Road, Xi'an, 710054, China.

Pu Yan, Institute for Arid Ecology and Environment (IAEE), Xinjiang University, No. 666 Shengli Road, Urumqi, Xinjiang, 830046, China.

Jonathan M. Yearsley, School of Biology and Environmental Science, Earth Institute, University College Dublin, Dublin, Dublin 4, Ireland.

Wang Yongqiang, Xinjiang Institute of Archaeology, Urumqi City, 830000, China.

Yanni Zhang, School of Safety Science and Technology, Xi'an University of Science and Technology, No. 58 Yanta Road, Xi'an, 710054, China.

Cao Zhanmin, School of Metallurgical and Ecological Engineering, University of Science and Technology, Beijing, 100083, China.

Jingyu Zhao, School of Safety Science and Technology, Xi'an University of Science and Technology, No. 58 Yanta Road, Xi'an, 710054, China.

Volume 5

Case Studies – Advances in Field and Laboratory Research

Upper Silesian Coal Basin, Poland

Douro Coalfield, Portugal

Scientists in the field collecting coal-fire samples and combustion data.

Loire Coal Basin, France

Jharia Coalfield, India

Photos: Tomasz Segit (Upper Silesian Coal Basin, Czerwionka-Leszczyny, Poland, 2017, Łukasz Kruszewski at a bituminous coal-waste pile, probing for H_2S with Dräger tubes prior to analysis with a portable FTIR GASMET DX4000 spectrometer), Irineu Dias (Douro Coalfield, Gondomar city, northwest Portugal, 2013, Joana Ribeiro at the São Pedro da Cova Mine, collecting a gas sample with a vacuum pump and Teflon® bag from the base of an unburnt part of an anthracite waste pile that burned elsewhere), Bernard Guy (Loire Coal Basin at La Ricamarie, France, 2014, Vincent Thiery collecting sulfur and efflorescent deposits at a gas vent in a burning anthracite coal-waste pile), Ram Karan (Jharia Coalfield, India, 2014, Varinder Sani about 5 m from a flaming Lower Permian bituminous (W-III grade) coal bed, possibly ignited by spontaneous combustion, at an open pit mine at Bharat Coking Coal Limited, Bokapahari, Dhanbad district, Jharkhand State. With audible crackling flames, Varinder could stand where she is for no longer than about a minute, moving away periodically and returning nearby to take a temperature measurement of 529 °C and some photographs. Attempts to extinguish the fire with water and a layer of soil failed).

Volume 5

Case Studies – Advances in Field and Laboratory Research

Professor Luo Zhenmin

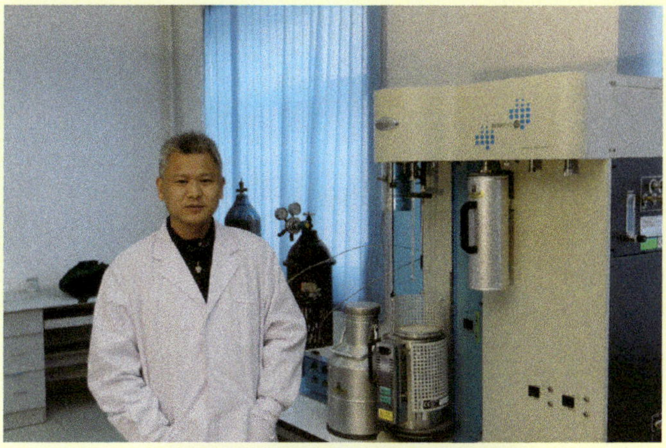

Professor Jun Deng

Scientists in the laboratory conducting coal-fires research.

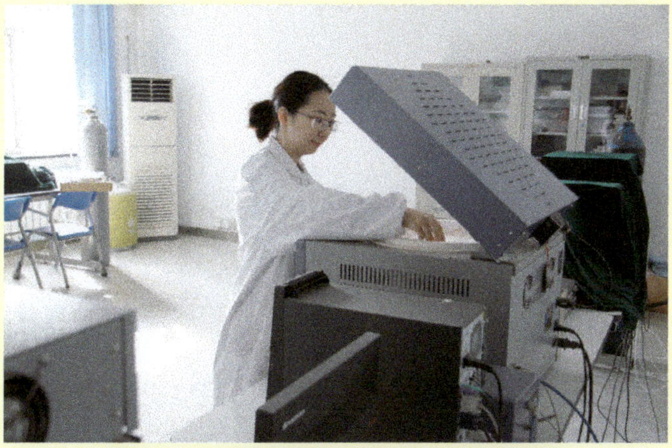

Professor Jingyu Zhao

Photos: Tao Wang (Xian University of Science and Technology, Xi'an, China, 2018, Professor Luo Zhenmin conducting gas explosion experiments with equipment she designed, Professor Jun Deng standing next to a physical and chemical adsorption instrument used to test the adsorption content of gas and the pore-size distribution of coal samples, Professor Jingyu Zhao conducting the proximate analysis of coal by measuring the moisture, ash, and volatile content).

Volume 5

Case Studies – Advances in Field and Laboratory Research

Yuqi Hu (left) and Eirik Christensen

Eirik Christensen

Scientists in the laboratory investigating smoldering peat.

Photos: Egle Rackauskaite (Hazelab, Department of Mechanical Engineering, Imperial College London, 2018, Yuqi Hu and Eirik Christensen using a self-designed gas emissions apparatus to study the relationship between the combustion dynamics of smoldering peat and its emissions, Eirik Christensen with a data logger that is recording temperatures from thermocouple probes inserted into smoldering peat. The temperatures are used to investigate the smoldering-propagation dynamics of the peat).

CHAPTER 1

The Earliest Known Uses of Coal as a Fuel: Paleolithic, Mesolithic, and Bronze Age Coal Fires

CHAPTER CONTENTS

Smelting ore to reduce lead (top left) and copper (top right). Coal from a mine (bottom right) is used as the fuel. Tin and lead smelting predate copper, which was smelted across China during the Bronze age. Bronze workings in China developed independently of outside influences. This metal signified wealth and social status. It is an alloy, usually of copper and tin, and may contain smaller amounts of other metals including nickel, zinc, manganese, and aluminum. It may also contain metalloids such as arsenic, silicon, and phosphorous. The components in bronze affect its physical properties, including toughness and hardness. The illustrations here are from China, AD 1637 (Sung, 1997).

Coal and Peat Fires: A Global Perspective
Edited by Glenn B. Stracher

https://doi.org/10.1016/B978-0-12-849885-9.00001-9

Photo from García (2017, p. 118).

1.1 The Earliest Known Uses of Burning Coal

Glenn B. Stracher
J. Marion Wampler
Li Jun
Wang Yongqiang

A shaft furnace from the first millennium BC, used for smelting copper in Hongfengshuiku, Daye County, Hubei Province, China. Hongfengshuiku is riddled with copper slag, bloomery iron, and technical ceramics.

Introduction

Coal has been mined for millennia, primarily because of the **chemical-potential energy** it contains. When this **fossil fuel** burns, its potential energy is converted into electromagnetic and **heat energy**. Ignition may be due to natural causes like lightning strikes, forest or brush fires, spontaneous combustion, or even volcanism. In addition, ignition may be linked to human activities. As such, ignition may be unintentional as is sometimes the case with mining and other activities. It may also be for an intentional nefarious purpose (arson). However, it is usually for beneficial purposes, namely, for use as a fuel (Stracher and Taylor, 2004; Stracher, 2010).

The **Industrial Revolution** that began in Great Britain about 1760 (Wikipedia, 2018A) and the amplified need for burning fossil fuels because of mass production launched innovations in steam-powered machinery such as railroad locomotives and steamships. Much is written about the historic use of fossil fuels and innovations in technology that began before and accelerated with the onset of the Industrial Revolution. There are fewer publications about the prehistoric use of coal as a fuel, and most publications about the **Bronze Age** use are about China. Although the ignition source (possibly woody vegetation) is questionable, this chapter chronicles documented locations of the oldest known uses of burning coal and briefly describes the periods during which those uses occurred.

Coal as a Fuel

Although the most common uses of coal today are to generate electricity and to manufacture steel, liquid fuel, and cement, it has many other uses. These include its use in the pharmaceutical, agricultural, toiletries, transportation, plastics, tar, and recreational industries (World Coal Association, 2018; National Geographic, 2018).

It is not known when coal was first mined and used as a fuel or how it was ignited as the need arose, although the first written accounts about it being a combustible rock and its use for metalwork were written by the Greek philosophers Aristotle, Theophrastus (Aristotle's student), and Pliny (Moore, 1922, p. 2; Caley and Richards, 1956). What is certain is that ***Homo*** (*Homo sapiens* and extinct species) has benefitted for millennia from coal by utilizing the heat energy produced during combustion for a multitude of purposes including cooking, warming their dwellings, socializing around a hearth, enhancing visibility in low light, physical protection from predators, firing pottery, modifying the mechanical properties of tools such as their toughness or hardness, and extractive metallurgy (i.e., **smelting**) (Stracher, 2007; Daemen, 2009; Gowlett, 2016; American Coal Foundation, 2018).

If readily available, coal could have been very useful as a fuel in communities where woody vegetation was sparse because of soil or climatic factors or because it was exhausted by activities such as the construction of lodgings and boats. Even if wood was abundant, coal may have been preferentially used, notwithstanding its higher ignition temperature, because it burns longer and produces more heat energy per unit of mass than wood (Bartok, 2003, p. 4).

Harvesting Potential Energy

The idea of harvesting the potential energy in coal for use as a fuel may have arisen when *Homo* observed a lightning strike, forest fire, or volcanic pyroclastics ignite a coal bed or when burning coal ignited by spontaneous combustion was observed. How *Homo* learned to purposefully ignite coal is not known. Before matches were invented in China, dry tinder-like fragments of woody vegetation, moss, or fungus were ignited by using a wooden stick as a hand drill, in a bow drill, in a fire plough, and so on (Wikipedia, 2018B) to convert the work done by friction into heat energy that then ignited the tinder. Alternatively, the tinder may have been ignited from sparks made by **percussion**, like striking together two stones such as flint against pyrite or marcasite or after metal was available, flint against a fire striker. Once ignited, the tinder might be hot enough to ignite coal. According to the writings of Tao (~AD 950), nonfriction matches were invented for cooking and heating by court women of the Northern Qi Dynasty in preparation for an enemy invasion in AD 577. The matches were pinewood sticks dipped in and impregnated with sulfur. They ignited quickly when touched to a preexisting flame and were used to light a fire elsewhere (MacDonald et al., 2008, p. 40). Tinder may have been placed in multiple places on a sheet or fragments of coal and quickly lit with the matches, causing the coal to ignite.

Paleolithic and Mesolithic Coal Fires

The **Paleolithic Period (Old Stone Age)** began with the earliest known use of stone tools including hammerstones and the biface. Depending on the reference, it occurred from about 3.3 to 2.6 million years ago and lasted until the end of the Pleistocene Epoch, when the last glacial period ended around 12,000 years ago (Groeneveld, 2017; Smithsonian Institution, 2018; Wikipedia, 2018C; International Commission on Stratigraphy, 2018). Based on the tools used and the materials used to make them such as stone, bone, or ivory, the Paleolithic is subdivided into the Lower or Early Paleolithic Period, Middle Paleolithic Period, and Upper or Late Paleolithic Period. The beginning and end of each of these subdivisions depend on geographic location (Groeneveld, 2017; Toth and Schick, 2007).

The **Lower Paleolithic Period**, also called the **Early Stone Age** in African archaeology, occurred from about 3.3–2.6 million to 300,000 years ago. Early *Homo* produced primitive tools including hammerstones, scrapers, and stone knives followed by stone hand axes, picks, and cleavers. There is some evidence for the use of fire by 1.5 million years ago or possibly earlier (Toth and Schick, 2007; Wikipedia, 2018C).

The **Middle Paleolithic Period**, depending on geographic location, occurred from about 250,000 to 30,000 years ago. New stone bifacial tools appeared such as the **bout-coupé** axe (Figure 1.1.1), leaf points, awls, and scrapers as did the first known cave paintings and widespread use of fire by *Homo*. Although *Homo* still lived in caves, these dwellings were partitioned for different activities (Ruebens and Wragg Sykes, 2016; Groeneveld, 2017; Smithsonian Institution, 2018; Hoffmann et al., 2018). The archaeological evidence for the use of fire includes burnt artifacts like bone, wood, and vegetation, and stone tools found in fire hearths along with baked sediment (Toth and Schick, 2007; Gowlett, 2016; Goldberg et al., 2017).

The **Upper Paleolithic Period** began about 50,000–40,000 years ago and lasted to about 10 000 years ago (Wikipedia, 2018C; Toth and Schick, 2007). This time is associated with ***Homo sapiens sapiens***; that is, *Homo sapiens* and extinct subspecies like *Homo neanderthalensis*. Stone tools continued to be produced and bows and arrows and spear throwers appeared. In addition, artifacts made of bone, ivory, and antler were produced as were musical instruments, architectural structures, paintings, sculptures, and engravings.

The **Mesolithic Period (Middle Stone Age)** followed the Paleolithic and preceded the **Neolithic Period (New Stone Age)**. Depending on the geographic location, the Mesolithic occurred between 12,000 and 7000 years ago whereas the Neolithic occurred between 10,200 and 2000 years ago. During the Mesolithic, *Homo sapiens* were hunters and gatherers although some domestication occurred. Farming and new technologies like pottery making developed during the Neolithic (Wikipedia, 2018D,E).

New stone tools for hunting and domestic use appeared during the Mesolithic. Many are illustrated in online resources such as the Museum of The Stone Age (2018). Mesolithic tools were typically made of chert with edges flattened by hand and they were sometimes reworked. They include scrapers to prepare animal hides for tents and clothing; a carpenter's tool called the **tranchet Adze** (Figure 1.1.2) that was used to make boats, housing, and fishing wharfs; and **microliths** (Figure 1.1.3), small tools diagnostic of the Mesolithic and beyond. Microliths were

Figure 1.1.1. A Middle Paleolithic bout-coupé biface made of chert; also called a flat-butted cordate or Paxton-type hand axe (Ruebens and Wragg Sykes, 2016). Such tools were bifacially worked and are roughly symmetrical along one dimension. They are cordiform in shape, resemble a triangle with rounded vertices, and have a straight (right side of the photo) or slightly convex edge. They were made by Neanderthal during the Mousterian, a Middle Paleolithic culture of Neanderthal that lasted from about 160,000 to 40,000 years ago (Shaw and Jameson, 1999, p. 408). This Neanderthal biface is as old as 40,000 years, measures 75 × 50 mm, and was collected from the surface of a field in Hampshire County, South Downs chalk hills, southeastern England. The soil here consists of about 0.6 m of clay mixed with chert and is underlain by chalk. Photo from Museum of The Stone Age (2018), courtesy of Richard Milton.

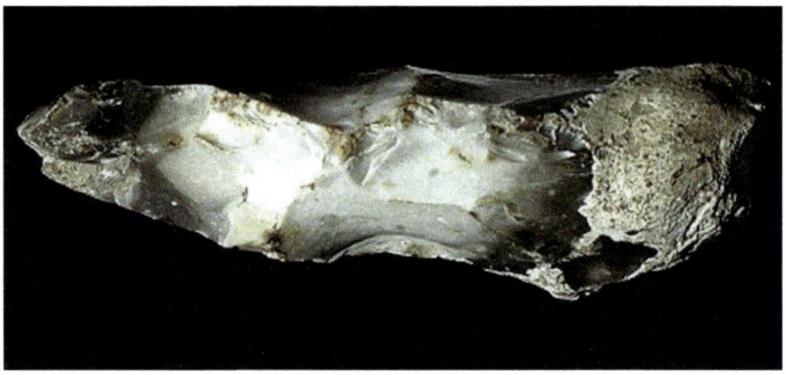

Figure 1.1.2. A Mesolithic chert tranchet adze (or adz). This is an axe head with a sharp and arched cutting edge that runs along its length and at a right angle to the handle. The tool was made by removing stone flakes, called tranchet flakes, parallel to the desired cutting edge of the tool. This sample was made about 8000 years ago, measures 151 × 48 mm, has a mass of 406 g, and was found on the surface of a field in Hampshire County, South Downs chalk hills, southeastern England. The soil here consists of about 0.6 m of clay mixed with chert and is underlain by chalk. Photo from Museum of The Stone Age (2018), courtesy of Richard Milton.

made from flakes of chert, quartz, or agate knapped off a larger piece and sometimes were reworked. They usually measured 1 cm or less in length and 0.5 cm or less in width but in some cases, were up to 2 cm wide and 5 cm long. They were attached to bone or wooden shafts with tree resin or twine for use as hand tools or to make arrows and spears for hunting. No more than two microliths were used per arrow but spears had as many as 6 to 18 (Wikipedia, 2018D; Museum of The Stone Age, 2018).

Southern France

The oldest known use of coal (lignite) as a fuel by *Homo* occurred at a Stone Age settlement at Causse du Larzac, a karst plateau in the south Massif Central of southern France. This settlement is **Les Canalettes** of Middle Paleolithic (**Mousterian**) age. Lignite was also burned much later at the nearby Mesolithic settlement of **Les Usclades** (Théry et al., 1996).

Figure 1.1.3. Mesolithic microliths reworked to be used as points. The one on the left measures 29 mm ×13 mm and has a mass of 1.8 g. The other one is 28 mm ×16 mm and its mass is 2.4 g. They are from Sussex, England and about 8000 years old. Photo from Museum of The Stone Age (2018), courtesy of Richard Milton.

Scanning electron microscope images of compressed cellular structures in what were thought to be charcoal fragments found at the settlements and laboratory replication of those structures with a mechanical press demonstrated that these samples are lignite. Outcrops of Jurassic lignite occur 7–15 m from the settlements. Their compressed structures appear analogous to those in the samples collected from the settlements, whereas other coal samples from the settlements came from outcrops elsewhere. Coal was possibly used as a fuel because of a wood shortage (Théry et al., 1996).

A thermoluminescence date for the lignite at Les Canalettes is 73,500 ± 6000 BP (Valladas et al., 1987; Meignen and Brugal, 2002). For different cultural settlements excavated at Les Usclades, the minimum radiocarbon date obtained by Michel Fontugne at the University of Versailles, France, is 8220 ± 70 BP and the maximum is 10,250 ± 80 BP (Théry et al., 1996).

Czech Silesia

The second oldest known use of coal as a fuel by humans was discovered during a 1952–1953 archaeological excavation of an approximately 177 m^2 area of the Landek Formation, an Upper Carboniferous sandstone, along the Oder (or Odra) River at **Ostrava-Petřkovice** in Czech Silesia, near the border with Poland. The Brno Archaeological Institute of the Czechoslovak Academy of Sciences uncovered a 30,000-year-old Paleolithic settlement where unburned coal, coke, gray ash, mammoth ivory, and fragments of charred and calcined animal bones were all found in association with six fire hearths. The incinerated bones include those of mammoth, horse, and reindeer. The hearths, fired to a red-color, were disrupted by solifluction but remain preserved in bowl-shaped depressions. They were found at the settlement and near **coal measures**, that is, Upper Carboniferous coal-bearing rocks. Stone artifacts, mostly made of flint, were found near the hearths and include microlith blades, scrapers, serrated blades, and assorted points (KLíama, 1956).

Bronze Age Coal Fires

China

The use of copper in China began during the Late Neolithic and extended into the **Chalcolithic Period** (**Copper Age**) that occurred in China from 3000–2000 BC (Birx, 2006, p. 570). The Chalcolithic Period was the transition time between the Neolithic and the Bronze Age. Overlapping with the Chalcolithic Period and depending on geographic location, the Bronze Age in Asia and the Middle East was from 3300 to 1200 BC (History, 2018). Carbon-14 dates matched with ceramics at archaeological sites reveal the use of copper and alloys from 3000 to 1500 BC at these sites, extending across northeastern Asia. In China, these sites range from the Xinjiang region in the northwest to the Central Plains (notably the Bronze Age settlement of Erlitou in the Yellow River valley of Henan Province), onward to Liaoning Province in the east, and as far north as Inner Mongolia (Linduff and Jianjun, 2008).

It was after the Xia Dynasty (2700–1556 BC) and during the time of the **Shang Dynasty** (**Yin Dynasty**; 1556–1046 BC) in the Yellow River valley in the Central Plain, that the **Early Bronze Age**, when tin was mixed with copper, developed across China (Figure 1.1.4). At that time, the physical properties of metal were well understood, and both the alloying process and metal casting were perfected (Linduff and Jianjun, 2008). The **Zhou Dynasty** (1046–256 BC) conquered and replaced the Shang Dynasty (Metropolitan Museum of Art, 2018). The Zhou Dynasty is the earliest Chinese dynasty corroborated from its own chronicles. The **Warring States Period** then occurred during the **Qin Dynasty** (221–206 BC) to create a unified China. This was followed by other dynasties. It was not until the **Song Dynasty** (AD 960–1279) that the bronze metallurgy of the Shang Dynasty was recognized by the inscriptions on bronze vessels created during the Shang Dynasty (Wikipedia, 2018F).

Ancient Use and Casting of Bronze in China Ancient cultures across China were interconnected and they used metals, symbolic of social status and wealth. Bronze workings in China developed independently of influences from other countries (Metropolitan Museum of Art, 2018). Along with gold, silver, brass, and copper, bronze was used to make items for personal adornment including rings, earrings, and nose rings. Copper and bronze were used to make vessels, chisel awls, parts of chariots, and weapons including knives. An important use for bronze was for sacraments that honored recently departed relatives as well as ancestors. Bronze artifacts were frequently included in burials (Metropolitan Museum of Art, 2018; Linduff and Jianjun, 2008).

In China, bronze was likely produced only by **piece-mold casting** until the end of the Shang Dynasty. In this method, a model is first made of the item to be cast in bronze and then clay is emplaced around the model to make a mold. The clay is then cut into sections as it is removed from the model. The pieces of clay are then reassembled and fired to make the mold into which bronze is poured for casting (Metropolitan Museum of Art, 2018).

Figure 1.1.4. Bronze artifacts from the Shang Dynasty of China. Left: Hou Mu Wu (Great Mother Wu) bronze ding, excavated in 1939 in Anyang city in Henan Province. The Emperor Zu Jia had this made for his mother, Wu. The 933 kg ding measures 112 × 79.2 cm along its open sides on the top (not visible here). It is 133 cm from the bottom of the legs to the top of the handles and the wall thickness is 6 cm. Right: Bronze fang zun (four-goat square vessel), excavated in Huangcai Town, Hunan Province, in 1938. Its maximum width is 52.4 cm and it is 58 cm high. Only people of high social status could acquire either of these kinds of bronze vessels. They were put on a table and filled with wine or food for different occasions such as a festival, anniversary, or government promotion. People would kneel by the table to worship their ancestors in gratitude. Both artifacts are now on display at the National Museum of China in Beijing. Photos from the personal collection of Li Jun.

Burning Coal in Bronze Age China One of the earliest known uses of coal occurred in China about 4000 BC, for embellishing the human ear (Robinson and Hsu, 2017, p. 9). However, except for the Stone Age fires in France and Silesia, the earliest known uses of coal as a fuel were found at archaeological sites near coal bed outcrops in northern and western China (Dodson et al., 2014).

Radiocarbon calendar dates (RCCDs) were obtained by Dodson et al. (2014) for human, pig, sheep, and cattle bones; millet seed; and charcoal from archaeological sites in Shaanxi province and Inner Mongolia. At these sites, coal fragments are mixed with human and domestic animal remains and in Inner Mongolia, also with pottery and bronze slag. The coal-slag association confirms the use of coal as a fuel for smelting. The earliest such use is indicated by RCCDs of 3496±132 BC and 3491±132 BC for charcoal from a Neolithic home (Zhang, 2013) excavated in 2010 at the **Xiahe Sitexi** at Xiahexi Village in Baishui County, Shaanxi province. Coal occurs with pottery-slag conglomerate at Zhukaiguo, where RCCDs of four of five bone samples indicate occupancy about 1900 BC.

More recently, excavations at the **Jiren Taigoukou Ruins** in Qialege'e Village in the Ili Valley of Nilka County in the Xinjiang Uygur Region of northwest China, near the present-day border with Kazakhstan, yielded a radiocarbon date of about 3600±35 years on animal bones found at the ruins (Ruan and Wang, 2017; Shen, 2014, Figure 1.1.5). Unburned coal, coal ash, ashcans, and coal-fire pits for protracting village fires were found at the site. The coal was burned as a fuel there because of the scarcity of woody plants. It was used for cooking and heating and likely to smelt copper, as suggested by copper tools found at the site (Chan, 2015).

By 1000 BC, coal used to smelt copper was mined at **Fushun**, on the Hun River in the Liaoning province of northeastern China (Chan, 2015; Kopp, 2018; Wikipedia, 2018G). The Eocene coal in Fushun is subbituminous to

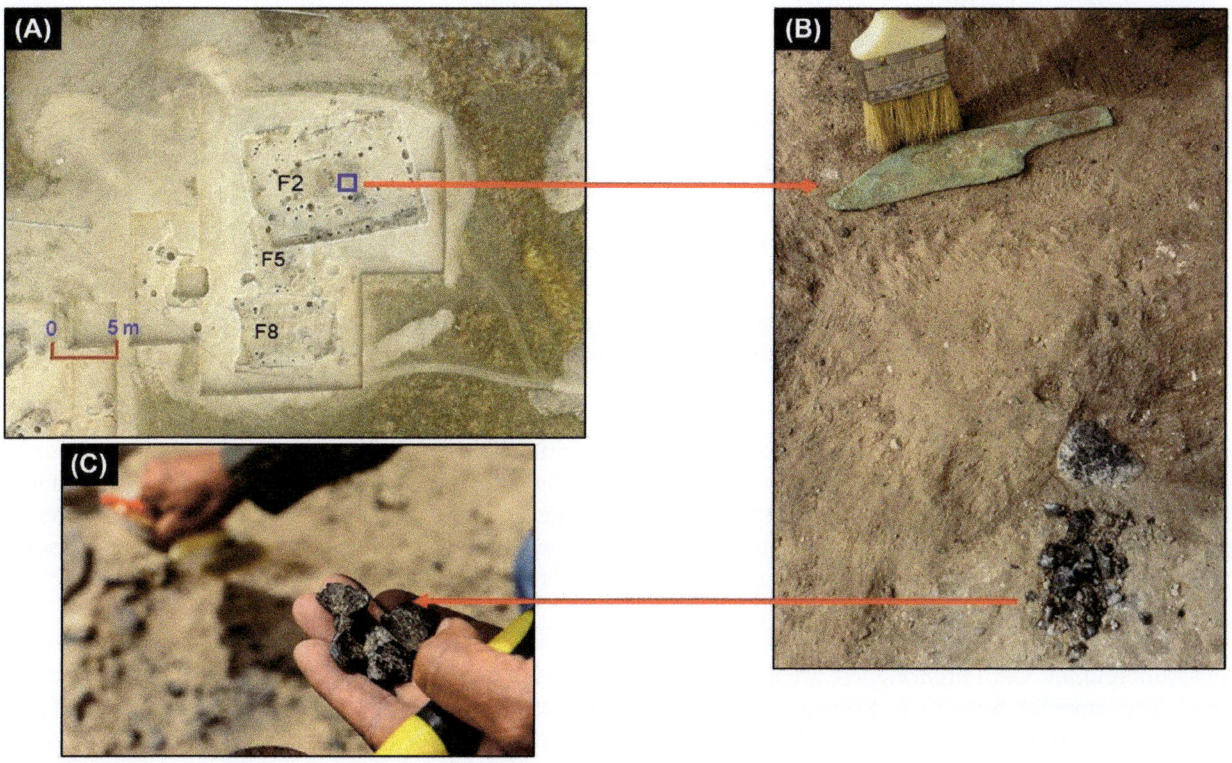

Figure 1.1.5. Bronze Age (3600±35 BP) settlement of Jiren Taigoukou in Nilka County, Xinjiang Uygur Region, northwest China. The ^{14}C date is for animal bone excavated at the site. The date was acquired at the Peking University mass accelerator lab in China and the Beta Analytic Testing Laboratory in Miami, Florida (Wang Yongqiang, Xinjiang Institute of Archaeology, personal communication). (A) Small-scale, residential-smelting operations, F2, F5, and F8, and open pits excavated by archaeologists from the Xinjiang Institute of Archaeology (Ruan and Wang, 2017). The ore was mined about 20 km away. Some pits were for wooden house posts; some were used for food storage; and others contained stones, animal bones, or pottery. (B) Copper blade excavated from the area outlined by the box in (A). (C) Coal at the excavation site was used as a fuel for smelting copper. Photos courtesy of Wang Yongqiang, Xinjiang Institute of Archaeology, Urumqi City, China.

bituminous in rank and was deposited in a fresh-water swamp (Johnson, 1990). Fushun contains several underground mines and is also the location of the **West-Open Pit Mine**, the largest such coal mine in Asia, where mining began in the 12th century and is currently winding down because of slope instability and a nearly exhausted supply of coal (Lei et al., 2015; Wikipedia, 2018G).

Fushun is known today in China by several other names including the **Capital of Coal** and **City of Coal**, and it is the home of the Fushun Mining Group Co., Ltd., a state-owned coal and coal bed methane company that processes coking and steam coal and methane (Johnson, 1990; Sui, 2001; Wikipedia, 2018G).

Wales

The two fuels most commonly used for **cremation** throughout human history are wood and coal. The first use of coal in Europe as a fuel for cremation occurred during the early Bronze Age in southern Wales (Brooks et al., 2011). Bronze Age human bones excavated there were found with coal fragments attached to them (Alberta Culture and Tourism, 2018). The cremated remains of Bronze Age humans, more common for that time than **inhumations** (noncremated burial), are found across Wales in pits and mounds along with amber, jet, ceramic beads, animal bones, flint scrapers, and vessels. Inhumations usually include skeletal remains accompanied by pottery; animal bones; and flint and metal artifacts including knives, scrapers, and arrow heads (Pettitt, 2015, p. 219–220).

The **Capel Eithin** cemetery near Gaerwin on the Welsh island of Anglesy contains evidence of Neolithic and Bronze Age activity as well as Roman and Early Christian activity during the Iron Age. Radiocarbon dates from this cemetery indicate that cremation had begun in Wales by 2300 BC. Cremated remains at the Bronze Age to Medieval cemetery at Tandderwen, near the town of Denbigh in northern Wales, are as old as 2000 BC, and inhumations there from the Bronze Age date from 2120 to 1880 BC (Pettitt, 2015, p. 218–220; Ancient Monuments, 2018; Coflein, 2018).

Additional Occurrences

Possible Uses of Coal for Smelting in Serbia and the Middle East

The oldest known bronze alloy containing a significant amount of tin (11.7% by weight) dates to about 4650 BC. It was found at a Neolithic archaeological site, that of the **Vinča culture**, in Plocnik, southern Serbia (Radivojević et al., 2013; Haemus, 2018). Named for the village of Vinča, 14 km downstream from Belgrade along the Danube River, the culture thrived from 6000 to 3000 BC in Yugoslavia and western Romania (Sutherland, 2015). The Vinča engaged in farming, exchanged artifacts with other communities, and were the first known culture to smelt copper. The symbols they inscribed on clay artifacts such as pottery may be an ancient form of written communication (Wikipedia, 2018H; Sutherland, 2015). Although the fuel source for smelting is undocumented, the Vinča alloy found in Plocnik is bronze foil, found on the floor of a dwelling near pottery vessels and next to a copper workshop. The foil may have been used to embellish the vessels for decorative purposes (Radivojević et al., 2013; Haemus, 2018).

In addition to the Vinča bronze, there are numerous occurrences of this and other metals used by ancient civilizations. The Israelites, Sumerians, Romans, Greeks, and Egyptians all smelted and used metals for a variety of purposes ranging from household items such as plates and vessels to weapons of war. Copper, lead, tin, mercury, gold, and iron were all used by these civilizations well before 2000 years ago (Encyclopaedia Judaica, 2008; Makin Metal Powders, 2018) but the fuel used for smelting during this time is also undocumented.

More About Europe

Coal and coal cinders in England, found among Roman ruins, reveal that the Romans were burning coal there soon after AD 43 (Kopp, 2018). Beginning around AD 100, the Romans burned coal to honor Minerva, the goddess of wisdom and military accomplishments. By AD 200, the Romans collected surficial deposits of coal from coalfields throughout England. They burned this coal to make iron weapons and to heat the baths in their military forts and in the dwellings of wealthy Romans. Often, the baths were heated with a hot air circulation system called a **hypocaust** (Speight, 2013, p. 129–130; National Geographic, 2018; Kopp, 2018; Alberta Culture and Tourism, 2018).

The Romans left England in AD 410, and from then until the end of the 12th century, no record exists of coal being burned in England, although it may have been. Around 1200, monk Reinier, who became prior of the Benedictine abbey of St. Jacques at Liège (Dury, 2018), a city on the Meuse River in eastern Belgium, was the first to record the mining of coal in Europe and its use by blacksmiths to smelt and forge metals. Limited use and trade of coal in Europe continued from the 13th century until the Industrial Revolution, when the demand for and use of coal as a fuel increased rapidly in Europe and abroad. Developments that contributed to this demand were the use of coke by the British iron founders Abraham Darby and John Wilkinson for manufacturing iron and the coal-powered steam engine invented by the Scottish mechanical engineer James Watt (Daemen, 2009; Curley, 2012, p. 90; Speight, 2013, p. 130; Kopp, 2018). As pointed out by García (2017), metallurgical analyses of iron artifacts can be used to determine if coal was the fuel used for smelting because of sulfur impurities in the iron from the coal.

The Americas

In northern Peru, anthracite coal was used for metallurgical processes from around 100 BC until at least the late 1800s. Carbon-14 dates of AD 1312–1438 reveal that a 25×70 m burn occurred a few meters from **Ciudadela** Tschundi in the Peruvian city of **Chan Chan** in the Moche Valley. This pre-Columbian city was the capital of the Kingdom of Chimor before the Incan conquest of 1470. The fire blackened and deteriorated adobe structures in this ciudadela, one of the city plazas for royalty in Chan Chan (Wikipedia, 2018I; Brooks et al., 2011). The chemical composition of burn ash and the presence of sulfur in the soil at Tschundi suggest that the burn fuel was coal for a fire that reached at least 1320°C and possibly was used for human cremation (Lechtman and Moseley, 1975; Brooks et al., 2011).

From the beginning of the 14th century and through the 16th, in what is now Mexico City, the Aztecs burned coal for heating, firing clay pottery, and cooking. They also used carved ornaments from it, including jewelry (Somervill, 2012, p. 11; Anderson, 2014).

From about the 13th century to the 17th, in what is now northern Arizona, the Hopi burnt coal that they strip- and auger-mined at coal-bearing mesas, including the Black Mesa and the Antelope Mesa. Hearths for the coal firing of pottery vessels occurred along mesa benches, and coal was burned for in-home domestic use. At Antelope Mesa, the Hopi burned an estimated minimum of 100,000 tons of coal in 300 years (Daemen, 2009; Bernardini, 2005, p. 132).

During his travels in the 13th century, the Venetian writer, explorer, and merchant Marco Polo observed the use of burning coal in China. Coal was not well known to most people in Europe and it was new to Marco Polo. His statements about its preferential use to wood as a fuel was recognized in prehistory and it applies even today: "It is a fact that all over the country of Cathay there is a kind of black stone existing in beds in the mountains, which they dig out and burn like firewood. If you supply the fire with them at night, and see that they are well kindled, you will find them still alight in the morning; and they make such capital fuel that no other is used throughout the country. It is true that they have plenty of wood also, but they do not burn it, because those stones burn better and cost less. Moreover, with that vast number of people, and the number of hot-baths that they maintain—for everyone has such a bath at least 3 times a week, and in winter if possible every day, whilst every nobleman and man of wealth has a private bath for his own use—the wood would not suffice for the purpose" (Yule, 1871, p. 395).

Peat is a precursor of coal. People living in the former USSR, Finland, Sweden, the Netherlands, Poland, Germany, England, and Ireland began using this fossil fuel at least 2000 years ago, as a substitute for firewood, for heating and cooking (Andriesse, 1988, p. 115). But the earliest recorded uses of peat as fuel is a story for another time.

Acknowledgments

The authors thank Professor Zeng Qiang, Institute for Arid Ecology and Environment, Xinjiang University, Urumqi City, China and David Larreina Garcia, Institute of Archaeology, University College London, England for their assistance in the acquisition of photos about China. We also thank Elsevier's Earth and Planetary Sciences staff for their review of and assistance with the publication of this work.

Important Terms

bout-coupé
Bronze Age
Capel Eithin
Capital of Coal
Chalcolithic Period (Copper Age)
Chan Chan
chemical potential energy
coal
coal measures
City of Coal
ciudadela
cremation
Early Bronze Age
fossil fuel
Fushun
heat energy
Homo
Homo sapiens sapiens
hypocaust
Industrial Revolution
inhumations
Jiren Taigoukou Ruins
Les Canalettes
Les Usclades
Lower Paleolithic (Early Stone Age)
Mesolithic Period (Middle Stone Age)
microliths
Middle Paleolithic Period
Mousterian
Neolithic Period (New Stone Age)
oracle bone script
Ostrava-Petřkovice
Paleolithic Period (Old Stone Age)
peat
percussion
piece-mold casting
Qin Dynasty
Shang Dynasty (Yin Dynasty)
Song Dynasty
smelting
technical ceramics
tranchet Adze
Vinča culture
Warring States Period
West-Open Pit Mine
Xiahe Site
Zhou Dynasty

References

Alberta Culture and Tourism, 2018. http://history.alberta.ca/energyheritage/coal/early-coal-history-to-1900/unearthing-ancient-mysteries/default.aspx (accessed March 2018).

American Coal Foundation, 2018. Coal Energy Facts. http://teachcoal.org (accessed February 2018).

Ancient Monuments, 2018. https://ancientmonuments.uk (accessed March 2018).

Anderson, C.L., 2014. The Coal Business in Wyoming. Wyoming State Historical Society Encyclopedia https://www.wyohistory.org/encyclopedia/coal-business-wyoming (accessed March 2018).

Andriesse, J.P., 1988. Nature and Management of Tropical Peat Soils, Food and Agricultural Organization of the United Nations. FOA Soils Bulletin 59, 165 p.

Bartok Jr., J.W., 2003. Heating with Wood and Coal, Natural Resource. Agricultural and Engineering Service (NRAES), Cornell University, Ithaca, New York. 69 p.

Bernardini, W., 2005. Hopi Oral Tradition and the Archaeology of Identity. University of Arizona Press, Tucson, Arizona. 220 p.

Birx, H.J. (Ed.), 2006. Encyclopedia of Anthropology, vol. 1. Sage Publications. 3128 p.

Brooks, W.E., Mora, C.G., Jackson, J.C., McGeehin, J.P., Hood, D.G., 2011. Coal and Ancient Man: cremation at the Tschudi burn, chan, chan, northern Peru. In: Stracher, G.B., Prakash, A., Sokol, E.V. (Eds.), Coal and Peat Fires: A Global Perspective, Volume 1: Coal – Geology and Combustion. Elsevier, pp. 63–81.

Caley, E.R., Richards, J.C., 1956. Theophrastus on Stones: Introduction, Greek Text, English Translation, and Commentary. Ohio State University. 238 p.

Chan, E., 2015. Remarkable Discovery Shows Humans Were Burning Coal as Fuel More than 3,500 Years Ago as World's Earliest Site of the Activity Is Unearthed in China. Daily Mail of the Associated Press Ltd. http://www.dailymail.co.uk/news/peoplesdaily/article-3202104 (accessed February 2018).

Coflein, 2018. Tandderwen, Cemetery. National Monuments Records of Wales. http://www.coflein.gov.uk/en/site/309402/details/tandderwen-cemetery (accessed March 2018).
Curley, R., 2012. Fossil Fuels, Energy: Past, Present, and Future, Britannica Educational Publishing. Rosen Educational Services, LLC, New York, New York. 160 p.
Daemen, J.J.K., 2009. Coal Industry, history of. In: Cleveland, C.J. (Ed.), Concise Encyclopedia of History of Energy, first ed. Elsevier, pp. 1–16.
Dodson, J., Xiaoqiang, L., Nan, S., Atahan, P., Xinying, Z., Hanbin, L., Keliang, Z., Songmei, H., Zemeng, Y., 2014. Use of coal in the bronze age of China. Holocene 24 (5), 525–530.
Dury, C., 2018. Reiner of St. James. In: Dunphy, G., Bratu, C. (Eds.), Encyclopedia of the Medieval Chronicle. https://doi.org/10.1163/2213-2139_emc_SIM_02165 (accessed March 2018).
Encyclopaedia Judaica, 2008. Metals & Mining. http://www.jewishvirtuallibrary.org/metals-mining (accessed March 2018).
García, D.L., 2017. Copper and Bloomery Iron Smelting in Central China: Technological Traditions in the Daye County, (Hubei) (Ph.D. thesis). University College London, England. 381 p.
Goldberg, P., Miller, C.E., Mentzer, S.M., 2017. Recognizing Fire in the Paleolithic Archaeological Record. Current Anthropology, 58, supplement 16, University of Chicago Press, Chicago, Illinois, pp. s175–s190.
Gowlett, J.A.J., 2016. The discovery of fire by humans: a long and convoluted process. Philos. Trans. R. Soc. B 371, 1–12. 20150164 rstb.royalsocietypublishing.org (accessed February 2018).
Groeneveld, E., 2017. Ancient History Encyclopedia, Paleolithic. https://www.ancient.eu/Paleolithic (accessed February 2018).
Haemus, 2018. 6,500-year Old Tin-bronze from Serbia. Center for scientific research and promotion of culture. http://haemus.org.mk/6500-year-old-tin-bronze-from-serbia (accessed March 2018).
History, 2018. Bronze Age. https://www.history.com/topics/bronze-age (accessed March 2018).
Hoffmann, D.L., Standish, C.D., García-Diez, M., Pettitt, P.B., Milton, J.A., Zilhão, J., Alcolea-González, J.J., Cantalejo-Duarte, P., Collado, H., de Balbín, R., Lorblanchet, M., Ramos-Muñoz, R., Weniger, G.-C., Pike, A.W.G., 2018. U-Th dating of carbonate crusts reveals Neanderthal origin of Iberian cave art. Science 359 (6378), 912–915.
International Commission on Stratigraphy, 2018. http://www.stratigraphy.org/ICSchart/ChronostratChart2017-02.pdf (accessed February 2018).
Johnson, E.A., 1990. Geology of the fushun coalfield, Liaoning province, People's Republic of China. Int. J. Coal Geol. 14 (3), 217–236.
KLíama, B., 1956. Coal in the ice age: the excavation of a palaeolithic settlement at Ostrava- Petřkovice in Silesia. Antiquity 30 (118), 98–101 (Cambridge University Press, United Kingdom).
Kopp, O.C., 2018. Encyclopedia Britannica, Coal: Fossil Fuel. https://www.britannica.com/science/coal-fossil-fuel (accessed February 2018).
Lechtman, H., Moseley, M.E., 1975. The scoria at Chan Chan: non-metallurgical deposits, Ñawpa Pacha. J. Andean Archaeol. 10–12, 135–170. Taylor and Francis. http://digitalassets.lib.berkeley.edu/anthpubs/ucb/text/nap010-12-009.pdf (accessed March 2018).
Lei, N., Zechuang, L., Min, Z., Lina, X., 2015. Deformation characteristics and mechanism of the landslide in west open-pit mine, fushun, China. Arabian J. Geosci. 8 (7), 4457–4468.
Linduff, K.M., Jianjun, M., 2008. Metallurgy in Ancient Eastern Asia: How Is it Studied? Where Is the Field Headed? Modeling Early Metallurgy: Old and New World Perspectives. British Museum. http://www.britishmuseum.org/pdf/Linduff%20Mei%20China.pdf (accessed March 2018).
MacDonald, G., Yow, Y.-S., Li, X., 2008. Innovation in China: The Dawning of the Asian Century. Adonis & Abbey Publishers Ltd, London. 267 p.
Makin Metal Powders, 2018. History of Bronze. Grimm Family of Companies. http://www.makin-metals.com (accessed March 2018).
Meignen, L., Brugal, J.-P., 2002. Territorial exploitation, technical traditions and environment in a mid-altitude context: the Canalettes rockshelter (Grands Causes, France). In: Conrad, N.J. (Ed.), Settlement Dynamics of the Middle Paleolithic and Middle Stone Age. Kerns Verlag, Tübingen, pp. 463–483.
Metropolitan Museum of Art, 2018. Shang and Zhou Dynasties: The Bronze Age of China. https://www.metmuseum.org/toah/hd/shzh/hd_shzh.htm (accessed March 2018).
Moore, E.S., 1922. Coal: Its Properties, Analysis, Classification, Geology, Extraction, Uses and Distribution. John Wiley & Sons, Inc. 462 p.
Museum of The Stone Age, 2018. Mesolithic Stone Tools. http://stoneagetools.co.uk/index.htm (accessed February 2018).
National Geographic, 2018. Coal. https://www.nationalgeographic.org/encyclopedia/coal (accessed February 2018).
Pettitt, R.G., 2015. Materiality in Early Bronze Age Wales (Ph.D. thesis). The University of Manchester, Faculty of Humanities, School of Arts, Languages and Cultures, Manchester, United Kingdom. 735 p.

Radivojević, M., Rehren, T., Kuzmanović-Cvetković, J., Jovanović, M., Northover, J.P., 2013. Tainted ores and the rise of tin bronzes in Eurasia, c. 6500 years ago. Antiquity 87, 1030–1045 (Cambridge University Press, United Kingdom).

Robinson, P.R., Hsu, C.S., 2017. Introduction to petroleum technology. In: Hsu, C.S., Robinson, P.R. (Eds.), Handbook of Petroleum Technology. Springer, pp. 1–81.

Ruan, Q., Wang, Y., 2017. The Jartai Pass Site in Nilka County, Xinjiang, Archaeology. Archaeological Magazine Press, Institute of Archaeology, Chinese Academy of Social Sciences, Beijing, China, pp. 57–70 (in Chinese).

Ruebens, K., Wragg Sykes, R.M., 2016. Spatio-temporal variation in late Middle Palaeolithic Neanderthal behaviour: British bout coupé handaxes as a case study. In: Di Modica, K., Pirson, S., Toussaint, M. (Eds.), Middle Palaeolithic in North-West Europe: Multidisciplinary Approaches. Quaternary International, 411, Part A., pp. 305–326.

Shaw, I., Jameson, R., 1999. A Dictionary of Archaeology. John Wiley & Sons. 736 p.

Shen, Q., 2014. Earliest Site of Coal Fuel Found in NW China, Chinese Archaeology. Institute of Archaeology, Chinese Academy of Social Sciences, Beijing, China. http://www.kaogu.cn/en/News/New_discoveries/2015/0814/51126.html (accessed March 2018).

Smithsonian Institution, 2018. Early Stone Age Tools. http://humanorigins.si.edu/evidence/behavior/stone-tools/early-stone-age-tools (accessed February 2018).

Somervill, B.A., 2012. The Story behind Coal. Heinemann-Raintree Books, Chicago, Illinois.

Speight, J.G., 2013. The Chemistry and Technology of Coal, third ed. CRC Press, Taylor & Francis Group. 845 p.

Stracher, G.B., Taylor, T.P., 2004. Coal fires burning out of control around the world: thermodynamic recipe for environmental catastrophe. In: Stracher, G.B. (Ed.), Coal Fires Burning Around the World: A Global Catastrophe. International Journal of Coal Geology. pp. 7–17.

Stracher, G.B., 2007. Coal fires burning around the world: opportunity for innovative and interdisciplinary research. GSA Today Geol. Soc. Am. 17 (11), 36–37.

Stracher, G.B., 2010. Coal fires: the rising global interest. Earth vol. 55 (9), 46–55.

Sui, Y., 2001. Investments Opportunities in Coal Mine Methane Projects in Fushun Mining Area. Fushun Mining Group Co., Ltd, United Stated Environmental Protection Agency and China Coalbed Methane Clearing House. 5 p. https://web.archive.org/web/20121022174631/http://www.epa.gov/cmop/docs/chinaprofile3.pdf (accessed February 2018).

Sung, Y.-H., 1997. Chinese Technology in the Seventeenth Century: T'ien-kung K'ai-Wu. 384 p. Dover Publications, Inc., Mineola, New York (Translated from the Chinese and annotated by E-Tu Zen and Shiou-Chuan Sun).

Sutherland, A., 2015. Mysterious Ancient Vinca Culture and its Undeciphered Script. Ancient Pages. http://www.ancientpages.com/?s=vinca (accessed March 2018).

Tao G., ~AD 950. Records of the Unworldly and Strange. National Library of China, Beijing. The original copy is lost but the library holds a handwritten copy of the original made in Jiangsu Province in AD 1572, during the Ming Dynasty, by the book collector Ye Sheng.

Théry, I., Gril, J., Vernet, J.L., Meignen, L., Maury, J., 1996. Coal used for fuel at two prehistoric sites in southern France: Les Canalettes (Mousterian) and Les Usclades (Mesolithic). J. Archaeol. Sci. 23, 509–512 (Academic Press, Elsevier).

Toth, N., Schick, K., 2007. In: Henke, W., Tattersall, I. (Eds.), Overview of Paleolithic Archaeology. Springer, pp. 1943–1963. https://doi.org/10.1007%2F978-3-540-33761-4_64 (accessed February 2018).

Valladas, H., Chadelle, J.P., Geneste, J.-M., Joron, J.L., Meignen, L., Texier, P.J., 1987. Datations par la thermoluminescence de gisements moustériens du Sad de la France. L'Anthropologie 91 (1), 211–226 (in French).

Wikipedia, 2018A. Industrial Revolution. https://en.wikipedia.org/wiki/Industrial_Revolution (accessed February 2018).

Wikipedia, 2018B. Fire Making. https://en.wikipedia.org/wiki/Fire_making (accessed February 2018).

Wikipedia, 2018C. Paleolithic. https://en.wikipedia.org/wiki/Paleolithic (accessed February 2018).

Wikipedia, 2018D. Mesolithic. https://en.wikipedia.org/wiki/Mesolithic (accessed February 2018).

Wikipedia, 2018E. Neolithic. https://en.wikipedia.org/wiki/Neolithic (accessed February 2018).

Wikipedia, 2018F. Shang Dynasty. https://en.wikipedia.org/wiki/Shang_Dynasty (accessed February 2018).

Wikipedia, 2018G. Fushun. https://en.wikipedia.org/wiki/Fushun (accessed February 2018).

Wikipedia, 2018H. Vinča Culture. https://en.wikipedia.org/wiki/Vinča_culture (assessed March 2018).

Wikipedia, 2018I. Chan Chan. https://en.wikipedia.org/wiki/Chan_Chan (accessed March 2018).

World Coal Association, 2018. Uses of Coal. https://www.worldcoal.org/coal/uses-coal (accessed February 2018).

Yule, H., 1871. The Book of Ser Marco Polo, the Venetian, Concerning the Kingdoms and Marvels of the East. 1, chapter XXX. John Murray, London, England. 409 p.

Zhang, P. (Y.K. Lee, Trans.). The Excavation of the House Remains of Yangshao Culture at Xiahe Site in Baishui County, Shaanxi, Chinese Archaeology 13, pp. 43–49, Walter de Gruter, Inc., 2013. http://www.kaogu.cn/uploads/soft/Chinese%20Archaeology/13/20160331house.pdf (accessed February 2018).

WWW Addresses: Additional Reading

(1) **American Coal Foundation: Timeline of Coal in the United States**
http://www.teachcoal.org/lessonplans/pdf/coal_timeline.pdf
(2) **Ancientcraft: Stone Age Paleolithic Stone Tools**
http://www.ancientcraft.co.uk/Tools/palaeolithic_tools.html
(3) **Art in China, by Craig Clunas**
Oxford University Press
http://www.oupcanada.com/catalog/9780199217342.html
(4) **Beginning of Metallurgy, by Katheryn Linduff, Han Rubin, Sun Shuyun**
Edwin Mellen Press
http://mellenpress.com/book/The-Beginnings-Of-Metallurgy-In-China/4156
(5) **China in the Early Bronze Age: Shang Civilization**
http://www.upenn.edu/pennpress/book/14210.html
(6) **Chinese Archaeology**
http://www.kaogu.cn/en
(7) **Coal: A Human History, by Barbara Freese**
Perseus Publishing
https://www.perseusbooks.com/?s=barbara+freese
(8) **Ding Vessel**
https://en.wikipedia.org/wiki/Ding_(vessel)
(9) **Four-Goal Square Zun**
https://en.wikipedia.org/wiki/Four-goat_Square_Zun
(10) **Fushun, China**
https://www.britannica.com/place/Fushun
(11) **Hopi and Coal Mining on Black Mesa, Arizona**
https://www.crossingworlds.org/hopi-and-coal-mining-on-black-mesa-arizona
(12) **Hypocaust**
https://en.wikipedia.org/wiki/Hypocaust
(13) **Iron and Steel in Ancient China**
E. J. Brill Publisher
https://brill.com/view/title/2665?qt-qt_product_details=0
(14) **Mesolithic Period**
http://academic.eb.com/levels/collegiate/article/Mesolithic-Period/52215
(15) **Neolithic Period**
  (a) **Encyclopedia Britannica**
  https://www.britannica.com/event/Neolithic-Period
  (b) **Wikipedia**
  https://en.wikipedia.org/wiki/Neolithic
(16) **Oracle Bone Script**
http://www.newworldencyclopedia.org/entry/Oracle_Bone_Script
(17) **The Coal Industry: 1600–1925**
http://spartacus-educational.com/CoalIndustry.htm
(18) **The Story of Coal**
https://grandemotte.wordpress.com/the-story-of-coal
(19) **The Vinča Culture: Ancient Wisdom**
http://www.ancient-wisdom.com/serbiavinca.htm
(20) **Yinxu, China**
https://en.wikipedia.org/wiki/Yinxu

CHAPTER 2

Coal-Fire Microarthropods From the Centralia, Pennsylvania and Healy, Alaska Mine Fires

An arthropod member of the order Coleoptera (beetle), the largest order of insects. The order name is derived from the Greek words *koleos* (sheath) and *ptera* (wings). Coleoptera have antennae, six legs, and most have two pairs of wings, including elytra (hard or soft forewings) that protect the abdomen and the rear wings (hindwings) used for flying. The order contains more species than any other, about 400,000, and comprises nearly 25% of all known life-forms (Wikipedia, 2017a). The specimen shown here is from the genus *Anotylus* (rove beetle). Its larvae and the adults feed on decomposing vegetation and dung (Hammond, 1976). *Photo courtesy of Michael Ferro, Louisiana State Arthropod Museum (Image ID 398), 2016.*

CHAPTER CONTENTS

Coal and Peat Fires: A Global Perspective
Edited by Glenn B. Stracher

https://doi.org/10.1016/B978-0-12-849885-9.00002-0

2.1 Arthropods and Coal Fires

Glenn B. Stracher
Jimmy Wedincamp
Breana Simmons
John P. Shields
Yelena White
Melissa A. Nolter
Anupma Prakash
Nancy Lindsley-Griffin (late)

Photo courtesy of the Virtual Fossil Museum (2017).

A springtail (subphylum Hexapoda, class Collembola) in amber (Pliocene to Pleistocene), Andes Mountains, Colombia. Usually less than 6 mm long (Meyer, 2016), Collembola inhabit grasslands and feed on fungi and decaying vegetation. They are discussed later in this chapter.

Introduction

Underground coal fires exhale combustion gases and particulate matter into the atmosphere from surficial vents and fissures, whereas combustion gases and particulates are exhaled directly into the atmosphere from coal beds burning at the surface. Live vegetation sometimes occurs on the surface immediately adjacent to the vents and fissures, sometimes several centimeters inside of them (Figure 2.1.1) or adjacent to **smoldering** (not flaming) coal beds burning at the surface.

It is not known why live vegetation habituates the environs of some vents and fissures as well as some outcrops of smoldering coal beds and not others, although this is likely a function of gas and soil temperature, gas chemistry, season of the year, and even climate. For example, at the gas vent illustrated in Figure 2.1.1, no sulfur was found. Moss adjacent to and growing several centimeters in the vent likely took advantage of the high concentration of carbon dioxide and water vapor in the coal-fire gas for **photosynthesis** (Stracher and Nolter, 2013a). However, less than a kilometer from the locations illustrated in Figure 2.1.1B and C, sulfur was found inside and immediately adjacent to a large gas vent, and all the ambient vegetation was dead and covered with sulfur (Figure 2.1.2).

This chapter, the first detailed coal-fires work of its kind, presents information about the phylum **Arthropoda** and the procedures for collecting and extracting **microarthropods** that potentially occur in vegetation at coal fires. We also present images of and a discussion about the following microarthropods: **mites**, **springtails**, **beetles**, **beetle larvae**, **thrips** (the name is both singular and plural), and **aphids**. They were extracted primarily from samples of vegetation collected at the **Centralia Mine Fire** in Centralia, Pennsylvania (Figure 2.1.1; Stracher et al., 2006) in June and December of 2003 and November of 2012. Some microarthropods were also extracted from a couple of samples collected at the **Healy Mine Fire** in Healy, Alaska (Figure 2.1.3; Prakash et al., 2013) in August of 2003. The information presented herein will be useful to researchers wishing to study the occurrence of arthropods at coal fires.

Frequently referred to by the **common** (nonscientific) **name arthropods**, the taxonomy and biology of the phylum Arthropoda, an implausibly diverse group of **invertebrates**, is discussed in detail in numerous publications, and there are many illustrations of their anatomical features on the Internet. Microarthropods, depending on the reference, are arthropods that are 5.0 mm in size or smaller (Martin and Davis, 2001, p. 2; Wikipedia, 2017f; Minor, 2017). The smallest known arthropod is *Stygotantulus stocki*, a parasitic crustacean only 94 μ long and about two-thirds as wide (Boxshall and Huys, 1989, p. 127).

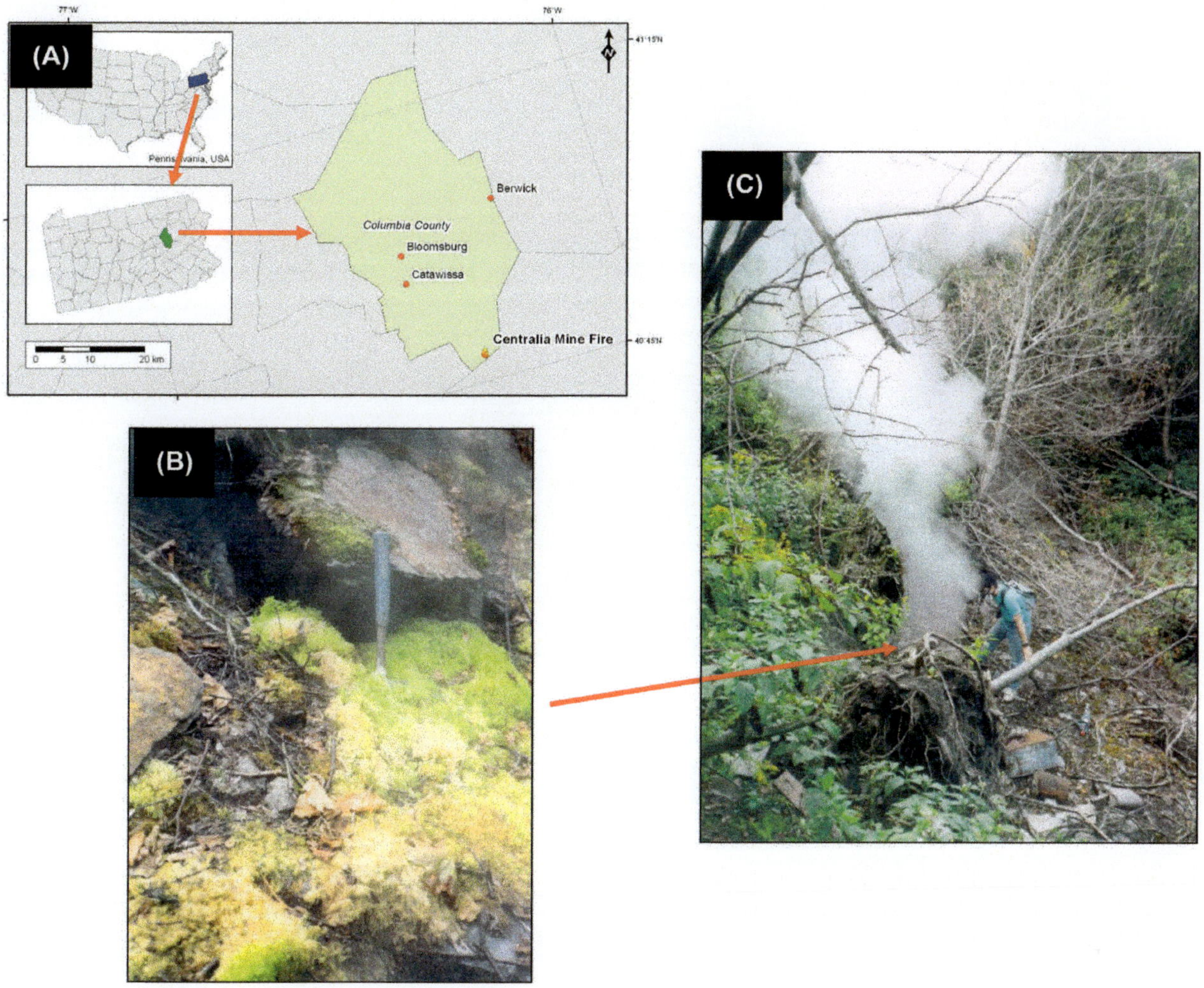

Figure 2.1.1. The Centralia Mine Fire in Columbia County, eastern Pennsylvania, started in 1962 when a landfill was ignited to reduce its volume and control rodents there. The landfill fire spread underground to the Pennsylvanian Buck Mountain anthracite coal bed (Stracher et al., 2006). (A) Location of the Centralia Mine Fire, still active today. (B) Moss growing adjacent to the outside of and for several centimeters into a coal-fire gas vent on the east side of Pennsylvania Route 61. The moss likely used CO_2 in the coal-fire gas for photosynthesis. (C) The gas vent and moss in (B) occurred at the base of a 20-m high hill that had a house (now demolished) on top. The vent and moss, in front of Stracher's right leg, are concealed by an uprooted tree. Stracher remembers seeing a male adult and at least one child adjacent to a car parked next to the house; exposed to the gas as it moved upslope, as seen here. While examining this vent, water vapor in the coal-fire gas condensed on Stracher's glasses. Map by Rudiger Gens, 2013 and photos by Janet L. Stracher, 1993: from Stracher and Nolter (2013a). Copyright © 2019 Glenn Blair Stracher. Published by Elsevier Inc. All rights reserved.

The Arthropods

The phylum Arthropoda, from the Greek *arthro* (joint) and *podos* (foot), comprises about 75%–85% of all known living and fossil members of the kingdom **Animalia** (Lane 1993; Levin and Smith, 2014, p. 138; Strauss, 2017; Wikipedia, 2017b; University of California Museum of Paleontology (UCMP), 2017).

Arthropods are adapted to life on land, in fresh and salt water, and in the air. They range in size from microscopic to that of king crabs with a 12-foot arm span, and they are found in ecosystems that vary from the deep sea to mountain tops (UCMP, 2017). Well-known to people are arthropod members of four subphyla. These include centipedes and millipedes (subphylum **Myriapoda**); spiders, mites, ticks, scorpions, and horseshoe crabs (subphylum **Chelicerata**); insects, springtails, protruans (coneheads), and diplurans (two-pronged bristle tails) (subphylum **Hexapoda**); and crayfish, shrimp, prawns, copepods, barnacles, woodlice, water fleas, crabs, and lobsters (subphylum **Crustacea**) (Heming, 2015; Chandra, 2017; Integrated Taxonomic Information System, 2016).

Figure 2.1.2. About 1 km from the location shown in Figure 2.1.1B and C, a gas vent in a sinkhole was found on Big Mine Run Road, just north of its intersection with Pennsylvania Route 61 in eastern Centralia. The sinkhole occurred because of ground subsidence caused by the volume reduction of coal as it burned underground. No living vegetation was seen here. The gas vent was encrusted with sulfur as were the ambient soil and dead vegetation. Photo by Glenn B. Stracher, 1992.

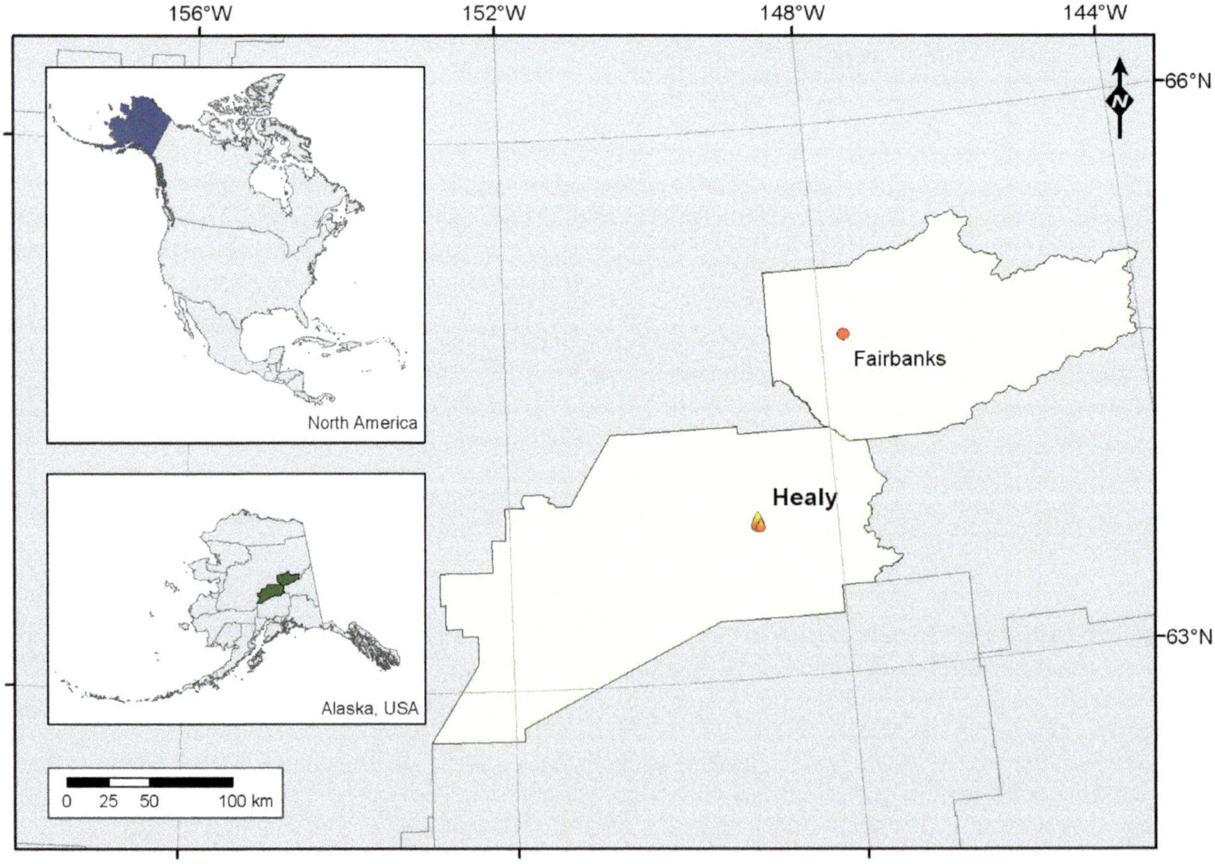

Figure 2.1.3. The Healy Mine Fire in Denali Borough, SW of Fairbanks in North Star Borough, Alaska. The Tertiary coal at Healy is low sulfur subbituminous C rank. No documentation exists about how and when the Healy coal-mine fire started, but residents of this area have known about coal fires in this region prior to coal mining that started in Healy in the early 1940s. During summer months, the solar heating of coal on south-facing slopes may lead to spontaneous combustion. Map by Rudiger Gens, 2017.

Over 1.2 million species of arthropods are identified (Wolfe et al., 2016); over 970,000 of these species are insects (including extinct insects). Insects comprise more than half of all known life forms. With an estimated 4–5 million living insect species, 80% of which are unidentified, as many as 95% of all insects that ever lived are thought to be extinct (Cranshaw and Redak, 2013, p. 2). Although most arthropods that ever lived are thought to be extinct, it is thought that millions of species living today have yet to be identified (Lane, 1993; UCMP, 2017). Thomas (1990) estimated that a conservative 6 to 9 million species of arthropods live in tropical forests and Lane (1993) suggested that up to 30 million species may exist today based on extrapolation from rain-forest studies.

The ancestors of the arthropods possibly appeared as soft-bodied **Edicaran fauna** (Precambrian, Neoproterozoic Era animals) about 555 million years ago (Glaessner, 1958; Lin et al., 2006; Edgecombe and Legg, 2014; UCMP, 2017). However, there are contending hypotheses about the ancestral **phylogeny** (evolutionary history) of arthropods, and it is unclear whether they evolved from a common or multiple unrelated ancestors (Heming, 2015; Wikipedia, 2017b). Some sources (Heming, 2015; Strauss, 2017; The University of Edinburgh, 2017) state that arthropods possibly evolved from ancestors of the phylum **Annelida** (segmented worms). Interestingly, Gould (1989) argued in his award-winning controversial book that most animals found in the **Burgess Shale** have no modern-day descendants. This shale is abundant in arthropods (Figure 2.1.4).

Arthropods appear with certainty during the Early Cambrian (e.g., trilobites). During the Middle Cambrian (c.510–505 million-years-ago), arthropods were prolific in the Burgess Shale, a famous **Lagerstätte**. A Lagerstätte (pl. Lagerstätten) is a deposit of diverse and/or well-preserved fossils (including **ichnofossils**, i.e., trace fossils) of organisms that did not live in situ, as many reef organisms do (Seilacher, 1970; Wikipedia, 2017c). The Burgess Shale was discovered in 1909 by Charles Walcott (Morris, 2009) in the Rocky Mountains of British Columbia's Yoho National Park, now a UNESCO world heritage site (UCMP, 2017). The shale was deposited by marine-turbidity currents along the base of the **Cathedral Escarpment**, a carbonate platform about 200 m high on the **Laurentian** (ancient North American) continental shelf, just south of the equator during the Cambrian (UCMP, 2017; Royal Ontario Museum, 2017). The Burgess Shale contains the oldest known (Cambrian, c.508 million-years-old) embryonic fossil-arthropod eggs that were **sclerotized**, that is, hardened by cross-linked, cuticle-protein molecules called **sclerotin** (Caron and Vannier, 2016).

Trilobites, members of the extinct arthropod class **Trilobita**, include the earliest known arthropods; there are over 20,000 species. They occur on every continent and were some of the first multicellular animals to live in pelagic and benthic shallow-marine environments around the world. Some species may have also lived in fresh water (Fortey, 2004; Kansas Geological Survey, 2017; Secher and Shugar, 2017). Well-preserved specimens with legs and antennae occur in the Burgess Shale (Smithsonian Institution, 2017). Their geologic range zone extends from the Early Cambrian (c.521 million-years-ago) to the Late Permian (c.240 million-years-ago); the cause of their extinction is unknown (Heming, 2015; Strauss, 2017; Secher and Shugar, 2017).

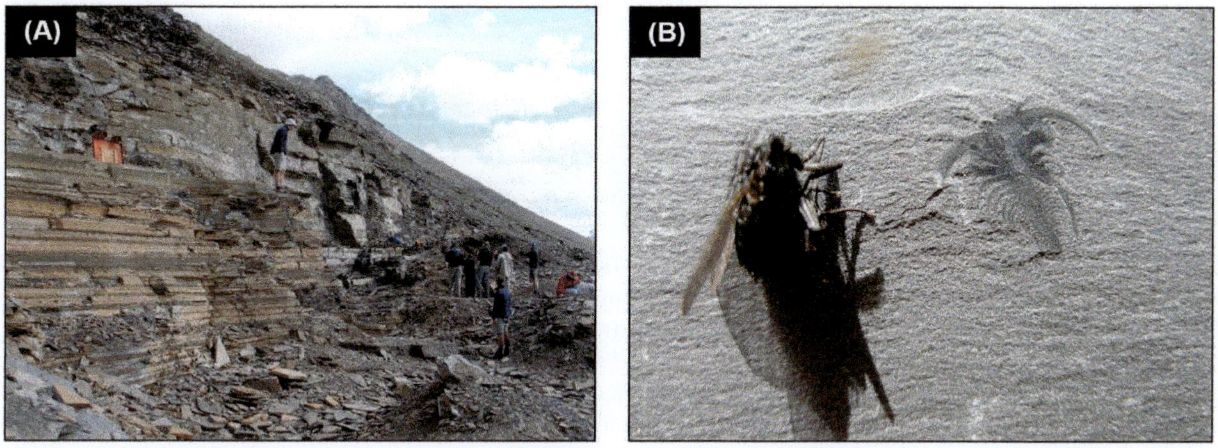

Figure 2.1.4. The Burgess Shale, part of the Middle Cambrian Stephen Formation, is in the Canadian Rocky Mountains; on Mt. Stephen in Yoho National Park, British Columbia, Canada. (A) The Walcott Quarry, named in honor of the Smithsonian Institution paleontologist Charles D. Walcott who discovered it in 1909. (B) *Marrella splendus* (a lace-crab ichnofossil) adjacent to a modern arthropod. Fossils of this extinct crab are the most abundant in the Burgess Shale. More than 15 000 specimens ranging in size from 0.25 – 1.91 cm were collected. *Marrella* may be the aquatic arthropod ancestor of chelicerates, crustaceans, and trilobites (Smithsonian Institution, 2017). Photos by Jacob Vinther, modified with permission of Palaeocast (2017).

Characteristics of Arthropods

External Features

All arthropods are invertebrate animals that share the following external **phenotypic** (observable) traits:

(1) Exoskeleton (external skeleton, also called a cuticle or shell) that is rigid, noncellular, deposited as two-layered plates called **sclerites**, and that cannot grow once the arthropod is completely formed (UCMP, 2017). The inner layer or **procuticle** is the thicker of the two. It protects and structurally supports the internal organs and has muscles attached to it. It is comprised mostly of **chitin** $(C_8H_{13}O_5N)_n$ and some sclerotin. Sometimes, the rigidity of the exoskeleton is increased with the deposition of calcium carbonate in the procuticle (Yeh, 2002; Hadley, 2008; Strauss, 2017; UCMP, 2017). The outer layer or **epicuticle** is comprised of a **wax-ester lipid** (chemically bonded fatty acid and fatty alcohol molecules) that repels and reduces water loss, as was the case with an insect collected from the location illustrated in Figure 2.1.5. The procuticle and epicuticle prevent desiccation (Hadley, 2008; Heming, 2015; Wikipedia, 2017d).

Chitin is an inelastic **polysaccharide** (a long-chain, polymeric carbohydrate) that is hydrophobic and insoluble in most organic solvents (Dutta et al., 2004, p. 21; Heming, 2015). Since it was first isolated from mushrooms in 1811 by the French chemist Henri Braconnot, chitin has been used in the biomedical, pharmaceutical, food, textile, and skin-care industries (Dutta et al., 2004, pp. 24–25; Park and Kim, 2010; Kohr, 2014; Singh and Dutta, 2016, p. 158). Because of its strength, chitin is used in the production of medical sutures (Nakajima et al., 1986; Dutta et al., 2004, p. 24).

Although the exoskeletons of trilobites now mineralized by carbonates and silica likely contained chitin (Teigler and Towe, 1974; Wilmot and Fallick, 1988; Klug et al., 2009), the oldest known occurrence of chitin (c.505 million years old) was found in 2013 in well-preserved fossil sponges (*Vauxia gracilenta*) in the Burgess Shale (Ehrlich et al., 2013). Many additional fossil occurrences of chitin occur, including those found in

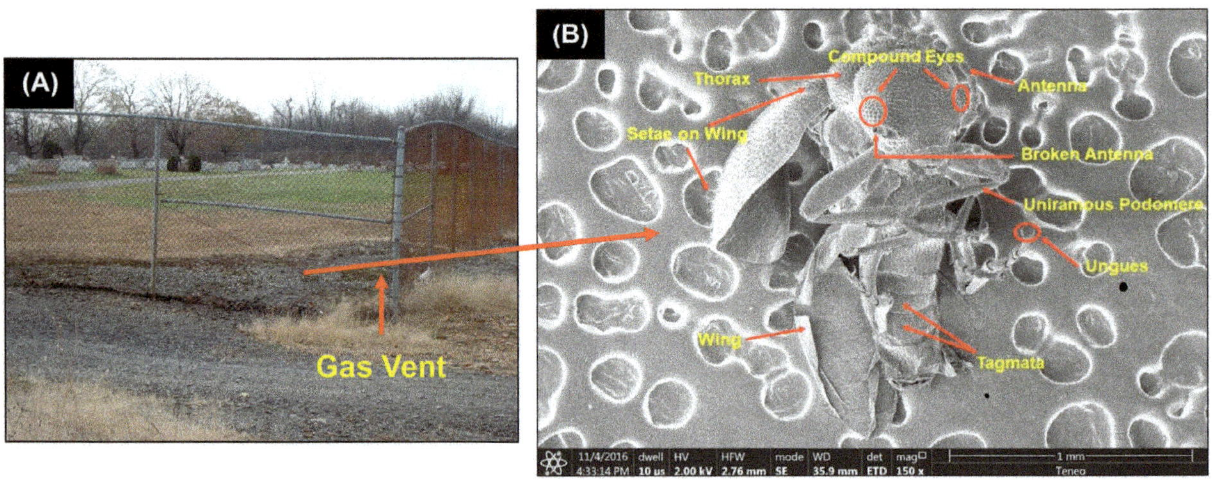

Figure 2.1.5. A November 12, 2012, sampling location (same as location 3 in Figure 2.1.13) for microarthropods at the Centralia Mine Fire in Pennsylvania. (A) St. Ignatius Cemetery on the west side of and adjacent to Pennsylvania Route 61, where microarthropods were collected from the green vegetation seen at the tail of the long *red arrow*. The temperature of the soil was 27°C. The distance between fence posts is about 3 m. (B) SEM image of a thrips, Hexapoda insects in the order Thysanoptera, collected from the vegetation. Springtails, beetle larvae, and mites were also found here (Table 2.1.1: 11/12/2012, Sample 3). The thrips has compound eyes with individual ommatidium (circular areas in the red ovals), visible on the head. The tagmata (metamere) are fused together on the abdomen, which is below the thorax, and the podomere of the uniramous legs (jointed appendages) are terminated with ungues. Setae are visible on one of the wings. The legs, always attached to the thorax, are visible. Also see Figure 2.1.14B. SEM notation in the black bar at the bottom of each photo: HV is high voltage; HFW is the horizontal field width; SE is secondary electrons; WD is the working distance from the lens to the object; det (detector) is EDT, an Everhardt-Thornley Detector; Teneo is the model of the SEM manufactured by FEI, now part of Thermo Fisher Scientific. Photo (A) by Melissa A. Nolter, 2012 and (B) by Yelena White and Glenn B. Stracher, 2016.

Oligocene insects (Stankiewicz et al., 1997); the oldest known occurrence (Middle Ordovician) of the extinct eurypterid, a sea scorpion (Lamsdell et al., 2015); and a Silurian eurypterid and Pennsylvanian scorpions (Gupta et al., 2010; Cody et al., 2011).

(2) **Molting** or **ecdysis** (shedding or sloughing) of the exoskeleton is necessary for the internal organs of an arthropod to increase in size as the arthropod grows or, as in the case of some insects, for it to change form, a process called **metamorphosis**. One well-known metamorphosis is the transformation of a caterpillar, the larva of the insect order **Lepidoptera** (also called **lepidopterans**), to a moth or butterfly.

Ecdysis triggering hormones (ETH), including mainly **ecdysterone** ($C_{27}H_{44}O_7$) (Dinan and LaFont, 2006; Heming, 2015), are responsible for molting. The separation of the exoskeleton from the cells beneath is called **apolysis**, and a new exoskeleton is secreted by **epidermal** (outer) muscle cells attached to the exoskeleton (Cranshaw and Redak, 2013; Strauss, 2017). Descriptions of ecdysis, including from the fossil record, are found in Roller et al. (2010) and Daley and Drage (2016). The topic is discussed in introductory biology textbooks and there are excellent YouTube videos on the Internet that illustrate different kinds of arthropods molting.

(3) **Tagma** (**metameres**) are multiple-hinged longitudinal body segments. Each individual-functioning body segment, called a **tagmata** (**metamere**), is fused to others by a process called **tagmosis** (Figure 2.1.5). Tagmata are a characteristic of all known fossil and living arthropods (Tobin and Duscheck, 2005, p. 487; Wikipedia, 2017b). The function, number, and anatomical names of tagma vary with taxa. Insects have three tagmata—head, **thorax**, and abdomen—but many spiders and crustaceans (Crustacea) have two, called a **cephalothorax** or **prosoma** (head plus chest) and abdomen (Myers, 2001; Wikipedia, 2017e; UCMP, 2017). In a trilobite, the **cephalon** (head), thorax, and **pygidium** (rump or tail) comprise the tagma (Levin and Smith, 2014, p. 140; Wikipedia, 2017e).

(4) **Jointed appendages** (also called outgrowths) are protrusions from the tagmata (Figure 2.1.5). These include mouthparts, legs, antennae, and reproductive organs. **Paired appendages** such as legs, antennae (only the Chelicerates lack antennae), and mouth parts occur on the body segments of an arthropod (Myers, 2001; Heming, 2015; Wikipedia, 2017b; The Canadian Encyclopedia, 2017). The legs of Myriapoda, Chelicerata, and Hexapoda are **uniramous**; that is, unbranched but segmented. In the Crustacea, they are **biramous**, meaning branched and segmented (Yeh, 2002; Nielsen, 2012, p. 244). The Trilobita had both uniramous and biramous legs (Hughes, 2003).

Regarding phylogeny, the uniramous morphology is thought to be derived from a biramous ancestry (Nielsen, 2012, p. 244). In addition, the ancestors of living arthropods likely had tagmata with paired appendages that served as limbs, but this feature was lost or occurs as a modified appendage found on the tagmata of living arthropods (Myers, 2001; Yeh, 2002; Wikipedia, 2017b). An example of a modified appendage is the **spinneret** of a spider located at the back of its abdomen, used to make silk (Yeh, 2002). Another example is that of **appendage dimorphism** such as the antennae that many male arthropods have for combating one another and for grasping females during mating (Fairbairn, 2013, p. 175).

(5) **Bilateral symmetry**; one half of the arthropod appears to be a mirror reflection of the other half when an imaginary plane is passed vertically from the front to the back and through the center of the arthropod (Cranshaw and Redak, 2013, Figure 2.1.5).

(6) **Compound eyes** and "simple eyes," called **ocelli** (sing. ocellus), occur in arthropods (Heming, 2015; Chandra, 2017). Some arthropods, including spiders and scorpions, only have ocelli. Scorpions have eight and spiders usually have eight but sometimes they have two, four, or six; in some cave spiders, there are none. Many crustaceans, including lobsters and crabs, have only one pair of compound eyes; each eye is on a movable eye appendage called a **stalk**. Insects have one to three ocelli and one pair of compound eyes (Cronin, 1986; Chandra, 2017; University of California Santa Barbara ScienceLine, 2017).

A compound eye (Figure 2.1.5) is a multilensed eye that contains optical receptors; each receptor is called an **ommatidium** (pl. ommatidia). Each ommatidium is covered by a **facet**, a transparent and polygonal, or dome-shaped **cornea** (lens). The facets are spread over the surface of a hemisphere and are usually hexagonal but may also be other polygonal shapes, with more than one shape in an ommatidia. An ommatidium contains photoreceptor (light-sensitive) and other cells that receive light transmitted through the facet. Each ommatidium is connected with others to an optic nerve that with a brain, makes vision possible. The ommatidia are oriented at different angles and produce a picture element of a precise area on an object, analogous to a pixel in a raster image. The picture elements formed by all the ommatidia (i.e., the ommatidia) produce an entire image that generally appears as light- and dark-colored spots. The greater the number of ommatidia, the higher the resolution of the image, and the fewer the number, the grainier the image (Chandra, 2017; Kimball, 2017a). The compound eyes in some arthropods can distinguish between colors. For example, honeybees are insects that can differentiate the wavelengths of white light (except red) visible to humans and in addition, ultraviolet light (Kimball, 2017a).

Compound eyes are useful in detecting motion over a wide range of angles (up to 360° on a stalk) because as an object moves, the ommatidia are visually stimulated at different positions of the object. These eyes are thus useful in helping an arthropod hunt or avoid predation. However, they are not useful for sharp-distant vision because each ommatidium is receptive to minute areas on an object, and the further away an object, the smaller those areas appear (Chandra, 2017; Kimball, 2017a).

A simple eye has one cornea at the surface through which light passes onto a layer of photoreceptor cells connected to an optic nerve that with a brain, makes vision possible. In many animals, including humans, the simple eye moves, and the curvature of the cornea is designed so that a crisp image is produced, even for objects afar. This characteristic of such a simple eye is called **accommodation** (Krapp, 2009). In arthropods, the simple eye, called an ocellus, also has one cornea, called an **ocellus lens**, but the eye cannot move. The design of the ocellus lens and its immobility permit photoreceptor cells together with the optic nerve and brain to generate an image but regardless of distance, the image is always blurry and shows little detail, like a photo taken with an out-of-focus camera lens. The primary function of ocelli is for orienting an arthropod in space by the transmission of variable light intensities through the ocelli, especially as the arthropod moves. Flying insects for example, typically have on their heads one to three ocelli used for orientation in space and even night vision by using starlight and light reflected from the surface of the moon (Chandra, 2017; Krapp, 2009). Although compound eyes can also provide flying arthropods with information about their orientation in space, there is a time delay in processing the transmission of light by ommatidia as such arthropods move. The ocelli are faster in permitting the arthropod to detect its orientation as it moves (Krapp, 2009).

(7) **Legs** and **wings** in the arthropods are used to propel them through the air, water, or over land. Movement is accomplished by the contraction and relaxation of muscle fibers, and this activity also allows individual tagmata to move relative to one another (The Canadian Encyclopedia, 2017).

Each segment of an arthropod leg used for aquatic or terrestrial movement is called a **podomere**, and claws on the end of a leg are called **ungues** (Angelini and Kaufman, 2005; Dunlop et al., 2013, Figure 2.1.5). Some adult insects have wings that permit them to fly (Figure 2.1.5). The wings are reinforced by veins, the pattern of which is useful for taxonomic classification. If an insect has two pairs of wings, one pair is called the **forewings** and the other is called the **hindwings**. If only one pair of wings occur, they are called forewings. It is not known why or with certainty how some insects evolved wings. Devonian (c.400 million-years-ago) fossils are wingless but by the Carboniferous (c.320 million-years-ago), fully functioning insect wings appear (Jockusch and Nagy, 1997; Wikipedia, 2017g).

Internal and Additional Features

Internal anatomical and additional features of arthropods are summarized here. Detailed descriptions are found in numerous publications about arthropods, including those in both the list of References and the links provided in the WWW Addresses: Additional Reading section at the end of this chapter.

(1) **Nervous system**, in which one or two nerve cords extend along the ventral (bottom) of the arthropod. Primitive arthropods have no brain but instead have **ganglia** (a mass of nerve cells) near their head or ganglia in their tagma. More advanced arthropods have two ventral nerve cords and a ring of nerves around the pharynx, all connected to a **cerebral ganglion** (brain) (Cranshaw and Redak, 2013; Myers, 2001; Strauss, 2017). The brain consists of two to three separate areas that serve different body functions (Edgecombe, 1998).

(2) **Open circulatory system** (no blood vessels) in which a tube-shaped muscle (the heart) on the dorsal side of the arthropod pumps **hemolymph** through the **hemocoel**. Although called blood when discussing invertebrate circulatory systems, hemolymph lacks the red blood cells that the vertebrates have. It contains mostly water and in addition, amino acids, ions, and other compounds (DeSalle, 2001). The hemocoel is the hemolymph-filled open part of the body cavity located between the internal organs (Raven and Johnson, 2001, p. 919; Cranshaw and Redak, 2013; Myers, 2001; Wikipedia, 2017b; The Canadian Encyclopedia, 2017). The hemolymph carries nutrients to and waste away from internal tissue; flows first from the heart to the anterior end of the arthropod, then toward the posterior end, and subsequently back to the heart through heart valves (Raven and Johnson, 2001, p. 919; Wikipedia, 2017b).

Because arthropods do not have the **respiratory pigments** of other animals, including vertebrates and numerous invertebrates, their hemolymph is usually colorless (Riggs, 2014; The Canadian Encyclopedia, 2017). Respiratory pigments are protein molecules that contain one or more metallic ions that combine reversibly with and enhance the transport of oxygen in blood (including hemolymph) cells (Starr, 2006, p. 595; Riggs, 2014). The pigments are tetrameric hemoglobins in the vertebrates and erythrocruorins (giant hemoglobins), hemocyanins, and hemerythrins in the invertebrates (Giardina et al., 1985; Riggs, 2014).

(3) Respiration (breathing) in animals is a gas-exchange process characterized by the intake of oxygen, its circulation to tissue within to produce energy, and the release of carbon dioxide in the opposite direction (Raven and Johnson, 2001, p. 1053). Oxygen and carbon dioxide as such are **respiratory gases**. Arthropods, unlike most animals, have no primary respiratory organ. Instead, respiration occurs by diffusion across the surface of the body and with or without the use of tracheae, gills, or book lungs (Raven and Johnson, 2001, p. 920; Yeh, 2002).

In most terrestrial arthropods, respiration commences with the intake of air through **spiracles**, apertures on various locations of the exoskeleton, that can be opened and closed as muscles relax or contract, respectively. Once air enters the spiracles, it travels through air canals called **tracheae** (sing. trachea). The tracheae branch into tubes called **tracheoles**, which transport oxygen passively or with muscle contractions through the body (Figure 2.1.6). The respiratory gases diffuse across cell membranes that are in contact with the tracheoles (Raven and Johnson, 2001, p. 920; Yeh, 2002).

Other terrestrial arthropods breathe with **book lungs** and with or without tracheae. A book lung is a respiratory organ comprised of stacked sheet-like structures (analogous to the pages of a book) around which air circulates. The plates are contained within an open abdominal-ventral chamber (Figure 2.1.7). Hemolymph passing over their surfaces exchanges respiratory gases with the plates (Oxford Reference, 2017). The plates absorb and expel oxygen through the chamber with muscular contractions. Scorpions have four pairs of book lungs and spiders up to two pairs (Raven and Johnson, 2001, p. 920; Encyclopedia Britannica, 2017a). Many Chelicerata, including some spiders, breathe solely with book lungs (Raven and Johnson, 2001, p. 920).

Aquatic arthropods breathe with **gills** or other aquatic respiratory devices (Jain, 2016). Gills are respiratory organs that absorb oxygen dissolved in water and release carbon dioxide into the water. They may or may not be enclosed in a gill chamber that has a gill cover over it. There are numerous kinds of gills, based on their shape and specific location on the body of an arthropod but in general, they are platy (**book gills**) and attached to appendages or tagma (Raven and Johnson, 2001, p. 920; Jain, 2016; Wikipedia, 2017h; The Canadian Encyclopedia, 2017).

(4) Feeding is accomplished with either mouth parts that consume (chew, suck, lap up) living or dead organic matter or by parasitizing other animals as, for example, blood-consuming insect fleas (subphylum Hexapoda, class **Insecta**) and ticks (subphylum Chelicerata, class **Arachnida**) do. The structure of the mouth and **alimentary canal** (digestive system) of the arthropods varies with diet (Drenkow, 2004; The Canadian Encyclopedia, 2017). All living species have a three-part alimentary canal that begins with a mouth at the front of the **foregut** (food consumption), ends with an anus at the end of the **hindgut** (water absorption and fecal formation), and includes a **midgut** (food digestion and absorption) between them (Yeh, 2002; Drenkow, 2004; Wikipedia, 2017i).

(5) Excretion of digestive waste occurs by different mechanisms in the arthropods. Muscles move food through the alimentary canal, where it undergoes extracellular digestion in the midgut. The products of digestion are then absorbed into and carried by hemolymph to the arthropod's tissue. **Metabolism** (cellular chemical reactions) of nutrients absorbed from hemolymph by cells in the tissue then generates water-soluble liquid waste. This waste is transferred from the cells to the hemolymph and eventually excreted (Wikipedia, 2017b,i). The nutritional requirements and metabolism of the arthropods is discussed by Stanley (1972, pp. 327–370).

In many terrestrial arthropods including the Hexapoda and Myriapoda, metabolic-liquid waste comprised of nitrogenous compounds including uric acid ($C_5H_4N_4O_3$), possibly guanine ($C_5H_5N_5O$), and electrolytic-water solutions diffuse from the hemolymph into **Malpighian tubules** located at the midgut and hindgut junction. These tubules are immersed in the hemolymph. Analogous to kidneys in their function, liquid waste from the

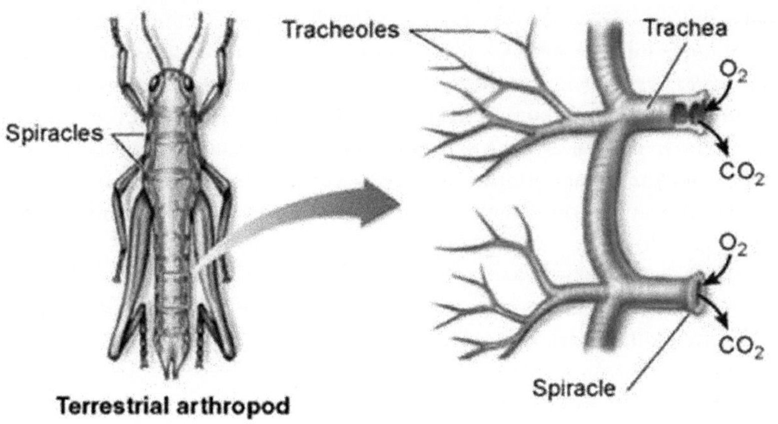

Figure 2.1.6. Spiracles, tracheae, tracheoles, and respiratory gases (O_2 and CO_2) in a terrestrial arthropod. These gases enter and leave the tracheoles during respiration. Since this arthropod has six legs, it is a member of the arthropod subphylum Hexapoda. Illustration with permission of Nazar (2017).

Figure 2.1.7. Book lungs occur within the pair of white-colored ventral chambers of this female jumping spider, *Thiodina puerpera*, a member of the arthropod class Arachnida in the subphylum Chelicerata. The specimen is about 11 mm long. Photo courtesy of Kaldari (2017); with modifications.

tubules moves into the hindgut (Figure 2.1.8). Once in the hindgut, liquid reabsorbed by the tubules results in the formation of solid compounds, including uric acid, that are eliminated from the anus as feces, possibly along with undigested food (Raven and Johnson, 2001, p. 920; Pacheco et al., 2014; Wikipedia, 2017b,j).

In addition to an anus, depending on the taxonomic group, arthropods use additional organs for excreting the liquid by-products of metabolism. For example, in many Chelicerata, including most Arachnida, a **coxal gland** is used for excreting urine at an opening at the base of the leg (Encyclopedia Britannica, 2017b). Many arthropods, including crustaceans belonging to the order **Decapoda** (10-footed arthropods including lobsters, crayfish, crab, and shrimp), excrete urine through an **antennal gland** (also called a **maxillary gland)** that opens to an external pore at the base of the antenna, where it meets the head (Roberts, 1971, pp. 221–222). In some crustaceans, the antennal gland is called a **green gland** because of its color. In aquatic arthropods, metabolic processes form toxic ammonia (NH_4) that is eliminated from the gills with a permeable membrane (Wikipedia, 2017b).

(6) **Reproduction** in most arthropod species occurs with **dioecious** individuals—those with either male or female reproductive organs, but not both. The sex organs of such species include paired testes or ovaries (**gonads**) connected to internal ducts that are open on the ventral surface at a species-dependent location. Fertilization is usually internal but may also be external (Myers, 2001; Encyclopedia Britannica, 2017c; Wikipedia, 2017b).

Terrestrial arthropods usually use **spermatophores** to reproduce by direct or indirect **internal fertilization** of the female sex cell, the **ovum** (pl. ova). A spermatophore is a waterproof protein capsule that contains motile spermatozoa inside (Allaby, 2003, p. 503; Amateur Entomologists' Society, 2017; Wikipedia, 2017b).

During direct-internal fertilization, a male and female arthropod copulate and a spermatophore is transferred into the female's genital opening through the male's reproductive organ, called an **aedeagus** (pl. aedeagi). Most Hexapoda reproduce this way (Encyclopedia Britannica, 2017c). Direct internal fertilization also occurs by the free transfer of sperm from the male aedeagus into the female's genital opening. This process occurs in some Hexapoda, Chelicerata, Myriapoda and in many Crustacea (Amateur Entomologists' Society, 2017; Encyclopedia Britannica, 2017c).

Indirect-internal fertilization occurs when a male (e.g., some mites and scorpions) first deposits a spermatophore on the ground, a leaf, and so on. Then following a ritual dance, the male maneuvers the female over to the spermatophore or she is chemically attracted to it, subsequently maneuvering it into her genital opening. Scorpions, spiders, and mites commonly reproduce this way, and this technique is used by some Hexapoda (Encyclopedia Britannica, 2017c).

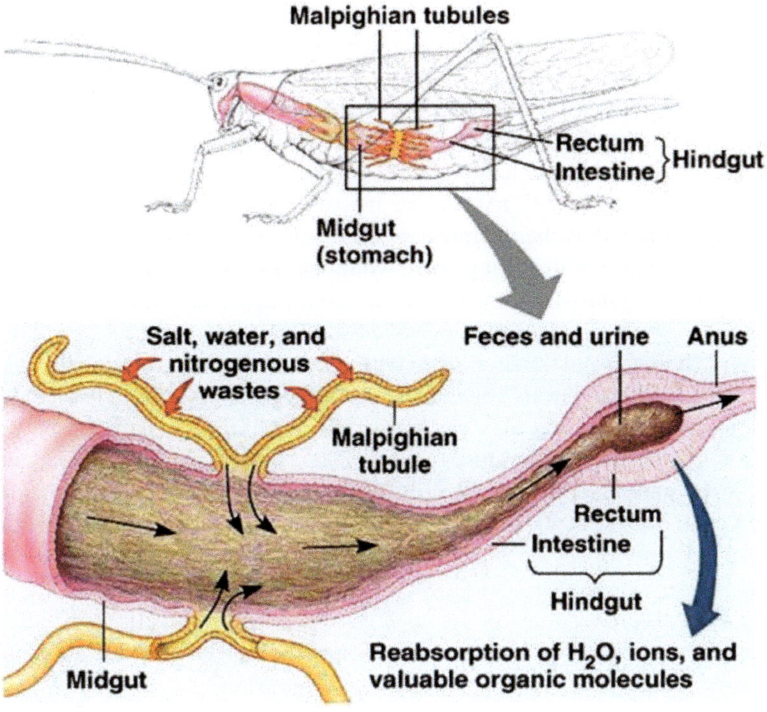

Figure 2.1.8. The arthropod digestive tract. Liquid waste comprised of nitrogenous compounds and electrolytic-water solutions diffuse from the hemolymph into Malpighian tubules immersed in the hemolymph. Liquid reabsorbed by the tubules results in the formation of solid compounds eliminated as feces, possibly along with undigested food, from the end of the rectum (anus). From Campbell, Neil A., Reece, Jane B., 2002. Biology, sixth ed. Reprinted by permission of Pearson Education, Inc., New York, New York, p. 943.

Most arthropods lay eggs. After hatching, depending on the species, the immature arthropod may anatomically be a small and modified version of its adult parent, but with different food requirements; or it may metamorphose from a larva to an adult (Myers, 2001; Yeh, 2002; The Canadian Encyclopedia, 2017). For example, some scorpions are **ovoviparous**, and give birth from eggs that hatch inside the female. The newborn scorpions (**scorplings**) are anatomically miniature and modified versions of a parent; they initially lack an exoskeleton. After five molts, over approximately 50 days, the scorplings become adults (ScorpionWorlds, 2017; Wikipedia, 2017b). As another example, many marine and freshwater Crustacea frequently pass through several larval stages before becoming adults. The number of stages is species-dependent and every stage is separated by a molt. After hatching, the most common first free-living larval stage in a Crustacea is called **nauplius** (pl. nauplii; other such first larval stages include zoea and megalopae in crabs and shrimp; see UCMP, 2017). Nauplius morphology is a defining characteristic of most Crustacea (Wikipedia, 2017k; Dahms et al., 2006). Nauplius larvae, identified in rocks as old as Early Cambrian, are characterized by an oval-, round-, or pear-shaped body with structures that include a median eye and in addition, three pair of appendages used for feeding, sensing, and swimming. These appendages become the mandibles and antenna in adult Crustacea (Martin et al., 2014, pp. 1–16; Wikipedia, 2017k).

Some of the Hexapoda, Crustacea, and Chelicerata reproduce by **parthenogenesis**, complex asexual reproduction involving the development of an embryo prior to birth (Wikipedia, 2017b; Pandian, 2016, p. 90; ScorpionWorlds, 2017). Finally, some Crustacea are either **simultaneous hermaphrodites** (also called **hermaphroditic**; contain male and female reproductive organs); **sequential hermaphrodites** (change sex at some point in their life); or as in the case of some shrimp and barnacles, are **protandric simultaneous hermaphrodites**, which first reproduce as males and then during maturation add female reproductive functions without discarding the male functions (Baeza, 2006; Yusa et al., 2013; Pandian, 2016, pp. 104–109; Wikipedia, 2017b).

(7) **Ecosystems** are profoundly influenced by arthropods because they link life forms on Earth by facilitating the consumption, decay, and recycling of organic matter in terrestrial and aqueous ecosystems (Cranshaw and Redak, 2013, p. 4; New Mexico State University Arthropod Museum (NMSUAM), 2017). Arthropods feed on plants as **phytophages**; on animals as parasites; and on dead or decaying plant or animal matter as **saprophages**; they consume seeds and feces. After feeding, they break down and excrete organic matter, freeing up nutrients for life on Earth. They are also a primary food source for other arthropods; a variety of animals including fish, amphibians, reptiles, birds, and mammals that include humans (Encyclopedia of Entomology, 2008, pp. 2876, 4238; Cranshaw and Redak, 2013, pp. 4–5). Some arthropods are consumed (e.g., flies) by carnivorous plants such as the pitcher plant. However, others (some ants) live in a **symbiotic** (mutually beneficial) relationship with carnivorous plants by living on and defending them from herbivores, while consuming prey and depositing nutrient-rich feces in the plant's pitcher or on its leaves (Hartmeyer, 1998). Any such mutually beneficial relationship between different kinds of organisms is called **symbiosis** and each organism is called a **symbiont**. Most terrestrial arthropods live in and biologically rework soil, a process called **bioturbation**. During this process the soil is churned, often by tunneling **social insects** like ants and termites, and its organic matter, including bacteria, eggs, and viruses, is redistributed as the chemical gradient of the soil is altered. Bioturbation is necessary for the retention of soil fertility and without it, plant growth would be lessened, thereby negatively affecting the link in the food chain between plants and animals (Meysman et al., 2006; Cranshaw and Redak, 2013, pp. 4–5).

Arthropods not only provide humans and other animals with food. They also provide humans with medicine; clothing; shellac; dyes; safeguard from dangerous organisms; criminal-investigatory evidence; specimens for research in physiology, genetics, and evolution; and they serve as trash disposals for plant waste, old wood, and manure. They have also inspired works of art and literature (Myers, 2001; NMSUAM, 2017). As one concrete example of their benefit to humans, insects are essential for the pollination of flowering plants and the production of vegetable and fruit seeds (Cranshaw and Redak, 2013, p. 3) that germinate into crops harvested, consumed as food, and used by people for medicinal purposes.

Despite their benefits, arthropods are responsible for cataclysmic famines, infectious diseases, and venomous bites that make people sick and cause millions of fatalities. They have infected livestock and destroyed building structures as well as unharvested and stored crops (Myers, 2001; Alexander, 2012, p. 399; NMSUAM, 2017). The Justinian Plague that started in 541 A.D., later followed by the Black Death (Great Plague) and then the Modern Plague were three major recorded pandemics caused by the bacterium *Yersinia pesti*. The bacteria can be transmitted to humans by bites from insect fleas that parasitized infected rats, ground squirrels, mice, and other mammals. It can also be transmitted from one person to another by droplets in an infected person's cough. Since the Justinian outbreak, these plagues have killed tens of millions of people, and outbreaks still occur across the world today (Center for Disease Control and Prevention, 2017).

Field Methods: Collecting Microarthropods

After collecting microarthropod-bearing vegetation at coal fires when first noticing plant matter growing adjacent to and inside gas vents, we realized that a more detailed collection protocol is needed to gather as much data as possible. Hence, we recommend the following eight-step procedure. However, depending on field circumstances and available equipment at the time of collection, all these steps may not be possible.

Coal fires are dangerous and adequate precautions should be used when sampling, such as probing the ground with a walking stick for dangerous subsidence and wearing a volatile organic compound (VOC) respirator that provides protection from both gas and particulate matter in smoke.

Collection Procedure

(1) Collect samples of vegetation and soil every 0.5 m along 3 m long linear transects; in a radial pattern, outward from a gas vent, fissure, or burning-coal bed exposed at the surface. Label each sample and record its GPS location. At least six transects are recommended for adequate ground surveillance.
(2) Record the physical characteristics of soil and rock along each transect, including the color and texture. Also, record as much additional data as possible including the ambient air temperature, humidity, and so on.
(3) Take photos of the entire collection locality, each transect, and the collection sites along each transect. Each photo should have a scale.
(4) Before collecting a sample, measure the temperature of the vegetation and soil at the surface and the temperature of the soil at 5 cm below the surface.
(5) When collecting samples of vegetation, take in situ photos and include a scale. If possible, collect the vegetation with some soil attached. Put each sample in a plastic bag.
(6) If possible, use a portable gas meter to measure select gas components such as CO_2 and CO within a mass of vegetation and in the soil at 5 cm below the surface of each sampling site. For additional gas components, a

sample of gas exhaled at the surface can be collected using stainless steel containers as illustrated and described in Stracher et al. (2005, 2008); Stracher (2011).

(7) At each collection site, collect a sample of soil at 5 cm below the surface, and put it in a plastic bag. A small hand shovel can be used for collecting. Wipe the shovel clean before collecting each sample. Again, label the sample and take photos with scales.

(8) Once the vegetation and soil samples are collected and put in sealed plastic bags, do not expose them to sunlight. Put them in a backpack or an insulated container (no ice). Send the samples to the lab for preparation, preferably within 48 h or less after collecting them.

If the samples must be stored overnight before sending them to the lab, keep them refrigerated above the freezing point of water.

Laboratory Extraction of Microarthropods

Collecting organisms from vegetation and soil samples is accomplished with a **Berlese funnel**, also known as a **Berlese trap** or **Tullgren funnel** (Mississippi Entomological Museum, 2017; Wikipedia, 2017l). There are numerous variations of these funnels; two varieties were used in the biology laboratory at East Georgia State College to extract microarthropods from samples of vegetation collected at the Centralia and Healy mine fires.

The Berlese funnel used in 2003 (Figure 2.1.9) for samples from Centralia and Healy included a plastic bucket with a glass beaker placed on the bottom. The beaker contained a 70% ethanol and 30% water solution used to preserve the microarthropods. A hemispherical-shaped metal strainer, with a wire mesh at the bottom, was inserted into the top of the bucket and one sample at a time was put on the mesh. As each sample sat on the mesh, a metal-hemispherical lamp, illuminated with a 50-W incandescent bulb at its center, was placed over the wire mesh for 24 h. During that time, heat energy from the bulb flushed microarthropods from the vegetation and they fell into the liquid in the

Figure 2.1.9. Jimmy Wedincamp at East Georgia State College using a Berlese-funnel in 2003 to extract microarthropods from samples collected at the Centralia, Pennsylvania and Healy, Alaska mine fires. Heat energy from a 50-W incandescent bulb shown in (A) and turned on for 24 h, flushed microarthropods from the sample shown in (B) into a glass beaker placed at the bottom of the bucket. The beaker contained a 70% ethanol and 30% water solution used for preservation. Liquid from the beaker was then poured into glass vials. Microarthropods from each vial were then identified with a Leica DM E compound microscope that was illuminated from below the stage. The beakers and sample holder, with a wire mesh at its bottom, were cleaned for each sample from which microarthropods were extracted. Photos by Glenn B. Stracher, 2003.

beaker. Subsequently, this liquid was poured into glass vials and then the microarthropods were identified using compound and stereo microscopes. Each beaker and the wire-mesh container were changed between each flushing.

Berlese funnels used for the Centralia samples collected in 2012 are shown in Figure 2.1.10. Each funnel consists of two glass halves that snap together. The bottom half has a wire mesh at the top, and a sample is placed on the mesh. A glass beaker containing a 70% ethanol and 30% water solution was placed on the bottom of this part of each funnel (Figure 2.1.10A and B). The Berlese funnels were then placed with others in a fume hood (Figure 2.1.10C) and the lights inside the hood were turned on high intensity for 24 h. As before, microarthropods flushed from the samples and into the beakers were then poured into glass vials and examined with compound and stereo microscopes.

Microscopy of Microarthropods

The microarthropods collected from Centralia and Healy were initially observed with both a binocular Leica DM E compound and an Olympus SZX16 stereo microscope. The samples were mounted with a drop of the solution used to preserve them on glass (Leica microscope) and cardboard micropaleontology slides (Olympus microscope). Although useful for taxonomic classification, this method did not reveal minute anatomical details for study and more precise classification. Consequently, the microarthropods observed with tabletop microscopes were processed

Figure 2.1.10. A Berlese funnel and fume hood used at East Georgia State College in 2012 to extract microarthropods from samples collected at the Centralia Mine Fire in Pennsylvania. (A) Berlese funnel. (B) Jimmy Wedincamp putting a sample on the wire mesh in the lower half of the funnel. A beaker containing 70% ethanol and 30% water used to preserve the microarthropods was placed on the bottom of this part of the funnel. (C) The top half of the funnel shown here was placed over the lower half, and the funnel was placed in the fume hood (behind Jimmy) with other Berlese samples. Three, 125-W heat lamps were turned on for 24 h in the fume hood. Heat energy from the lights flushed microarthropods into the beakers and the liquid was then poured into glass vials in preparation for study. Photos by Glenn B. Stracher, 2012.

for and preferentially imaged with a **scanning electron microscope (SEM)**, and representative samples are presented in this chapter.

Processing of Microarthropods for SEM Imaging

When air dried, microarthropods without hard exoskeletons (e.g., Collembola) would collapse as they desiccate. They would then be deformed and not useful for SEM imaging. If they are taken out of the solution used to preserve them, coated with an inert metal by using a sputter coater, and placed into the SEM sample chamber, they would also collapse and deform. This is because they would desiccate from the high vacuum in the chamber. Consequently, microarthropods must be chemically processed for maximizing the resolution of their details with an SEM. This process involves rehydrating and then dehydrating the samples, as described in the following. Additional details are found in Palade (1952), Sabatini et al. (1963), Nation (1983), and Heraty and Hawks (1998).

Rehydration

The microarthropods were removed from the preservative solution with a fine-pointed artist's paintbrush. They were then rehydrated every 15 min by reducing the ethanol content of the preservative with the addition of distilled-deionized water in increments of 25%. Once rehydrated so that they were in 100% water, the water was removed with a pipette and then a 2% **glutaraldehyde ($C_5H_8O_2$)** fixative in a 0.1 M **phosphate-buffered saline solution** (PBS) was added to the samples for 1 h at room temperature. This mixture was then removed with a pipette, taking care to not remove any microarthropods as well, and replaced with PBS. This was repeated three times to ensure that excess glutaraldehyde was removed. The PBS used to replace the glutaraldehyde/PBS mixture was then removed with a pipette and replaced for 1 h with a secondary fixative of 1% **osmium tetroxide (OsO_4)** in 0.1 M PBS. This mixture was next removed with a pipette and replaced with distilled-deionized water. Water was used to avoid the precipitation of salts during ethanol dehydration, discussed in the following.

Chemical fixatives such as the aldehyde glutaraldehyde and osmium tetroxide are chemicals used with a buffer that has a similar osmolarity to the tissue being fixed. The fixatives stabilize tissue by preserving cellular structure and promoting tissue cohesion (Huang and Yeung, 2015, pp. 23–43; Hughes, 2017).

Dehydration

Following rehydration, the samples in water were reduced every 15 min with the addition of ethanol in increments of 25%. Once the samples were in 100% ethanol, the ethanol was reduced every 15 min with **Hexamethyldisilizane ($C_6H_{19}NSi_2$)** (HDMS), increased in 25% increments. After the samples were in 100% HMDS, they were kept in a loosely covered container to avoid dust and contamination and completely air dried for 12 h in a running fume hood. HMDS was used to avoid the collapse of delicate structures associated with the elevated pressure of conventional critical point drying. Note that the samples cannot be put from water into HDMS because HDMS and water react explosively. When the samples were completely dry, they were placed on double-sided, electrically conductive carbon tape attached to SEM sample stubs. Next, the samples were coated with an approximately 15-nm-thick layer of vacuum-evaporated chromium. The samples were imaged on an FEI Teneo Field-Emission SEM (FE-SEM) with both scattered electron (SE) and backscattered electron (BSE) detectors.

Images of Coal-Fire Microarthropods

Representative images of microarthropods found at the Centralia and Healy mine fires are presented next. Select anatomical structures are labeled in the images. Details about these arthropods and their anatomy can be found in Romoser and Stoffolano (1998), Elzinga (2003), and Triplehorn and Johnson (2005).

Samples of vegetation containing microarthropods illustrated in this chapter were collected by three coauthors before two entomologists highly recommended that the collection protocol stated earlier be used for future research. Table 2.1.1 records sample data for the Centralia and Healy collection sites.

Table 2.1.1
Collection localities for the extraction of microarthropods from coal fires.

Date Collected (Month/Day/Year) and Sample Number	GPS Coordinates (Latitude and Longitude)	Altitude (Meters)	Material Collected	Temperature[‖] (°C)	Microarthropods Identified	Archetypal Microarthropod Chapter Figure
		Centralia Mine Fire, Pennsylvania				
6/22/2003 Sample 1 (Field Location 2A)	40°47.803′ N 076°20.549′ W Pennsylvania Route 61*	447.4	Moss + Gas Sample	63.7 Gas Vent	Springtails	2.1.11
6/22/2003 Sample 1 (Field Location 2B)	40°47.803′ N 076°20.549′ W Pennsylvania Route 61*	447.4	Moss	69.4 Gas Vent	Springtails	2.1.11
12/29/2003 Sample 1	40°47.782′ N 076°20.580′ W Death Valley[†]	436.8	Moss	65.6 Gas Vent	Springtails	2.1.11
12/29/2003 Sample 2	40°48.060′ N 076°20.493′ W Cemetery Front[‡]	468.2	Moss	76.7 Gas Vent	Springtails	2.1.11
12/29/2003 Sample 3	40°48.047′ N 076°20.643′ W Cemetery Front[‡]	468.2	Moss	65.6 Moss	Springtails and Mites	2.1.11, 2.1.12, 2.1.14
11/12/2012 Sample 1	40°48.047′ N 076°20.643′ W Inside Cemetery[§]	468.2	Moss	49 Gas Vent	Springtails	2.1.11
11/12/2012 Sample 2	40°48.047′ N 076°20.643′ W Inside Cemetery[§]	468.2	Moss	35 Moss	Springtails and Mites	2.1.11, 2.1.12, 2.1.14
11/12/2012 Sample 3	40°48.047′ N 076°20.643′ W Inside Cemetery[§]	468.2	Unidentified Vegetation	27 Soil	Springtails, Mites, Thrips, Beetles, Beetle Larvae	2.1.5, 2.1.11, 2.1.12, 2.1.14
11/12/2012 Sample 4	40°48.047′ N 076°20.643′ W Inside Cemetery[§]	468.2	Grass	24 Soil	Springtails and Mites	2.1.11, 2.1.12, 2.1.14
11/12/2012 Sample 5	40°48.047′ N 076°20.643′ W Inside Cemetery[§]	468.2	Unidentified Vegetation	15 Soil	Springtails, Mites, Thrips, Beetles, Beetle Larvae	2.1.5, 2.1.12, 2.1.14
		Healy Mine Fire, Alaska				
8/23/2003 Sample V2-P1	63°51.375′ N 148°50.810′ W	~409	Unidentified Burnt Vegetation	67 Gas Vent	Springtails	2.1.17
8/23/2003 Sample V2-P3	63°51.375′ N 148°50.810′ W	~409	Unidentified Vegetation	12 Soil	Aphids	2.1.17

*Samples from a foot down and inside a crack in an abandoned section of Pennsylvania Route 61 (PA 61). See Figure 2.1.11.
[†]An abandoned strip-mining cut engulfed in toxic fumes at the time of sampling (Stracher et al., 2006, Figure 8A); west of PA 61. See Figure 2.1.12A.
[‡]Samples from in front of St. Ignatius Cemetery, on the west side of PA 61. See Figure 2.1.12B.
[§]Samples from the inside fence corner of St. Ignatius Cemetery, on the west side of PA 61. GPS coordinates nearly the same as the Cemetery Front. See Figures 2.1.12B and 2.1.13.
[‖]Gas Vent designates the gas temperature at about 5 cm inside the vent. The other temperatures are those of samples of vegetation or soil collected near a gas vent (see text).

SEM images of springtails and aphids are not presented in this chapter but instead, photos of these Hexapoda, originally taken with a Leica DM E compound microscope (equipped with a Nikon Coolpix 990 camera) and an Olympus SZX16 microscope (equipped with a built-in camera) are shown. This is because when the springtails and aphids were observed with the Olympus microscope just prior to preparation for SEM imaging, they inadvertently dried out in the lab and did not rehydrate as well as anticipated, resulting in less than desirable SEM images.

Centralia Mine Fire, Pennsylvania

The anthracite coal at Centralia (Figure 2.1.1) formed during the Pennsylvanian Period (Upper Carboniferous). A mine fire started there has burned since May 1962. The fire started when trash was ignited in an abandoned strip-mining cut in an unregulated dump on the south limb of the Centralia syncline. The trash was ignited to eliminate and reduce the volume of rodents in the dump. The trash fire then spread into mining tunnels under Centralia (Stracher et al., 2006).

On June 22, 2003, plant moss was collected adjacent to a gas vent, about a foot down inside a crack in an abandoned section of Pennsylvania Route 61 (PA 61) in Centralia. The temperature at the entrance to the gas vent varied from 63.7 to 69.4°C over a 5-min interval on a breezy day. The only microarthropods found in June 2003 were springtails (Table 2.1.1: 6/22/2003, Sample 1; Figure 2.1.1).

Springtails

Springtails (Figure 2.1.11C and D) are members of the subphylum Hexapoda and class Collembola. They first appeared in the Devonian (Penney and Jepson, 2014, p. 16), and about 6500 species have been identified. They are wingless, and most are microarthropods because they are less than 5 mm long, but some can grow longer than microarthropods; up to 10 mm in length. They live on every continent, even in deserts and on snow. Some species live in fresh water and others in the marine littoral zone.

Most species of springtails feed on decomposing vegetation and fungi hyphae. Others feed on living plants, human corpses, zooplankton (e.g., rotiers), roundworms (nematodes), and other springtails. Although some are pests and cause crop damage, most are beneficial. As decomposers, they add plant nutrients to soil and some species keep plant fungal diseases in check (Hopkins, 1997, pp. 2–5). Springtail colors include brown, white, black, gray, yellow, red, orange, purple, and cobalt blue; in addition, they can be multicolored (University of Minnesota Extension, 2017; Penn State College of Agricultural Sciences, 2017a).

On December 29, 2003, microarthropods were collected from moss next to a gas vent at the top of a smoke-filled, abandoned strip-mining cut (Death Valley) adjacent to the west side of PA 61 (right side of Figures 2.1.11A and 2.1.12A). As Table 2.1.1 shows, the temperature at the entrance to the gas vent was 65.6°C and the only microarthropods found by the Death Valley sampling site were springtails. The table also shows that on this same day, microarthropods were collected from two different locations at Cemetery Front, the area in front of the fence around St. Ignatius Cemetery (Figure 2.1.12B), north of and adjacent to the west side of a drivable section of PA 61. Springtails were collected from moss adjacent to a gas vent at Cemetery Front, and the gas temperature at about 5 cm inside the vent was 76.7°C. Several meters to the south of this vent, mites (Figure 2.1.12C) and springtails were collected from moss where the temperature was 65.6°C.

Mites

Mites (Figures 2.1.12C and 2.1.14A) are members of the subphylum Chelicerata and a subclass of Arachnida called **Acari** (or **Acarina**). The subclass Acari includes mites and ticks, and the study of this subclass is called **acarology**. Acari first appeared in the Devonian. Mites occupy terrestrial (land), fresh and salt water, cave, and thermal spring ecosystems (Wikipedia, 2017m). About 48,200 species of mites have been identified (Halliday et al., 1997, p. 193). They feed on both living and dead organic matter; some are decomposers whereas others parasitize animals and plants. Mites have pear-shaped bodies with stiff, hair-like fringes called **setae** (sing. seta). Most mites are less than 1 mm in size (ticks are larger) and most have four pairs of legs but some have one, two, and three pairs. They have no wings; it is difficult to discern segmentation between the head, thorax, and abdomen; and they have no eyes, except possibly **eye spots** that can detect the direction and intensity of light. Mites have a pair of **pedipalps** that use

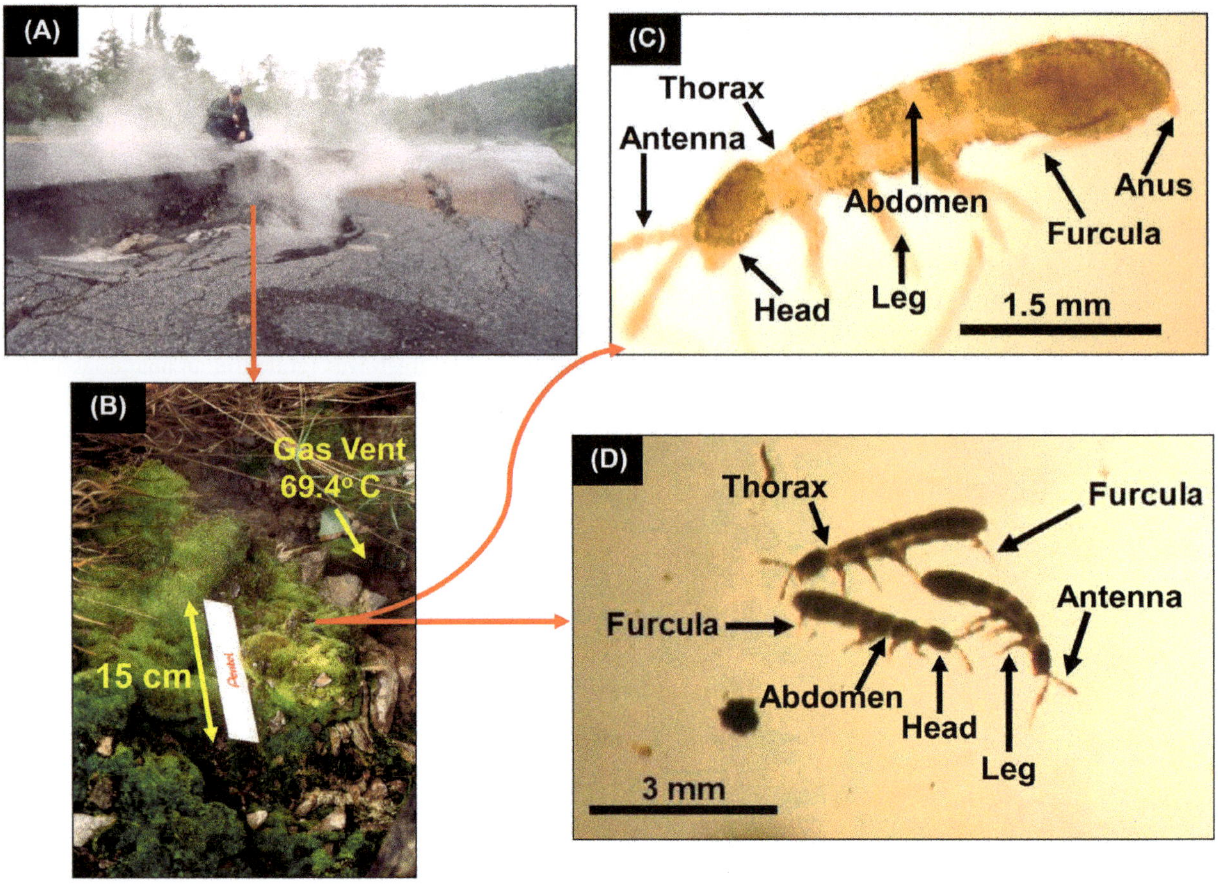

Figure 2.1.11. Microarthropods collected from the Centralia Mine Fire in Pennsylvania on June 22, 2003. (A) Collection locality on an abandoned section of Pennsylvania Route 61 (PA 61). (B) A gas vent with moss growing adjacent to it, about a foot below the surface in a crack in PA 61. The gas was exhaled in pulses every few seconds, and this photo was taken between pulses. The gas temperature at about 5 cm inside the vent varied from 63.7°C to 69.4°C (Table 2.1.1: 6/22/2003, Sample 1) over a 5-min interval on a breezy day. Vegetation utilizes CO_2 from coal combustion for photosynthesis. (C) and (D) Leica DM E compound microscope images (substage illumination, Nikon Coolpix 990 camera) of order Entomobryomorpha springtails, Hexapoda in the class Collembola, extracted with a Berlese funnel from the moss growing adjacent to the gas vent in (B). The abdomen, labeled in the photo, includes all the tagmata behind the thorax. The **furcula**, one of the defining characteristics of springtails, is a tail-like appendage at the end of the abdomen and when folded underneath, is used to spring away from a threat. Photos (A) and (B) by Janet L. Stracher, 2003 and (C) and (D) by Jimmy Wedincamp, 2003.

chemical sensors to canvass their surroundings by touch and smell (Wikipedia, 2017m,n; Saito, 2010, pp. 1–3). Images of mites on the Internet reveal that their colors include brown, red, orange, green, cream, and yellow; in addition, they can be multicolored.

On November 12, 2012, microarthropods were collected from five locations inside St. Ignatius Cemetery. The first location was adjacent to a gas vent and the other four were about every 0.5 m to the south of the vent (Figure 2.1.13). As Table 2.1.1 shows, the microarthropods found are springtails (e.g., Figure 2.1.11), mites, thrips, beetles, and beetle larvae (Figure 2.1.14). The table also shows that the temperatures varied from 49 to 15°C from the gas vent to the furthest collection site, respectively, and that springtails occurred at every location. Figures 2.1.15 and 2.1.16 record the distribution of microarthropods from the collection sites (Figure 2.1.13) as a function of temperature.

Figures 2.1.15 and 2.1.16 record the distribution of two of the three main groups of springtails found in the samples collected from St. Ignatius Cemetery on November of 2012. These members of the class Collembola identified with a Leica DM E compound microscope, but not differentially illustrated in this chapter, are the orders **Poduromorpha** and **Entomobryomorpha**. The springtail order **Symphypleona** was not found. All three orders can be identified by body shape. Poduromorpha have oval-shaped and plump bodies with short legs. Entomobryomorpha, the most common springtails, have slim and elongated bodies, a pronounced furcula, and either short or long legs.

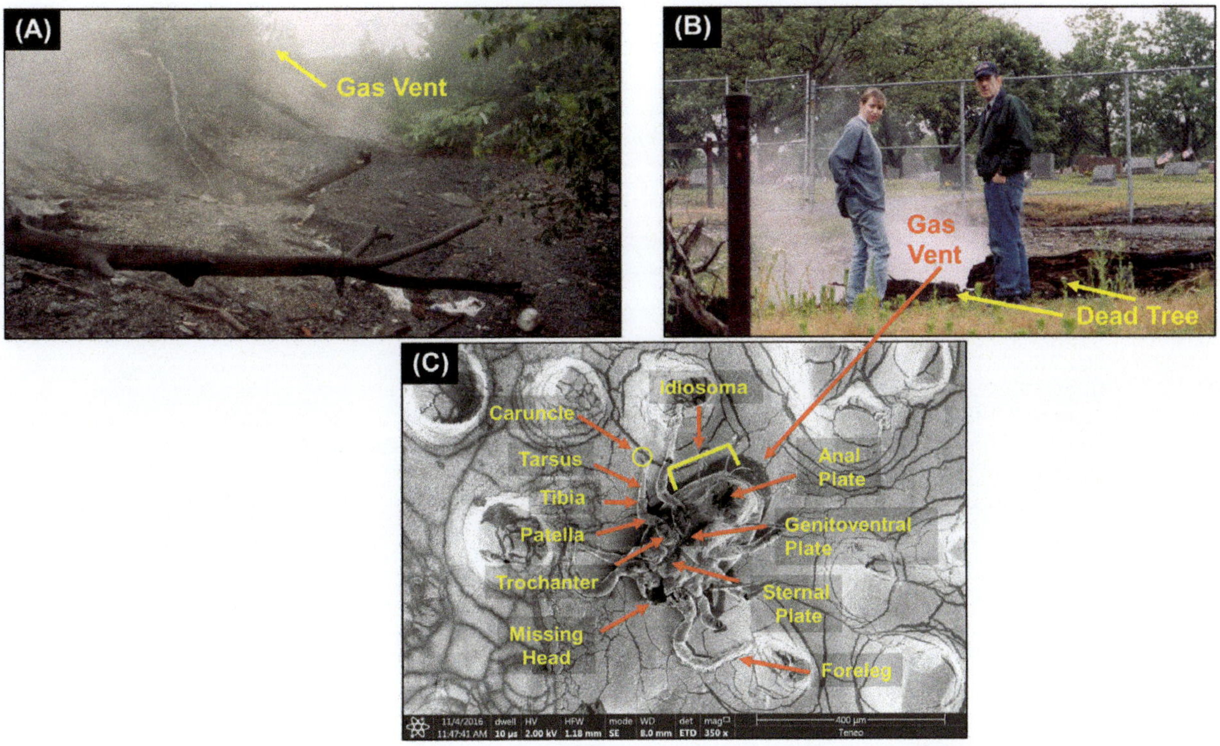

Figure 2.1.12. December 29, 2003, sampling locations for microarthropods at the Centralia Mine Fire in Pennsylvania. (A) An abandoned strip-mining cut (Death Valley) immediately west of abandoned Pennsylvania Route 61 (off to the right side of Figure 2.1.11A). Only springtails (e.g., Figure 2.1.11) were found in vegetation adjacent to a gas vent at the top of Death Valley, where the gas temperature at about 5 cm inside the vent was 65.6°C. (B) Melissa A. Nolter and Glenn B. Stracher at Cemetery Front, outside the fence enclosing St. Ignatius Cemetery. The gas vent is behind the logs and not visible here. Springtails were extracted from a moss sample adjacent to the gas vent. The temperature at about 5 cm inside the vent was 76.7°C. (C) Mites (SEM ventral image, possibly of an Oribatida), Chelicerata in the class Arachnida and subclass Acari, in addition to springtails were found in a sample of moss from Cemetery Front; several meters to the south of the gas vent in (B). The temperature of the moss was 65.6°C. The anatomical structures illustrated here include the **idiosoma**, the posterior end of a mite, and the **caruncle**, a sucker that male mites have at the end of their legs for holding a female during copulation. Leg segments include the trochanter and patella. Also see Table 2.1.1 (12/29/2003, Samples 1–3; and Figure 2.1.14A). SEM notation in the black bar at the bottom of the photo: HV is high voltage; HFW is the horizontal field width; SE is secondary electrons; WD is the working distance from the lens to the object; det (detector) is EDT, an Everhardt-Thornley Detector; Teneo is the model of the SEM manufactured by FEI, now part of Thermo Fisher Scientific. Photos (A) and (B) by Janet L. Stracher, 2003 and (C) by Yelena White and Glenn B. Stracher, 2016.

Symphypleona have nearly spherical bodies and long antennae (Wikipedia, 2017o). Images of springtail body shapes and identification schematics are available on the Internet.

Also recorded in Figures 2.1.15 and 2.1.16 is the distribution of two orders of mites, also not differentially illustrated here, found in the November 2012 samples from St. Ignatius Cemetery. Mites belonging to the order **Mesostigmata** have long legs, setae on their bodies, and one pair of spiracles between the third and fourth pair of legs. Most live in or on soil and soil litter and on plants, where they are predatory and feed on Collembola and plant-feeding pests including spider mites and nematodes. Others are parasitic bloodsuckers and can cause dermatitis in rodents, birds, humans, and other animals (Fain, 1984, pp. 92–93; Koehler, 1999; Wikipedia, 2017m; American Association of Veterinary Parasitologists, 2017). Mites in the order **Oribatida** are also called **moss mites** and **beetle mites**. They are oval-shaped, appear leathery, and have a sclerotized **integument** (outer surface). Oribatida live in soil and on plants and tree trunks. They are important decomposers and consume dead and living plants, fungi, lichens, and carrion. Although not parasitic, some are predatory and feed on living microorganisms including nematodes, while others are tapeworm hosts and may transmit the worms to **ruminants**. Ruminants are herbivorous mammals, including cattle, goats, and deer, with a specialized stomach for plant fermentation and digestion (Rockett, 1980; Wissuwa et al., 2013; Encyclopedia Britannica, 2017d; Wikipedia, 2017m; Merriam Webster Dictionary, 2017; New World Encyclopedia, 2017). Oribatida have a symbiotic relationship with carrion beetles (family **Silphidae**). The mites climb onto and ride on the back of these beetles to carrion, such as a dead

Figure 2.1.13. November 12, 2012, sampling locations for microarthropods at the Centralia Mine Fire in Pennsylvania. The samples were collected inside St. Ignatius Cemetery, west of and adjacent to a travelable section of Pennsylvania Route 61. Plant samples were collected at five different locations with the first one adjacent to a gas vent and the others about every 0.5 m south of the vent; in a slightly zigzag pattern (as plants occurred) in the direction of the yellow arrow. The numbers 1–5 designate the sampling locations and the sample numbers in Table 2.1.1. The distance between the fence posts is about 3 m. Photo by Melissa A. Nolter, 2012; with modifications by Glenn B. Stracher, 2016.

mouse, and eat the maggots and eggs deposited by flies, thereby reducing the competition for food with beetle larvae deposited on the carrion (National Park Service, 2017). Images of mites and identification schematics are available on the Internet.

Thrips

Thrips (Figures 2.1.5 and 2.1.14B), also called thunderflies or thunderbugs, are members of the subphylum Hexapoda, class Insecta, and order **Thysanoptera** (meaning fringed wing). Thrips first appeared in the Triassic (Penney and Jepson, 2014, p. 122), and about 5000 species have been identified. Different species inhabit tropical to arctic ecosystems. Some cause extensive crop damage. They feed on plants, insects, and other thrips, including members of their own species (Kirk, 1995, pp. 21–29). They have a bilaterally symmetric epicuticle that is a wax-ester-lipid sclerite that repels water on the outside and prevents desiccation from within. Adult thrips are needle-like in appearance (long and slender); less than 1 mm long; have setae on the edges of two pair of wings; and are white, yellow, dark brown, black, or translucent (University of California Agriculture and Natural Resources, 2017).

Beetles

Beetles (Figure 2.1.14C) are members of the subphylum Hexapoda, class Insecta, and order **Coleoptera**. They first appeared in the Lower Triassic, and there are about 400,000 species (Wikipedia, 2017a). They consume any type of organic matter including dead organics in any state of decay. The process of consuming fungi, called **fungivory** (also called **mycophagy**), is common among beetles, and they frequently use dung for food and in which to lay

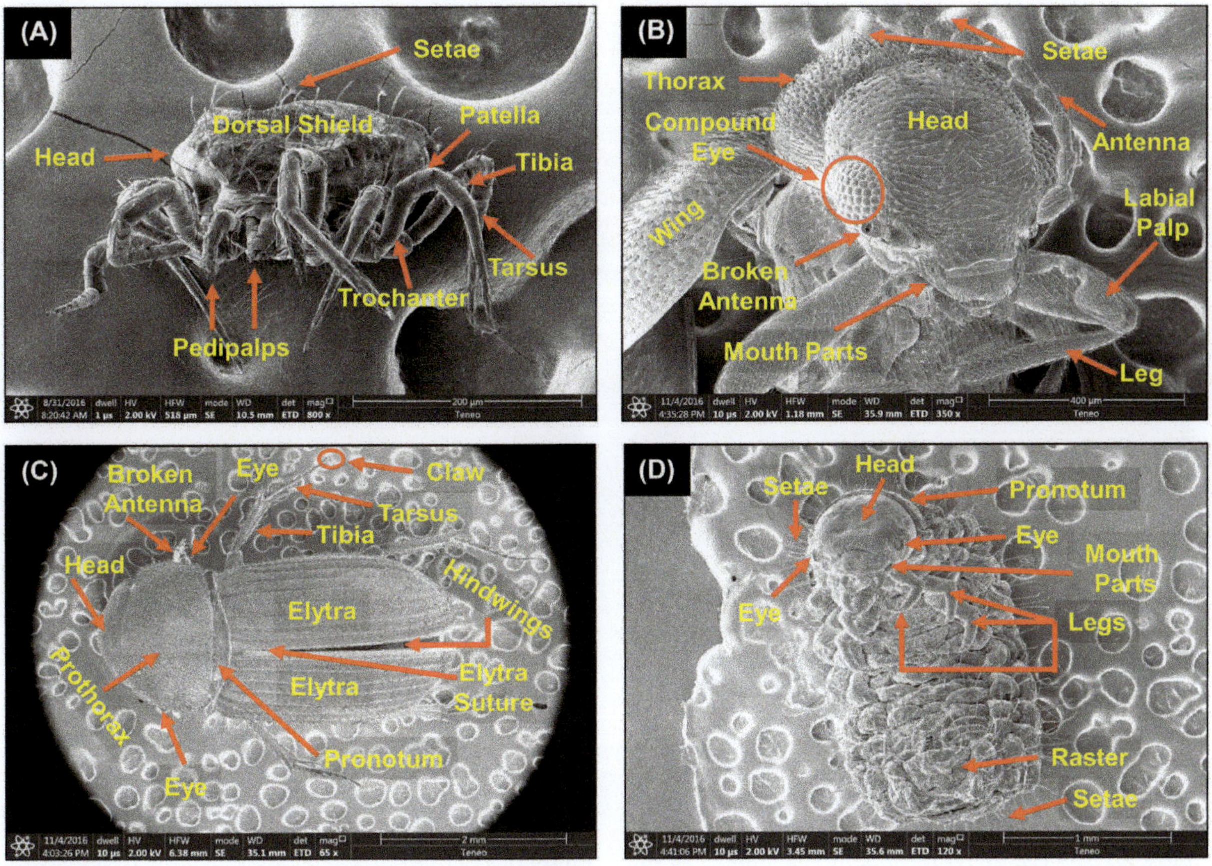

Figure 2.1.14. Microarthropods collected on November 12, 2012, from five locations in St. Ignatius Cemetery at the Centralia Mine Fire in Pennsylvania (Figure 2.1.13). Table 2.1.1 lists those found and shows that the temperatures at these locations varied from 49°C to 15°C from the gas vent to the furthest collection site, respectively. The microarthropods include springtails (e.g., Figure 2.1.11C and D), and only they occurred at all five locations. (A) Mite (location 3 in Figure 2.1.13; also see Figure 2.1.12C). Pedipalps are used to explore the environment. Leg segments include the trochanter and patella. (B) Thrips (location 3 in Figure 2.1.13; also see Figure 2.1.5). The left **labial palp** is a mouth part used for feeding. (C) The adult beetle illustrated here, a Hexapoda insect member of the order Coleoptera, exemplifies one found at the furthest sampling location from the gas vent (location 5 in Figure 2.1.13). Additional such beetles were found at location 3. Anatomical structures pictured here include four of the six legs, all attached to the thorax. The other two are broken off the **prothorax** (front thorax) or bent underneath it. The **pronotum**, also called the **prothoracic shield**, is a hard plate that covers part of the prothorax. The **elytra** are hard or soft forewings attached to the thorax. They are used to protect the underlying hindwings used by species that fly (not all do) when the elytra are raised. The elytra are attached to the flexible, articulated back part of the thorax underneath called the **pterothorax** (comprised of the fused **mesothorax** and **metathorax**) (Wikipedia, 2017a). The hindwings beneath the elytra are barely visible here; near the back of the pterothorax. (D) Image of a beetle larva (location 3 in Figure 2.1.13), also called a grub. Beetle larvae were also found at location 5. An inverted Y, characteristic of beetle larvae, is subtly visible at the arrowhead that points to the grub's head. Also visible is the **raster**, located at the bottom of the last abdominal segment and in front of a grub's anus. It contains the collection of spines, setae, and the tissue between them (Penn State College of Agricultural Sciences, 2017b). The pattern of spines is used for taxonomic classification (University of Georgia Extension, 2017). SEM notation in the black bar at the bottom of the photo: HV is high voltage; HFW is the horizontal field width; SE is secondary electrons; WD is the working distance from the lens to the object; det (detector) is EDT, an Everhardt-Thornley Detector; Teneo is the model of the SEM manufactured by FEI, now part of Thermo Fisher Scientific. Photos by Yelena White and Glenn B. Stracher, 2016.

their eggs. Beetles habituate diverse environments including external and internal plant (e.g., inside trees) and animal remains, fresh water, and marine intertidal zones. They have antennae, three pairs of legs, most have two pairs of wings, and they commonly use camouflage to avoid predation by mimicking the color of the substrate they occupy. Images of beetles on the Internet reveal that their colors include black, brown, tan, orange, yellow, blue, and green; in addition, they can be multicolored.

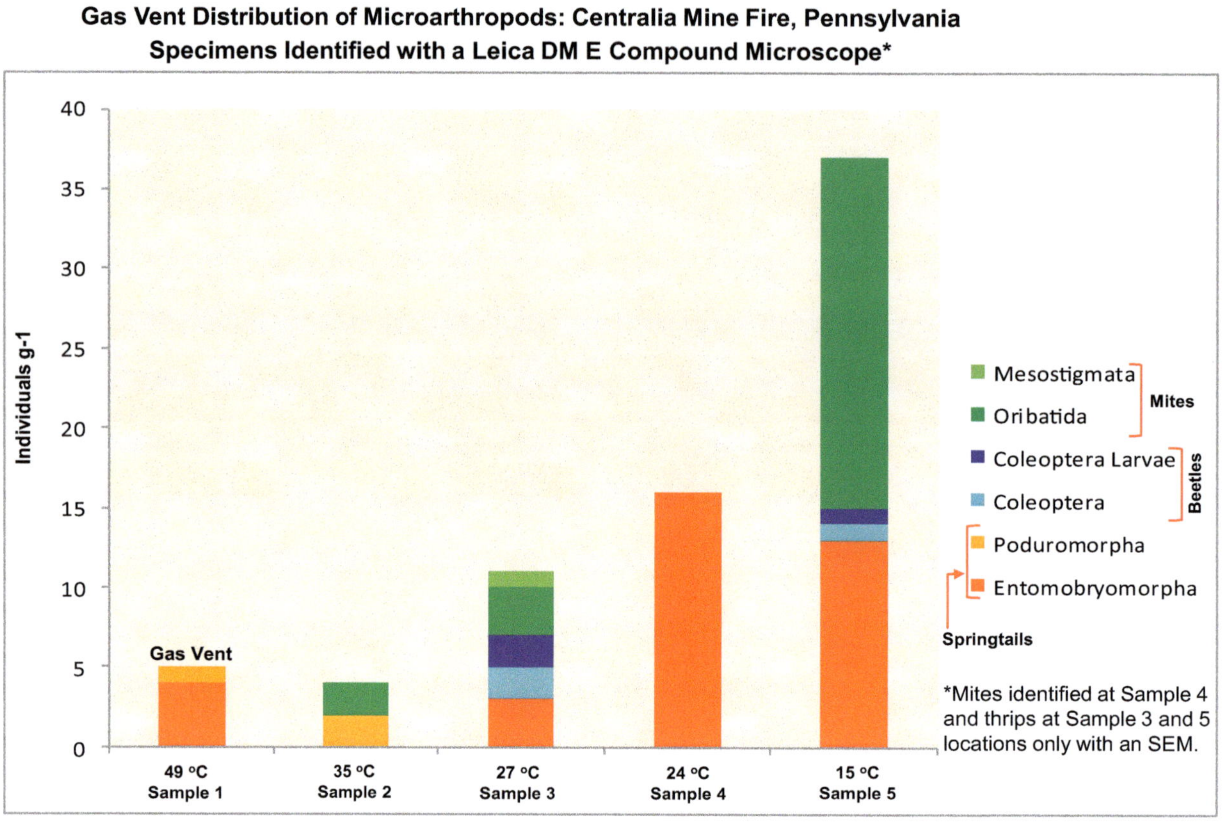

Figure 2.1.15. Bar graph illustrating the number of microarthropods per gm of sample as a function of temperature for Samples 1–5 collected on November 12, 2012, at St. Ignatius Cemetery in Centralia, Pennsylvania (Figure 2.1.13; Table 2.1.1). The microarthropods were identified and counted using a Leica DM E compound microscope before preparation for SEM imaging. Only springtails (e.g., Figure 2.1.11C and D) occurred immediately adjacent to the gas vent; at the highest temperature (49°C; Table 2.1.1: 11/12/2012, Sample 1). The springtail orders Poduromorpha and Entomobryomorpha and the mite orders Mesostigmata and Oribatida are discussed in the text. Mites were found in the Sample 4 and thrips in the Sample 3 and 5 locations (Table 2.1.1) with the SEM, but they are not recorded on the graph because they were not identified with the Leica DM E microscope. Also see Figure 2.1.16. Figure by Breana Simmons and Glenn B. Stracher, 2017.

Some beetles are pests and destroy crops including potatoes, cotton, coconuts, and stored grains, in addition to trees, dried animal skins, and leather. However, some are beneficial and are used for crop and garden-weed pest control (Gullan and Cranston, 2010, pp. 487–488; Wikipedia, 2017a). They are as tiny as the featherwing beetle and as big as the adult goliath beetle (weighing as much as 100 gm) with lengths of about 0.325 μm–11 cm, respectively (Wikipedia, 2017a).

Beetle Larvae

Beetle larvae (Figure 2.1.14D), also called **grubs**, are Hexapoda insect members of the order Coleoptera. Some grubs have legs, and those that do not are said to be **apodus**. Grubs are elongate and cylindrical or flattened and they may be curled into a c-shape. They have clearly discernible mouthparts and a hardened head (often brown) with an inverted Y extending toward the top (BugwoodWiki, 2017).

A larva is the second life cycle of an insect. The first is the egg from which it hatched. Some larvae feed on the food source in which they hatched while others are predatory or feed externally on vegetation (Wikipedia, 2017a). Once an egg hatches, the larva undergoes various **instars**, the developmental stages of metamorphism between insect molts. The last instar, which is the third life cycle between the larva and adult cycles, is the **pupal stage**. During this time, the insect is a **pupa** (pl. pupae), a nonfeeding and relatively immobile insect in a cocoon. The pupa develops the wings and other features of a sexually mature adult before it leaves its cocoon. The adult insect (**imago**) is the fourth life cycle, called the **imaginal stage** (Wikipedia, 2017a; Arizona State University School of Life Sciences, 2017; Encyclopedia of Life, 2017).

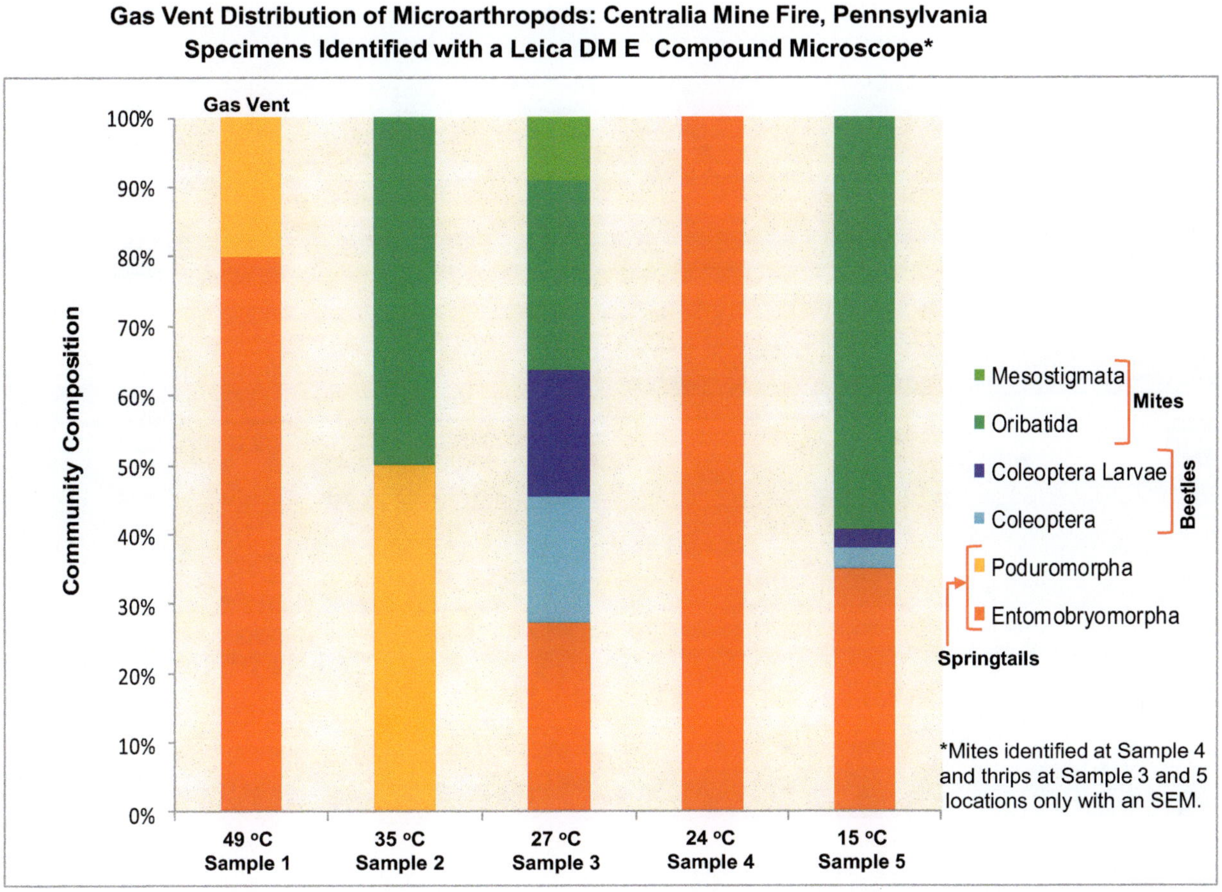

Figure 2.1.16. Bar graph illustrating the percent composition of each type of microarthropod relative to all other microarthropods as a function of temperature for Samples 1–5 collected on November 12, 2012, at St. Ignatius Cemetery in Centralia, Pennsylvania (Figure 2.1.13; Table 2.1.1). The microarthropods were identified and counted using a Leica DM E compound microscope before preparation for SEM imaging. Only springtails (e.g., Figure 2.1.11C and D) occurred immediately adjacent to the gas vent; at the highest temperature (49°C; Table 2.1.1: 11/12/2012, Sample 1). The springtail orders Poduromorpha and Entomobryomorpha and the mite orders Mesostigmata and Oribatida are discussed in the text. Mites were found in the Sample 4 and thrips in the Sample 3 and 5 locations (Table 2.1.1) with the SEM, but they are not recorded on the graph because they were not identified with the Leica DM E microscope. Also see Figure 2.1.15. Figure by Breana Simmons and Glenn B. Stracher, 2017.

Healy Mine Fire, Alaska

The subbituminous C rank coal at Healy (Figure 2.1.3) formed during the Tertiary Period. There is no written record stating how and when the Healy Mine Fire started. One possibility is spontaneous combustion caused by the solar heating of coal rubble and outcrops exposed on south-facing slopes during long summer days. Coal fires are known to have occurred near Healy before coal mining began there in the early 1940s (Prakash et al., 2013).

On August 23, 2003, microarthropods were collected from vegetation at two locations at the Healy Mine Fire (Figure 2.1.17; Table 2.1.1: 8/23/2003). Springtails (e.g., Figure 2.1.11) were extracted from charred vegetation collected adjacent to a gas vent where the temperature at about 5 cm inside the vent was 67°C. Aphids (Figure 2.1.18) were extracted from vegetation about 2 m downslope from the gas vent, and the soil temperature there was 12°C.

Aphids

Aphids are members of the subphylum Hexapoda, class Insecta, order **Hemiptera** (formerly order **Homoptera**) and the family **Aphididae**. Also called plant lice, blackflies, greenflies, whiteflies, and ant cows, they occur worldwide and about 4400 species have been identified. Aphids are white, green, yellow, black, brown, or multicolored

Figure 2.1.17. August 23, 2003, sampling locations for microarthropods at the Healy Mine Fire in Alaska. (A) Two gas vents, V1 and V2–P1, at the Healy fire. No vegetation occurred adjacent to V1. Charred vegetation, from which springtails (e.g., Figure 2.1.11) were extracted, occurred adjacent to vent V2–P1. The temperature at about 5 cm inside V2–P1 was 67°C. Field assistant John Styers is at the base of the slope. (B) Charred vegetation, adjacent to gas vent V2–P1, from which springtails were extracted. (C) and (D) Vegetation sample V2–P3, about 2 m downslope from gas vent V2–P1. Aphids (Figure 2.1.18) were extracted from this vegetation. The soil temperature here was 12°C. Also see Table 2.1.1 (8/23/2003). Photos by Anupma Prakash, 2003.

and are 1–10 mm in length. Most have compound eyes, and they have pear-shaped bodies with straw-like mouthparts, called **stylets**, used for feeding by sucking up the juices (sap) of leaves, flowers, or plant stems (American Orchid Society, 2017; Missouri Department of Conservation, 2017; Wikipedia, 2017p).

Aphids may reproduce sexually or unlike most insects, by a form of asexual reproduction called parthenogenesis, giving birth by **viviparity** (i.e., to live young). Some species produce winged offspring that will fly to other plants when a host plant becomes crowded or wanes. Because aphids breed quickly and in great numbers, they do a lot of damage to garden plants and farming crops, leaving them with mottled, yellowed, and curled leaves. Pesticides are used to eradicate them (Missouri Department of Conservation, 2017; Townsend, 2017; Wikipedia, 2017p; Encyclopedia Britannica, 2017e).

All aphids have a pair of **cornicles**, also called **siphuncules** or **siphunculi** (sing. siphuncule) (Figure 2.1.18). These are a pair of hollow, backward-pointing tubes used for defense. They are located on the last dorsal abdominal segment of the abdomen. When under attack, muscles at the base of the cornicles contract and secrete fluids, including an alarm **pheromone (trans-beta-farnesene)** that hardens quickly into a solid plate called **cornicle wax**. The wax is used for defense since it is distasteful to many predators including beetles, wasps, parasites, and parasitoids. It also warns other aphids, within detection distance, of danger (Heathcote, 1988; Wu et al., 2010; Insect Identification Key, 2017; Amateur Entomologists' Society, 2017; Missouri Department of Conservation, 2017; Townsend, 2017). Pheromones are chemicals released by an organism to communicate with members of its own species (Kimball, 2017b). Another anatomical structure in aphids is the **cauda**, a conical protrusion above the rectal orifice. It secrets **honeydew**, a fluid comprised of excess plant sugars used as food by ants and other insects that may protect aphids from predation. A fungus called **sooty mold** grows on honeydew deposited on branches and leaves, turning it black. When objects such as automobiles are parked under trees infested with aphids, honeydew that falls from the aphid's cauda leaves spots on glass and other surfaces (Missouri Department of Conservation, 2017; Wikipedia, 2017p; Townsend, 2017).

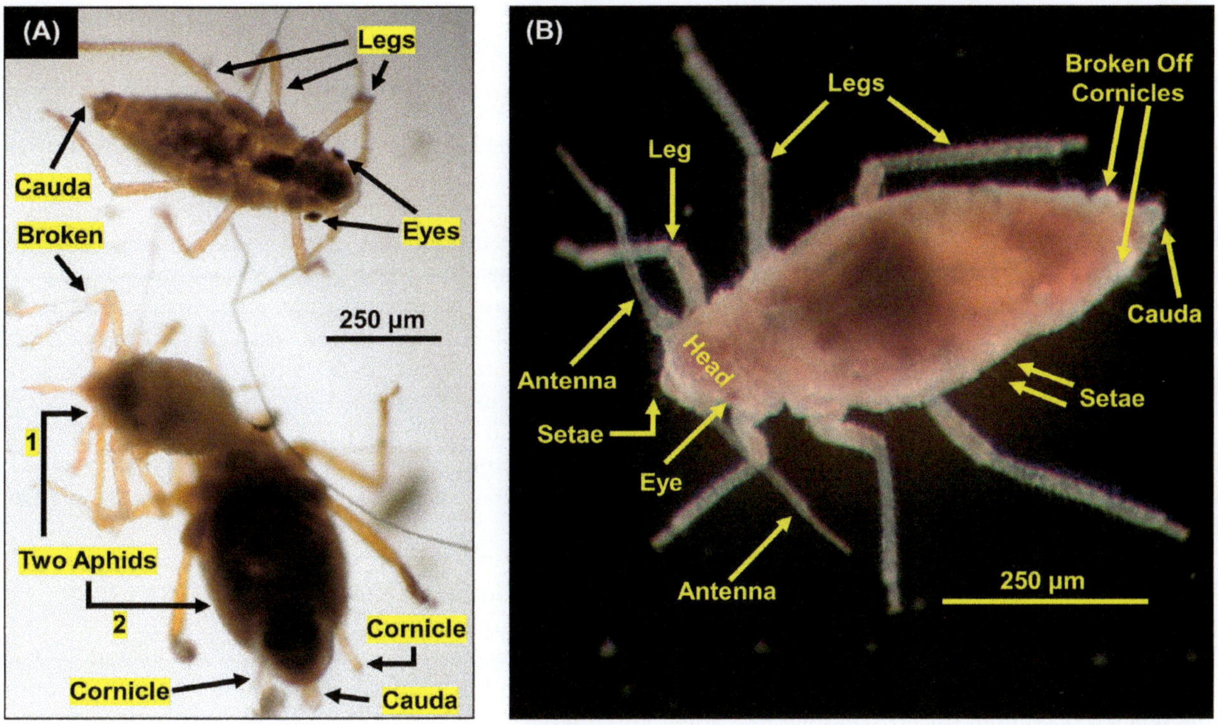

Figure 2.1.18. Aphids from the Healy Mine Fire in Alaska. These Hexapoda insect members of the order Hemiptera (formerly order Homoptera) and family Aphididae were extracted from vegetation sample V2–P3 (Table 2.1.1: 8/23/2003), about 2 m downslope from gas vent V2–P1 illustrated in Figure 2.1.17. The soil temperature at V2–P3 was 12°C. (A) Image of three aphids acquired with a Leica DM E compound microscope equipped with substage illumination and a Nikon Coolpix 990 camera. There are two aphids at the bottom of the image. One of these aphids is partially beneath the other; some legs of aphid 1 overlap those of aphid 2, as seen above the label "Two Aphids" in the bottom left-side of the figure. There are also legs that are stuck together and broken off from at least one of the aphids or an unidentified microarthropod. Aphids have a fused thorax and abdomen. The cornicles secrete chemicals for defense and to warn other aphids of predation. The cauda secrets honeydew, a fluid comprised of plant sugars that some insects feed on. (B) Image of an aphid acquired with an Olympus SZX 16 stereo microscope equipped with overhead and oblique lighting and a built-in camera. The cauda is present, but the cornicles are broken off. Photo (A) by Jimmy Wedincamp, 2003 and (B) by John Shields, 2016.

Discussion

The occurrence of microarthropods at coal fires was first reported by Stracher et al. (2013) and Simmons et al. (2013). This chapter is the first study in which a detailed description of the arthropods is followed by the identification, description, and illustration of microarthropods as a dynamic component of the food web at or near active coal-fire gas vents. Field and laboratory studies by other authors about surface- and soil-dwelling organisms focus on population densities before and after forest or grassland fires, the heat tolerance of microarthropods exposed to such fires, and recovery rates after a fire. Depending on the study, springtails and mites present prior to burning exhibited a variable tolerance to heat as well as variable post-fire recovery rates (Barratt et al., 2006; Malmström, 2008; Hutchins et al., 2011; Gongalsky et al., 2012). A study conducted by Van Dooremalen et al. (2013) of five surface-dwelling and four soil-dwelling species of springtails revealed that the surface-dwelling species were more heat tolerant than the soil-dwelling.

Gas vents and fissures that emit water vapor and CO_2 during coal combustion can be auspicious environments for the growth of vegetation adjacent to or near these features of many coal fires. As shown in this study, the vegetation and its proximity to gas exhaled into the atmosphere create a living habitat for select microarthropods. Such fauna may be opportunists because of a decrease in competition from other organisms that are more sensitive to coal-fire gas, combustion-related temperatures, or both. In addition, the gas and elevated temperatures at or near gas vents appear to create an oasis for select microarthropods by warding off larger predators that would consume the vegetation. Yet, an ongoing mystery is the ability of the fauna identified to survive in an oxygen-depleted and CO_2- rich environment when for respiration, they need to intake oxygen and release CO_2. The gas exhaled and collected from some of the vents typically includes up to 45 compounds and some, including toluene (C_7H_8), benzene(C_6H_6), and CO, are toxic to humans and other animals (Stracher and Nolter, 2013b). The microarthropods we identified can tolerate these toxins, but it is unknown for how long.

Unfortunately, it was not possible to collect additional samples at the Healy Mine Fire. However, the plot illustrated in Figure 2.1.15 for the November 12, 2012, Centralia samples reveals that the abundance of microarthropods increases with distance from the gas vent. Table 2.1.1 shows that springtails were found at all sampling locations except for Healy Sample V2–P3, collected about 2 m downslope from gas vent V2–P1. This table and Figures 2.1.15 and 2.1.16 show that only springtails occurred adjacent to gas vents, the highest temperature sampling locations, suggesting that of all the microarthropods identified, springtails are the most tolerant of coal-fire gas and the higher temperatures. Figures 2.1.15 and 2.1.16 also illustrate that where Sample 2 was collected in November, about 0.5 m from the gas vent (Sample 1), only springtails and mites occurred in the vegetation, suggesting that these microarthropods have a better tolerance than others to the coal-fire gas and temperature near the gas vent. These figures also reveal some unexplained phenomenon. One such example is the disappearance at the Sample 2 location of Entomobryomorpha and the appearance of Poduromorpha and then at the Sample 3–5 locations, the disappearance of Poduromorpha and reappearance of Entomobryomorpha. As another example, there are less Oribatida in Sample 3 than Sample 5.

Mites were identified from all the November collecting locations in Centralia (Table 2.1.1) except at the Sample 1 gas vent. They are not recorded for Sample 4 in Figures 2.1.15 or 2.1.16 and neither are thrips for Samples 3 and 5 because although identified with the SEM, they were not identified with the Leica DM E microscope. The beetles appear only in Samples 3 and 5 and are less numerous in Sample 5, where the Oribatida reach a maximum for all sampling locations. This suggests that the beetles are less competitive for food as the number of Oribatida increase, or more likely, that the Oribatida picked up and carried on the backs of the beetles to the Sample 5 location outnumbered the beetles that migrated there. The second proposal is notably reasonable because there are typically numerous Oribatida carried on the back of a beetle as it moves while foraging for food.

Additional research about microarthropods at coal fires is necessary to firmly establish the relationship between these fauna; coal-fire gas chemistry; and gas, soil, and vegetation temperatures at and radially outward from gas vents or other hot spots at coal fires. The characteristics of arthropods and collection procedures described earlier as well as the study presented here will facilitate the acquisition and interpretation of the data needed to establish such relationships.

Acknowledgments

The authors thank Janet L. Stracher, Joseph T. Nolter, and John Styers for their assistance with field work. Rudiger Gens created the maps of the Centralia and Healy mine fires. Most of the images of the microarthropods were acquired at the Electron Microscopy Laboratory at the University of Georgia. We also thank the Earth and Planetary Sciences team at Elsevier for their review of the manuscript. This work was supported by East Georgia State College and the Stracher-Griffin Fund (#241100 A000401).

Important Terms

accommodation
Acari
Acarina
acarology
aedeagus
alimentary canal
lepidopterans
Malpighian tubules
maxillary gland
Mesostigmata
mesothorax
metabolism

Animalia
Annelida
antennal gland
Aphididae
aphids
apodus
appendage dimorphism
apolysis
Arachnida
Arthropoda
arthropods
beetle larvae (sing. larva)
beetle mites
beetles
Berlese funnel
Berlese trap
bilateral symmetry
bioturbation
biramous
book gills
book lungs
Burgess Shale
caruncle
Cathedral Escarpment
cauda
Centralia Mine Fire
cephalon
cephalothorax
cerebral ganglion
Chelicerata
chemical fixatives
chitin
Coleoptera
cornicles (sing. cornicle)
cornicle wax
coxal gland
common name
compound eyes
cornea
Crustacea
Decapoda
dioecious
ecosystems
ecdysis
ecdysterone ($C_{27}H_{44}O_7$)
metamorphosis
metamere
metameres
metathorax
microarthropods
midgut
mites
molting
moss mites
mycophagy
Myriapoda
nauplius (pl. nauplii)
nervous system
ocelli (sing. ocellus)
ocellus lens
ommatidium (pl. ommatidia)
open circulatory system
Oribatida
osmium tetroxide (OsO_4)
ovoviparous
ovum (pl. ova)
paired appendages
parthenogenesis
pedipalps (sing. pedipalp)
phenotypic
pheromone
phosphate-buffered saline solution
photosynthesis
phylogeny
phytophages (sing. phytophage)
podomere
Poduromorpha
polysaccharide
procuticle
Pronotum
prosoma
protandric simultaneous hermaphrodites (sing. hermaphrodite)
prothoracic shield
prothorax
pterothorax
pupa (pl. pupae)
pupal stage
pygidium
raster
reproduction

Edicaran fauna
elytra
Entomobryomorpha
epicuticle
epidermal
excretion
exoskeleton
eye spots
facet
feeding
foregut
forewings
fungivory
furcula
ganglia
gills
glutaraldehyde ($C_5H_8O_2$)
gonads
green gland
grubs (sing. grub)
Healy Mine Fire
Hemiptera
hemocoel
hemolymph
hermaphroditic
hexamethyldisilizane ($C_6H_{19}NSi_2$)
Hexapoda
hindgut
hindwings
Homoptera
honeydew
ichnofossil
idiosoma
imaginal stage
imago
Insecta
instars
integument
internal fertilization
invertebrates
jointed appendages
labial palp
Lagerstätte (pl. Lagerstätten)
Laurentian
legs
Lepidoptera
respiration
respiratory gases
respiratory pigments
ruminants (sing. ruminant)
saprophages (sing. saprophage)
scanning electron microscope (SEM)
sclerites (sing. sclerite)
sclerotin
sclerotized
scorplings (sing. scorpling)
sequential hermaphrodites (sing. hermaphrodite)
setae (sing. seta)
Silphidae
simultaneous hermaphrodites (sing. hermaphrodite)
siphuncules (sing. siphuncle)
siphunculi (sing. siphuncule)
smoldering
social insects
sooty mold
spermatophores
spinneret
spiracles (sing. spiracle)
springtails
stalk
stylets
symbiont
symbiosis
symbiotic
Symphypleona
tagma
tagmata
tagmosis
thorax
thrips
Thysanoptera
tracheae (sing. trachea)
tracheoles (sing. tracheole)
trans-beta-farnesene
Trilobita
Tullgren funnel
ungues
uniramous
viviparity
wax-ester lipid
wings

References

Alexander, J.O., 2012. Arthropods and Human Skin. Springer, 424 p.

Allaby, M., 2003. A Dictionary of Zoology. Oxford University Press, 597 p.

Amateur Entomologists' Society, 2017. https://www.amentsoc.org/insects/glossary/terms/aedeagus (accessed June 2017).

American Association of Veterinary Parasitologists, 2017. http://www.aavp.org/wiki/arthropods/arachnids/mesostigmata (accessed October 2017).

American Orchid Society, 2017. Aphids. http://www.aos.org/orchids/orchid-pests-diseases/aphids.aspx (accessed October 2017).

Angelini, D.R., Kaufman, T.C., 2005. Insect appendages and comparative ontogenetics. Dev. Biol. 286 (1), 57–77.

Arizona State University School of Life Sciences, 2017. Beetle Reproduction, Ask a Biologist. https://askabiologist.asu.edu/home (accessed October 2017).

Baeza, J.A., 2006. Testing three models on the adaptive significance of protandric simultaneous hermaphroditism in a marine shrimp. Evolut. Int. J. Organ. Evolut. 60 (9), 1840–1850.

Barratt, B.I.P., Tozer, P.A., Wiedemer, R.L., Ferguson, C.M., Johnstone, P.D., 2006. Effect of fire on microarthropods in New Zealand indigenous grassland. Rangel. Ecol. Manag. 59 (4), 383–391 Elsevier.

Boxshall, G.A., Huys, R., 1989. New Tantulocarid, *Stygotantulus stocki*, parasitic on Harpacticoid copepods, with an analysis of the phylogenetic relationships within the Maxillopoda. J. Crustac. Biol. 9, 126–140.

BugwoodWiki, 2017. Coleoptera – Beetles. https://wiki.bugwood.org/Coleoptera (accessed October 2017).

Caron, J.-B., Vannier, J., 2016. *Waptia* and the diversification of brood care in early arthropods. Curr. Biol. 26 (1), 69–74.

Center for Disease Control and Prevention, 2017. https://www.cdc.gov/plague/history/index.html (accessed June 2017).

Chandra, G., 2017. Vision in Arthropoda, Zoology and Entomology. IASZoology.com. http://www.iaszoology.com/arthropod-eye-vision (accessed October 2017).

Cody, G.D., Gupta, N.S., Briggs, D.E.G., Kilcoyne, A.L.D., Summons, R.E., Kenig, F., Plotnick, R.E., Scott, A.C., 2011. Molecular signature of chitin-protein complex in Paleozoic arthropods. Geology 39, 255–258.

Cranshaw, W., Redak, R., 2013. Bugs Rule!: An Introduction to the World of Insects. Princeton University Press, 496 p.

Cronin, T.W., 1986. J. Crustac. Biol. 6 (1), 1–23.

Dahms, H.-U., Fornshell, J.A., Fornshell, B.J., 2006. Key for the identification of crustacean nauplii. Org. Divers. Evol. 6 (1), 47–56, Elsevier.

Daley, A.C., Drage, H.B., 2016. The fossil record of ecdysis, and trends in the moulting behaviour of trilobites. Arthropod Struct. Dev. 45 (2), 71–96.

DeSalle, R., 2001. How Is Bug Blood Different from Our Own? Scientific American. https://www.scientificamerican.com/article/how-is-bug-blood-differen (accessed June 2017).

Dinan, L., Lafont, R., 2006. Effects and applications of arthropod steroid hormones (ecdysteroids) in mammals. J. Endocrinol. 191, 1–8.

Drenkow, D., 2004. The Diversity of the World of Life: Arthropods. http://douglasdrenkow.com/theworldoflife/systematics/arthropods.htm (accessed June 2017).

Dunlop, J.A., Scholtz, G., Selden, P.A., 2013. Water-to-land transition. In: Minelli, A., Boxshall, G., Fusco, G. (Eds.), Arthropod Biology and Evolution, Molecules, Development, Morphology. Springer, pp. 417–440.

Dutta, P.K., Dutta, J., Tripathi, V.S., 2004. Chitin and Chitosan: chemistry, properties and applications. J. Sci. Ind. Res. 63, 20–31.

Edgecombe, G.D., 1998. Arthropod Fossils and Phylogeny. Columbia University Press. 347 p.

Edgecombe, G.D., Legg, D.A., 2014. Origin and evolution of arthropods, paleontology, 57 (Part 3), 457–468. http://onlinelibrary.wiley.com/doi/10.1111/pala.12105/pdf. (accessed October 2017).

Ehrlich, H., Rigby, J.K., Botting, J.P., Tsurkan, M.V., Werner, C., Schwille, P., PetrášEk, Z., Pisera, A., Simon, P., Sivkov, V.N., Vyalikh, D.V., Molodtsov, S.L., Kurek, D., Kammer, M., Hunoldt, S., Born, R., Stawski, D., Steinhof, A., Bazhenov, V.V., Geisler, T., 2013. Discovery of 505-million-year old chitin in the basal demosponge *Vauxia gracilenta*. Sci. Rep. 3 (3497), 1–24. http://www.ncbi.nlm.nih.gov/pubmed/24336573 (accessed October 2017).

Elzinga, R.J., 2003. Fundamentals of Entomology, sixth ed. Prentice Hall, Upper Saddle River, New Jersey. 512 p.

Encyclopedia Britannica, 2017a. Book Lung. https://www.britannica.com/science/book-lung (accessed May 2017).

Encyclopedia Britannica, 2017b. Coxal Gland. https://www.britannica.com/science/coxal-gland (accessed June 2017).

Encyclopedia Britannica, 2017c. Arthropods. https://www.britannica.com/animal/arthropod (accessed June 2017).
Encyclopedia Britannica, 2017d. Mite. https://www.britannica.com/animal/oribatid-mite (accessed October 2017).
Encyclopedia Britannica, 2017e. Aphid. https://www.britannica.com/animal/aphid (accessed October 2017).
Capinera, J.L. (Ed.), 2008. Encyclopedia of Entomology. Springer, 4346 p.
Encyclopedia of Life, 2017. Coleoptera. http://eol.org (accessed October 2017).
Fain, A., 1984. Diseases of Man. In: Nutting, W.B. (Ed.), Mammalian Diseases and Arachnids: Volume II, Medico-Veterinary, Laboratory, and Wildlife Diseases, and Control. CRC Press. 304 p.
Fairbairn, D.J., 2013. Odd Couples: Extraordinary Differences between the Sexes in the Animal Kingdom. Princeton University Press. 312 p.
Fortey, R., 2004. The lifestyles of the trilobites. 92 (5), 446–453.
Giardina, B., Coletta, M., Zolla, L., Brunori, M., 1985. Oxygen transport proteins: a unitarian view based on thermodynamic kinetic and stereochemical considerations. In: Lamy, J., Truchot, J.-P., Gilles, R. (Eds.), Respiratory Pigments in Animals. Springer, pp. 125–140.
Glaessner, M.F., 1958. New fossils from the base of the Cambrian in south Australia. Trans. Roy. Soc. S. Aust. 81, 185–188.
Gongalsky, K., Malmström, A., Zaitsev, A., Shakhab, S., Bengtsson, J., Persson, T., 2012. Do burned areas recover from inside? An experiment with soil fauna in a heterogeneous landscape. Appl. Soil Ecol. 59, 73–86.
Gould, S.J., 1989. Wonderful Life: The Burgess Shale and the Nature of History, Norton. 347 p.
Gullan, P.J., Cranston, P.S., 2010. The Insects: An Outline of Entomology, fourth ed. Wiley-Blackwell. 565 p.
Gupta, N.S., Kilcoyne, A.L.D., Briggs, D.E.G., Summons, R.E., Cody, D.G., 2010. Identification of Chitin Preserved in a Pennsylvanian Age Fossil Scorpion Cuticle, Lunar and Planetary Institute, Astrobiology Science Conference, League City, Texas. Abstract No. 5423. http://www.lpi.usra.edu/meetings/abscicon2010/pdf/5423.pdf (accessed October 2017).
Hadley, N.F., 2008. Cuticular lipids of terrestrial plants and arthropods: a comparison of their structure, composition, and waterproofing function. Biol. Rev. 56 (1), 23–47.
Halliday, R.B., Oconnor, B.M., Baker, A.S., 1997. Global Diversity of Mites. In: Raven, P.H. (Ed.), Nature and Human Society: The Quest for a Sustainable World. National Academy Press, Washington, D.C., pp. 192–203.
Hammond, P.M., 1976. A review of the genus Anotylus C.G. Thomson: (Coleoptera: Staphylinidae). Bull. Br. Mus. Nat. Hist. Entomol. 33, 139–187.
Hartmeyer, S., 1998. Carnivory in *Byblis* revisited II: the phenomenon of symbiosis in insect trapping plants, carnivorous plant newsletter. Int. Carniv. Plant Soc. 27 (4), 110–113.
Heathcote, G.D., 1988. A Defense Mechanism of Aphids: Can We Turn it against Them? Suffolk Natural History, the Suffolk Naturalists' Society, vol. 24. Suffolk Biodiversity Information Service, Ipswich, Suffolk, United Kingdom, pp. 35–38.
Heming, B.S., 2015. Arthropoda, Canadian Encyclopedia. http://www.thecanadianencyclopedia.ca/en (accessed October 2017).
Heraty, J., Hawks, D., 1998. Hexamethyldisilazane – a chemical alternative for drying insects. Entomol. News 109 (5), 369–374.
Hopkins, S.P., 1997. Biology of the Springtails (Insecta: Collembola). Oxford University Press. 340 p.
Huang, B.Q., Yeung, E.C., 2015. Chemical and physical fixation of cells and tissues: an overview. In: Yeung, E.C.T., Stasolla, C., Sumner, M.J., Huang, B.Q. (Eds.), Plant Microtechniques and Protocols. Springer. 576 p.
Hughes, N.C., 2003. Trilobite tagmosis and body patterning from morphological and developmental perspectives. Integr. Comp. Biol. 43 (1), 185–206.
Hughes, L.C., 2017. Biological Electron Microscopy. http://www.biologicalelectronmicroscopy.com (accessed August 2017).
Hutchins, M.W., Reynolds, B.C., Patch, S.P., 2011. Prescribed fire and the abundance of soil microarthropods in Northeast Georgia. SE. Nat. 10 (3), 489–500 Eagle Hill Institute.
Insect Identification Key, 2017. http://www.knowyourinsects.org/Aphids.html (accessed October 2017).
Smithsonian Institution, 2017. Department of Paleobiology, the Burgess Shale. http://paleobiology.si.edu/index.html (accessed February 2017).
Integrated Taxonomic Information System, 2016. http://www.itis.gov (accessed October 2017).
Jain, K., 2016. Forms of Respiration Among Arthropods, Biology Discussion. http://www.biologydiscussion.com/invertebrate-zoology/arthropods/forms-of-respiration-among-arthropods/33558 (accessed May 2017).
Jockusch, E.L., Nagy, L.M., 1997. Insect evolution: how did insect wings originate? Curr. Biol. 7 (6), R358–R361.
Kaldari, R., 2017. Wikimedia Commons, Photos of Spiders by Ryan Kaldari. https://commons.wikimedia.org/wiki/Category:Photos_of_spiders_by_Ryan_Kaldari (accessed May 2017).

Kansas Geological Survey, 2017. Fossil Trilobites. http://www.kgs.ku.edu/Extension/fossils/trilobite.html (accessed October 2017).
Kimball, J.W., 2017a. Kimball's Biology Pages, the Compound Eye. The Saylor Foundation. http://www.biology-pages.info/C/CompoundEye.html. http://www.saylor.org (accessed October 2017).
Kimball, J.W., 2017b. Kimball's Biology Pages, Pheromones. The Saylor Foundation. http://www.biology-pages.info/P/Pheromones.html (accessed October 2017).
Kirk, W.D.J., 1995. Feeding behavior and nutritional requirements. In: Parker, B.L., Skinner, M., Lewis, T. (Eds.), Thrips Biology and Management. Springer, Boston, Massachusetts, pp. 21–29.
Klug, C., Sculz, H., De Baets, K., 2009. Red devonian trilobites with green eyes from Morocco and the silicification of the trilobite exoskeleton. Acta Palaeontol. Pol. 54 (1), 117–123 Institute of Paleobiology, Polish Academy of Sciences.
Koehler, H.H., 1999. Predatory mites (Gamasina, Mesostigmata). Agric. Ecosyst. Environ. 74 (1–3), 395–410.
Kohr, E., 2014. Chitin, second ed. Elsevier. 158 p.
Krapp, H.G., 2009. Ocelli. Curr. Biol. 19 (11), R435–R437.
Lamsdell, J.C., Briggs, D.E.G., Liu, H.P., Witzke, B.J., McKay, R.M., 2015. The oldest described eurypterid: a giant Middle Ordovician (Darriwilian) megalograptid from the Winneshiek Lagerstätte of Iowa. BMC Evol. Biol. 15 (169), 1–31.
Lane, R.P., 1993. Introduction to the arthropods. In: Lane, R.P., Crosskey, R.W. (Eds.), Medical Insects and Arachnids. Springer Science + Business Media Dordrecht, pp. 30–46.
Levin, H.L., Smith, M.S., 2014. Laboratory Studies in Earth History, tenth ed. McGraw Hill Education. 283 p.
Lin, J.P., Gon, S.M., Gehling, J.G., Babcock, L.E., Zhao, Y.L., Zhang, X.L., Hu, S.X., Yuan, J.L., Yu, M.Y., Peng, J., 2006. A Parvancorina-like arthropod from the Cambrian of South China. Hist. Biol. 18, 33–45.
Louisiana State Arthropod Museum, 2016. http://www.lsuinsects.org (accessed July 2016).
Malmström, A., 2008. Temperature tolerance in soil microarthropods: simulation of forest-fire heating in the laboratory. Pedobiologia 51 (5-6), 419–426 Elsevier.
Martin, J.W., Davis, G.E., 2001. An updated classification of the recent Crustacea, natural history museum of Los Angeles county. Science 39, 124.
Martin, J.W., Olesen, J., Høeg, J.T., 2014. Introduction (p. 1-7) and The Crustacean Nauplius (p. 8–16). In: Martin, J.W., Olesen, J., Høeg, J.T. (Eds.), Atlas of Crustacean Larvae. Johns Hopkins University Press, Baltimore, Maryland. 369 p.
Merriam Webster Dictionary, 2017. Oribatid. https://www.merriam-webster.com/dictionary/oribatid (accessed October 2017).
Meyer, J., 2016. General Entomology, Collembola. North Carolina State University. https://www.cals.ncsu.edu/course/ent425/index.html (accessed October 2017).
Meysman, F.J.R., Middelburg, J.J., Help, C.H.R., 2006. Bioturbation: a fresh look at Darwin's last idea. Trends Ecol. Evol. 21 (12), 688–695.
Minor, M., 2017. Soil Mites and Other Animals. Ecology Group, Massey University. http://www.massey.ac.nz/~maminor/mites.html (accessed February 2017).
Mississippi Entomological Museum, 2017. http://mississippientomologicalmuseum.org.msstate.edu (accessed August 2017).
Missouri Department of Conservation, 2017. Aphids. https://nature.mdc.mo.gov/discover-nature/field-guide/aphids (accessed October 2017).
Morris, S.C., 2009. Walcott, the Burgess Shale and rumours of a post-Darwinian world. Curr. Biol. 19 (20), R927–R931.
Myers, P., 2001. Arthropoda. University of Michigan Museum of Zoology. http://animaldiversity.org/accounts/Arthropoda (accessed October 2017).
Nakajima, M., Atsumi, K., Kifune, K., Miura, K., Kanamaru, H., 1986. Chitin is an effective material for sutures, National center for Biotechnology Information. Jpn. J. Surg. 16 (6), 418–424.
Nation, J.L., 1983. A new method using hexamethyldisilazane for preparation of soft insect tissues for scanning electron microscopy. Stain Technol. 58 (6), 347–351.
National Park Service, 2017. A Symbiotic Relationship - Carrion Beetles and Mites, Bandelier National Monument, New Mexico. https://www.nps.gov/band/learn/nature/symbiotic.htm (accessed November 2017).
Nazar, M.Z., 2017. Respiratory System of Insects. Department of Agriculture, University of Sargodha, Punjab, Pakistan. https://www.slideshare.net/MuhammadZeeshanNazar/respiratory-system-of-insects-ppt (accessedMay 2017).
New Mexico State University Arthropod Museum, 2017. https://arthropods.nmsu.edu (accessed June 2017).
New World Encyclopedia, 2017. Ruminant. http://www.newworldencyclopedia.org/entry/Ruminant (accessed November 2017).

Nielsen, C., 2012. Animal Evolution: Interrelationships of the Living Phyla. Oxford University Press. 402 p.

Oxford Reference, 2017. http://www.oxfordreference.com (accessed May 2017).

Pacheco, C.A., Alevi, K.C.C., Ravazi, A., de Azeredo Oliveira, M.T.V., 2014. Review: Malpighian tubule, an essential organ for insects. Entomol. Ornithol. Herpetol. 3 (2), 1–3. https://repositorio.unesp.br/bitstream/handle/11449/122591/ISSN2161-0983-2014-03-02.pdf?sequence=1 (accessed June 2017).

Palade, G.E., 1952. A study of fixation for electron microscopy. J. Exp. Med. 95 (3), 285–298 plates 19-23.

Palaeocast, 2017. Episode 48: The Burgess Shale. http://www.palaeocast.com (accessed February 2017).

Pandian, T.J., 2016. Series on Reproduction and Development in Aquatic Invertebrates, Volume 1: Reproduction and Development in Crustacea. CRC Press (Taylor and Francis Group). 315 p.

Park, B.K., Kim, M.-M., 2010. Application of chitin and its derivatives in biological medicine. Int. J. Mol. Sci. 11 (12), 5152–5164.

Penn State College of Agricultural Sciences, 2017a. Insect Advice from Extension: Springtails. http://ento.psu.edu/extension/factsheets/springtails (accessed October 2017).

Penn State College of Agricultural Sciences, 2017b. Insect Advice from Extension: White Grubs in Home Lawns. http://ento.psu.edu/extension/factsheets/white-grubs-lawns (accessed October 2017).

Penney, D., Jepson, J.E., 2014. Fossil Insects: An Introduction to Palaeoentomology. SIRI Scientific Press, Manchester, United Kingdom. 224 p.

Prakash, A., Gens, R., Styers, J., Stracher, G.B., 2013. Coal Fires of the United States: Alaska. In: Stracher, G.B., Prakash, A., Sokol, E.V. (Eds.), Coal and Peat Fires: A Global Perspective, Volume 2: Photographs and Multimedia Tours. Elsevier, pp. 451–455.

Raven, P.H., Johnson, G.B., 2001. Biology, sixth ed. McGraw Hill. 1344 p.

Riggs, A.F., 2014. Respiratory Pigments (Invertebrate). AccessScience, McGraw-Hill Education. https://doi.org/10.1036/1097-8542.583500 (accessed May 2017).

Roberts, M.B.V., 1971. Biology: A Functional Approach, fourth ed. 695 p.

Rockett, C.L., 1980. Nematode predation by oribatid mites (Acari: Oribatida). Int. J. Acarol 6 (3), 219–224.

Roller, L., Žitňanová, I., Dai, L., Šimo, L., Park, Y., Satake, H., Tanaka, Y., Adams, M.E., Žitňan, D., 2010. The ecdysis triggering hormone signaling in arthropods. Peptides 31 (3), 429–441.

Romoser, W.S., Stoffolano Jr., J.G., 1998. The Science of Entomology, fourth ed. McGraw-Hill, Boston, Massachusetts. 605 p.

Royal Ontario Museum, 2017. The Burgess Shale. http://www.rom.on.ca/en#/gallery/recent (accessed February 2017).

Sabatini, D.D., Bensch, K., Barrnett, R.J., 1963. Cytochemistry and electron microscopy: the preservation of cellular ultrastructure and enzymatic activity by aldehyde fixation. J. Cell Biol. 17, 19–58.

Saito, Y., 2010. Plant Mites and Sociality: Diversity and Evolution. Springer, Tokyo, Japan. 187 p.

ScorpionWorlds, 2017. Scorpion Facts, BioExpedition. http://www.scorpionworlds.comhttp://www.bioexpedition.com (accessed June 2017).

Secher, A., Shugar, M.A., 2017. Paleontology: Trilobite Website, American Museum of Natural History. https://research.amnh.org/paleontology/trilobite-website (accessed October 2017).

Seilacher, A., 1970. Begriff und Bedeutung der Fossil-Lagerstätten, Neues Jahrbuch fur Geologie und Paläontologie, Monatshefte (in German), pp. 34–39.

Simmons, Breana, Wedincamp, Jimmy, Barnes, Melissa, A., Stracher, G.B., 2013. Soil arthropods associated with continuous coal fires in Centralia. In: Pennsylvania, Soil Ecology Society 14th Biennial Meeting (June 11-14), Camden, New Jersey, SES Program, p. 8.

Singh, B.K., Dutta, P.K., 2016. Chitan, Chitosan, and silk fibroan electrospun nanofibrous scaffolds: a prospective approach for regenerative medicine. In: Dutta, P.K. (Ed.), Chitin and Chitosan for Regenerative Medicine. Springer, pp. 151–192.

Stankiewicz, B.A., Derek, E.G., Briggs, R.P., Evershed, M.B., Flannery, M.W., 1997. Preservation of chitin in 25-million-year-old fossils. Science 276 (5318), 1541–1543.

Stanley, M.S.M., 1972. Cultivation of arthropod cells. In: Rothblat, G.H., Cristofalo, V.J. (Eds.), Growth, Nutrition, and Metabolism of Cells in Culture, vol. II. Academic Press. 459 p.

Starr, C., 2006. Biology: Concepts and Applications, sixth ed. Thomson -Brooks/Cole. 857 p.

Stracher, G.B., 2011. Gas vent mineralization and coal combustion. In: Stracher, G.B., Prakash, A., Sokol, E.V. (Eds.), Coal and Peat Fires: A Global Perspective, Volume 1: Coal – Geology and Combustion. Elsevier, pp. 135–153.

Stracher, G.B., Nolter, M.A., 2013a. Coal fires of the United States: Pennsylvania. In: Stracher, G.B., Prakash, A., Sokol, E.V. (Eds.), Coal and Peat Fires: A Global Perspective, Volume 2: Photographs and Multimedia Tours. Elsevier, p. 498–509.

Stracher, G.B., Nolter, M.A., 2013b. Coal fires of the United States: Pennsylvania. In: Stracher, G.B., Prakash, A., Sokol, E.V. (Eds.), Coal and Peat Fires: A Global Perspective, Volume 2: Photographs and Multimedia Tours, Companion Materials: Data Repository. http://booksite.elsevier.com/9780444594129.

Stracher, G.B., Tabet, D.E., Anderson, P.B., Pone, J.D.N., 2005. Utah's state rock and the emery coalfield: geology, mining history, and natural burning coal beds. In: Pederson, J., Dehler, C.M. (Eds.), Interior Western United States, Geological Society of America Field Guide 6, pp. 199–210.

Stracher, G.B., Nolter, M.A., Schroeder, P., McCormack, J., Vice, D.H., 2006. The great Centralia mine fire: a natural laboratory for the study of coal-mine fires. In: Pazzaglia, F.J. (Ed.), Excursions in Geology and History: Field Trips in the Middle Atlantic States, Geological Society of America Field Guide 8, pp. 33–45.

Stracher, G.B., Lindsley-Griffin, N., Griffin, J.R., Renner, S., Schroeder, P., Viellenave, J.H., Masalehdani, M.N.-N., Kuenzer, C., 2008. Revisiting the south cañon number 1 coal mine fire during a geologic excursion from denver to glenwood springs, Colorado. In: Raynolds, R.G. (Ed.), Roaming the Rocky Mountains and Environs: Geological Field Trips, Geological Society of America Field Guide 10, pp. 101–110.

Stracher, G.B., Lindsley-Griffin, N., Simmons, B., Wedincamp, J., Griffin, J.R., Barnes, M.A., 2013. Newly discovered arthropods at gas vents and fissures: Centralia mine fire, eastern Pennsylvania. Geol. Soc. Am. Abstr. Progr. 45 (7), 651.

Strauss, B., 2017. 10 Facts about Arthropods, ThoughtCo. https://www.thoughtco.com/facts-about-arthropods-4069412 (accessed October 2017).

Teigler, D.J., Towe, K.M., 1974. Microstructure and composition of the trilobite exoskeleton. Fossils Strata 4 (4), 137–149.

The Canadian Encyclopedia, 2017. http://www.thecanadianencyclopedia.ca/en/article/arthropoda (accessed March 2017).

The University of Edinburgh, 2017. The Natural History Collections, Phylum Arthropoda. http://www.nhc.ed.ac.uk/index.php?page=24.25.312.330 (accessed October 2017).

Thomas, C.D., 1990. Fewer species. Nature 347, 237.

Tobin, A.J., Duscheck, J., 2005. Asking about Life, third ed. Thompson: Brooks/Cole. 879 p.

Townsend, L., 2017. Aphids, ENTFACT – 103. University of Kentucky Entomology. https://entomology.ca.uky.edu/ef103 (accessed October 2017).

Triplehorn, C.A., Johnson, N.F., 2005. Borror and DeLong's Introduction to the Study of Insects, seventh ed. Thomson Brooks/Cole, Belmont, California. 888 p.

University of California Agriculture and Natural Resources, 2017. Statewide Integrated Pest Management Program: Thrips. http://ipm.ucanr.edu/PMG/PESTNOTES/pn7429.html (accessed October 2017).

University of California Museum of Paleontology, 2017. Introduction to the Arthropoda. http://www.ucmp.berkeley.edu/index.php (accessed October 2017).

University of California Santa Barbara ScienceLine, 2017. http://scienceline.ucsb.edu/index.html (accessed October 2017).

University of Georgia Extension, 2017. White Grub Pests of Turfgrass, Circular 940. http://extension.uga.edu/publications/detail.html?number=C940 (accessed October 2017).

University of Minnesota Extension, 2017. Insects: Springtails. https://www.extension.umn.edu/garden/insects/find/springtails (accessed October 2017).

Van Dooremalen, C.V., Berg, M.P., Ellers, J., 2013. Acclimation responses to temperature vary with vertical stratification: Implications for vulnerability of soil-dwelling species to extreme temperature events. Global Change Biol. 19 (3), 975–984 John Wiley & Sons Ltd.

Virtual Fossil Museum, 2017. http://www.fossilmuseum.net (accessed May 2017).

Wikipedia, 2017a. Beetle. https://en.wikipedia.org/wiki/Beetle (accessed October 2017).

Wikipedia, 2017b. Arthropod. https://en.wikipedia.org/wiki/Arthropod (accessed October 2017).

Wikipedia, 2017c. Lagerstätte. https://en.wikipedia.org/wiki/Lagerst%C3%A4tte (accessed October 2017).

Wikipedia, 2017d. Wax Ester. https://en.wikipedia.org/wiki/Wax_ester (accessed October 2017).

Wikipedia, 2017e. Tagma (Biology). https://en.wikipedia.org/wiki/Tagma_(biology) (accessed October 2017).

Wikipedia, 2017f. Microarthropods. http://dictionary.sensagent.com/microarthropods/en-en (accessed October 2017).

Wikipedia, 2017g. Insect Wing. https://en.wikipedia.org/wiki/Insect_wing (accessed October 2017).

Wikipedia, 2017h. Gill. https://en.wikipedia.org/wiki/Gill (accessed October 2017).

Wikipedia, 2017i. Extracellular Digestion. https://en.wikipedia.org/wiki/Extracellular_digestion (accessed October 2017).

Wikipedia, 2017j. Malpighian Tubule System. https://en.wikipedia.org/wiki/Malpighian_tubule_system (accessed October 2017).

Wikipedia, 2017k. Crustacean Larvae. https://en.wikipedia.org/wiki/Crustacean_larvae (accessed October 2017).

Wikipedia, 2017l. Tullgren Funnel. https://en.wikipedia.org/wiki/Tullgren_funnel (accessed October 2017).

Wikipedia, 2017m. Mite. https://en.wikipedia.org/wiki/Mite (accessed October 2017).

Wikipedia, 2017n. Acari. https://en.wikipedia.org/wiki/Acari (accessed October 2017).
Wikipedia, 2017o. Springtail. https://en.wikipedia.org/wiki/Springtail (accessed October 2017).
Wikipedia, 2017p. Aphid. https://en.wikipedia.org/wiki/Aphid (accessed October 2017).
Wilmot, N.V., Fallick, A.E., 1988. Optical mineralogy of trilobite exoskeletons. Paleontology 32 (2), 297–304.
Wissuwa, J., Salamon, J.-A., Frank, T., 2013. Oribatida (Acari) in grassy arable fallows are more affected by soil properties than habitat age and plant species. Eur. J. Soil Biol. 59, 8–14.
Wolfe, J.M., Daley, A.C., Legg, D.A., Edgecombe, G.D., 2016. Fossil calibrations for the arthropod tree of life. Earth Sci. Rev. 160, 43–110.
Wu, G.-M., Boivin, G., Brodeur, J., Giraldeau, L.-A., Outreman, Y., 2010. Altruistic defense behaviours in aphids. BMC Evol. Biol. 10, 19. https://bmcevolbiol.biomedcentral.com/articles/10.1186/1471-2148-10-19 (accessed October 2017).
Yeh, J., 2002. Arthropoda. The Gale Group, Inc. Encyclopedia.com. http://www.encyclopedia.com/topic/Arthropoda.aspx (accessed October 2017).
Yusa, Y., Takemura, M., Sawada, K., Yamaquchi, S., 2013. Diverse, continuous, and plastic sexual systems in barnacles. Integr. Comp. Biol. 53 (4), 701–712.

WWW Addresses: Additional Reading

(1) American Arachnological Society
http://www.americanarachnology.org

(2) Arthropods

- **(a) Animal Diversity Web**
  http://animaldiversity.org
- **(b) Arthropod Vision**
  - **(i) Vision in Arthropoda**
    http://www.iaszoology.com/arthropod-eye-vision
  - **(ii) Arthropod Eye**
    https://en.wikipedia.org/wiki/Arthropod_eye
- **(c) Arthropod Structure and Development**
  http://www.journals.elsevier.com/arthropod-structure-and-development
- **(d) Clemson University Arthropod Collection**
  http://www.clemson.edu/cafls/departments/esps/database/museum
- **(e) The Earth Life Web**
  http://www.earthlife.net/inverts/arthropoda.html
- **(f) University of Arkansas Arthropod Museum**
  http://www.uark.edu/ua/arthmuse/museum.html

(3) Berkeley Lab's Advanced Light Source Finds Big Surprise in Paleozoic Scorpion Fossil
http://newscenter.lbl.gov/2011/02/28/scorpion-fossil

(4) Burgess Shale

- **(a) Wikipedia**
  https://en.wikipedia.org/wiki/Burgess_Shale
- **(b) Palaeocast**
  http://www.palaeocast.com/episode-48-the-burgess-shale

(5) Chelicerata

- **(a) The Virtual Fossil Museum**
  http://www.fossilmuseum.net
- **(b) Beutel, R.G., Friedrich F., Ge, S-Quin, Yang, Xing-Ke, 2014**, Insect Morphology and Phylogeny: A Textbook for Students of Entomology, Walter de Gruyter GMBH, Berlin, 515 p.
- **(c) Thorp, J.H. and Rogers, D.C., 2011,** Field Guide to Freshwater Invertebrates of North America, Mites and Spiders: Subphylum Chelicerata, Class Arachnida, Chapter 14, pp. 121–126.

(6) Coleoptera
http://www.coleoptera.org

(7) C.P. Gillette Museum of Arthropod Diversity
http://bspm.agsci.colostate.edu/01-2/gillette-museum

(8) Crustacea

- **(a) The Earthlife Web**
  http://www.earthlife.net/inverts/crustacea.html

(b) **Conway, D.V.P., 2012**, Marine Zooplankton of Southern Britain, Part 2: Arachnida, Pycnogonida, Cladocera, Facetotecta, Cirripedia and Copepoda, in John, A.W.G. ed., Marine Zooplankton of Southern Britain, Marine Biological Association of the United Kingdom Occasional Publications No 26, Plymouth, United Kingdom, 163 p.
(c) **Martin, J.W., Olesen, J., Høeg, J.T., 2014,** Atlas of Crustacean Larvae, Johns Hopkins University Press, Baltimore, Maryland, 370 p.

(9) **Diseases from Arthropods**
(a) **Berger, S., 2017,** Arthropod Infections: Global Status: 2017 edition, Gideon Informatics Inc., 139 p.
(b) **Journal of Arthropod-Borne Diseases,** Iranian Society of Medical Entomology http://jad.tums.ac.ir/index.php/jad
(c) **Marcondes, C.B.,** 2016, Arthropod Borne Diseases, Springer, 645 p.
(d) **Wikel, S., Aksoy, S., Dimopoulos, G., 2017,** Arthropod Vector: Controller of Disease Transmission, Volume 1: Vector Microbiome and Innate Immunity of Arthropods, Academic Press, 270 p.
(e) **Wikel, S., Aksoy, S., Dimopoulos, G., 2017,** Arthropod Vector: Controller of Disease Transmission, Volume 2: Vector Saliva-Host-Pathogen Interactions, Volume 2, Academic Press, 332 p.

(10) **Entomological Society of America**
http://www.entsoc.org

(11) **Hexapoda**
(a) **The Virtual Fossil Museum**
http://www.fossilmuseum.net
(b) **Tree of Life Web Project**
http://tolweb.org
(c) **Thorp, J.H. and Rogers, D.C., 2011,** Field Guide to Freshwater Invertebrates of North America, Introduction to Insects and their Near Relatives: Subphylum Hexapoda, Chapter 19, pp. 169–178.

(12) **Myriapoda**
(a) **Australian Museum**
https://australianmuseum.net.au
(b) **Roberts, R.G., 2014,** What Goes “99-Thump?” Plos Biology, V. 12 (November)
http://journals.plos.org/plosbiology/article?id=10.1371/journal.pbio.1002006

(13) **The Field Museum**
https://www.fieldmuseum.org/node/5046

(14) **The Smithsonian National Museum of Natural History**
http://naturalhistory.si.edu

(15) **The Virtual Fossil Museum**
http://www.fossilmuseum.net/index.htm

(16) **Trilobita**
(a) **A Guide to the Order of Trilobites**
https://www.trilobites.info
(b) **Fossilguy.com: Celebrating the Richness of Paleontology through Fossil Hunting**
http://www.fossilguy.com
(c) **New World Encyclopedia**
http://www.newworldencyclopedia.org

CHAPTER 3

Coal Fires of Northeastern Pennsylvania

CHAPTER CONTENTS

The Laurel Run Fire in Luzerne County, Pennsylvania. In 1915, an underground anthracite fire started in Laurel Run at the Red Ash Coal Mine when a coal miner left his lit-carbide lamp suspended from a timber support. Because there was no watchman, the fire was not spotted during the weekend. It was noticed the following week. Sand and concrete were used to block the entry of air into the mine and the mine's owners thought the fire was under control. However, by 1921 the fire spread and the company used barriers to prevent the spread of the fire so that mining could continue. The mine was closed in 1957 and the fire was left to burn. Thereafter, it started threatening homes in Laurel Run and Wilkes-Barre, including the home of Congressman Daniel Flood. The fire also threatened interstate highway I-81 and the mainline of a railroad. The portions of the fire threatening Laurel Run and Wilkes-Barre were extinguished, possibly by being dug out, but a portion of the fire was left to burn because it was not considered a threat. Harold Aurand, Jr. is seen here standing at an active Laurel Run fire site adjacent to a road (not visible) leading into a trailer court. *Photo by Daniel H. Vice, 2011.*

Coal and Peat Fires: A Global Perspective
Edited by Glenn B. Stracher

https://doi.org/10.1016/B978-0-12-849885-9.00003-2

3.1 Coal Fires of Northeastern Pennsylvania

Harold J. Aurand, Jr.
Melissa A. Nolter
Daniel H. Vice

Photo by Harold Aurand, Jr., 2004.

A warning sign at the Centralia Mine Fire in Pennsylvania. The sign was removed when residents complained that it attracted trespassers rather than deterred them from investigating the area.

Introduction

The **anthracite** coal fields of northeastern **Pennsylvania** (Figure 3.1.1) cover an area of 1400 mi^2 (2253.08 km^2) of the Central Appalachian Mountain section of the Valley and Ridge Province of Pennsylvania; however, due to erosion, only 439 mi^2 (706.50 km^2) of the area still contain coal (Faill and Nickelsen, 1999). The anthracite coal has been preserved in a series of complex synclinoria, which form four main fields, the southern, eastern middle, western middle, and northern, with several smaller pockets of anthracite and semianthracite in outlying areas (Figure 3.1.1). The general trend of each field is northeast to southwest, with each field being a complex series of secondary basins (Eggleston et al., 1999). The coal beds within each field can be thought of as a series of nested bowls with only the rim being exposed.

The coal beds occur in the **Pottsville and Llewellyn Formations** (Pennsylvanian) in northeastern Pennsylvania and are similar in age to the coals of western Pennsylvania in that they represent deposition in the Pennsylvanian sediments of the Appalachian Basin.

The Pottsville and Llewellyn Formations consist predominantly of conglomerate, sandstone, siltstone, claystone, and shale. The Pottsville Formation contains a distinctive conglomerate, which has a framework of large, white quartz pebbles and dark-grayish, sand-rich matrix that has considerable organic matter. Much of the shale and/or claystone in the Llewellyn Formation is dark gray in color.

It is generally understood that coal fires can start through human activity, spontaneous combustion, or surface fires (Kim and Chaiken, 1993). Spontaneous combustion has never been clearly defined, but is usually understood to be the tendency of a coal bed or stockpile to heat to the point where it starts burning (Uludag, 2007; Cao et al., 2007; Nelson and Chen, 2007). Generally, coals that spontaneously combust (Figure 3.1.2) have a low rank and high sulfur/sulfide mineral content (Cao et al., 2007). Anthracite coal in Pennsylvania does not spontaneously combust because of its high rank (92%–98% fixed carbon) and its low sulfur content (0.3%–1.2%) (Eggleston et al., 1999). There is also little evidence of surface fires igniting underground anthracite coal deposits. All the mine fires in the anthracite region of Pennsylvania where an original cause has been determined have been started by human activity, whether purposeful or accidental.

Early Mine Fires

Because of the Pennsylvania anthracite region's proximity to the major markets of Philadelphia and New York City, large-scale mining started there earlier than in other parts of America with coal deposits. Production increased

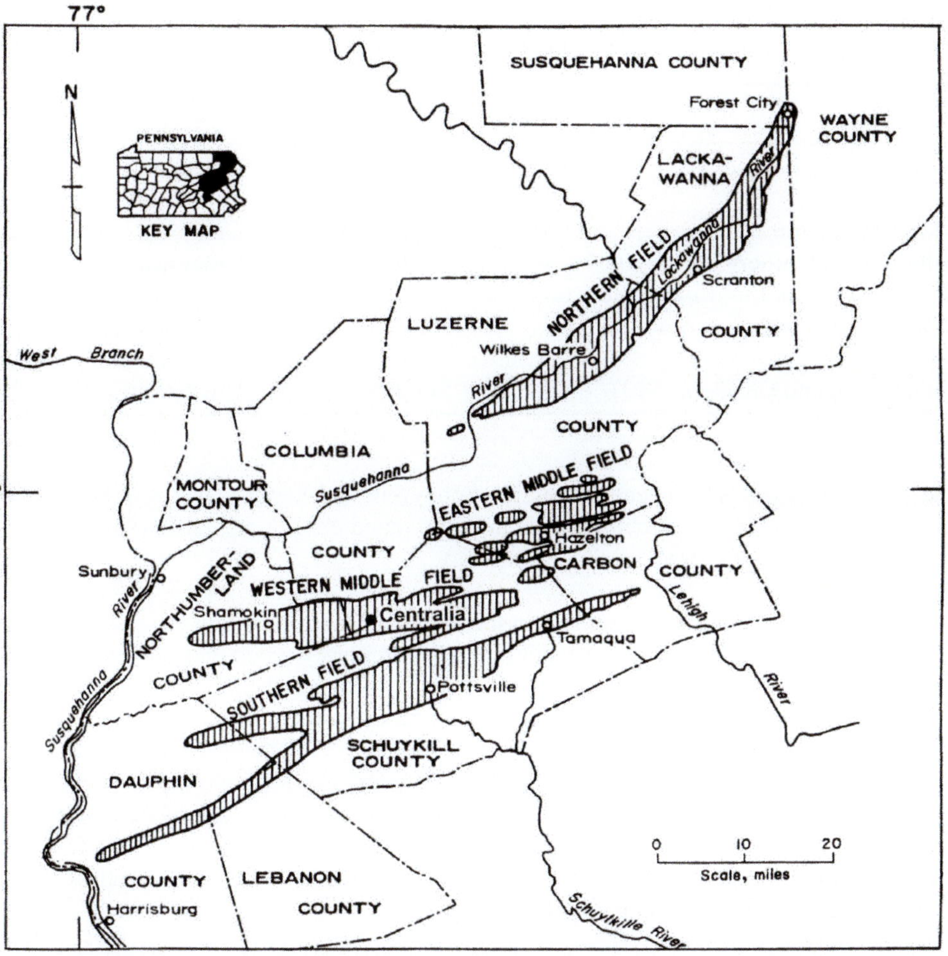

Figure 3.1.1. Map of the southern, western, eastern middle, and northern anthracite fields in eastern Pennsylvania. The famous Centralia Mine Fire is in the western middle field, Columbia County. From Stracher and Nolter (2013); Courtesy of Timothy Altares, Pennsylvania Department of Environmental Protection, Harrisburg, PA.

Figure 3.1.2. One of the last active burn areas at the Cemetery Front of the Centralia Mine Fire. Note the growth of young trees. Photo by Daniel H. Vice, 2014.

Table 3.1.1
Early coal fires in Pennsylvania.

Name and Reference	Field	Cause
Greenwood Colliery (Detterline, 1968; Scherer, 1976)	Southern	Untended warming fire
	Southern	Spark from lamp on January 1, 1877
Summit Hill (Mazaika, 2007)	Southern	Unknown
Heckscherville Anchor Colliery (Blasé, 1997)	Southern	Unknown
Cameron Colliery at Shamokin (Lindemuth, 2008)	Western middle	Unknown
Laurel Run (Randolph, 2002; Ashmead, 1922)	Northern	Untended carbide lamp left in mine timber in 1915
Peach Mountain (Kim and Chaiken, 1993)	Southern	Unknown

rapidly in the 1820s as canals were opened along the Schuylkill and Lehigh-Delaware River systems. This early start to mining meant anthracite production significantly predated environmental laws regulating how businesses could use the land and much of current mining technology.

Among the causes of early mine fires were those associated with the production process itself. Some early mines used furnaces to drive ventilation. Heated air would go up a temporary wood and canvas flue, drawing cool, fresh air into the mine. Furnace accidents occasionally ignited nearby coal supplies. Flues caught fire from rising embers and collapsed into mines. Miners lit their workplaces with oil or carbide lamps. If left hanging on mine supports they could, and sometimes did, lead to larger fires.

Another potential trouble spot was mining company philanthropy. In many anthracite communities, poor people were allowed and even encouraged to forage for waste coal at certain designated sites. During the winter months, it was natural for such foragers to build small warming fires. There are accounts of these fires falling into air shafts, or in other ways spreading into active mine workings.

Unfortunately, few generalizations can be made about these early anthracite fires. Prior to World War II, the state of Pennsylvania viewed mine fires as economic, rather than environmental or public health problems. The main concern was that unless the fire could be put out or somehow contained, affected mines would close and workers would lose their jobs. As owners of the land and mineral rights, the mining companies were responsible for responding to fires. As an illustration, Kim and Chaiken (1993) have pointed out that prior to 1949 no federal or state agency even bothered to collect information about coal fires in abandoned mine lands.

Some fires from this early period were covered by local newspapers or show up in town histories, but locating such sources is very much a hit-or-miss endeavor. Few of the newspapers have been digitized, and academic libraries rarely bother holding sources perceived as only having antiquarian interest. Thus, while we have found evidence of some fires (Table 3.1.1), we cannot claim that they represent either all the fires or even a representative sampling.

The Changing Anthracite Industry

The anthracite industry began to decline in the 1920s, when alternative fuels like oil and natural gas began to take over the home heating market, while coked bituminous coal replaced anthracite in furnaces smelting iron and steel. There was a brief revival in the anthracite fields during World War II, but then the industry went into a period of near collapse. During these decades when the anthracite industry was trying to adapt to a shrinking market, several changes took place that would affect the future of mine fires. First, as the market contracted, mining companies shifted from an almost complete reliance on underground mining to a mix of underground and surface mines. Second, the Pennsylvania state and federal governments began to pass legislation regulating mining on environmental grounds. Finally, some anthracite mining companies began to go out of business. Coal fires that had once been put out or contained underground were now left free to burn (e.g., Laurel Run).

The anthracite fields of Pennsylvania faced America's most serious problem with abandoned coal lands because this was the first area to develop large-scale coal mining, and the industry had continued for well over a century, so there was a great deal of land that had been marred by mining. By the time reclamation legislation was passed, the companies that had done much of the mining no longer existed, or lacked the financial wherewithal to restore the land. Consequently, large amounts of abandoned coal mine land occur in Northeastern Pennsylvania. Even the large railroads, with their extensive landholdings, could do little more than provide token security forces to guard their land. Further, as mining declined, communities in the anthracite region lost population and the tax base necessary to deal with abandoned mine lands.

Large areas of unrestored land became open to trespassing by all-terrain vehicles (e.g., Excelsior), and stripping pits started being used as impromptu or formal garbage dumps (e.g., **Centralia**). The result was an increased number of underground and coal waste fires starting at the very time the companies and local communities had less ability to fight them. Kim and Chaiken (1993) noted that the most common type of fire in coal mines had started in trash within old strip mining pits. These sorts of statements are only possible because the government has begun to collect statistics on mine fires.

The anthracite region's experience with mine fires in this period of change was mixed. The government had assumed the mining companies would be primarily responsible for fire suppression and land reclamation. They knew that occasionally a mining company might be bankrupt and unable to fulfill its obligations, so some money was set aside for those instances. In Pennsylvania's anthracite region, though, abandoned lands, or mining companies facing severe economic distress, were the norm. Pennsylvania often found itself having to prioritize, focusing on fires that immediately threatened major population centers or infrastructure, while other fires were left for later. As the years went by, this approach, while sensible, became less politically acceptable. In the years after World War II, most male residents of the anthracite region had first-hand experience working in the mines. They understood the risks of mine fires and other potential threats of working underground. Today, with only a small part of the population having mining experience, there is more often panic, and an unwillingness to wait one's turn in line. Individual fires are discussed in the Appendix.

The Future

The current coal mine fires have largely been contained, such as Palo Alto (Figure 3.1.3), and are no longer a threat to lives or property. Centralia, the most notorious coal mine fire, appears to be going out (Figure 3.1.2). Looking

Figure 3.1.3. Site of the Palo Alto coal fire showing the isolation trench. Photo by Daniel H. Vice, 2009.

Figure 3.1.4. McAdoo cogeneration plant. Photo by Daniel H. Vice, 2010.

forward, Pennsylvania's anthracite region should expect a steady decline in the number of mine fires. Active mining has stabilized at a very low level, which will limit the possibility of work accidents causing fires. Garbage fires in abandoned mine workings should also decrease because of contained reclamation of coal lands by the federal government and the state of Pennsylvania.

In addition, considerable reclamation is being done by private industry such as the cogeneration power plants that are burning renewable or waste energy resources (Inners et al., 1996). These **cogeneration plants** are using a new technology, known as a **fluidized bed combustion chamber**, to burn coal waste and generate electricity (Figure 3.1.4). Burning the waste, locally called culm, removes surface pollution and limits the possibility of waste-pile runoff polluting nearby streams. The ash left over from when culm is burned has proven useful in filling in surface mining pits (Aurand and Vice, 2015). **Northampton Fuel Supply**, just to use one example, has spent the past 11 years removing and burning a 700-ft-high pile of culm outside Nanticoke (Jackson, 2016). As cogeneration plants burn waste piles and use the resulting ash to fill abandoned surface mines, and as the limited federal funds for abandoned mine land reclamation take effect, there will be less possibilities for accidental garbage fires to burn into underground mine workings.

Summary and Conclusion

The history of coal fires in northeastern Pennsylvania has been divided into three unequal times: pre-1950, 1950 to 2010, and 2010 to the present. These periods have been based on the way coal fires were treated. Before approximately 1950, coal fires were considered the responsibility of the mine owner and no statistics were collected on occurrences. Information on the number and history of these coal fires was limited and has been obtained from local histories and similar sources. From 1950 to 2010, many coal fires occurred in stripping areas that had not been reclaimed and were largely unsupervised. The federal and Pennsylvania state governments were forced to step in and control the coal fires. By 2010 most of the coal fires had been contained or put out. Based on the continued reclamation of stripping areas and the general decrease in coal mining in northeastern Pennsylvania, the number of coal fires will continue to decline in the future.

Appendix: Coal Fires

Although coal-mine fires are known to have occurred since the start of mining in northeastern Pennsylvania (Table 3.1.1), there is not a wealth of published research about the early coal fires from the start of mining in Pennsylvania to about 1950.

Carbondale, Centralia, and Other Middle Period Coal Fires

The coal mine fires in the middle period from approximately 1950 to 2000 illustrate the shift from company efforts of control in order the save a valuable resource to the first efforts of government control because of threats to public safety (Table 3.1.2). The two best known of these fires are the **Carbondale** and Centralia, but many more fires are known from the middle period in part because of the lists available in U.S. Bureau of Mines publications (e.g., Kim and Chaiken, 1993). Many of these middle period fires started in trash that was being dumped in old stripping pits, as can be seen from Table 3.1.2.

Recent Fires

Several recent coal mine fires have been discovered since 2010 (Table 3.1.3). The fires of Palo Alto, Dolph, and Excelsior are described in the following to illustrate the change in reactions for both the community and government after about 2000. This change is often blamed on how the government handled the notorious Centralia mine fire. For most of the state, government policy in the 1950s and 1960s made sense. Faced with several large fires at

Table 3.1.2
Middle Period coal fires in Pennsylvania.

Fire and Reference	Field	Cause
Carbondale (Munley, 1998; Kim and Chaiken, 1993)	Northern	Started in trash dump in old stripping pit in late 1940s
Kulpmont (Kim and Chaiken, 1993)	Western middle	Unknown
Mt. Carmel (Kim and Chaiken, 1993)	Western middle	Unknown
Shamokin (Kim and Chaiken, 1993)	Western middle	Unknown
Cedar Avenue in Scranton (Dierks et al., 1971)	Northern	Unknown
Tower City (Kim and Chaiken, 1993)	Southern	Unknown
Shenandoah (Kim and Chaiken, 1993)	Western middle	Unknown
Coal Run (Kim and Chaiken, 1993)	Western middle	Unknown
Centralia (Nolter and Vice, 2004; Kim and Chaiken, 1993)	Western middle	Started in trash dump in old stripping pit in 1962
Enyon Street (Kim and Chaiken, 1993)	Northern	Unknown
Hazleton (Kim and Chaiken, 1993)	Eastern middle	Unknown
Kehley Run (Kim and Chaiken, 1993)	Western middle	Unknown
Throop (Dierks et al., 1971)	Northern	Observed in 1966 and was believed to have started in garbage in an old stripping pit that was used as a garbage dump
Warrior Run (Kim and Chaiken, 1993)	Northern	Unknown
Swoyersville (Kim and Chaiken, 1993)	Northern	Unknown
Eddy Creek (Kim and Chaiken, 1993)	Northern	Unknown
Forestville (Malenka, 1985)	Southern	Started in trash in old stripping pit in 1982
Hughestown (Kim and Chaiken, 1993)	Northern	Unknown
Sugar Notch (Kim and Chaiken, 1993)	Northern	Unknown
Larksville (Kim and Chaiken, 1993)	Northern	Unknown
Maffett (Kim and Chaiken, 1993)	Northern	Unknown

Table 3.1.3
Recent coal fires in Pennsylvania.

Fire and Reference	Field	Cause
Mildred (Sullivan Review, 2004)	Western northern	Uncertain but believed to have started in trash dumped at an old mine
Palo Alto (Pottsville Republican-Herald, 2008, 2009)	Southern	Uncertain but believed to be hunter's fire built on a **culm bank** (waste pile) in December of 2008
Dolph or Olyphant	Northern	Unknown origin in 2004 in abandoned underground mine workings
Coal Township or Excelsior (Shamokin New-Item, 2006)	Western middle	Fire (possibly arson) in 2005 in a leaf-based mulch on reclaimed mine lands that caught coal waste on fire
Simpson	Northern	Unknown origin on a culm bank in 2014

the same time, the government prioritized, using limited funds to control the fires that threatened people's lives or major infrastructure. The Centralia fire did neither, thus it received less attention. By the time the state was ready to focus on Centralia, what had been a small fire had become a major one, expanding on four separate fire fronts. The town, despite 20 years of nominal fire control efforts, had to be destroyed. Today, when new mine fires are discovered, the affected communities immediately seek out the media and invoke the name of Centralia. Fearful of bad publicity, the federal and state governments respond swiftly (Nolter and Vice, 2004).

Palo Alto is located on the southern margin of the Southern Anthracite Field close to the boundary between the Llewellyn and Pottsville Formations. Coal beds in the area are essentially vertical (Levine and Eggleston, 1992). In December of 2008, the community became aware of a coal fire burning just outside their borough limits in East Norwegian Township. The fire had probably started in July, but because the site was isolated, about 200 yards from any homes or well-travelled roads, it hadn't been noticed until a local hunter, John Ketner, stumbled across it. Ketner notified the East Norwegian Township supervisors, who, after examining the site, decided "this was way beyond the scope of any municipality," and contacted the Office of Surface Mining (Pottsville Republican, 2008).

John Mack, Office of Surface Mining project manager, was at the site within 2 days. After heat guns registered a ground temperature of 178°F, exploratory boreholes were drilled to determine the extent of the fire. Results showed that what was burning was an 8-ft-long ×70-ft-wide teardrop-shaped deposit of buried coal waste. The local water table was only 60 ft below the surface, so there was little chance of the fire spreading downward into lower strata (Pottsville Republican, 2008).

Once the extent of the fire was known, Mack said the next step was to "design a trench to isolate the fire." By spring the trench was in place, with two or three acres effectively isolated (Figure 3.1.3). Although Office of Surface Mining Engineer Dave Philbin admitted, "there's a lot of burnable material there," everyone agreed that the fire had been cut off from spreading, and would not pose a threat to the community of Palo Alto. Later the fire could be entirely dug up and extinguished or left to burn (Pottsville Republican, 2008).

Olyphant lies on the southern limb of the Northern Anthracite Field. In the summer of 2004 an underground mine fire broke out in the abandoned mine workings south of town. Eventually it spread through the mine workings to encompass an area of seven acres. As in the case of other recent fires, the government's response was swift. A 609.6 m-long trench, up to 39.62 m deep, was dug to isolate the fire. During the digging of the trench, when temperatures at one area of coal and rocks were warmer than the blasting contractor would tolerate, a sprinkler system was set up to water the affected site until it cooled. This allowed the blasting contractor to set his charges where they were initially scheduled, and prevented the delays that would have accompanied redesigning and extending the trench around the hot area. The entire trenched area was subsequently fenced off.

Excelsior (Figure 3.1.5) is in the Western Middle Anthracite Field, just outside Shamokin. Elwood Swank, owner of the Split Vein Coal Company in Paxinos, leased land near Excelsior from Reading Anthracite. He had a permit from Pennsylvania's Department of Environmental Protection (DEP) to conduct surface mining on the area, and bring in refuse for reclamation processes. Included in the refuse was a leaf-based mulch approved by the department. In December 2005, the compost caught fire. Most thought it had been set intentionally by people riding on

Figure 3.1.5. Area of the Excelsior coal fire. The reddish areas on the hillside are the locations where shale had been heated to temperatures of 750°C or more by the fire. Photo by Daniel H. Vice, 2009.

all-terrain vehicles. Fumes from the fire drifted toward the city of Shamokin, and soon coal waste on the site had caught fire as well (Gilger, 2005; Sciccitano, 2006a; Long, 2006a).

The DEP was soon on the site. Swank was ordered to have a backhoe and bulldozer work 8 h a day to remove the burning refuse and push it into a pond. (A nearby stream was already dead from acid mine drainage, so runoffs from the pond could not make things worse.) Workers from Reading Anthracite helped with the clean-up efforts. By January the DEP's Tom Rathburn could report that the western end of the refuse pile was extinguished, and they were working toward the eastern end. By summer the fire appeared to be out, but then it restarted. The DEP again suspected arson, and fire mitigation efforts were delayed until State Police could conduct what turned out to be an inconclusive investigation. Swank was now ordered to remove all the compost, and again he set to work (Sciccitano, 2006a; Sciccitano, 2006b; News-Item Staff, 2006a,b; Long, 2006b,c; Dorman, 2006; Wheary, 2006; News-Item Staff, 2006c). In 2007 the DEP realized the fire had burned beneath the refuse into a coal bed. This led to the fire being reclassified as a mine rather than coal refuse fire. Efforts to put the fire out were redoubled. When the authors visited the site in July of 2010, there was no evidence of a continuing fire (Figure 3.1.5).

Since the summary article on coal fires in northeastern Pennsylvania was prepared, some other fires have been identified. In some cases, the fires had existed for some time and were only recently brought to the attention of the state. In other cases, these were apparently new coal fires.

The four recent coal fires near Mildred, Palo Alto, Olyphant, and Excelsior have several similarities. In all four cases, as soon as the communities discovered the fires help was sought from the Office of Surface Mining and/or Pennsylvania's DEP. Representatives from those agencies were on site almost at once; plans were put in place to extinguish or control the fires and were promptly implemented. Although the Excelsior fire proved harder to contain than anticipated, possibly because of the actions of arsonists, few would argue that the threats weren't handled well. There was very little of the community infighting and resulting political unresponsiveness predicted by the Couch and Kroll-Smith model.

Ironically, one reason for the effective handling of these fires might be Centralia itself. The Dolph, Palo Alto, and Excelsior mine fires received extensive press coverage, and Centralia seemed to be very much on people's minds. The comparison flowed both ways. For example, in the case of Palo Alto, nervous citizens, fearful for their community, cited Centralia to encourage action. In the December 31, 2008 issue of the *Pottsville Republican*, Lee Dalton was quoted as saying, "I've heard about Centralia. That's the first thing I think of. It's been burning since June and nobody said anything to us? I have a problem with that." On the other side government officials mentioned Centralia to explain what the Palo Alto fire was not. For example, when State Representative Neal Goodman and Congressman Tim Holden visited the fire, they were quoted as saying, "it's far enough away from people's homes that it's not a danger. It is not a Centralia-type fire." Palo Alto borough council member Gerald Richter was quoted in the *Pottsville Republican* on January 7, 2009 saying: "This is not another Centralia". He made similar comments to the paper on January 13 and January 19. Overall, out of the 10 articles the *Pottsville Republican* devoted to the fire between December 20, 2008 and May 5, 2009, seven of them, in one way or another, referenced Centralia (Pottsville Republican, 2008).

The fire outside Excelsior showed a similar pattern. Szymanski (2006) asked in the *News-Item* of Shamokin, Pennsylvania, "Is Excelsior another Centralia in the making? That was just one of the concerns brought up at Thursday's Coal Township Commissioners meeting by residents worried about a compost pile that continues to burn near Excelsior." The township commissioners and DEP had a fiery debate in the paper over who approved of the leaf-based mulch Swank had used, and whether the public had had too small of an oversight role. According to the *News-Item Staff* (2006a), "The name Centralia will forever be synonymous with government inaction at all levels. The fire spread while some ignored the problem or when the fire was addressed ineffectively. Surely this area will never allow this to happen again." Essentially, the memory of the Centralia fire causes communities to respond to coal fires differently than in the past, and the perceived failure of the government (whether rightly or wrongly assigned) in the case of Centralia encourages the state bureaucracy to respond to fires in a more robust way (Szymanski, 2006; Rathbun, 2006; Fetterman et al., 2006; Gilger, 2006; News Item Staff, 2006a; Sciccitano, 2006c; Szymanski and Heintzelman, 2007; Wolf, 2007; Sciccitano, 2007a,b).

A second factor that may explain the new ways communities respond to mine fires as compared to what happened in Centralia in the 1960s is the passage of time. In the 1960s most residents of the anthracite coal fields had experience with the mining industry, and were not surprised by its inherent dangers. Today that is no longer the case. When the Palo Alto fire broke out, borough councilman Charles Dries was quoted as saying, "The kicker here is there hasn't been any mining in our town for probably 75 years." (Later callers to the newspaper suggested 40 years might be more accurate.) Either way, only the oldest residents of Palo Alto would have had much experience with mine problems. This would have made the fire less familiar and more alarming. Centralia, though, has achieved something of a cult status as a tourist destination. Everyone has seen it, and knows what a mine fire can do, and reacts accordingly. As the authors conducted research on this and other papers, and have talked to people who live near coal fires, almost everyone has told us to go see Centralia. It has come to epitomize coal fires in a way no other fire has.

Acknowledgments

The authors would like to thank project editor Glenn B. Stracher for his long years of support as they have conducted their research. Mike Korb has helped create a sense of community among the geologists in northeastern Pennsylvania. Jessica A. Mazaika did the initial research on the Summit Hill mine fire as a class project. The library staffs of Penn State-Schuylkill and Penn State-Hazleton campuses both stepped up at various times with assistance. Additional research was done at the Luzerne County Historical Society and Osterhout Free Public Library, both in Wilkes-Barre, Pennsylvania and also provided research help.

Important Terms

anthracite
Carbondale
Centralia
cogeneration plants
culm bank
fluidized bed combustion chamber
Northampton Fuel Supply
Pennsylvania
Pottsville and Llewellyn Formations

References

Ashmead, D.C., May 11, 1922. Red Ash Company's Mine Fire, Thought to Be Slushed Out, Blazes up, Threatening Nearby Communities: Coal Age, pp. 769–770.

Aurand Jr., H., Vice, D.H., 2015. Co-generation plants: from environmental impact to resource availability. In: Guidebook for the 80th Annual Conference of Pennsylvania Geologists, pp. 39–44.

Blasé, F., March 29–30, 1997. Heckscherville (Anchor) Colliery: The Pottsville Republican & Evening Herald, p. 5.

Cao, D., Fan, X., Guan, H., Wu, C., Shi, X., Jia, Y., 2007. Geological models of spontaneous combustion in the Wuda Coalfield, Inner Mongolia, China. In: Stracher, G.B. (Ed.), Geology of Coal Fires: Case Studies from Around the World, Geological Society of America Reviews in Engineering Geology XVIII, pp. 23–30.

Detterline Jr., M.E., 1968. Tamaqua Tales, Kutztown Publishing Company, Kutztown, pp. 54–57.

Dierks, H.A., Whaite, R.H., Harvey, A.H., 1971. Three Mine Fire Control Projects in Northeastern Pennsylvania: U.S. Bureau of Mines Information Circular. 8524, 53 p.

Dorman, I., November 20, 2006. It's Time to Do Something about the Stench. News-Item, Shamokin, Pennsylvania, p. 4.

Eggleston, J.R., Kehn, T.M., Wood Jr., G.H., 1999. Anthracite. In: Schultz, C.H. (Ed.), The Geology of Pennsylvania, Pennsylvania Geological Survey and Pittsburgh Geological Society. Harrisburg, Pennsylvania, pp. 459–469.

Faill, R.T., Nickelsen, R.P., 1999. Appalachian mountain section of the Valley and Ridge Province. In: Schultz, C.H. (Ed.), The Geology of Pennsylvania. Pennsylvania Geological Survey Special Publication 1, pp. 269–285.

Fetterman, C., Saintor, A., Welsh, G., Zalor, G., 2006. Coal Township Commissioners Rap DEP. The News-Item, Shamokin, Pennsylvania, p. 4.

Gilger, M., December 15, 2005. Mystery Solved: Burning Compost Pile at Reclamation Site Is Source of Foul Odor. News-Item, Shamokin, Pennsylvania, pp. 1–7.

Gilger, M., November 10, 2006. Coal Township Sick of Stench: Businessman Promises to Rent Bus to Take Their Fight to the State Capital. The News-Item, Shamokin, Pennsylvania, pp. 1–7.

Inners, J.D., Edmunds, W.E., Laregina, J.A., 1996. Independent steam-electric power plants in the anthracite region, Northeastern Pennsylvania: site geology, coal-refuse bank utilization, and environmental benefits. In: Chiang, S.-H. (Ed.). Chiang, S.-H. (Ed.), Coal Energy and Environment, 13th Annual International Pittsburgh Coal Conference Proceedings, vol. 2. University of Pittsburgh, Center for Energy Research, pp. 813–818.

Jackson, K., October 30, 2016. New Uses for Coal Waste in Pennsylvania, Lebanon Daily News. The Washington Times, p. 13A. http://www.washingtontimes.com/news/2016/oct/29/new-uses-for-waste-coal-in-pennsylvania (accessed July 2017).

Kim, A.G., Chaiken, R.F., 1993. Fires in Abandoned Coal Mines and Waste Banks: U.S. Bureau of Mines, Information Circular. 9352, 58 p.

Levine, J.R., Eggleston, J.R., 1992. In: The Anthracite Basins of Pennsylvania, US: Guidebook for Joint Meeting of the International Committee for Coal and Organic Petrology (44th Meeting) and the Society for Organic Petrology (9th Meeting), U.S. Geological Survey Open-file Report # 92-568, 72 p.

Lindemuth, J.R., June 15, 2008. Colliery fire took months to extinguish. Republican & Herald, Pottsville. Sunday, p. C-3.

Long, K., August 17, 2006a. Cause of Compost Fire Unknown: DEP Looks to Have Burning Smelly Heap Extinguished in a Week. The News-Item, Shamokin, Pennsylvania, pp. 1–6.

Long, K., September 1, 2006b. Fire Fight Goes on: DEP "Burning Compost Pile Could Be Out in Weeks. The News-Item, Shamokin, Pennsylvania, pp. 1–7.

Long, K., October 5, 2006c. The Smell of Progress. The News-Item, Shamokin, Pennsylvania, pp. 1–7.

Malenka, W.T., 1985. Final Report: Forestville Mine Fire, Office of Surface Mining. Project No. PA 83-004, 3 p.

Mazaika, J.A., 2007. The Burning Mines of Summit Hill. Unpublished paper for a project at Penn State Hazleton, 9 p.

Munley, K.P., 1998. Mine fire: the West side Carbondale mine fire, 1947–1961. In: Proceedings of the Tenth Annual Conference on the History of Northeastern Pennsylvania. Luzerne County Community College, pp. 67–85.

Nelson, M.I., Chen, X.D., 2007. Survey of experimental work on the self-heating and spontaneous combustion of coal. In: Stracher, G.B. (Ed.), Geology of Coal Fires: Case Studies from Around the World, Geological Society of America Reviews in Engineering Geology XVII, pp. 13–83.

News Item Staff, December 9, 2006c. DEP: Time's up on Fire: Gilberton Coal Ordered to Start Work in Five Days, Finish in 120. The News-Item, Shamokin, Pennsylvania, pp. 1–7.

News-Item Staff, August 12, 2006a. It's Time to Solve Compost Problem once and for All. The News Item, Shamokin, Pennsylvania, p. 4.

News-Item Staff, August 15, 2006b. Stench Returns: DEP, Police Set Meeting in Excelsior to Address Noxious Compost Fire. The News-Item, Shamokin, Pennsylvania, pp. 1–7.

Nolter, M.A., Vice, D.H., 2004. Looking back at the Centralia coal fire: a synopsis of its present status. In: Stracher, G.B. (Ed.), Coal Fires Burning Around the World: A Global Catastrophe, International Journal of Coal Geology, vol. 59, pp. 99–106. Issue 1–2.

Pottsville Republican, 2008. Palo Alto Mine Fire, Pottsville, Pennsylvania. Palo Alto Vertical File in the County Collection. Ciletti Memorial Library, Penn State Schuylkill Campus, The Pennsylvania State University, Schuylkill Haven, Pennsylvania. December 30–May 5, 2009.

Randolph, A., 2002. Overview of the Laurel Run Mine Fire, Borough of Laurel Run, Luzerne County, Pennsylvania. Luzerne County Historical Society, Wilkes-Barre. 7 p.

Rathbun, T., October 18, 2006. DEP Offers its Take on Stench Complaints. The News-Item, Shamokin, Pennsylvania, p. 4.

Scherer, P., 1976 exact date unknown. Tamaqua tragedies of the 1850s, The County Collection. Penn State Hazelton Library (non-circulating) 16 p.

Sciccitano, E., January 11, 2006a. Burning Question: DEP Smoldering Refuse Doesn't Threaten Coal Vein. The News-Item, Shamokin, Pennsylvania, pp. 1–7.

Sciccitano, E., January 22, 2006b. Coal Refuse Fire "Shrinking": Bulldozing, Dousing, Continues; DEP Visits Regularly. The News-Item, Shamokin, Pennsylvania, p. 2.

Sciccitano, E., December 14, 2006c. Threats Bring Action: DEP Prodded by Planned Bus Trip to Voice Worries over Compost Fire. The News-Item, Shamokin, Pennsylvania, pp. 1–7.

Sciccitano, E., January 12, 2007a. Burning Compost "Mine Fire" but DEP Doesn't See This as Threat to Nearby Properties. The News-Item, Shamokin, Pennsylvania, pp. 1–7.

Sciccitano, E., February 11, 2007b. CO Readings High at Coal Refuse Site but Toxic Gas Doesn't Pose Risk DEP Says. The News-Item, Shamokin, Pennsylvania, pp. 1–10.

Stracher, G.B., Nolter, M.A., 2013. Coal fires of the United States: Pennsylvania. In: Stracher, G.B., Prakash, A., Sokol, E.V. (Eds.), Coal and Peat Fires: A Global Perspective, Volume 2: Photographs and Multimedia Tours, pp. 498–509.

Sullivan Review, June 3, 2004. Subsurface Mine Fire Located between Mildred and Lopez, p. 1 Sullivan County, Pennsylvania.

Szymanski, M., Heintzelman, A., January 14, 2007. DEP: There's No Mine Fire: Nearby Property Owners Shouldn't Be Concerned Official Adds. The News-Item, Shamokin, Pennsylvania, pp. 1–5.

Szymanski, M., August 11, 2006. Another Centralia in Making? Excelsior Residents Fear Consequences of Burning Compost Pile. The News-Item, Shamokin, Pennsylvania, pp. 1–5.

Uludag, S., 2007. The spontaneous combustion index and its application: past, present, and future. In: Stracher, G.B. (Ed.), Geology of Coal Fires: Case Studies from Around the World, Geological Society of America Reviews in Engineering Geology XVIII, pp. 15–22.

Wheary, R., December 3, 2006. Coal Company Posts Blame for Compost Fire: Lessee Says Blaze Would Be Out if Landowner Wouldn't Interfere. The News-Item, Shamokin, Pennsylvania, pp. 1–6.

Wolf, L., January 14, 2007. Resident Needs to Stand up and Scream. The News-Item, Shamokin, Pennsylvania, p. 6.

CHAPTER 4

The Summit Hill Coal-Mine Fire, Pennsylvania

Beginning in 1827, a switchback gravity railroad was used to transport anthracite from the coal mines around Summit Hill, Pennsylvania to the docks along the Lehigh River at Mauch Chunk (whose name was changed to Jim Thorpe in the 1950s) (Mauch Chunk Historical Society, 2017; Wikipedia, 2017). A coal car was loaded with anthracite in Summit Hill, and then the coal was transported by boat to New York City and Philadelphia. Empty coal cars were pulled back to the mines by mules. The gravity railroad was also used for tourism, and one of the main attractions for tourists was the mine fire at Summit Hill, where lunch was served to sightseers. *Photo by Daniel H. Vice, 2017.*

CHAPTER CONTENTS

Coal and Peat Fires: A Global Perspective
Edited by Glenn B. Stracher

https://doi.org/10.1016/B978-0-12-849885-9.00004-4

Photo by Daniel H. Vice, 2017.

4.1 The Summit Hill Mine Fire

Daniel H. Vice
Harold J. Aurand, Jr.
Nicole M. Andel

Coal was transported to Jim Thorpe (formerly Mauch Chunk) along the Lehigh River by a gravity railroad from the mines around Summit Hill. The coal was then loaded onto boats and delivered to New York City and Philadelphia.

Introduction

Because of their location near the urban markets of Philadelphia and New York City, the **anthracite** coal fields of Northeastern **Pennsylvania** developed into America's earliest center of coal production. Canals along the Lehigh, Schuylkill, Delaware, and Susquehanna rivers provided miners with access to the market in the 1820s. This was long before laws were in place to regulate mining. Problems like mine fires, which grew as the industry expanded, weren't viewed as threats to public health and safety, but rather as natural results of the mining process to be handled by the companies producing the coal.

One result of this lack of government oversight is that we have little information on these early mine fires. Occasionally they were covered in the day's newspapers or written about later by local antiquarians, but such mentions tended to be brief. This has made it nearly impossible for researchers to look at trends of fire occurrence or the techniques used in fighting the fires. The **Burning Mines Fire** at the **Summit Hill Mine**, of the **Lehigh Coal and Navigation Company**, is the exception. Because of the fire's unique location near the tourist town of **Mauch Chunk** (since renamed Jim Thorpe), its progress was covered extensively in the press and travelers' accounts. It is the best documented early anthracite fire.

Summit Hill is a coal mining town east of Tamaqua on the southeastern limb of the **Southern Anthracite Coalfield** (Figure 4.1.1). Commercial mining began in the 1820s, when canals along the Lehigh and Delaware rivers gave the area access to the Philadelphia market. Along with mining came environmental problems, including mine fires. Although some coals can reportedly start burning spontaneously, anthracite, with its high carbon content, cannot. Human intervention, whether purposeful or accidental, is necessary to start an anthracite fire. We are not aware of any anthracite fires that predate the mining industry.

We believe the Summit Hill Mine fire was one of the earliest major coal fires in the anthracite region. Proving that assertion, though, is difficult. Prior to World War II, coal-mine fires in Pennsylvania were not seen as public health hazards, requiring state intervention, but as economic problems, to be handled or not as the coal companies that owned the land saw fit. Without state intervention and oversight, records of the fires, and how they progressed, are scarce. Most anthracite mining companies went out of business long ago. Their corporate records were often scattered or destroyed. Sometimes a fire would attract the interest of the local press or antiquarian writers, but many did not achieve such notoriety, or have not yet been uncovered by researchers (Aurand et al., 2015). The Summit Hill Mine fire is better documented than most because it became a tourist attraction (Figure 4.1.2). What we have tried to do is gather together the pieces of information available and create a story of an early Pennsylvania mine fire, including both how the fire was fought and its overall economic implications.

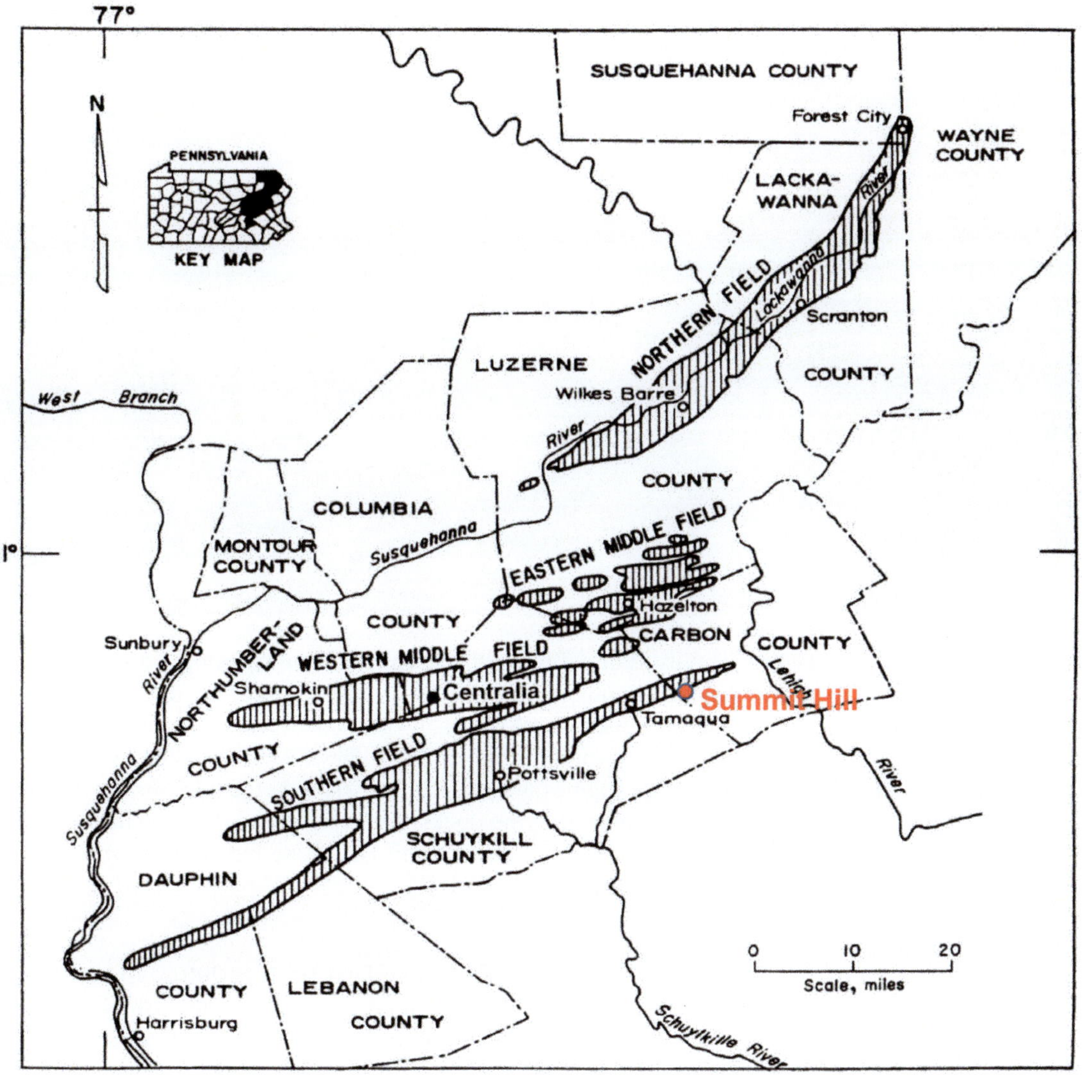

Figure 4.1.1. Map of the southern, western, eastern middle, and northern anthracite fields in eastern Pennsylvania. The famous Centralia Mine Fire is in the western middle field, Columbia County. From Stracher and Nolter (2013); Courtesy of Timothy Altares, Pennsylvania Department of Environmental Protection, Harrisburg, Pennsylvania.

History and Development

In February of 1791, Philip Ginder discovered shiny black stones near the roots of an upturned tree on the summit of Sharp Mountain. Ginder took the strange rocks to his neighbor, Colonel Jacob Weiss. Weiss had been Deputy Quartermaster General during the Revolutionary War, and had contacts extending from George Washington to members of the Philadelphia merchant community. Weiss had moved to the area at the end of the Revolution, buying a piece of land along the Lehigh River from the Moravian Church (Hydro, 2002). He took Ginder's discovery to Philadelphia to have it assessed. When informed that it was anthracite coal, he met with his cousin, Michael Hillegas (former treasurer of the United States), his brother-in-law and publisher Charles Cist, and noted land speculator and Comptroller-General of Pennsylvania, John Nicholson to discuss commercially developing the site. They knew the area around Sharp Mountain was largely unsettled forest. The local wood supply easily met the needs for heating and cooking. Philadelphia, though, with its growing population, was in the early stages of a fuel crisis. Together, Weiss and his associates created two joint-stock companies, the Lehigh Coal Mine Company and Lehigh Navigation Company, to mine the anthracite and ship it to market (Knies, 2001).

When the first shipments arrived in Philadelphia, wealthy residents saw an opportunity for investment. Large sections of land were purchased along the Nesquehoning Creek. Others invested in the now-merged Lehigh

Figure 4.1.2. In 1827, a switchback gravity railroad opened in Pennsylvania, carrying anthracite coal from the mines around Summit Hill to docks at Mauch Chunk, about nine miles to the west along the Lehigh River. Gravity was used to transport ore cars downhill to the docks where the coal was loaded onto canal and riverboats for transport to New York City and Philadelphia, Pennsylvania. Once emptied, the cars were pulled by mules back to the mines. Years later, gravity and stationary steam engines were used. Tourists rode the train on days when the mines were closed. In 1872 a shorter route for shipping coal to the docks was created and the switchback was used only for tourism. One of the main attractions on the route was the Summit Hill Mine Fire in Summit Hill. The switchback closed in 1932. Photo from Wikimedia Commons, 2017.

Navigation and Coal Company, allowing that entity to take up more land (Hydro, 2002). At the same time, the investors discovered two problems with the operation. First, the public had to be taught to use anthracite coal (Hydro, 2002). While anthracite gives off a tremendous amount of heat, it can be difficult to ignite. The Philadelphia Water Works, for example, purchased an early load of anthracite coal, and wound up using it to gravel their walkways (Niehoff, 1995). Special furnaces and stoves or grates for burning anthracite domestically had to be introduced before it could be widely used.

A second problem was how to transport the coal from the top of Sharp Mountain, where a monument was erected in 1941 to mark the site where Grinder first discovered coal (Knies, 2001), to the Lehigh River, over eight miles away. The Lehigh Coal and Navigation Company erected a three-part system. To move the coal off the mountain they originally constructed a stone turnpike, although everyone assumed it would later be replaced by a railroad (Knies, 2001). Along the shores of the Lehigh River coal storage and boat loading facilities were developed. Finally, a system of slack water navigation, using short canals and dammed pools in the Lehigh River, brought the coal down to the confluence with the Delaware River.

For purposes of this chapter, the railroad is the most important part. Originally Lehigh Coal and Navigation used gravity for motive force. Several cars of coal would be strung together, followed by a car loaded with mules. Gravity would pull the loaded train down to the Lehigh River, then the mules would be unloaded and they would pull the empty cars back to the top. Later, two stationary steam engines were built to replace the mules. The workers who ran the gravity railroad noticed that the public was interested in their work. They received permission from Lehigh Coal and Navigation to run passenger trains on days when the mines weren't operating. In 1872 the Hauto Tunnel was completed through Sharp Mountain, and suddenly the gravity railroad wasn't needed for coal at all—it became a pure tourist train (Niehoff, 1995).

Tourism was a viable option because of what was at the other end of the railroad. While Sharp Mountain and nearby Panther Valley had large deposits of coal, the area immediately around the loading docks in Mauch Chunk did not. Instead there was the Lehigh River gorge and several steep mountains. Sometimes called the "Switzerland of America," Mauch Chunk was second only to Niagara Falls as an American tourist attraction in the 1800s. Key

Figure 4.1.3. Buck Mountain coal in vertical beds in the south limb of the southern anthracite coal field along Route 61 in Pottsville, Pennsylvania. Also see Figure 4.1.1. Photo by Daniel H. Vice, 2017.

attractions were Glenn Onoko Falls and the gravity railroad (Niehoff, 1995). One of the goals of the people of Summit Hill was to convince some of those tourists riding the railroad to get off and spend a few hours at the top of the mountain. Hopefully, they would patronize local shops and restaurants, or perhaps even stay the night. One attraction at the top of the hill was an ice cave, where ice remained throughout the summer. Then, in 1874, Summit Hill began advertising a new attraction, the Burning Mines Fire (Knies, 2001).

The Burning Mines Fire

On February 15, 1859, an active mine fire was discovered in some abandoned mine workings on the Mammoth vein (Anonymous, 1910). Like many other coal beds in the area, the Mammoth was vertical to slightly overturned (Figure 4.1.3). This orientation favored small, individual operators who could come in, lease the land, and focus on the most accessible coal. Large mining operators sought easier to access deposits (Wallace, 1988). Work at the fire site had begun in 1850 and was abandoned shortly thereafter.

The exact cause of the fire is unknown. One story claims that some boys, who were playing on the slope, built a fire to warm themselves and neglected to put it out, allowing it to spread into the workings. Another story focuses on the mine ventilation system. In the early days of anthracite mining one method of drawing fresh air into the mines was to put a large stove or furnace at the bottom of a ventilation shaft. As the furnace heated the air, it would rise to the surface, drawing fresh, outside air in to take its place. Some believe an accident with a stove or furnace started the mine fire (Brenckman, 1913). Finally, the Number Nine Wash Shanty Museum in Lansford, Pennsylvania claims a wooden mine car, carrying a newly raked pile of ashes from a stove, was mistakenly shunted into an old disused passageway. The mine car eventually ignited and the fire spread to the timbers and coal and the passageway.

Regardless of its origin, by 1874, the Burning Mines Fire had become an impressive site. Tourists, particularly those unfamiliar with mining operations, were willing to get off the gravity railroad and take a look, even if it meant having to kill some time shopping or eating in Summit Hill while waiting for the next train.

Dealing with the Fire

While the mining operation that gave rise to the Burning Mines Fire was defunct, it was still a threat to other nearby mines. Gases from underground fires could spread into working mines, causing suffocation. Worse, most mining in the anthracite region started near the surface and gradually moved further and further underground. Mining deeper drove up the production costs. Seeing their profits shrink, many mine owners began to look at the coal around the Burning Mines Fire as a lost opportunity. Putting out the fire might give easy access to cheaper coal deposits.

Putting out the Summit Hill fire, though, would not be an easy process. Usually, the easiest ways to put out a mine fire is to flood the mine with water or cut off the fire's access to air. Due to cave-ins and fissures in the rock around the Mammoth vein, there was no way to cut off the air supply or prevent water from running away from the fire (Anonymous, 1910). In the 1860s a plan was devised to dig a great open cut, with sloping sides, across the path of the fire in the hopes than when it reached the break in the coal seam, it would burn out. The cost of digging the cut proved prohibitive, and no effort was made to carry it out (Anonymous, 1910).

Years later an attempt was made to block the fire by drilling bore holes and smothering the fire with culm, coal dirt, and water. The fire burned through the barrier (Brenckman, 1913). It was only after 1900, over 40 years after the start of the fire, that W.A. Lathop, President of the Lehigh Coal and Navigation Company, came up with the plan that would permanently end the Summit Hill fire. Lathop's plan called for a 274-m (900 ft)-long clay barrier built from the surface down to the water line. The clay would be supported by two concrete walls, 18 m (60 ft) in height and 3.7 m (12 ft) wide. Each concrete wall had "windows" built in at regular intervals. When wet clay and sediment were poured in, they would seep through on both sides. The material would fill in any voids, preventing the fire from advancing. Lathrop's plan worked, and by 1910 the fire was extinguished (Anonymous, 1910).

Lathrop's plan had required 50,000 cubic yards of clay and 8000 cubic yards of concrete to finish the barrier. Ten miles of timber were used in the shafts and open cut, all of which came from North Carolina. One of the holes bored had 24,000 tons of silt poured into it. It is estimated the total cost of the total cost of the plan was $250,000. Had the fire continued, it might have ultimately followed the Mammoth vein into the Panther Valley, up past the town of Lansford. Such a fire would have threatened 400 million tons of coal (Anonymous, 1910). Of course, losing the Burning Mines Fire hurt the Summit Hill tourist industry, but the accidental burning of the palatial Glen Onoko Hotel probably hurt more. By the 1930s the gravity railroad was torn up and sold for scrap.

Discussion

The area around Summit Hill, Pennsylvania was both a center of the early anthracite coal industry, and thanks to the gravity railroad, an early Victorian tourism site. That rare combination guaranteed that the Summit Hill Mine fire would receive a lot more attention than other early mine fires. At the time, the state and federal governments considered mine fires to be economic problems, which should be solved by the mining companies that owned the land. There was no reason for government monitoring or record-keeping of fire sites. For the mining companies, whether and how to fight the fires was decided by comparing the costs of controlling the fire to the expected revenue that might be gained if mining continued. The Summit Hill fire was more complex because the tourist economy had to be factored into those calculations. Regardless, it is the most completely documented account of how a major anthracite fire was dealt with that we have found to date.

Acknowledgments

This chapter originally started as a class project by Jessica A. Mazaika, a student in one of the senior author's classes. She received help from her father, David Mazaika Sr., who used his personal experience to provide background information on mining and the anthracite fields of Pennsylvania. All the authors wish to thank the personnel at the Number Nine Wash Shanty in Lansford for providing maps and showing their old artifacts. Michael Korb, who works in mine reclamation, provided connections to local historians and websites. Finally, we wish to give special thanks to Lee Mantz of the Summit Hill Historical Society.

Important Terms

Anthracite
Burning Mines Fire
Lehigh Coal and Navigation Company
Mauch Chunk
Pennsylvania
Southern Anthracite Coalfield
Summit Hill
Summit Hill Mine
Summit Hill Mine fire

References

Anonymous, January 30, 1910. Check Mine Fire after 51 Years: $250,000,000 Worth of Coal Burned in that Time in Working Near Summit Hill. Penn, New York Times, p. 6.

Aurand Jr., H., Nolter, M.A., Vice, D.H., 2015. Summary of coal fires in Northeastern Pennsylvania. In: Guidebook for 80th Annual Field Conference of Pennsylvania Geologists, pp. 59–74.

Brenckman, F.C., 1913. History of Carbon County, Nungesser J.J., Pennsylvania, Harrisburg, p. 626.

Hydro Jr., V., 2002. The Mauch Chunk Switchback: America's Pioneer Railroad. Easton, Pennsylvania Canal History and Technology Press, p. 358.

Knies, M., 2001. Coal on the Lehigh 1790–1827: Beginnings and Growth of the Anthracite Industry in Carbon County, Pennsylvania. Easton, Pennsylvania Canal History and Technology Press, p. 83.

Mauch Chunk Historical Society, 2017. http://www.mauchchunkhistory.com. (accessed November 2017).

Niehoff Jr., W.H., 1995. The Reincarnation of the Switchback Gravity Railroad: The Story of the Switchback and the Switchback Operating Model, p. 110. Self-published.

Stracher, G.B., Nolter, M.A., 2013. Coal fires of the United States: Pennsylvania. In: Stracher, G.B., Prakash, A., Sokol, E.V. (Eds.), Coal and Peat Fires: A Global Perspective, Volume 2: Photographs and Multimedia Tours, pp. 498–509.

Wallace, A.F.C., 1988. St. Clair: A Nineteenth Century Coal Town's Experience with a Disaster Prone Industry. Cornell University Press, Ithaca, p. 519.

Wikimedia Commons, 2017. File: Jefferson Plane Looking up.jpg. https://commons.wikimedia.org/wiki/File:Jefferson_plane_looking_up.jpg. (accessed November 2017).

Wikipedia, 2017. Jim Thorpe, Pennsylvania. https://en.wikipedia.org/wiki/Jim_Thorpe_Pennsylvania. (accessed November 2017).

CHAPTER 5

Modern and Ancient Coal Fires in the Powder River Basin, Wyoming and Montana

A coal train passing through a railroad cut in clinker. The coal storage silos from the Antelope Mine, the southernmost active coal mine in the Powder River basin, are visible in the distance. The clinker formed during prehistoric combustion of the Anderson coal bed, probably within the past few hundred-thousand years. This low-sulfur, subbituminous coal is from the uppermost Fort Union Formation of the Paleocene age. It likely ignited naturally due to range fires or spontaneous combustion and baked the overlying sandstone, siltstone, and shale layers to produce the reddish clinker. The Anderson coal bed is being mined today at the Antelope Mine and transported by rail to power plants in several states where the coal is burned to produce electricity. Crushed clinker is also being used as railroad ballast below the tracks. The view looks northwest toward the mine from Converse County Road 37 along Logan Draw, 36 km southeast of the town of Wright, Wyoming. *Photo from Heffern and Coates (2004); with permission of Elsevier.*

CHAPTER CONTENTS

Coal and Peat Fires: A Global Perspective
Edited by Glenn B. Stracher

https://doi.org/10.1016/B978-0-12-849885-9.00005-6

5.1 Coal Fires and Clinker

Edward L. Heffern
Catherine A. Riihimaki
Peter W. Reiners

Photo by Edward L. Heffern, 2008.

An active coal fire near the Tongue River, north of Sheridan, Wyoming. The fissures were caused by the slumping of soil and rock due to the volume reduction of a coal bed burning underground. The baked rock, called clinker, is reddish in color.

Introduction

Hundreds of coal fires burn today in the Powder River basin (PRB) of northeast Wyoming and southeast Montana. Some consume coal in abandoned underground mines or burn in highwalls and spoil piles at active surface mines, while others naturally ignite on coal outcrops from range fires and spontaneous combustion. But they are only part of the picture. Natural coal fires also have occurred in the geologic past, as coal beds exposed by erosion dewatered and ignited. These ancient fires left a record of **clinker** on hilltops and ridges throughout the landscape. The reddish, brick-to-lava-like clinker rock formed when fires in coal outcrops burned back into hillsides and baked and melted the overlying sandstone and shale. Clinker outcrops cover 3812 km^2 (1472 mi^2) of the basin (Heffern et al., 1993, 2013) and represent natural burning of tens of billions of tons of coal. Isotopic ages of zircon grains in baked sandstones reveal that coal has burned to form clinker throughout the past few million years (Reiners et al., 2011; Heffern et al., 2007).

This chapter focuses on the eastern part of the basin in Wyoming, where 40.5% of the coal mined in the United States in 2015 was extracted (Energy Information Administration, 2016). The text and pictures in this chapter accompany a map (Figure 5.1.1) that shows ancient and modern coal fires in an area of about 7000 km^2 (2700 mi^2) in the vicinity of the towns of Gillette and Wright. Much of this chapter is summarized or adapted from earlier publications (Heffern et al., 2007, 2013; Riihimaki and Reiners, 2012; Reiners et al., 2011; Riihimaki et al., 2009; Heffern and Coates, 2004) that discussed coal fires and clinker in the entire PRB.

Causes of Coal Fires

Heat energy and air are necessary for coal to ignite. Coal surfaces must be exposed to the air and the coal must be above the water table (i.e., not saturated). The natural process of erosion brings coal above the water table as coal beds are exposed by streams that carve down through the nearly flat-lying strata of the PRB, by erosion of hillsides adjacent to streams, and by landslides. Surface mining creates highwalls that expose fresh coal seams to air and subsidence of mined-out rooms between pillars in underground mines allows air to infiltrate. Once air is available, the heat energy produced by chemical reactions, range fires, or lightning strikes can trigger fires in coal beds. The highly volatile subbituminous coal in the PRB is particularly prone to spontaneous combustion caused by oxidation and heat energy of wetting reactions (Lyman and Volkmer, 2001; Kim and Chaiken, 1993). Weathered coal outcrops are less likely to spontaneously combust, as the coal is already degassed and oxidized.

The coal beds in the PRB are nearly horizontal, and when a fire starts it first spreads along the outcrop. With time, it burns deeper into the hillside, causing the overlying rocks to progressively subside into the burned-out void left by the loss of the coal. The resulting fracturing allows air to enter and gas to escape, feeding the fire. The fire goes out naturally when the overburden becomes so thick that fractures from the collapse fail to reach the surface to draw in more air, when the fire burns down to the water table in the coal, or when surface runoff fills the fractures with

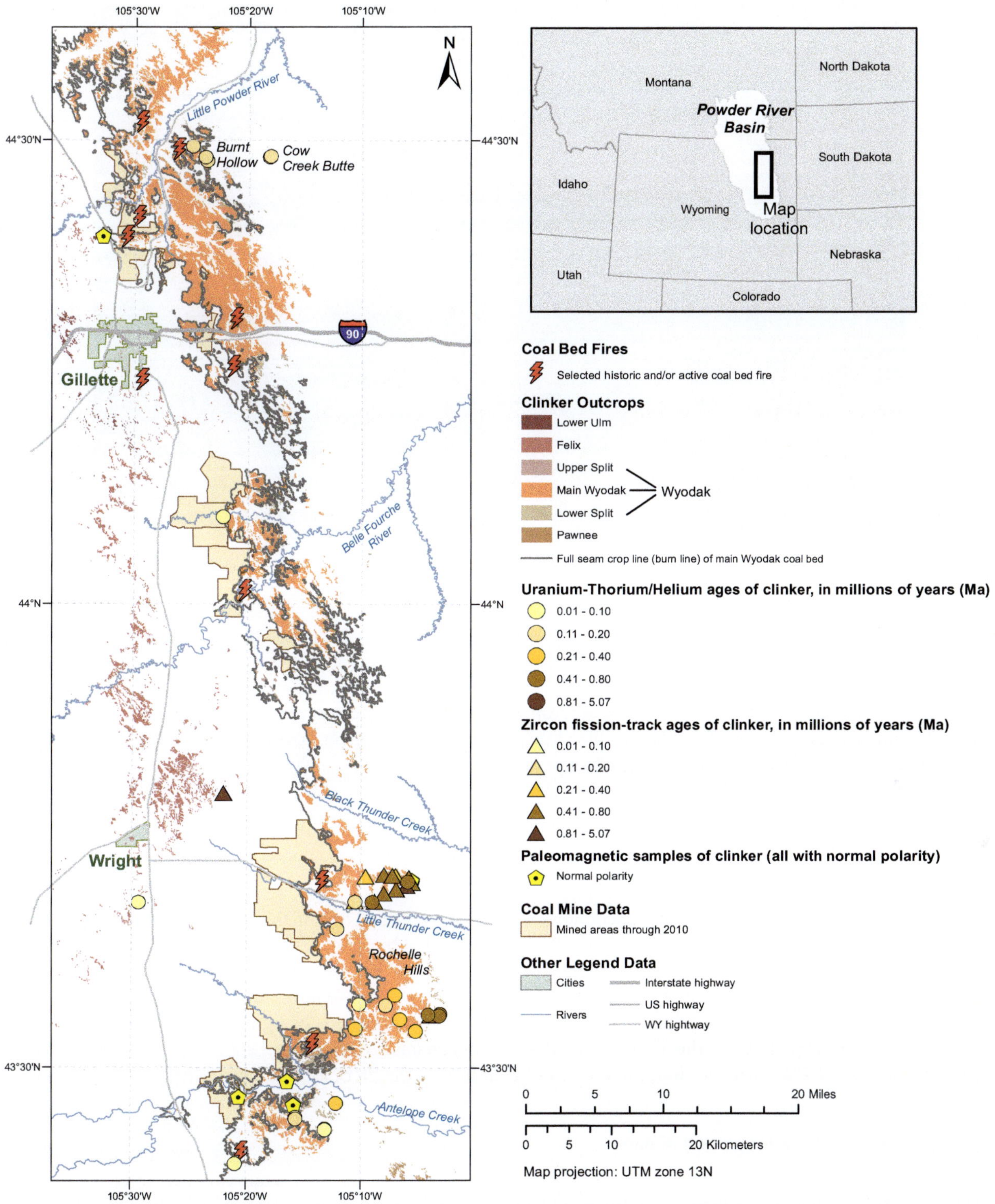

Figure 5.1.1. Modern and ancient coal fires in the vicinity of Gillette and Wright, Wyoming. Known historic and active coal fire locations are shown, along with clinker outcrops and their ages. Areas mined for coal through 2010, as well as the crop line of the main Wyodak coal bed, are included in the map. Figure compiled by Catherine A. Riihimaki and Edward L. Heffern.

Figure 5.1.2. Burn line exposed in the highwall of a surface coal mine in eastern Powder River basin. Photo by Edward L. Heffern, 1993.

sediment and chokes off the fire. Toward the outcrop, coal thickness tapers to zero as increasingly lower parts of the coal bed are burned. Figure 5.1.2 shows the structure of a typical clinker-coal interface. The contact between clinker and unburned coal is called the **burn line** or burn front.

Modern Coal Fires

The explorers Lewis and Clark had many encounters with natural coal bed fires in their journeys along the Missouri and Yellowstone rivers two centuries ago. They correctly associated burning coal beds with the burnt hills of clinker in the northern Great Plains. Clark named one river the Redstone River, because of all the cobbles of reddish clinker in that riverbed where it joined the Yellowstone (Thwaites, 1969). We now call that river the Powder River, which "derives its name from the sulphurous vapors rising from burning beds of lignite in its vicinity," according to an early military expedition in the Yellowstone River drainage basin (Raynolds, 1868, p. 8). The Powder River lent its name to the PRB. There are four main types of coal fires in the PRB, as explained next.

Fires in Abandoned Underground Mines

These are most common in areas mined in the late 1800s and early 1900s before large surface mines became dominant, such as the underground mines along the Tongue River valley north of Sheridan, Wyoming, northwest of the area in Figure 5.1.1. As the rooms between pillars collapse, sinkholes allow air to enter. Coal fires sporadically ignite in the sinkholes and occasionally produce explosions (Dunrud and Osterwald, 1980).

Fires in Highwalls and Spoil Piles

Most surface mining in the PRB occurs in a north–south trending belt east of the towns of Gillette and Wright in northeast Wyoming (Figure 5.1.1). Surface mines in this region have regular programs to extinguish the many fires that start spontaneously in the exposed coal. Fires occasionally occur in surface mines that were abandoned prior to the reclamation laws or where reclamation was not fully successful, as in Figure 5.1.3.

Figure 5.1.3. Underground burning of coal causes steam and vapor to rise from fissures at partially reclaimed Canfield coal mine, Section 14, T41N, R70W, Rochelle Hills, Wyoming. From Heffern and Coates (2004); with permission of Elsevier.

Fires in Natural Coal Outcrops

Outcrops of fresh coal in gullies, stream banks, and highwalls of landslides are especially prone to spontaneous combustion. Also, in the semiarid climate of the PRB, wild range and forest fires triggered by lightning or humans regularly ignite coal beds. Many of these fires were started by burning juniper and pine trees that were rooted in coal beds, such as in Figure 5.1.4.

Low-Level Smoldering of Coal Beds

These are most apparent in winter when the temperature difference is greatest and may be revealed as bare patches of ground or ice-encrusted vents where the snow cover has melted. Rising steam or green vegetation in an otherwise snow-covered area may also indicate smoldering.

Table 5.1.1 lists 11 historic and/or active fires shown in Figure 5.1.1. These include both outcrop and mine fires. Some, such as the Burning Coal Draw fire, were burning at least since Anglo settlers arrived in the late 1800s (Russell and Smith, 1951). Others, such as the Canfield and Andrew Ditto mine fires, resulted from surface and underground mining in the early 1900s. Still others, such as the Rourke fire in 2002, were ignited by range fires (Heffern and Coates, 2004). These 11 fires are by no means all the active or historic coal fires in the map area. Heffern et al. (2013) documented 81 known fires in the entire PRB, of which 53 are outcrop fires, 24 are mine fires, and 4 are spoil fires. These 11 fires are a representative sample of what can be gleaned from the literature (Dobbin and Barnett, 1927; Shellenberger and Donner, 1979; Goodson and Associates, 1986; Kim and Chaiken, 1993). In the Wyoming PRB, the Wyoming Department of Environmental Quality's Abandoned Mine Division has an ongoing program to reclaim and monitor coal fires. The now-disbanded U.S. Bureau of Mines conducted 22 fire control projects in the Wyoming PRB between 1949 and 1977. The U.S. Bureau of Land Management and the Campbell County Fire Department have extinguished other coal fires that ignited from range fires. Success of these efforts has varied and some fires still smolder, such as at the Canfield mine (Figure 5.1.3).

Figure 5.1.4. Smoldering fire in coal bed north of Gillette, Wyoming. July 2002 range fires ignited coal fires in 60 locations. At this location, a range fire burned trees rooted in a coal bed, which caught fire. Photo by Edward L. Heffern, 2002.

Table 5.1.1
A Sample of 11 active and/or historic coal fires near Gillette and Wright, Wyoming.*

Name of Fire	Size	Type of Fire	Year of Origin	Year of Control Project
Moyer Gulch	2 ha (5 acres)	Mine fire	1885, set by horse thief; burning 1915	1935, CCC†; 1950, USBM‡
Little Thunder Basin (Burning Coal Draw)	360 × 180 m; 20 m deep	Outcrop fire	Burning since 1870s or earlier	1951, USBM
Laur	90 × 120 m	Mine fire	Burning since 1939 or earlier	1939, CCC; 1950, USBM
Canfield and Canfield No. 2	340 × 180 m; 30 m deep	Mine fire	Burning since 1925	1950, USBM; 1964, USBM; still smoldering
Robertson	120 × 30 m	Outcrop fire	Unknown; as long as locals recall	1964, USBM
Rourke	3 small fires	Outcrop fire	July 2002 range fire	2002, CCFD§
T7 Ranch	Unknown	Outcrop fire	Burning since 1800s; visited in 1911	Unknown
Rawhide School	<100 m^2	Outcrop fire	Unknown; smoldering in 1986	Unknown
E. W. Record	<10 m^2	Spoil fire	Unknown; smoldering in 1986	Unknown
Andrew Ditto Mine	Unknown	Mine fire	Burning before 1927	Unknown
D. C. Holler	<10 m^2	Mine fire	Unknown; smoldering in 1986	Unknown

*Modified from Heffern et al. (2013); with permission of the Montana Bureau of Mines and Geology.
†*CCC*, Civilian Conservation Corps.
‡*USBM*, U.S. Bureau of Mines.
§*CCFD*, Campbell County Fire Department.

Engle et al. (2012) measured gas emissions from three coal fires in the Tongue River valley north of Sheridan, Wyoming, two on natural coal outcrops and one at an abandoned coal mine. They found that the fires produced locally dangerous levels of carbon monoxide, carbon dioxide, hydrogen sulfide, benzene, and elemental mercury, among other gases. Total CO_2 fluxes ranged from 0.9 to 780 mg/s/m^2. Tar and coal fire mineral samples were also analyzed.

Ancient Coal Fires

Clinker—rock that has been baked, welded, or melted by the burning of coal beds—provides a record of coal fires in the geologic past. Clinker, hardened by heating, forms resistant reddish layers that cap plateaus, hilltops, and escarpments in the landscape of the PRB (Figure 5.1.5).

The clinker is also highly fractured, which allows much of the rainfall and snowmelt to infiltrate rather than run off the surface and erode the outcrop. The unbaked overlying and underlying rocks are eroded more rapidly, leaving the clinker standing in relief (Coates and Heffern, 2000). In the Rochelle Hills of eastern Wyoming, natural burning of the Wyodak coal zone has left an eastward-facing escarpment 100–200 m (300–700 ft) high, capped by a layer of reddish clinker 20–50 m (60–150 ft) thick (Figure 5.1.5). Clinker fragments are abundant in landslides below these escarpments. It is, essentially, a landscape formed by fire.

The block diagram in Figure 5.1.6 shows the structure of a typical clinker hillside. Note the progressive collapse of the sandstone and shale layers in the overburden as the coal fire advances into the hillside. As the burn front retreats from the outcrop, the overburden sags into the void left by the burned coal, leaving a chaotic mass of jumbled rocks. Slump blocks are separated by tension cracks filled with welded breccia and **paralava** (Figure 5.1.7) that indicate areas of intense heating. Rock types in the clinker reflect the rock types in the overburden, and have varying degrees of hardness depending on the degree of heating. Baked sandstone retains its grain structure but hardens to a bricklike texture, while baked shale layers may become welded together to form a rock called **porcellanite**.

Figure 5.1.5. East-facing escarpment capped by clinker formed by ancient burning of Wyodak coal bed, Section 12, T41N, R69W, Keyton Canyon, Rochelle Hills, Wyoming. Photo by Edward L. Heffern, 2002.

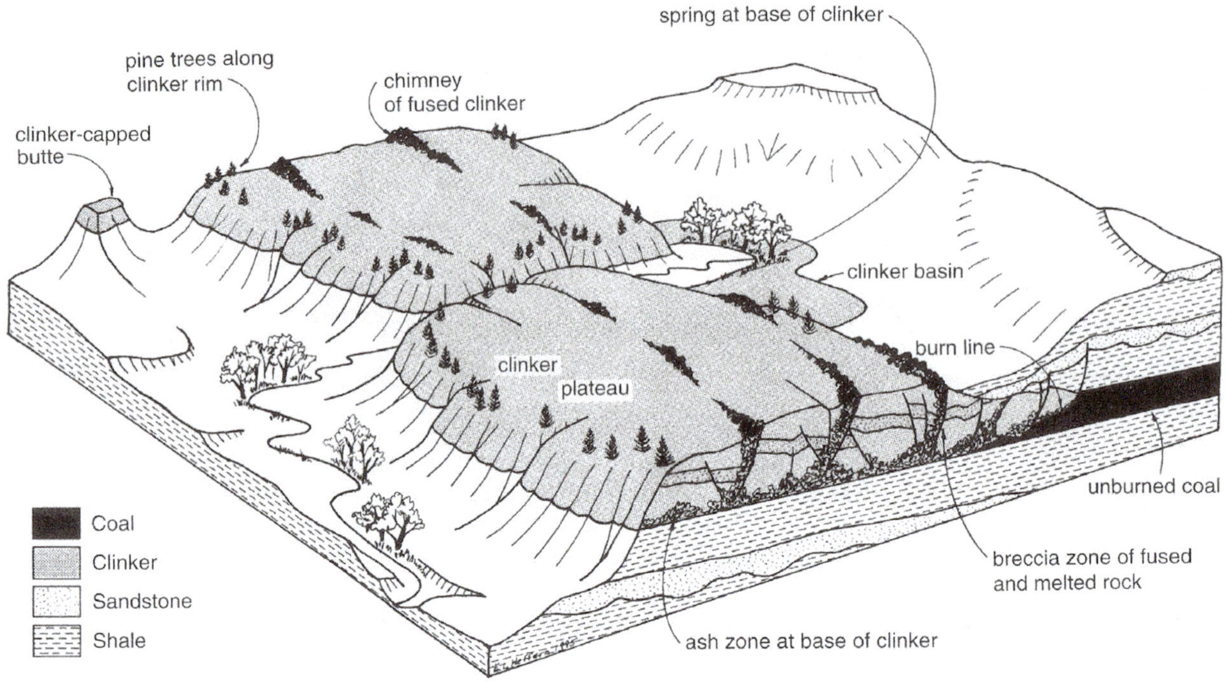

Figure 5.1.6. Block diagram showing the structure of a typical hillside capped by clinker. As the coal fire retreats from the outcrop, overburden sags into the void left by the burned coal. Behind the burn line, unburned coal remains. From Heffern and Coates (1997); with permission of the Wyoming State Geological Survey.

Figure 5.1.7. Welded breccia of orange baked siltstone and black paralava in clinker quarry of surface coal mine, eastern Powder River basin, Wyoming. Photo by Edward L. Heffern, 2002.

Figure 5.1.8. Peter W. Reiners examining ash zone at base of clinker formed by burning of the lower Wyodak (Canyon) coal bed, Section 12, T40N, R70W, Antelope Creek valley, Wyoming. Photo by Edward L. Heffern, 2006.

Paralava forms where rocks were heated enough to melt. **Chimneys** of **welded breccia** and paralava are very hard and resistant to erosion and may stand up to 8 m (25 ft) above the surrounding landscape. At the base of the clinker is a layer, up to a meter thick, of light-tan **ash** and glassy, greenish **scoria**, formed by unburned minerals left over from the original coal bed, which may have been tens of meters thick (Figure 5.1.8). As most of the heat energy of burning is conducted upward, strata more than half a meter below the base of the original coal bed are unbaked (Coates and Heffern, 2000).

Extent of Clinker

Outcrops of clinker cover 3812 km^2 (1472 mi^2) of the PRB, of which 2833 km^2 (1094 mi^2) are in Montana and 979 km^2 (378 mi^2) are in Wyoming (Heffern et al., 2013). In the area of Figure 5.1.1, clinker outcrops produced by prehistoric burning of coal beds cover 466 km^2 (180 mi^2)—7% of the total map area. The clinker averages 15–30 m (50–100 ft) thick but can be as much as 50 m (160 ft) thick. Most (425 km^2 (164 mi^2)) of the clinker in the map area was produced by ancient burning of the Wyodak (Wyodak-Anderson) coal zone in the Tongue River Member of the Paleocene Fort Union Formation (Luppens et al., 2008, 2015; Flores and Bader, 1999). The clinker of the Wyodak coal zone is just east of the 12 active mines on the eastern flank of the PRB (Figure 5.1.9) that produce from this coal zone west of the burn line (Figure 5.1.1). These 12 mines produced 329.6 million metric tons (363.4 million short tons) of coal in 2015. This was 40.5% of total U.S. production (Energy Information Administration, 2016) and included 8 of the top 10 producing mines in the United States. As of 2010, the mines shown in Figure 5.1.1 removed coal from an area of 314 km^2 (121 mi^2) out of a total mined area of 425 km^2 (164 mi^2) in the entire PRB (Heffern et al., 2013). As of 2015, these mines removed 9.5 billion short tons of coal (Energy Information Administration, 2016, 2015, and 2014; Luppens et al., 2015).

Figure 5.1.9. Clinker in foreground of North Antelope/Rochelle coal mine, which produced 109 million short tons in 2015, the most of any coal mine in the United States. The mine is just west of the Wyodak clinker burn line. From Heffern and Coates (2004); with permission of Elsevier.

In the overlying Eocene Wasatch formation, 36 km^2 (14 mi^2) of clinker shown in Figure 5.1.1 were produced by the burning of the Felix coal zone, 150–270 m (500–900 ft) above the Wyodak coal (Luppens et al., 2015, 2008) and 10–20 km (6–12 mi) west of the Wyodak outcrop (Figure 5.1.1). The Felix clinker forms the red hills around the town of Wright. Figure 5.1.1 also shows clinker from the Lower Ulm bed (stratigraphically, the highest coal in the map area), which caps hills such as Antelope Butte just southwest of Gillette and clinker from lower coal beds in the Fort Union Formation in the southeast corner of the map.

Clinker is commonly 1–3 times as thick as the coal bed that burned, as shown on many cross sections in coal mining plans on public file with the Land Quality Division of the Wyoming Department of Environmental Quality. Most of the eastern PRB mines are producing from coal beds 15 to over 30 m (50 to over 100 ft) thick. If we assume that the coal beds that burned to produce the clinker in the area of Figure 5.1.1 averaged 15 m (50 ft) thick, then the 466 km^2 (180 mi^2) of clinker in place records the prehistoric burning of 9 billion metric tons (10 billion short tons) of coal—about as much as the 9.5 billion short tons of coal mined over the past century in the eastern PRB (Luppens et al., 2015, plus U.S. Energy Information Administration production updates).

These clinker outcrops are only remnants of a much larger volume of clinker that has been removed by erosion (Heffern and Coates, 2004). As erosion deepened and widened valleys, successively lower coal beds were exposed, ignited, and burned back into the hillsides. Clinker fragments were carried downslope and downstream and are found in talus slopes, landslides, alluvial fans, and the alluvium and gravel terraces of the tributaries of the Missouri River that drain the PRB (Figure 5.1.10). The clinker remaining in its original position may be an order of magnitude less than the volume of clinker produced over time, so the amount of coal burned by natural processes in the map area may be closer to one hundred billion tons.

Figure 5.1.10. Catherine A. Riihimaki and students standing on alluvial fan of clinker eroded from top of Cow Creek Butte (Section 17, T52N, R70W) northeast of Gillette, Wyoming. Very little clinker remains on the top of the butte. Photo by Edward L. Heffern, 2006.

From Heffern and Coates (2004); with permission of Elsevier.

5.2 Geologic History of Clinker in the Eastern Powder River Basin

A chimney of welded breccia and paralava surrounded by and enclosed in clinker from the Wyodak coal zone, Rochelle Hills, Wyoming. A person is standing adjacent to the bottom right side of the 8 m (25 ft) high chimney.

Methods of Dating

Clinker has been dated by three different methods (Heffern et al., 2007): (1) uranium-thorium/helium ratios of zircon grains in baked sandstones (**ZHe ages**), (2) fission-track counts of zircon grains in baked sandstones (**ZFT ages**), and (3) **paleomagnetic** orientations of magnetite, hematite, and goethite in paralava.

The first two methods use thermochronometry (Reiners, 2005; Naeser, 1979) to provide absolute ages. When zircon grains in the overlying sediments are baked, they are annealed, and ZHe and ZFT ages show the age of cooling. The only helium or fission tracks in the zircon grains formed after the coal bed burned to produce the baked sandstone. Because the clinker is very young compared to other types of rock that are typically dated by this technique, fission tracks are sparse and ZFT dates have a large uncertainty. The ZHe dates confirm the fission-track ages in general, but are more precise (with about a 10% level of uncertainty for samples of this type and age), and many more ZHe ages have been obtained for PRB clinker than ZFT ages. The third method is a coarse dating technique that can provide relative age—if the magnetic polarity is reversed, the clinker burned before the most recent magnetic reversal at 0.78 million years ago (Jones et al., 1984).

Age of Clinker

Clinker formed as the ancestral tributaries of the Yellowstone and Missouri rivers excavated the sedimentary fill in the PRB and exposed the coal beds there. From that time to the present, coal beds have been burning from natural causes—not as a single large fire, but as many fires, large and small, separated in space and time (Heffern et al., 2007).

Detrital thermochronometry of zircon grains in an alluvial deposit capping a strath terrace west of Colstrip, Montana shows that coal beds in the PRB have burned to form clinker at least as far back as 4 million years ago (Ma), in the Pliocene Epoch. There is also a 5.5 Ma ZHe date but it may be influenced by partial resetting during clinker formation (Reiners et al., 2011). The maximum age of the terrace deposit appears to be around 2.6 Ma, near the onset of Pleistocene glaciation (Reiners et al., 2011).

The oldest in situ clinker samples dated thus far in the PRB are two ZFT samples from a clinker capping the Little Wolf Mountains southwest of Colstrip, Montana. They are dated at 2.9 ± 0.9 Ma and 2.2 ± 0.8 Ma. All other samples are 1.5 Ma or younger (Heffern et al., 2007).

ZHe ages of 86 in situ clinker samples, collected across the basin, range from 0.007 to 1.1 Ma (Reiners et al., 2011; Riihimaki et al., 2009). The age distribution shows two temporal patterns: (1) a bias toward younger ages because of erosion of older clinker, and (2) periodic occurrence of coal fires likely corresponding with different climates (Riihimaki and Reiners, 2012). The ages preferentially cluster during times in which the eccentricity of the earth's orbit was high, possibly driven by increased seasonality in the region causing increased erosion rates and coal exhumation. These times usually, but not always, correlate with interglacial periods (Figure 5.2.1).

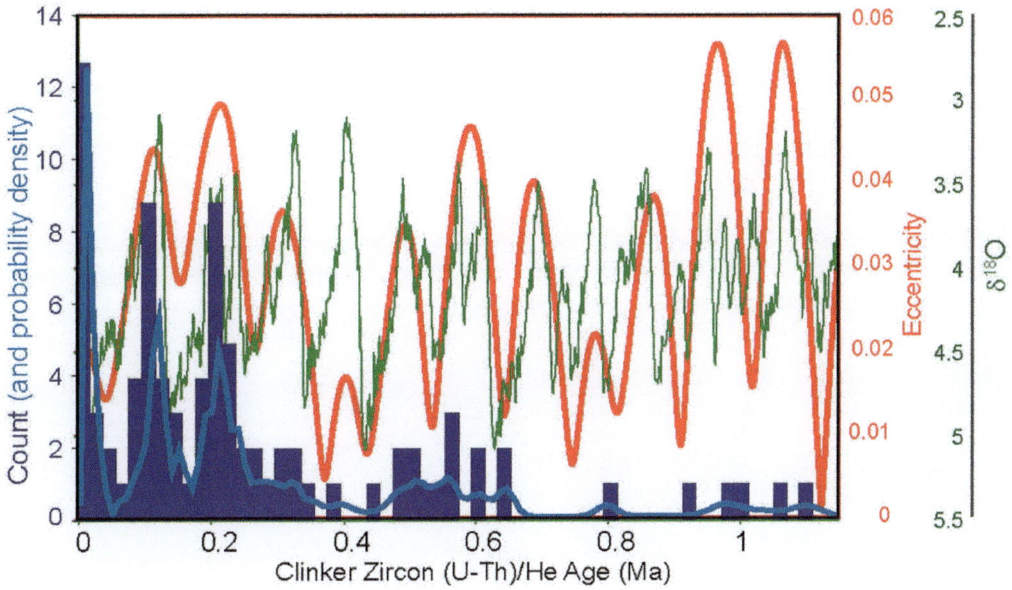

Figure 5.2.1. Histogram and probability density (blue bars and trend) of zircon He ages from 86 in situ clinker samples in the Powder River basin. Also shown are $\delta^{18}O$ patterns of benthic foraminifera (representing glacial/interglacial cycles) and orbital eccentricity from 1.2 Ma to the present. Clinker ages are clustered in peaks corresponding to the last three interglacials, two of which correspond to periods of high eccentricity (and maximum precessional variation in solar insolation). From Reiners et al. (2011); with permission of the Geological Society of America.

In the eastern PRB of Wyoming shown in Figure 5.1.1, ZHe ages reveal that coal has burned to form clinker as early as 0.615 Ma and as recently as 0.010 Ma, with an average uncertainty of ±10%. Less precise ZFT ages are as old as 0.9±0.6 Ma. A total of 28 ZHe and 19 ZFT ages have been derived for clinker samples collected in the area illustrated in Figure 5.1.1 (Reiners et al., 2011; Heffern et al., 2007; Coates and Naeser, 1984). Five paleomagnetic samples in the map area all had normal polarity (Jones et al., 1984). All sample locations and relative ages are shown on Figure 5.1.1.

Most of the ZHe and ZFT ages in Figure 5.1.1 are from outcrops of the Wyodak clinker in the Rochelle Hills east of the Jacobs Ranch, Black Thunder, North Antelope/Rochelle, and Antelope coal mines southeast of the town of Wright (Figure 5.2.2). On both headlands, north and south of Little Thunder Creek, the oldest clinker is more than 500,000 years old and is farthest east on the tip of the east-facing escarpment, which stands about 200 m (700 ft) above the local drainage (Figure 5.2.3). The Wyodak coal bed that formed the clinker dips 9–11 m/km (50–60 ft/mi) toward the west, and the clinker ages systematically decrease in that direction to 10,000 years old, as shown by cross section A-A' in Figure 5.2.2. Ten kilometers (6 mi) to the west of the headland, where clinker meets unburned coal at the present-day burn line, a natural coal fire at Burning Coal Draw was extinguished in the early 20th century (Russell and Smith, 1951), and mine operations extract shallow coal that has not yet burned (Figure 5.1.9).

Figure 5.1.1 also shows relative ages from the Wyodak clinker escarpment in the Burnt Hollow and Weston Hills recreation areas northeast of Gillette. In these areas, the Wyodak clinker is much more fine-grained, with baked shales and siltstones dominant, than in the Rochelle Hills east of the coal mines to the south, where there are many areas of baked sandstones. However, we were able to obtain ZHe ages of clinker samples from four outcrops of the Wyodak clinker as well as from detrital clinker boulders in two alluvial fans descending from the top of Cow Creek Butte (Figure 5.1.10). The ZHe ages of burning were all within the range of 0.10–0.24 Ma, considerably younger than many of the ages in the Rochelle Hills in the southeast corner of the map area. In this northern area, it appears that the 150-m (500-ft) high escarpment of the Cow Creek Wall has eroded since the clinker capping the butte was formed about 0.2 Ma. While additional sampling could possibly discover older ages, it is also possible that the fine-grained sediments in this area may be less resistant, erode more rapidly, and therefore result in younger clinker ages.

One ZHe and three ZFT ages have been derived for the clinker of the Felix coal zone that caps hilltops near Wright (Figure 5.1.1). The oldest ZFT age of 0.9±0.6 Ma has a large range of error and is farthest east. Here again, the younger ages (<0.3 Ma) are further west near the present burn line of the Felix coal, which gently dips to the west (Heffern et al., 2007). This pattern (albeit with few data points) makes sense if the Felix burn line and escarpments are retreating to the west over geologic time, as is the case with the Wyodak coal zone. The one ZHe sample is very young (0.024±0.001 Ma) and near the burn line as well.

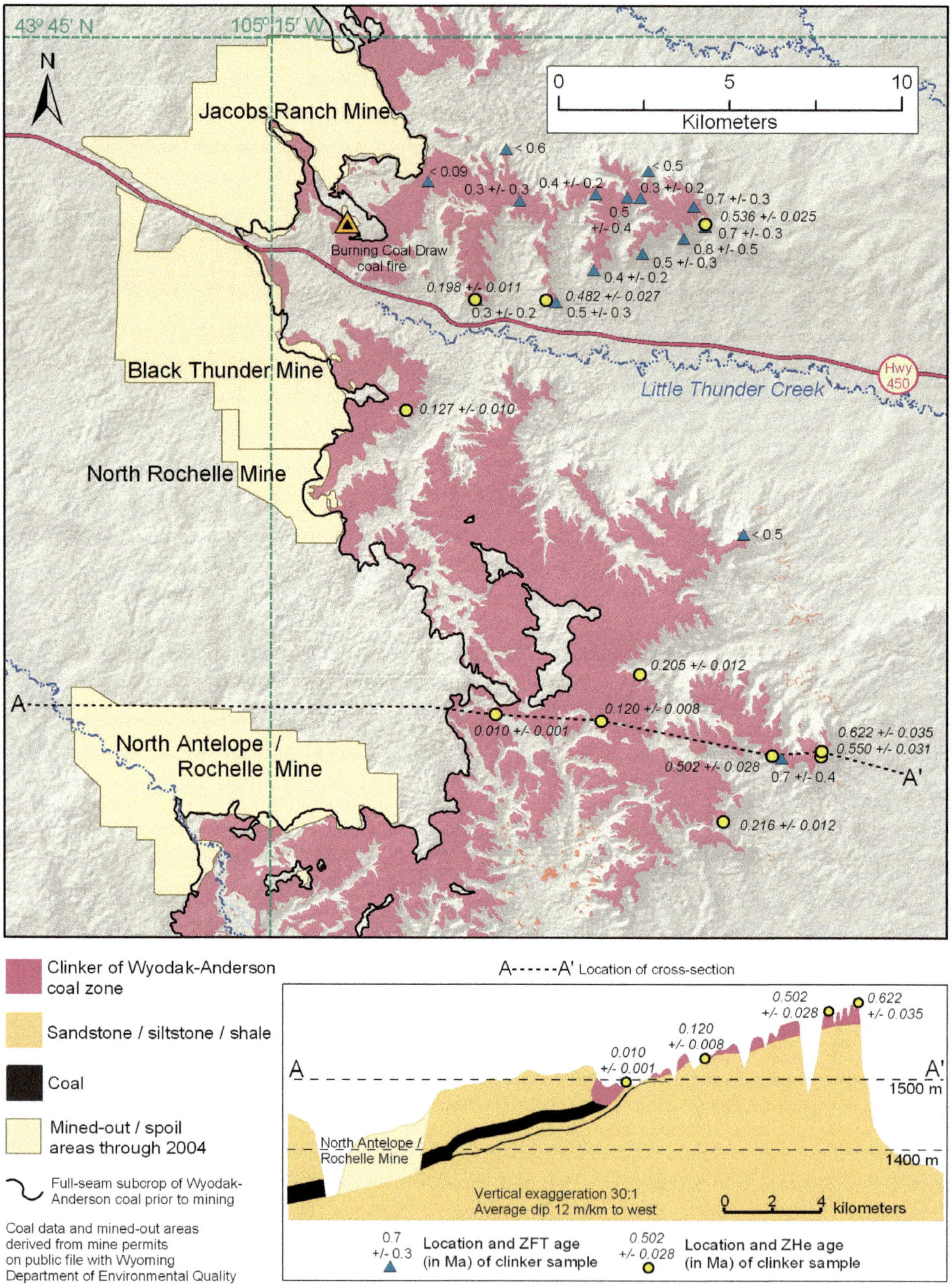

Figure 5.2.2. Clinker outcrops, sample ages, and coal mines, Rochelle Hills, Wyoming. *ZHe*, Zircon (U–Th)/He age; *ZFT*, Zircon fission-track age. From Heffern et al. (2007); with permission of the Geological Society of America.

Figure 5.2.3. Edward L. Heffern standing by clinker outcrop with oldest ZHe date in area of Figure 5.2.2 (0.615±0.035 Ma), Section 5, T41N, R68W, Rochelle Hills, Wyoming. Photo by Peter W. Reiners, 2002.

Discussion

In the Rochelle Hills, the Wyodak coal gently dips to the west at 0.5 degrees or less and the landscape drains to the east. Consequently, the plateaus and ridges of clinker that extend eastward from the burn line rise higher toward the east above the streams and above the local water table, and the older ZHe and ZFT ages are further to the east. Regression analyses of ZHe ages for transects of the Wyodak clinker east of Wright give an apparent vertical erosion rate of 210 m (690 ft) per million years (Myr) for the escarpment north of Little Thunder Creek and an apparent rate of 250 m (820 ft)/Myr for the escarpment south of Little Thunder Creek (Heffern et al., 2007), as the burn line retreated downdip to the west at a rate of 10,000–14,000 m (6–9 mi)/Myr (Figure 5.2.4). Vertical erosion rates in the area of Cow Creek Butte northeast of Gillette appear to be somewhat faster.

Coal beds in the PRB have been exposed and have burned naturally since at least the Pliocene. Burning progresses in a systematic pattern as rivers carve valleys into the stack of sediments and remove the sedimentary fill of the basin, exposing stratigraphically lower coal beds (Heffern et al., 2007; Coates and Heffern, 2000). The exhumation of the landscape and burning of coal beds appear to be sensitive to fluctuations in climate, with coal fires more common at times of high solar insolation, high eccentricity, and increased seasonality, and during interglacial periods (Riihimaki and Reiners, 2012; Reiners et al., 2011; Riihimaki et al., 2009). The burning of a thick coal bed generally produces higher temperatures over a longer time than burning of a thin bed, so that the clinker produced is thicker, harder, and more resistant. The sheets of clinker produced by major coal beds are more durable and extensive than the clinker produced by thin coal beds, and the oldest clinker preserved in place is the thick clinker capping the higher divides (Figure 5.2.5). Thin clinker rims along valley sides are generally young. In addition, detrital clinker in gravel terraces and alluvial fans, derived from clinker bodies that have eroded away in the distant past, can in some cases provide older dates of burning than in situ clinker outcrops. The radioisotopic ages collected from clinker in the PRB provide a glimpse of the geologic history of the region over the past several million years. These ages reveal systematic spatial patterns that make sense when we consider how landforms develop in a basin undergoing exhumation.

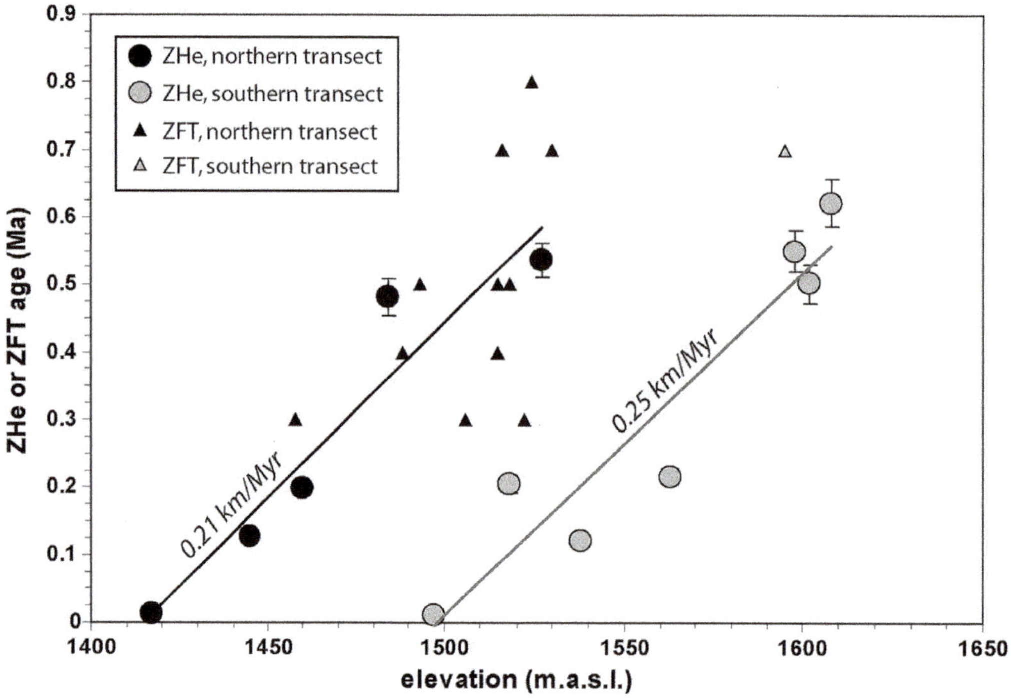

Figure 5.2.4. Zircon ZHe and ZFT ages versus sample elevation in Wyodak clinker, Rochelle Hills, Wyoming (see Figure 5.2.2). *Circles*, ZHe ages; *triangles*, ZFT ages; *black symbols*, transect north of Little Thunder Creek; *gray symbols*, transect south of Little Thunder Creek. Error bars are 2δ. *Solid lines* are regressions through ZHe data for each transect, yielding apparent vertical erosion rates of 0.21 ± 0.01 km/Myr and 0.25 ± 0.01 km/Myr for the northern and southern transects, respectively (*a.s.l.*, above sea level). From Heffern et al. (2007); with permission of the Geological Society of America.

Figure 5.2.5. Clinker formed as rivers carved into the stack of sediments in the PRB and ignited coal beds. Here in the Tongue River Valley of southeast Montana, looking west toward the town of Ashland, U. S. Highway 212 slices through an outcrop of clinker created by burning of the Knobloch coal bed 152,000 years ago (Heffern et al., 2013; Reiners et al., 2011). West of the Tongue River, which occupies the valley bottom, the Knobloch clinker also caps the tree-covered bench on the opposite side of the valley. The high plateau on the horizon to the west is capped by clinker of the Wyodak-Anderson coal zone, 300–400 m stratigraphically above the Knobloch clinker. This high clinker is as much as 1.07 million years old. Photo by Edward L. Heffern, 2016.

Acknowledgments

Field work for Drs. Reiners and Riihimaki and for ZHe analyses were funded by the National Science Foundation from grants EAR-0518818 and EAR-0518754. The authors thank Don Coates and Charles Naeser, both retired from the U.S. Geological Survey, for discussions on their earlier ZFT work in the PRB. Gerald Queen and Steve Hannan of the U.S. Bureau of Land Management's Buffalo Field Office provided information on modern coal fires in northeast Wyoming. Dr. Bill Reiners of the University of Wyoming; Camille Jones, Zoe Ruge, Jessica Scheik, and Kaia Davis of Bryn Mawr College; Jason Whiteman of the Northern Cheyenne Tribe; and Paul F. Gore helped us collect clinker samples. Pam Kania and the late Ken Kania provided hospitality when we did field work near Ashland, Montana. We benefitted from discussions with Susan Vuke of the Montana Bureau of Mines and Geology about clinker mapping in southeastern Montana. Stefan Nicolescu of the University of Arizona provided assistance with ZHe dating. Larry Neasloney of the U.S. Bureau of Land Management provided helpful advice on preparing the ArcGIS map in Figure 5.1.1.

Important Terms

Ash	Fine-grained minerals at the base of a clinker layer, typically of a light tan color and loosely cemented. This ash layer commonly marks the base of the burned coal bed. The ash is formed by nonorganic inert materials in the coal bed and minerals precipitated by groundwater passing down through the porous clinker.
Burn line	The contact between clinker and unburned coal or overburden. The burn line at depth commonly penetrates farther into the hillside than the surface expression of the burn line. The term "burn front" usually refers to the advancing edge of an actively burning coal fire.
Chimney	A vertical dike or mound of welded breccia and paralava. Chimneys mark the former location of fissures through which hot gases from a coal fire traveled upward and ignited to fuse and melt the surrounding rocks. On the surface of a clinker outcrop, chimneys can stand 1–8 m (3–25 ft) above areas of less intensely baked rock due to their resistance to erosion.
Clinker	Rock that has been altered by heating from a coal fire. It includes a variety of thermally metamorphosed or melted lithologies, varying from baked sandstones, siltstones, and shales, to fused shales (porcellanite), to welded breccias, paralava, and ash, all resulting from the burning of coal and heating of nearby sediments. Clinker is commonly reddish in color where iron in the rocks has been oxidized, but may be black or gray where heating occurred in reducing conditions.
Detrital thermochronometry	The analysis of decay of the radioactive elements in mineral grains (such as zircon and apatite) that are preserved in sediments. The ratio between concentrations of parent elements and products of radioactive decay can provide a date when the mineral cooled through a closure temperature, such as the date of a prehistoric coal fire.
Paleomagnetic dating	The analysis of the orientation of magnetic grains in rocks. The pattern of normal and reverse polarity orientations can bracket the time when the grain was formed.
Paralava	Rock that has been melted by a coal fire. The texture of the rock can be ropy to vesicular to rough and jagged. Paralava is most commonly found in chimneys and near the base of the original coal bed, indicating locations where hot gases heated rock past the melting point. The black color of much paralava indicates melting in reducing conditions. Greenish glass may be present, especially near the base of clinker.
Porcellanite	Shale and siltstone that has been fused into a fine-grained porcelain-like rock rich in feldspar. Heat energy and dehydration converted the clay to feldspar.
Scoria	A term often used for clinker. However, in the context of coal fires, scoria refers to a rough form of paralava or welded breccia rather than the full suite of pyrometamorphic rocks.

Welded breccia	Shattered blocks of shale, siltstone, and sandstone whose edges have been fused together. Commonly, voids and fractures in shattered rock are intruded or coated by paralava. In clinker, welded breccias are most commonly found in chimneys and above the ash zone, where collapse along tension cracks and heating is most intense.
ZFT (zircon fission-track) dating	A method of thermochronometry where a thin section of a zircon grain is prepared, the uranium concentration is determined, and the number of fission tracks is counted over a specific area. During spontaneous fission of uranium in a zircon crystal, particles smash through the crystal lattice, leaving tracks that can be revealed by etching and counted under a microscope. The ratio between the density of fission tracks and the uranium concentration in a crystal provides an estimate of how long the tracks have been accumulating (Naeser, 1979). When sandstone is baked by burning of a coal bed, the fission tracks in the detrital zircon grains are annealed. The only fission tracks in the zircon grains were produced after the clinker was formed. The density of fission tracks indicates age of the clinker—the older the clinker, the more fission tracks. Typical uncertainties produced by this method are ±50%.
ZHe (zircon (uranium-thorium)/helium) dating	A method of thermochronometry where the ratios between the radioactive parent elements (uranium and thorium) and the alpha decay product (helium) in a zircon grain are measured to determine how long ago the zircon grain cooled below the closure temperature. In the case of clinker, the zircon grains in baked sandstones are reset by the burning of the coal bed, which completely degasses all the helium produced before the burning occurred. The helium present was only produced after the coal bed burned and the clinker that was formed cooled below the closure temperature of 375–450°C (Reiners, 2005). Typical uncertainties of ages obtained by this method are ±10%.

References

Coates, D.A., Heffern, E.L., 2000. Origin and geomorphology of clinker in the Powder River Basin, Wyoming and Montana. In: Miller, R. (Ed.), Coal Bed Methane and Tertiary Geology of the Powder River Basin: Wyoming Geological Association, 50th Annual Field Conference Guidebook, pp. 211–229.

Coates, D.A., Naeser, C.W., 1984. Map Showing Fission-track Ages of Clinker in the Rochelle Hills, Southern Campbell and Weston Counties. U.S. Geological Survey Miscellaneous Investigations Map I-1462, Wyoming. Scale 1:50,000.

Dobbin, C.E., Barnett, V.H., 1927. The Gillette Coal Field, Northeastern Wyoming. U.S. Geological Survey Bulletin 796-A, pp. 1–64.

Dunrud, C.R., Osterwald, F.W., 1980. Effects of coal mine subsidence in the Sheridan, Wyoming, area. In: U.S. Geological Survey Professional Paper 1164 49 p.

Energy Information Administration, 2016, 2015, and 2014. Annual Coal Reports for 2015, 2014 and 2013. U.S. Department of Energy. http://www.eia.gov/coal/annual (accessed December 2017).

Engle, M.A., Radke, L.F., Heffern, E.L., O'Keefe, J.M.K., Hower, J.C., Smeltzer, C.D., Hower, J.M., Olea, R.A., Eatwell, R.J., Blake, D.R., Emsbo-Mattingly, S.D., Stout, S.A., Queen, G., Aggen, K.L., Kolker, A., Prakash, A., Henke, K.R., Stracher, G.B., Schroeder, P.A., Román- Colón, Y., ter Schure, A., 2012. Gas emissions, minerals, and tars associated with three coal fires, Powder River Basin, USA. Sci. Total Environ. 420, 146–159.

Flores, R.M., Bader, L.R., 1999. Fort union coal in the Powder River Basin, Wyoming and Montana – a synthesis. In: Fort Union Coal Assessment Team (Ed.), 1999 Resource Assessment of Selected Tertiary Coal Beds and Zones in the Northern Rocky Mountains and Great Plains Region: U.S. Geological Survey Professional Paper 1625-A Chapter PS, 75 p., CD-ROM.

Goodson and Associates, 1986. Phase 1 Report of Investigation – Wyoming Mine Fire Control Study. (Unpublished report submitted to State of Wyoming Department of Environmental Quality, Land Quality Division).

Heffern, E.L., Coates, D.A., 1997. Clinker – its occurrence, uses, and effects on coal mining in the Powder River Basin. In: Jones, R.W., Harris, R.E. (Eds.), Proceedings of the 32nd Annual Forum on the Geology of Industrial Minerals. Wyoming State Geological Survey, Public Information Circular 38, pp. 151–165.

Heffern, E.L., Coates, D.A., 2004. Geologic history of natural coal-bed fires, Powder River Basin, USA. In: Stracher, G.B. (Ed.), Coal Fires Burning Around the World – A Global Catastrophe, pp. 25–47 Int. J. Coal Geol. 59 (1–2).

Heffern, E.L., Coates, D.A., Whiteman, J., Ellis, M.S., 1993. Geologic Map Showing Distribution of Clinker in the Tertiary Fort Union and Wasatch Formations, Northern Powder River Basin. U.S. Geological Survey Coal Investigations Map C-142, Montana.

Heffern, E.L., Reiners, P.W., Naeser, C.W., Coates, D.A., 2007. Geochronology of clinker and implications for evolution of the Powder River Basin landscape, Wyoming and Montana. In: Stracher, G.B. (Ed.), Geology of Coal Fires: Case Studies from Around the World, pp. 155–175 Geological Society of America Reviews in Engineering Geology XVIII.

Heffern, E.L., Reiners, P.W., Riihimaki, C.A., 2013. Clinker Distribution and Age in the Powder River Structural Basin. Montana Bureau of Mines and Geology. Geologic Map 64.

Jones, A.H., Geissman, J.W., Coates, D.A., 1984. Clinker deposits, Powder River Basin, Wyoming and Montana – a new source of high-fidelity paleomagnetic data for the Quaternary. Geophys. Res. Lett. 11 (12), 1231–1234.

Kim, A.G., Chaiken, R.F., 1993. Fires in Abandoned Coal Mines and Waste Banks. U.S. Bureau of Mines Information Circular 9352. 58 p.

Luppens, J.A., Scott, D.C., Haacke, J.E., Osmonson, L.M., Rohrbacher, T.J., Ellis, M.S., 2008. Assessment of coal geology, resources, and reserves in the Gillette coalfield, Powder River Basin, Wyoming. In: U.S. Geological Survey Open-File Report 2008-1202, p. 38 (p. plus figures and tables).

Luppens, J.A., Scott, D.C., Haacke, J.E., Osmonson, L.M., Pierce, P.E., 2015. Coal geology and assessment of coal resources and reserves in the Powder River Basin, Wyoming and Montana. In: U.S. Geological Survey Professional Paper 1809 218 p. https://doi.org/10.3133/pp1809 (accessed December 2017).

Lyman, R.M., Volkmer, J.E., 2001. Pyrophoricity (Spontaneous Combustion) of Powder River Basin Coals – Considerations for Coal-bed Methane Development. Wyoming State Geological Survey, pp. 18–22. Wyoming Geo-Notes no. 69.

Naeser, C.W., 1979. Fission-track dating and geologic annealing of fission tracks. In: Jager, E., Hunsaker, J.C. (Eds.), Lectures in Isotope Geology. Springer-Verlag, Berlin, pp. 154–169.

Raynolds, W.F., 1868. Report on the Exploration of the Yellowstone River. U.S. Government Printing Office, U.S. Secretary of War.

Reiners, P.W., Riihimaki, C.A., Heffern, E.L., 2011. Clinker geochronology, Plio-Pleistocene glaciation, and landscape evolution in the northern Rockies. GSA Today 21 (7), 4–9.

Reiners, P.W., 2005. Zircon (U-Th)/He thermochronometry. Rev. Mineral. Geochem. 58, 151–179.

Riihimaki, C.A., Reiners, P.W., 2012. Empirical evidence of climate's role in Rocky Mountain landscape evolution. J. Geophys. Res. 117, 1–15. https://doi.org/10.1029/2011JF002137. F02007.

Riihimaki, C.A., Reiners, P.W., Heffern, E.L., March 2009. Climate control on Quaternary coal fires and landscape evolution, Powder River Basin, Wyoming and Montana. Geology 37 (3), 255–258.

Russell, H.W., Smith, J.H., 1951. Control of Coal Crop Fire in Little Thunder Basin, Hilight, Campbell County. U.S. Bureau of Mines, Wyoming (unpublished final report, 38 p.).

Shellenberger, F.H., Donner, D.L., 1979. Controlling Fires in Unmined Coal Deposits and Abandoned Coal Mines in the Western United States and Alaska. U.S. Bureau of Mines, Denver Research Center. Progress Report 10047, 90 p.

Thwaites, R.G. (Ed.), 1969. Original Journals of the Lewis and Clark Expedition. Arno Press, New York, pp. 1804–1806. 5 volumes.

CHAPTER 6

The Occurrence and Use of Coal Cinders in Washington State

The Landsburg coal-waste pile was on fire at SE 253rd Street (also called Morris Mine Connection Road) in the Black Diamond–Green River coal basin, Ravensdale, near the Seattle–Tacoma metropolitan area. The fire was discovered in January of 2003 and extinguished within a week. Some vehicles are visible through the smoke, near and to the right of the center of the photo. *Photo by William Kombol, 2003.*

CHAPTER CONTENTS

Coal and Peat Fires: A Global Perspective
Edited by Glenn B. Stracher

https://doi.org/10.1016/B978-0-12-849885-9.00006-8

Photo by William Kombol, January 2003.

6.1 The Occurrence and Use of Coal Cinders in Washington State

William Kombol
Daniel H. Vice

A fire in the SE 253rd Street coal-slag pile in the Black Diamond–Green River coal basin near the Seattle–Tacoma metropolitan area.

Introduction

Coal **clinker** or **cinders** form when **coal-waste piles** burn, often because of **spontaneous combustion**. The cinders are a valuable resource for landscaping and other uses including walking paths, running tracks, and as a lightweight fill (Palmer Coking Coal Company, 2014).

Kim and Chaiken (1993) noted that approximately 25% of the material removed from a coal mine is rejected as **coal waste** and stored in coal-waste piles, also commonly referred to as **boney piles**, **gob piles**, and **spoils piles**.

Coal-waste piles in many areas of Washington State contain significant amounts of **slack coal**; that is, coal dust and small chunks of coal. The slack coal was discarded along with shale, carbonaceous shale, and other waste rock brought to the surface with coal during underground mining. Only the more valuable stoker-size, nut-size, and lump-size coal were sent to market. Coal from the McKay seam in and around Ravensdale, Black Diamond, and Franklin, and other coal seams from the Black Diamond–Green River coal basin (Figure 6.1.1) were prone to spontaneous combustion that ignited large coal-waste piles. Once the coal waste ignited, it often burned for years before consuming all the available fuel. Spontaneous combustion was not limited to the Black Diamond–Green River coal basin but also included the Rosyln area in eastern Washington (Figure 6.1.1), where Rector (1963) noted clinker or cinders in coal-waste piles there.

The Use of Cinders

Some coal-waste piles that burned for years were later mined for cinders (also known as **red dog**, **cintrex**, **scoria**, and **ash**) (Figures 6.1.2 and 6.1.3). The cinders were first used for road construction and then for high-school running tracks and various landscaping projects (Kombol, 2014a,b). The red cinders are associated with coal seams that contained pyrite in which the iron oxidized to ferrous oxide (Fe_2O_3) (Figure 6.1.4). Gray or ashen-colored cinders also formed during combustion and represent less oxidation than the red-colored cinders. Regardless of color, the cinders are used for landscaping (Figure 6.1.5) and running tracks. The red-colored cinders are more popular for running tracks because of their brighter color.

Lightweight cinders are also used for fill projects in soft, boglike soils where lightweight material is required. For example, the Bellefield Office Park in south Bellevue was built in a wetlands area, and cinders were used as a road base to keep the soft soil there from subsiding, as would have happened if heavy sand and gravel or quarry rock were used.

Palmer Coking Coal Company (Palmer) has mined and sold red cinders since 1953 from the coal-waste pile at Mine #11 (Morgan Slope Mine), Mine #14, and the Lawson Mine, all in Black Diamond, Washington. Palmer acquired

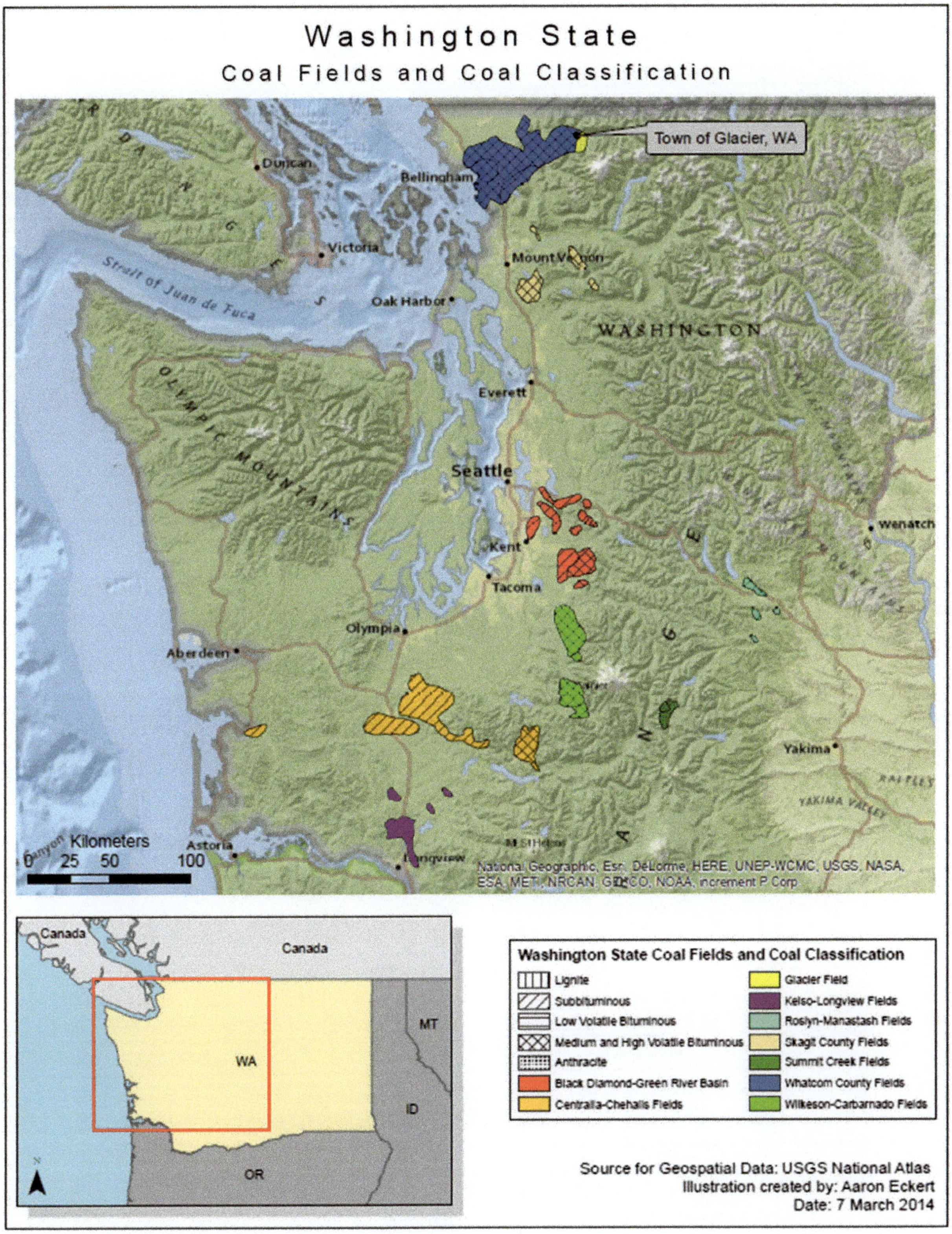

Figure 6.1.1. Map showing coal deposits in Washington State. The Black Diamond–Green River coal deposits are shown in red; the Roslyn deposits are in purple. From Vice et al., 2015.

Figure 6.1.2. Aerial view looking south at Mine #11 in Black Diamond, Washington, showing an industrial yard and the mining of red cinders from a coal-waste pile that burned. Photo by William Kombol, June of 1990.

Figure 6.1.3. Photo of a cinder pile at Mine #14 and nearby homes in Black Diamond, Washington. Photo by William Kombol, May of 1990.

Figure 6.1.4. Cinders used for running tracks; sold at the Palmer Coking Coal Company's retail sales yard at 31407 Highway 169, Black Diamond, Washington. Photo by William Kombol, August of 2010.

Figure 6.1.5. An example of Palmer Coking Coal's three-quarter-inch cinders used for landscaping. This is also at the company's retail sales yard (Figure 6.1.4). Photo by William Kombol, October 2014.

these properties in 1953 from the Pacific Coast Coal Company, which had owned and operated these underground coal mines. The first cinders that Palmer mined from coal-waste piles were used to build the Black Diamond–Ravensdale Road (originally known as the Red Top Road). Palmer has mined out most of the available red-cinder stockpiles and now only about 4000–5000 cubic yards remain.

Acknowledgments

The authors wish to thank Glenn B. Stracher, East Georgia State College, Swainsboro, for his review of the manuscript.

Important Terms

ash
boney piles
cintrex
clinker
cinders
coal waste
coal-waste piles
gob piles
red dog
scoria
slack coal
spoils piles
spontaneous combustion

References

Kim, A.G., Chaiken, R.F., 1993. Fires in Abandoned Coal Mines and Waste Banks: US Bureau of Mines. Information Circular. 9352, 58 p.

Kombol, W., 2014a. When Coal Was King: Voice of the Valley, Maple Valley, Weekly, p. 2 May 20th.

Kombol, W., 2014b. When Coal Was King: Voice of the Valley, Maple Valley, Weekly, p. 5 May 13th.

Palmer Coking Coal Company, 2014. http://www.palmercc.com/redcinders.htm (accessed November 2014).

Rector, R.J., 1963. Geologic and Mineral Evaluation of the Swauk-teanaway Area. Kittitas County, Washington: Unpublished report prepared for the Geology Division of the Northern Pacific Railroad Company, 30 pp. This document can be accessed at the Washington Division of Geology and Earth Resources, 1111 Washington St., SE, P. O. Box 47007, Olympia, Washington, 98504.

Vice, D., Stracher, G.B., Eckert, A., 2015. Coal fires of the Pacific Northwest, USA. In: Stracher, G.B., Prakash, A., Sokol, E.V. (Eds.), Coal and Peat Fires: A Global Perspective, vol. 3, pp. 671–680.

CHAPTER 7

The Spontaneous Combustion of Coal-Mine Waste and Stream Effects in the El Bierzo Coalfield, Spain

A self-burning coal-waste pile (right) in the Fabero anthracite mine. The mine is in the El Bierzo coalfield, Spain. Active combustion is occurring on the right side of the waste pile, seen here in Arroyo Galladas. *Photo by Joana Ribeiro, 2012.*

CHAPTER CONTENTS

Coal and Peat Fires: A Global Perspective
Edited by Glenn B. Stracher

https://doi.org/10.1016/B978-0-12-849885-9.00007-X

Photo by Joana Ribeiro, 2012.

7.1 The El Bierzo Coalfield, Spain

Joana Ribeiro
Conchi O. Ania
Isabel Suárez-Ruiz
Deolinda Flores

The effect of combustion in the waste pile in the Fabero anthracite mine is visible as red-colored oxidized waste in Arroyo Galladas.

Introduction

Mining activities in general cause environmental disturbances in the areas where they are carried out. Some of the environmental issues related to **coal mining** are like those related to mining in general, whereas others are specific to coal mining. The environmental impacts and human health problems caused by coal have long been a cause for concern, probably since its first use as a fuel in China (Suárez-Ruiz and Crelling, 2008).

The most important environmental issues relating to coal mining are changes in landscape, land exploitation, topography, and hydrographic systems; land instability and subsidence; soil erosion; the production and disposal of solid wastes; the propagation development of coal-related fires; the generation of **acid mine drainage**; surface and groundwater pollution; soil and sediment pollution; and atmospheric pollution due to the dispersion of solid particles and gas emissions from coal-related fires and their impact on biodiversity and ecosystems (Bell et al., 2001; Younger, 2004; Suárez-Ruiz and Crelling, 2008; Bian et al., 2010). The disposal of coal mine wastes by itself may lead to the atmospheric dispersion of solid particles, spontaneous combustion, landslides and mass movements, the mobilization of solid materials, leaching, and mining acid drainage.

The magnitude and severity of these environmental problems depend on many factors, such as coal composition, geological setting, local hydrology, climate, topography, mining methods, and local regulations, among others (Suárez-Ruiz and Crelling, 2008).

This chapter provides an overview of the environmental impact of mining acid drainage and spontaneous combustion processes occurring in coal-waste piles (CWPs) located in the north of Spain, resulting from the exploitation of coal deposits in the Cantabrian mountain range.

Mine-Acid Drainage

One of the most significant environmental concerns associated with coal mining is the generation of acid mine drainage, which is a common pathway for the environmental contamination of soils and waters (Younger, 2004). Mine-acid drainage (MAD) is the run-off produced when water comes into contact with exposed rocks containing sulfur-bearing minerals that react with water and air to form sulfuric acid, dissolved Fe and Al, and other chemical elements (Blowes et al., 2004). The acidic run-off promotes the dissolution of **heavy metals** such as Cu, Pb, and Hg, which pollute soils, sediments, and ground and surface waters after discharge into natural **streams** (Drever, 1997).

Coal mining also promotes the exposure of sulfides (primarily pyrite) in coal and associated rocks to oxygen and moisture. The stoichiometry of the oxidation of the metallic sulfides has been extensively described in the literature. It basically leads to the formation of sulfuric acid and/or the release of trace elements that may precipitate in oxygenated surface water bodies and concentrate in stream sediments (Blowes et al., 2004). The resulting acidic waters adversely impact ecosystems in watersheds downstream of both active and abandoned waste piles (Orem and Finkelman, 2003; Drever, 1997). Environmental concerns associated with MAD include, generally, the loading of dissolved contaminants, the sedimentation of chemical precipitates, soil erosion, and loss of biodiversity in contact with waters with high metal loads (Williams et al., 2002; Younger, 2004). The waters that drain from many coal mines are strongly acidic because of the oxidation of pyrite in the coal (Drever, 1997). However, in coal mining, acidic drainage is predominantly associated with waste piles containing pyritic shales (Younger, 2004). When streams become contaminated by MAD, living organisms die and the precipitation of ferric hydroxide occurs over long distances (Drever, 1997).

Spontaneous Combustion of Coal Waste

Coal-related fires can affect unmined coal seams, underground and open-cut coal mines, coal stockpiles, and CWPs. The most representative environmental impacts of spontaneous combustion of CWPs include the emission of harmful gases (CO, CO_2, H_2S, SO_x, CH_4, VOCs, PAHs, and phenols) and particulate matter into the atmosphere, the mobilization and leaching of hazardous elements, and the sublimation of coal fire gas minerals.

Besides being a loss of energy resources, coal-related fires cause a significant amount of environmental and human health problems, and therefore not surprisingly, coal-related fires occurring all over the world have been extensively reported in the literature (Stracher and Taylor, 2004; Stracher, 2007; Stracher et al., 2011, 2013, 2015, and references therein). Most case studies focus on the environmental and human health impacts of coal fires (Finkelman, 2004; Stracher and Taylor, 2004; Pone et al., 2007; Querol et al., 2008; Zhao et al., 2008; Ribeiro et al., 2010a,b; Finkelman and Stracher, 2011; Querol et al., 2011; Suárez-Ruiz et al., 2012; Ribeiro et al., 2013; Ribeiro et al., 2016a), whereas the effects of spontaneous combustion processes in associated streams has been scarcely investigated.

Case Study

The coal-bearing deposits in the Cantabrian Range, from the Late Carboniferous (Pennsylvanian) age, contain about 70% of the coal resources in Spain. The Stephanian B-C coal-bearing series are widespread in the south of the Cantabrian Mountains, unconformably overlying basement units from the Precambrian to the Carboniferous age, with a thickness between 3000 and 3500 m (Suárez-Ruiz and Jiménez, 2004; Colmenero et al., 2008).

El Bierzo is the westernmost and largest Stephanian coalfield (310 km^2) in the Cantabrian Range that has been exploited since the middle of the 19th century (Hernández-Mendoza et al., 2013). The mining activities were carried out in three sectors—Torre-Bembibre, Fabero-Matarrosa, and Toreno-Valdesamario—and altogether about 70 coal seams have been mined (Colmenero et al., 2008). The coal of the El Bierzo coalfield has a rank ranging from bituminous A to anthracite C, B, A (vitrinite reflectance between 1.87% and 5.25%), a low-to-high ash yield, and vitrinite as its major organic constituent (>96%) (Colmenero et al., 2008).

The environmental issues associated with coal mining in El Bierzo have been addressed in several studies (Alberruche et al., 2008; CIUDEN-CIEMAT, 2010; de la Losa et al., 2010; Rodríguez Gómez et al., 2010; Hernández-Mendoza et al., 2013). The intense coal-mining activity in El Bierzo has generated 571 waste dumping zones, covering an area of 52.51 km^2 over the original soil. The mining waste was dumped over a period of many years and was exposed to natural alterations that include physical, chemical, and biological processes. These alterations have been the cause of MAD and the spontaneous combustion of CWPs, both of which are responsible for environmental impacts on the air, waters, soils, and vegetal and animal life of the surrounding areas.

Three CWPs originating from the mining activities in the region of Fabero (Fabero-Matarrosa sector) of El Bierzo have been studied: Arroyo Galladas, Arroyo Mourin, and Fabero (Figure 7.1.1). Arroyo Galladas and Fabero CWPs are prone to spontaneous combustion while Arroyo Mourin CWP is not. All three CWPs have been investigated in depth with the aim of characterizing the mineralogical, petrological, and geochemical composition of the waste materials, the changes induced by the combustion processes, and the related potential environmental problems (Ribeiro et al., 2014; Ribeiro et al., 2016a; b).

Figure 7.1.1. Location of the studied area. Adapted from http://siams.chminosil.es/visorg/ (accessed 2017).

The effects of coal mining on ground and surface waters in some areas of El Bierzo have already been investigated (de la Losa et al., 2010; Rodríguez Gómez et al., 2010). However, these studies did not include a complete characterization of the composition of the surface water in streams related to the previously mentioned CWPs (Arroyo Galladas, Arroyo Mourin, and Fabero), and dealt only briefly with the self-burning CWP.

The main rivers crossing El Bierzo are (from east to west) Cúa, Sil, Noceda, Boeza, and Tremor. They all are north–south oriented and characterized by an abundant and constant flow. The physicochemical parameters of the river waters are quite diverse due to the variable geological materials and structural complexity of the sediments they cross (Alberruche et al., 2008; CIUDEN-CIEMAT, 2010). Apart from this, most of the rivers in this region have undergone degradation due to the mining activities, and they show significant concentrations of sulfates arising from the dissolution of precipitates in the CWP during rainy periods (such as gypsum, epsomite, melanterite, and rozenite; and Ca, Mg, and Fe hydrated salts), and high concentration of iron oxides in the sediments (Alberruche et al., 2008; CIUDEN-CIEMAT, 2010; de la Losa et al., 2010; Rodríguez Gómez et al., 2010).

The Cúa River is in the north area of the El Bierzo municipality (León, Spain) and in the hydrographic basin of Minõ-Sil (Figures 7.1.1 and 7.1.2), which embraces the villages of Peranzos, Fabero, Vega de Espinareda, Araganza, Cacabelos, Carracedelo, and Villadecanes-Toral de los Vados. The Cúa River flows into the Sil River, which is the main affluent of the Minõ River (that crosses northwest Spain, constitutes a natural border with Portugal, and finally flows into the Atlantic Ocean), and has a total length of 62 km and an average flow of 6.57 m^3/s (CIUDEN-CIEMAT, 2010).

In the inventory of mining structures and potential sources of contamination associated with the mining activities in El Bierzo, some areas have been identified as priority concern for environmental restoration, the Cúa River area being one of them (Alberruche et al., 2008). The mining area of Fabero encompasses 7 km of the Cúa River, which circulates very close to the areas occupied by the Arroyo Galladas and Arroyo Mourin CWPs (Figure 7.1.3). In these areas, the Galladas, Mourin, and Lamocos streams are tributaries of the Cúa River (Figure 7.1.3). In general, the Cúa riverine water system is of a high quality but critical environmental problems have been identified in the valley of the Galladas tributary, which is occupied (about 413 ha) by the Arroyo Galladas CWP in the mining area of Fabero. These are associated with acid mine drainage and waste pile self-burning issues (CIUDEN-CIEMAT, 2010). The water flow of the Galladas tributary was channeled under the Arroyo Galladas CWP and very acidic waters (pH 3) are produced during intense rainy periods. Others parameters relevant to these waters such as their electrical conductivity (c. 1456 μS/cm) and their sulfate (c. 990 ppm), aluminum (c. 33.5 ppm), and iron contents (c. 11.7 ppm) have also been reported (CIUDEN-CIEMAT, 2010).

Figure 7.1.2. Minõ-Sil hydrographic basin (17080 km^2 in Portugal and Spain). Adapted from http://siams.chminosil.es/visorg/ (accessed 2017).

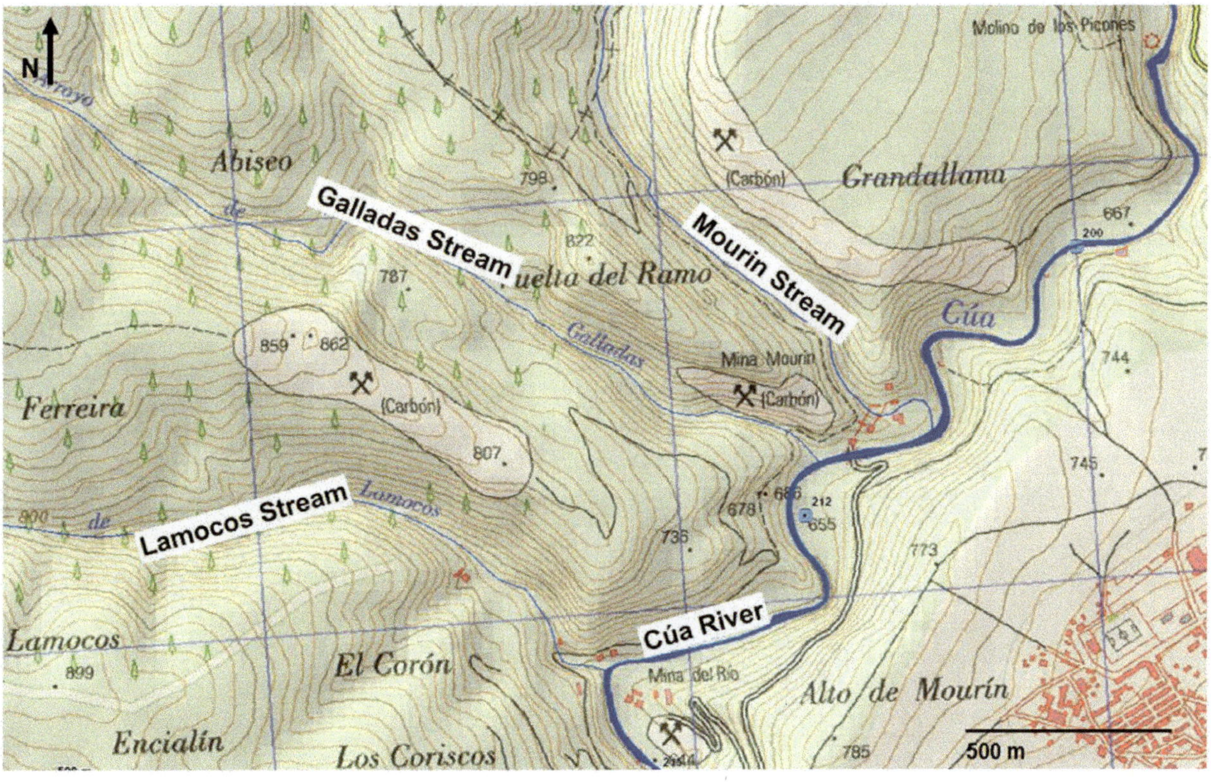

Figure 7.1.3. Tributaries of the Cúa River in the mining area of Fabero where the Arroyo Galladas (GWP) and the Arroyo Mourin (MWP) CWP are located. Adapted from http://siams.chminosil.es/visorg/ (accessed 2017).

Considering what has been said so far, the present study aims to investigate the influence of CWPs (including the acid drainage and spontaneous combustion processes) on streams from the **riverine system** in the mining region of Fabero. The objectives of the study are to (1) determine the physicochemical parameters of water samples from the streams, (2) determine the elemental composition of water samples from the streams, and (3) identify the source of potential contaminants and their potential effects on the streams in the riverine system. In addition, this work is expected to contribute to an assessment of the nature and extent of the water pollution of the riverine system in the mining area of El Bierzo, and to scientific knowledge about the **environmental effects** of coal-related fires occurring worldwide.

Photo by Isabel Suárez-Ruiz, 2012.

7.2 Materials and Methods

Water samples 4A and 4B, discussed in the text, were taken here from the Lamocos tributary of the Cúa River in the El Bierzo coalfield. Electrical conductivity, pH, and temperature measurements were made here.

Sampling

To investigate the effects of **coal-waste disposal** and the spontaneous combustion of coal waste along the riverine system of the mining region of Fabero, water samples from the Cúa River and its tributaries were collected (Galladas, Mourin, and Lamocos tributaries) (Figure 7.1.3). A total of six sites were sampled in two different periods of the year: a wet and a dry season (May and December, respectively). Figure 7.2.1 shows the sampled locations.

Two locations were sampled along the Galladas tributary, upstream (sampling area 1) and downstream (sampling area 2) of the deposit of **self-burning coal waste**. Comparison of the data collected from the water samples at these locations is intended to establish the changes induced in the waters that percolate through the materials from the

Figure 7.2.1. Location of the sampling areas 1 to 6 investigated in this study; the circles indicate the location of Arroyo Galladas and Mourin waste piles, GWP and MWP, respectively. Adapted from http://siams.chminosil.es/visorg/ (accessed 2017).

Figure 7.2.2. Field aspects of sampling area 1, where Galladas tributary was sampled upstream of the self-burning coal-waste deposit in the dry season (A) and in the wet season (B).

Figure 7.2.3. Sampling area 2: (A) Exit of the water flow channeled through Arroyo Galladas CWP. (B) Junction of the Galladas tributary with the Cúa River. The width of the Galladas tributary in the images is about 3 m.

Arroyo Galladas CWP and are subsequently discharged into the Cúa River. Figure 7.2.2 shows some field aspects of the sampling area in the Galladas tributary located upstream of the Arroyo Galladas CWP (sampling area 1) corresponding to the dry season (Figure 7.2.2A) and wet season (Figure 7.2.2B) campaigns. This location can be expected to be the less affected by the CWP formed due to the mining activities, since it is upstream. On the other hand, Figure 7.2.3A shows images of sampling area 2, corresponding to the water flow channeled under the Arroyo Galladas CWP. Figure 7.2.3B shows the junction of the Galladas Tributary with the Cúa River.

The Cúa River was sampled (sampling area 3) after the Mourin tributary discharge and before the Arroyo Galladas CWP; that is, before the discharge of the Galladas tributary (as can be observed in Figure 7.2.1). Data obtained from the analysis of the water samples taken from this sampling area allow the influence of the mining infrastructures upstream (at least the Arroyo Mourin CWP) to be identified as well as the changes caused by the contribution of the Galladas tributary that percolates through the Arroyo Galladas CWP (via a comparison of sampling areas 3 and 2). Figure 7.2.4 shows some field aspects of the area where the water samples from the Cúa River were collected.

Sample 6 was collected from the Mourin tributary, immediately downstream of the Arroyo Mourin CWP (Figures 7.2.5A and 7.2.5B, in the wet and dry seasons, respectively), allowing the water parameters attributable to the circulation of water through unburned-coal waste to be identified. Figures 7.2.5C and 7.2.5D show the junction of the Mourin tributary with the Cúa River in the dry and wet seasons, respectively.

The field observations reveal that the Lamocos tributary is about 5 m near the areas where a coal seam and past mining infrastructures were identified, downstream of water-sampling area 4 and upstream of water-sampling area 5. It is believed that the water of the Lamocos tributary may be affected by its circulation through these geological

Figure 7.2.4. Field aspects of sampling area 3, corresponding to the Cúa River in the dry (A) and wet season (B). The width of the Cúa River in the images is about 8–10 m.

Figure 7.2.5. Sampling area 6. (A) and (B) illustrate the Mourin tributary, immediately after the CWP in the dry and wet seasons, respectively. (C) and (D) illustrate the junction of the Mourin tributary with the Cúa River, in the dry and wet season, respectively. The width of the Mourin tributary in the images is about 4–5 m.

materials, since significant amounts of sulfur on the exposed and weathered surface of the coal seam were detected near the Lamocos tributary, as can be seen in the field images in Figure 7.2.6.

Field aspects of sampling areas 4 and 5 are illustrated in Figures 7.2.7 and 7.2.8. The brownish color characteristic of ferric hydroxides and other precipitates was clearly observed downstream of the Lamocos tributary (sampling area 4) as opposed to the upstream sampling area (Figure 7.2.8). Therefore, it can be expected that the latter will be less affected by the CWP.

Figure 7.2.6. Previous coal-mine infrastructure (left) and a close-up of the outlined coal seam. The mine is close to the Lamocos tributary. The yellow color is sulfur.

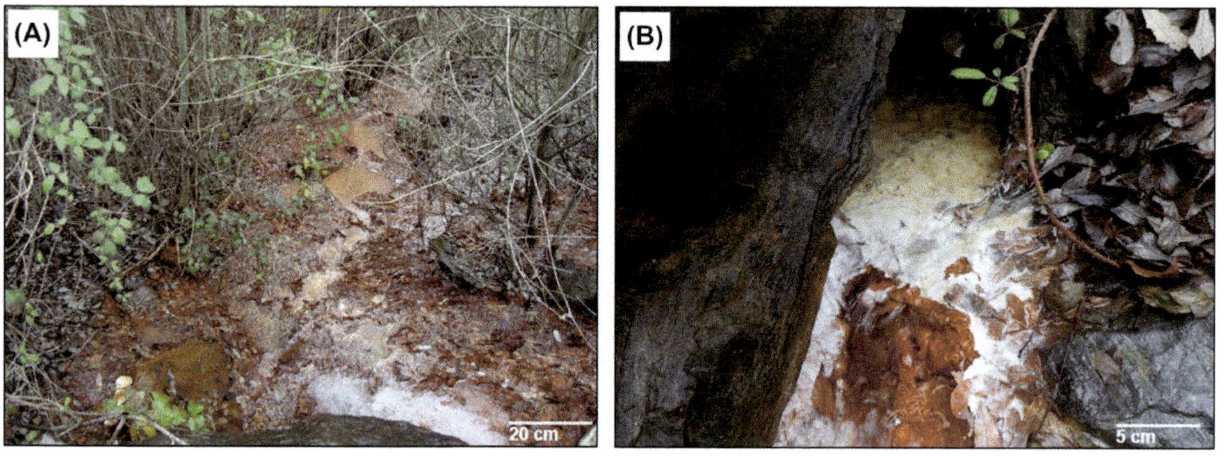

Figure 7.2.7. Field aspects of sampling area 4 corresponding to area downstream of the Lamocos tributary.

Figure 7.2.8. Sampling area 5, upstream in the Lamocos tributary. The width of the Lamocos tributary in the image is about 1 m.

Table 7.2.1
Samples collected during the dry season (May) and the wet season (December).

Samples		Month of Collection	Coordinates		Stream	Observations
1	A	May	42°46′36.51″N	6°38′52.21″W	Galladas tributary	Not influenced by GWP*
	B	December				
2	A	May	42°46′24.79″N	6°38′24.52″W	Galladas tributary	Influenced by GWP
	B	December				
3	A	May	42°46′25.38″N	6°38′22.52″W	Cúa River	Influenced by MWP†
	B	December				
4	A	May	42°46′09.78″N	6°38′45.88″W	Lamocos tributary	Influenced by old mining infrastructures
	B	December				
5	A	May	42°46′14.11″N	6°38′51.70″W	Lamocos tributary	Not influenced by old mining infrastructures
	B	December				
6	A	May	42°46′30.36″N	6°38′10.15″W	Mourin tributary	Influenced by MWP
	B	December				

*GWP, Arroyo Galladas CWP.
†MWP, Arroyo Mourin CWP.
CWP, coal waste piles.

Table 7.2.1 contains a summary of the sampling areas explored in this study during the wet and dry seasons (May and December, respectively), together with the associated nomenclature used in this work. Each water sample was collected in 500 mL screw-capped pyrex borosilicate glass bottles, carried to the laboratory, and immediately stored in a refrigerator at 4°C for 24 h in preparation for the analysis. For longer storage, the samples were stabilized with 2 M nitric acid (for 48 h) and frozen at −20°C.

Analytical Methods

For the pH and ionic conductivity measurements, a multimeter (SevenMulti, Mettler Toledo) apparatus equipped with a glass pH electrode and a conductivity probe was used. Measurements were taken at ambient temperature on the as-collected samples (before storage or acidic stabilization).

The total carbon (TC), inorganic carbon (IC), organic carbon (TOC), and total nitrogen (TN) contents were determined by combustion catalytic oxidation on a Shimadzu VCSH TOC analyzer. Briefly, the samples were heated at 680°C in an oxygen-rich environment in a combustion tube filled with a platinum catalyst, and the carbon dioxide generated upon oxidation was detected using an infrared gas analyzer (for the TC content). Additionally, by subjecting the sample to a sparing process, the inorganic carbon content in the sample was converted to carbon dioxide and detected as IC. The TOC content was then calculated by subtracting the IC from the TC values measured. Furthermore, the nonpurgeable organic carbon (NPOC) was also detected, allowing the large errors commonly associated with the calculation of TOC in samples with a significant amount of inorganic carbon to be avoided. The method consists of purging the IC from the sample before any carbon measurement is performed (by sparing the sample in the presence of a small amount of acid so as to convert the IC to carbon dioxide before removing it). The remaining organic matter in the sample is also oxidized to CO_2 and quantified as NPOC. In this approach, TOC is the sum of purgeable organic carbon (POC) and NPOC. An analysis of the difference between TOC and NPOC also gives an indication of the contribution of organic molecules that can be purged with the IC in the purge step. The amount of carbonates CO_3^{2-} + HCO^{3-} was recalculated from the IC content (expressed in mg C/g). The TN content was also measured by means of catalytic combustion at 720°C, which converts nitrogen monoxide and nitrogen dioxide species to NO_2 by reaction with oxygen/ozone, and then detected by chemiluminescence.

The concentration of chloride, nitrates, and sulfate anions was determined by ion chromatography on a Metrohm apparatus (940 Professional IC Vario) equipped with a conductivity detector and a 250 × 4.0 mm anion column (Metrosep A Supp 7–250, 40°C, 0.7 mL/min of 3.6 m M sodium carbonate as mobile phase).

The trace elements, including rare-earth elements, were analyzed by Acme Analytical Laboratories (Vancouver, Canada), using inductively coupled plasma mass spectrometry (ICP-MS).

Photo by Joana Ribeiro, 2012.

7.3 Physicochemical Parameters

The Cúa River flowing to the right of a coal-waste deposit. The waste pile on the left is about 20 m high. The direction of water flow is from the top of the image to the bottom and off to the right.

Physicochemical Parameters

The results for the physicochemical parameters obtained from the water samples (i.e., pH, ionic conductivity, forms of carbon, and TN) are compiled in Table 7.3.1.

The pH of the studied samples varied between 3.5 and 6.5 in the dry season and between 2.8 and 6.9 in the wet season (Table 7.3.1). As can be seen, only in the samples corresponding to sampling areas 2 (downstream of the Galladas tributary) and 3 (Cúa River before the discharge of Galladas tributary) displayed acidity. The rest of the sampling areas showed slightly acidic values (below a pH value of 7) characteristic of water samples in real environments. Sampling area 6 (downstream of the Mourin tributary) showed slightly lower values around 5 units for both seasons. The pH variation between the seasons was very low for all the samples with the exception of the area downstream of the Galladas tributary (sampling area 2), where the water was found to be more acidic in winter. The pH values evidenced the production of acid mine drainage as being the result of the circulation of water from the Galladas tributary through the materials of the Arroyo Galladas CWP (values of sampling area 2), and in the Cúa River due to the impact of the Arroyo Mourin CWP (values of sampling areas 3 and 6).

The ionic conductivity values of the waters were high for sampling areas 2, 3, and 4, as illustrated in Figure 7.3.1. These values correspond to the samples collected after the circulation of water through the Arroyo Galladas CWP (2A and 2B), to samples from the Cúa River (3A and 3B) and to samples collected from the Lamocos tributary, after

Table 7.3.1
Physicochemical parameters and data for the water samples.

Samples	1A	1B	2A	2B	3A	3B	4A	4B	5A	5B	6A	6B
pH	6.4	6.6	3.5	2.8	6.5	3.8	6.3	6.4	6.4	6.9	5.1	5.0
Cond. (μs/cm)	20	13	665	2900	163	795	1055	1304	37	21.0	32	21
TC (mg C/L)	1.6	4.2	6.2	13.7	2.9	5.0	15.5	35.4	1.8	4.8	0.5	2.3
IC (mg C/L)	0.8	0.8	5.7	12.4	2.4	3.1	14.6	34.0	0.9	0.8	0.1	0.4
TOC (mg C/L)	0.8	3.4	0.5	1.1	0.5	1.9	0.9	1.4	0.9	4.0	0.4	1.9
NPOC (mg C/L)	0.9	3.3	0.7	1.2	0.5	1.9	0.9	1.2	0.7	4.1	0.3	1.8
TN (mgN/L)	0.3	0.1	1.0	0.6	0.3	0.6	0.5	0.1	0.7	0.1	0.3	0.1

C/L, carbon per liter; Cond., ionic conductivity; IC, inorganic carbon; NPOC, nonpurgeable organic carbon; TC, total carbon; TN, total nitrogen; TOC, total organic carbon.

the circulation of water through certain mining infrastructures (4A and 4B). The high values of conductivity in samples 2A and 2B can be associated with their acidic pH, and therefore, to the production of acid drainage in the Arroyo CWP, which led to the dissolution of metals, thereby enhancing the ionic conductivity. The same causes may explain the conductivity values in samples 3A and 3B, since the Cúa River receives water from some tributaries that are fed by water that has circulated through mining infrastructures present in the region. On the other hand, the high ionic conductivity of samples 4A and 4B is associated with almost neutral pH values. It should therefore be attributed to the precipitation of trace elements such as ferric-hydroxides, sulfates, oxyhydroxysulfates, and others. This would be in agreement with the brownish precipitates observed in the sampling area (Figure 7.2.7). The pH and conductivity values of the Galladas tributary after percolation through coal waste that spontaneously combusted are in accordance with those reported in the inventory of mining structures and potential sources of contamination associated with the mining activities in the El Bierzo coalfield (Alberruche et al., 2008).

Figure 7.3.2 illustrates the TC, TOC, and IC contents of all the samples studied. As can be seen, the TC content is more abundant in the samples collected in the wet season, with samples 2 and 4 showing the highest values (above 5 mg C/L). Inorganic carbon is the most abundant in samples 2, 3, 4, and 5A while organic carbon is most abundant in samples 1, 5B, and 6. TN content varies between 0.1 and 1 mg N/L, with the highest values corresponding to the dry season.

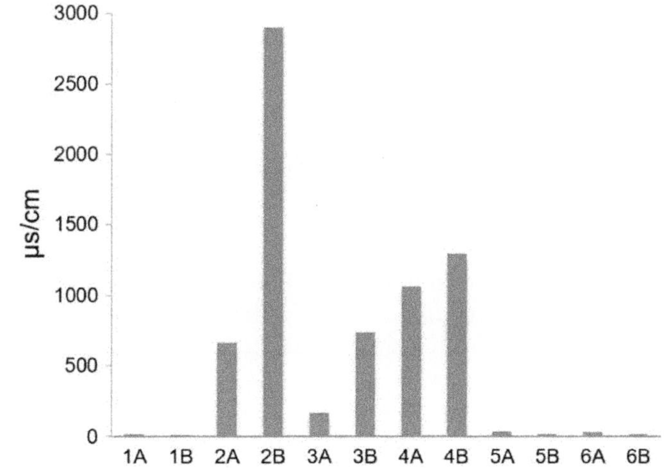

Figure 7.3.1. Ionic conductivity of the water samples. The sample numbers are labeled on the horizontal axis.

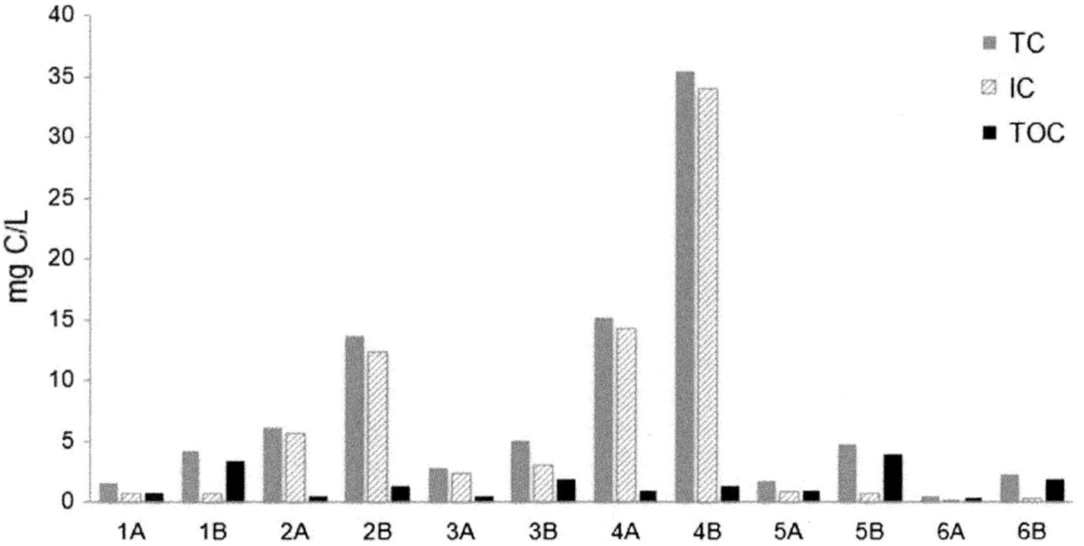

Figure 7.3.2. Total carbon, inorganic carbon, and total organic carbon in the water samples measured in mg C/L. The sample numbers are labeled on the horizontal axis.

Table 7.3.2 shows the concentration of the main anions detected in the streams, namely chlorides, nitrates, sulfates, and carbonates (recalculated from the IC content). As can be seen, SO_4^{2-} is the dominant anion in almost all the samples (with the exception of samples 1A and 1B) with minor concentrations of Cl^- and NO^{3-}.

The trends of the different anions in the samples studied can be more clearly appreciated in Figures 7.3.3 and 7.3.4. The concentration of nitrates varies between 0.2 and 5.2 ppm, the values being higher, as a general rule, in the samples collected in the wet season (Table 7.3.2 and Figure 7.3.3). The concentrations of nitrates in samples 1 and 2 are very similar, which indicates that the concentration of this anion is not affected by the circulation of water through burnt-coal waste. The concentrations of chloride anions are also similar among the studied samples, with the exception of 2B and 3B, whose higher values may be attributed to the fact that these samples were collected in the wet season.

The concentration of sulfates displayed a great variability, with higher values in samples collected in the wet season. Concentrations higher than 500 ppm were reported for samples 2A, 2B, 4A, and 4B, about 2–3 times larger than values measured in sampling area 3 (i.e., 100–200 ppm), and exceeding considerably the values measured for the rest of the sampling areas (Table 7.3.2 and Figure 7.3.4). The high values found in samples 2A, 2B, 4A, and 4B are attributed to the circulation of water from the Galladas and Lamocos tributaries through coal waste and mining infrastructures and also to the dissolution of precipitates promoted by raining events, principally in the wet season. The concentration of CO_3^{2-} follows a similar trend, showing higher values for the samples collected in the wet season and in samples 2A, 2B, 4A, and 4B.

The Piper diagram (Piper, 1944) has been used to infer the hydrochemical facies of the studied waters. Hydrochemical facies are distinct zones with cation and anion concentration categories, and were calculated using Qualigraph software to generate a trilinear Piper diagram (Figure 7.3.5). This is a useful tool to reveal the analogies, dissimilarities, and different types of waters, and it explains the variations of cation and anion concentrations. The

Table 7.3.2
Concentrations in parts per million (ppm) of chloride, nitrate, sulfate, and carbonate anions in the water samples.*

	1A	1B	2A	2B	3A	3B	4A	4B	5A	5B	6A	6B
Cl^-	3.1	2.9	3.4	8.1	2.8	9.2	2.1	2.7	3.3	3.4	2.6	2.7
NO^{3-}	3.1	3.2	3.2	4.2	2.9	5.2	0.2	1.2	1.2	1.2	2.5	3.1
SO_4^{2-}	3.4	3.4	495	853	106	176	715	951	10.7	11.3	13.4	14.8
CO_3^{2-*}	4.0	4.0	28.5	62.0	12.0	15.5	73.0	170	4.5	4.0	0.5	2.0

*Recalculated from the inorganic content (IC).

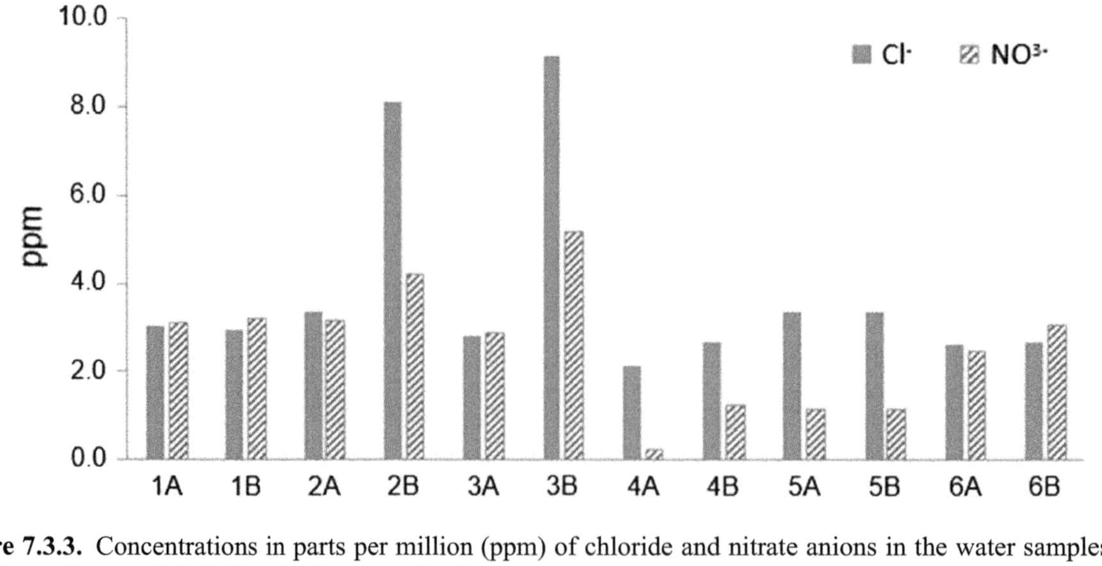

Figure 7.3.3. Concentrations in parts per million (ppm) of chloride and nitrate anions in the water samples. The sample numbers are labeled on the horizontal axis.

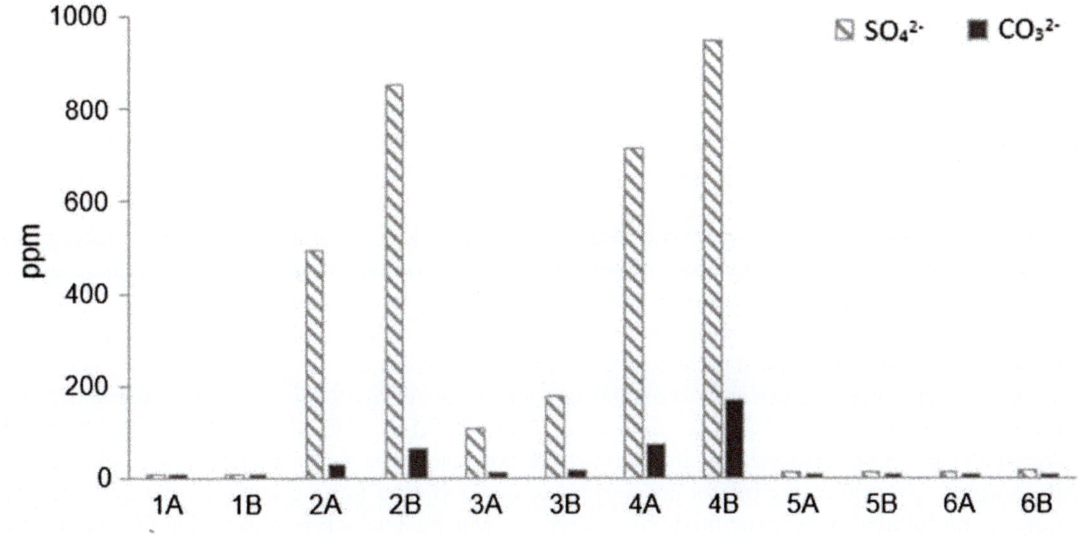

Figure 7.3.4. Concentrations in parts per million (ppm) of sulfate and carbonate anions in the water samples. The sample numbers are labeled on the horizontal axis.

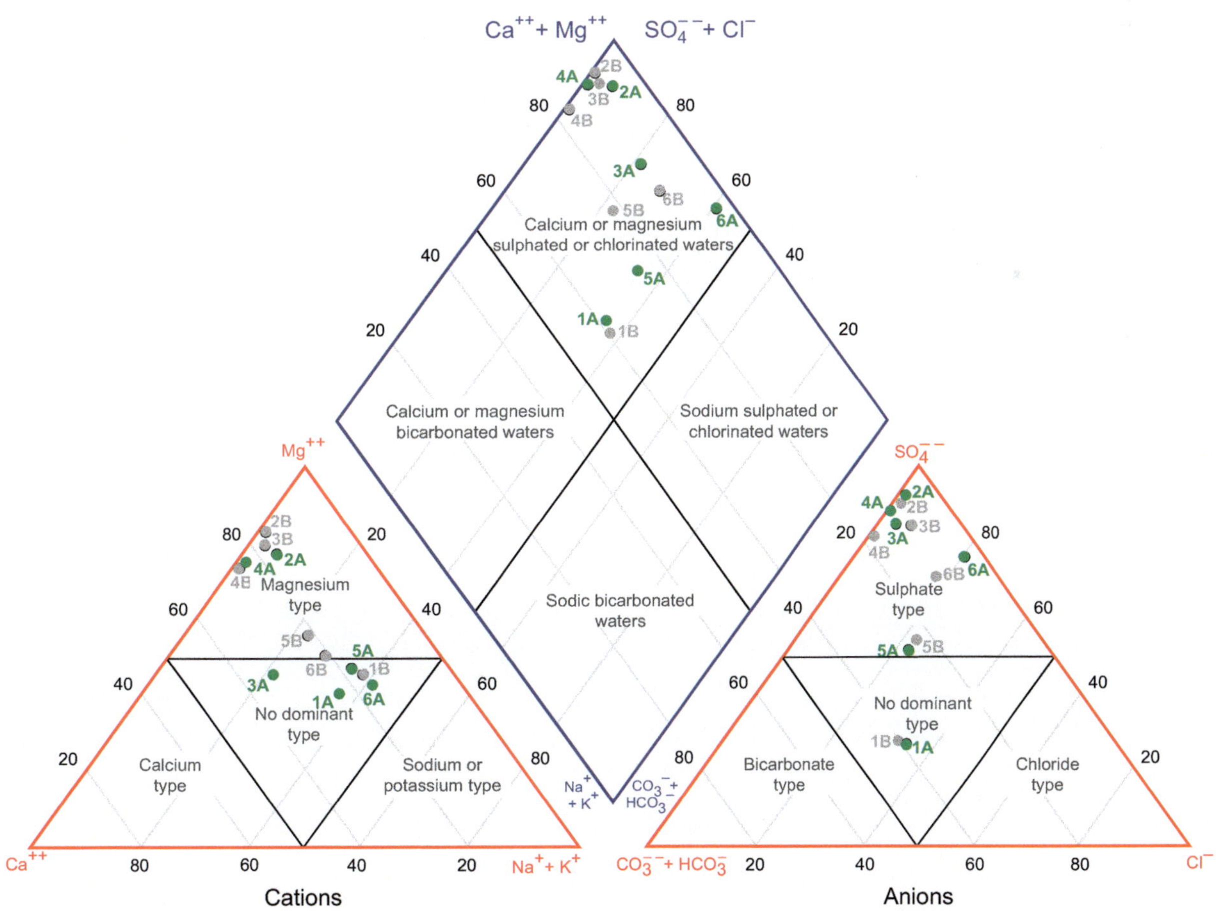

Figure 7.3.5. Piper diagram of the water samples (1A, 1B—Galladas tributary, upstream. 2A, 2B—Galladas tributary, downstream. 3A, 3B—Cúa River after the discharge of the Mourin tributary, before the Arroyo Galladas CWP. 4A, 4B—downstream of the Lamocos tributary. 5A, 5B—upstream of the Lamocos tributary. 6A, 6B—downstream of the Mourin tributary).

different studied waters exhibit a low variability in terms of hydrochemical facies, ranging from Ca - Mg - SO_4^{2-} - Cl^- waters to nondominant ones (Figure 7.3.5).

From the analysis of the diamond-shaped areas in the Piper diagram, useful information about the chemical relationships among the water samples and the sources of pollution can be learned. As we can see in Figure 7.3.5, sampling areas 1 (Galladas tributary, upstream) and 5 (upstream of the Lamocos tributary) are classified as nondominant waters with lower contents of cations. Thus the ions reflect the composition of the water bodies themselves before any pollution (e.g., Ca - SO_4^{2-} type waters result most likely from the dissolution of calcite, dolomite, and pyrite). In contrast, the rest of the studied water bodies corresponding to the sampling areas affected by the mining activities are classified as multiple cations and anion waters according to the Piper diagram $\left(a - Mg - SO_4^{2-} + Cl^-\right)$. Such multiple cation and anion water types reflect the contribution of the acid drainage from the CWP to the water. As seen, the concentration of ions is lower for sampling area 6 (downstream of the Mourin tributary), indicating a lower environmental impact of the Arroyo Mourin CWP to the water stream, as compared to the Arroyo Galladas CWP (sampling areas 2 and 4). Moreover, samples 2, 3, and 4 (regardless the season campaign) tend to be clustered, presenting a more noticeable magnesium and sulfated character. On the other hand, samples 1, 5, and 6 (regardless the season campaigns) tend to be grouped, although more dispersed between magnesium sulfated waters and no dominant type of waters. The similarities noticed for the studied samples in the Piper diagram agree with the similarities noticed for other physicochemical parameters shown in Tables 7.3.1 and 7.3.2.

Photo by Joana Ribeiro, 2012.

7.4 Minor and Trace Element Concentrations

The Arroyo Galladas coal-waste pile. The waste is heterogeneous and contains shale, arenites, and coal. The particles sizes range from a few μm to about 5 cm.

Minor and Trace Element Concentrations

The chemical composition of the waters was determined by ICP-MS, the data for which are presented in Table 7.4.1. The results obtained for sample 2B need to be analyzed with caution since quantification of many elements was not possible due to unusually high levels of interference from other elements. In any case, and as in the case of sample 2A, this sample displayed a large concentration of many elements, accurate quantification of which was not possible with the methods applied.

As a general trend, the contents of Ca, Cl, K, Mg, Na, and S predominate in the chemical composition of water samples, with higher concentrations being found in samples 2A, 2B, 3A, 3B, 4A, and 4B. Aluminum, Fe, and Mg were also found to be present in quite high concentrations in some samples. For all the samples, the following elements were also detected in moderate concentrations: Al, Ba, Ca, Co, Cu, K, Li, Mg, Mn, Na, Ni, Rb, S, Si, Sr, and Zn.

The sum of the trace elements in the samples from both sampling campaigns, wet and dry seasons, was calculated, the results of which are shown in Figure 7.4.1. Samples 2A, 2B, 3A, 3B, 4A, and 4B show the highest concentrations (i.e., $\sum$ trace elements > 100 ppm) while samples 5A, 5B, 6A, 6B, 1A, and 1B present correspondingly low element concentration values (i.e., $\sum$ trace elements < 100 ppm). These trends are in good agreement with expectations based on the preliminary physicochemical parameters (i.e., surface pH, ionic conductivity) indicating that sampling areas 1 (upstream of the Arroyo Galladas CWP) and 5 (upstream of the old mining infrastructures) are less influenced by the circulation of waters through coal waste and other mining infrastructures. The low concentrations in sampling area 6 could be related to the fact that the Arroyo Mourin CWP is relatively small and the area in contact with the Mourin tributary is restricted. On the other hand, the very high concentrations of the sum of trace element in sampling areas 2 and 4 may be related to the fact that they were collected downstream of the circulation of waters through coal waste and/or mining infrastructures. Similarly, in the case of sampling area 3, the high concentrations of trace elements may be linked to the fact that the Cúa River collects more tributary discharge and has a higher flow than its sampled tributaries.

Considering seasonal effects, it should be noted that sampling areas 1 (upstream of the Galladas tributary), 5 (upstream of the Lamocos tributary), and 6 (downstream of the Mourin tributary) show a higher element concentration in the dry season while sampling areas 2 (downstream of the Galladas tributary), 3 (the Cúa River before the discharge of the Galladas tributary), and 4 (downstream of the Lamocos tributary) follow the oppose trend, presenting higher concentrations of trace elements in the wet season.

However, the differences between the two campaigns in sampling areas 4, 5, and 6 are small, and cannot be considered significant (i.e., the ratio between the sums of the trace element concentrations for the wet and dry seasons is less than 2). On the other hand, in the samples collected in areas 2 and 3, the concentration of trace elements is 10- and 5-fold higher in the wet than in the dry season, respectively. This can be attributed to the fact that sampling

Table 7.4.1
Concentrations of trace elements in water samples collected in May (dry season - samples 1A to 6A) and December (wet season - samples 1B to 6B).

Sample	1A	1B	2A	2B	3A	3B	4A	4B	5A	5B	6A	6B
						ppm (parts per million)						
Ca	1.0	0.52	17.1	144	13.7	33.8	73.3	115	1.4	1.3	0.94	0.75
Cl	28.0	1.0	20.0	n.a.*	15.0	1.0	5.0	<DL†	5.0	1.0	4.0	<DL
K	1.2	0.37	3.9	1.0	12.5	1.5	1.7	1.3	2.9	0.35	2.8	0.26
Mg	1.1	0.90	48.6	470	11.6	93.9	144	210	2.3	2.1	1.5	1.1
Na	1.1	1.2	4.9	9.0	2.8	5.0	3.6	4.4	1.4	1.4	1.1	1.0
S	2.0	<DL	117	1188	24.0	250	234	348	4.0	4.0	4.0	4.0
Si	1.8	1.5	4.8	n.a.	2.8	8.1	3.5	4.5	2.1	2.0	2.2	1.7
						ppb (parts per billion)						
Ag	<DL	<DL	0.19	<DL	0.49	<DL	<DL	<DL	0.16	<DL	<DL	<DL
Al	25.0	78.0	13,486	211,182	21.0	47,076	16.0	3281	24.0	79.0	297	298
As	0.60	<DL	<DL	<DL	<DL	0.50	<DL	<DL	<DL	<DL	<DL	<DL
B	<DL	<DL	36.0	<DL	<DL	8.0	10.0	12.0	<DL	<DL	<DL	6.0
Ba	1.4	1.7	8.4	<DL	2.6	13.0	17.1	14.8	1.6	2.2	2.4	3.3
Be	<DL	<DL	3.0	n.a.	0.06	5.0	<DL	0.65	0.09	<DL	0.12	<DL
Bi	<DL	<DL	<DL	<DL	<DL	<DL	<DL	<DL	<DL	<DL	<DL	0.08
Br	14.0	22.0	12.0	n.a.	10.0	11.0	9.0	9.0	10.0	17.0	<DL	5.0
Cd	<DL	<DL	2.8	<DL	<DL	5.4	0.93	0.79	<DL	<DL	0.08	0.07
Ce	<DL	0.07	9.8	n.a.	<DL	25.9	0.13	4.4	<DL	0.09	0.07	0.17
Co	0.04	1.3	166	1597	0.50	297	67.6	86.1	0.04	0.90	3.5	3.6
Cr	<DL	<DL	0.70	<DL	<DL	2.6	<DL	<DL	<DL	<DL	<DL	<DL
Cs	<DL	<DL	0.44	n.a.	0.01	0.27	0.07	0.09	<DL	<DL	<DL	<DL
Cu	3.4	2.1	32.0	121	1.6	41.4	5.8	14.8	0.70	3.3	2.1	1.7
Fe	<DL	126	2185	95,897	<DL	1770	<DL	6691	19.0	300	16.0	230
Ga	<DL	<DL	0.09	n.a.	<DL	0.11	<DL	<DL	<DL	<DL	<DL	<DL
Hg	<DL	<DL	<DL	n.a.	<DL	<DL	<DL	<DL	<DL	<DL	<DL	<DL
Li	2.0	1.6	222	1751	4.3	394	38.1	56.1	3.9	3.7	5.4	4.0
Mn	0.44	16.3	7065	68,472	34.1	12,004	2651	3653	1.3	24.0	20.8	26.8
Mo	<DL	<DL	<DL	<DL	<DL	0.10	<DL	0.10	<DL	0.20	<DL	0.20
Ni	2.0	3.2	297	3660	3.0	618	134	132	2.7	5.0	9.1	7.9
P	12.0	<DL	14.0	<DL	14.0	28.0	<DL	<DL	<DL	17.0	<DL	<DL
Pb	<DL	0.30	3.4	<DL	1.0	3.2	0.20	0.40	0.20	0.70	1.2	0.70
Rb	0.21	0.30	2.6	n.a.	0.67	3.8	1.1	1.7	0.33	0.33	0.24	0.32
Sb	<DL	<DL	<DL	<DL	0.08	0.05	0.06	0.05	<DL	<DL	<DL	<DL
Sc	<DL	<DL	3.00	n.a.	1.0	1.0	<DL	<DL	<DL	<DL	<DL	<DL
Se	<DL	<DL	<DL	n.a.	<DL	<DL	<DL	<DL	<DL	<DL	<DL	<DL
Sn	<DL	<DL	<DL	n.a.	<DL	<DL	<DL	<DL	<DL	0.07	<DL	<DL
Sr	2.6	2.1	39.5	112	86.3	69.7	210	296	4.3	3.8	4.5	4.6
Th	<DL	<DL	0.08	<DL	<DL	0.36	<DL	0.07	<DL	<DL	<DL	<DL
Tl	<DL	<DL	0.06	n.a.	<DL	0.06	<DL	0.02	<DL	<DL	<DL	<DL
U	<DL	<DL	1.6	<DL	<DL	2.8	0.03	0.40	<DL	<DL	<DL	0.02
V	5.7	<DL	4.3	<DL	1.2	0.30	0.30	<DL	0.60	<DL	<DL	<DL
W	<DL	0.02	0.02	<DL	<DL	<DL	<DL	<DL	<DL	<DL	0.04	<DL
Y	0.04	0.04	26.9	n.a.	0.02	67.6	0.27	5.1	<DL	0.05	0.11	0.17
Zn	14.8	8.7	589	5190	17.4	1086	196	183	17.1	12.3	35.5	15.2
Zr	0.02	0.12	0.07	n.a.	<DL	0.15	0.04	0.09	<DL	0.13	<DL	0.14
∑REE‡	0.02	0.14	47.7	n.a.	0.01	117	0.39	14.7	<DL	0.19	0.16	0.39

*n.a., data not available.
†<DL, below detection limit.
‡∑REE, sum of rare earth elements.

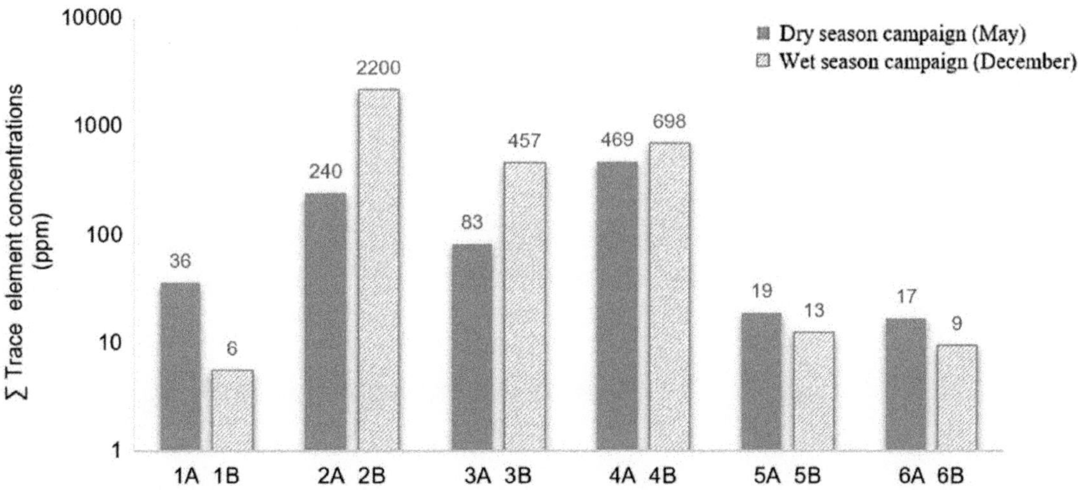

Figure 7.4.1. Sum of trace element concentrations in parts per million (ppm) presented in Table 7.4.1 in the study samples. The vertical axis is a logarithmic scale because of the large range in values. The sample numbers are labeled on the horizontal axis.

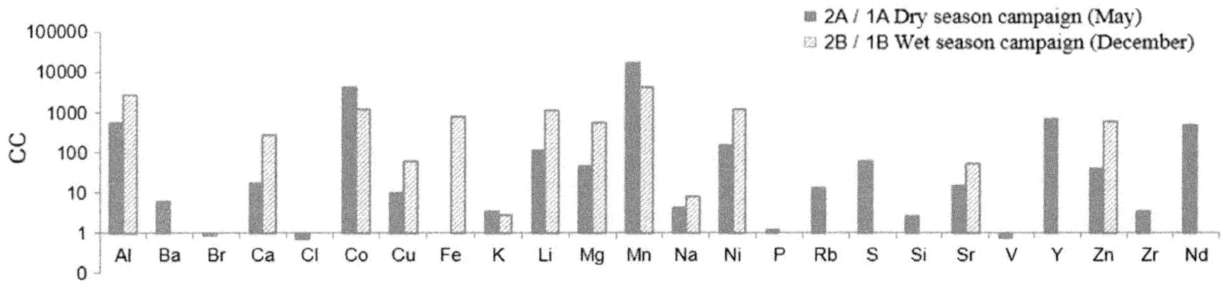

Figure 7.4.2. Concentration coefficients (CC) of elements in samples 2A and 2B compared to samples 1A and 1B.

was carried out during the rainy season, when the rate of water infiltration through the CWP was at its maximum. In the case of sampling area 1 (upstream of the Galladas tributary before it passes through the Arroyo Galladas CWP), which is hardly affected by the mining activities, the concentration of trace elements was about six times higher in May than in December, probably due to the self-dilution effect that is prevalent in the wet season.

Data from sampling areas 1 and 2 were compared to identify the effects that the water percolating through the Arroyo Galladas CWP material had on the Galladas tributary, the Cúa River, and the riverine system of the Fabero region of northwest Spain. The ratios between the element concentrations in samples 2A and 2B (downstream of the Galladas tributary) and in samples 1A and 1B (upstream of the Galladas tributary) allow the trace element concentration coefficients (CCs) to be determined. Based on the results illustrated in Figure 7.4.2, it is possible to identify the elements that show higher concentrations after the water has circulated through the Arroyo Galladas CWP (CC>1).

As Figure 7.4.2 shows, the elements Al, Ca, Co, Cu, K, Li, Mg, Mn, Na, Ni, Sr, and Zn are present in higher concentrations in water samples collected downstream (sampling area 2) of the Arroyo Galladas CWP in both May (dry season) and December (wet season) campaigns compared to samples 1A and 1B. The elements Ba, P, Rb, S, Si, and Y are present in large quantities only in samples collected in May, while Fe presents its highest concentration only in December. Of the elements just mentioned the presence of Al, Ca, Co, Fe, Li, Mg, Mn, Ni, Y, Zn, and Nd is more than 100 times greater (CC>100) in water after circulation through the CWP (in at least one of the samples); Cu, Rb, S, and Sr have a CC between 10 and 100, and Ba, K, Na, P, Si, and Zr have a CC between 1 and 10 (in at least one of the samples). The elements Br, Cl, and V show slightly higher concentrations in the samples upstream of the Arroyo Galladas CWP (CC<1).

The higher concentrations of K, Mg, Na, S, Si, Al, B, Ba, Cd, Ce, Co, Cu, Fe, Li, Mn, Ni, Pb, Rb, Sc, Sr, U, Y, Zn, and rare earth elements (REEs) in samples taken from the Cúa River (samples 3A and 3B) may be a consequence

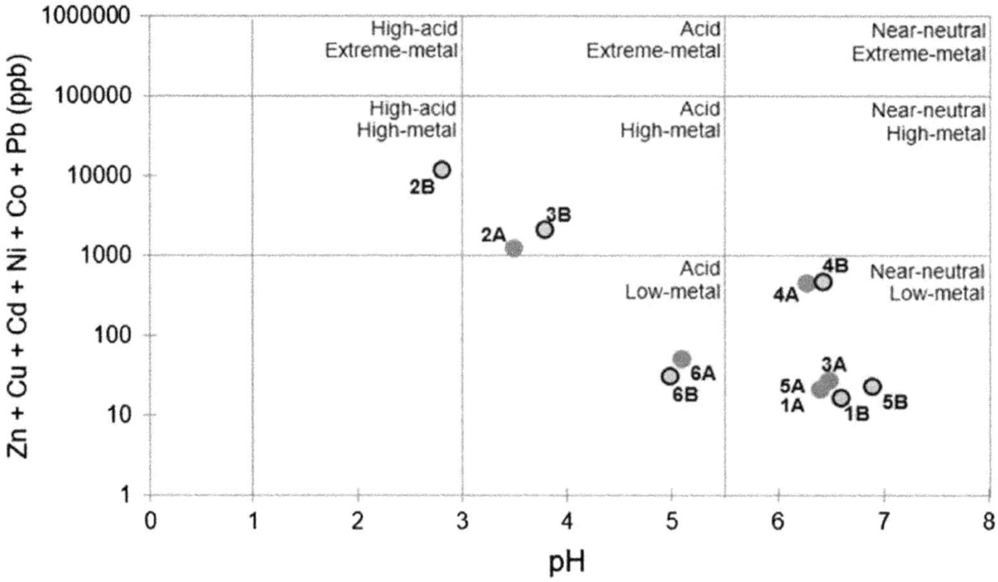

Figure 7.4.3. Ficklin diagram of concentrations in parts per billion (ppb) versus pH of the water samples (numbered circles) studied.

of the high concentration of these elements in water samples taken downstream of the Arroyo Galladas CWP (samples 2A and 2B) that are discharged into the Cúa River. However, most of these elements are also present in high concentrations in samples 4A and 4B, indicating that the concentrations of these elements may result from the circulation of water through coal waste and other mining infrastructures. Only the elements Al, Cd, Co, Cu, Mn, U, Y, and Zn have high concentrations in samples 2 and 3, which can be attributed to changes induced in waste materials due to spontaneous combustion in the Arroyo Galladas CWP.

Using the Ficklin diagram (Ficklin et al., 1992) depicted in Figure 7.4.3, and the pH and the sum of some metals, the water samples can be classified as follows: 2B is a high-acid, high-metal water sample; 2A and 3B are acid high-metal water samples; 6A and 6B are acid low-metal water samples; and the remaining are classified as almost neutral low-metals (Figure 7.4.3). The Ficklin diagram also shows that the production of MAD promotes the dissolution of heavy metals.

7.5 Surface Water Quality

Photo by Joana Ribeiro, 2012.

Junction of the Mourin tributary with the Cúa River. The Mourin tributary is about 4–5 m wide here. Along 200 m of its length, it flows around the Mourin waste pile.

Surface Water Quality

To evaluate the quality of the water in streams in the region of Fabero, the data obtained were compared with previous studies related to groundwater quality in the El Bierzo area, studies on mining acid drainage waters in other coal-mining regions, and the guideline values established by water quality legislation. The data used for comparison are compiled in Table 7.5.1. First, a comparison based on water quality data available in the literature on groundwater and few surface water samples collected relatively near the mining area of Fabero and corresponding to shallow wells and springs was made (de la Losa et al., 2010). Second, a comparison was made using data from waters discharged from CWPs resulting from coal-mining activities in England and Norway (Banks et al., 1997). In respect of legislation, comparisons were made with (1) Spanish legislation based on the European Directives (RD 817, 2015) on the standards for the environmental quality of surface waters, with maximum admissible concentrations of inorganic elements and other substances; and (2) Canadian Environmental Quality Guidelines that provide science-based goals regarding the quality of aquatic and terrestrial ecosystems (Canadian Council of Ministers of the Environment (CCME), 1999) and including freshwater quality guidelines to prevent anthropogenic pollution from chemical inputs or changes to physical components.

An analysis of the results of the comparison with data from previous studies related to water quality in the El Bierzo (de la Losa et al., 2010) revealed that the surface waters taken from the region of Fabero have higher concentrations of chlorides and Pb in almost all the samples as well as higher concentrations of sulfates, Al, Ca, K, Mg, Cd, Co, Cu, Fe, Mn, and Zn in sampling areas 2 (downstream of the Galladas tributary), 3 (from the Cúa River before the discharge of the Galladas tributary), and 4 (downstream of the Lamocos tributary). In addition, the elements Al, Fe, and Zn also have higher concentrations in some of samples 5 and 6. The pH of the studied waters was also more acidic in sampling areas 2 and 6 and sample 3B.

A comparison of the drainage waters from CWP in England and Norway (Banks et al., 1997) indicates that in the region of Fabero, the surface waters show pH values with identical ranges and concentrations to most of those analyzed and in both studies, are lower than or within the same range of concentrations as those found by Banks et al. (1997). In addition, as expected, samples 2A, 2B, 3A, and 3B are analogous to the water investigated by Banks et al. (1997). On the other hand, the elements Al, Mn, and Zn in samples 2B and 3B have higher concentrations (or values that are very close in the case of Zn in sample 3B).

In view of the Spanish legislation relating to the quality of surface waters (RD 817, 2015), which is in accordance with European directives, the studied waters present concentrations of Cd, Co, Ni, and Zn that are above the established limits in samples 2A, 2B (corresponding to Galladas tributary downstream of the after Arroyo Galladas CWP), 3B (in the Cúa River before the discharge of the Galladas tributary), and samples 4A, and 4B (downstream of the Lamocos tributary, after the water has passed through a coal seam and mining infrastructures). According to Canadian Environmental Quality Guidelines (CCME, 1999), the Cd, Fe, and Mn contents exceed the limits established in samples 2A, 2B, 3B, 4A, and 4B (except Fe in sample 4A) where the pH is more acidic in most cases (particularly in samples 2A, 2B, 3B, 6A, and 6B).

These comparisons indicate the outstanding presence of certain metallic species (i.e., Cd, Co, Cu, Fe, Mn, Ni, Pb, and Zn) and their likely environmental impact on the riverine system in the Fabero mining area, and consequently in the hydrographic Mino-Sil system.

Table 7.5.1
Data from the water samples in this study and from previous studies with legislation, and international guidelines for comparison.

Samples	1A	1B	2A	2B	3A	3B	4A	4B	5A	5B	6A	6B		de la Losa et al. (2010)	Banks et al. (1997)	RD 817/2015	CCME (1999)
Water	Surface Water												Shallow Well	Spring Water (n=3)	CWP Drainage Waters	Spanish Guidelines	Canadian Guidelines
pH	6.4	6.6	3.5	2.8	6.5	3.8	6.3	6.4	6.4	6.9	5.1	5.0	6.7	5.4–7.4	2.7–5.5	n.a.[*]	6.5–9
ppm (parts per million)																	
Cl^-	3.1	2.9	3.4	8.1	2.8	9.2	2.1	2.7	3.3	3.4	2.6	2.7	2	1–3	4.5–7	n.a.	n.a.
SO_4^{2-}	3.4	3.4	495	853	106	176	715	951	10.7	11.3	13.4	14.8	396	9–86	77–1358	n.a.	n.a.
Ca	1.0	0.52	17.1	144	13.7	33.8	73.3	115	1.4	1.3	0.94	0.75	71	2–15	n.a.	n.a.	n.a.
K	1.2	0.37	3.9	1.0	12.5	1.5	1.7	1.3	2.9	0.35	2.8	0.26	3	0	n.a.	n.a.	n.a.
Mg	1.1	0.90	48.6	470	11.6	93.9	144	210	2.3	2.1	1.5	1.1	62	1–17	n.a.	n.a.	n.a.
Na	1.1	1.2	4.9	9.0	2.8	5.0	3.6	4.4	1.4	1.4	1.1	1.0	11	2–6	n.a.	n.a.	n.a.
ppb (parts per billion)																	
Al	25.0	78.0	13,486	211,182	21.0	47,076	16.0	3281	24.0	79.0	297	298	53	0–382	970–27500	n.a.	n.a.
As	0.60	<DL[†]	<DL	<DL	<DL	0.50	<DL	<DL	<DL	<DL	<DL	<DL	0.8	0	n.a.	50	5
B	<DL	<DL	36.0	<DL	<DL	8.0	10.0	12.0	<DL	<DL	<DL	6.0	n.a.	n.a.	n.a.	n.a.	1500
Ba	1.4	1.7	8.4	<DL	2.6	13.0	17.1	14.8	1.6	2.2	2.4	3.3	25.1	1.1–5.4	n.a.	n.a.	n.a.
Be	<DL	<DL	3.0	n.a.	0.06	5.0	<DL	0.65	0.09	<DL	0.12	<DL	n.a.	n.a.	n.a.	n.a.	n.a.
Cd	<DL	<DL	2.8	<DL	<DL	5.4	0.93	0.79	<DL	<DL	0.08	0.07	n.a.	n.a.	n.a.	<0.08–0.25[‡]	0.09
Co	0.04	1.3	166	1597	0.50	297	67.6	86.1	0.04	0.90	3.5	3.6	0	0–3.5	n.a.	5–120[*]	n.a.
Cr	<DL	<DL	0.70	<DL	<DL	2.6	<DL	<DL	<DL	<DL	<DL	<DL	n.a.	n.a.	n.a.	50	1 $Cr_{(IV)}$–8.9 $Cr_{(III)}$
Cu	3.4	2.1	32.0	121	1.6	41.4	5.8	14.8	0.70	3.3	2.1	1.7	2.9	0	<7-168	n.a.	n.a.
Fe	<DL	126	2185	95,897	<DL	1770	<DL	6691	19.0	300	16.0	230	240	0	1,5000–287,000	n.a.	300
Hg	<DL	<DL	<DL	n.a.	<DL	<DL	<DL	<DL	<DL	<DL	<DL	<DL	n.a.	n.a.	n.a.	n.a.	0.026
Mo	<DL	<DL	<DL	<DL	<DL	0.10	<DL	0.10	<DL	0.20	<DL	0.20	n.a.	n.a.	n.a.	n.a.	73
Mn	0.44	16.3	7065	68,472	34.1	12,004	2651	3653	1.3	24.0	20.8	26.8	7.3	2.18–31.6	400–5200	n.a.	n.a.
Ni	2.0	3.2	297	3660	3.0	618	134	132	2.7	5.0	9.1	7.9	3.1	0–7.13	n.a.	20	n.a.
Pb	<DL	0.30	3.4	<DL	1.0	3.2	0.20	0.40	0.20	0.70	1.2	0.70	0.0	0.0	n.a.	7.2	n.a.
Se	<DL	<DL	<DL	n.a.	<DL	<DL	<DL	<DL	<DL	<DL	<DL	<DL	n.a.	n.a.	n.a.	1	1
Th	<DL	<DL	0.08	<DL	<DL	0.36	<DL	0.07	<DL	<DL	<DL	<DL	n.a.	n.a.	n.a.	n.a.	0.8
U	<DL	<DL	1.6	<DL	<DL	2.8	0.03	0.40	<DL	<DL	<DL	0.02	n.a.	n.a.	n.a.	n.a.	15
Zn	14.8	8.7	589	5190	17.4	1086	196	183	17.1	12.3	35.5	15.2	18.3	5.1–15.3	50–1300	30–500[‡]	30

[*]n.a., data not available.
[†]<DL, below detection limit.
[‡]Values variable with the $CaCO_3$ content.

7.6 Sources of Contamination

Photo by Joana Ribeiro, 2012.

The Mourin coal waste and the Mourin tributary (right). The waste pile on the left is about 8 m high.

Potential Sources of Contamination

In order to identify the potential source of the predominant elements detected in the water samples, the chemical composition of the surface waters was compared with the chemical composition of the materials that constitute the Arroyo Galladas and Arroyo Mourin CWPs through which the tributaries pass before being discharged into the Cúa River. In addition, the Arroyo Galladas CWP is currently undergoing spontaneous combustion. A comparison of the drainage waters with the solid material from the Arroyo Galladas waste pile will therefore highlight the environmental impact of the spontaneous combustion of CWP on bodies of water. Table 7.6.1 presents the data obtained from the studied water samples together with the data from coal waste for comparison (Ribeiro et al., 2016b).

As expected, the concentration of elements in the surface waters is lower than in the coal waste, except for Mn, Ni, and Zn detected in sample 2B, where the Mn and Ni concentrations are higher than the minimum values in the Arroyo Galladas CWP materials and the Zn concentration is similar. This evidences that the percolation of waters through the Arroyo Galladas CWP promotes the dissolution of toxic metals (specifically Mn, Ni, Zn) under acidic conditions by means of solid-water interactions, leading to high concentrations of these metallic species in the water bodies of the riverine system in the Fabero mining region. These elements (along with another 22, namely Ag, As, B, Ba, Be, Cd, Cl, Co, Cr, Cu, F, Hg, Mo, P, Pb, Sb, Se, Sn, Tl, Th, U, and V) have been identified by Finkelman (1995) as environmentally sensitive elements in coal, and therefore worthy of attention to mitigate and/or resolve the environmental problems affecting the El Bierzo mining area.

Investigation on the enrichment/depletion behavior of chemical elements in burned and unburned solids from the Fabero and Arroyo Galladas CWP also revealed that Fe, K, Na, S, Ti, Nb, Re, Ta, Te, and W are enriched in burned materials, while Ca, Mg, P, Cd, Cu, Mn, Zn, and REE are depleted (Ribeiro et al., 2016b). The high concentration of some elements in water samples collected downstream of the Arroyo Galladas CWP (corresponding to sampling area 2) compared to samples collected upstream of the coal waste (sampling area 1) may be attributed to the percolation of water through the coal waste and solids/water interaction under acidic conditions. The depletion of certain trace elements in the coal waste because of self-combustion in the Arroyo Galladas CWP, therefore, may promote the enrichment of these elements in the drainage waters. This would explain the concentrations of Ca, Cu, Mg, Mn, P, and Zn detected in the water samples collected downstream of the CWP.

Table 7.6.1
Chemical composition of surface-water and waste-material samples collected from the Arroyo Galladas and Mourin CWP (GWP and MWP, respectively) (Ribeiro et al., 2016b).

	1A	1B	2A	2B	3A	3B	4A	4B	5A	5B	6A	6B	GWP (n=15) Min	GWP (n=15) Max	MWP
	Surface Water Samples												Waste Materials		
	ppm (parts per million)														
Ca	1.0	0.52	17.1	144	13.7	33.8	73.3	115	1.4	1.3	0.94	0.75	200	4000	2000
Cl	28.0	1.0	20.0	n.a.*	15.0	1.0	5.0	<DL†	5.0	1.0	4.0	<DL	n.a.	n.a.	n.a.
K	1.2	0.37	3.9	1.0	12.5	1.5	1.7	1.3	2.9	0.35	2.8	0.26	9000	32,000	23,000
Mg	1.1	0.90	48.6	470	11.6	93.9	144	210	2.3	2.1	1.5	1.1	1000	5000	5000
Na	1.1	1.2	4.9	9.0	2.8	5.0	3.6	4.4	1.4	1.4	1.1	1.0	2000	5000	4000
S	2.0	<DL	117	1188	24.0	250	234	348	4.0	4.0	4.0	4.0	4000	100,000	4000
Si	1.8	1.5	4.8	n.a.	2.8	8.1	3.5	4.5	2.1	2.0	2.2	1.7	n.a.	n.a.	n.a.
	ppb (parts per billion)														
Ag	<DL	<DL	0.19	<DL	0.49	<DL	<DL	<DL	0.16	<DL	<DL	<DL	0	400	100
Al	25.0	78.0	13,486	211,182	21.0	47,076	16.0	3281	24.0	79.0	297	298	3,300,000	9,800,000	81,000,000
As	0.60	<DL	<DL	<DL	<DL	0.50	<DL	<DL	<DL	<DL	<DL	<DL	11,800	90,900	25,600
B	<DL	<DL	36.0	<DL	<DL	8.0	10.0	12.0	<DL	<DL	<DL	6.0	n.a.	n.a.	n.a.
Ba	1.4	1.7	8.4	<DL	2.6	13.0	17.1	14.8	1.6	2.2	2.4	3.3	11,000	713,000	566,000
Be	<DL	<DL	3.0	n.a.	0.06	5.0	<DL	0.65	0.09	<DL	0.12	<DL	2000	4000	2000
Bi	<DL	<DL	<DL	<DL	<DL	<DL	<DL	<DL	<DL	<DL	<DL	0.08	0	1000	300
Br	14.0	22.0	12.0	n.a.	10.0	11.0	9.0	9.0	10.0	17.0	<DL	5.0	n.a.	n.a.	n.a.
Cd	<DL	<DL	2.8	<DL	<DL	5.4	0.93	0.79	<DL	<DL	0.08	0.07	20	200	200
Ce	<DL	0.07	9.8	n.a.	<DL	25.9	0.13	4.4	<DL	0.09	0.07	0.17	n.a.	n.a.	n.a.
Co	0.04	1.3	166	1597	0.50	297	67.6	86.1	0.04	0.90	3.5	3.6	700	32,700	6700
Cr	<DL	<DL	0.70	<DL	<DL	2.6	<DL	<DL	<DL	<DL	<DL	<DL	34,000	111,000	88,000
Cs	<DL	<DL	0.44	n.a.	0.01	0.27	0.07	0.09	<DL	<DL	<DL	<DL	5000	16,400	9500
Cu	3.4	2.1	32.0	121	1.6	41.4	5.8	14.8	0.70	3.3	2.1	1.7	13,200	63,600	29,000
Fe	<DL	126	2185	95,897	<DL	1770	<DL	6691	19.0	300	16.0	230	900,000	7,400,000	3,200,000
Ga	<DL	<DL	0.09	n.a.	<DL	0.11	<DL	<DL	<DL	<DL	<DL	<DL	8900	31,400	21,000
Hg	<DL	<DL	<DL	n.a.	<DL	<DL	<DL	<DL	<DL	<DL	<DL	<DL	n.a.	n.a.	n.a.
Li	2.0	1.6	222	1751	4.3	394	38.1	56.1	3.9	3.7	5.4	4.0	15,600	108,000	107,000

Mn	0.44	16.3	7065	68,472	34.1	12,004	2651	3653	1.3	24.0	20.8	26.8	18,000	308,000	121,000
Mo	<DL	<DL	<DL	<DL	<DL	0.10	<DL	0.10	<DL	0.20	<DL	0.20	600	8000	4300
Ni	2.0	3.2	297	3660	3.0	618	134	132	2.7	5.0	9.1	7.9	3400	109,000	25,900
P	12.0	<DL	14.0	<DL	14.0	28.0	<DL	<DL	<DL	17.0	<DL	<DL	10,000	130,000	80,000
Pb	<DL	0.30	3.4	<DL	1.0	3.2	0.20	0.40	0.20	0.70	1.2	0.70	11,600	99,600	32,900
Rb	0.21	0.30	2.6	n.a.	0.67	3.8	1.1	1.7	0.33	0.33	0.24	0.32	55,900	159,000	128,000
Sb	<DL	<DL	<DL	<DL	0.08	0.05	0.06	0.05	<DL	<DL	<DL	<DL	2000	18,600	3000
Sc	<DL	<DL	3.00	n.a.	1.0	1.0	<DL	<DL	<DL	<DL	<DL	<DL	5800	18,200	14,700
Se	<DL	<DL	<DL	n.a.	<DL	<DL	<DL	<DL	<DL	<DL	<DL	<DL	600	57,600	2100
Sn	<DL	<DL	<DL	n.a.	<DL	<DL	<DL	<DL	<DL	0.07	<DL	<DL	1400	4800	3900
Sr	2.6	2.1	39.5	112	86.3	69.7	210	296	4.3	3.8	4.5	4.6	42,000	168,000	101,000
Th	<DL	<DL	0.08	<DL	<DL	0.36	<DL	0.07	<DL	<DL	<DL	<DL	6800	21,200	12,600
Tl	<DL	<DL	0.06	n.a.	<DL	0.06	<DL	0.02	<DL	<DL	<DL	<DL	400	1200	900
U	<DL	<DL	1.6	<DL	<DL	2.8	0.03	0.40	<DL	<DL	<DL	0.02	1700	4400	3300
V	5.7	<DL	4.3	<DL	1.2	0.30	0.30	<DL	0.60	<DL	<DL	<DL	56,000	145,000	118,000
W	<DL	0.02	0.02	<DL	<DL	<DL	<DL	<DL	<DL	<DL	0.04	<DL	800	3100	900
Y	0.04	0.04	26.9	n.a.	0.02	67.6	0.27	5.1	<DL	0.05	0.11	0.17	5300	14,700	8700
Zn	14.8	8.7	589	5190	17.4	1086	196	183	17.1	12.3	35.5	15.2	6500	134,000	42,800
Zr	0.02	0.12	0.07	n.a.	<DL	0.15	0.04	0.09	<DL	0.13	<DL	0.14	35,700	91,300	76,600
ΣREE‡	0.02	0.14	47.7	n.a.	0.01	117	0.39	14.7	<DL	0.19	0.16	0.39	71,500	204,000	134,000

*n.a., data not available.
†<DL, below detection limit.
‡ΣREE, sum of rare earth elements.
CWP, coal waste piles.

Conclusions

The production of acid mine drainage and spontaneous combustion processes related to coal-waste deposits resulting from past coal-mining activities in the El Bierzo coalfield were identified. The investigation of the effects of the coal waste on the riverine system in the region of Fabero was the main goal of this study. For this purpose, water samples from the Cúa River and its tributaries in the region (Galladas, Mourin, and Lamocos tributaries) were analyzed to determine their physicochemical parameters.

The pH and conductivity values in waters evidence that MAD occurs as the result of the circulation of waters through the waste material deposits, which promotes the dissolution of metals. The high ionic conductivity of samples with near-neutral pH values are probably due to the precipitation of trace elements such as ferric-hydroxides, sulfates, oxy-hydroxysulfates, and others.

The concentration of sulfates varies greatly with higher values being found in water samples collected downstream of the coal waste and mining infrastructures due to the circulation of water through these materials and mining infrastructures and to the dissolution of precipitates promoted by rainy events, principally in the wet season.

The studied waters exhibited a low variability in terms of hydrogeochemical facies, ranging from multiple cations/anions waters to nondominant ones. The Piper diagram also revealed useful information about the chemical relationship of the water samples and the sources of pollution. Upstream water bodies were classified as nondominant waters with low contents of cations, the composition reflecting the water bodies before any pollution. On the other hand, downstream water samples were classified as multiple cations and anions waters $(Ca - Mg - SO_4^{2-} - Cl^-)$, and the composition reflected the contribution of the acid drainage from the CWP.

From the physicochemical parameters and the chemical composition of the waters, the results indicate that those collected in areas upstream of the coal waste and old mining infrastructures are less affected by the circulation of waters through those materials. On the other hand, the samples collected downstream of the coal waste and old mining infrastructures exhibit the highest trace element concentration (including some toxic metals), a lower pH, a higher conductivity and sulfate concentration as the result of the circulation of waters through these materials.

Regarding seasonal effects, it is to be noted that the samples obtained from downstream areas of the coal-mining deposits and mining infrastructures have higher concentrations of trace elements in the wet season. This may be related to the fact that sampling was carried out during the rainy season, when the rate of water infiltration through the CWP was at its highest.

By comparing the chemical composition of the studied samples with data from previous studies related to groundwater quality in El Bierzo area, studies on mining acid drainage associated with other coal-mining regions, and guideline values, it is possible to evaluate the quality of the water in the riverine system. The results point to the presence of significant concentrations of certain metallic elements of environmental concern (i.e., Cd, Co, Cu, Fe, Mn, Ni, Pb, and Zn).

To identify the potential source of the predominant elements detected in the water samples, their chemical composition was compared with the chemical composition of the CWP materials that percolated through the analyzed waters. This comparison indicates that the percolation of the waters through the Arroyo Galladas CWP caused the dissolution of toxic metals (specifically Mn, Ni, and Zn) under acidic conditions, giving rise to high concentrations of these metallic species in the water capable of having a serious environmental impact.

Acknowledgments

Joana Ribeiro has benefited from a postdoctoral scholarship awarded by the Portuguese Science and Technology Foundation, FCT – Fundação para a Ciência e Tecnologia (grant SFRH/BPD/76230/2011). The authors also acknowledge the funding provided by the Institute of Earth Sciences (ICT), under contracts UID/GEO/04683/2013 with FCT (the Portuguese Science and Technology Foundation), and COMPETE POCI-01-0145-FEDER-007690 and the Spanish MINECO (grant CTM2014-56670-R). The authors would like to thank Dionisio Luis García from Instituto Nacional del Carbón (INCAR-CSIC) for his help in collecting the samples.

Important Terms

acid mine drainage
coal mining
coal-waste disposal
environmental effects
heavy metals
riverine system
self-burning coal waste
streams

References

Alberruche, E., Gómez, V.R., Vadillo, L., 2008. Evaluatión del Impacto y Restauración Ambiental de la Cuenca Minera de El Bierzo. Impacto Ambiental de la Minería de Carbón em Ecosistemas Acuáticos. Instituto Geológico y Minero de Espanã. 466 p.

Banks, D., Younger, P.L., Arnesen, R.-T., Iversen, E.R., Banks, S.B., 1997. Mine-water chemistry: the good, the bad and the ugly. Environ. Geol. 32 (3), 157–174.

Bell, F.G., Bullock, S.E.T., Halbich, T.F.J., Lindsay, P., 2001. Environmental impacts associated with an abandoned mine in the Witbank Coalfield, South Africa. Int. J. Coal Geol. 45, 195–216.

Bian, Z., Inyang, H.I., Daniels, J.L., Otto, F., Struthers, S., 2010. Environmental issues from coal mining and their solutions. Min. Sci. Technol. 20, 215–223.

Blowes, D.W., Ptacek, C.J., Jambor, J.L., Weisener, C.G., 2004. The geochemistry of acid mine drainage. In: Holland, H.D., Turekian, K.K., Mackenzie, F.T. (Eds.), Treatise on Geochemistry, vol. 9. Elsevier, Amsterdam, pp. 149–204.

Canadian Council of Ministers of the Environment - CCME, 1999. Canadian Environmental Quality Guidelines.

CIUDEN-CIEMAT, 2010. Plan Director para la Restauración Edafopaisajística y Recuperación Ambiental de los espacios degradados por la minería de carbón en la comarca de El Bierzo (García Álvarez, A., Coord.). CIUDEN y CIEMAT, Ponferrada, España, 276 p.

Colmenero, J.R., Suárez-Ruiz, I., Fernández-Suárez, J., Barba, P., Llorens, T., 2008. Genesis and rank distribution of upper carboniferous coal basins in the Cantabrian Mountains, northern Spain. Int. J. Coal Geol. 76, 187–204.

de la Losa, A., Moreno, L., Núñez, I., 2010. Calidad química de las aguas subterráneas en una zona de actividad minera (Cuenca del Bierzo-León). Bol. Geol. Min. 121 (1), 103–122.

Drever, J.I., 1997. The Geochemistry of Natural Waters. Surface and Groundwater Environments, third ed. Prentice Hall. 436 p.

Ficklin, W.H., Plumlee, G.S., Smith, K.S., McHugh, J.B., 1992. Geochemical classification of mine drainages and natural drainages in mineralized areas. In: Proceedings of the 7th International Symposium on WaterRock Interaction, Balkema, Rotterdam, pp. 381–384.

Finkelman, R.B., Stracher, G.B., 2011. Environmental and health impacts of coal fires. In: Stracher, G.B., Prakash, A., Sokol, E.V. (Eds.), Coal and Peat Fires: A Global Perspective, vol. 7. Elsevier, Amsterdam, pp. 115–125.

Finkelman, R.B., 1995. Modes of occurrence of environmentally-sensitive trace elements of coal. In: Swaine, D.J., Goodarzi, F. (Eds.), Environmental Aspects of Trace Elements of Coal. Kluwer Academic Publishers, Netherlands, pp. 24–50.

Finkelman, R.B., 2004. Potential health impacts of burning coal beds and waste banks. Int. J. Coal Geol. 51, 19–24.

Hernández-Mendoza, H., Mejuto, M., Cardona, A.I., García-Álvarez, A., Millán, R., Yllera, A., 2013. Optimization and validation of a method for heavy metals quantification in soil samples by inductively coupled plasma sector field mass spectrometry (ICP-SFMS). Am. J. Anal. Chem. 4, 9–15.

Orem, W.H., Finkelman, R.B., 2003. Coal formation and geochemistry. In: Holland, H.D., Turekian, K.K., Mackenzie, F.T. (Eds.), Treatise on Geochemistry, vol. 7. Elsevier, Amsterdam, pp. 191–222.

Piper, A.M., 1944. A graphic procedure in the geochemical interpretation of water analyses. Am. Geophys. Union Trans. 25, 914–923.

Pone, J.D.N., Hein, K.A.A., Stracher, G.B., Annegarn, H.J., Finkelman, R.B., Blake, D.R., McCormack, J.K., Schroeder, P., 2007. The spontaneous combustion of coal and its byproducts in the Witbank and Sasolburg coalfields of South Africa. Int. J. Coal Geol. 72, 124–140.

Querol, X., Izquierdo, M., Monfort, E., Alvarez, E., Font, O., Moreno, T., Alastuey, A., Zhuang, X., Lu, W., Wang, Y., 2008. Environmental characterization of burnt coal gangue banks at Yangquan, Shanxi Province, China. Int. J. Coal Geol. 75, 93–104.

Querol, X., Zhuang, X., Font, O., Izquierdo, M., Alastuey, A., Castro, I., van Drooge, B.L., Moreno, T., Grimalt, J.O., Elvira, J., Cabañas, M., Bartroli, R., Hower, J.C., Ayora, C., Plana, F., López-Solar, A., 2011. Influence of soil cover on reducing the environmental impact of spontaneous coal combustion in coal waste gobs: a review and new experimental data. Int. J. Coal Geol. 85, 2–22.

Real Decreto 817/2015, 2015. Criterios de seguimiento y evaluación del estado de las aguas superficiales y las normas de calidad ambiental. Ministério de Agricultura, Alimentacion y Medio Ambiente. Boletin Oficial del Estado. ISSN: 0212–033X. 96 p.

Ribeiro, J., Ferreira da Silva, E., Flores, D., 2010a. Burning of coal-waste piles from Douro Coalfield (Portugal): petrological, geochemical and mineralogical characterization. Int. J. Coal Geol. 81, 359–372.

Ribeiro, J., Flores, D., Ward, C., Silva, L.F.O., 2010b. Identification of nanominerals and nanoparticles in burning coal-waste piles from Portugal. Sci. Total Environ. 408, 6032–6041.

Ribeiro, J., Taffarel, S.R., Sampaio, C.H., Flores, D., Silva, L.F.O., 2013. Mineral speciation and fate of some hazardous contaminants in coal waste pile from anthracite mining in Portugal. Int. J. Coal Geol. 109–110, 15–23.

Ribeiro, J., Suárez-Ruiz, I., Ward, C., Flores, D., 2014. Mineralogy of thermally altered coal mining residues in self-burning coal-waste piles in Portugal and Spain. Comunicações Geológicas 101, 187–190.

Ribeiro, J., Suárez-Ruiz, I., Flores, D., 2016a. Geochemistry of self-burning coal mining residues from El Bierzo Coalfield (NW Spain): environmental implications. Int. J. Coal Geol. 159, 155–168.

Ribeiro, J., Suárez-Ruiz, I., Ward, C.R., Flores, D., 2016b. Petrography and mineralogy of self-burning coal wastes from anthracite mining in the El Bierzo Coalfield (NW Spain). Int. J. Coal Geol. 154–155, 92–106.

Rodríguez Gómez, V., Vadillo Fernández, L., Lacal Guzmán, M., Alberruche del Campo, E., Herrero Barrero, T., de la Losa Román, A., 2010. Atenuación natural y calidad ambiental de los ríos mineros de la Cuenca Carbonífera del Bierzo (León). Boletín Geológico y Minero 121 (1), 89–102.

Stracher, G.B., Taylor, T.P., 2004. Coal fires burning out of control around the world: thermodynamic recipe for environmental catastrophe. Int. J. Coal Geol. 59, 7–17.

Stracher, G.B., Prakash, A., Sokol, E.V. (Eds.), 2011. Coal and Peat Fires: A Global Perspective Coal - Geology and Combustion Volume 1. Elsevier, Amsterdam (380 p.).

Stracher, G.B., Prakash, A., Sokol, E.V. (Eds.), 2013. Coal and Peat Fires: A Global Perspective, Photographs and Multimedia Tours Volume 2. Elsevier, Amsterdam (584 p.).

Stracher, G.B., Prakash, A., Sokol, E.V. (Eds.), 2015. Coal and Peat Fires: A Global Perspective, Case Studies — Coal Fires Volume 3. Elsevier, Amsterdam (816 p.).

Stracher, G.B. (Ed.), 2007. Geology of Coal Fires: Case Studies from Around the World. Reviews in Engineering Geology XVIII (278 p.).

Suárez-Ruiz, I., Crelling, J.C. (Eds.), 2008. Applied Coal Petrology. The Role of Petrology in Coal Utilization. Elsevier (388 p.).

Suárez-Ruiz, I., Jiménez, A., 2004. Coal facies studies in Spain. Int. J. Coal Geol. 58, 31–39.

Suárez-Ruiz, I., Flores, D., Mendonça Filho, J.G., Hackley, P.C., 2012. Review and update of the applications of organic petrology: Part 2, Geological and Multidisciplinary Applications. Int. J. Coal Geol. 98, 73–94.

Williams, D.J., Bigham, J.M., Cravotta III, C.A., Traina, S.J., Andersen, J.E., Lyon, J.G., 2002. Assessing mine drainage pH from color and spectral reflectance of chemical precipitates. Appl. Geochem. 17, 1273–1286.

Younger, P.L., 2004. Environmental impacts of coal mining and associated wastes: a geochemical perspective. In: Gieré, R., Stille, P. (Eds.), Energy, Waste and the Environment: A Geochemical Perspective. Geological Society, vol. 236. Special Publications, London, pp. 169–209.

Zhao, Y., Zhang, J., Chou, C.-L., Li, Y., Wang, Z., Ge, Y., Zheng, C., 2008. Trace element emissions from spontaneous combustion of gob piles in coal mines, Shanxi, China. Int. J. Coal Geol. 73, 52–62.

CHAPTER 8

Analyzing the Status of Thermal Events in Longwall Coal Mine Gobs

Smoke from combustion in abandoned coal mine gob piles in Datong, China. Surface cracks caused by underground mine subsidence contribute to the circulation of air both underground and through the gob piles. *Photo by Jianwei Cheng, 2017.*

CHAPTER CONTENTS

Coal and Peat Fires: A Global Perspective
Edited by Glenn B. Stracher

https://doi.org/10.1016/B978-0-12-849885-9.00008-1

8.1 Thermal Accidents in Underground Coal Mines

Jianwei Cheng

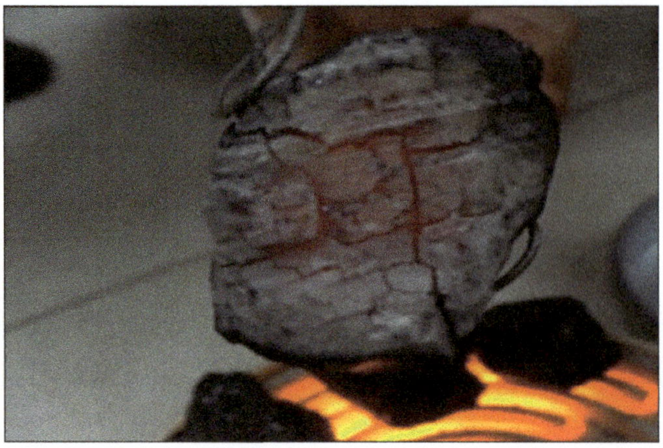

Photo by Jianwei Cheng, 2014.

Lignite Coal obtained from the Datong coal field in China was burnt to 700°C on electrical coils as a laboratory experiment for students. Afterward, the coal was cooled with water, producing a lot of water vapor.

Introduction

Fire in underground mines has long been a concern for workers. A mine fire produces gases and heat that can be carried by airflow traveling through the confined and connected mine ventilation system. The liberated gases can be poisonous or explosive and the released heat can disturb the ventilation system. Worst of all, an uncontrolled mine fire could ignite the available fuels (e.g., methane, coal dust, wood) on its propagation path and probably result in a severe mine disaster. Most deaths arising from mine fires and explosions are caused not by burning or blast effects, but by the inhalation of toxic gases, especially carbon monoxide. There are two major differences between underground fires and those that occur in surface structures. The first concern is the long distance, often several kilometers, for the miners to travel in passageways that may be smoke-filled. Second, the ventilation routes are bounded by the confines of the airways and workings, causing closely coupled interactions between the airflows and behavior of the fire (McPherson, 1993). Unlike fire accidents on the surface, the limited numbers of evacuation routes in an underground coal mine may slow down the evacuation process, and cause worse damage and a larger threat to miners' lives.

Understanding the hazards and risk of mining gobs, especially when a thermal event is identified, is very important for the safety of coal-mine workers. In this chapter, gob is waste rock that fills part of the mine from which coal was removed. The purpose of this chapter is to (1) present the **spontaneous combustion** characteristics of gob piles, (2) review **fire ratios** typically used to characterize the stage of a thermal event in a sealed coal mine, and (3) provide an overview of gas explosibility analysis along with **mine-gas distribution** patterns that occur with changing ventilation parameters.

Mining Hazards

A mine fire can occur at any time and impose hazards to the health and safety of miners. Statistics from the Mine Safety and Health Administration (MSHA) indicate that 137 reportable fires occurred in underground coal, metal, and nonmetal mines from 1991 to 2000 in the United States. Those fires caused 2 fatalities and 34 injuries. The statistic data combined by Alexander (2006, p.38) based on some published statistical analysis of coal mine fires has shown that there were still about 10 reportable fires in 2001 even though the number of reportable fires have decreased from 1950 to 2001 as shown in Figure 8.1.1.

On the other hand, mine gas explosions present the most feared hazards in the coal industry worldwide. In 2009, a total of 157 gas explosions were responsible for 755 fatalities in Chinese coal mines. This number represented 28.7% of all Chinese coal mine fatalities (Huang, 2010). Two explosions in the Pike River mine disaster began on November 19, 2010 in New Zealand, killing 29 miners. This mine accident ranks as New Zealand's worst mining disaster since 43 men died at Ralph's Mine in Huntly in 1914 (Wikipedia, 2010). On March 21, 2011, a serious gas explosion in Pakistan's Sorange mine killed 43 miners (CNN, 2011). In U.S. coal mining history, gas explosions are also considered the most dangerous hazard. The Monongah mine disaster in Monongah, West Virginia that

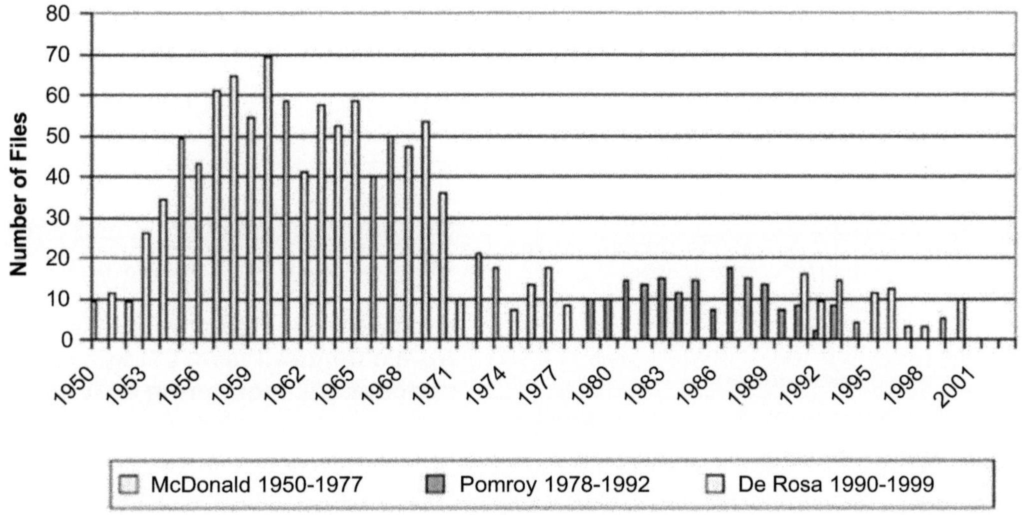

Figure 8.1.1. Underground coal mine fires 1950–2001. From Alexander (2006).

Table 8.1.1
Coal mine disasters due to gas explosions (1970–2010).*

No	Year	Day	Mine Name	Locations in the United States†	Type	Deaths
1	2010	4–5	Upper Big Branch Mine	Montcoal, WV	Explosion	29
2	2006	5–20	Darby Mine No. 1	Holmes Mill, KY	Explosion	5
3	2006	1–2	Sago Mine	Tallmansville, WV	Explosion	12
4	2001	9–23	No. 5 Mine	Brookwood, AL	Explosion	13
5	1992	12–7	No. 3 Mine	Wise, VA	Explosion	8
6	1989	9–13	William Station No. 9 Mine	Sullivan, KY	Explosion	10
7	1983	6–21	McClure No. 1 Mine	Dickinson, VA	Explosion	7
8	1982	1–20	RHF No. 1 Mine	Floyd, KY	Explosion	7
9	1981	12–8	Grundy Mining Co. No. 21 Mine	Marion, TN	Explosion	13
10	1981	12–7	Adkins Coal No. 11 Mine	Knott, KY	Explosion	8
11	1981	4–15	Dutch Creek No. 1	Redstone, CO	Explosion	15
12	1980	11–7	Ferrell No. 17	Boone, WV	Explosion	5
13	1976	3–9	Scotia Mine	Oven Fork, KY	Explosion	26
14	1972	12–16	Itmann No. 3 Mine	Wyoming, WV	Explosion	5
15	1970	12–30	Nos. 15 and 16 Mines	Hyden, KY	Explosion	38

*Data from Mine Safety and Health Administration (MSHA) (2017), and the Centers for Disease Control and Prevention (CDC) (2017).
†WV, KY, AL, VA, TN, and CO are the states of West Virginia, Kentucky, Alabama, Virginia, Tennessee, and Colorado, respectively.

occurred on December 6, 1907 has been described as the worst mining disaster in American history. The lives of 362 workers including children were lost in this underground explosion.

Statistics from the MSHA show underground coal mine disasters due to gas explosions from 1970 to 2010 (Table 8.1.1). More than half of the disasters resulted in 10 or more deaths per accident. Due to inaccessibility at a sealed gob mine for safety reasons, it is nearly impossible to implement direct measurements of the overall atmospheric composition in the mine. The gas composition in a sealed mine changes with time, especially during the first few weeks or months after the mine is sealed (Zipf and Mohamed, 2010). Most explosions occur when the atmospheric composition at the mine changes during the critical period (McPherson, 1993). Hence, the atmospheric composition at a coal mine should be carefully monitored and well controlled. Explosion and fire-control mitigation procedures may include pumping water and injecting CO_2/N_2 along **longwall gob** to control coal oxidation. Barometric pressure can be problematic in explosion control in that during falling pressure, there can be an increased gas-emission rate that may cause explosive-gas leakage flow from sealed areas into the atmosphere at an active mine (Mitchell, 1996).

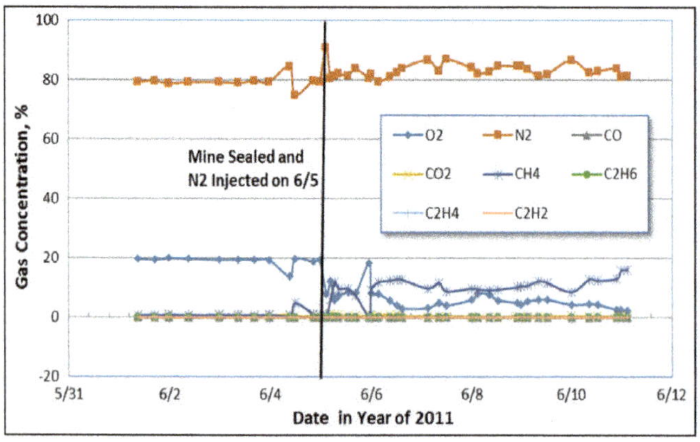

Figure by Jianwei Cheng (2012); from an unpublished report to an anonymous Chinese coal company.

8.2 Spontaneous Combustion of Longwall Gob

Gas compositions before and after implementing control measures in an anthracite coal mine in northwestern China. Gas samples were taken from gob from a longwall face; using a telemetric atmospheric monitoring system.

Introduction

To effectively control the spontaneous combustion of coal, it is necessary to detect the locations where the coal is susceptible to high oxidation rates and exothermic reactions that generate enough heat energy to ignite the coal. Meanwhile, there are certain rules to follow in detecting these locations. To be specific, if the four conditions (stockpiled coal or coal waste, atmospheric exposure, accumulation of heat energy, and time) for spontaneous combustion are met, the coal will likely ignite. Based on production practices and the rule of airflow pattern in the longwall working section, it can be known that the coal spontaneous combustion is more likely to occur in the oxidation zone in the gob, two roadways and two lines (air intake roadway and air return roadway, longwall setup and recovery lines), and the top-coal caving region in mine tunnels. However, it does not mean that other regions and areas would be excluded from spontaneous combustion.

Three Zones in Coal-Mine Gob

Air leakage occurs in the interior of a gob pile from the working face of both pillar and nonpillar mining. The distribution and rate of air leakage is influenced by roof lithology, roof falling, the rate of mining the working face, and the rate of rock compaction in the gob (Hartman et al., 1997). The Air leakage flowrate within the range of ~25–35 m behind the longwall working face is relatively large but it is small within the range of ~30–60 m. Once reaching to ~50–70 m behind the face, the air leakage flowrate is very small and it can not be detected far more than 100 m. Polish experts (Brann, 1994) divided the gob area behind the working face into three zones from the perspective of coal spontaneous combustion: nonspontaneous combustion, spontaneous combustion, and suffocation zones.

Nonspontaneous Combustion Zone

This refers to Zone I, shown in Figure 8.2.1. This zone is closely adjacent to the longwall working face. Although the residual coal accumulates in this zone, no heat storage condition is met due to the strong air leakage and large space among the roof falling rocks. Besides, the duration time of exposure between the superficial coal layer and air is still quite short. All these make coal spontaneous combustion unlikely to occur. The width of this zone is about 5 m calculated from the center of the longwall working face.

Spontaneous Combustion Zone

This refers to Zone II, shown in Figure 8.2.1. As shown, it covers the range of 25–60 m extending from nonspontaneous combustion zone to the inside gob. Since the roof falling rocks are gradually compacted firmly, the space among the rocks becomes smaller, which in turn increases the airflow resistance. Consequently, the air leakage

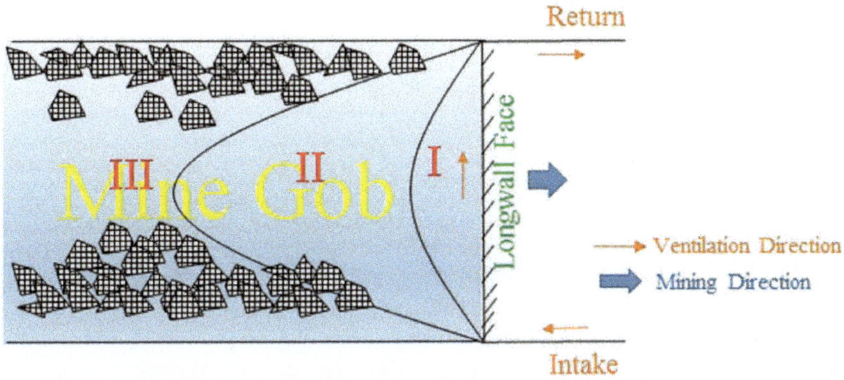

Figure 8.2.1. Distribution of spontaneous combustion zone in coal-mine gob. Figure by Jianwei Cheng, 2017.

becomes weak and falls into a transition flow regime. The remaining coal in the gob is oxidized gradually as time goes by. The heat generated in the process of oxidization is accumulated, leading to spontaneous combustion. That is why this zone is named the spontaneous combustion zone. The width of this zone is influenced by roof lithology, the size and compaction degree of roof falling rock, and the pressure difference between the two endpoints of the working face.

Suffocation Zone

This refers to Zone III, shown in Figure 8.2.1. This zone is located behind the spontaneous combustion zone. In this zone, the roof falling rocks are compacted firmly and there is virtually no air leakage, which leads to the decline of oxygen concentration. The spontaneous combustion of the residual coal developed in Zone II will suffocate due to lack of oxygen if falling into Zone III. This explains why this zone is named the suffocation zone. In addition, the heat generated and accumulated in Zone II will be gradually conducted away by the rock, thus decreasing the coal temperature to the normal state.

The location of the three zones moves as the working face advances. The more slowly the spontaneous combustion zone moves, the wider this zone is; the longer the residual coal stays in this zone, the more likely it is for spontaneous combustion to occur. Therefore, to prevent spontaneous combustion, it is important to speed up the advance of spontaneous combustion zone and to reduce its width. The only way to speed up the advance of the spontaneous combustion zone is to increase the longwall face mining production speed. One method to control its width is to reduce the air resistance in the working face or pressure difference between the two endpoints; the other methods are to fill the pores and cavities with slurry in mine gob, to inject water to promote the formation of regenerated roof, and to improve the air leakage resistance in the gob.

In these three zones, the characteristic parameters of the spontaneous combustion zone (Zone II) are most noteworthy in terms of spontaneous combustion prevention, which are advance rate, width, and porosity. Although there is no universally acknowledged division of the zones in a gob pile, two divisions are commonly used (Wang, 2007).

Gob Zones and Air-Flow Velocity

This method is conducted by simulating mining conditions in the laboratory. On-site measurements still have a long way to go because of the difficulties in establishing measurement points in the gob, the precision of measurement instruments, and the unpredictability of air leakage directions through the gob.

Air-flow velocity (gob air leakage) and the three zones mentioned previously are related as follows (Yang et al., 2000).

1. The flow velocity of the nonspontaneous combustion zone: over 0.24 m/min.
2. The flow velocity of the spontaneous combustion zone: 0.24 ~ 0.1 m/min.
3. The flow velocity of the suffocation zone: below 0.1 m/min.

The reason for this division lies in critical velocity, which is the maximum velocity necessary to maintain coal oxidization and the accumulation of heat energy necessary for spontaneous combustion. Research (Wang, 2007) has shown that when the air-flow velocity is less than 0.1 m/min or over 0.24 m/min, oxidization and the accumulation of heat energy in the gob pile decrease, reducing the chance of spontaneous combustion. When the air-flow velocity is less than 0.1 m/min, the lack of oxygen inhibits oxidation, and when the air-flow velocity is over 0.24 m/min, heat energy does not accumulate as it would otherwise in the gob pile.

Gob Zones and Oxygen Concentration

The oxygen concentration in a gob pile, measured in the field, and the three zones mentioned earlier are related as follows (Zhang, 2000).

1. The oxygen concentration of nonspontaneous combustion zone: over 18% or 19%.
2. The oxygen concentration of spontaneous combustion zone: 8/9% ~ 18/19%.
3. The oxygen concentration of suffocation zone: below 8% or 10%.

The oxygen concentration is taken as the standard to distinguish between nonspontaneous combustion zone and spontaneous combustion zone. Nonspontaneous combustion zone has sufficient oxygen supply conditions, but due to the air leakage in this zone, the heat generated in the oxidization process is dispersed by the air flow. Consequently, the oxidization process proceeds very slowly. It should be noted that oxygen concentration cannot be taken as the exclusive index to divide the nonspontaneous combustion zone and spontaneous combustion zone. Usually, other on-site factors also need to be considered. The fundamental reason for nonspontaneous combustion in this zone is not because the oxygen concentration goes beyond a special value, but because the heat generated in the oxidization process is dispersed by the too-strong air leakage. That is also why this zone is also called the cooling zone or heat dissipation zone. Therefore, it is better to take critical velocity as the primary index to define the transition from heat dissipation to heat accumulation in the oxidization process. Both the nonspontaneous combustion zone and spontaneous combustion zone can be classified as oxidization zone. In contrast, critical oxygen concentration is used to divide the oxidization zone and suffocation zone, because suffocation, in nature, is caused by a subcritical oxygen concentration. The critical oxygen concentration is generally known as 8%.

The decisive factor of coal spontaneous combustion is the supply of oxygen. The measurement results of oxygen concentration in the gob can be taken as the standard to divide the three zones of oxidization and spontaneous combustion. The application of a **tube bundle monitoring system** (Zipf et al., 2013) makes it feasible. This method is the most widely used and at the same time is the most effective at present. By means of sampling tubes preinstalled in the gob and gas chromatograph, the gas component in the gob is monitored and analyzed. The key to this method lies in the setup of measuring points in the gob. There are two ways to set up the measuring points: one is to set up the measuring points along the air intake roadway and air return roadway; the other is to set up the measuring points along the gob dip. The first way is easy to operate, but has great limitations in monitoring parameters, thus failing to represent the overall distribution of the gas component in the gob. The second way, in contrast, implies a huge amount of work in the setup of measuring points, but the parameters monitored can present a whole picture of the actual gas distribution in the gob.

In addition, it is not appropriate to take temperature as the main index for the division of the three zones, because the temperature in the whole gob does not necessarily rise to a certain value. The residual coal in the oxidization zone has the conditions and possibility of spontaneous combustion, but does not necessarily self-ignite and rise in temperature. Even if the temperature in the gob does not rise, it does not mean that the three zones do not exist. What's more, the suffocation zone can be formed by oxygen consumption in the oxidization process, where the temperature does not rise noticeably. Therefore, temperature can only be taken as a complementary standard for the division of the three zones under proper circumstances. If oxidization has occurred and heat has been generated in the gob, and the temperature monitoring can measure out the changing of gob temperature, the temperature change can be taken as the important evidence for the division of the three zones; if there is no oxidization and no heat generated, the temperature monitoring results cannot be taken as the major factor.

8.3 Fire Ratios and Indicators

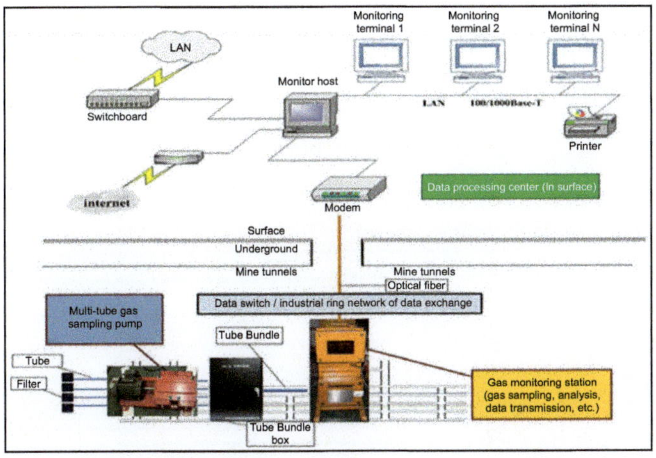

From Guan (2016) with modifications, reprinted with permission from Wangli Guan.

An example of a Chinese tube bundle system for monitoring coal-fire gases underground.

Introduction

Concealed thermal events in longwall gob present safety hazards for underground coal miners and could greatly interrupt the mining production. The causes for such thermal events could range from coal oxidation at slightly above ambient temperature to smoldering fires due to spontaneous combustion of the broken coal left in the gobs. Locating the hot spots is the key step for planning and implementing mitigation measures to bring such thermal events under control. If part of a mine or even the entire mine has to be sealed, understanding the status of the atmosphere in the sealed mine is important for the decision of subsequent actions.

The gas composition obtained through the tube bundle atmospheric monitoring system is information of great value. Fire ratios proposed by various researchers and derived from the gas composition data are useful in determining the fire status and even approximating the source locations. However, the applicability of each of them should be carefully evaluated. Properly combined use of different fire ratios could enhance the certainty of analysis.

Graham's Ratio

Graham's Ratio (ICO) is used to evaluate the status of a mine fire event and is defined by Eq. (8.3.1) (Vutukuri and Lama, 1986), in which CO, N_2, Ar, and O_2 are the volume percents of these gases. The term $0.265(N_2+Ar)-O_2$ is an indicator of oxygen deficiency (Graham, 1914).

$$\mathrm{ICO} = \frac{100 \times \mathrm{CO}}{0.265\,(\mathrm{N_2} + \mathrm{Ar})\ - \mathrm{O_2}} \tag{8.3.1}$$

The resulting ICO values can be used in the following two ways. If an individual ICO value is obtained, a value larger than 0.5 indicates an active fire while a value smaller than 0.2 shows that the fire is under control. When multiple ICO values have been determined, an ICO rise over time indicates the existence of fire and temperature increases, whereas a decreasing trend shows the oxygen has been consumed and the temperatures fall to near ambient levels. A level and stable ICO trend is indicative that fire is under control.

Figure 8.3.1 shows the derived ICO ratio from the gas monitoring data in the longwall panel. Three abnormal data points have been deleted from the figure due to the irrational sudden increase in O_2 level after the panel has been sealed. Using the derived ICO ratio, the earliest indication of a fire was observed at the monitoring point 4 in the **tail entry** (also called a **tailgate entry**), used as a ventilation airway. The other points showed ICO > 0.5 nearly 20 h after the fire event but with much short durations (<3 h). About 45 h after the fire event and 31 h after sealing the panel, the ICO ratio at point 4 was reduced below 0.5. It took varying lengths of time for the ICO at the monitoring points to reach value below 0.2. Point 3, located far back in the gob was the last point to get down to ICO < 0.2. Since the determined ICO ratios at all monitoring points prior to the fire event were essentially zero, the sole source of CO gas at the monitoring points was the flame fire observed at the longwall face.

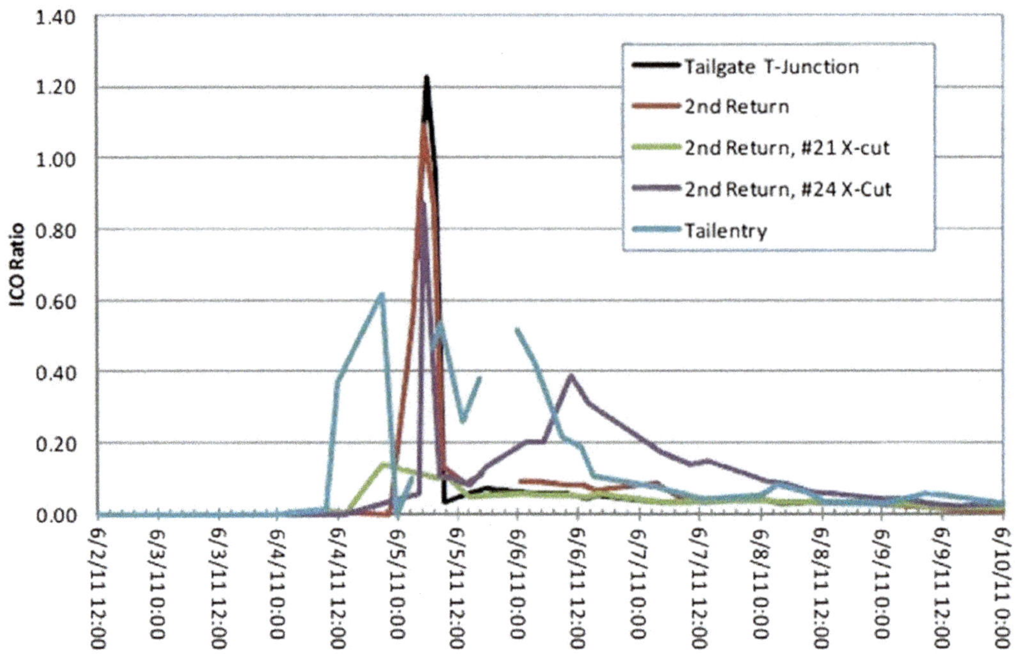

Figure 8.3.1. Graham's Ratio versus date and time. A T-Junction is where one mining tunnel crosses the other, forming the shape of the letter T. Figure by Jianwei Cheng (2012).

Trickett's Ratio

Trickett's Ratio (TR), also called the **Jones-Trickett Ratio** (Jones and Trickett, 1954), indicates the status of a fire and what is burning. It is based on the principle that the number of molecules of matter consumed in a fire is proportional to the number of molecules or the volume of gas produced at constant temperature. The TR is defined by Eq. (8.3.2) in which CO_2, CO, N_2, O_2, Ar, and H_2 are in volume percents.

$$TR = \frac{CO_2 + 0.75CO - 0.25H_2}{0.265\,(N_2 + Ar) - O_2} \tag{8.3.2}$$

The implications of the TR are shown in Table 8.3.1.

The resulting TRs are shown in Figure 8.3.2. Irregular development patterns have been observed at the monitoring point in the second return, especially before the fire event. The irregular spikes are normally associated with the sudden increase of O_2 level at the monitoring point. Similar problems occurred at point 4 in the tail entry but with lower spike values. If these two sets of TR data are ignored, the peak TR value at points 1 and 4, about 0.3, occurring at the time when the panel was sealed, show that coal burning was not involved in the thermal event.

Relative Intensity

Relative Intensity (RI) is used by the power industry to determine fuel efficiency and the access or deficiency of oxygen necessary for combustion. RI measures the relationship between the air required to burn a unit mass of fuel like coal, the oxygen consumed, and the temperature of combustion. RI is determined by the following equation (Mitchell, 1990).

$$RI = \left(\frac{1 - 0.0383 \times O_2}{N_2} \times ICO \times 10^4\right) \tag{8.3.3}$$

where N_2 is in volume percent. Adapting the RI to mine fire analysis, RI is used to determine the proximity of the sampling location to the fire or the intensity of the fire. If the RI increases, the fire is either growing or moving

Table 8.3.1
The implications of Trickett's Ratio (TR).

TR Value	Implication
<0.4	Fire probably out
0.4–0.5	Methane probably burning
0.5–1.0	Coal, oil, or conveyor belting probably burning
0.8–1.6	Wood probably burning
>1.6	Not normally possible in a coal mine fire; sample is suspicious

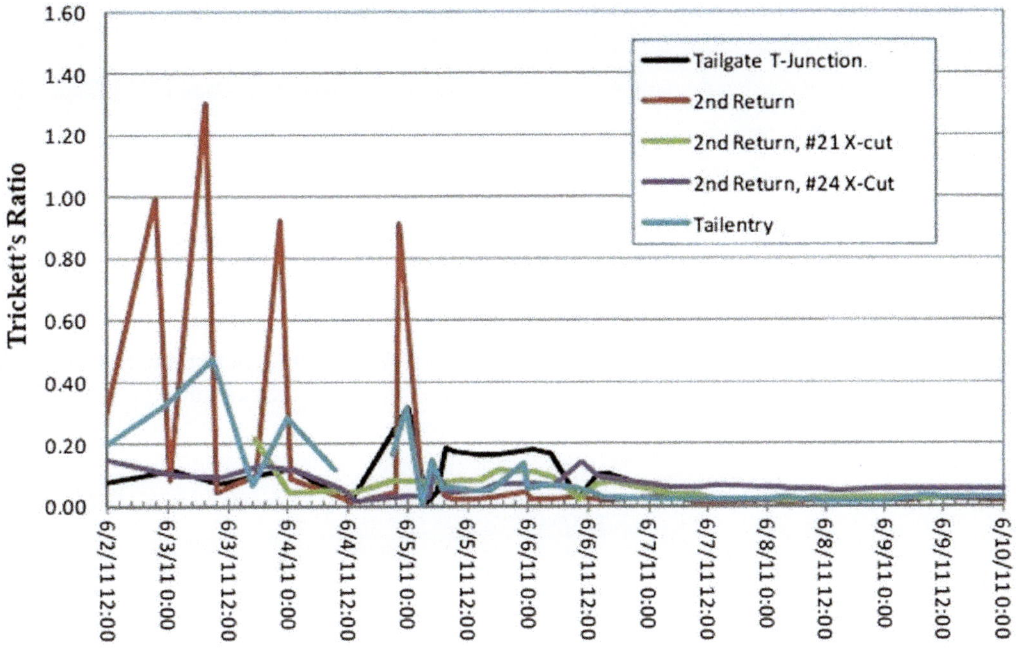

Figure 8.3.2. Trickett's Ratio versus date and time. Figure by Jianwei Cheng, 2011.

closer to the sampling location. Conversely, the decreasing RI indicates a fire is being throttled or moving away from the sampling location.

Figure 8.3.3 shows the resulting RI at the five monitoring points. Since RI is a function of ICO, their distributions are like those of ICO shown in Figure 8.3.1. The peak RI values at the points decrease as the distances between the fire location and the monitoring points increase. All points show decreasing RI after June 6, indicating that the thermal effect caused by the fire event disappeared.

Litton Ratio

Litton (1986) develop a method to monitor the sealed-mine atmosphere. It was designed to reduce the potential for re-ignition when a mine or a section is reopened and oxygen is reintroduced. In this method, only four gases, O_2, CO, CH_4, and ethane (C_2H_6), which are independent of the injected N_2 and CO_2, are required. It requires a residual gas (Rg) in volume percent be determined first using the following equation:

$$R_g = 100.0 - 4.774 \times O_2 - CH_4 - C_2H_6 \tag{8.3.4}$$

where all gases are in volume percent.

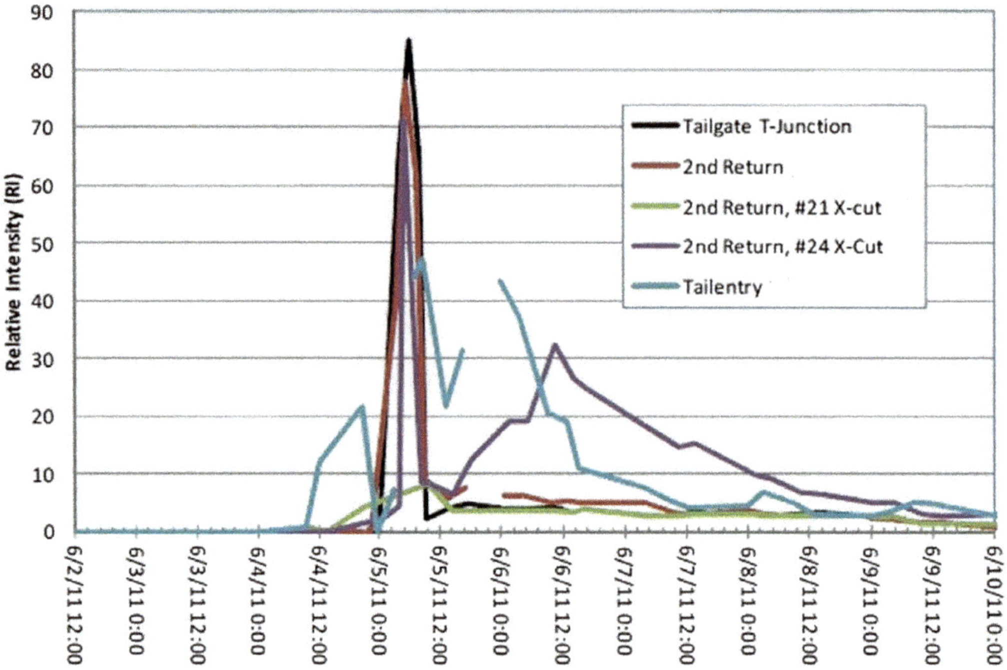

Figure 8.3.3. Relative intensity versus date and time. Figure by Jianwei Cheng, 2011.

Based on the concentration of oxygen in the atmosphere, the **Litton Ratio** (LR) is determined by Eq. (8.3.5). In this equation, the CO concentration is in ppm and the O_2 concentration is in volume percent.

$$LR = \begin{cases} \frac{1}{3}\frac{CO}{(R_g)^{3/2}\sqrt{O_2}} & \text{if } O_2>1\% \\ \frac{1}{3}\frac{CO\times O_2}{(R_g)^{3/2}} & \text{if } O_2<1\% \end{cases} \quad (8.3.5)$$

If the ratio is greater than 1, smoldering combustion or above-ambient temperature oxidation is occurring. If it is equal to or less than 1, it is possible that ambient temperature has been reached. Equilibrium exists only if LR stabilizes at a value less than one. Litton also recommends that the minimum time for the LR to remain stabilized at less than 1 is 30 days.

The derived LRs for the monitoring points are plotted in Figure 8.3.4. Since even the peak LRs are less than 1, it indicates that no smoldering fire or above temperature oxidation of broken coal in the gob as the suspicious cause for the previous abnormal CO event was involved in the fire event.

Hydrocarbon Ratio

The **Hydrocarbon (HC) Ratio** method (Justin and Kim, 1988) was developed since low molecular weight hydrocarbon gases desorbed from coal in direct proportion to increasing temperature. Among them, methane is readily liberated at ambient temperatures; the progression of gas desorption due to heating is carbon dioxide, carbon monoxide, hydrogen, ethylene (C_2H_4), propylene (C_3H_6), and acetylene (C_2H_2). Hydrocarbon concentrations increase with rising temperatures (Justin and Kim, 1988). The HC (R_1) is defined by Eq. (8.3.6). In this equation, THC is the total hydrocarbon concentration, including hydrogen, in ppm, CH_4 is the methane concentration in ppm, and c is the constant 0.01 ppm.

$$R_1 = \frac{1.01\,(THC) - CH_4}{THC + c} \times 1000 \quad (8.3.6)$$

The HC is zero when no hydrocarbons are detected, 10 when methane is the only measured hydrocarbon, and about 1010 at the upper limit. For bituminous coal, R_1 values are closely dependent on temperature. Because of this,

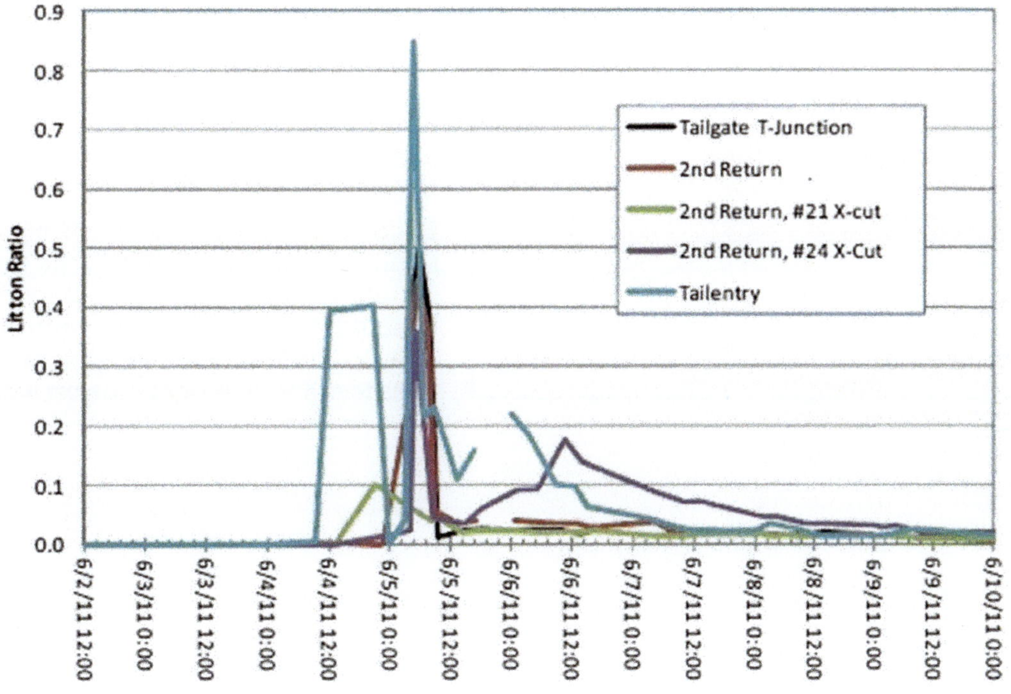

Figure 8.3.4. Litton Ratio versus date and time. Figure by Jianwei Cheng, 2011.

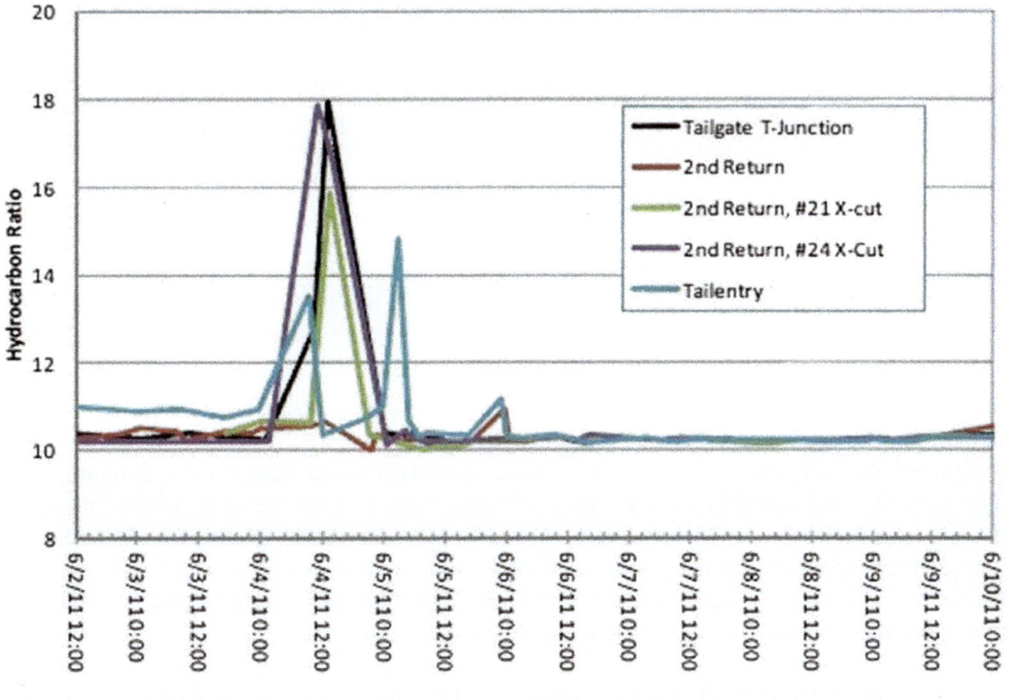

Figure 8.3.5. Hydrocarbon ratio. Figure by Jianwei Cheng, 2011.

limits have been derived to aid in determining the temperature of coal being analyzed. R_1 values from 0 to 50 indicate that normal oxidation of coal is occurring. Values from 50 to 100 indicate possible elevated temperatures. Above 100, coal is undergoing elevated temperature oxidation.

Figure 8.3.5 shows the derived HCs at the monitoring points inside the longwall panel. Since the peak values are still considerably less than 50, it shows that the fire event was neither originated from nor caused coal above-normal temperature oxidation.

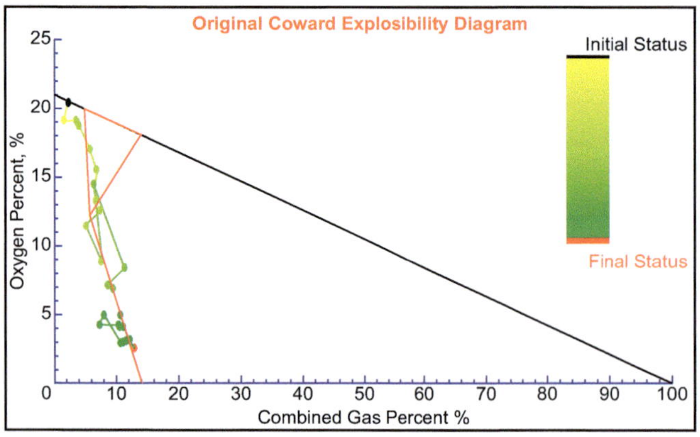

Figure by Jianwei Cheng (2012).

8.4 Explosibility Analysis of Mining Atmospheres

Because coal-mine gob in a Chinese mine spontaneously combusted, the longwall panel was sealed and mine operators injected N_2 into the gob to control the fire. The Coward Explosibility Diagram is used to record the risk of explosion with time.

Introduction

In a sealed coal mine, the atmosphere includes (1) gases emitted from coal seams including CH_4, CO_2, and H_2. Methane is the No. 1 combustible gas in the underground atmosphere and is also the most dangerous in the underground coal mining. Methane has an explosive range between 5% and 15% and the concentration of 9.5% is the most dangerous scenario due to complete combustion of the air-methane mixture. Other underground gases are mainly by-products of the coal formation process. (2) Products of chemical reactions in underground mines including those formed during the low temperature oxidation of coal, coal combustion, and gas explosions. Most of them are the alkane (CnH_{2n+2}), alkene (CnH_{2n}), or alkyne (CnH_{2n-2}) series of hydrocarbon gases. (3) Atmospheric gases, which refer to the gases that can be found in an ambient air sample. Almost 99% of them consist of nitrogen and oxygen; the remains are a very small quantity of argon and carbon dioxide. (4) Other inert gas, N_2 or CO_2, injected into the sealed volume to extinguish the mine fire events or minimize the risk of potential explosions. Generally, the so-called mine gas is a mixture of these gases. The following sections discuss the methods to determine mine-gas explosibility.

Tertiary Diagrams

Tertiary diagrams are used to represent different combinations of three gases. Dwyer et al. (2003) introduced the diagram to determine the explosibility of the fuel-air-inert mixtures. Figure 8.4.1A shows a tertiary diagram with a mixture of hydrogen, oxygen, and nitrogen (all in volume percent). The top vertex of the diagram represents 100% hydrogen and the base of the triangle represents 0% hydrogen, The same thing for oxygen and nitrogen. Therefore, the diagram can be used to represent any ratio of hydrogen, oxygen, and nitrogen at ambient conditions. Using line A as an example, it represents 30% hydrogen. It is the same thing for line B (30% oxygen) or line C (30% nitrogen). Accordingly, the example point D on line A consists of 30% hydrogen, 27% oxygen, and 43% nitrogen; and point E can be considered pure air (0% hydrogen, 21% oxygen, and 79% nitrogen).

Figure 8.4.1B illustrates the tertiary diagram for a mixture of hydrogen, oxygen, and nitrogen including the flammability envelope for ambient conditions. If the gas sample is inside the envelope, it indicates explosibility. The explosions may take place once a flame is provided. Line F is drawn from air (point E in Figure 8.4.1A) to 100% hydrogen and thus any mixture of hydrogen and air alone must lie on line F. Two points of intersections that are formed by line F and the flammability envelope is the lower flammability limit (LFL) and upper flammability limit for hydrogen in air, respectively (Dwyer et al., 2003).

It can be apparently seen that the biggest drawback of the tertiary diagram is that it can only deal with the gas mixture of three gases at most. However, like the previous statement, there are many combustible gases existing in the real mine atmosphere. Hence, the tertiary diagram is not a very perfect or accurate method to determine the mine-gas explosibility due to the limitation of gas species. In addition, generating the flammability envelope in the

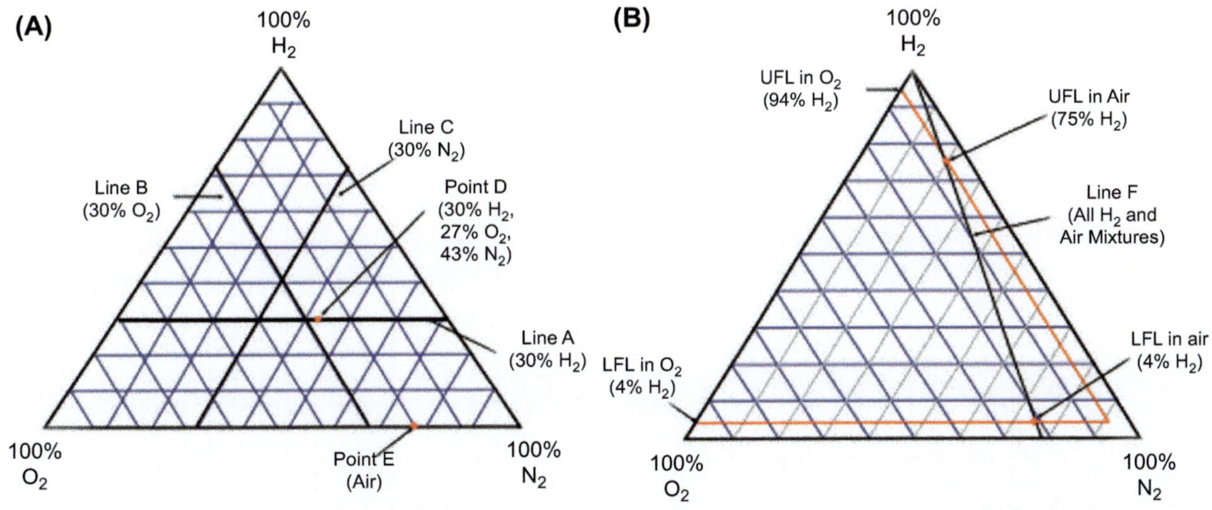

Figure 8.4.1. Tertiary diagram for hydrogen, oxygen, and nitrogen. (A) Without flammability envelope, (B) with flammability envelope. *LFL*, lower flammability limit; *UFL*, upper flammability limit. Reprinted, with permission, from Dwyer, J., Hansel, J.G., Pilips, T., 2003. Temperature influence on the flammability limits of heat treating atmospheres. In: Dahotre, N., Gaster, R., Hill, R., Popoola, O. (Eds.), Proceedings of the 22nd Heat Treating Society Conference and the 2nd International Surface Engineering Congress, Indianapolis, Indiana, pp. 24–28. Copyright 2003 by Air Products & Chemicals, Inc., all rights reserved.

diagram is also time-consuming and complicated work, and many experiments or theoretical calculations should be done to identify the envelope prior to use.

Revised Le Chatelier's Rule

The flammability limits of gases that are a mixture of several combustible gases can be derived with the help of Le Chatelier's rule (Mashuga and Crowl, 2000). This rule assumes that the individual flammable compositions neither react with others nor have any catalytic influence upon each other.

For mixtures of gases with air and inert diluents, the influence of the ratio of inert gas to the flammable gas on the flammability limit must be considered (Greuer, 1974). Figure 8.4.2 shows these limits for hydrogen, carbon monoxide, and methane diluted with nitrogen or carbon dioxide when mixed with air. The upper and lower limits of mixtures of complex gases with inert gases are dissected into simpler mixtures, each of which contains only one flammable gas and mixtures are determined from graphs like Figure 8.4.2. If $V_1, V_2, V_3, \ldots V_n$ are the volume percent of the mixtures and $N_1, N_2, N_3, \ldots N_n$ are the flammability limits (either upper or lower limit) for each mixture, the flammability limit, L (either upper or lower limit), of the total mixture of gases (when mixed with air) can be calculated based on Le Chatelier's rule:

$$L = \frac{100}{\frac{V_1}{N_1} + \frac{V_2}{N_2} + \cdots + \frac{V_n}{N_n}} \tag{8.4.1}$$

This method was used with sufficient accuracy for mixtures of hydrogen, carbon monoxide and methane with nitrogen, and carbon dioxide, and can therefore be used for mine fire gases (Greuer, 1974).

Maximum Allowable Oxygen Analysis

The quantity of oxygen required for flame propagation; therefore, the maximum allowable oxygen (MAO) depends on the relative quantities of methane, carbon monoxide, and hydrogen in a mine-fire area. When oxygen concentration in a mixture is below the specified maximum oxygen value, the mixture is either nonexplosive or explosive when mixed with air (Zabetaksi et al., 1959). Hence, the **MAO analysis** could be used to analyze the mine atmosphere status.

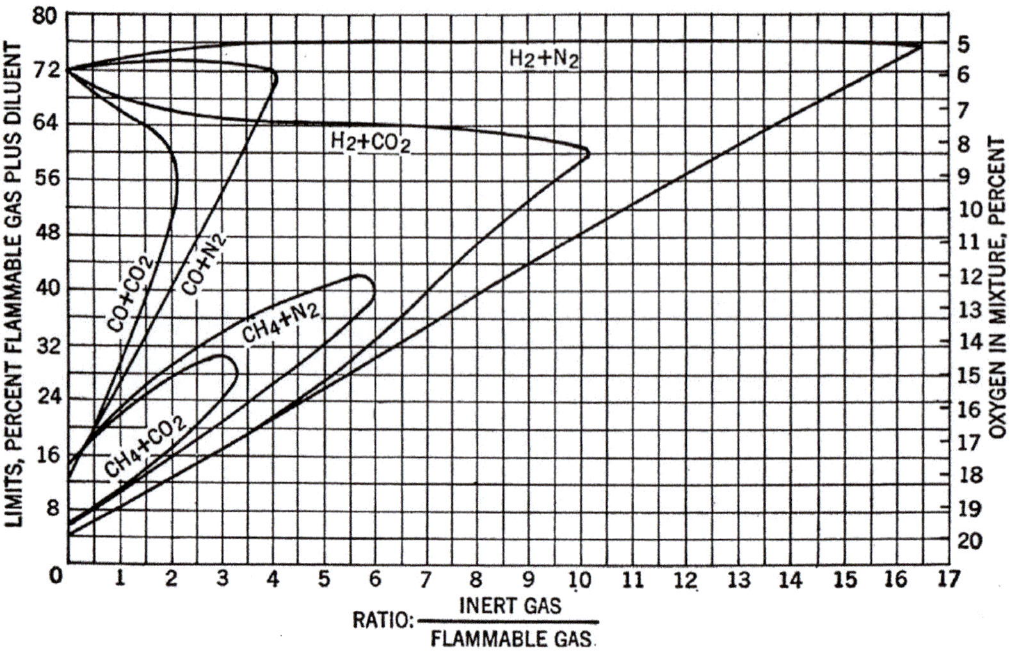

Figure 8.4.2. Limits of flammability of hydrogen, carbon monoxide, and methane containing variable amounts of carbon dioxide and nitrogen versus the ratio of inert gas to flammable gas. From Greuer (1974).

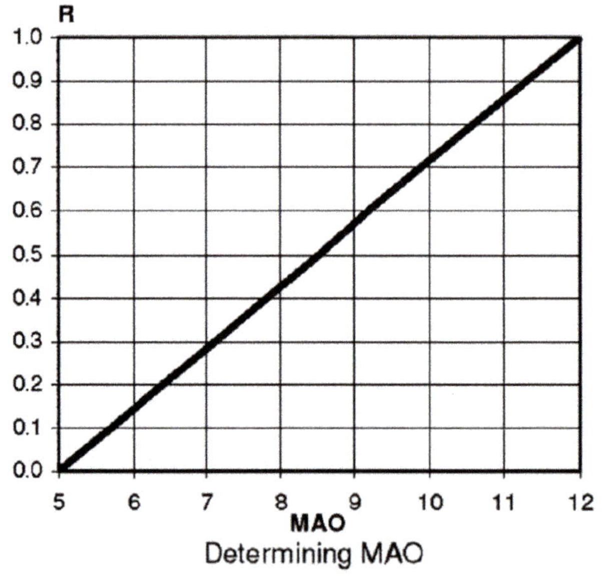

Figure 8.4.3. Graphically determine MAO using R. From Timko and Derick (2006).

To use this method, the ration R should be calculated using the following equation:

$$R = \frac{V_{CH_4}}{V_{CH_4} + V_{H_2} + V_{CO}} \quad (8.4.2)$$

where V_{CH_4}, V_{H_2}, and V_{CO} are the volume percents for CH_4, H_2, and CO, respectively. The MAO-value can then be determined based on the corresponding R-value, referring to Figure 8.4.3 or with the following equation:

$$MAO = 5.0 + 7.0R \quad (8.4.3)$$

If the actual measured oxygen concentration is less than the MAO value, the atmosphere cannot be ignited in its present state. Conversely, it may be explosive. The MAO analysis provides an easy and simple way to judge the status of the sealed atmosphere. But like previously discussed, the drawback of this method is that it is hard to tell if the atmosphere is explosive if the measured oxygen value is greater than the MAO value. Therefore, when only using the MAO value, mine operators should pay more careful attention on the results because the nonexplosive atmosphere may become explosive when mixed with air. Under such conditions, it is highly suggested that other explosibility methods should be used to conduct a comprehensive analysis.

Another alternative method to calculate the MAO value was calibrated by two Chinese researchers. Their equation considers more combustible gases. Accordingly, the MAO result is more accurate (Lu and Han, 1991) and is calculated as follows using the volume percent of gases:

$$MAO = \frac{12 \times V_{CH_4} + 6 \times V_{CO} + 5 \times (V_{H_2} + V_{C_mH_n})}{V_T} \tag{8.4.4}$$

where V_T is the total percentage of combustible gas and C_mH_n stands for various hydrocarbon gases.

U.S. Bureau of Mines Explosibility Diagram

A simplified graphical method for determining the explosibility of mining atmospheres from atmospheric composition data was presented by Zabetaksi et al. (1959). This method uses a U.S. Bureau of Mines (USBM) diagram referred to herein as a **USBM Explosibility Diagram**. It is a popular method widely used in the U.S. mining industry. Figure 8.4.4 shows the diagram developed by Zabetaksi. This diagram shows that a methane-air-inert gas mixture falls into one of three categories: (1) explosive, (2) explosive when mixed with air, or (3) nonexplosive, depending on the percentage of methane and the percentage of effective inert. (Kissell, 2006). The effective inert can be obtained from excess nitrogen and carbon dioxide volume percentages. The arrows in the diagrams represent how the composition point moves if more methane, air, or inert gas is added. It is a simple method to determine the explosibility of the mixture gas. The USBM method to determine mine-gas explosibility follows these steps (Zabetaksi et al., 1959):

1. Analyze the mine atmosphere; determine volume percents of carbon dioxide (CO_2), oxygen (O_2), carbon monoxide (CO), methane (CH_4), hydrogen (H_2), and nitrogen (N_2).

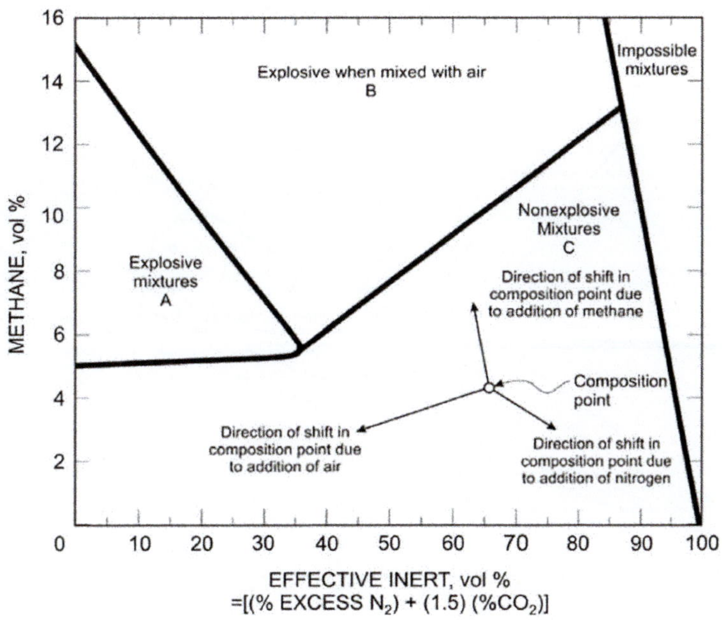

Figure 8.4.4. Explosibility diagram for methane-nitrogen-air mixtures at 800°F and 0.1 MPa (1 bar) atmospheric pressure. From Zabetaksi et al. (1959).

2. Calculate the volume percents of excess nitrogen (E_{N_2}) and effective inert (E_{Inert}) gases as shown in Eqs. (8.4.5) and (8.4.6):

$$E_{N_2} = N_2 \text{ in sample} - N_2 \text{ in normal air} \tag{8.4.5}$$

where N_2 in sample = 100%–($CO_2 + O_2 + CO + CH_4 + H_2$) and N_2 in atmosphere air = $3.8 \times O_2$ in the sample. The factor of 3.8 is the volumetric ratio of N_2 to O_2 in air (Encyclopedia Britannica, 2017).

$$E_{Inert} = E_{N_2} + 1.5 \times CO_2 \text{ in sample} \tag{8.4.6}$$

3. Calculate the percent of effective combustible gases ($E_{Combustible}$) as follows:

$$E_{Combustible} = CH_4 + 1.25 \times H_2 + 0.4 \times CO \tag{8.4.7}$$

The factors of 1.25 and 0.4 are, respectively, the ratios of the lower explosive limit (LEL) of CH_4 to that of H_2 and CO.

4. Calculate the R-value to determine the explosibility curve in Figure 8.4.5.

$$R = \frac{CH_4}{CH_4 + H_2 + CO} \tag{8.4.8}$$

Based on the chosen explosibility triangle and the position of the **gas-mixture point** (coordinated by the percent effective combustible and the effective inert), the explosibility status can be identified.

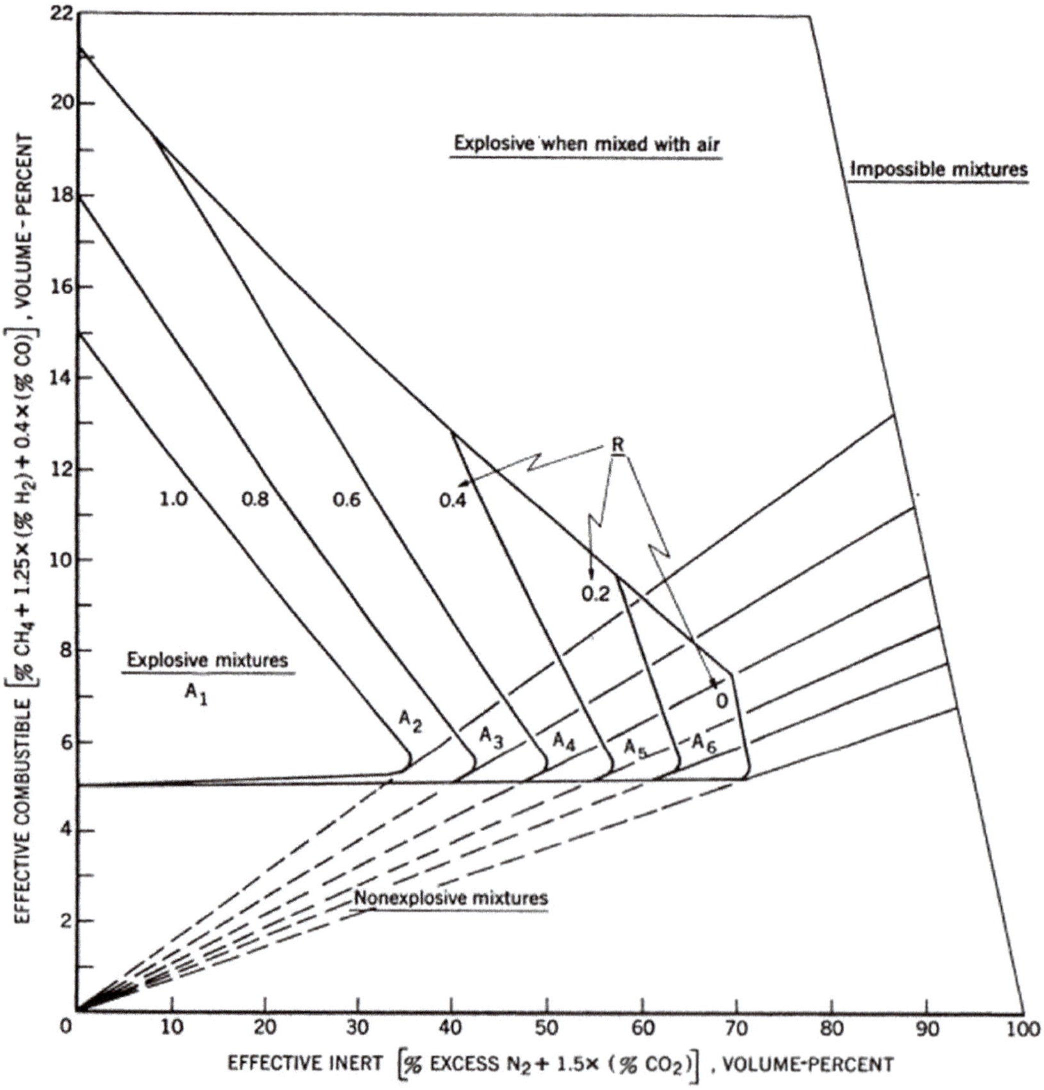

Figure 8.4.5. Explosibility diagram for methane-hydrogen-nitrogen-carbon monoxide-carbon dioxide-air mixtures at 800°F and 0.1 MPa (1 bar) atmospheric pressure. From Zabetaksi et al. (1959).

Coward Explosibility Triangle

The **Coward Explosibility Triangle** (Coward and Jones, 1952) is considered a fast and easy way to determine the explosibility of the mixture of air and combustible gases. It considers only three combustible gases (CH_4, CO, and H_2). When the three combustible gases mix with normal air, the explosibility of the mixture depends on the volume percentages of the combustible gases and oxygen. Figure 8.4.6 shows the individual explosibility triangles for the three combustible gases. Each of the explosibility triangles is defined by three characteristic points, the lower explosive and upper explosive limits (LEL and upper explosive limit [UEL], respectively), and the **nose limit** (**nose**; i.e., the minimum oxygen concentration to support an explosion) (Coward and Jones, 1952). The characteristic points for the three combustible gases are listed in Table 8.4.1. The explosibility triangle for a combustible gas divides the plot area into five distinctive zones. Taking CO as an example in Figure 8.4.6, the zone located above line AB is a zone in which it is impossible to form an air–gas mixture. The area under line AE is the nonexplosive zone, also known as the absolute safety zone. Triangle COD is the zone with the potential for explosion. The air–gas mixture in triangle AOC is nonexplosive. However, it could become explosive if more CO is infused. Zone ODBE is a nonexplosive zone but could become explosive as more fresh air is introduced.

The procedure to generate the Coward Explosibility Triangle for the combustible gases CH_4, CO, and H_2 is described as follows (McPherson, 1993):

1. Determine the total combustible volume percent, V_T, using the volume percentages of the three combustible gases:

$$V_T = V_{CH_4} + V_{CO} + V_{H_2} \tag{8.4.9}$$

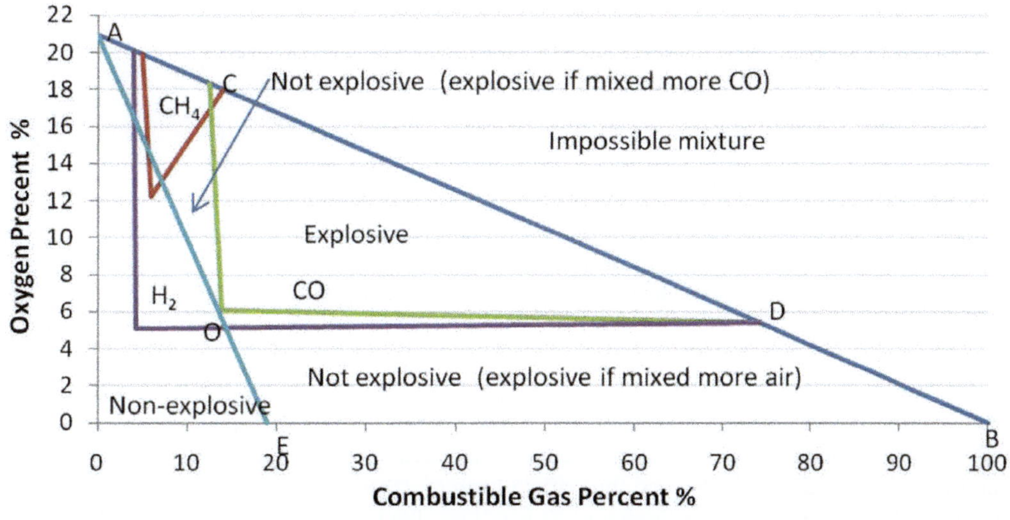

Figure 8.4.6. Coward explosibility triangles for methane, carbon monoxide, and hydrogen. Figure by Jianwei Cheng (2012).

Table 8.4.1
Gas vertices of explosion triangles.

	Flammable Limits		Nose Limits†		Gaseous Nitrogen Needed to Make an Extinctive Mixture: (m^3 of Nitrogen per m^3 of Combustible Gas)
Gas*	Lower	Upper	Gas	Oxygen	
Methane (CH_4)	5.00	14.00	5.90	12.20	6.07
Hydrogen (H_2)	4.00	74.20	4.30	5.10	4.13
Carbon monoxide (CO)	12.50	74.20	13.8	6.10	16.59

*Gas percent by volume.
†Nose limit: The minimum concentration to support an explosion.

2. Determine the gas flammability. Le Chatelier's rule in the form of Eq. (8.4.10) is used to determine the upper, lower, and nose flammability of mixed gases. To apply this equation for the lower explosibility limit of the mixture (L_{mix}), the LFLs of the three gases are substituted in the places of L_1, L_2, and L_3. The same can be done for the upper nose limits.

$$\frac{V_T}{L_{mix}} = \frac{V_{CH_4}}{L_{CH_4}} + \frac{V_{CO}}{L_{CO}} + \frac{V_{H_2}}{L_{H_2}} \quad (8.4.10)$$

3. Determine the required excess nitrogen.
An effective way to render an air–gas mixture as nonexplosive is to inject an excessive volume of pure nitrogen (N_{ex}) into the mixture as shown in Eq. (8.4.11). In this equation, L_n is the nose flammability of the mixed gases, and N^+ is the volume percent of excess pure nitrogen gas needed to CH_4, CO, or H_2 gas extinctive by themselves in air. It is obtained by experimentally adding it to flammable gases. Table 8.4.1 also presents the excess nitrogen if the combustible content consisted of one gas only.

$$N_{ex} = \frac{L_n}{V_T}\{N^+_{CH_4}V_{CH_4} + N^+_{CO}V_{CO} + N^+_{H_2}V_{H_2}\} \quad (8.4.11)$$

4. Determine the volume percent of oxygen, O_n, at the nose limit:

$$O_n = 0.2093\,(100 - N_{ex} - L_n) \quad (8.4.12)$$

Using the data obtained from these equations, the explosibility triangle for the air–gas mixture can be constructed, and the actual mixture point (defined by the volume percent of oxygen and total combustible gases) can be plotted on the same diagram. The relative position between the explosibility triangle and the state point shows the explosibility status of the air–gas mixture at the current state and the potential when condition changes.

Kukuczka Method

The Polish scientist Kukuczka (Kukuczka, 1982; Yu, 1992) analyzed the composition of coal-mine gas and developed the **Kukuczka Method**, a unique method to determine the explosibility of gas. This method converts combustible gases by volume percent into a single gas (CH_4) via a mathematical transformation to calibrate the position of a gas-mixture point. By plotting a CH_4 explosibility triangle (described in the following) and the gas-mixture point on the same diagram, the relative position between the triangle and the point shows the explosibility of the gas mixture at its current state. The procedure used by Kukuczka is as follows:

1. Determine the CH_4 explosibility triangle.
In the coordinate system of the Kukuczka explosibility diagram, the volume percent concentration of oxygen and methane are plotted on the y and x axes, respectively. Different from the Coward Explosibility Triangle, the explosibility triangle used by Kukuczka is fixed, and defined by three characteristic points, the LEL (5.00, 19.88), and UEL limit (15.00, 17.79), and the critical limit (5.18, 9.47) for methane.
2. Determine the influence factor, α, for the flammability limit due to the volume percent of the gases:

$$\alpha = \frac{CO_2 - 0.03}{CO_2 + N_2 - 3.778 \times O_2} \quad (8.4.13)$$

3. Determine the total combustible volume percentage:

$$T = CH_4 + CO + H_2 + C_2H_2 + C_2H_4 + C_2H_6 + C_3H_6 + C_3H_8 \quad (8.4.14)$$

In the next three equations, V_i is the volumetric percent of a combustible gas and a_i, b_i, c_i, d_i, e_i, and f_i are coefficients of combustible gases whose values are listed in Table 8.4.2.

4. Determine the influence factor, β_i, for each combustible gas ($\beta_1, \beta_2 \ldots \beta_n$) due to the carbon dioxide and volume percent of oxygen:

$$\beta_i = \frac{20.93 - (O_2 + 0.2093T)}{a_i - b_i\alpha} \quad (8.4.15)$$

If the value of β_i is over 1, it must be reset as 1.

Table 8.4.2
Values of gas coefficients.*

Gas		Coefficients									
Name	Formula	a_i	b_i	c_i	d_i	e_i	f_i	c_i'	d_i'	e_i'	f_i'
Methane	CH_4	10.376	3.016	0	1	0	−0.78	0	0	1	−2.852
Hydrogen	H_2	14.918	3.533	4.643	0.140	0.010	−0.107	5.401	0.116	0.698	−2.435
Carbon monoxide	CO	13.039	3.396	3.117	0.161	0.007	−0.400	3.622	0.133	0.797	−2.619
Ethylene	C_2H_4	14.269	3.526	4.121	0.385	0.009	−0.216	4.849	0.072	0.729	−2.519
Ethane	C_2H_6	11.872	2.909	1.937	1.052	0.005	−0.724	2.233	−0.037	0.875	−2.391
Propene	C_3H_6	12.869	3.383	2.934	1.098	0.006	−0.429	3.442	−0.061	0.808	−2.637
Propane	C_3H_8	12.105	3.294	2.164	1.382	0.005	−0.538	2.537	−0.110	0.858	−2.710
Acetylene	C_2H_2	15.308	3.577	4.901	0.127	0.011	−0.045	5.719	0.115	0.680	−2.415

*From Kukuczka (1982). The coefficients are dimensionless.

5. Determine the position of the actual gas-mixture point as follows:
   a. The vertical Y coordinate calculated with the volume percent of oxygen is:

$$Y=\sum_{i=1}^{n}\left(c_i'+d_i'T+e_i'O_2+\alpha\beta_i f_i'\right)\times\frac{V_i}{T} \tag{8.4.16}$$

   b. The horizontal X coordinate calculated with the volume percent of oxygen is:

$$X=\sum_{i=1}^{n}(c_i+d_iT+e_iO_2+\alpha\beta_i f_i)\times\frac{V_i}{T} \tag{8.4.17}$$

   The relative position between the explosibility triangle and the gas-mixture point shows the explosibility status of the air–gas mixture in its current state and the potential explosibility when conditions change, such as mixing with fresh air.

Mine-Gas Distribution Patterns

Many gas-explosion accidents have occurred at the corner of the underground longwall working face, where methane levels often exceed the LEL despite the use of ventilation, or in mine gobs (Zhang, 2000). To control the accumulation of coal-mine gas such as methane, the migration and distribution of coal gas is of high priority to mining researchers (Hu et al., 2008, 2009). A basic strategy of gas control in underground is to increase the ventilation airflow for sweeping the longwall working face and diluting the gas concentration (Cheng et al., 2016). But this also results in the leakage of an amount of air into the mine gob area, which may change the gas concentration (Mishra et al., 2016; Toraño et al., 2009). It was suggested that the working faces would be safe if 80%–94% of the total mine emissions in gob is removed (Curl, 1978; Schatzel et al., 1992). As overlying strata collapse and underground mining gob is mixed with these fractured rocks and accumulated toxic gases, the gob is extremely dangerous and inaccessible (Stoltz et al., 2006; Prosser and Oswald, 2006). It is impossible for researchers to perform any direct measurements of the gas concentration in an actual gob. To better study the migration and concentration distribution of coal gas, it is necessary to make a scaled physical model of mine gob that catches the main features of the real situation (Lunarzewski, 1998; Karacan et al., 2007). It is expected to help researchers understand the impacts of ventilation strategy on gas distribution in mine gob and in caved zone, which is very meaningful for preventing a gas explosion.

Longwall Mining Gob Experiment

The Datong Mine

The designed physical model is based on a gas-filled coal mine, the Datong #1 mine located in Chongqing, China. The field ventilation records show that the air velocities in intake and return roadways are 1.24–2.61 and

1.82–3.98 m/s, respectively. The exhaust system is applied as the ventilation system. The most recent records show that the **gas release rate (GRR)** fluctuated with the time from 25.41 to 60.86 m^3/min, which could be instructive for setting a reasonable simulated GRR in experiments.

Figure 8.4.7 shows the geological column of the overburden strata. A total of nine measuring levels are set in the overlying strata. In addition, another two that are close to top of the gob and floor are set in the gob separately, which are not shown in Figure 8.4.7 but are partially zoomed in Figure 8.4.8. The layout of 12 measuring points at each measuring level above the gob is the same as that at levels 1 and 2, shown in Figure 8.4.8.

The Experimental Model

A photo of the experimental model and a design blueprint are shown in Figure 8.4.9. Two PVC tubes are placed on both sides to form a U-type ventilation system, and a gas release channel is set on the right side. As shown within the yellow frame, the lower part of the physical model, which looks like a container, is composed of a steel frame surrounded by eight steel plates. In order to show the strata deformation clearly, the two front plates are replaced with glass ones later.

The strata seams are simulated by means of different combinations of limestone, sands, and calcium carbonate mixed with water. Hence the strength of prototype rocks could be well represented. The very beginning step in this experimental process is to use the mechanical hydraulic pressure system to exert a loading pressure (the ground pressure) coming from strata seams over such simulated ones. Thus deformed strata occur and the bottom strata would collapse to a great amount of porous material that would fill the gob area. This process well represents the characteristics of a real gob. Then the whole model is perfectly sealed. A special gas sampling and measuring system is built to take the gas samples at the two levels within the gob and the other nine levels in overlying strata.

Air Velocity and Gas Release Rates

To develop relatively stable gas concentration distribution patterns that are meaningful under preset similarity ratio modeling, it is necessary to note that the ventilation condition or gas release in the gob area is always

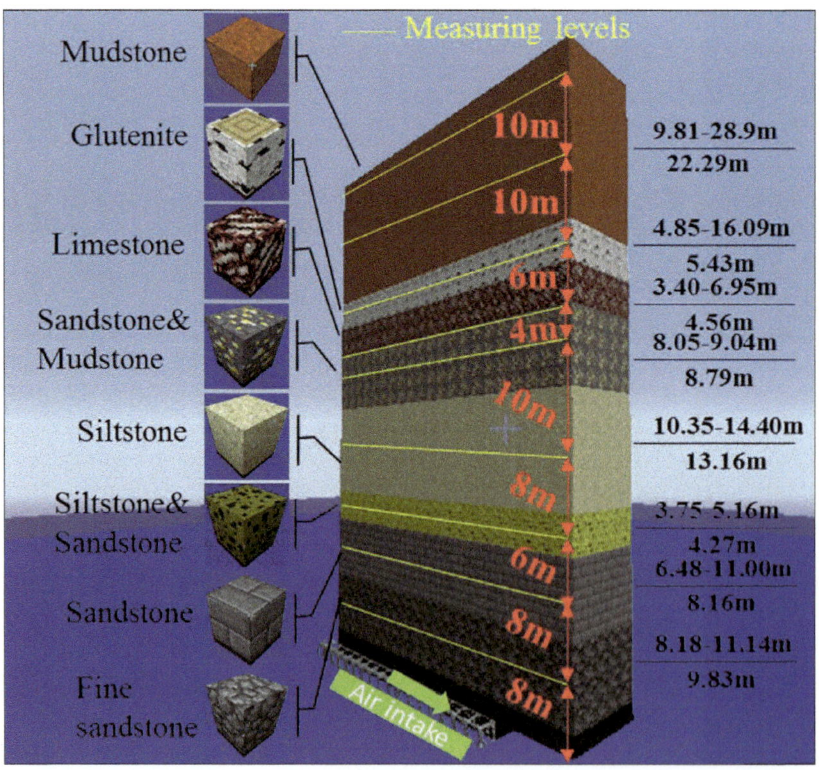

Figure 8.4.7. Coal seam overlying strata and measuring levels. Figure by Jianwei Cheng, 2016.

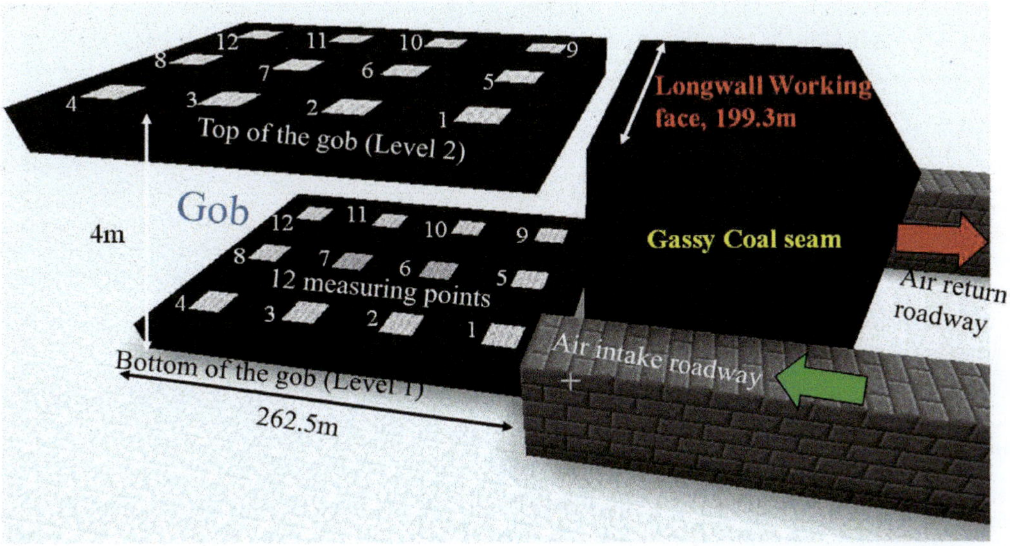

Figure 8.4.8. Model of the locations of gas measuring points in the gob of a coal mine. There are 12 points in the two levels. Gas volume concentrations could be obtained from 24 points, thereby providing a good representation of gas distribution in the gob. Figure by Jianwei Cheng, 2016.

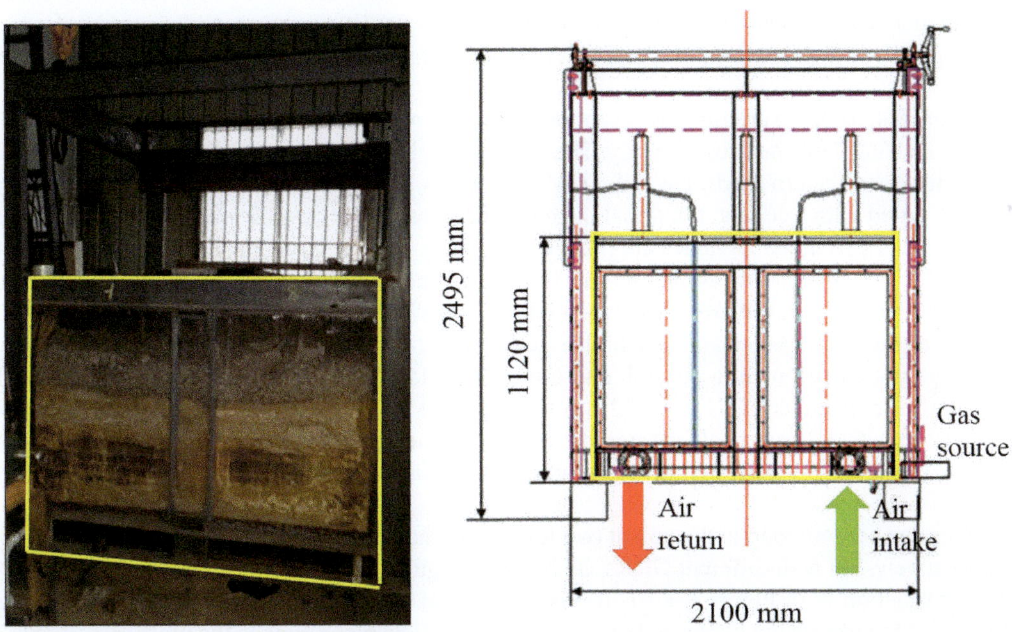

Figure 8.4.9. Physical model and design of the experiment. Photo and Figure by Jianwei Cheng, 2014.

changing with the time, which could be verified by the field records. Meanwhile, the combination of air velocities and GRRs varies in any given time. Hence, considering the representativeness of the combination of two parameters and the similar proportional principles, four different GRRs are set in the experiment: 0.17, 0.33, 0.50, and 0.67 m^3/min in the model; that correspond, respectively, to 16.67, 33.33, 50, and 66.67 m^3/min in an actual mining operation (Table 8.4.3). In addition, three **combination modes (CMs)** of air velocity in the air intake way and return way, in accordance with the field testing of air velocity, are proposed and shown in Table 8.4.3.

In conclusion, a total of 12 groups of data can be taken from 12 × 11 sampling points from the model and overlying strata under the combinations of four different GRRs and three modes of ventilation air quantities.

Table 8.4.3
Air velocity combinations in coal-mine air intakes and air returns.

	Air Velocity (m/s)	
Air Velocity Combination Modes*	Air Intake	Air Return
CM 1	1	2
CM 2	2	4
CM 3	2.5	3.5

*As discussed in the text, three combination modes (CMs) for intake and return air velocities from the working face of an underground coal mine were designed to simulate the gas distribution pattern in the mining gob and the overlying strata.

Results and Discussion

Gas Distribution in the Gob

As mentioned before, the gas concentration on the 11 levels (2 in the gob and 9 in the overlying strata) can be obtained under 12 ventilation patterns (combinations of 4 GRRs and 3 modes of air velocity in intake and return way). Figure 8.4.10 represents a comprehensive summary of all the data obtained in the experiment with a column chart. It shows how gas distributes in the gob under a certain pattern. The two bottom figures in Figure 8.4.10A–D present the gas concentration data from the roof level (level 2) and the floor level (level 1) as shown in Figure 8.4.8, with the black/white columns showing the difference value (D-value) between these two levels. As an example, when the concentration at a measurement point on level 2 is greater than the lower corresponding one on level 1, D-value is positive and represented by a black solid column in the black/white column chart at the same point and vice versa.

As shown in Figure 8.4.10, there are four groups of data, namely A, B, C, and D, in different conditions of GRR at 16.67, 33.33, 50, and 66.67 m^3/min, respectively. The colors (red, blue, and green) of the columns present different air velocities in intake and return roadways as for CM 1, 2, and 3. It also can be seen that the gas concentration on different levels is varied with a general trend that the gas concentrations on level 2 are greater than those on level 1. For example, in case of CM 3, the D-values are positive and huge; and the gas concentration is significantly greater than that of the other two air velocity combinations. However, a noteworthy exception is that the red and blue columns are apparently higher than the green one when GRR is 16.67 m^3/min. This indicates that, in most GRR cases (16.67, 33.33, 50 m^3/min), CM 3 could make a higher gas concentration compared with the other two combinations, which means more gas emission from the gob and more explosion risks. In other words, once the GRR is set at 16.67, 33.33, or 50 m^3/min, either CM 1 or CM 2 is suitable for gas control purposes. In contrast, when GRR is large enough (66.67 m^3/min), CM 3 could hold most of the gas within the gob area, with only a small part of gas escaping to working faces. It is therefore more helpful to mitigate the explosion risk compared with the other two CMs.

The gas-concentration contour map in the gob at two levels with heights of 0 m and 4 m close to the bottom and top of the gob, respectively, but with different GRRs, is shown in Figure 8.4.11. In this figure, A, B, C, and D represent the same kind of situation as in Figure 8.4.10. To visibly present the explosion risk for roof and floor levels, five colors are employed to indicate the likelihood of a gas explosion. The black zone is the most explosively hazardous area in the gob because the gas concentration at these zones is very close to 9%, which amounts to the strongest explosion intensity concentration for methane. Red represents the explosion range. Yellow represents the areas at which methane concentrations are very close to UEL or LEL. The areas shaded in green and light green represent the gas concentration either above the UEL or below the LEL. The concentration could fall in the gas explosion range at any time due to the fluctuation of ventilation and GRR.

The patterns of gas concentrations on the floor (level 1) and roof (level 2) of the bottom of the gob are shown in Figure 8.4.11. As indicated, the space close to the top of the gob has a higher explosion risk with a larger explosive/potential explosive area, which can also be seen in Figure 8.4.10. The gas concentration distribution in the gob area is greatly influenced by the ventilation and gas release from the residual coal pieces and rock in the gob area, which makes it necessary to understand the integrative effect of ventilation and gas release on gas distribution in underground mines so as to choose a suitable ventilation plan. Sometime, simply increasing air quantities in both air intake way and return way may not be helpful for gas control. For example, the left two figures of CM 1 in Figure 8.4.10C are more effective for gas control than increasing both air velocities as in CM 3. Through analyzing

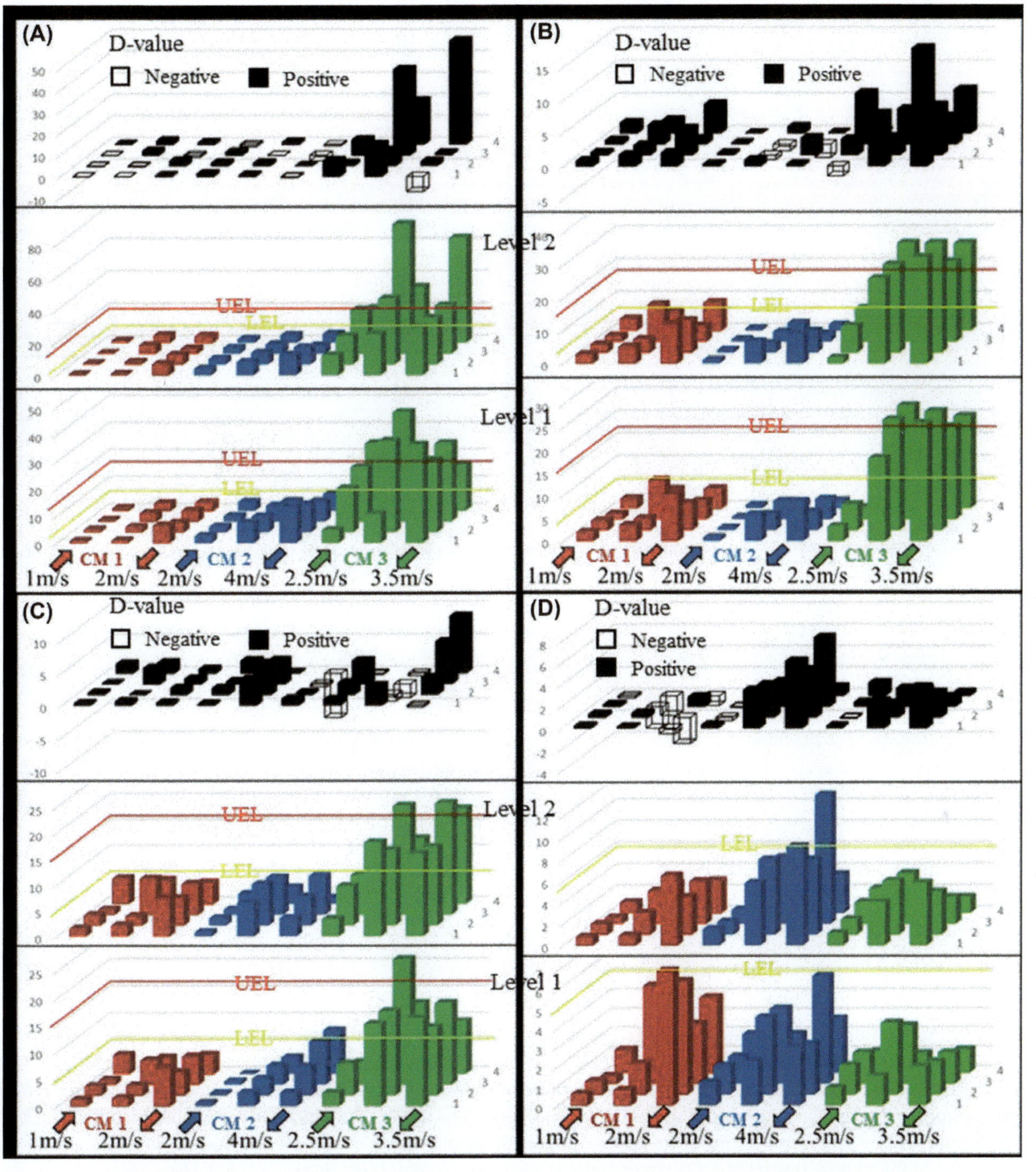

Figure 8.4.10. Gas concentration and distribution chart. Figure by Jianwei Cheng, 2016.

Figure 8.4.11, it can be drawn that CM 1 is the best ventilation strategy for gas control when the GRR is 16.67 m^3/min or 33.33 m^3/min, while CM 2 is suitable when the GRR reaches 50 m^3/min. CM 3 should be adopted when GRR is 66.67 m^3/min, because contrary to diluting the gas in the gob by ventilation, these air quantities just keep the gas in the gob at a very high concentration, which makes an explosion impossible. However, CM 1 is not suitable for gas control when GRR are 50 and 66.67 m^3/min, because it diffuses the gob gas and enlarges the explosive area; similarly, setting ventilation as CM 3 is not a good choice when GRR is 33.33 m^3/min.

Gas Distribution in Overlying Strata

The influence of the rate of airflow and GRR on gas distribution patterns in strata overlying coal-mine gob were studied in nine levels over the top of the gob. Figure 8.4.12 shows the basic layout of numbered measuring points.

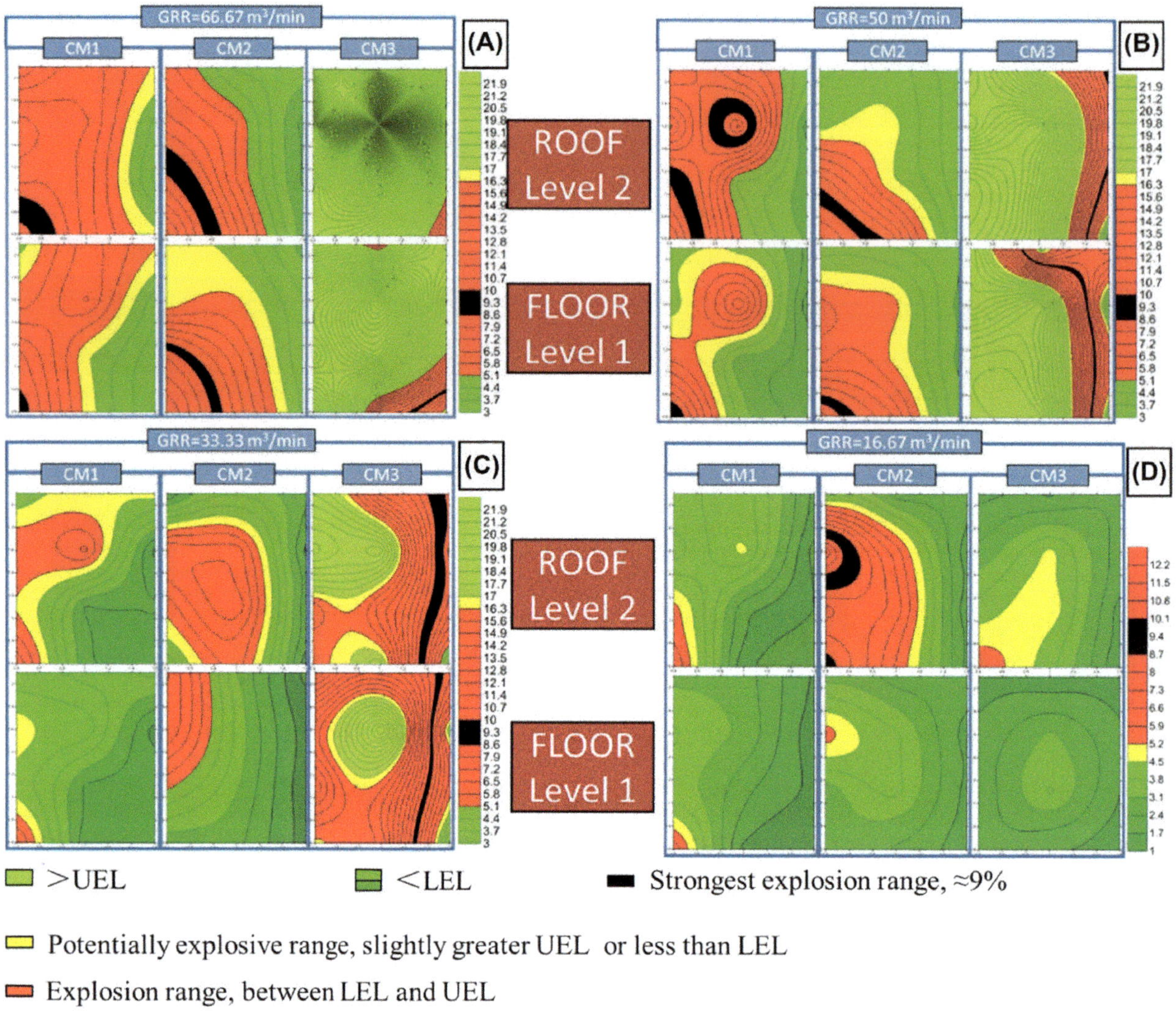

Figure 8.4.11. Gas concentration distribution at the top and bottom of mine gob; different ventilation modes. Figure by Jianwei Cheng, 2016.

Figures 8.4.12B and C show the top and side views of the measured locations within. Gas samples were taken from 108 points set in 9 levels over the gob. Figure 8.4.12A shows the tubes used for gas sampling. Figures 8.4.13–8.4.15 illustrate gas concentrations with different combinations of CM 1, CM 2, CM 3, and four GRRs. In other words, the gas concentration patterns in both the gob and the overburden are studied with the GRR changing under CM 1, CM 2, and CM 3. All 11 measuring levels are arranged along the x axis by height over the bottom of the gob and the points in the yellow frames are gas concentration data from two gob levels (Figure 8.4.10).

Figure 8.4.13 shows that with the ventilation air velocity of CM 1, more and more of a gas explosive zone arises in the gob area with GRR increasing, while the gas concentration in levels above 30 m remains low and stable. An enlightening phenomenon is that when GRR is set at 50 m^3/min, some part (8.3%) of measuring level, about 8 m above the bottom of the gob in overlying strata, is explosively hazardous.

Figure 8.4.14 shows that when the air velocities in the air intake and return pathways are doubled compared with the previous case, and GRR is set at 33.33 m^3/min, point 11 close to the air return way on all levels is explosive because the gas data measured there are between LEL and UEL. From Figure 8.4.14D, it can be concluded that under this GRR and CM 2, the overlying strata (8–22 m off the bottom of the gob) contain explosive zones of various sizes, because the gas concentration data concentrate between the limits of gas explosion. The gas concentration pattern in the gob and the overlying strata is very different from the previous two cases when air quantities in the air intake/return way are set at CM 3. From Figure 8.4.15, it must be noted that the gas explosive limits in B, C, and D are all

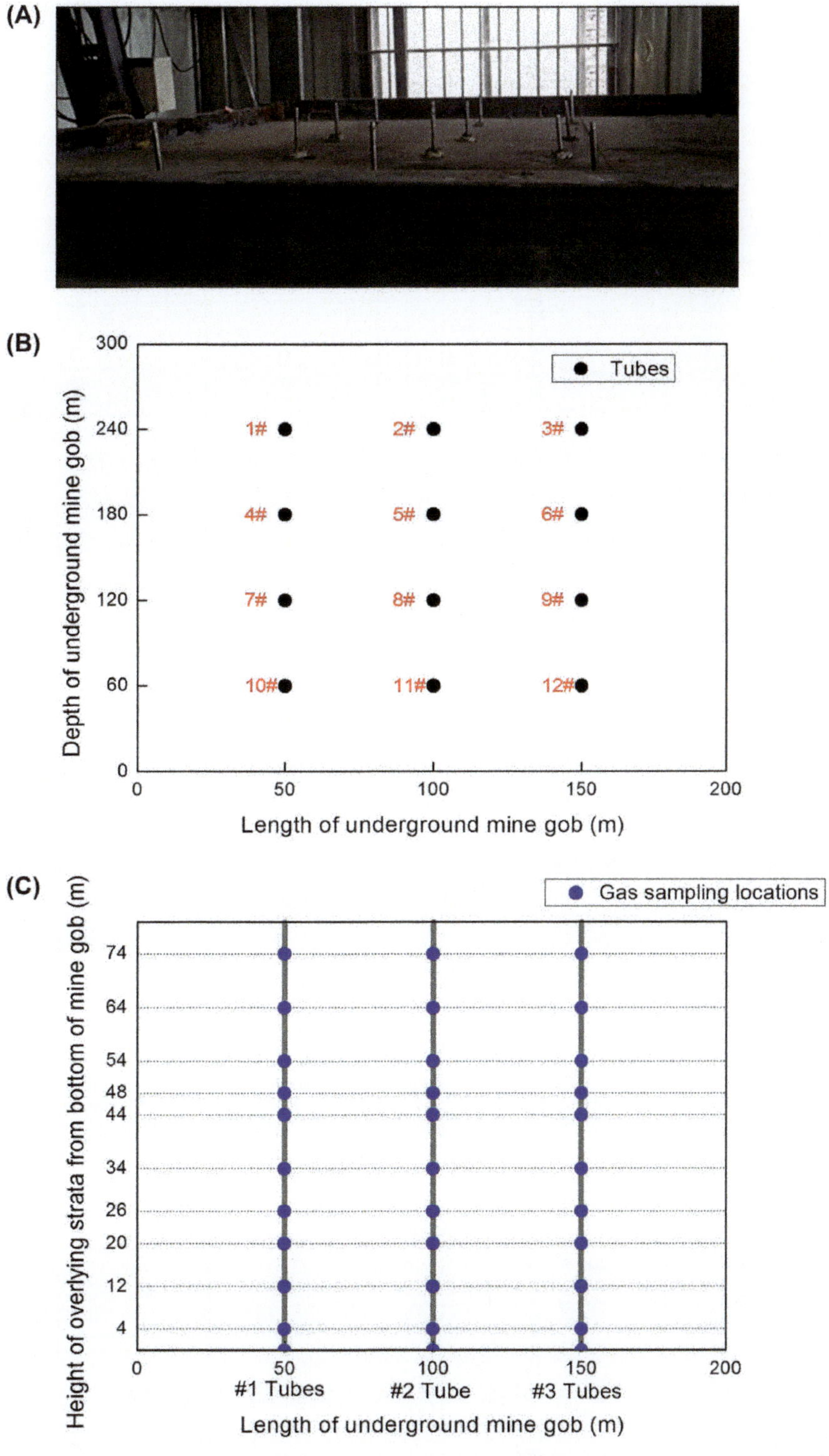

Figure 8.4.12. Schematic arrangement of tubes used for gas sampling in a gob pile. (A) The actual tubes used for gas sampling. (B) Top view of tubes emplaced in the gob. (C) Another view showing an arrangement for the locations of sampling holes. Figure by Jianwei Cheng, 2016

narrower than that in A because the gas concentration at most measuring points in most levels are large, which makes the vertical axis show a larger range of numbers. This explains the shrinkage of the gas explosive limits that represent the same range as in the other figures. From Figure 8.4.15B, when GRR is set at 33.33 m^3/min, about 75% of the area of 8 m level is explosively hazardous and about 17% of the area of 16 and 22 m levels are potential for an explosion. When GRR is increased to 50 and 66.67 m^3/min, even part of the area of 30 m level is explosive.

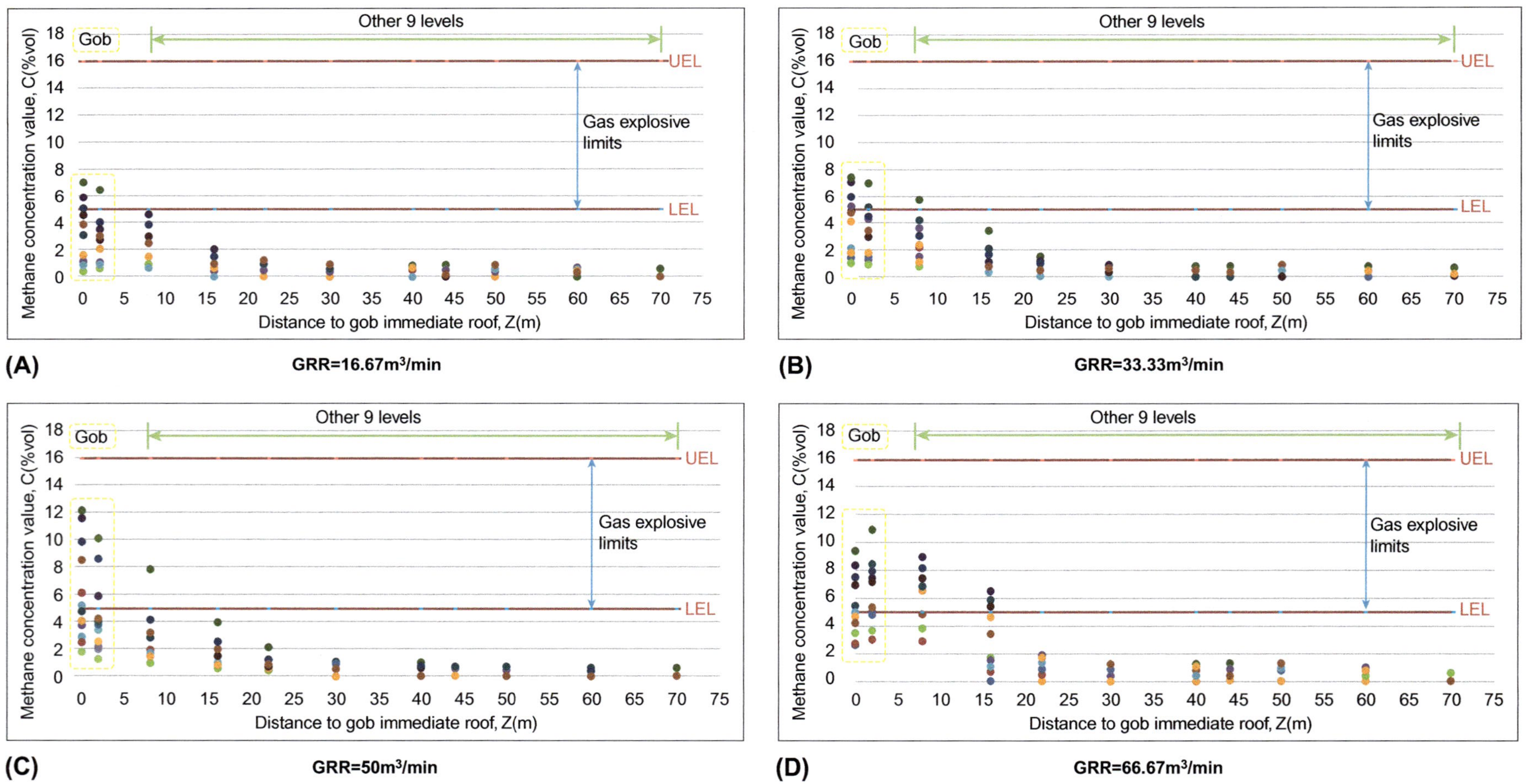

Figure 8.4.13. Scatter diagram of methane concentrations versus methane release rates when air velocities in the air intake and return roadways are 1 m/s and 2 m/s, respectively (CM 1). Figure by Jianwei Cheng, 2016.

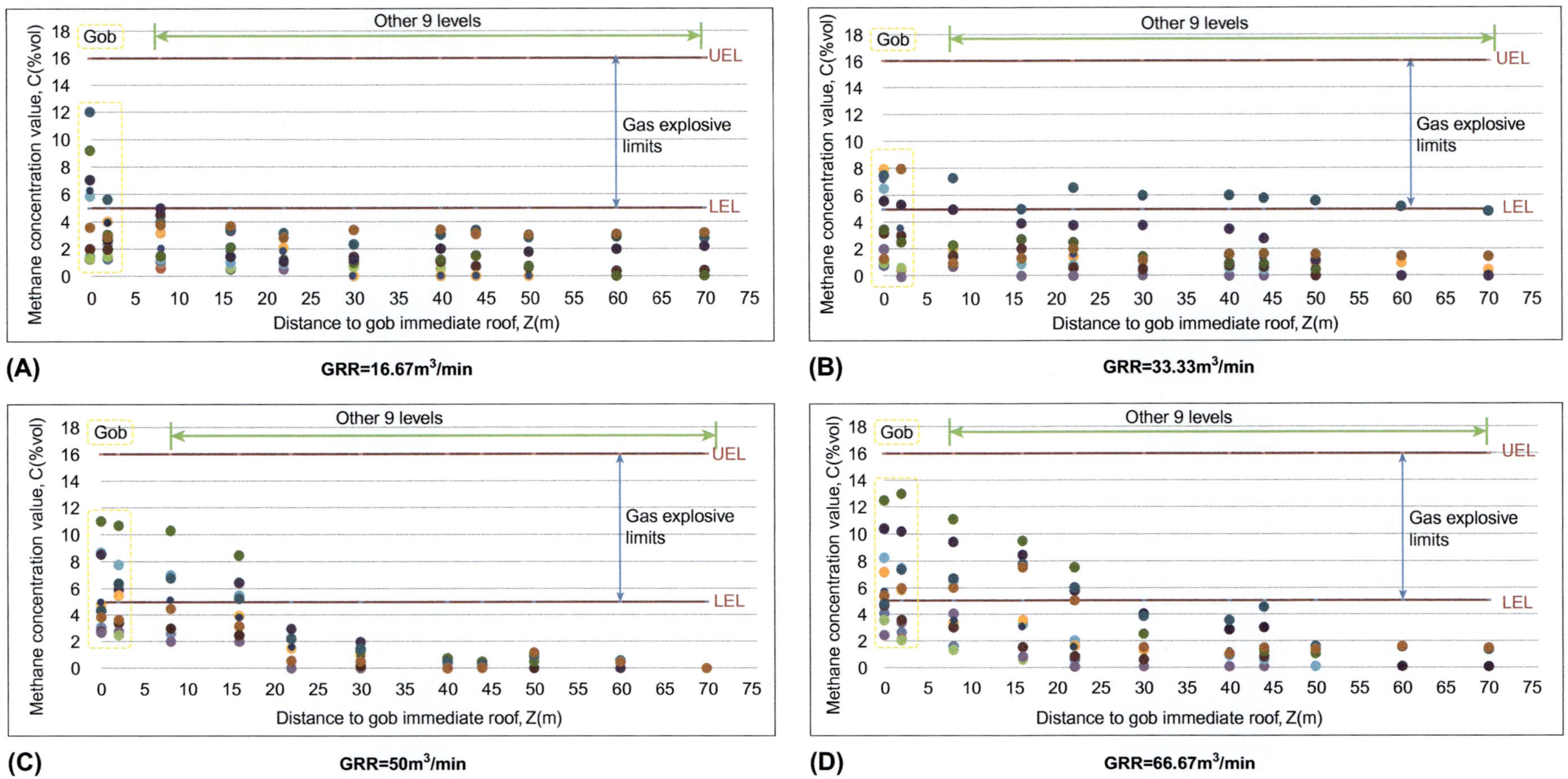

Figure 8.4.14. Scatter diagram of methane concentrations versus methane release rates when air velocities in the air intake and return roadways are 2 m/s and 4 m/s, respectively (CM 2). Figure by Jianwei Cheng, 2016.

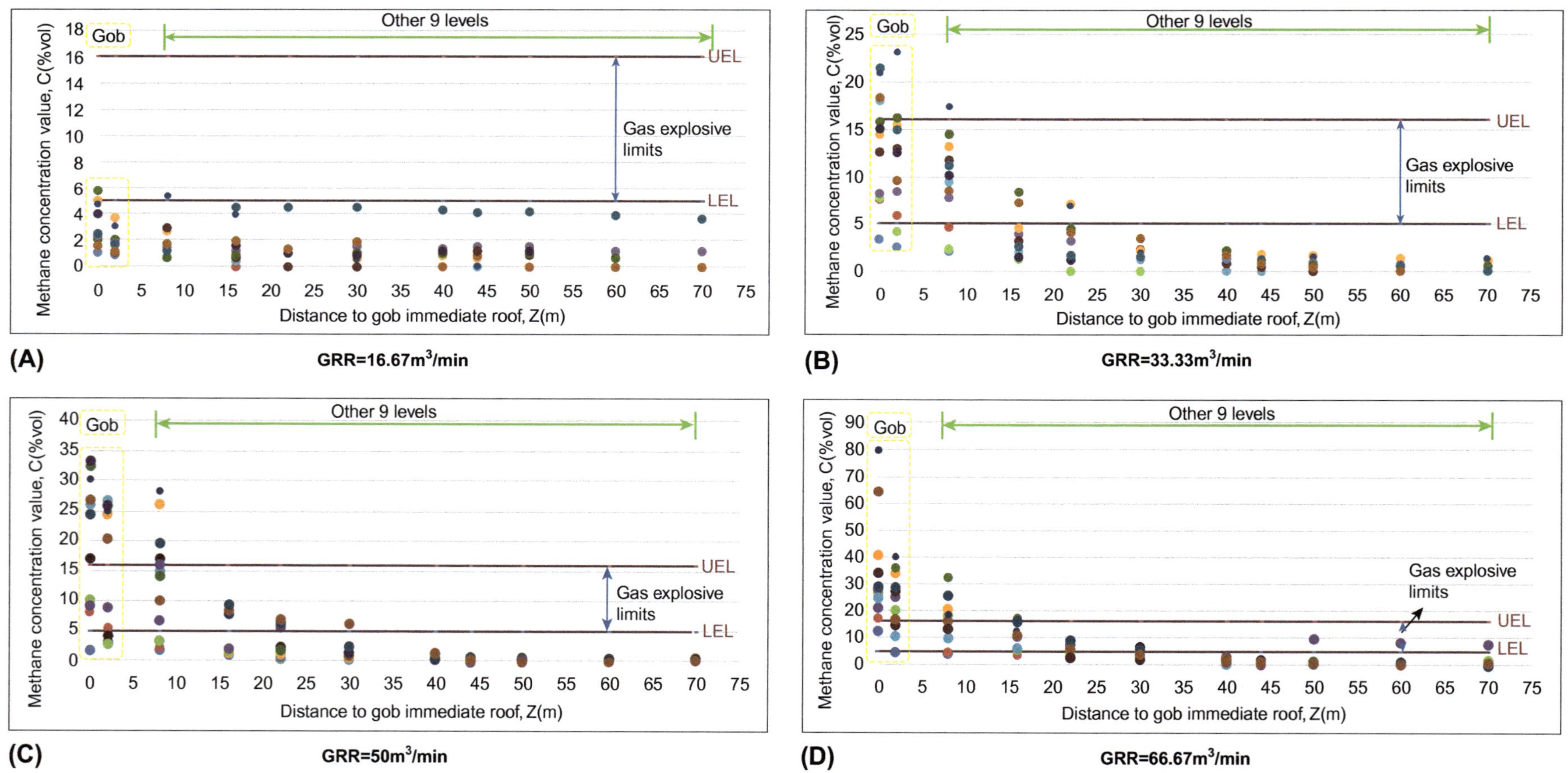

Figure 8.4.15. Scatter diagram of methane concentrations versus methane release rates when air velocities in the air intake and return roadways are 2.5 m/s and 3.5 m/s, respectively (CM 3). Figure by Jianwei Cheng, 2016.

Acknowledgments

Partial field work and analyses were funded by the National Science Foundation of China (Grant No. 51304203). The author thanks Professor Emeritus Glenn B. Stracher, East Georgia State College, University System of Georgia, Swainsboro, Georgia, USA, for his review of this chapter.

Important Terms

Combustion mode (CM)
Coward Explosibility Triangle
fire ratios
gas-mixture point
gas release rate (GRR)
Graham's Ratio
Hydrocarbon (HC) Ratio
Jones-Trickett Ratio
Kukuczka Method
Litton Ratio
Longwall gob
Maximum Allowable Oxygen (MAO)
mine-gas distribution
nonspontaneous combustion zone
nose limit (nose)
Relative Intensity
spontaneous combustion
spontaneous combustion zone
suffocation zone
tail entry
tailgate entry
tertiary diagrams
Trickett's Ratio (TR)
tube bundle monitoring system
USBM Explosibility Diagram

References

Alexander, D.W., 2006. The Expected Value of Catastrophes in Underground Coal Mines 1987-2001 (Ph.D. dissertation). West Virginia University. 238 p.

Brann, M., 1994. Modelling of endogenous fire in lumps or in fissures (in polish), Central Mining Institute. In: XXVI Krakowska Konferencja Naukowo-Techniczna Przeróbki Kopalin – A Scientific-technological Conference, September 7–9, Ustroń. Jaxa Publishing, Kraków, Poland.

Centers for Disease Control and Prevention (CDC), 2017. The National Institute for Occupational Safety and Health (NIOSH). https://www.cdc.gov/niosh/mining/statistics/content/allminingdisasters.html (accessed July 2017).

Cheng, J., Li, S., Zhang, F., Zhao, C., Yang, S., Ghosh, A., 2016. CFD modelling of ventilation optimization for improving mine safety in longwall working faces. J. Loss Prev. Process. Ind. 40, 285–297.

Cheng, J., 2012. Comprehensive and Integrated Model for an Atmospheric Status in Sealed Underground Mine Areas (Ph.D. dissertation). West Virginia University, Morgantown, West Virginia. 163 p.

CNN (Cable News Network), 2011. http://articles.cnn.com/2011-03-22/world/pakistan.mineblast_1_methane-gas-explosion-mine-rescuers?_s¼PM:WORLD (accessed March 2011).

Coward, H.F., Jones, G.W., 1952. Limits of Flammability of Gases and Vapors, U.S. Bureau of Mines, Bulletin 503. US Government Printing Office. 153 p.

Curl, S.J., 1978. Methane Prediction in Coal Mines, IEA Coal Resources Report ICTIS/TR 04, 79 p.

Dwyer, J., Hansel, J.G., Pilips, T., 2003. Temperature influence on the flammability limits of heat treating atmospheres. In: Dahotre, N., Gaster, R., Hill, R., Popoola, O. (Eds.), Proceedings of the 22nd Heat Treating Society Conference and the 2nd International Surface Engineering Congress, Indianapolis, Indiana, pp. 24–28.

Encyclopedia Britannica, 2017. https://www.britannica.com/science/air (accessed August 2017).

Graham, J.I., 1914. Adsorption of oxygen by coal. V Trans. Inst. Min. Eng. XLVIII, 521.

Greuer, R.E., 1974. Study of Mine Fire Fighting Using Inert Gases, U.S. Bureau of Mines Contract Report No. S0231075. 212 p.

Guan, W., 2016. Mine Firefighting Practices in Shendong Coal Field, China (M.S. thesis). China University of Mining and Technology, Xuzhou, China, p. 63.

Hartman, H.L., Mutmansky, J.M., Ramani, R.V., Wang, Y.J., 1997. Mine Ventilation and Air Conditioning, third ed. John Wiley & Sons, Inc. 730 p.

Hu, G.Z., Wang, H.T., Tan, H.X., Fan, X.G., Yuan, Z.G., 2008. Gas seepage equation of deep mined coal seams and its application. J. China Univ. Min. Technol. 18, 483–487 Xuzhou, China.

Hu, G.Z., Wang, H.T., Fan, X.G., Yuan, Z.G., Hong, S., 2009. Mathematical model of coalbed gas flow with Klinkenberg effects in multi-physical fields and its analytic solution. Transport Porous Media 76, 407–420.

Huang, S., 2010. China Coal Outlook 2010. China Coal Industry Publishing House, Beijing, China (Number of pages?).

Jones, J.H., Trickett, J.C., 1954. Some observations on the examination of gases resulting from explosions in collieries. Trans. Inst. Min. Eng. 114, 768–791 London.

Justin, T.R., Kim, A.G., 1988. Mine Fire Diagnostics to Locate and Monitor Abandoned Mine Fires, USBM Information Circular IC 9184, pp. 348–355.

Karacan, C.Ö., Esterhuizen, G.S., Schatzel, S.J., Diamond, W.P., 2007. Reservoir simulation- based modeling for characterizing longwall methane emissions and gob gas vent hole production. Int. J. Coal Geol. 71, 225–245.

Kissell, F.N., 2006. Handbook for Methane Control in Mining, National Institute for Occupational Safety and Health Document, IC 9486. 180 p.

Kukuczka, M., 1982. A new method for determining explosibility of complex gas mixtures. Mechanizacja I Automatuzacja Gornictwa 164 (11), 36–39.

Litton, C.D., 1986. Gas Equilibrium in Sealed Coal Mines, U.S. Bureau of Mines Report of Investigation RI 9093. 13 p.

Lu, L., Han, X., 1991. Handbook of Coal Mine Safety. Coal Mining Industry Press (in Chinese), Beijing. 163 p.

Lunarzewski, L., 1998. Gas emission prediction and recovery in underground coal mines. Int. J. Coal Geol. 35, 117–145.

Mashuga, C.V., Crowl, D.A., 2000. Derivation of Le Chatelier's mixing rule for flammable limits. Process Saf. Prog. 19, 112–117.

McPherson, M.J., 1993. Subsurface Ventilation and Environmental Engineering. Chapman & Hall, New York. 905 p.

Mine Safety and Health Administration (MSHA), 2017. United States Department of Labor. http://www.msha.gov (accessed July 2017).

Mishra, D.P., Kumar, P., Panigrahi, D.C., 2016. Dispersion of methane in tailgate of a retreating longwall mine: a computational fluid dynamics study. Environ. Earth Sci. 75, 1–10.

Mitchell, D.W., 1990. Mine Fires. Maclean Hunter, Chicago, IL. 167 pp.

Mitchell, D.W., 1996. Mine Fires Prevention, Detection, Fighting [Book]. Intertec Publishing Inc, Chicago, IL. 211p.

Prosser, B.S., Oswald, N.L., 2006. Ventilation surveying and modeling of longwall bleeder and gob areas. In: Mutmansky, J.M., Ramani, R.V. (Eds.), The 11th U.S./North American Mine Ventilation Symposium, Pennsylvania State University, University Park, Pennsylvania. CRC Press, pp. 587–590. https://www.crcpress.com/11th-USNorth-American-Mine-Ventilation-Symposium-2006-Proceedings-of-the/Mutmansky-Ramani/p/book/9780415401487 (accessed July 2017).

Schatzel, S.J., Garcia, F., McCall, F.E., 1992. Methane sources and emissions on two longwall panels of a Virginia coal mine. In: 9th Annual International Pittsburgh Coal Conference, Pittsburgh, Pennsylvania, pp. 991–998.

Stoltz, R.T., Francart, W.J., Adair, L., Lewis, J., 2006. Sealing a recent United States coal mine longwall gob fire. In: Mutmansky, J.M., Ramani, R.V. (Eds.), The 11th U.S./North American Mine Ventilation Symposium, Pennsylvania State University, University Park, Pennsylvania. CRC Press, pp. 331–335.

Timko, R.J., Derick, R.L., 2006. Methods to Determine the Status of Mine Atmospheres – an Overview. National Institute for Occupational Safety and Health Publications. https://www.cdc.gov/niosh/mining/UserFiles/works/pdfs/mtdts.pdf (accessed on July 2017).

Toraño, J., Torno, S., Menendez, M., Gent, M., Velasco, J., 2009. Models of methane behavior in auxiliary ventilation of underground coal mining. Int. J. Coal Geol. 80, 35–43.

Vutukuri, V.S., Lama, R.D., 1986. Environmental Engineering in Mines. Cambridge University Press, London. 520 p.

Wang, D., 2007. Mine Ventilation and Safety (In Chinese). China University of Mining and Technology Press, Xuzhou, China. 330 p.

Wikipedia, 2010. Pike River Mine Disaster. http://en.wikipedia.org/wiki/Pike_River_Mine_disaster (accessed February 2012).

Yang, S., Zhang, R., Di, Z., 2000. Study on three zones distribution law of spontaneous combustion in gob of a fully mechanized mining face. J. China Univ. Min. Technol. 29, 93–96 Xuzhou, China.

Yu, Q., 1992. Prevention of Coal Mine Methane (In Chinese). China University of Mining and Technology Press, Xuzhou, China. 150 p.

Zabetaksi, M.G., Stahl, R.W., Watson, H.A., 1959. Determining the Explosibilty of Mine Atmospheres, U.S. Bureau of Mines, Bulletin IC7901. 20 p.

Zhang, G., 2000. Ventilation and Safety Engineering (In Chinese). China University of Mining and Technology Press, Xuzhou, China. 415 p.

Zipf, R.K., Mohamed, K.M., 2010. In: Hardcastle, S., McKinnon, D. (Eds.), Composition Change Model for Sealed Atmosphere in Coal Mines, in the 13th United States/North American Mine Ventilation Symposium. Laurentian University, Sudbury, Ontario, pp. 493–500.

Zipf, R.K., Marchewka, W., Mohamed, K., Addis, J., Karnack, F., 2013. Tube bundle system for monitoring of coal mine atmosphere. Min. Eng. 65 (5), 57–63.

WWW Addresses: Additional Reading

(1) China State Administration of Coal Mine Safety
http://www.chinasafety.gov.cn/newpage/mkaj/index.htm (in Chinese)

(2) Coal Mining Method: Energy and Minerals Field Institute Summary
http://emfi.mines.edu/emfi2011/Coal%20Mining%20Methods%20-%20EMFI%20Summary.pdf

(3) Jianwei Cheng's Fundamentals Mine Dust and Control, China University of Mining and Technology, Xuzhou, China
http://www.mineteaching.cn

(4) Jianwei Cheng's Safety Engineering, China University of Mining and Technology, Xuzhou, China
http://www.cheng-jianwei.com

(5) National Institute for Occupational Safety and Health (NIOSH) Mining Topic: Fire Control and Suppression
https://www.cdc.gov/niosh/mining/topics/firecontrolsuppression.html

(6) The National Institute for Occupational Safety and Health (NIOSH) Mining Topic: Explosion Prevention
https://www.cdc.gov/niosh/mining/topics/Explosions.html

CHAPTER 9

Gases Generated During the Low-Temperature Oxidation and Pyrolysis of Coal and the Effects on Methane-Air Flammable Limits

A gas-explosion experiment in a box sealed by polytetrafluoroethylene film. The box is 2.5 m on each side. *Photo by Fangming Cheng, 2012.*

CHAPTER CONTENTS

Coal and Peat Fires: A Global Perspective
Edited by Glenn B. Stracher

https://doi.org/10.1016/B978-0-12-849885-9.00009-3

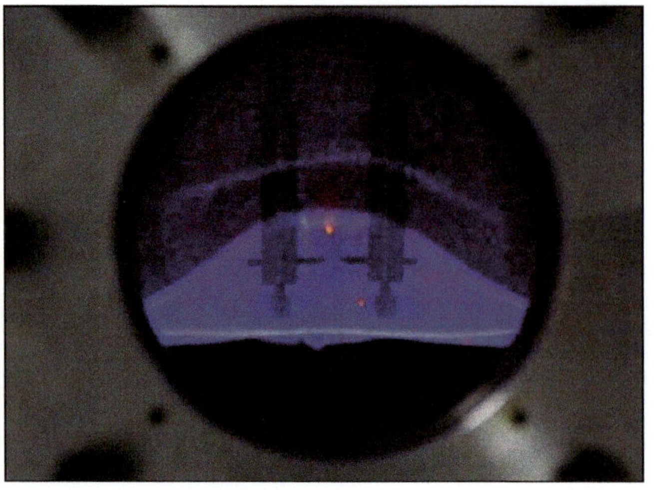

Photo by Fangming Cheng, 2014.

9.1 Gases Generated During Low-Temperature Oxidation and Pyrolysis

Zhenmin Luo
Tao Wang
Fangming Cheng

Methane flame propagation in a 20 L spherical explosion chamber. The diameter of the chamber is 33.68 cm.

Introduction

Multiple combustible gases in a coal fire area are mainly composed of gases from coal seams, gases produced during the processes of **coal oxidation** and **coal pyrolysis** in low temperature. The coal seam gases are mainly methane (CH_4), carbon dioxide (CO), ethane (C_2H_6), propane (C_3H_8), hydrogen (H_2), sulfur dioxide (SO_2), and hydrogen sulfide (H_2S), among others. Among them, the CH_4 accounts for more than 90%, while the content of other combustible gases is less. At ambient temperature and pressure, the gas absorption and desorption keep balance in coal surface, but this balance will be broken and gases are released when crush and **spontaneous combustion** occurs.

In the process of coal oxidation in low temperature, the coal reacts with the oxygen and produces CO, CO_2, and H_2O. At the initial process of coal oxidation, absorbed gases such as CH_4, CO_2, CO, N_2, and others in the coal surface release with temperature rise. Moreover, coal structures crack with the rising of **coal temperature**, releasing gases as CH_4, C_2H_6, C_3H_8, C_2H_4, C_2H_2, and such. Researchers at the Xian University of Science and Technology used a large-scale experiment platform to study the spontaneous combustion of coal, namely XK-III, XK-IV, and ZRM-15, to deal with coal samples (experimental maximum temperature is 500°C). Experiment results indicate that the coal samples generate CO in the low-temperature oxidation and ignition stages. To the general coal samples, the quantity of the carbon monoxide is less than 2% beneath the temperature of 300°C. Before oxidation, CH_4 can be detected in many coal samples, and its release continuously rises with the temperature rise. When the coal temperature reaches a certain point, other hydrocarbon gases are produced. The experimental results also show that the maximum content of hydrocarbon gases besides CH_4 is C_2H_6 released in the process of oxidation and spontaneous combustion, and its quantity is less than 1% when the temperature is lower than 300°C. Li Zenghua and Lin Baiquan from China University of Mining and Technology measured H_2 produced in the process of coal oxidation by hydrogen measurement. The experimental results show that H_2 can be detected when the temperature rises to 200–300°C, and the H_2 concentration increases exponentially as temperatures rise. In addition, H_2 concentration can rise up to 0.49% when the coal temperature rises to 280°C, and the lower the coal rank, the lower the temperature of H_2 release, and the greater the content of H_2 generated. Although the amount of H_2 is slight in the process of coal oxidation and spontaneous combustion, H_2 is an important explosion hazard. Its mere existence also has a significant impact on explosive properties of other combustible gases.

In addition, the coal can also produce an amount of H_2, CO, and CH_4 when it explodes for the first time in the coal mine. Sometimes the content of H_2 can be up to 4%, the lower explosive limit. Due to the lower ignition temperature of H_2 and CO, which increases the mixed gas explosion hazard, newly generated gas, original gas, and coal dust mixing together after the first explosion constitutes a lower explosion limit, and this mixed gas will explode again when the gases have exploded once. The above situation often occurs, which greatly increases the difficulty of coal mine rescue.

Experimental Conditions

The **oil bath temperature-programmed system** for studying the spontaneous combustion of coal was adopted to test the regularity of gas generation during oxidation. This system is composed of four parts, including oil bath temperature control system, gas supply system, cooling system, and the gas collection and analysis system; the structure is shown in Figure 9.1.1. Four coal samples of different ranks were chosen for the experiment: Baelapukilia, Tangkou, Dongshan, and Baijiao. Air flow rate is 120 mL/min, the rate of temperature rise is 0.3 °C/min, and the gases were identified and analyzed by SP-3430 gas chromatograph.

Gas Production Rates

During the oxidation process, coal and oxygen reacted, consuming oxygen, and produced CO. By consuming oxygen, the **concentration of oxygen** in the test tube decreases along the airflow direction, and the concentration of CO increases. CO production rate of coal is proportional to the oxygen consumption rate.

$$V_{co}(T)=\frac{C_{O2}}{C^{0}_{O2}}\cdot V^{0}_{co}(T) \quad (9.1.1)$$

where $V_{CO}(T)$ is CO production rate (mol/(cm^3.s)) and V^0_{CO} is the standard CO production rate of 21% oxygen concentration (mol/(cm^3.s)).

By the Arrhenius we can infer the oxygen concentration at any point in the furnace to be:

$$C=C_i\cdot e^{-\frac{V^{0}_{O2}(T)\cdot S}{Q\cdot C^{0}_{O_2}}\cdot(z-z_i)} \quad (9.1.2)$$

where C_i is the certain oxygen concentration of one point in the furnace.

$$dC_{CO}=\frac{V_{CO}(T)}{n}d\tau,\ d\tau=\frac{dz}{u},\ u=\frac{Q}{S\cdot n} \quad (9.1.3)$$

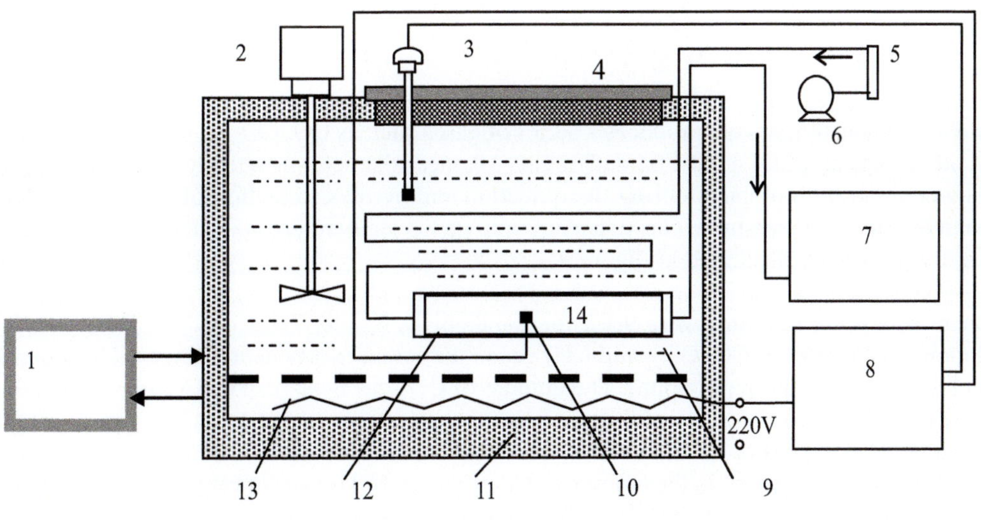

Figure 9.1.1. An oil bath temperature-programmed system. 1: Cooling system, 2: stirrers, 3: thermocouple, 4: sealing cover, 5: flowmeter, 6: air compressor, 7: gas chromatograph, 8: temperature control instrument, 9: oil bath, 10: temperature sensor, 11: thermal barrier, 12: sample container, 13: heating cord, and 14: coal sample. Figure by Yanni Zhang, 2010.

The oxygen concentration at unknown point is C_1, the distance to the inlet is Z_1, the oxygen concentration at a certain point is C, and the distance from the inlet is Z_2. Put into Eq. (9.1.3) and integrating,

$$C_{co}^{2} - C_{co}^{1} = \int_{1}^{2} \frac{V_{CO}(T)}{u \cdot n} dz = \int_{1}^{2} \frac{S}{Q} \cdot \frac{C \cdot V_{CO}^{0}(T)}{C^{0}_{O_2}} dz = \frac{S \cdot V_{CO}^{0}(T)}{Q \cdot C^{0}_{O_2}} \int_{1}^{2} C_1 e^{\frac{V_{O_2}^{0}(T) \cdot S}{Q \cdot C^{0}_{O_2}} \cdot (z - z_1)} dz \tag{9.1.4}$$

From Eq. (9.1.4), the standard oxygen concentration production rate of CO is

$$V_{CO}^{0}(T) = \frac{V_{O_2}^{0}(T) \cdot \left(C_{co}^{2} - C_{co}^{1} \right)}{C_1 \cdot \left[1 - e^{\frac{V_{O_2}^{0}(T) \cdot S \cdot (z_2 - z_1)}{Q \cdot C^{0}_{O_2}}} \right]} \tag{9.1.5}$$

Similarly, the equation of production rate of CO_2 and CH_4 are as follows,

$$V_{CO_2}^{0}(T) = \frac{V_0(T) \cdot \left(C_{CO_2}^{2} - C_{CO_2}^{1} \right)}{C_0 \cdot \left[1 - e^{\frac{V_0(T) \cdot s \cdot n \cdot (z_2 - z_1)}{Q \cdot C_0}} \right]} \tag{9.1.6}$$

$$V_{CH_4}^{0}(T) = \frac{V_0(T) \cdot \left(C_{CH_4}^{2} - C_{CH_4}^{1} \right)}{C_0 \cdot \left[1 - e^{\frac{V_0(T) \cdot s \cdot n \cdot (z_2 - z_1)}{Q \cdot C_0}} \right]} \tag{9.1.7}$$

where $V_{co_2}^{0}$ is the production rate of CO_2, and $V_{CH_4}^{0}$ is that of CH_4.

Gas Generation During Oxidation and Pyrolysis

The spontaneous combustion of coal produces a series of gases such as CO, CO_2, H_2O, H_2, CH_4, C_2H_4, C_2H_2, NO, SO_2, among others. CO and CO_2 are the oxidation gas. Alkanes, alkenes, and acetylene are the gases generated by pyrolysis. The concentration of gases during the oxidation and pyrolysis reaction of coal with different rank and different **particle size** were measured. Gas concentration versus temperature was analyzed by mixed particle size coal samples; the graphs are illustrated in Figure 9.1.2.

The experiment results are seen in Figure 9.1.2 for Barakpuriya lignite, Tangkou coal, Dongshan lean-coal, and Baijiao anthracite that produced CO, CO_2, and CH_4 gas in the initial experimental stages. The amount of CO produced by each coal is low, less than 10 ppm. The amount of CO_2 gas is large. The CO_2 gas produced by Tangkou coal, Dongshan lean-coal, and Baijiao anthracite are nearly 600 ppm. Lignite's CO_2 gas production is nearly 2000 ppm; anthracite's CH_4 gas production is 200 ppm, and the concentration of other three coal samples are very low. With the rise of coal temperature, the amount of CO, CO_2, and CH_4 gas concentration increases exponentially. When the coal temperature is close to 100°C, C_2H_6 gas appeared in Barakpuriya lignite, Tangkou coal, and Dongshan lean-coal, while Baijiao anthracite has C_2H_6 gas at 70°C. C_2H_4 gas emerged at different temperatures of the different rank of coal samples. Barakpuriya lignite at 108°C, Tangkou at 110.5°C, 144°C as Dongshan lean-coal, and Baijiao anthracite is not measured for C_2H_4.

According to the experimental results, it can be inferred that the types and order of the gases produced by different coal ranks in the process of low-temperature oxidation are basically the same, but the temperature of appearance and the

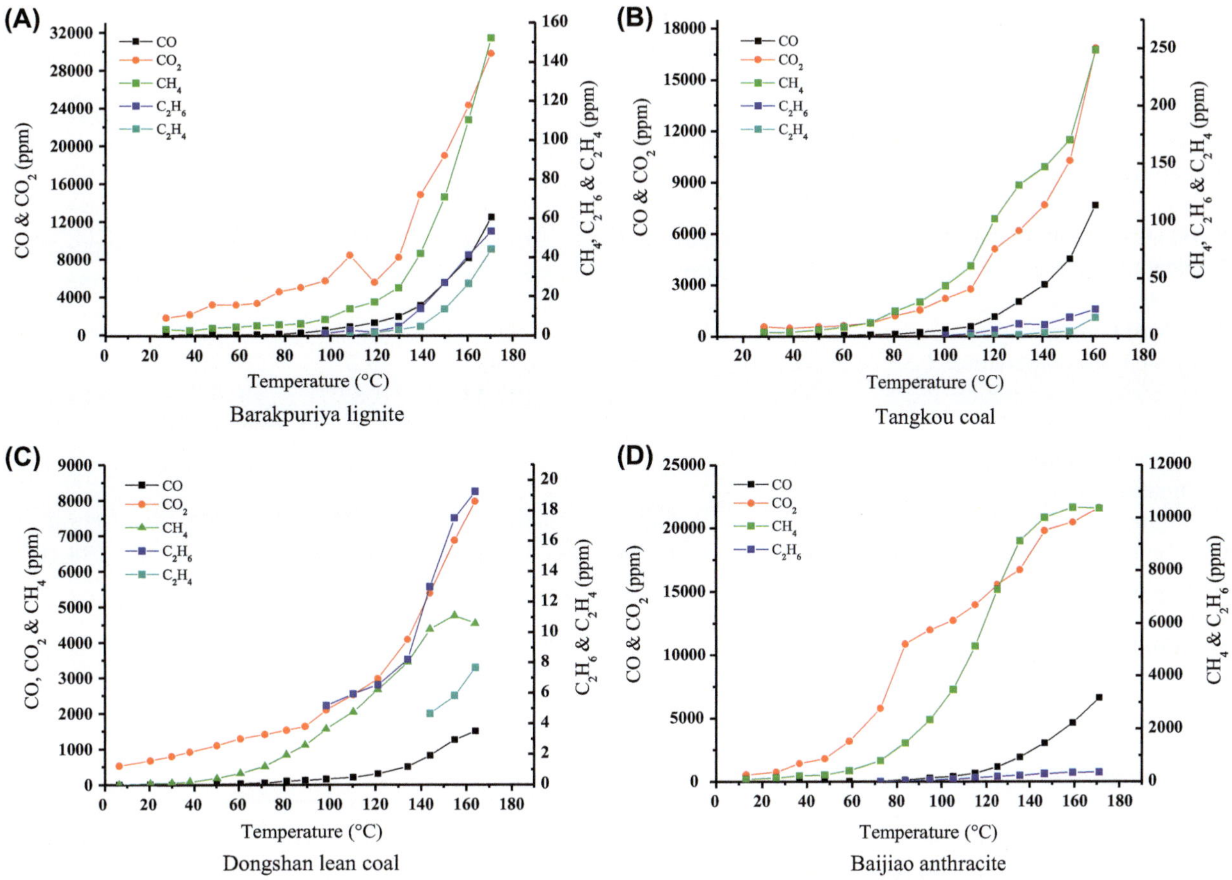

Figure 9.1.2. Graphs of gas concentration versus temperature. Figures by Tao Wang, 2016.

amount of the same gas are different. This is mainly caused by the same kind of **functional groups** in different ranks of coal having different activities, reflecting differences in the microchemical structure of the different ranks of coal.

Infrared spectra (IR) and oxidation simulation of coal samples show that the main **organic structure of coal** is composed of the aromatic ring with various side chains connected by bridge bonds such as methane, ethylidene, and ether bond. The monolithic structure is a polymer with the dimensional network. During the low-temperature oxidation, the nonaromatic structure side chains and bridge bonds in coal molecules react with oxygen to release CO and CO_2. In the high-temperature oxidation stage, the structures of aromatic rings start to break gradually and react with oxygen to generate CH_4, C_2H_6, C_2H_4 alkanes, and olefins gas. Therefore, CO and CO_2 gas are the symbols of the coal low-temperature oxidation stage, while alkanes such as CH_4, C_2H_6, and C_2H_4 are the high-temperature oxidation symbols. Gas production in different stages can reflect the oxidation degree of coal and internal structure.

During the initial stage of the experiment, CO, CO_2, and CH_4 gases were detected in all samples. A small quantity of CO gas is produced for the coal samples, which are exposed to the air in the process of crush, and the active groups in coal molecular react with oxygen on the surface of coal. The larger amount of CO_2 gas and lesser amount of CH_4 gas are not generated by coal oxidation and pyrolysis but adsorption in the coal seams of the mining area. The desorption of those gases occurs when the temperature rises. The CO_2 content of Barakleah lignite is the highest, and the CH_4 content of Baijiao anthracite is the highest. C_2H_6 gas is detected in the Baijiao anthracite at a relatively low temperature of 70°C. This means that C_2H_6 gas also lies in Baijiao coal. The generation of C_2H_4 gas in the temperature-programmed experiments indicates that the side chains of coal molecules begin to break down and participate in the oxidation, and the spontaneous combustion of coal enters an accelerated oxidation stage. While the Baijiao anthracite sample does not produce C_2H_4 during the whole oxidation process, for the high rank of Baijiao anthracite and the production temperature of C_2H_4 it may be close to 200°C, and the generation amount of C_2H_4 is too small to detect.

Figure 9.1.3 shows the graph of the production rate of CO versus temperature. The characteristic temperatures of gases appearance and coal samples are shown in Table 9.1.1.

Table 9.1.1
The temperatures at which gases were released in coal samples.

Coal Sample	1#	2#	3#	4#	5#	6#
Barapukulia Coal Samples						
CO mutation temperature/°C	57.95	60.51	68.46	68.21	67.44	58.46
C_2H_4 appear temperature/°C	109.23	112.05	108.97	109.74	109.23	108.46
Critical temperature/°C	55–70					
Crack temperature/°C	105–110					
Tangkou Coal Samples						
CO mutation temperature/°C	66.15	69.49	67.44	60.0	70.77	60.0
C_2H_4 appear temperature/°C	105.64	111.54	107.44	108.97	110.77	110.51
Critical temperature/°C	60–70					
Crack temperature/°C	105–110					
Dongshan Coal Samples						
CO mutation temperature/°C	66.67	68.21	66.15	62.82	68.72	71.54
C_2H_4 appear temperature/°C	138.46	137.95	137.18	133.08	139.74	134.10
Critical temperature/°C	65–70					
Crack temperature/°C	130–140					
Baijiao Coal Samples						
CO mutation temperature/°C	67.95	76.9	82.8	90.7	87.9	72.56
C_2H_4 appear temperature/°C	–	–	–	–	–	–
Critical temperature/°C	70–90					
Crack temperature/°C	–					

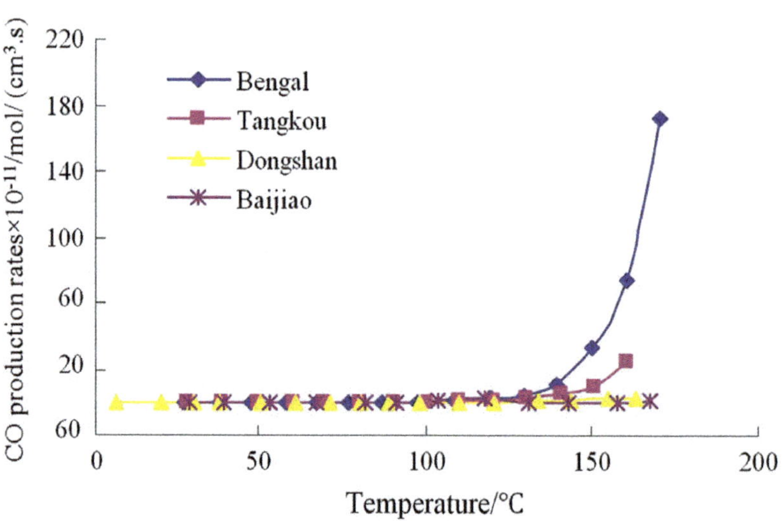

Figure 9.1.3. Graph of CO production rate versus temperature. Figure by Tao Wang, 2016.

It can be seen from Figures 9.1.2 and 9.1.3 that the CO production rate increases as the coal temperature rises. The higher the temperature reached, the higher the rate followed. At the same temperature, the CO production rate decreases with the ranks of coal increases.

As shown in Table 9.1.1, the macro behavior of the spontaneous combustion of coal is the change of gas concentration at a certain temperature. Different microchemical structure in different ranks of coal leads to a diversity of macrogenerated gas types and amount. The critical characteristic temperatures for the spontaneous combustion of coal can be determined by the changes of CO concentration. When the coal temperature exceeds the critical temperature and the cracking temperature, the bridge bond, alkyl side chain, and oxygen-contained functional groups in the coal molecule begin to break and participate in the oxidation and release in the form of gases. Thus another gas, C_2H_4, appears, and the curve of CO concentration is sharply increased. Therefore, the characteristic temperatures for the spontaneous combustion of coal can be determined by the temperature at which C_2H_4 appears and is supplemented by the drastic change of CO concentration.

Oxidation and Spontaneous Combustion Gases

The **coal-oxygen reaction** is mainly divided into three parts: physical adsorption on the surface of coal, chemisorption on the surface of coal, and the chemical reaction in the part active structure of chemisorption. Infer the chemisorption of coal and oxygen, chemical reactions, and thermal behaviors that follow.

Due to the complexity of coal structure, the coal–oxygen composite chemical reaction is not a single structure and one-step reaction, but a multistep hybrid reaction of a variety of surface-active structures. According to the reaction process and products, it is inferred that the chemical reaction can be divided into three steps; each step is accompanied by the change of bond energy. In coal surface molecules, functional groups react with oxygen. The chemisorption that occurs is

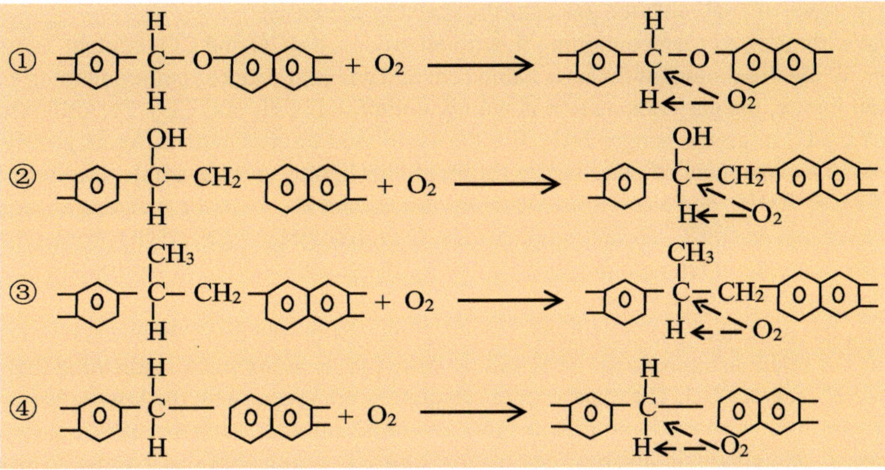

In these reactions,  and 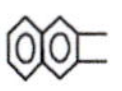 represent two different structural units and O_2 indicates oxygen adsorption.

A part of **α-carbon atoms** (atoms that are directly linked to functional groups) and oxygen, which formed chemic adsorption, have a larger molecular energy than the activation energy required for the chemical bond rearrangement. This will lead to:

(1) The mechanism of CO and CO_2 generation. Taking the active group –O–CH_3 as an example, the reaction process is divided into three steps, the chemical adsorption process, the dehydration reaction, and the gas production process, as follows:

(a) The chemical adsorption process:

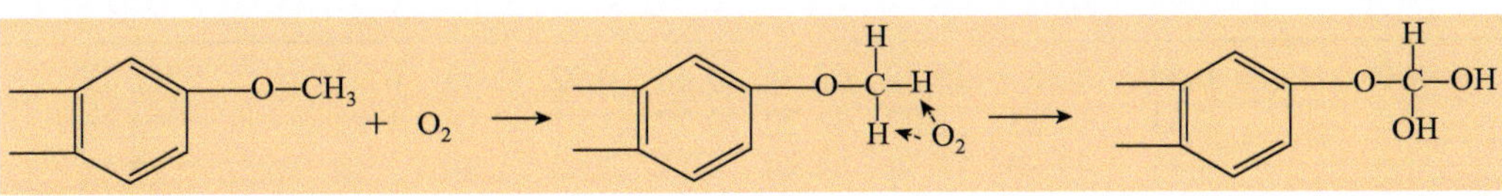

(b) The dehydration reaction:

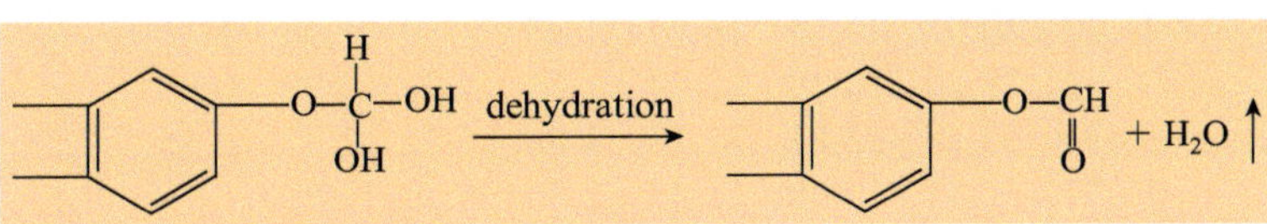

(c) The gas production process:

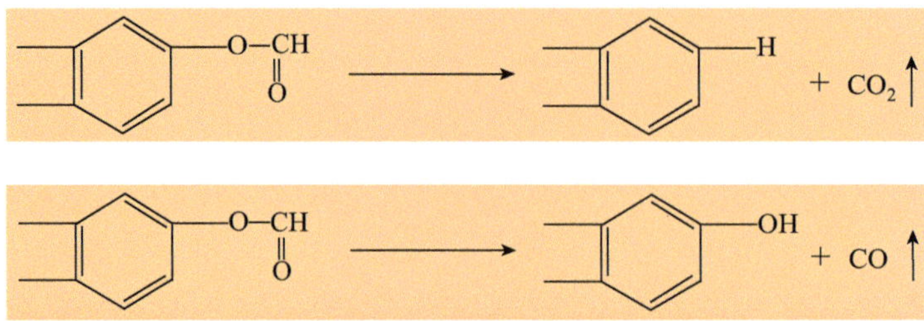

(2) The mechanism of hydrocarbon gas generation.
  (a) With the increase of coal rank, the gas calorific quantity increased.
  For lignite with more oxygen-containing functional groups, the CO and CO_2 content as pyrolysis products are higher. While the content of the long flame coal and gas coal with oxygen-containing functional groups is less than lignite, the gas composition is mainly composed of H_2 and CH_4. For different coal ranks, the content of C_2–C_4 in gases changes little. For the same kind of coal, with different pyrolysis temperature, the content of a constituent of gases is also different. For lignite, as an example, the constituent of gases is CO and CO_2 at 400°C, and also a small amount of CH_4 and H_2. With the increase of temperature, the H_2 content increased. At 600°C, H_2 increased rapidly. On the contrary, CO and CO_2 content decreased gradually with the increase of temperature. Before 550°C, the content of CH_4 is increased with the temperature rise but decreased after 550° (Table 9.1.2).
  (b) The production of methane is associated with –CH and its production rate increased when –CH content increased. The relationship between –CH and tar from pyrolysis are linear, indicating that formation of tar relates to the crack of –CH. The experimental results are consistent with the low-temperature tar constituent; there are little aromatic hydrocarbons in low-temperature tar, which mostly consists of unsaturated hydrocarbons, straight-chain alkanes, and cycloalkanes. It is assumed that tar and hydrocarbon gases of small molecules, such as methane, are generated by the –CH cracking.
  (c) From the chemical reaction mechanism of coal pyrolysis, citing multiple equation models. Coal pyrolysis is regarded as a parallel reaction of coal functional groups. Linking the precipitation process of a single pyrolysis product with the radical reaction of free radicals put forward a new pyrolysis model and deduced the formation mechanism of different pyrolysis products.

Table 9.1.2
Characteristics of the pyrolysis of coal-fire gases.

Coal Samples	Temperature (°C)	Composition of Coal Gases (Volume %)									Calorific Value (MJ/m^3)	Density (kg/cm^3)
		H_2	CH_2	CO	CO_2	C_2H_4	C_2H_6	C_3H_6	C_3H_8	CH_4		
Lignite	400	0.44	5.38	25.57	67.38	0.48	0.92	1.23	0.92	–	11.09	1.66
	450	0.81	10.68	21.07	61.22	0.77	1.85	2.18	1.41	–	12.26	1.64
	500	3.43	22.50	23.58	33.89	3.13	5.83	1.80	1.86	0.56	22.08	1.34
	550	9.21	25.72	24.51	29.03	2.67	5.24	1.96	1.28	0.37	23.06	1.25
	600	19.04	24.07	22.56	26.04	2.33	3.58	1.18	0.74	0.43	20.75	1.11
	650	34.15	22.95	12.20	22.27	2.68	3.41	1.12	0.79	0.42	18.90	0.98
Long-flame coal	550	9.92	33.26	24.31	21.60	2.10	5.38	1.46	1.97	–	23.73	1.12
	600	19.43	33.56	20.16	18.09	1.73	4.53	1.13	1.37	–	22.75	0.98
	650	31.27	30.02	13.64	16.85	1.83	4.19	1.00	1.21	–	22.54	0.90
	700	37.32	28.68	14.19	12.19	1.65	4.01	0.98	0.98	–	21.09	0.82
Gas coal	550	8.67	32.18	23.81	21.13	1.80	6.22	2.37	3.45	0.27	26.07	1.16
	600	19.37	37.51	14.40	15.43	2.84	5.30	1.89	3.10	0.16	26.98	0.96
	650	29.56	38.20	11.72	11.01	1.52	3.70	1.59	1.63	0.07	25.65	0.82
	700	36.92	30.76	12.75	10.98	2.96	4.50	0.94	0.19	–	23.59	0.81

The mechanism of CH_4 generation is:

① 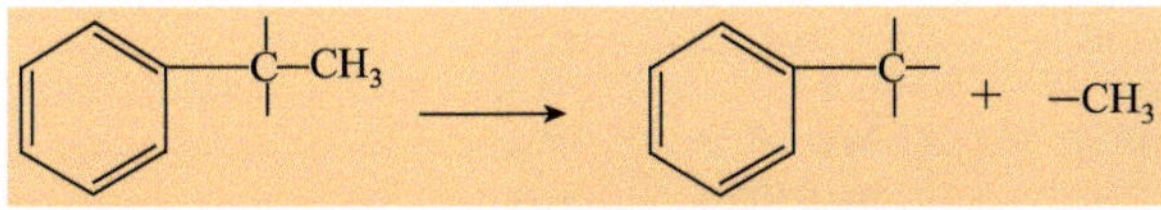

② 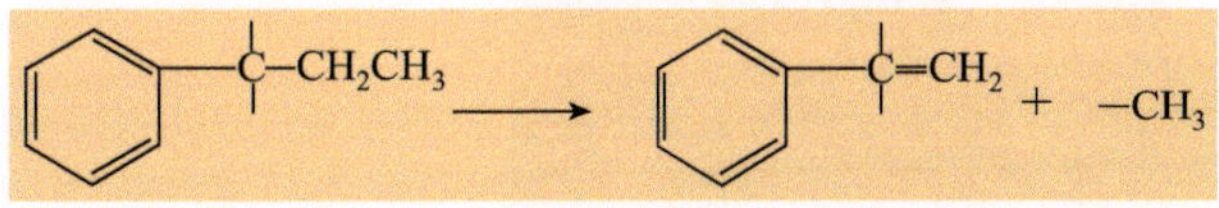

③ 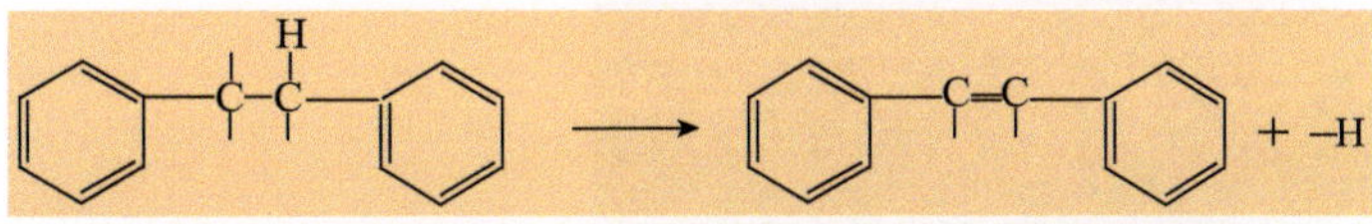

④ 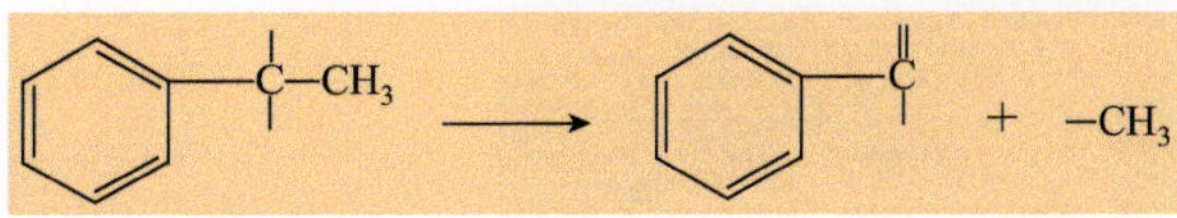

The mechanism of C_2H_6 generation is:

① 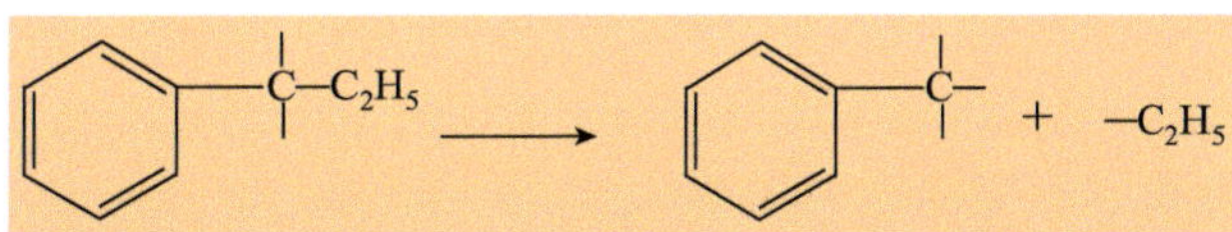

② 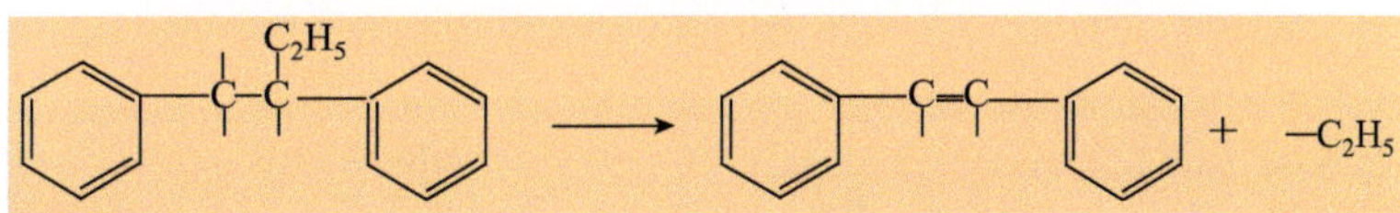

③ 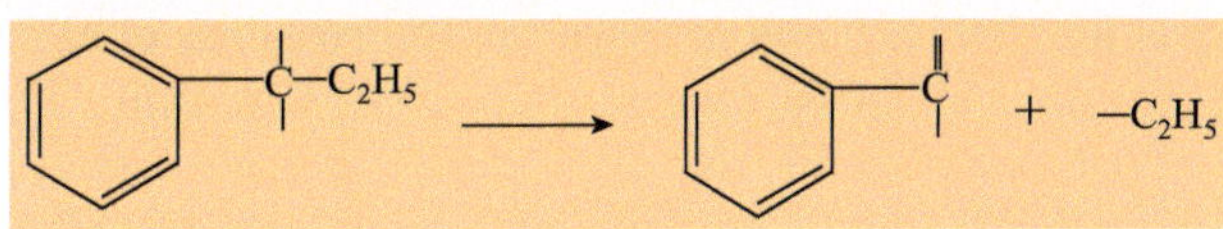

The mechanism of C_2H_4 generation is:

① 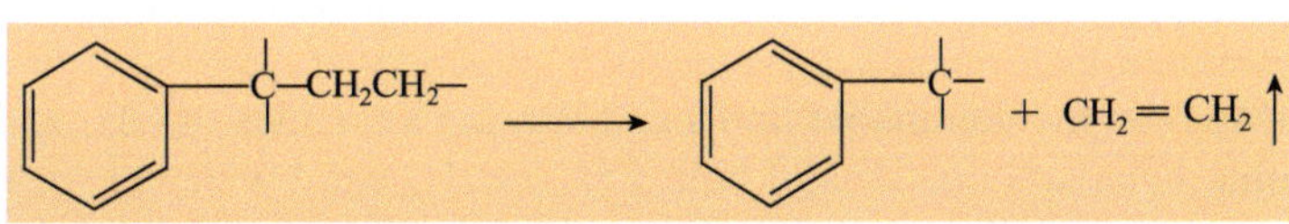

Discussion

The types and sequence of gases produced by different ranks of coal in a low-temperature oxidation process are basically the same, but the difference in the microstructure of the coal and the occurrence and amount of the same gaseous product are different. CO is one of the important gas products in low-temperature oxidation, and the amount increases when the temperature rises exponentially. In the low-temperature oxidation stage, nonaromatic structure side chains and bridge bonds in coal molecules are attacked by oxygen, which shows the results of the generation of CO and CO_2 as macrobehavior. In a high-temperature oxidation stage, the structure of the coal's molecule is splitting and cracking, and the macroperformance is the generation of alkanes and olefins, such as CH_4 and C_2H_6.

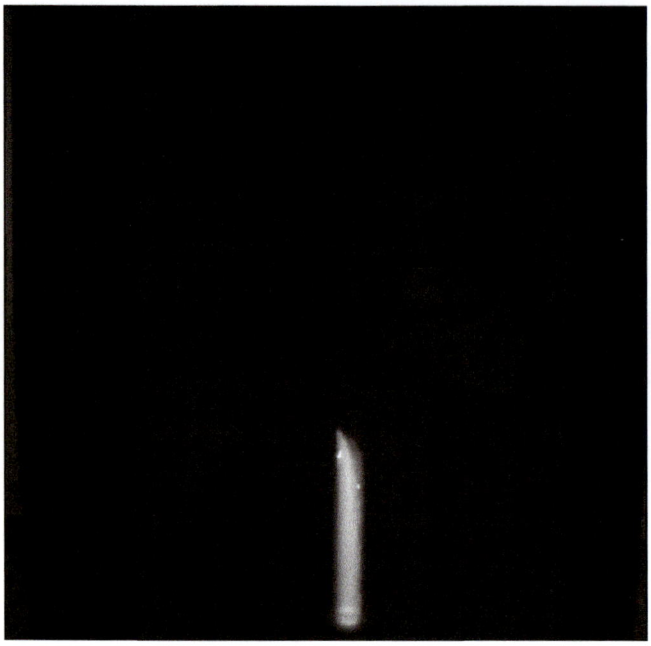

Photo by Fangming Cheng, 2015.

9.2 Gas Effects on Methane-Air Flammable Limits

A methane-air flame propagates upward during an experiment in a silver-colored cylinder. The cylinder is 6 cm wide and 140 cm high.

Introduction

In the mining industry, gas explosions are another leading disaster besides coal fires, causing tremendous casualties and loss of property. In past decades, many scholars, concerned with the safety of CH_4, undertook abundant experiments focusing on **explosion characteristics** of CH_4–air mixtures, like blast pressure evolution (Zhang et al., 2014; Zhang and Ma, 2015; Huang et al., 2012), flame propagation (Fairweather et al., 1999; Huang et al., 2012; Liu et al., 2015), explosion venting (Bao et al., 2016; Ferrara et al., 2006; Guo et al., 2016; Kasmani et al., 2013; Shao et al., 2016), and the influences of obstacles on CH_4 explosion (Fairweather et al., 1999; Kindracki et al., 2007; Na'inna et al., 2015; Oh et al., 2001). All of these studies revealed the features and hazardous effects of the CH_4–air explosion in various situations.

For CH_4–air **flammable limits**, Hustad and Sønju (1988) studied the lower flammable limits of CH_4, C_4H_{10}, H_2, CO, and their mixtures at elevated temperatures up to 450°C in an upward propagation chamber. Results showed that flammable limits of each gas presented linear decrease with the increasing temperature. Shebeko et al. (1995) conducted a group of systematic experiments to evaluate the influences of elevated initial temperature (up to 250°C) and pressure (up to 4.0 Mpa) on the flammable limits and burning velocities of CH_4 and H_2 mixtures. The experiment results indicated that flammable regions of both CH_4 and H_2 were widened with temperature increase, but narrowed with pressure rise. Cashdollar et al. (2000) tested the factors of chamber size, ignition source kind, and initial turbulence condition on the flammability of CH_4, C_3H_8, and H_2. Takahashi et al. (2003) measured the effect of chamber shape and size on flammability limits of CH_4 and C_3H_8. They reported that the quenching effect of the cylinder vessel wall, heated gas accumulation at vessel ceiling, unburnt gas heating, and self-heating of the incipient flame were determined factors affecting the flammable limits. Furthermore, a number of mathematic models to predicate flammable limits under various conditions have been proposed in past years (Cheng and Yang, 2011; Hansel et al., 1992; Liaw and Chen, 2016; Molnarne and Schröder, 2011; Shebeko et al., 2002; Takahashi et al., 2015; Zhao et al., 2010).

At a coal fire, gases generated from the coal oxidation and pyrolysis at low temperature will blend with CH_4, forming an explosive mixture. From this point, attention to the effects of other combustible gases on CH_4 explosion should also be paid; at present, there have been little research published about this issue. Deng et al. (2015) reported a series of experiments on the explosion of CO–CH_4–air mixtures that were conducted in a 20-L spherical tank and also simulated by FLACS. The results demonstrated that both the upper and lower flammable limits of CH_4 were decreased by CO. Also, the existence of CO intensified the explosion of CH_4 under stoichiometric concentration but posed a slightly negative effect on that of CH_4 above stoichiometric concentration. Salzano et al. (2012)

performed tests of H_2–CH_4 explosion characteristics in stoichiometric air using a 5 L cylindrical chamber. The laminar burning velocities of the mixtures were calculated by a **Le Chatelier's Principle**-like formula. Experimental results indicated the existence of hydrogen does not regularly influence stoichiometric CH_4 explosion, whereas Li et al. (2012) found that H_2 would increase the maximum explosion pressure and pressure rise rate of H_2–CH_4–air mixtures in oxygen-rich condition. Similar effects were observed by Ma et al. (2014). It is worth noting that Shen et al. (2016) measured the explosion parameters of CH_4–C_3H_6 mixtures in a 20 L spherical vessel. Results manifested C_3H_6 could promote the explosion severity and enhance explosion risk of the CH_4–air system.

Experiment Conditions

A photo of the flammable test set-up is illustrated in Figure 9.2.1 and the experimental results are listed in Table 9.2.1. The upward propagation cylindrical equipment was designed according to ISO 10156 (2010). It mainly consists of an **explosion cylinder**, a gas mixing system, an electric ignition system, and a pressure gauge. The explosion cylinder, with an inner diameter of 60 mm and length of 1400 mm, was made of quartz. A relief valve with the size of 25 mm was located at the bottom of the cylinder. The compound gas used in experiments with an accuracy of 0.1% was mixed by the partial pressure method. A pair of electric igniting probes was mounted 100 mm above the bottom of the cylinder.

Figure 9.2.1. Flammable limits test device. Photo by Fangming Cheng, 2015.

Table 9.2.1
Experimental tests scenarios.

Gas	Volume Fraction	Test
C_2H_6	0%–2.0%	Flammable limits
C_2H_4	0%–2.0%	Flammable limits
CO	0%–2.0%	Flammable limits
H_2	0%–2.0%	Flammable limits

All the experiments were conducted in the room temperature and ambient pressure with relative humidity range of 52%–73%. The gases of C_2H_6, C_2H_4, CO, H_2, and CH_4 with purity above 99.9% were produced by Hengli. Co., Ltd. (Xi'an, China).

Flammable Limits

The flammable limits and the explosive risk degree of C_2H_6, C_2H_4, CO, H_2, and CH_4 were listed in Table 9.2.2. The explosion risk value F, indicating the relative explosion risk of flammable gases, was defined as (Han, 1995):

$$F = \left[(L_1 \cdot L_2)^{1/2} - L_1\right] / (L_1 \cdot L_2)^{1/2} \tag{9.2.1}$$

where L_1 is the lower flammable limit and L_2 is the upper flammable limit. Eq. (9.2.1) shows that the F value was distributed between 0 and 1. A larger F value indicates a greater explosive risk for one type of gas. The explosive risk of gases used in this work is in turn that $CH_4 < C_2H_6 < CO < C_2H_4 < H_2$.

Table 9.2.2
The UFL, LFL, and corresponding explosion risk value of the tested gases.

	Experimental Result		
Gas	UFL (%)	LFL (%)	Explosion Risk Value F
CH_4	16.15	5.15	0.44
C_2H_6	12.95	2.65	0.55
C_2H_4	32.05	2.55	0.73
CO	73.35	11.05	0.61
H_2	74.65	3.85	0.77

The upper flammable limit (UFL) and lower flammable limit (LFL) of CH_4 in five situations are depicted in Figure 9.2.2. As showed in the picture, both the UFL and LFL declined with the increment of C_2H_6, C_2H_4, CO, H_2, and their compounds. The decline range differed due to the differences in gas species. But the flammable range presented an increasing trend. Generally, the effect of single gas on the LFL of CH_4 was greater than the UFL. For example, with the **volume fraction** of C_2H_6, C_2H_4, CO, and H_2 increased from 0% to 2% respectively, the LFL declined by 68%, 79.6%, 25.4%, and 64%, correspondingly, while the UFL diminished merely by 15.5%, 5.6%, 3.7%, and 5.6%. Meanwhile, the flammable range of CH_4 extended to 12%, 14.2%, 11.7%, 13.4% separately from 11%. In general, CO manifested a smaller influence on the LFL of CH_4 compared to other gases. The compound exerted a bigger effect on decreasing the UFL of CH_4 than each of its components. And what's more, the flammable range of CH_4 distributed between 11% and 12% with the volume fraction of the compound increased to 2% from 0%.

Overall, in the same volume fraction, the gas with a lower LFL itself led to a greater reduction on the LFL of CH_4, in accordance with Le Chatelier's Principle (Mashuga and Crowl, 2000). It's worth noting that in most industrial areas CH_4 was thought to be explosive beyond the volume fraction of 5%, which value was set as an alarm point, whereas in this work, the LFL of CH_4 was proved to be diminished greatly when some other flammable gases existed. Thus it is necessary to lower the monitor alarm threshold of CH_4 and to take appropriate measures in case of the potential explosion hazard when facing a similar situation.

Explosive Risk Value F

Figure 9.2.3 showed the explosive risk value F, which was calculated according to Eq. (9.2.1) and Figure 9.2.2. The explosive value F of CH_4 increased due to the increment of C_2H_6, C_2H_4, CO, and H_2. For these four kinds of flammable gases, C_2H_4 imposed the greatest effect on the explosive risk of CH_4. As the volume fraction of C_2H_4 rose to

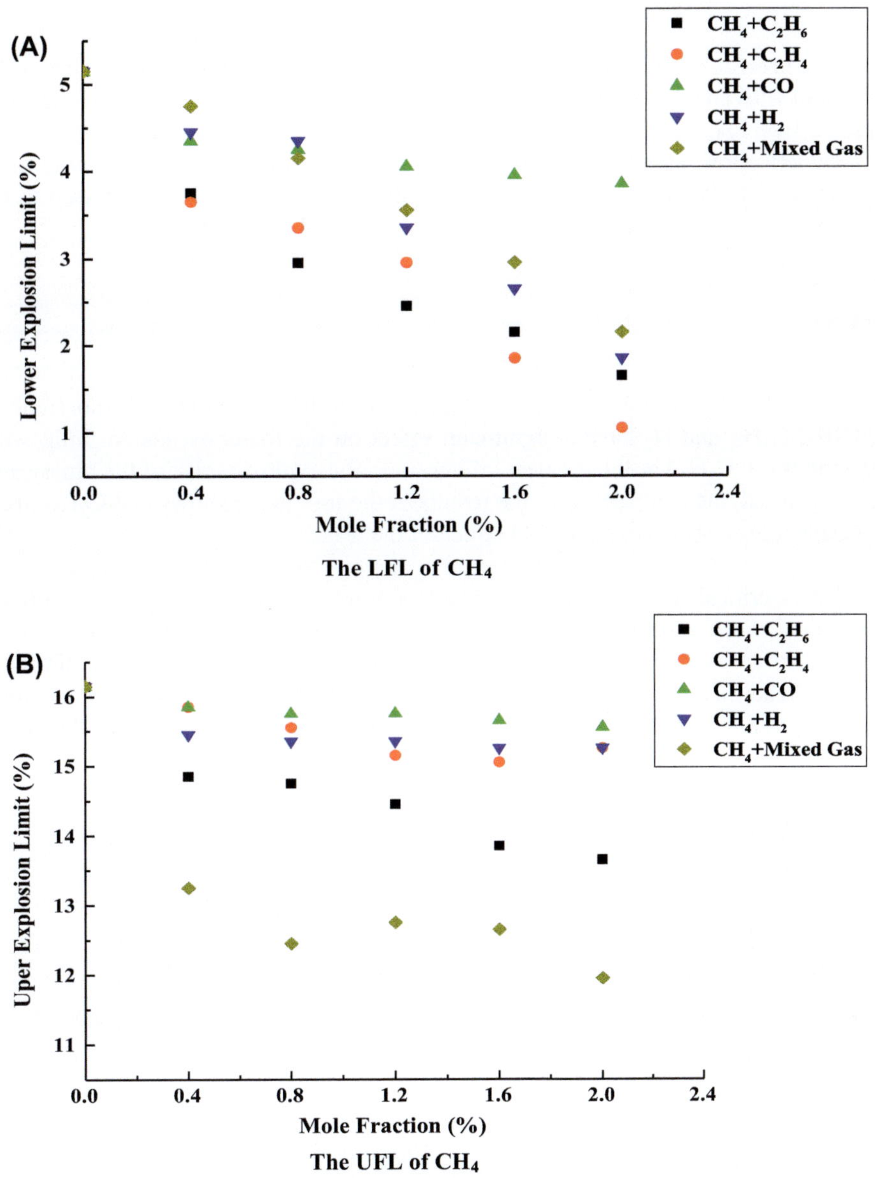

Figure 9.2.2. The LFL and UFL of CH_4 versus volume fractions of added gases. Figures by Tao Wang, 2016.

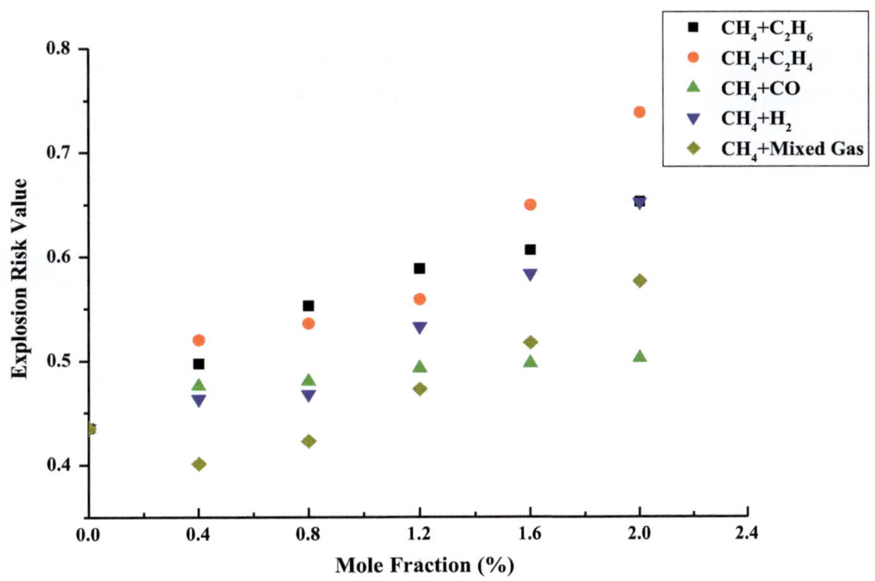

Figure 9.2.3. The explosion risk value F of CH_4 versus volume fractions of added gases. Figure by Tao Wang, 2016.

2%, the F value of CH_4 grew to 0.74 from 0.44 by an increment of 68%. Then the influences of C_2H_6 and H_2 followed that of C_2H_4, both with an increment of 48%. CO exerted little influence on the UFL and LFL of CH_4 leading to the F-value increase by 11%, which was the least. It was notable that the existence of the compound made the CH_4 explosive risk value F show a tendency to diminish first and then rise. For example, with the volume fraction of the compound increased to 0.4%, the F decreased to 0.40 from 0.44; it finally increased to 0.58, with a total increment of 32% when the volume fraction of the compound was enhanced to 2%.

Discussion

Organic flammable gases have more influence on the explosion limit of methane than inorganic flammable gases do. Furthermore, C_2H_6, C_2H_4, and H_2 have a significant effect on the lower explosion limit, while C_2H_6 greatly affects upper explosion limit. C_2H_4 has the greatest influence on the explosive risk of the binary gas mixture, while CO has less effect. In general, the comparison of the results of the four gases shows that C_2H_4 affects the explosion limit most, C_2H_6 affects it less, H_2 is next, and CO affects it the least.

C_2H_6, C_2H_4, and H_2 have a critical effect on the lower explosion limit of the binary gas mixture containing CH_4, which leads to the decline of the lower explosion limit of the gas mixture, although CO results in an elevated lower explosion limit. Compared with other gases, the C_2H_6 gas has an unsteady effect on the upper explosion limit of the gas mixture. There are almost same effects on the explosion limit of a gas mixture by either adding C_2H_4 or adding C_2H_4; influences of either adding C_2H_6 or adding CO are nearly equal, but C_2H_4 and H_2 have more influence than C_2H_6 and CO do.

Acknowledgments

The authors thank the financial support from the National Natural Science Foundation of China (Grant No. 51674193), the Key Science and Technology Program of Shaanxi Province (Grant No. 2015SF280), and the Doctoral Fund of Xi'an University of Science and Technology (Grant No. 2013QDJ048).

Important Terms

α-carbon atoms
coal oxidation
coal-oxygen reaction
coal pyrolysis
coal temperature
concentration of oxygen
explosion characteristics
explosion cylinder
flammable limits
functional groups
infrared spectra (IR)
Le Chatelier's Principle
oil bath temperature-programmed system
organic structure of coal
particle size
spontaneous combustion
volume fraction

References

Bao, Q., Fang, Q., Zhang, Y., Chen, L., Yang, S., Li, Z., 2016. Effects of gas concentration and venting pressure on over pressure transients during vented explosion of methane-air mixtures. Fuel 175, 40–48.
Cashdollar, K.L., Zlochower, I.A., Green, G.M., Thomas, R.A., Hertzberg, M., 2000. Flammability of methane, propane, and hydrogen gases. J. Loss Prev. Process. Ind. 13, 327–340.
Cheng, J.W., Yang, S.Q., 2011. Improved coward explosive triangle for determining explosibility of mixture gas. Process Saf. Environ. Prot. 89, 89–94.
Deng, J., Cheng, F., Song, Y., Luo, Z., Zhang, Y., 2015. Experimental and simulation studies on the influence of carbon monoxide on explosion characteristics of methane. J. Loss Prev. Process. Ind. 36, 45–53.

Fairweather, M., Hargrave, G.K., Ibrahim, S.S., Walker, D.G., 1999. Studies of premixed flame propagation in explosion tubes. Combust. Flame 116, 504–518.

Ferrara, G., Di Benedetto, A., Salzano, E., Russo, G., 2006. CFD analysis of gas explosions vented through relief pipes. J. Hazard Mater. 137, 654–665.

Guo, J., Wang, C., Li, Q., Chen, D., 2016. Effect of the vent burst pressure on explosion venting of rich methane-air mixtures in a cylindrical vessel. J. Loss Prev. Process. Ind. 40, 82–88.

Han, M., 1995. New method to evaluate gas explosion risk value. Low. Temp. Spec. Gas 2, 57–60. (In Chinese).

Hansel, J.G., Mitchell, J.W., Klotz, H.C., 1992. Predicting and controlling flammability of multiple fuel and multiple inert mixtures. Plant Oper. Prog. 11, 213–217.

Huang, Z., Liu, Z., Chen, S., Zhang, Y., Zhang, Y., 2012. Numerical simulation and study on the transmission law of flame and pressure wave of pipeline gas explosion. Saf. Sci. 50, 806–810.

Hustad, J.E., Sonju, O.K., 1988. Experimental studies of lower flammability limits of gases and mixtures of gases at elevated temperatures. Combust. Flame 71, 283–294.

ISO 10156, 2010. Gases and Gas Mixtures-Determination of Fire Potential and Oxidizing Ability for the Selection of Cylinder Valve Outlets (ISO 10156:2010), pp. 1–34.

Kasmani, R.M., Andrews, G.E., Phylaktou, H.N., 2013. Experimental study on vented gas explosion in a cylindrical vessel with a vent duct. Process Saf. Environ. Prot. 91, 245–252.

Kindracki, J., Kobiera, A., Rarata, G., Wolanski, P., 2007. Influence of ignition position and obstacles on explosion development in methane-air mixture in closed vessels. J. Loss Prev. Process. Ind. 20, 551–561.

Li, Q., Lin, B., Dai, H., Zhao, S., 2012. Explosion characteristics of H_2/CH_4/air and CH_4/coal dust/air mixtures. Powder Technol. 229, 222–228.

Liaw, H.J., Chen, K.Y., 2016. A model for predicting temperature effect on flammability limits. Fuel 178, 179–187.

Liu, Q., Zhang, Y., Niu, F., Li, L., 2015. Study on the flame propagation and gas explosion in propane/air mixtures. Fuel 140, 677–684.

Ma, Q., Zhang, Q., Chen, J., Huang, Y., Shi, Y., 2014. Effects of hydrogen on combustion characteristics of methane in air. Int. J. Hydrogen Energy 39, 11291–11298.

Mashuga, C.V., Crowl, D.A., 2000. Derivation of Le Chatelier's mixing rule for flammable limits. Process Saf. Prog. 19, 112–117.

Molnarne, M., Schroder, V., 2011. Comparison of calculated data for the flammability and the oxidation potential according to ISO 10156 with experimentally determined values. J. Loss Prev. Process. Ind. 24, 900–907.

Na'inna, A.M., Somuano, G.B., Phylaktou, H.N., Andrews, G.E., 2015. Flame acceleration in tube explosions with up to three flat-bar obstacles with variable obstacle separation distance. J. Loss Prev. Process. Ind. 38, 119–124.

Oh, K.H., Kim, H., Kim, J.B., Lee, S.E., 2001. A study on the obstacle-induced variation of the gas explosion characteristics. J. Loss Prev. Process. Ind. 14, 597–602.

Salzano, E., Cammarota, F., Di Benedetto, A., Di Sarli, V., 2012. Explosion behavior of hydrogen-methane/air mixtures. J. Loss Prev. Process. Ind. 25, 443–447.

Shao, H., Jiang, S., Wu, Z., Zhang, W., Wang, K., 2016. Application and effect of negative pressure chambers on pipeline explosion venting. J. Loss Prev. Process. Ind. 41, 8–17.

Shebeko, Y.N., Tsarichenko, S.G., Korolchenko, A.Y., Trunev, A.V., Navzenya, V.Y., Papkov, S.N., Zaitzev, A.A., 1995. Burning velocities and flammability limits of gaseous mixtures at elevated temperatures and pressures. Combust. Flame 102, 427–437.

Shebeko, Y.N., Fan, W., Bolodian, I.A., Navzenya, V.Y., 2002. An analytical evaluation of flammability limits of gaseous mixtures of combustible-oxidizer-diluent. Fire Saf. J. 37, 549–568.

Shen, X., Zhang, B., Zhang, X., Xiu, G., 2016. Explosion characteristics of methane-ethane mixtures in air. J. Loss Prev. Process. Ind. https://doi.org/10.1016/j.jlp.2016.11.012.

Takahashi, A., Urano, Y., Tokuhashi, K., Kondo, S., 2003. Effect of vessel size and shape on experimental flammability limits of gases. J. Hazard Mater. 105, 27–37.

Takahashi, S., Ebisawa, T., Bhattacharjee, S., Ihara, T., 2015. Simplified model for predicting difference between flammability limits of a thin material in normal gravity and microgravity environments. Proc. Combust. Inst. 35, 2535–2543.

Zhang, Q., Ma, Q.J., 2015. Dynamic pressure induced by a methane-air explosion in a coal mine. Process Saf. Environ. Prot. 93, 233–239.

Zhang, B., Zhao, W., Wang, W., Zhang, X., 2014. Pressure characteristics and dynamic response of coal mine refuge chamber with underground gas explosion. J. Loss Prev. Process. Ind. 30, 37–46.

Zhao, F., Rogers, W.J., Mannan, M.S., 2010. Calculated flame temperature (CFT) modeling of fuel mixture lower flammability limits. J. Hazard Mater. 174, 416–423.

CHAPTER 10

Determination of the Characteristics of Coal-Spontaneous Combustion and the Danger Zone

Coal fire mountain at the bituminous Daquanhu Coal Mine in Urumqi, Xinjiang, China. The mountain is about 5 m high, and there is coal inside that extends below its base. Coal once at the surface ignited spontaneously about 30 years ago. All attempts to extinguish the fire failed, including pouring a slurry of water and sand onto the burning-coal gangue. The sand poured onto the gangue now forms the outside shell of the mountain. The road is about 3 m wide. *Photo by Jingyu Zhao, 2017.*

CHAPTER CONTENTS

Coal and Peat Fires: A Global Perspective
Edited by Glenn B. Stracher

https://doi.org/10.1016/B978-0-12-849885-9.00010-X

Photo by Jingyu Zhao, 2017.

10.1 Coal-Spontaneous Combustion Analysis

Jun Deng
Jingyu Zhao
Yang Xiao
Yanni Zhang
Caiping Wang

Coal-spontaneous combustion monitoring system in the School of Safety Science and Technology, Xian University of Science and Technology, Xi'an, China.

Introduction

Coal-spontaneous combustion is a phenomenon in which coal catches fire without igniting. It is a complicated process involving a physical and chemical reaction, which is caused by heat energy release from a coal–oxygen reaction. When the heat energy storage reaches a certain level, the process of coal oxidation is promoted. With the increase in the oxidation degree, coal-spontaneous combustion eventually occurs. The frequent occurrence of coal-spontaneous combustion is one of the critical factors that restrict the development of the coal industry in China. Coal-spontaneous combustion causes severe harm to the economy, the ecological environment, and social development.

The most evident phenomenon of coal-spontaneous combustion is the release of gasses and heat energy. During coal-spontaneous combustion, pungent gasses are diffused, which makes it difficult for people to breathe, and can lead to suffocation and even death. The gasses released from coal-spontaneous combustion can be divided into carbon–oxygen gasses and carbon–hydrogen gasses, such as carbon monoxide (CO), carbon dioxide (CO_2), methane (CH_4), ethylene (C_2H_4), and ethane (C_2H_6), and are used as an indicator to characterize the degree of coal-spontaneous combustion (as CO and C_2H_4). With the release of gaseous products during spontaneous combustion, the oxygen content continuously declines. The characteristic parameter that reveals the changes in oxygen content is known as the oxygen consumption rate.

Moreover, heat energy is accumulated when gasses are generated. The ability of coal to produce heat energy spontaneously reflects its intrinsic nature, which is called heat energy release intensity. The heat energy release of coal is extremely complex due to the diversity of the coal–oxygen reaction, mineral ingredients, and coal–rock components.

When the conditions of residual coal thickness, particle size, and oxygen are met, coal-spontaneous combustion occurs in the mined-out area. Therefore, it is crucial to probe the characteristic parameters such as the generation of gasses, heat energy release intensity, and the limit parameters (limiting thickness of residual coal, limiting oxygen concentration, and limiting air leakage intensity), which could help determine the danger zone of the mined-out area.

Coal Samples

Coal samples were obtained from the Hebi coal mine, Henan Province, the People's Republic of China. The coal rank is meager-lean coal, which has a low content of volatile matter, moisture, and ash. The samples were not subjected to water injection, spraying, or other treatments and were, instead, sealed and packed with multilayer plastic and nylon bags prior to being transported to the laboratory. The coal quality test is provided in Table 10.1.1.

Table 10.1.1
Proximate and elemental analysis of coal.

Sample	M_{ad}/%	A_{ad}/%	V_{ad}/%	C_{daf}/%	H_{daf}/%	O_{daf}/%	N_{daf}/%	S_{daf}/%
Hebi	1.64	12.71	14.23	89.14	4.65	4.37	1.58	0.26

ad and daf designate air-dried and ash-free basis, respectively. In the coal, M_{ad}% is the percent of moisture, A_{ad}/% is the percent of ash, V_{ad}/% is the percent of volatiles, C_{daf}/% is the percent of carbon, H_{daf}/% is the percent of hydrogen, O_{daf}/% is the percent of oxygen, N_{daf}/% is the percent of nitrogen content, and S_{daf}/% is the percent of sulfur.

Experimental Apparatus

A low loading capacity is not conducive to coal-spontaneous combustion. To more closely resemble the actual situation and achieve the spontaneous combustion of coal, a 15-t spontaneous combustion experimental furnace was constructed to simulate the process. Other systems were also constructed to conduct the experiments, including an air supply system, a gas sample collection and analysis system, as well as an automatic temperature measurement and control system. The structural diagram of the spontaneous combustion test furnace is provided in Figure 10.1.1. Thermistor probes were installed inside the furnace to measure coal temperature, and tubes were arranged in the central axis of the furnace for the collection of gas samples. The air was obtained using an oil-free air compressor, and the humidity and temperature of the airflow into the coal body were controlled by a humidity and temperature control box. Gas samples were collected manually or automatically by using beam tubes. A gas chromatograph was employed to analyze the gas composition and concentration, and the results were stored. A remote data acquisition and control module was used for the automatic temperature measurement and control system. The outer-layer temperature and volume of the air supply were monitored in real time by the monitoring system. The parameters of the experimental furnace are listed as follows:

1. The inner diameter of the furnace was 2800 mm, and the outer diameter was 3800 mm. The height of coal loading was 2200 mm, and the largest coal loading capacity was 15 t.
2. Buffer layers of 100 mm were added to the top and bottom of the furnace to achieve a uniform airflow over the coal body.
3. Four coal loading ports with a diameter of 600 mm and one smoke vent with a diameter of 50 mm at the top of the furnace were added.
4. The spacing between the layers of the monitoring point was 200 mm, and the radial spacings were 200, 400, and 600 mm. Twelve layers were added where the monitoring point was arranged. Moreover, each of the two buffer layers set at the top and bottom of the furnace had one monitoring point to monitor the ambient temperature in the furnace.

The parameters and accuracy of automatic detection and control devices are demonstrated as follows:

1. Temperature accuracy: 0.5°C
2. Temperature detection time: 1–30 min
3. Temperature difference controlled by water: 1–5°C
4. Gas detection accuracy: 1 ppm
5. Maximum air supply volume of the oil-free air compressor: $2.5 \times 1010\,mm^3/h$

The coal sample was placed in the furnace after being pulverized into pieces. Coal samples of mixed particle size were used. The range of particle sizes and the experimental conditions are provided in Tables 10.1.2 and 10.1.3.

The initial temperature of the experiment was 24.6°C. When the temperature was below 50°C, the heat energy release intensity was low, and the volume of air supply affected the heat accumulation. A higher volume of air supply accelerates heat loss. Therefore, a lower volume of air supply, $1.5 \times 10^8\,mm^3/h$, was used to adjust the coal-spontaneous combustion experiment. With a slight change in the volume of air supply, the heating rate decreased sharply, indicating that the coal had burned. The air supply was stopped to cool down the temperature.

When the experiment was conducted, the air was compressed by an oil-free air compressor and then passed into a humidity and temperature control box. A flow meter controlled the volume of air that was passed into the humidity

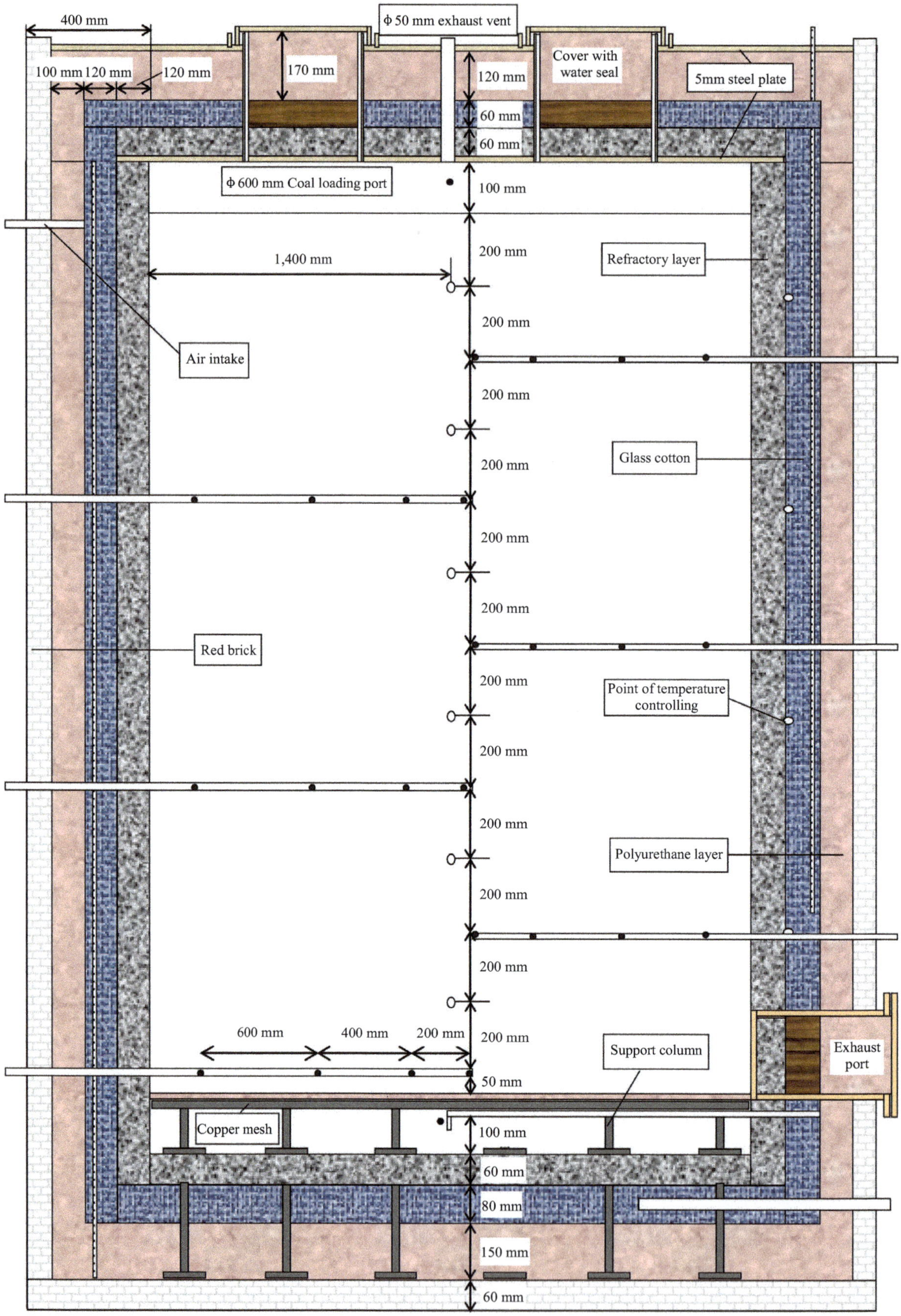

Figure 10.1.1. Structure of experimental furnace for studying coal-spontaneous combustion. Figure by Deng et al. (2015).

Table 10.1.2
Coal particle sizes.*

Particle Size	+10 mm	+7 mm, −10 mm	+5 mm, −7 mm	+3 mm, −5 mm	+3 mm, −0.9 mm	−0.9 mm
Frequency (%)	9.78	11.08	8.26	14.68	18.38	37.81

*The sign "+" indicates that the coal sample did not pass through the screen, and "–" indicates that the sample passed through the screen.

and temperature control box. After the humidity and temperature were controlled, the air was passed over the coal body at the bottom of the furnace through the water layer, thus ensuring the same temperature for air and water. The air was then vented using an exhaust at the top of the furnace. A control machine monitored the temperature of the coal body, water layer, temperature control box, and the ambient temperature in the furnace in real time at the monitoring points. By adjusting the power of the electrical heat tube, the temperatures of water and air were monitored to ensure that the coal body was not heated. The maximum temperature difference between the coal body and water layer was not allowed to exceed 2°C, thus reducing the heat energy loss. Moreover, the heat energy removal by airflow was calculated using the difference in the ambient temperature at the top and bottom of the furnace. The gas samples at different measuring points were collected to analyze the gas composition and concentration through gas chromatography. The varying trends of temperature and gasses during the entire process were studied. The flow program is schematized in Figure 10.1.2.

Table 10.1.3
Experimental conditions.

Reference Particle Size d_{50} (mm)	3.32
Height (mm)	1420
Mass (g)	1.51×10^7
Volume (mm^3)	1.61×10^{10}
Lump coal density (g/mm^3)	1.40×10^{-3}
Bulk density (g/mm^3)	0.99×10^{-3}
Porosity	0.24
Air volume (mm^3/h)	1.5×10^8
Initial temperature (°C)	24.6

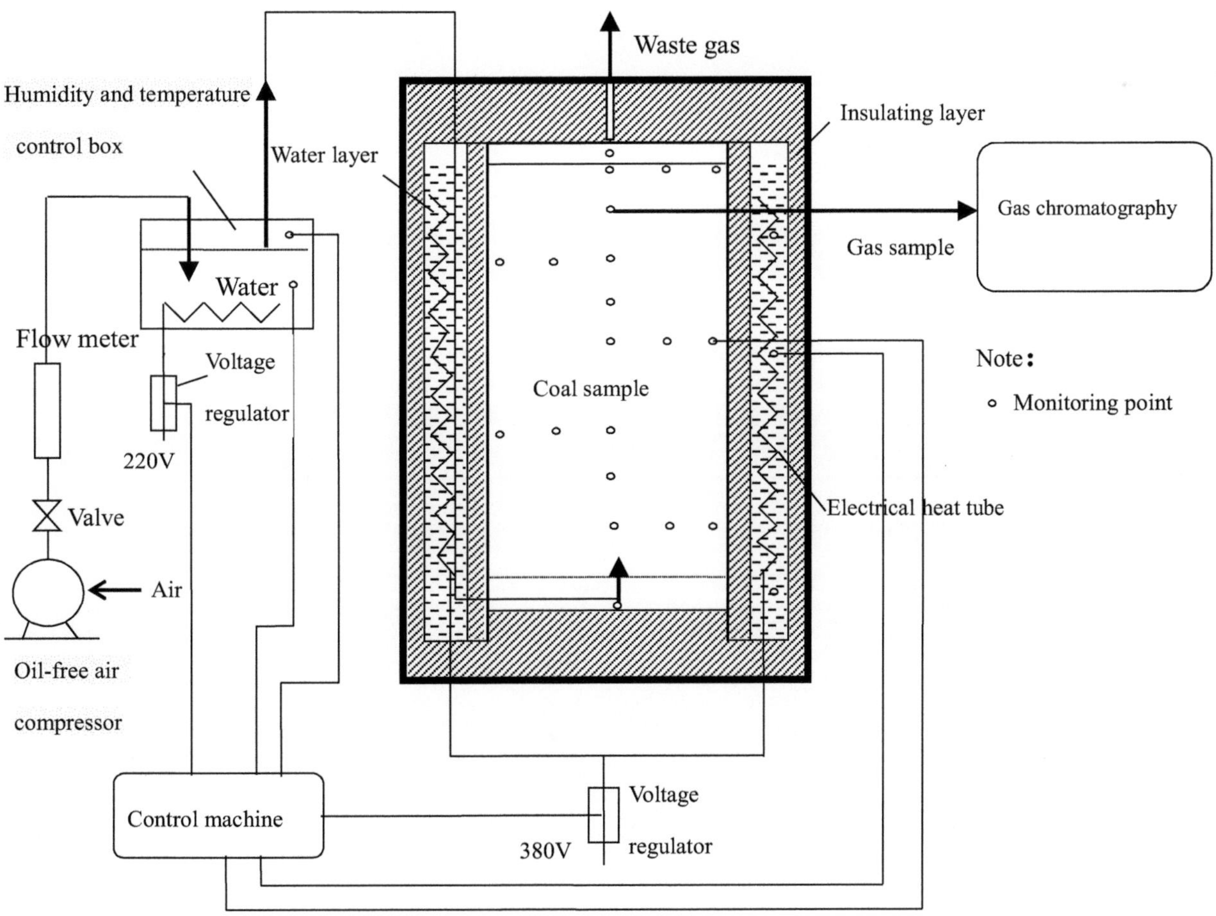

Figure 10.1.2. Flow program experiment for the spontaneous combustion of coal. Figure by Deng et al. (2015).

Gas Concentration

Typically, CO, CO_2, alkanes, alkenes, and alkynes are released during coal oxidation. Among them, CO is the most pronounced emission gas during oxidation. Figure 10.1.3 presents the relationship between the CO concentration and temperature of the coal samples. CO, which appears at 40°C, can be used as an indicator gas, because it is not absorbed in the surface (the initial temperature of the experiment was 24.6°C). With an increase in temperature, the CO concentration gradually increases to 120°C. After that, the CO concentration rapidly increases and continues to increase until the end of the oxidation.

CO_2 and CH_4 gasses are the maximum emissions during the oxidation process. Coal adsorbs CO_2 and CH_4 gasses from the air on its surface before oxidation. Figure 10.1.4 reveals that the CO_2 content increased with an increase in temperature in a linear fashion initially, which indicates that the generation of CO_2 is the complete reaction, and the concentration of CO_2 is higher than that of CO. CH_4 concentration was stable before 50°C and increased to its peak value at 140°C (Figure 10.1.5).

The contents of C_2H_6 and C_2H_4 gasses were very low (Figures 10.1.6 and 10.1.7); the highest concentrations were 1200 ppm and 30 ppm, respectively, and the concentrations increased with an increase in temperature. Moreover, C_2H_4 gas emerged at 115°C, which evidenced that the crack reaction in the coal molecule commenced.

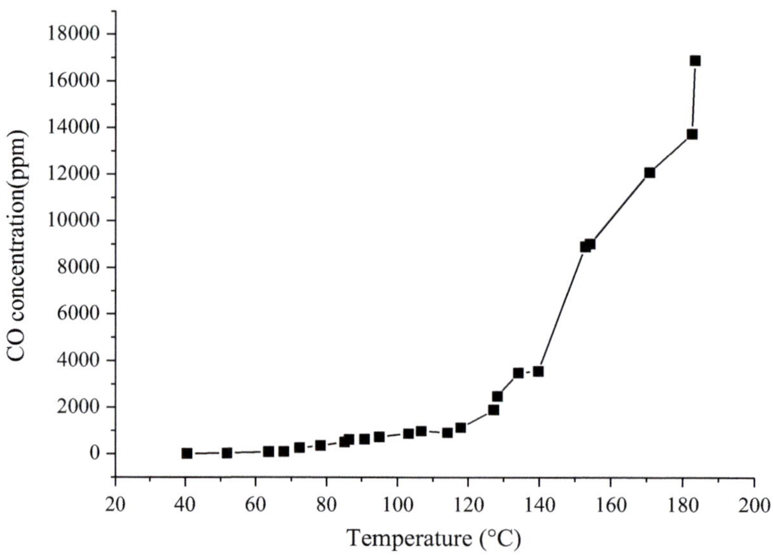

Figure 10.1.3. CO concentration versus temperature. Figure by Jingyu Zhao, 2017.

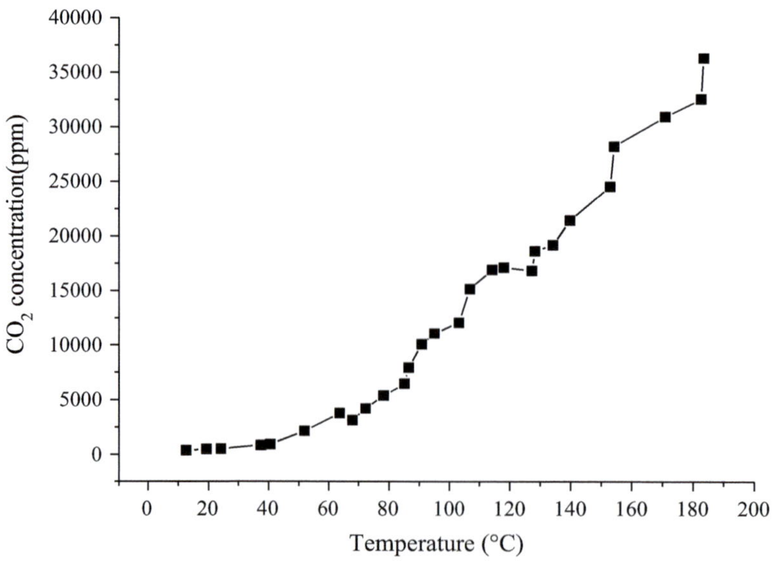

Figure 10.1.4. CO_2 concentration versus temperature. Figure by Jingyu Zhao, 2017.

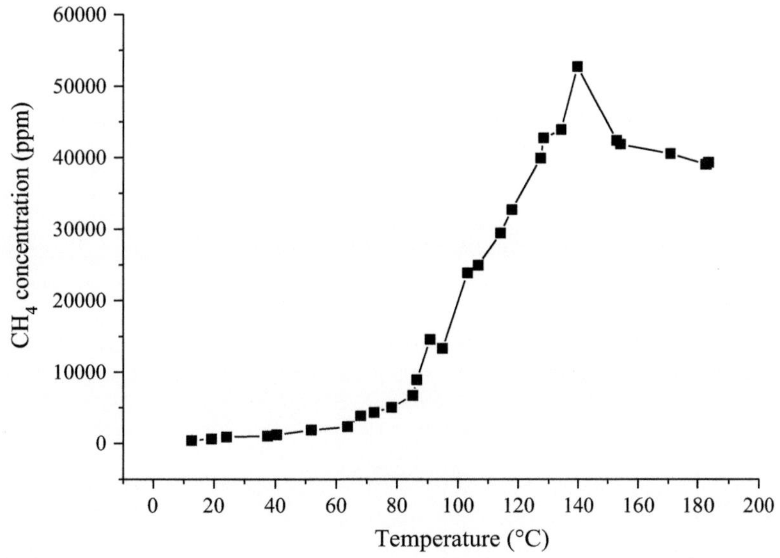

Figure 10.1.5. CH_4 concentration versus temperature. Figure by Jingyu Zhao, 2017.

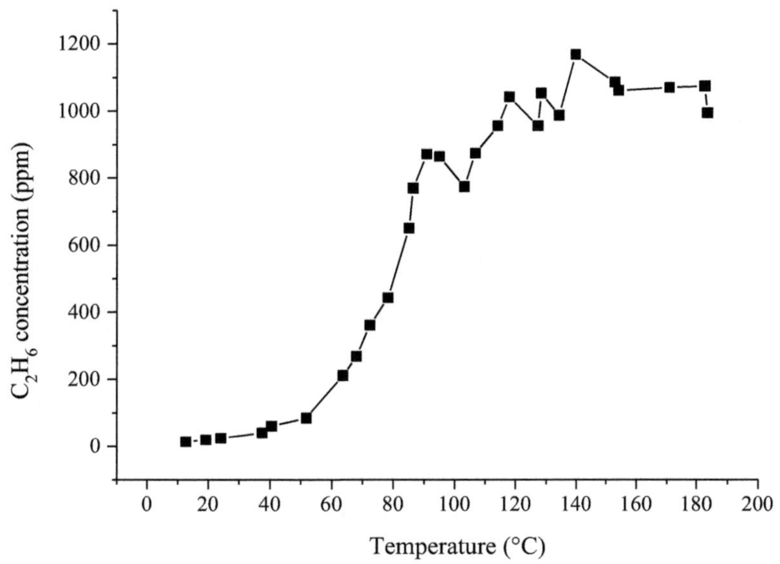

Figure 10.1.6. C_2H_6 concentration versus temperature. Figure by Jingyu Zhao, 2017.

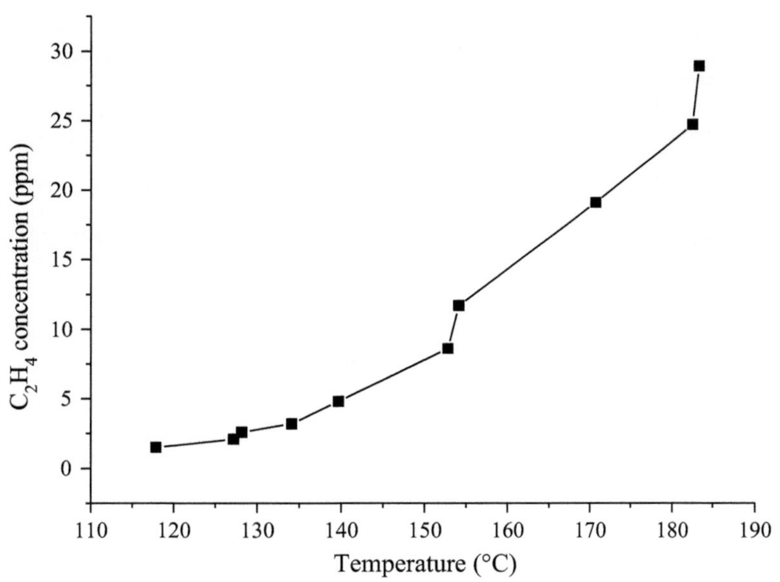

Figure 10.1.7. C_2H_4 concentration versus temperature. Figure by Jingyu Zhao, 2017.

Oxygen Consumption Rate

The oxygen consumption rate is a crucial characteristic parameter of spontaneous combustion and can be obtained as follows:

$$dC / d\tau = -V(T), d\tau = dx / \bar{Q} \tag{10.1.1}$$

where oxygen concentration at a certain point is denoted as C (ppm); V(T) is the oxygen consumption rate (mol/m^3·s); τ is the reaction time (s); and $\bar{Q}$ is the average airflow of a plane with coal in the reactor (cm^3/s). Hence,

$$\bar{Q}dC / dx = -V(T) \tag{10.1.2}$$

The oxygen consumption rate is proportional to the oxygen concentration; thus, Eq. (10.1.2) can be written as follows:

$$dC = -V_0(T) \times \frac{C}{\bar{Q}C_0} dx \tag{10.1.3}$$

where C_0 is the standard oxygen concentration (21%); $V_0(T)$ is the oxygen consumption rate of coal in air (mol/(cm^3·s)).

Integrating the two sides of Eq. (10.1.3) leads to Eq. (10.1.4):

$$V_0(T) = \frac{QC_0}{S(z_2 - z_1)} \ln \frac{C_1}{C_2} \tag{10.1.4}$$

where z is the distance to the air intake in centimeters and S is the oven cross-sectional area in square centimeters.

Figure 10.1.8 provides the oxygen consumption rate of the coal sample. It is evinced that the oxygen consumption rate increases with an increase in temperature and an increase in gas emission. The higher the gas emission, the higher the oxygen consumed. At the beginning of oxidation, the adsorbed gas release leads to slight changes in

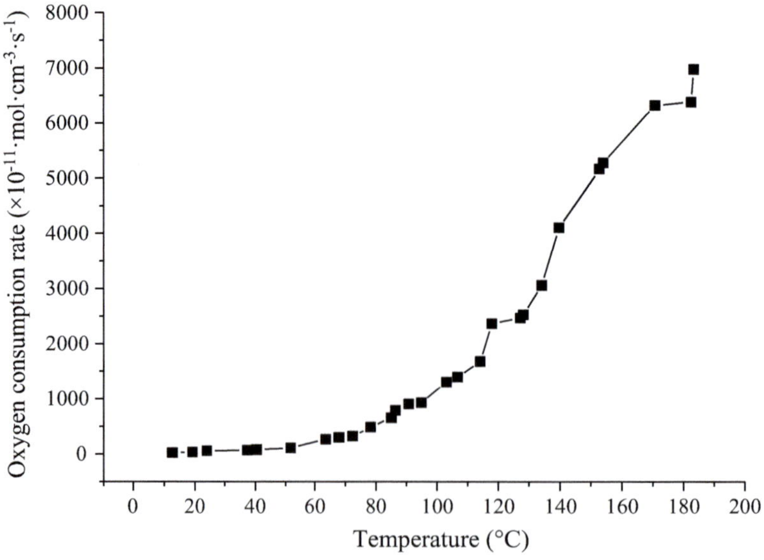

Figure 10.1.8. Oxygen consumption versus temperature. Figure by Jingyu Zhao, 2017.

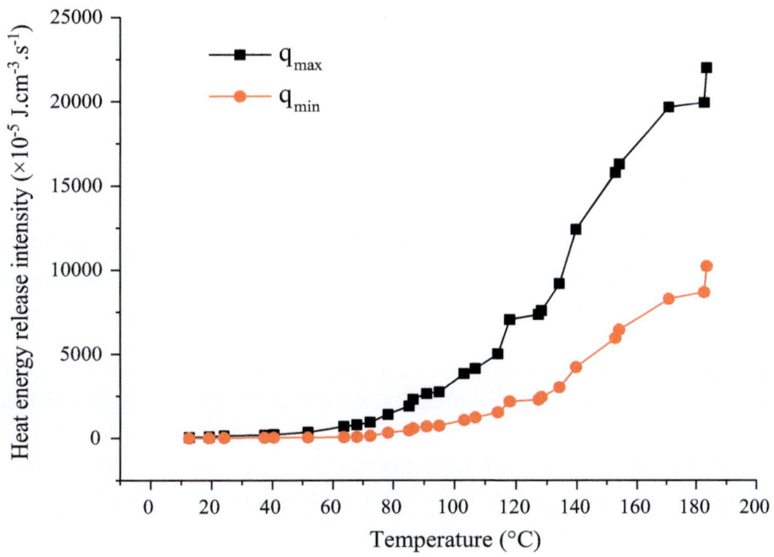

Figure 10.1.9. Heat energy release intensity versus temperature. Figure by Jingyu Zhao, 2017.

oxygen concentration. After the temperature reaches 50°C, the oxygen consumption rate sharply increases because the compound in the coal reacts with oxygen.

Heat Energy Release

The heat energy release intensity reflects the exothermicity of coal. The heat energy is derived from the coal–oxygen reaction. Xu, J.C. (2001) derived the equation of heat energy release and proved the actual heat energy release intensity is between the maximum and minimum value, as displayed in Eqs. (10.1.5) and (10.1.6):

$$q_{max}(T) = \frac{V^0_{co}(T)}{V^0_{co}(T) + V^0_{co_2}(T)} V_0(T)\Delta H^{co} + \frac{V^0_{co_2}(T)}{V^0_{co}(T) + V^0_{co_2}(T)} V_0(T)\Delta H^{co_2} \quad (10.1.5)$$

$$q_{min}(T) = \Delta H^r[V_0(T) - V^0_{co}(T) - V^0_{co_2}(T)] + \Delta H^{co} V^0_{co}(T) + \Delta H^{co_2} V^0_{co_2}(T) \quad (10.1.6)$$

where $q_{max}(T)$ is the maximum heat energy release rate (J/(cm^3 s)) and $q_{min}(T)$ is the minimum heat energy release rate (J/(cm^3 s)); V^0_{co} (T) is the standard CO production rate (mol/(cm^3 s)); $V^0_{CO_2}$ (T) is the standard CO_2 production rate (mol/(cm^3 s)); ΔH^{co_2} and ΔH^{co} are the reaction heat energy rates of CO_2 and CO (kJ/mol), respectively; and $\Delta H^r = 58.8$ kJ/mol.

The minimum and maximum heat energy release intensity rates exhibit the same trend with an increase in temperature (Figure 10.1.9). The intensity increases gradually before 50°C and begins to release much more heat energy after this point. The maximum heat energy release intensity is twice as high as the minimum heat energy release intensity.

Limiting Parameters: Mined-Out Area

The limiting parameters include limiting the thickness of residual coal, limiting the oxygen concentration, and limiting the air leakage intensity. Based on the energy conservation principle, coal-spontaneous combustion occurs when the heat release in the mined-out areas is larger than the heat dissipation from the roof, floor, and airflow. Eq. (10.1.7) is expressed as follows:

$$\mathrm{div}[\lambda_e \mathrm{grad}(T_m)] + q_0(T) - \mathrm{div}(n\rho_g c_g \overline{U} T_m) \geq 0 \tag{10.1.7}$$

where ϱ_g, c_g represents the air density (g/cm^3) and heat capacity (J/(g °C)) of the working face; $q_0(T)$ is the heat release intensity (J/(cm^3 s)); λ_e is the thermal conductivity of the residual coal (J/(s cm)); and $\overline{U}$ is the average air leakage velocity in the mined-out areas (cm/s).

Thus there are three conditions for coal-spontaneous combustion in the mined-out area. First, the spontaneous combustion of residual coal in the mined-out area must have sufficient thickness so that the heat energy released by the oxidation of coal can be accumulated. Second, it must have sufficient oxygen concentration to ensure the residual coal produces heat energy and increases the temperature. Third, it must exhibit a suitable air leakage intensity to avoid the heat energy being consumed by the airflow. Coal-spontaneous combustion can occur only if these three conditions are satisfied synchronously.

Considering the mined-out area as an infinite plane, heat energy dissipation is conducted through rocks, which exhibit a slight and unidimensional air leakage intensity. Coal temperature is well-distributed; thus Eq. (10.1.7) can be simplified as Eq. (10.1.8):

$$-\rho_g \cdot c_g \cdot \overline{Q} \cdot \frac{\partial T_m}{\partial x} + \lambda_e \frac{\partial^2 T_m}{\partial z^2} + q_0(\overline{T}_m) > 0 \tag{10.1.8}$$

where

$$\frac{\partial^2 T_m}{\partial z^2} \approx -\frac{2\times(T_m - T_y)}{(h/2)^2}, \quad \frac{\partial T_m}{\partial x} \approx \frac{T_m - T_g}{x}, \quad \overline{T}_m = \frac{1}{2}(T_m + T_y) \tag{10.1.9}$$

By substituting Eq. (10.1.9) into Eq. (10.1.8), we obtain the prerequisite of coal heating,

$$h > \sqrt{\frac{8\times(T_m - T_y)\cdot\lambda_e}{q_0(\overline{T}_m) - \rho_g \cdot c_g \cdot \overline{Q} \cdot (T_m - T_g)/x}} = h_{min} \tag{10.1.10}$$

where h is the thickness of the residual coal (cm); h_{min} is the limiting thickness of residual coal (cm); T_m is the highest temperature point in coal (°C); $\overline{T}_m$ is the average temperature in coal (°C); T_y is the temperature of rock (20°C); T_g is the temperature of airflow (°C); $q_0(\overline{T}_m)$ is the heat energy release intensity under the temperature of $\overline{T}_m$(J/(cm^3 s)); $\overline{Q}$ is air leakage intensity (cm/s); and x is the distance between the mined-out area and the working face (cm).

From Eq. (10.1.10), it is known that h_{min} relates to the temperature, heat energy release intensity, and distance between the mined-out area and working face. Moreover, coal will not spontaneously combust when $h \leq h_{min}$.

As for the Hebi coal sample, T_m was 25°C, T_g was the temperature of airflow (°C), T_y was 20°C, λ_e was 0.92×10^{-3}(J/s cm), ϱ_g was 1.29×10^{-3} (g/cm^3), and C_g was 1.01 (J/(g °C)); the value of $q_0(\overline{T}_m)$ is provided in Table 10.1.4.

From the data in Table 10.1.4 and Eq. (10.1.10), the limiting thickness of residual coal can be calculated as presented in Table 10.1.5. Based on this calculation, the temperature of residual coal is 50°C when the limiting thickness reaches the highest level. Spontaneous combustion is not possible when the thickness of the residual coal is lower than 0.72 m.

The heat energy release intensity is proportional to the oxygen concentration as presented in Eq. (10.1.11):

$$q(T_m) = \frac{C_{O_2}}{C^0_{O_2}} q_0(T_m) \tag{10.1.11}$$

Table 10.1.4
Heat energy release intensity at different temperatures in the mined-out area.

Temperature (°C)	30	40	50	60	70	80	90	100	110	120	130	140
$q_0(\overline{T}_m) \times 10^{-5}$ J/cm³·s	1.89	2.81	4.18	6.22	9.24	13.74	20.42	30.35	45.12	67.07	99.69	148.19

Table 10.1.5
Minimum thickness of residual coal (m).

$\overline{Q}$ (cm/s)* Temperature (°C)	0.005	0.01	0.015	0.02	0.03	0.04	0.05	0.06	0.08	0.2
30	0.47	0.49	0.51	0.53	0.57	0.61	0.66	0.70	0.80	1.51
40	0.68	0.72	0.76	0.80	0.88	0.98	1.08	1.18	1.40	2.91
50	0.72	0.76	0.81	0.86	0.95	1.06	1.17	1.29	1.54	3.25
60	0.70	0.74	0.78	0.82	0.92	1.02	1.12	1.23	1.47	3.06
70	0.65	0.68	0.72	0.76	0.83	0.92	1.01	1.10	1.30	2.67
80	0.59	0.61	0.64	0.67	0.74	1.80	0.88	0.95	1.11	2.22
90	0.52	0.54	0.56	0.59	0.64	0.69	0.75	0.80	0.93	1.79
100	0.16	0.17	0.49	0.51	0.55	0.59	0.63	0.67	0.76	1.42
110	0.40	0.41	0.42	0.44	0.46	0.49	0.52	0.56	0.63	1.11
120	0.34	0.35	0.36	0.37	0.39	0.41	0.44	0.46	0.51	0.86
130	0.29	0.30	0.31	0.32	0.33	0.35	0.36	0.38	0.42	0.67
140	0.25	0.26	0.26	0.27	0.28	0.29	0.30	0.31	0.34	0.52
150	0.21	0.22	0.22	0.23	0.23	0.24	0.25	0.26	0.28	0.40
160	0.18	0.18	0.19	0.19	0.20	0.20	0.21	0.21	0.23	0.32

*The air leakage into the coal mine, used to calculate the minimum thickness of residual coal.

where $C^0_{O_2}$ represents standard air oxygen concentration, 21% by volume; C_{O_2} is the actual oxygen concentration in %; $q(T_m)$ denotes heat energy release intensity with the standard oxygen concentration, J/(s cm³); $q_0(T_m)$ denotes heat energy release intensity with the actual oxygen concentration, J/(s cm³). Substituting Eq. (10.1.11) into Eqs. (10.1.7) and (10.1.12) can be expressed as

$$C_{O_2} > \frac{-C^0_{O_2}}{q_0(\overline{T}_m)}[\mathrm{div}(\lambda_e \mathrm{grad} T_m) - \mathrm{div}(\rho_g c_g \overline{Q} T_m)] = C_{min} \qquad (10.1.12)$$

C_{min} is the limiting oxygen concentration. The heat produced by the oxidation of coal is less than consumption when $C \le C_{min}$, and it is difficult for the coal temperature to increase at this condition.

The mined-out area can be simplified as dimensional heat energy transfer in an infinite plate; thus, Eq. (10.1.12) can be modified into Eq. (10.1.13):

$$C_{min} = \frac{C^0_{O_2}}{q_0(\overline{T}_m)}[\frac{8 \times \lambda_e (T_m - T_y)}{h^2} + \rho_g c_g \overline{Q} \cdot \frac{2 \times (T_m - T_g)}{x}] \qquad (10.1.13)$$

It can be inferred from Eq. (10.1.13) that C_{min} changes with the coal temperature, air leakage intensity, working distance, and thickness of residual coal. According to Eq. (10.1.13), the limiting oxygen concentration at different thicknesses and temperatures of residual coal can be obtained when the airflow enthalpy is disregarded (Table 10.1.6). In this experiment, the limiting oxygen concentration of Hebi was 20.27%.

Table 10.1.6
Parameters used with Eq. (10.1.13) to calculate the limiting oxygen concentration.

		Thickness of Residual coal (m)							
Temperature (°C)	$q_0(T)\times 10^5$* (J/(s·cm³))	0.7 m	0.8 m	1.0 m	1.2 m	1.4 m	1.6 m	1.8 m	2.0 m
30	1.8	8.96	6.88	4.42	3.09	2.28	1.75	1.39	1.13
40	2.81	18.08	13.88	8.92	6.23	4.60	3.54	2.81	2.88
50	4.18	20.27	15.56	10.01	6.98	5.15	3.96	3.15	2.56
60	6.22	19.09	14.56	9.42	6.58	4.85	3.73	2.96	2.41
70	9.24	16.52	12.67	8.15	5.69	4.20	3.23	2.56	2.09
80	13.74	13.58	10.42	6.70	4.68	3.45	2.66	2.11	1.72
90	20.42	10.80	8.29	5.33	3.72	2.74	2.11	1.68	1.36
100	30.35	8.38	6.43	4.14	2.89	2.13	1.64	1.30	1.06
110	45.12	6.39	4.90	3.15	2.20	1.62	1.25	0.99	0.81
120	67.07	4.80	3.69	2.37	1.65	1.22	0.94	0.75	0.61
130	99.69	3.57	2.74	1.76	1.23	0.91	0.70	0.55	0.45
140	148.19	2.63	2.02	1.30	0.91	0.67	0.51	0.41	0.33
150	220.28	1.92	1.48	0.95	0.66	0.49	0.38	0.30	0.24
160	327.43	1.40	1.07	0.69	0.48	0.36	0.27	0.22	0.08

*$qo(T)\times 10^5$ is the intensity of heat energy release. It is used to calculate the oxygen concentration and the air leakage intensity.

Under the condition of a sufficient oxygen supply as well as a residual coal thickness higher than h_{min}, unidimensional airflow, and constant flow rate, Eq. (10.1.7) can be simplified as follows:

$$\lambda_e \frac{\partial^2 T_m}{\partial z^2} + q_0(\overline{T}_m) - n\rho_g c_g U \frac{\partial T_m}{\partial z} > 0$$

$$U < \frac{q_0(\overline{T}_m)}{n\rho_g c_g \frac{\partial T_m}{\partial x}} + \frac{\lambda_e \frac{\partial^2 T_m}{\partial z^2}}{n\rho_g c_g \frac{\partial T_m}{\partial z}} = U_{max} \tag{10.1.14}$$

and $\overline{Q}_{max} = Un$. Thus,

$$\overline{Q} < \frac{q_0(\overline{T}_m)}{\rho_g c_g \frac{\partial T_m}{\partial x}} + \frac{\lambda_m \frac{\partial^2 T_m}{\partial z^2}}{\rho_g c_g \frac{\partial T_m}{\partial z}} = \overline{Q}_{max} \tag{10.1.15}$$

$$\overline{Q}_{max} = x \cdot \frac{q_0(\overline{T}_m) - 8\lambda_e(T_m - T_y)/h^2}{\rho_g \cdot c_g \cdot (T_m - T_g)} \tag{10.1.16}$$

The coal temperature does not increase when $\overline{Q} \geq \overline{Q}_{max}$, and $\overline{Q}_{max}$ denotes the limiting air leakage intensity, (cm/s). The greater the intensity of the air leakage is, the more heat energy dissipation. $\overline{Q}_{max}$ is related to the coal temperature, working distance, and thickness of residual coal. According to Eq. (10.1.16), the limiting air leakage intensity can be calculated at different thicknesses and temperatures of residual coal (Table 10.1.7). The limiting air leakage intensity of the Hebi coal was 0.22 cm/s.

Table 10.1.7
Limiting air leakage intensity at different thickness and residual-coal temperatures.

Temperature (°C)	$q_0(T) \times 10^5$* (J/(s·cm³))	Thickness of Residual Coal (m)							
		0.7m	0.8m	1.0m	1.2m	1.4m	1.6m	1.8m	2.0m
30.00	1.89	5.92	7.92	11.58	15	18.28	21.47	24.59	27.68
40.00	2.81	0.77	2.03	4.22	6.17	7.97	9.69	11.34	12.96
50.00	4.18	0.22	1.40	3.44	5.23	6.88	8.44	9.94	11.39
60.00	6.22	0.50	1.72	3.84	5.71	7.44	9.07	10.65	12.19
70.00	9.24	1.25	2.57	4.91	6.99	8.93	10.78	12.58	14.33
80.00	13.74	2.44	3.94	6.62	9.05	11.33	13.52	15.66	17.75
90.00	20.42	4.18	5.93	9.10	12.02	14.80	17.49	20.12	22.71
100.00	30.35	6.62	8.72	12.59	16.21	19.69	23.08	26.40	29.69
110.00	45.12	10.03	12.61	17.45	22.04	26.49	30.85	35.15	39.41
120.00	67.07	14.75	18.01	24.20	30.14	35.94	41.65	47.30	52.91
130.00	99.69	21.32	25.51	33.58	41.40	49.08	56.66	64.19	71.67
140.00	148.19	30.47	35.97	46.65	57.08	67.37	77.57	87.71	97.81
150.00	220.28	43.24	50.57	64.90	78.98	92.92	106.77	120.56	134.31
160.00	327.43	61.13	71.01	90.46	109.65	128.70	147.66	166.56	185.42

*$q_0(T) \times 10^5$ is the intensity of heat energy release. It is used to calculate the oxygen concentration and the air leakage intensity.

Discussion

Based on the experiment of coal-spontaneous combustion, the spontaneous combustion processes of coal from 24.6°C to 190°C were simulated. The experimental scenarios simulated the conditions of an actual environment, such as coal thermal storage and air supply, where abundant burnt loose coals result in spontaneous combustion.

The characteristic parameters of gas concentrations, oxygen consumption rate, and heat energy release intensity increased with an increase in temperature. Gas concentrations increased gradually at the early stage, but drastically later. The oxygen consumption rate and heat energy release intensity increased with the increase in gas concentration.

When the coal temperature was 50°C, the limiting thickness of residual coal reached its peak value at different air leakage intensities. The limiting air leakage intensity increased with the increase in thickness. By contrast, the limiting oxygen concentration decreased with the increase in the thickness of residual coal.

Photo by Yang Xiao, 2015.

10.2 The Danger Zone: Determining Coal-Spontaneous Combustion

A surface crack caused by spontaneous combustion in an open-pit coal mine at the Daquanhu mine in Urumqi, Xinjiang, China. The fire started about 30 years ago and all attempts to extinguish it have failed. The width of the crack is about 10 cm.

The 3006-Working Face in the Hebi Coal Mine

The Hebi coal mine is located in the north of Henan Province, in the People's Republic of China (Figure 10.2.1). The 3006-working face of the Hebi coal mine is located in the middle of the mining area. The main roof of this working face is composed of granule sandstone with a height of 3.04 m. The immediate roof is covered by iron pyrite, with a thickness of 4.85 m. The structure of the working face is complicated, and the oblique faults are well developed. The occurrences of coal seam in the working face are different, the structure is relatively stable with a dip angle between 15 and 23 degrees.

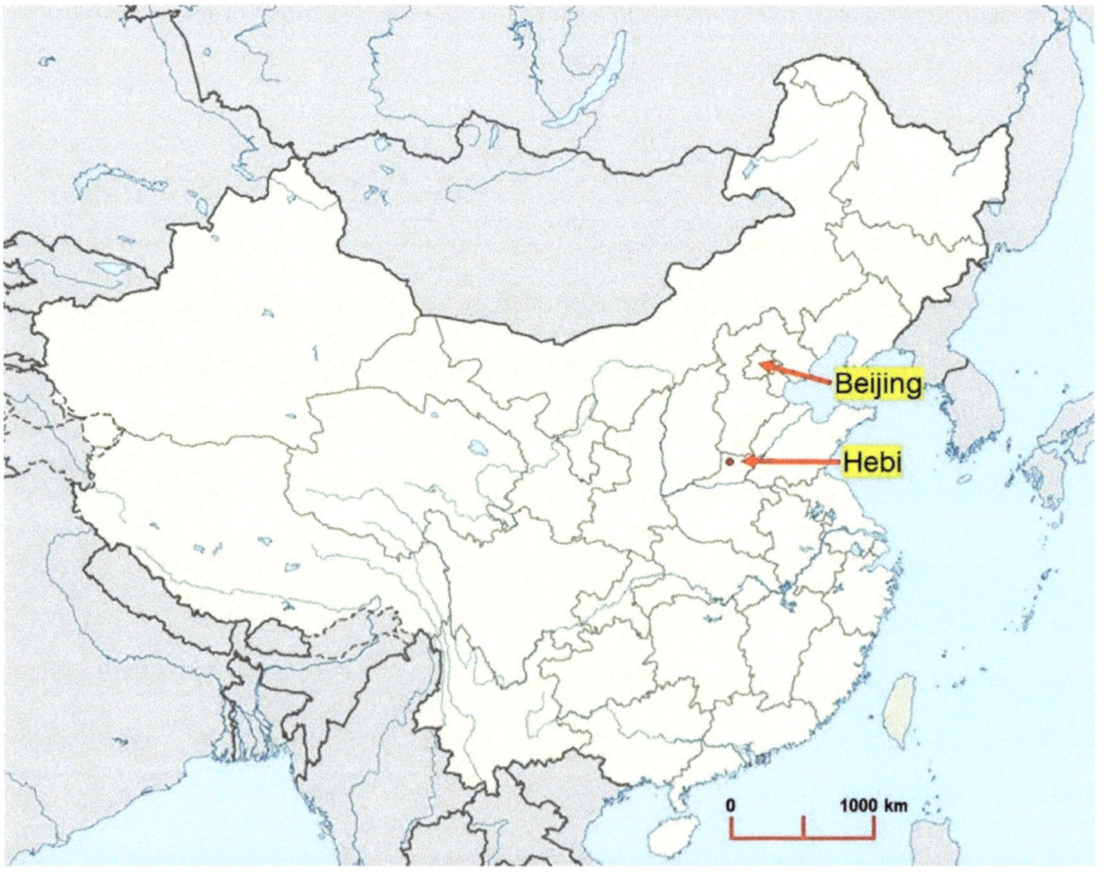

Figure 10.2.1. Location of the Hebi coal mine in China. Modified from Uwe Dedering, 2018.

The mining length of the 3006-working face was 1000 m, the incline was 140 m long, the thickness of the coal seam was 6.2–8.1 m, the average coal thickness was 7.15 m, and the hardness was 1.0. Moreover, the immediate roof was vulnerable to collapsing. Longwall mining on the strike and retreating were adopted, and fully mechanized coal mining was employed with the caving of the roof method to conduct the fieldwork. The air supply for the working face was 1400 m^3/min.

In Situ Observation

The working face was observed through a buried pipeline in the mined-out area to test temperature and gasses to obtain the changes in the temperature and gas concentration with the working face advancing in the mined-out area. Moreover, the airflow rate in the working face was measured at the same time. The instruments employed to test those parameters included a wind watch, stopwatch, portable gas determination system, gas phase chromatograph (specially manufactured for coal mines), and an underground small vacuum pump.

The vacuum pumping method and buried thermal resistance method were employed to observe the gas composition and temperature in the mined-out area. The measurement range was 120 m away from the working surface. The probes were set in the mined-out area every 30 m, and three probes were kept in the inner upper and lower side in the mind-out area.

A 1.5-inch steel tube was adopted as a buried pipeline and penetrated in three beam tubes of different colors with diameters of 8 mm. Each beam tube was used for a certain gas measurement point, and probes were raised to 0.5 m to prevent the accumulation of water. The temperature-measuring probe was embedded in the pipe parallel with the beam tube inlet (as shown in Figures 10.2.2 and 10.2.3).

Danger Zone Determination

The necessary condition for the experimental spontaneous combustion of residual coal in the mined-out area was

$$(h > h_{min}) \cap (C_{O_2} > C_{min}) \cap (\overline{Q} < \overline{Q}_{max}) \tag{10.2.1}$$

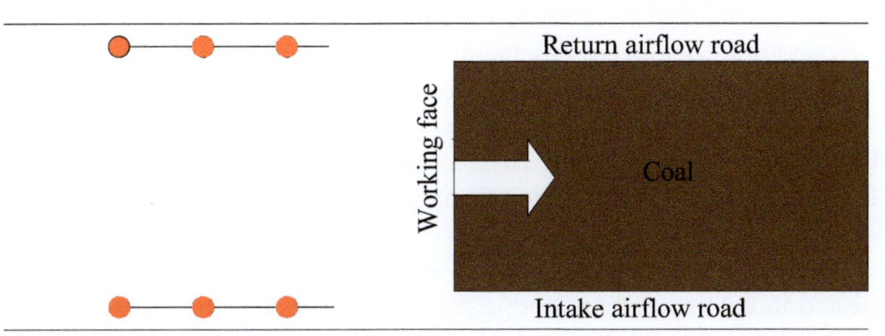

Figure 10.2.2. Observation points of a pipeline buried in a mined-out area. Figure by Yang Xiao, 2017.

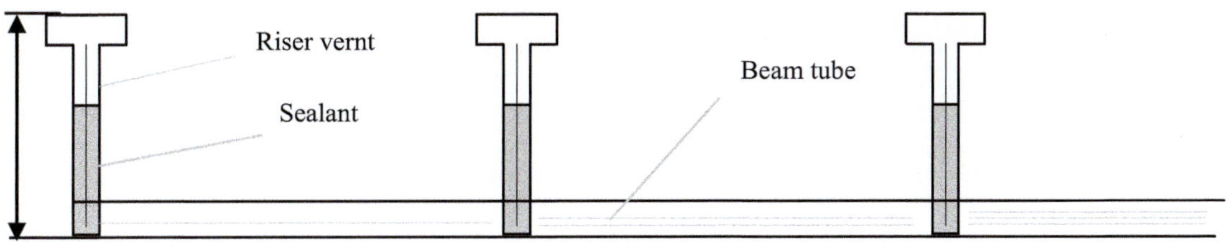

Figure 10.2.3. Observation probes of a tube buried in a mined-out area. Figure by Yang Xiao, 2017.

This indicated that if the area in the mined-out area met the necessary condition, spontaneous combustion would occur. Limiting thickness of the residual coal, limiting oxygen concentration, and limiting air leakage intensity codetermine the danger of coal-spontaneous combustion. Nevertheless, due to the advance of the working face, the actual condition of a certain point in the mined-out area during production was changed; thus, Eq. (10.2.1) was a necessary condition, but not the sufficient condition. The conditions also change because of the state of air leakage, oxygen distribution, and temperature heating in the mined-out area. The high temperature region was dynamically changed in the mined-out area.

Additionally, the danger zone of coal-spontaneous combustion must have sufficient time to maintain the condition and must not change during a period, as described in Eq. (10.2.2):

$$\tau > \tau_{min} \tag{10.2.2}$$

where τ is the coal-spontaneous combustion time (d) and τ_{min} is the minimum coal-spontaneous combustion period (d).

The limiting advance velocity is equal to the value of the distance from the heat dissipation area to the heat quenching zone divided by the minimum coal-spontaneous combustion period. Spontaneous combustion occurs when the advance velocity of the working face is less than the limiting advance velocity. The equation is as follows:

$$v < \left(v_{min} = \frac{L_{max}}{\tau_{min}} \right) \tag{10.2.3}$$

where v represents the advance velocity of the working face (m/d); v_{min} represents the limiting advance velocity of the working face (m/d); and L_{max} is the maximum width of the oxidation zone (m).

In summary, the necessary and sufficient condition for the spontaneous combustion of residual coal in the mined-out area is

$$(h > h_{min}) \cap (C_{O_2} > C_{min}) \cap (\overline{Q} < \overline{Q}_{max}) \cap (v < v_{min} \tag{10.2.4}$$

According to the actual conditions, the isogram ichnography and section plan of the thickness of residual coal in the mined-out area were determined, as depicted in Figure 10.2.4.

According to Eqs. (10.1.10), (10.1.13), and (10.1.16), the isogram curves of oxygen concentration distribution and air leakage intensity in the mined-out area were drawn, as provided in Figures 10.2.5 and 10.2.6.

According to Eqs. (10.1.10), (10.1.13), and (10.1.16), T_y and T_g from the on-site measurement as well as $q_0(T)$ and air leakage intensity tested in the experiment were used to calculate the three-dimensional extreme values h_{min}, C_{min}, and $\overline{Q}_{max}$ under the conditions of different thicknesses of residual coal, temperature, and distance from the working face. Based on the three-dimensional extreme values, the maximum value at a certain temperature was determined and can be expressed as follows:

$$\left\{ \begin{array}{l} h_{min}(\overline{Q}_i) = \max\left[h_{min}(T_1, x_1, \overline{Q}_i), h_{min}(T_2, x_2, \overline{Q}_i) \cdots h_{min}(T_i, x_i, \overline{Q}_i) \cdots \right] \\ C_{min}(h_i) = \max\left[C_{min}(T_i, x_1, h_1, \overline{Q}_1), C_{min}(T_i, x_2, h_2, \overline{Q}_2) \cdots C_{min}(T_i, x_i, h_i, \overline{Q}_i) \cdots \right] \\ \overline{Q}_{max}(h_i) = \min\left[\overline{Q}_{max}(T_i, x_1, h_1), \overline{Q}_{max}(T_i, x_2, h_2) \cdots \overline{Q}_{max}(T_i, x_i, h_i) \cdots \right] \end{array} \right\} \tag{10.2.5}$$

The results are provided in Table 10.2.1.

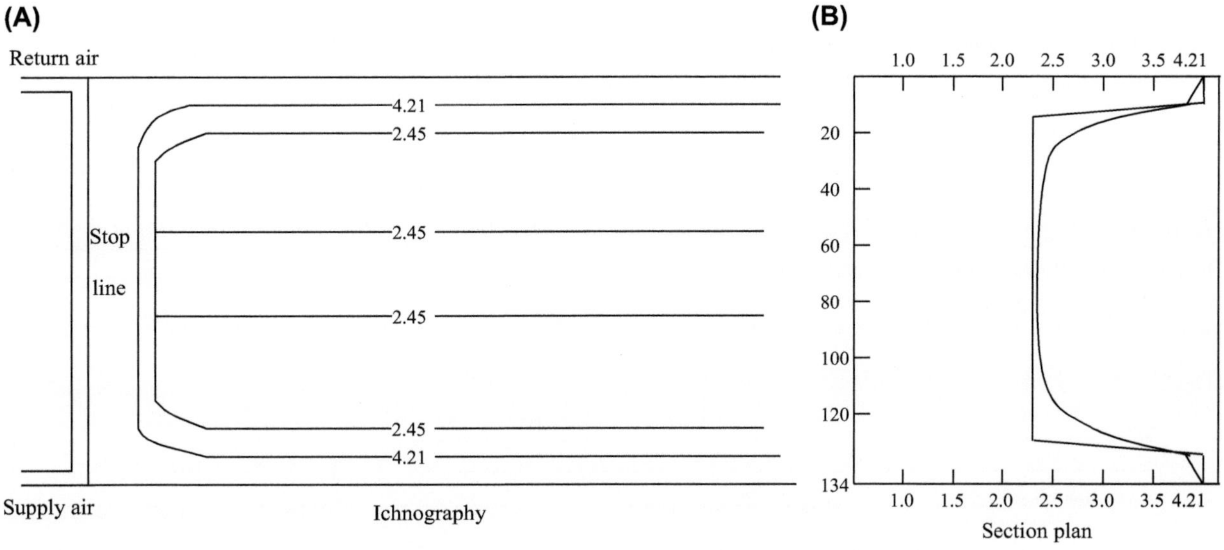

Figure 10.2.4. Isogram of the thickness (m) of residual coal in a mined-out area. Figure by Yang Xiao, 2017.

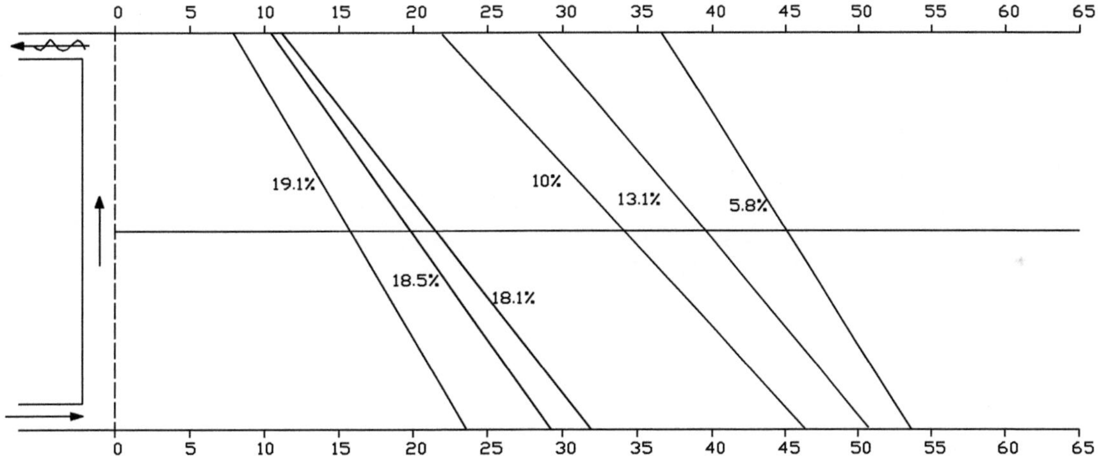

Figure 10.2.5. Isogram of oxygen concentration (%) in a mined-out area. Figure by Yang Xiao, 2017.

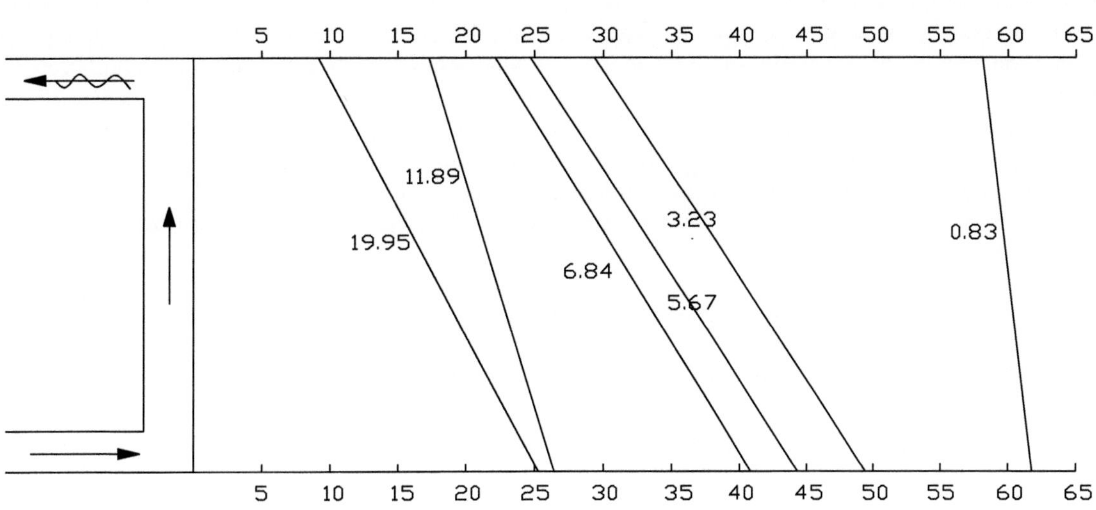

Figure 10.2.6. Isogram of air leakage intensity (cm/s) in a mined-out area. Figure by Yang Xiao, 2017.

Table 10.2.1
Limiting oxygen concentration and air leakage intensity as a function of residual-coal thickness in the 3006-working face.

h_{min} (m)*	0.7	0.8	1.0	1.2	1.4	1.6	1.8	2.0	3.0	4.0	5.0	6.0
C_{min} (%)†	20.27	15.56	10.01	6.98	5.15	3.96	3.15	2.56	1.16	0.67	0.44	0.31
$\overline{Q}_{max}$($\times 10^{-2}$cm/s)	‡0.22	1.40	3.44	5.23	6.88	8.44	9.94	11.39	18.34	25.04	31.34	38.18

*h_{min} (m) is the minimum thickness of residual coal.
†C_{min} (%) is the limiting oxygen concentration.
‡$\overline{Q}_{max}$($\times 10^{-2}$cm/s) is the limiting air leakage intensity.

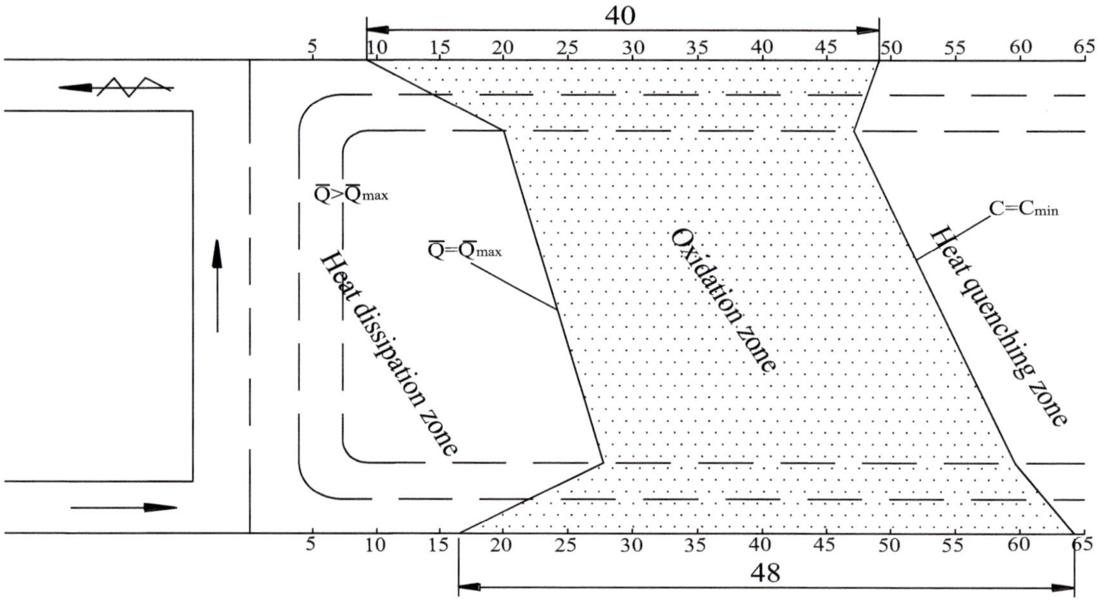

Figure 10.2.7. Determination of three zones in the mined-out area of the 3006-coal seam. Figure by Jingyu Zhao, 2017.

Three Zone Determination: Mined-Out Area

The three zones in the mined-out area were determined by drawing the isogram of the thickness of residual, oxygen concentration, and air leakage intensity in one figure. h_{min}, C_{min}, and $\overline{Q}_{max}$ were marked on Figure 10.2.7 according to Eq. (10.2.5), and they were then connected with a line. Therefore, three zones could be divided on the figure: the heat dissipation zone, oxidation zone, and heat quenching zone. Additionally, the width of the oxidation zone (L_x) was obtained.

From Figure 10.2.7 and the actual calculation, the range of the oxidation zone was determined to be 16–64 m in the air inlet side and 9–49 m in the air return side of the mined-out area. The maximum width was 48 m. The oxidation zone in the mined-out area of the 3006-coal seam was distributed on the two sides and the thick residual coal was distributed in the middle of the mined-out area.

Determination of Danger Zone: Mined-Out Area

According to the width of the oxidation zone, the maximum width of the oxidation heating zone was determined as $L_{max} = \max\{L\}$. Then, the limiting advance velocity of the working face (v_{min}) was calculated based on Eq. (10.2.7), which was compared with the actual advance velocity (v) to monitor the spontaneous combustion risk in the mined-out area. Finally, the minimum width ($L_{min} = v \cdot \tau_{min}$) was obtained using the actual advance velocity in the working face and the spontaneous combustion period, which is called the coal-spontaneous combustion danger zone in the mined-out area. It is equal to the oxidation zone and is depicted in Figure 10.2.7.

The maximum distance of the oxidation zone in the mined-out area was 48 m, and the equivalent of the oxygen concentration was 9.13%. The similarity ratio coefficient (κ) of the experimental oxygen concentration and the on-site oxygen concentration was calculated as follows:

$$\kappa = \frac{C^0_{O_2}}{\overline{C_{O_2}}} = \frac{21}{9.13} = 2.29 \tag{10.2.6}$$

where $\overline{CO_2}$ is the percent equivalent of the oxygen concentration in the mined-out area.

The minimum coal-spontaneous combustion period in the mined-out area was 46 days. Therefore, the limiting advance velocity in the mined-out area of 3006-working face was as follows:

$$v_{min} = \frac{L_{max}}{\tau_{min} \cdot \kappa} = \frac{48}{46 \times 2.29} = 0.46\text{m/d} \tag{10.2.7}$$

Thus there was no danger of coal-spontaneous combustion when the advance velocity was higher than 0.46 m/d. On the contrary, the risk of spontaneous combustion was inevitable.

This confirmed that the daily average advance velocity was 0.76 m/d.

The width of the actual safety oxidation zone (L_0) was computed as follows:

$$L_0 = \tau_{min} \times v_0 \times \kappa = 46 \times 0.76 \times 2.29 = 80.06\,\text{m} \tag{10.2.8}$$

The width of the coal-spontaneous combustion danger zone in the mined-out area, which was less than 80 m, is not expected to cause a disaster.

Discussion

The working face can be divided into three zones: heat dissipation zone, oxidation zone, and heat quenching zone. The oxidation zone was the coal-spontaneous combustion danger zone in the mined-out area. The parameters of limiting advance velocity of the working face and the width of oxidation zone determined the danger zone.

According to the experiments and actual observations, the 3006 working face in the mined-out area of the Hebi coal mine captures coal-spontaneous combustion at the condition in which the residual coal thickness was higher than 0.72 m, oxygen concentration was greater than 20.27%, air leakage intensity was lower than 0.22 cm/s, advance velocity was higher than 0.46 m/d, and the width of the oxidation zone was greater than 80.06 m.

Acknowledgments

Research was conducted at the Hebi coal mine and experimental analyses were completed at the Xi'an University of Science and Technology. The authors are grateful for the support for this project, provided by the National Natural Science Foundation of China (No. 51134019, No. 51674191, and No. 51504187), the China Postdoctoral Science Foundation (No. 2016M590963) and the Xi'an University of Science and Technology Doctoral Science Foundation (No. 2017QDJ057). Part of this chapter was reproduced with the permission of Elsevier (Deng et al., 2015).

References

Deng, J., Xiao, Y., Li, Q.W., Lu, J.H., Wen, H., 2015. Experimental studies of spontaneous combustion and anaerobic cooling of coal. Fuel 157, 261–269.

Xu, J.C., 2001. Determination Theory of Coal-spontaneous Combustion Zone. China Coal Industry Publishing House, Beijing. 288 p.

Additional Reading

Baris, K., Kizgut, S., Didari, V., 2012. Low-temperature oxidation of some Turkish coals. Fuel 93, 423–432.

Beamish, B.B., Barakat, M.A., George, J.D.S., 2001. Spontaneous-combustion propensity of New Zealand coals under adiabatic conditions. Int. J. Coal Geol. 45, 217–224.

Carras, J.N., Day, S.J., Saghafi, A., Williams, D.J., 2009. Greenhouse gas emissions from low-temperature oxidation and spontaneous combustion at open-cut coal mines in Australia. Int. J. Coal Geol. 78, 161–168.

Da Costa, J.C.D., Prasad, P., Pagan, R.J., 2004. Modern environmental improvement pathways for the coal power generation industry in Australia. Process Saf. Environ. Protection 82, 191–199.

Deng, J., Xu, J.C., Li, L., Zhang, X.H., Wen, H., 1999a. Investigation on the relation of the rate of oxygen consumption and the size of coal sample. J. Xi'an Jiaot. Univ. 33, 106–117.

Deng, J., Xu, J.C., Zhang, X.H., Wen, H., Li, L., 1999b. Investigation on the relation of coal low temperature oxidation and the size of coal sample. Coal 8, 13–25.

Deng, J., Zhao, J.Y., Zhang, Y.N., Huang, A.C., Liu, X.R., Zhai, X.W., Wang, C.P., 2016. Thermal analysis of spontaneous combustion behavior of partially oxidized coal. Process Saf. Environ. Protection 104, 218–224.

García-Torrenta, J., Ramírez-Gómezb, Á., Querol-Aragóna, E., Grima-Olmedoa, C., Medic-Pejica, L., 2012. Determination of the risk of self-ignition of coals and biomass materials. J. Hazard Mater. 213–214, 230–235.

Gürdal, G., Hoşgörmez, H., Özcan, D., Li, X., Liu, H., Song, W., 2015. The properties of Çan Basin coals (Çanakkale–Turkey): spontaneous combustion and combustion by-products. Int. J. Coal Geol. 138, 1–15.

Jia, R.H., Harris, G.H., Fuerstenau, D.W., 2000. An improved class of universal collectors for the flotation of oxidized andror low-rank coal. Int. J. Miner. Process. 58, 99–118.

Joshi, K., Raghavan, V., Rangwala, A.S., 2013. Effect of weathering of coal and organic dusts on their spontaneous ignition. Fire Technol. 49, 843–856.

Kuenzer, C., Zhang, J., Sun, Y., Jia, Y., Dech, S., 2012. Coal fires revisited: the Wuda coal field in the aftermath of extensive coal fire research and accelerating extinguishing activities. Int. J. Coal Geol. 102, 75–86.

Li, L., Jiang, D., Beamish, B.B., 2010. Calculation of ignition times under adiabatic conditions by activation energy. J. China Coal Soc. 35, 802–815.

Lu, W., Wang, D., Zhong, X., Zhou, F., 2006. Tendency of spontaneous combustion of coal based on activation energy. J. China Univ. Min. Technol. 35, 201–215.

Mahidin, Usui, H., Ishikawa, S., Hamdani, 2002. The evaluation of spontaneous combustion characteristics and properties of raw and upgraded Indonesian low rank coals. Coal Prep. 22, 81–91.

Moghtaderi, B., Dlugogorski, B.Z., Kennedy, E.M., 2000. Effects of wind flow on self-heating characteristics of coal stockpiles. Process Saf. Environ. Protection 78, 445–453.

Nordon, P., Young, B.C., Bainbridge, N.W., 1979. The rate of oxidation of char and coal in relation to their tendency to self-heat. Fuel 58, 443–449.

Nugroho, Y.S., Mclntosh, A.C., Gibbs, B.M., 2000. Low-temperature oxidation of single and blended coals. Fuel 79, 1951–1961.

Perdochova, M., Derychova, K., Veznikova, H., Bernatik, A., Pitt, M., 2015. The influence of oxygen concentration on the composition of gaseous products occurring during the self-heating of coal and wood sawdust. Process Saf. Environ. Protection 94, 463–470.

Song, Z.Y., Kuenzer, C., Zhu, H.Q., Zhang, Z., Jia, Y.R., Sun, Y.L., Zhang, J.Z., 2015. Analysis of coal fire dynamics in the Wuda syncline impacted by fire-fighting activities based on in-situ observations and Landsat-8 remote sensing data. Int. J. Coal Geol. 141–142, 91–102.

Spears, D.A., Tewalt, S.J., 2009. The geochemistry of environmentally important trace elements in UK coals, with special reference to the Parkgate coal in the Yorkshire-Nottinghamshire Coalfield, UK. International J. Coal Geol. 80, 157–166.

Stracher, G.B., Taylor, T.P., 2004. Coal fires burning out of control around the world: thermodynamic recipe for environmental catastrophe. In: Stracher, Glenn, B. (Eds.), Coal Fires Burning Around the World, pp. 7–17 International Journal of Coal Geology, 59, issues 1-2.

Taku Ide, S., Orr Jr., F.M., 2011. Comparison of methods to estimate the rate of CO_2 emissions and coal consumption from a coal fire near Durango, CO. Int. J. Coal Geol. 86, 95–107.
Tang, Y.L., 2015. Sources of underground CO: crushing and ambient temperature oxidation of coal. J. Loss Prev. Process. Ind. 38, 50–57.
Wang, H., Dlugogorski, B.Z., Kennedy, E.M., 1998. Low-temperature oxidation of coal at elevated pressures. J. Loss Prev. Process. Ind. 11, 373–381.
Wang, H., Dlugogorski, B.Z., Kennedy, E.M., 2003a. Analysis of the mechanism of the low-temperature oxidation of coal. Combust. Flame 134, 107–117.
Wang, H., Dlugogorski, B.Z., Kennedy, E.M., 2003b. Coal oxidation at low temperatures: oxygen consumption, oxidation products, reaction mechanism and kinetic modeling. Prog. Energy Combust. Sci. 29, 487–513.
Warmuzinski, K., 2008. Harnessing methane emissions from coal mining. Process Saf. Environ. Protection 86, 315–320.
Wen, H., Xu, J.C., Ge, L.M., Deng, J., 2001. Test technology and numerical analysis of coal-spontaneous combustion. J. Univ. Sci. Technol. 6, 499–501.
Wen, H., Zhang, F.Y., Jin, Y.F., Liu, W.Y., 2011. Experiment research on effect of sulfur on characteristic parameters of coal-spontaneous combustion. Saf. Coal Mine 42, 5–7.
Xie, J., Xue, S., Cheng, W.M., Wang, G., 2011. Early detection of spontaneous combustion of coal in underground coal mines with development of an ethylene enriching system. Int. J. Coal Geol. 85, 123–127.
Zhang, Y.T., Zhang, X.C., Tien, J.C., Li, Y.Q., 2011. Evaluation of the propensity for coal-spontaneous combustion based on catastrophe theory. J. Coal Sci. Eng. 17, 265–269.
Zhong, M.H., Fu, T., 2008. Illegal mining could revive Xinjiang's coalfield fires. Nature 451, 16.
Zhou, Y.G., Li, P., Ni, S.A., Qian, H.P., Zhao, H., 2014. Effect of oxygen volume fraction on induction of coal-spontaneous combustion. J. Chin. Soc. Power Eng. 34, 421–426.

CHAPTER 11

Quantification of the Environmental Impact of Coal Fires: Xinjiang Province, China

A fire, about 50 m deep, in Jurassic coal at the Songshutou fire site, Urumqi, China. The fire started in the early 2000s during illegal mining of a 40 m thick outcrop. Mining was forbidden here by the local administration because of the potential for fire and environmental issues including air pollution, damage to vegetation, and contamination of nearby surface and ground water with heavy metals and other harmful matter. The fire was extinguished in 2007 by surface stripping, injecting water into boreholes for cooling followed by slurry injection for reducing the flow of oxygen, and then burial with soil. From Nie et al. (2017) with modifications, reprinted with permission of the Journal of China Coal Society. *Photo by Zeng Qiang, 2006.*

CHAPTER CONTENTS

Coal and Peat Fires: A Global Perspective
Edited by Glenn B. Stracher

https://doi.org/10.1016/B978-0-12-849885-9.00011-1

Photo by Zeng Qiang, 2010.

11.1 Coal Fire Pollutants

Zeng Qiang
Pu Yan
Nie Jing
Cao Zhanmin

Mirabilite crystals ($Na_2SO_4 \cdot 10H_2O$) nucleated from bituminous coal-fire gas at the Heshituoluogai No. 5 fire site, Hoboksar County, Xinjiang Province, China. The fire started spontaneously during mining in the 1990s and was extinguished in 2016 using the same techniques described on the previous page.

Introduction

Underground coal fire (UCF) is a natural disaster associated with coal mining. Globally, UCFs are widely reported in all major coal-producing countries (Chatterjee, 2006; Kuenzer et al., 2007; Stracher and Taylor, 2004, 2011; Pone et al., 2007; Nolter and Vice, 2004; Finkelman, 2004; Zeng, 2012), causing large economic losses and severe ecological impacts. In a UCF event, coal–oxygen reactions produce heat and generate reaction products. The resultant products are ultimately responsible for the environmental impacts of UCFs and thus have been the targets of quantitative studies. Hower et al. (2009, 2011) monitored pollutant concentrations (e.g., CO, CO_2, Hg, H_2S) in gases released from UCFs. O'Keefe et al. (2010, 2011) measured CO, CO_2, and Hg emissions from UCFs in the field. Engle et al. (2011, 2012, 2013) measured greenhouse gas emissions from UCFs by ground-based and airborne surveying techniques. Liang et al. (2014) analyzed Hg concentrations in fumes released from a UCF in Inner Mongolia (China). Zeng et al. (2016, 2014) investigated the distribution of heavy metals in soils at two UCF zones for the first time. Fu (2002) characterized the emission of polycyclic aromatic hydrocarbons during coal combustion. Feng (2004) studied the thermodynamic and kinetic behaviors of typical trace pollutant elements in coal (i.e., Hg, Cd, As, Pb, Cr, Ni) during combustion. Xu et al. (2005) compared the effects of experimental conditions on the combustion characteristics of coal samples and found heating rate to be the dominant influencing factor. Kong (2005) studied the emission of NO and SO_2 from coal combustion in a circulating fluidized bed. However, no study has reported the patterns of product formation during coal–oxygen reactions under conditions relevant to UCFs.

The study of this part characterized the production of typical pollutants under conditions like those experienced during UCFs, by combining programmed heating–oxidation, thermogravimetric analyses (TGA), and chemistry thermodynamic simulations (FactSage). The goal of this study was to quantify the production of typical pollutants during UCFs, which will aid assessments of UCF impact on local environment.

Analytical Methods

Coal samples were collected from the Beitashan UCF zone, in the East Junggar coalfield (Xinjiang, Province, China). The sampling site (90°56′26″ E, 44°50′59″ N) was 178 km northeast of Qitai, Changji, Xinjiang. In the Beitashan UCF zone, seam B_6 (thickness: 27.97 m, dip: 7°) has been combusting at a depth of 17–18 m over an area of 150,888 m^2, causing up to 80,000 tons loss of coal annually. Table 11.1.1 lists the physicochemical characteristics of the samples collected from the zone.

Programmed Heating–Oxidation

Coal samples were treated by programmed heating–oxidation using a temperature controller, an adiabatic heating furnace, a gas chromatography (GC) station, gas pipes, a temperature monitoring system, and required accessories.

Table 11.1.1
Characteristics of noncaking bituminous boal seam B_6 at the Beitashan coal fire.

Coal seam	Properties (%)				Element (%)					Calorific value (MJ/Kg)
	Fixed carbon	Moisture	Volatile matter	Ash	C	H	O	N	S	
B_6	59.30	9.98	28.30	2.42	69.47	3.75	13.59	0.67	0.12	26.68

Each sample (100 g) was loaded into the furnace; dry air was set to flow through the furnace at 80 mL min^{-1} and it was heated to 500°C in a progressive manner (room temperature to 50°C in 15 min, 50°C–75°C at 1.67°C min^{-1}, 75°C–100°C at 1.67°C min^{-1}, and 100°C–500°C at 1°C min^{-1}). After reaching each temperature point (50°C, 75°C, 100°C, and 500°C), the temperature was maintained for 5 min and then the gas was collected from the outlet of the oven. GC analyses were performed to characterize the gases (e.g., O_2, CO, C_nH_m) present. The temperatures of the furnace and coal samples were continuously monitored until a coal sample's temperature exceeded that of the furnace, indicating coal burning, defined as the critical temperature.

Thermogravimetric Analysis

Coal samples were analyzed by TGA to characterize their mass loss with the increases in temperature. Samples were loaded in a Hitachi STA 7300 TGA station and heated from 30°C to 1000°C in a dry air atmosphere (flow rate: 300 mL min^{-1}). TGA and differential TGA curves were recorded.

Thermodynamic Simulations

Coal–oxygen reactions were simulated with FactSage (Thermfact, CRCT, Montreal, QC, Canada). In modeling, minimization of Gibbs free energy was used as the criterion for equilibration and the composition of the reaction system at equilibrium was determined by the Lagrange multiplier method.

Results and Discussion

The critical temperature and the concentrations of O_2, CO, and C_nH_m during low-temperature oxidation were monitored and analyzed by programmed heating–oxidation. Table 11.1.2 summarizes the results.

TGA and differential TGA curves of the samples (Figure 11.1.1) showed a series of features during heating: moisture loss–oxygen absorption (T_1), thermal decomposition (T_2), ignition (T_3), maximum combustion rate (T_4), and burning-out (T_5). Table 11.1.3 lists the characteristic temperatures for these events.

Coal combustion in a coal-fire zone is a complicated process. Combustion reaction kinetic parameters are key to understanding coal combustion and the basis for investigating pollutant formation at the microscopic level. According to theory, coal combustion may involve the following coal–oxygen reactions:

$$\text{(a) } C+O_2 \rightarrow CO_2+q_a$$

$$\text{(b) } C+1/2O_2 \rightarrow CO+q_b$$

$$\text{(c) } C+CO_2 \rightarrow 2CO+q_c$$

$$\text{(d) } CO+1/2O_2 \rightarrow CO_2+q_d$$

Of these, reactions (a), (b), and (c) are solid phase exothermic reactions at the coal surface and reaction (d) is a gas-phase endothermic reaction; the letter q (with subscripts) represents the heat of the reaction. During coal combustion, reactions (a) and (b) are the leading reactions.

Table 11.1.2
Results of the programmed heating–oxidation of noncaking bituminous B_6 coal samples.

	Sample temperature (°C)	Oven temperature (°C)	O_2 (%)	CO (ppm)	CH_4 (ppm)	CO_2 (ppm)	C_2H_4 (ppm)	C_2H_6 (ppm)	C_2H_2 (ppm)	C_3H_8 (ppm)
B_6	11.2	12.0	21.09	0.00	0.00	617.21	0.00	0.00	0.00	0.00
	36.8	43.9	20.92	0.00	0.00	804.70	0.00	0.00	0.00	0.00
	60.2	69.7	20.84	0.00	0.00	873.68	0.00	0.00	0.00	0.00
	85.1	95.2	20.73	94.67	0.00	1240.70	0.00	0.00	0.00	0.00
	107.4	117.0	20.31	442.84	0.32	1957.30	0.00	0.00	0.00	0.00
	123.5	128.4	20.13	472.40	0.34	2554.11	0.00	0.00	0.00	0.00
	141.0	143.5	19.53	1236.88	0.61	3775.32	0.00	0.00	0.00	0.00
	157.2	156.5	18.79	5016.91	5.61	7398.13	0.00	0.00	0.00	0.00
	173.5	168.6	16.44	17283.33	6.80	16556.05	9.53	3.41	4.50	10.19
	246.1	175.2	3.16	65550.80	3377.60	42475.28	214.28	286.76	7.19	200.30
	226.5	158.0	3.97	249836.58	2695.64	174504.92	168.99	224.34	5.01	139.59
	216.6	165.5	3.44	190575.21	1673.44	185998.94	145.26	167.05	11.12	104.69

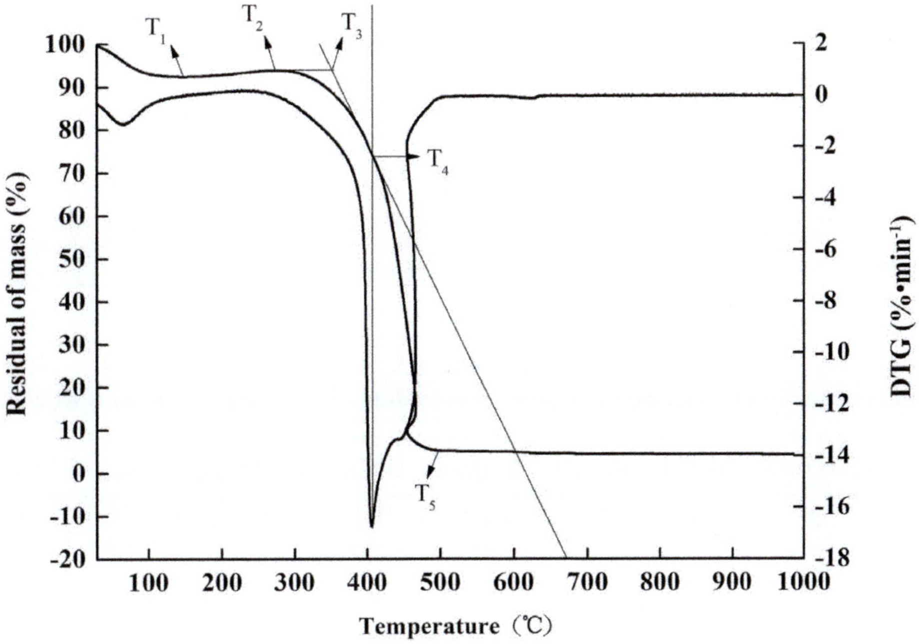

Figure 11.1.1. TGA and differential TGA curves of the coal samples.

Table 11.1.3
Characteristic thermogravimetric temperatures of the noncaking bituminous B_6 coal samples.

T_1 (°C)	T_2 (°C)	T_3 (°C)	T_4 (°C)	T_5 (°C)
146.97	274.69	351.10	406.54	505.32

It was established that the rate constant for a coal combustion reaction follows the Arrhenius equation:

$$k=A\exp^{(-E/RT)} \tag{11.1.1}$$

where k is the rate constant; A is the frequency factor; E (kJ mol^{-1}) is the activation energy for the reaction; T (K) is the absolute temperature; and R is the universal gas constant (8.314 J K^{-1} mol^{-1}).

The activation energy (E) is a critical factor that determines the rate of a reaction and how easily it occurs. It is also an important thermal kinetic parameter that characterizes the coal–oxygen reaction process. This parameter can be calculated using multiple methods, such as the frequently used Freeman–Carroll, Flynn–Wall–Ozawa, and Coats–Redfern methods. In the present study, the activation energies for the coal–oxygen reactions were calculated by the Coats–Redfern method. In the calculation, coal combustion was treated as a first-order reaction and the conversion efficiency of coal expressed as:

$$\alpha=\frac{m_0-m_t}{m_0-m_\infty} \tag{11.1.2}$$

where α is the conversion efficiency of the coal sample (i.e., the proportion of mass converted to products); m_0 is the mass of the sample at the start of the experiment; m_t is the mass at time t; and m_∞ is the mass of the sample at the end of the experiment. The reaction rate was expressed as:

$$\frac{d\alpha}{dt}=k(1-\alpha) \tag{11.1.3}$$

where k is the rate constant and t represents time.

Table 11.1.4
Kinetic parameters for coal–oxygen reactions in noncaking bituminous B_6 coal samples.

Coal seam B6	Temperature range (°C)	Regression equation	Activation energy (kJ/mol)	Frequency factor	Coefficient of correlation
	30–146.97	y=−4975.01x+2.2618	41.36	4.78×10^{2}	−0.93
	147.05–274.69	y=−11517.06x+10.4560	95.75	4.00×10^{6}	−0.95
	274.77–505.32	y=−16294.80x+9.8924	135.47	3.22×10^{6}	−0.96

Table 11.1.5
Reactant configurations selected for FactSage modeling, noncaking bituminous B_6 coal seam.

C (mol)	O_2 (mol)	H (mol)	S (mol)	N (mol)	As (mol)	Hg (mol)	Cr (mol)	Cd (mol)	Ni (mol)	Si (mol)
26.84	24.77	17.70	0.09	0.30	6.67×10^{-6}	3.42×10^{-6}	4.13×10^{-4}	8.90×10^{-7}	1.70×10^{-5}	0.01

Thermal analysis kinetics indicates that, for reactions of different orders (n = 1 or n ≠ 1), the Coats–Redfern equation is written as:

$$n = 1, \ln\left[\frac{1-(1-\alpha)^{1-n}}{T^2(1-n)}\right] = \ln\left[\frac{AR}{\beta E}\left(1-\frac{2RT}{E}\right)\right] - \frac{E}{RT} \tag{11.1.4}$$

$$n \neq 1, \ln\left[\frac{-\ln(1-\alpha)}{T^2}\right] = \ln\left[\frac{AR}{\beta E}\left(1-\frac{2RT}{E}\right)\right] - \frac{E}{RT} \tag{11.1.5}$$

Generally, because 2RT/E is close to zero, 1 − 2RT/E approaches 1. Under this condition, Eq. (11.1.4) can be written as:

$$\ln\left[\frac{-\ln(1-\alpha)}{T^2}\right] = -\frac{E}{RT} \tag{11.1.6}$$

Eqs. (11.1.2), (11.1.3), and (11.1.6) were used to calculate kinetic parameters for the coal–oxygen reactions in this study as follows: ln $[-\ln(1-\alpha)/T^2]$ was plotted against 1/T and the slope of the regression line was the activation energy. Table 11.1.4 summarizes the results.

FactSage Modeling

According to the analyses of the coal samples and the data from the programmed heating–oxidation experiments, parameters were selected for the FactSage simulation (Table 11.1.5). Four reaction systems (C+O; C+O+As; C+O+Hg; C+O+H+S+N+Hg) were constructed and the results are depicted in Figures 11.1.2–11.1.5.

Figures 11.1.6 and 11.1.7 depict the changes in O_2 and CO concentrations with temperature during programmed heating–oxidation; with increasing temperature, the O_2 concentrations decreased.

In TGA (Table 11.1.3), the samples exhibited a series of features: moisture loss–oxygen absorption (T_1: 146.97°C), thermal decomposition (T_2: 274.69°C), ignition (T_3: 351.10°C), maximum combustion rate (T_4: 406.54°C), and final burning-out (T_5: 505.32°C). Thermodynamic calculations (Table 11.1.4) revealed that the activation energy

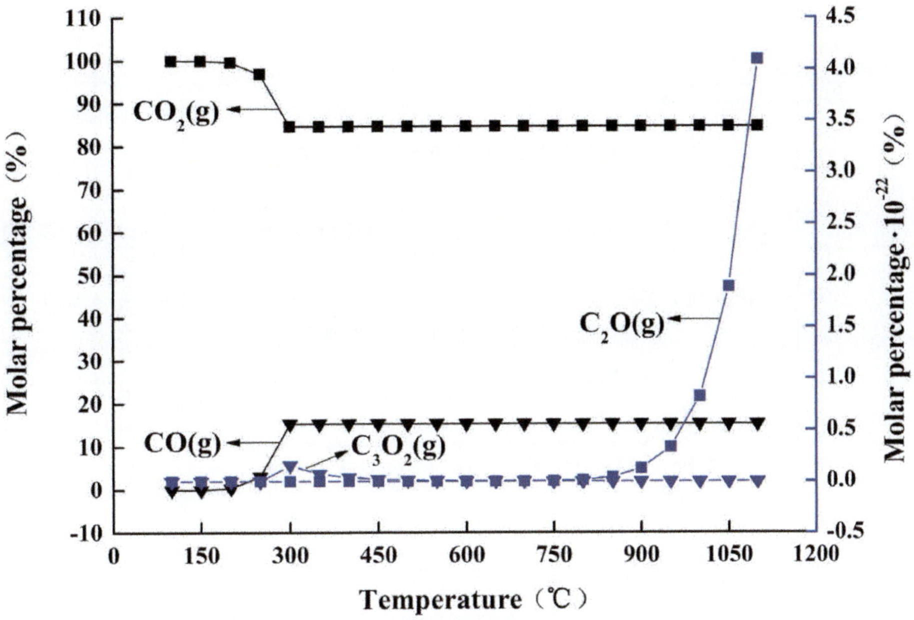

Figure 11.1.2. Molar percentages versus temperature in the C and O system.

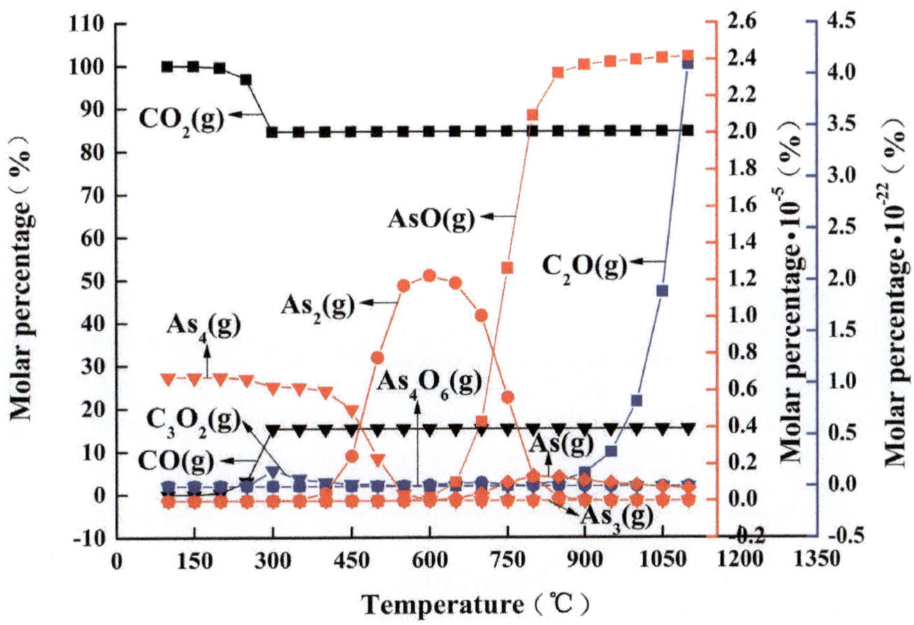

Figure 11.1.3. Molar percentages versus temperature in the C, O, and As system.

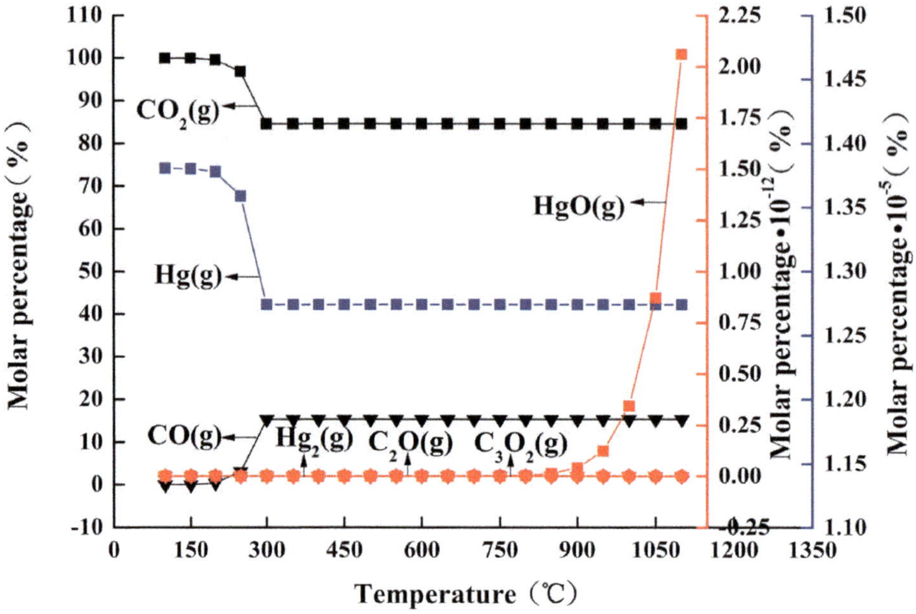

Figure 11.1.4. Molar percentages versus temperature in the C, O, and Hg system.

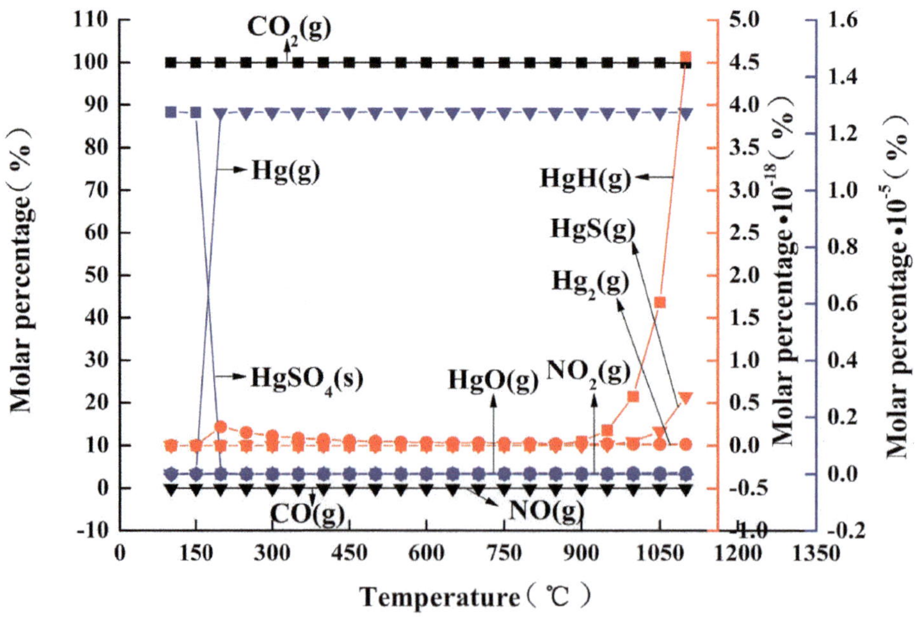

Figure 11.1.5. Molar percentages versus temperature in the C, O, H, S, N, and Hg system.

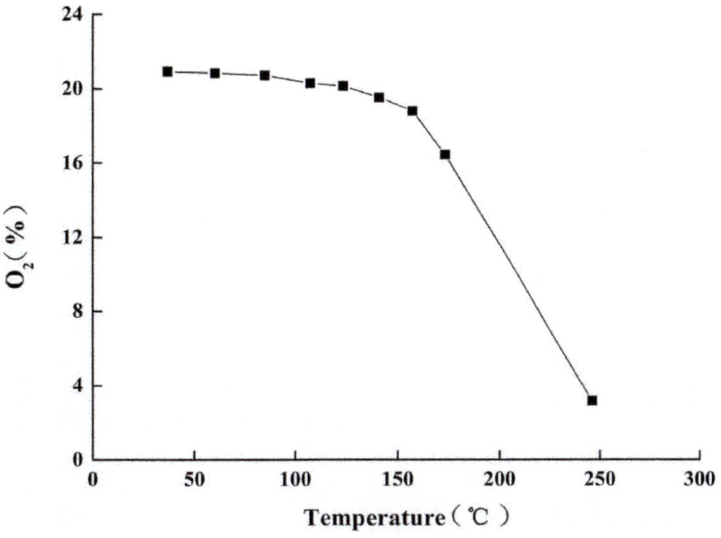

Figure 11.1.6. O_2 concentration as a function of temperature.

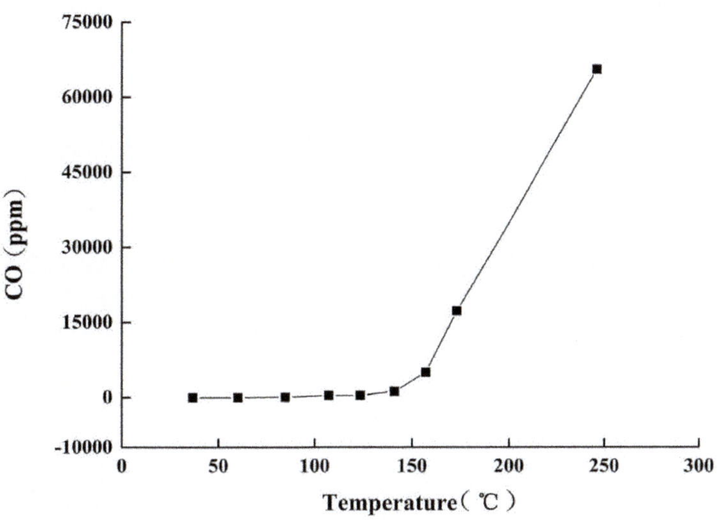

Figure 11.1.7. CO concentration as a function of temperature.

for the temperature range from ambient temperature to T_1 was 41.36 kJ mol^{-1}; for T_1-T_2 the activation energy was 95.75 kJ mol^{-1}, and for T_2-T_5 it was 135.47 kJ mol^{-1}. Based on the activation energy and frequency factor for each stage, the rate constants for the three ranges were expressed by Eqs. (11.1.7)–(11.1.9):

$$k = 4.00\times10^{2}\exp\left(-\frac{95.75}{8.314\times(t+273.15)}\right) \tag{11.1.7}$$

$$k = 4.78\times10^{2}\exp\left(-\frac{41.36}{8.314\times(t+273.15)}\right) \tag{11.1.8}$$

$$k = 3.22\times10^{6}\exp\left(-\frac{135.47}{8.314\times(t+273.15)}\right) \tag{11.1.9}$$

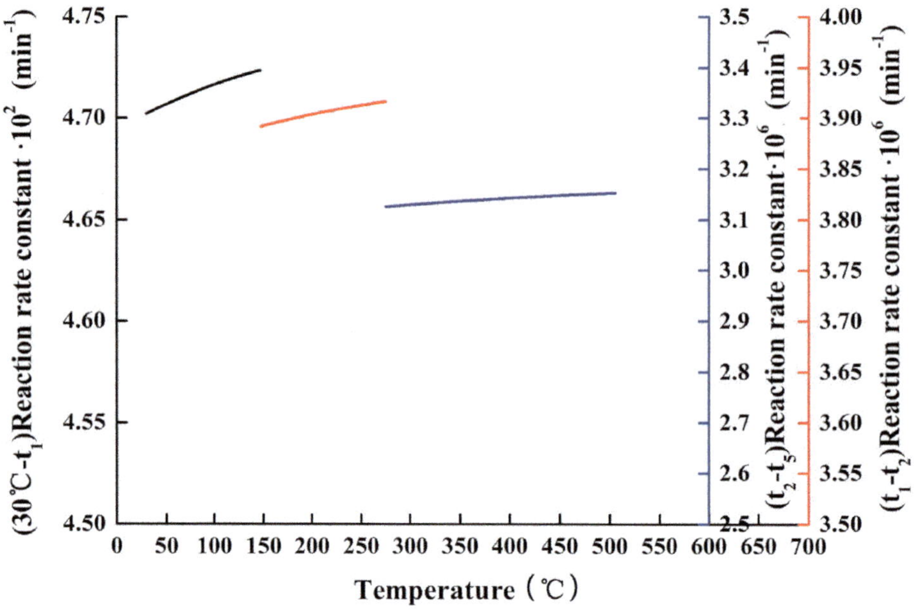

Figure 11.1.8. Reaction rate constants as a function of temperature.

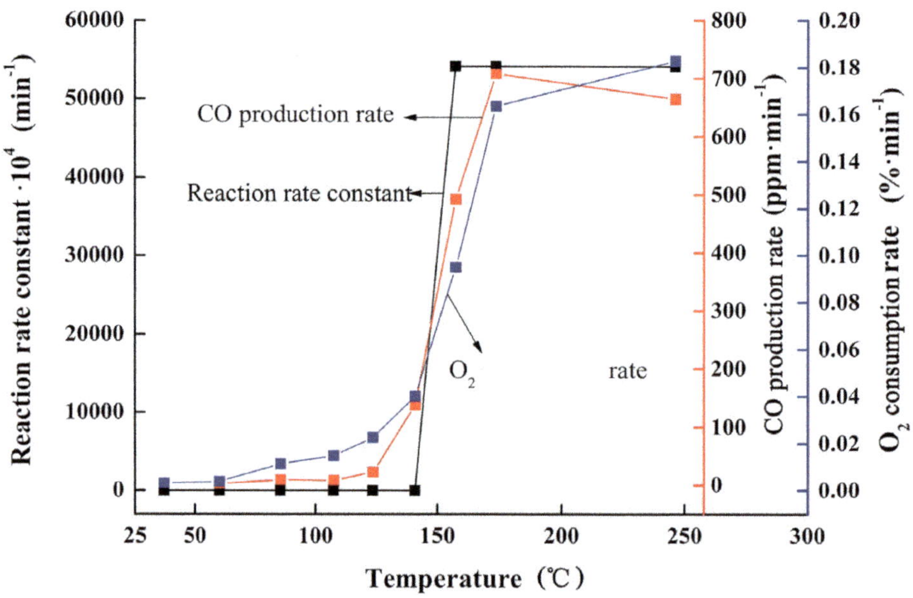

Figure 11.1.9. Reaction rate constant, CO production, and O_2 consumption rates as a function of temperature.

Figure 11.1.8 depicts the rate constants calculated from these equations. Additionally, from the data in Table 11.1.2 and Eqs. 11.1.7–11.1.9, the changes in the rate constant, oxygen consumption rate, and CO formation rate with temperature were calculated (Figure 11.1.9), and they all varied in generally similar patterns.

The FactSage simulation for the system C+O (Figure 11.1.2) predicted that the products consisted primarily of CO_2 and CO. At ≥800°C, trace C_2O (g) also formed. For the system C+O+As (Figure 11.1.3), the products also primarily consisted of CO_2 and CO, with the formation of trace As_4 starting at 100°C and trace As_2 starting at 350°C. The formation of As_4 and As_2 fluctuated with increasing temperature, showing a valley and a peak, respectively. For the system C+O+Hg (Figure 11.1.4), CO_2 and CO remained the major products, and Hg (g) decreased with increasing temperature, reaching a steady concentration at 300°C. Starting from 800°C, trace HgO (g) formed and it proportionally increased with further increases in temperature. The system C+O+H+S+N+Hg (Figure 11.1.5) produced predominantly CO_2, in addition to trace Hg (g) and $HgSO_4$ (s). From 950°C, trace HgS (g) and HgH (g) started to form and these increased with further increases in temperature.

Photo by Zeng Qiang, 2014.

11.2 Migration of Heavy Metals in Soil Affected by Coal Fires

Soil profile sampled from the Daquanhu bituminous underground coal mine fire, Uruqmi city, Xinjiang, Province, China. This fire started by spontaneous combustion in the 1970s during mining of the B_7 and B_8 coal seam.

Introduction

Underground coal fires not only consume massive coal resources but also have a serious impact on the local ecological environment. The transfer of pollutants within the soil, atmosphere, or aquatic environment affected by a coal fire should be the core topic of quantitative research into the environmental impact of coal fires, but no relevant research has been reported in the literature.

The present study of this part is based on previous research into the spatial distribution of heavy metals in soil above coal fires. This is the first study to examine the characteristics of adsorption and desorption of arsenic (As) in coal fire–affected soil, as well as to explore the transport pattern of this typical heavy metal in soil using soil column leaching tests. The research findings are expected to provide a basis for scientific evaluation of coal fires' environmental impact and guide environmental control in coal fire zones.

The Daquanhu coal fire is located 8 km west of Urumqi, and the coordinates of its geographic center are 43°47′07″ N, 87°24′00″ E. This area features rolling hills that range from 962 to 1080 m in altitude. It has an arid continental climate with hot summers and cold winters. The highest surface temperature of the coal fire is 121°C. The annual average wind speed is 2.8 m/s, and the maximum wind speed reaches 26.7 m/s. Lying within the northern limb of the Xishan anticline, the coal seams in this area strike roughly east–west and dip 52°–62°. Burning coal seams include the 7 m-thick B_7 and 5 m-thick B_8, and the depth of the fire is between 40 and 144 m. The coal fire consumes an estimated 226.6 kilotons of coal annually.

Analytical Methods

Soil Sampling

The tested soil samples were collected at a location with a temperature anomaly in the Daquanhu coal fire area (49°47′68″ N, 87°26′13″ E). The soil was sampled at intervals of 10 cm, with layer one to layer six at the sampling depth ranging from 0 to 60 cm. After being air-dried and having debris removed, the soil samples were filtered through a 2-mm sieve. Then, 50 g of soil was extracted from each soil layer using the method of coning and quartering, and the 50 g samples were filtered through a 0.149 mm sieve for use in chemical tests.

Experimental Apparatus and Reagents

The types of experimental apparatus used in this study include an ST-X101A thermostatic oscillator manufactured by Shinetek Instruments (Beijing) Co., Ltd.; 80-2 electric centrifuge produced by the Jintan Medical Instrument Factory, Jiangsu Province; PF6 nondispersive atomic fluorescence spectrometer; and pHS-2CA acidometer made by Shanghai Dazhong Analytical Instrument Factory. The reagents used are arsenic trioxide, potassium iodide, ascorbic acid, thiourea, sodium hydroxide, hydrochloric acid (guaranteed reagent), potassium hydroxide, and potassium borohydride. Conventional analytical methods in soil science and agricultural chemistry were used to test the soil samples. The pH values of the samples were determined by measuring electric potential. Organic content was measured through potassium dichromate heating and oxidation. The mechanical compositions of the samples were analyzed by sieve and pipette analysis. The carbonate contents of the soil samples were determined by titration. A mixed solution of $HCl–HNO_3–HClO_4$ was used to heat and digest the soil samples, and then the As concentration in the digestion fluid was measured using the nondispersive atomic fluorescence spectrometer.

Isothermal Adsorption and Desorption

The soil that passed through the 0.194 mm sieve was placed in a number of plastic centrifuge tubes, each of which held 0.5 g. The soil samples in the tubes were divided into nine treatment groups, each with three duplicate soil samples. Then As solutions at different concentrations (0.1 mg/L, 0.2 mg/L, 0.5 mg/L, 1.0 mg/L, 2.0 mg/L, 2.5 mg/L, 5 mg/L, 10 mg/L, and 20 mg/L) were separately added to the soil samples, one concentration for one treatment group. After that, the tubes were put into the oscillator for 24 h of thermostatic oscillation, and the three samples in each of the nine treatment groups were oscillated at 25°C, 50°C, and 70°C, respectively. Then the tubes were taken out and put in an LD5-2A centrifuge for 30 min of centrifugation at 4000 r/min. Immediately, after the centrifugation, the atomic fluorescence spectrometer was used to measure the resulting As (III) concentrations in the samples. To serve as a control, samples of As (III) solutions that contained no soil underwent the same oscillation, centrifugation, and fluorescence spectrometer measurement. The differences between the As (III) concentrations measured before and after the adsorption experiment were assumed to represent the amounts of As adsorbed by the soil. Twenty-four hours after the adsorption process, all the balanced solutions were taken out of the centrifuge tubes, and the As-containing soil samples left in the tubes were later used in the desorption experiment. During the desorption experiment, 20 mL of a sodium hydroxide solution (0.01 mol/L) was added to each of the centrifuge tubes with As-containing soil samples (treatment groups) or blank samples of As (III) solutions (control group). The tubes were then placed in the oscillator for 24-h thermostatic oscillation at either 25°C, 50°C, or 70°C. Next, the tubes were put in the LD5-2A centrifuge for 30 min of centrifugation at 4000 r/min. The resulting As (III) concentrations in the samples were then measured using the atomic fluorescence spectrometer.

Soil Column Construction

The six soil layers collected at the sampling site were separately put into six identical PVC pipes (diameter $\Phi = 7$ cm and length = 25 cm) to create a single-layer soil column for each layer. To analyze the actual heavy metal transfer through the soil profile at the sampling site, the soil layers were put layer by layer into a PVC pipe ($\Phi = 10$ cm and length = 65 cm) to build an integrated soil column, and they were stacked in the same order as observed in situ. During loading, the soil in each pipe was compacted into a column of a specified height in order to ensure that its bulk density and water content equaled or approximated the in-situ values and that the soil particles were evenly distributed in the column.

Leachate Setting

The study area is in Xishan, Urumqi, where the average annual rainfall is around 300 mL. An estimated 60% of local rainfall enters the soil every year, while the rest of the rainfall produces surface runoff. In this study, the total leachate volume was set at 2 L, approximately nine times the average annual rainfall. Dilute hydrochloric acid was added to the leachate to render it slightly acidic.

Leaching Test

Endogenous As was added to the soil samples in this test. At the beginning, an aqueous solution of a known As concentration was added to the bottom of each column, so that the solution could move up by capillary action to saturate the soil. Then the leachate was injected into the top of each column at a rate of around 0.1 mL/min. Fixed amounts of leachate were gathered from the bottom of the columns regularly, and the heavy metal concentrations in the leachates were measured. After the leaching test was completed, the soil columns were taken out of the pipes, and each column was then divided into six sublayers, whose depth ranges were 0 cm, 0–2 cm, 2–5 cm, 5–10 cm, 10–15 cm, and 15–20 cm. Three subsamples were collected from each layer and analyzed for their heavy metal contents.

The accumulated amount of a certain heavy metal released into a leachate and the release rate can be calculated using the following equations:

$$A = \frac{\sum_{i=1}^{n} C_i \times V}{m} \quad (11.2.1)$$

$$B = \frac{A}{S} \times 100\% \quad (11.2.2)$$

where A is the accumulated amount of the heavy metal released from the soil under simulated rainfall (μg/kg), including the heavy metal contents of both polluted and natural soils. Ci is the concentration of this heavy metal in the leachate sample, V is the leachate volume (set at 0.2 L in this test), M is the mass of the tested soil, B is the release rate of the heavy metal from the soil column, and S is the initial content of the heavy metal in the soil column (μg/kg).

Results and Discussion

Table 11.2.1 summarizes the physical and chemical properties of the soil samples collected from the coal fire. The experimental results show that the soil samples' pH levels were between 7.03 and 8.26, the organic matter content ranged from 11.44% to 18.17%, the carbonate content varied between 6.93% and 7.35%, and the total As content ranged from 13.25 to 17.45 mg/kg.

Adsorption Characteristics

Figures 11.2.1–11.2.3 show the adsorption isotherms of As with the tested soils at 25°C, 50°C, and 70°C, respectively. Figure 11.2.4 illustrates the relationship between soil temperature and adsorption at an initial As concentration of 20 mg/L for each of the six soil layers. Figures 11.2.1–11.2.3 show that when the As concentration was below 2.5 mg/L, the adsorption isotherms climbed rapidly at 25°C and 50°C, and slowly at 70°C. At all three temperatures, the adsorption amount varied slightly among soil layers. As the As concentration increased from 2.5 mg/L

Table 11.2.1
Properties of soil samples collected from the Dauqanhu coal fire, Urumqi city, Xinjiang Province, China.

Soil layer (cm)	Sand (%)	Silt (%)	Clay (%)	pH	Organic matter (%)	$CaCO_3$ (%)	Total As (mg/kg)
I	32.05	54.73	23.22	7.03	18.17	6.93	16.23
II	24.38	55.58	20.04	7.33	15.85	7.17	15.19
III	35.66	52.18	12.16	7.66	12.98	7.27	16.97
IV	28.80	60.17	11.03	7.94	12.37	7.35	17.45
V	23.23	57.20	19.57	7.77	11.52	7.16	14.85
VI	30.85	55.62	13.53	8.26	11.44	7.22	13.25

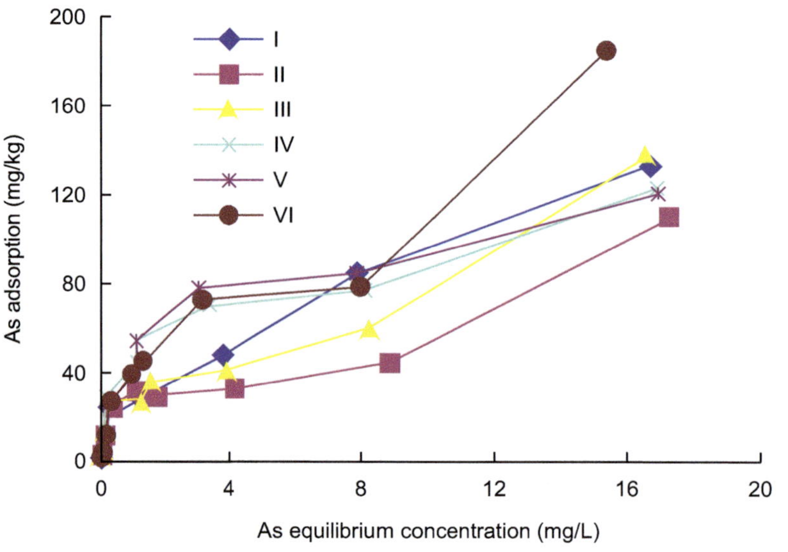

Figure 11.2.1. As adsorption as a function of the As adsorption-equilibrium concentration at 25°C.

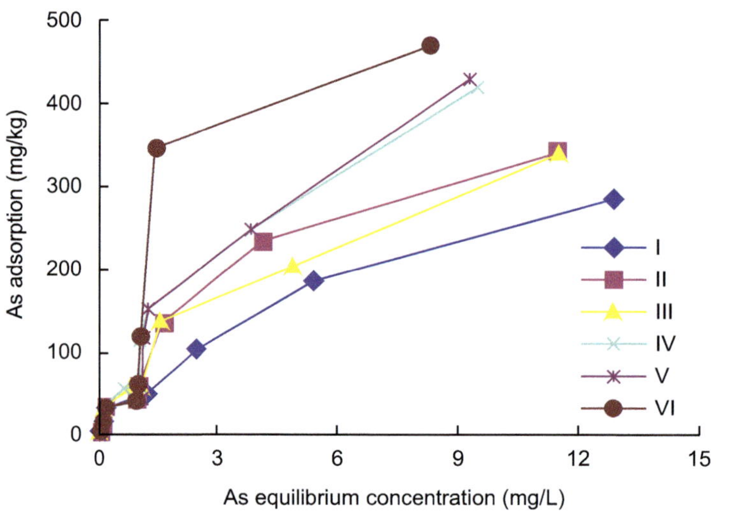

Figure 11.2.2. As adsorption as a function of the As adsorption-equilibrium concentration at 50°C.

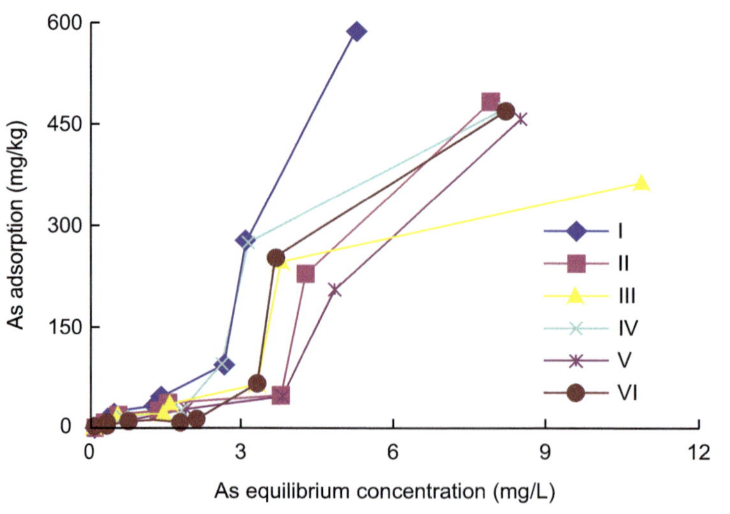

Figure 11.2.3. As adsorption as a function of the As adsorption-equilibrium concentration at 70°C.

to 20 mg/L, the upward trends in the adsorption isotherms at all three temperatures changed significantly, and all layers experienced continuous increases in the amount adsorbed. At 25°C, as the As concentration increased from 2.5 mg/L to 20 mg/L, the amount of As adsorbed by layer I continuously grew, first at a high rate and then at a relatively low rate, while adsorption in the other soil layers was relatively slow at first and then accelerated. At an initial As concentration of 20 mg/L, layer VI demonstrated the highest amount of adsorption, at 185.5 mg/kg, while layer II showed the lowest amount of adsorption, at 110.1 mg/kg. At 50°C, As was adsorbed by each layer at a high rate at first and then at a lower rate. At a concentration of 20 mg/L, the highest amount of adsorption occurred in layer VI at 469.6 mg/kg, while the lowest value, 285.5 mg/kg, occurred in layer I. At 70°C, the amount of adsorption in each layer grew with increasing As concentration, and the adsorption rate was high at the beginning before decreasing. At an initial As concentration of 20 mg/L, the maximum amount of adsorption occurred in soil layer I at 589.2 mg/kg, and the minimum amount occurred in layer III at 365.1 mg/kg. Figure 11.2.4 reveals that the amount of As adsorbed by each layer increased as the temperature rose.

Desorption Characteristics

Previous research has suggested that NaOH and phosphates are the most efficient substances for desorption of As from soil. Therefore, a 0.01 mol/L NaOH solution was used as the desorbent in this experiment. Figures 11.2.5 through 11.2.7 illustrate the relationships between the amounts of As adsorption and desorption obtained by fitting. Figure 11.2.8 shows the variation in the amount of As desorbed from each soil layer at a concentration of 20 mg/L.

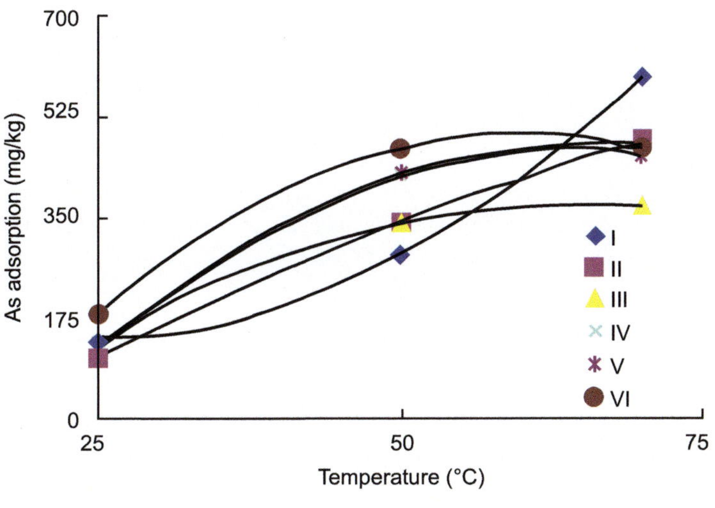

Figure 11.2.4. As adsorption as a function of temperature.

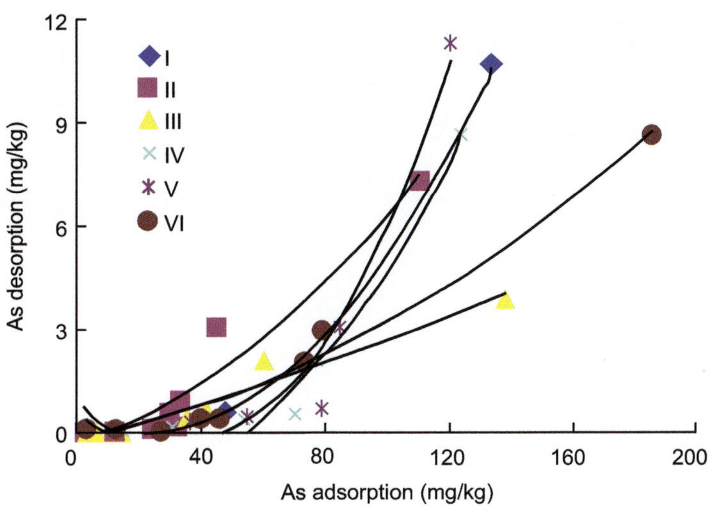

Figure 11.2.5. As desorption as a function of As adsorption, at 25°C.

As Figures 11.2.5 through 11.2.7 show, the relationships between the amounts of As adsorption and desorption can be properly fitted by quadratic functions ($y=ax^2+bx+c$). The values of R^2 range between 0.9372 and 0.9988, demonstrating a good correlation between adsorption and desorption and confirming the validity of the results of fitting. The amount of As desorbed from the tested soil increased as the amount adsorbed increased. This may have occurred because in the region of low adsorption, specific adsorption of As is the dominant adsorption process occurring in the soil, and the desorbent could hardly substitute the specifically adsorbed As. As the amount of adsorption increased, the number of specific adsorption sites on the surface of soil particles gradually decreased, while the nonspecific adsorption increased. Figure 11.2.8 shows that the amount of As desorbed from layers I, II, III, and VI increased with rising temperatures. As the temperature increased from 25°C to 50°C, layers V and IV experienced an increase in the amount of desorption, followed by a decrease when the temperature rose from 50°C to 70°C.

Table 11.2.2 shows the As concentrations in the leachates collected from the single-layer soil columns. Table 11.2.3 shows the As contents at different depths of the single-layer soil columns. Table 11.2.4 presents the As contents at different depths of the integrated soil column.

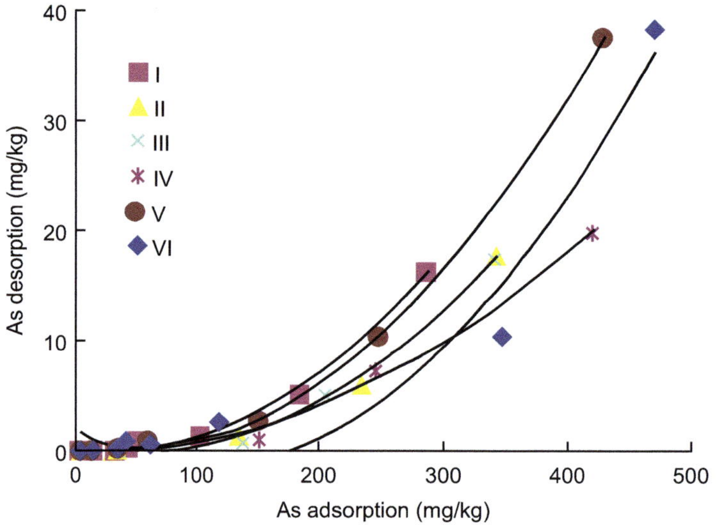

Figure 11.2.6. As desorption as a function of adsorption, at 50°C.

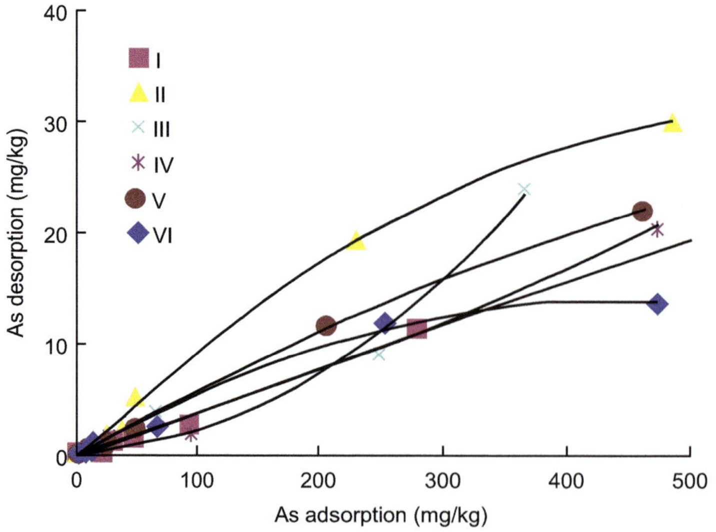

Figure 11.2.7. As desorption as a function of As adsorption, at 70°C.

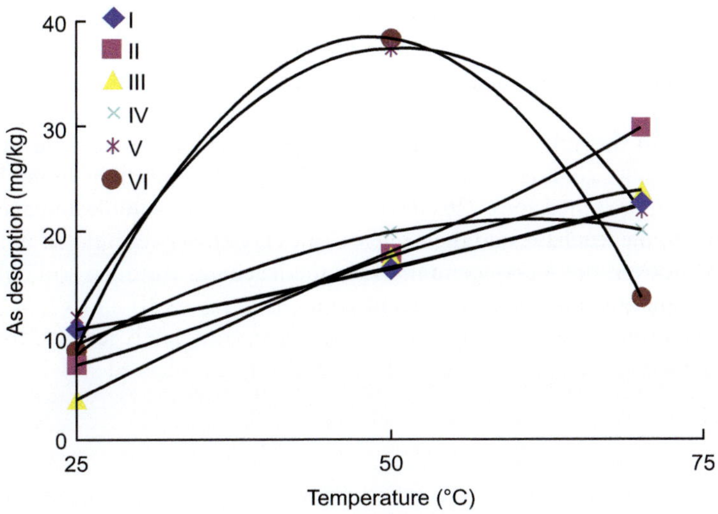

Figure 11.2.8. As desorption as a function of temperature.

Table 11.2.2
As concentrations in leachates from single-layer soil columns.

Leachate volume (mL)	I (μg/L)	II (μg/L)	III (μg/L)	IV (μg/L)	V (μg/L)	VI (μg/L)
0	1258.632	1095.815	359.338	856.723	1168.728	264.104
200	1169.244	987.063	211.141	804.629	1053.944	190.976
400	1067.268	948.542	199.206	638.141	803.592	160.016
600	906.048	850.833	166.516	554.596	768.76	124.08
800	750.324	562.074	138.985	441.833	455.3496	93.44
1000	483.996	423.0532	115.906	230.164	349.783	66.624
1200	331.632	264.2692	72.033	187.068	259.392	56.288
1400	167.844	129.228	52.848	89.526	95.907	53.04
1600	111.714	73.444	25.572	51.732	66.43	45.892
1800	54.096	46.1831	16.194	28.014	37.478	19.698
2000	44.505	38.982	15.594	27.534	37.058	19.551

Table 11.2.3
As contents at different depths of single-layer soil columns.

Depth (cm)	I (mol/m^3)	II (mol/m^3)	III (mol/m^3)	IV (mol/m^3)	V (mol/m^3)	VI (mol/m^3)
2	0.3063	0.2997	0.3358	0.3227	0.3199	0.3423
5	0.3197	0.3137	0.3463	0.3399	0.3265	0.3589
10	0.3445	0.3383	0.3787	0.3621	0.3508	0.3896
15	0.3782	0.3697	0.4059	0.3907	0.3793	0.4234
20	0.3932	0.4055	0.4438	0.4306	0.4107	0.4575

Table 11.2.4
As contents at different depths in an integrated soil column.

Depth (cm)	5	10	15	20	25	30	35	40	45	50	55	60
Concentration (mol/m^3)	0.40	0.42	0.43	0.43	0.44	0.44	0.45	0.45	0.45	0.46	0.46	0.46

Data from Tables 11.2.2, 11.2.3, and 11.2.4 were plotted in Figures 11.2.9–11.2.11, respectively, to depict the relationship between As concentrations and leachate volume for single-layer soil columns, as well as the relationships between As content and depth for both the single-layer and integrated soil columns.

Figure 11.2.9 indicates that all the soil layers released As rapidly when the leachate volume was smaller than 1500 mL; however, as the leachate volume further increased, the As concentrations in the leachate gradually stabilized, and the As release rate approached zero. This occurred because water-soluble and exchangeable As in the soil leach easily from the soil into the leachate, and their release rates are closely correlated with the percolation rate of the leachate in the soil. Moreover, the As concentrations in the leachates from the soil columns in layers I, II, IV, and V were substantially higher than those from the soil columns in layers III and VI. When the leachate volume exceeded 1500 mL, the easily soluble and exchangeable As had been completely leached from the soil columns, and the remaining small amounts of Ca–As, Fe–As, and Al–As tended to be released slowly.

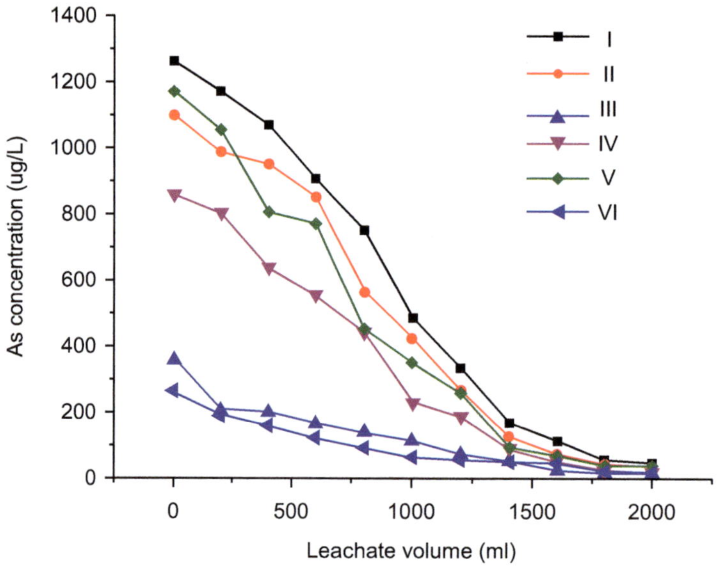

Figure 11.2.9. As concentration as a function of leaching volume.

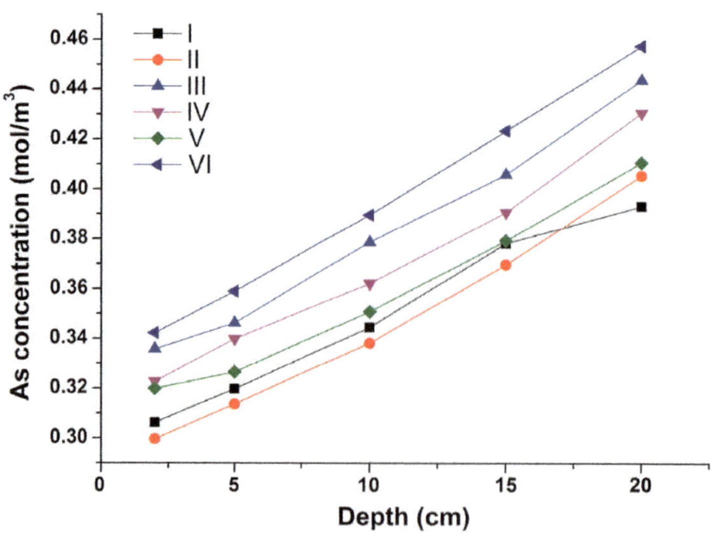

Figure 11.2.10. As concentration as a function of soil depth.

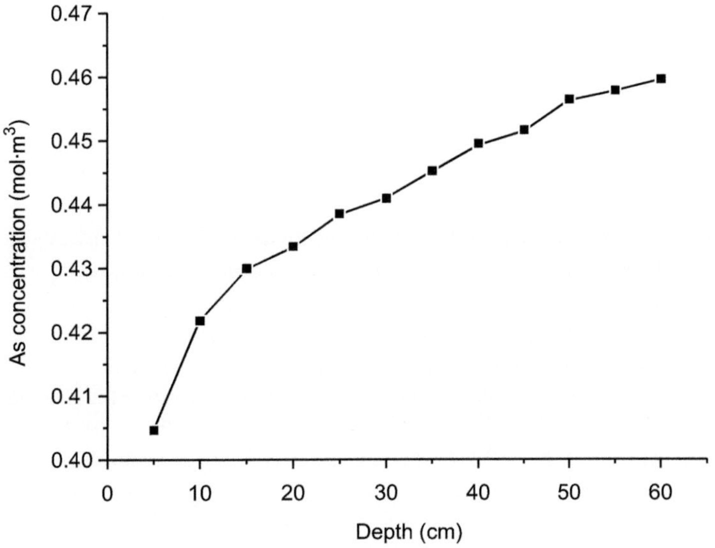

Figure 11.2.11. As concentration as a function of depth in the integrated soil column.

Table 11.2.5
Accumulated As release and release rates.

Soil layer	I	II	III	IV	V	VI
Accumulated As release (μg/kg)	3051.14	2540.20	710.66	1828.25	2230.07	607.77
Release rate (%)	4.0025	3.3784	0.9233	2.3606	2.9794	0.8297

Table 11.2.6
Analysis of correlation between accumulated As release and soil properties.

Parameter	Sand	Silt	Clay	pH	Organic matter	$CaCO_3$	Total As	Bulk density	Adsorption capacity K_F
Release	−0.495	0.256	0.827*	−0.782	0.722	−0.667	0.198	0.788	−0.513

*Significant correlation if $P>.05$.

Figure 11.2.10 shows that the As content of each single-layer soil column increased with increasing depth, suggesting that the amount of leached As decreased with increasing depth. Additionally, after the leaching process, the single soil layers were ranked by As content as follows: VI>III>IV>V>I>II.

Figure 11.2.11 shows that the integrated soil column experienced an increase in As content as the depth increased, with the most marked increase occurring in the topsoil. This suggests that the leaching process leached more As from the topsoil than from the other soil layers.

Accumulated Amount and Rate of Arsenic Release

Table 11.2.5 lists the accumulated amounts of As released from the soil columns and the corresponding release rates. The soil layers were ranked in descending order of release rate as follows: I, II, V, IV, III, and VI. This order reflects the relationship between release rate and As adsorption capacity of the soil layers: A soil layer with a higher As adsorption capacity tends to release As at a lower rate.

The statistical software SPSS was used to analyze correlations between the accumulated amount of As released and various soil properties including sand, silt, and clay percentages; pH levels; organic matter content; $CaCO_3$ content; total As; bulk density; and the adsorption capacity K_F (Table 11.2.6). The results show that, compared to other parameters, the bulk density and contents of clay and organic matter were more strongly correlated with the accumulated As release. Further analysis of the soil media's pore structure is needed to determine the reasons behind these findings.

Conclusions

The generation of typical pollutants from a UCF was studied by programmed heating–oxidation, TGA, and a FactSage simulation for the first time. The gaseous pollutants predominantly comprised CO and CO_2 in addition to trace C_nH_m (g). Moreover, the products included two major greenhouse gases: CO_2 and CH_4. With increasing temperatures, the coal–oxygen reactions accelerated, yielding products at higher rates. The reaction constant of the coal–oxygen reactions, oxygen consumption rates, and CO formation rates varied in similar patterns. The FactSage simulations suggested that temperature affected the composition of the products. The simulated products were primarily CO_2 and CO, and the concentrations of other trace products varied with temperature. Further studies are required to characterize the formation of solid products in coal–oxygen reactions and their accumulation characteristics under UCF conditions.

This study also was the first to examine the characteristics of As adsorption and desorption in soil affected by a coal fire. The As content of the soil samples collected in Daquanhu coal fire zone ranged from 13.25 to 17.45 mg/kg. Both the amounts of adsorption and desorption increased with increasing temperature, but the former increased faster than the latter. Since this study focused on soil samples from only one site, the experimental data only reflect the characteristics of As adsorption and desorption by the soil layers at this site. Further research is required to systematically explore how the combination of temperature and soil properties affects the soil's adsorption and desorption behavior.

The results of the leaching test performed on single-layer soil columns show that, under the application of endogenous As, all the soil layers released As rapidly during the early period of leaching. In the later period, the release rate gradually approached zero, as indicated by the stable As concentrations in the leachates. In each single-layer soil column, the As content increased with depth, indicating that the amount of As removed by the leaching process increased as the depth increased. By the end of the leaching process, the soil layers were ranked in descending order of As content as follows: VI > III > IV > V > I > II. The data from the leaching test conducted on the integrated soil column also demonstrate that the As content increased with increasing depth. This trend was especially marked in the top soil, suggesting that the leaching process transported more As through the top soil than through the other soil layers.

Acknowledgments

This study was supported by the Nature Science Foundation of China (NSFC) with grant no. 51374182.

References

Chatterjee, R.S., 2006. Coal fire mapping from satellite thermal IR data-a case example in Jharia Coalfield, Jharkhand. ISPRS J. Photogrammetry Remote Sens. 60, 113–128.

Engle, M.A., Olea, R.A., O'Keefe, J.M.K., Hower, J.C., Geboy, N.J., 2013. Direct estimation of diffuse gaseous emissions from coal fires: current methods and future directions. Int. J. Coal Geol. 112, 164–172.

Engle, M.A., Radke, L.F., Heffern, E.L., O'Keefe, J.M.K., Hower, J.C., Smeltzer, C.D., Hower, J.M., Olea, R.A., Eatwell, R.J., Blake, D.R., Emsbo-Mattingly, S.D., Stout, S.A., Queen, G., Aggen, K.L., Kolker, A., Parakash, A., Henke, K.R., Stracher, G.B., Schroeder, P.A., Román-Colón, Y., Schure, A., 2012. Gas emissions, minerals, and tars associated with three coal fires, Powder River Basin, USA. Sci. Total Environ. 420, 146–159.

Engle, M.A., Radke, L.F., Heffern, E.L., O'Keefe, J.M.K., Smeltzer, C.D., Hower, J.C., Hower, J.M., Prakash, A., Kolker, A., Eatwell, R.J., Schure, A., Queen, G., Aggen, K., Stracher, G.B., Henke, K.R., Olea, R.A., Colón, Y.R., 2011. Quantifying greenhouse gas emissions from coal fires using airborne and ground-based methods. Int. J. Coal Geol. 88, 147–151.

Feng, R., 2004. Study and Comparison of Chemical Thermodynamic and Kinetic Calculation of Representative Trace Elements in Coal Combustion Process (M.S. Thesis). Huazhong University of Science and Technology, Wuhan, Hubei province, China (in Chinese).
Finkelman, R.B., 2004. Potential health impacts of burning coal beds and waste banks. Int. J. Coal Geol. 59 (1–2), 9–24.
Fu, G., 2002. Study on the Characteristics of Organic PAHs Pollutants Emission from Coal Combustion (M.S. Thesis). Zhejiang University, Hangzhou, Zhejiang province, China (in Chinese).
Hower, J.C., Henke, K., O'Keefe, J.M.K., Engle, M.K., Blake, D., Stracher, G.B., 2009. The Tiptop coal-mine fire, Kentucky: Preliminary investigation of the measurement of mercury and other hazardous gases from coal-fire gas vents. Int. J. Coal Geol. 80, 63–67.
Hower, J.C., O'Keefe, J.M.K., Henke, K.R., Bagherieh, A., 2011. Time series analysis of CO concentrations from an Eastern Kentucky coal fire. Int. J. Coal Geol. 88, 227–231.
Kong, D.H., 2005. Study on the Emission of Pollutants from Coal Combustion of Circulating Fluidized Bed (M.S. Thesis). Shandong University, Shandong province, China (in Chinese).
Kuenzer, C., Zhang, J., Tetzlaff, A., Dijk, P.V., Voigt, S., Mehl, H., Wagner, W., 2007. Uncontrolled coal fires and their environmental impacts: investigating two arid mining regions in north-central China. Appl. Geogr. 27, 42–62.
Liang, Y.C., Liang, H.D., Zhu, S.Q., 2014. Mercury emission from coal seam fire at Wuda, Inner Mongolia, China. Atmos. Environ. 83, 176–184.
Nie, J., Zeng, Q., Pu, Y., Jin, S., 2017. Migration of typical heavy metal As in the soil of underground coal fire. J. China Coal Soc. 42 (2), 527–537 (in Chinese).
Nolter, M.A., Vice, D.H., 2004. Looking back at the Centralia coal fire: a synopsis of its present status. Int. J. Coal Geol. 59 (1–2), 99–106.
O'Keefe, J.M.K., Henke, K.R., Hower, J.C., Engle, M.A., Stracher, G.B., Stucker, J.D., Drew, J.W., Staggs, W.D., Murray, T.M., Hammond III, M.L., Adkins, K.D., Mullins, B.J., Lemley, E.D., 2010. CO_2, CO, and Hg emissions from Truman Shepherd and Ruth Mullins coal fires, eastern Kentucky, USA. Sci. Total Environ. 408, 1628–1633.
O'Keefe, J.M.K., Neace, E.R., Lemley, E.W., Hower, J.C., Henke, K.R., Copley, G., Hatch, R.S., Satterwhite, A.B., Blake, D.R., 2011. Old Smokey coal fire, Floyd County, Kentucky: estimates of gaseous emission rates. Int. J. Coal Geol. 87, 150–156.
Pone, J.D.N., Hein, K.A.A., Stracher, G.B., Annegarn, H.J., Finkleman, R.B., Blake, D.R., McCormack, J.K., Schroeder, P., 2007. The spontaneous combustion of coal and its by- products in the Witbank and Sasolburg coalfields of South Africa. Int. J. Coal Geol. 72, 124–140.
Stracher, G.B., Taylor, T.P., 2011. The effects of global coal fires. In: Stracher, G.B., Prakash, A., Sokol, E.V. (Eds.). Stracher, G.B., Prakash, A., Sokol, E.V. (Eds.), Coal and Peat Fires: A Global Perspective, vol. 1. Elsevier, pp. 101–114.
Stracher, G.B., Taylor, T.P., 2004. Coal fires burning out of control around the world: thermodynamic recipe for environmental catastrophe. In: Stracher, G.B. (Ed.), Coal Fires Burning Around the World: A Global CatastropheInt. J. Coal Geol., vol. 59, pp. 7–17 (1–2).
Xu, C.F., Zhang, P.Y., Xia, M., Sun, X.X., 2005. Effect of the experiment condition on performance of coal combustion. J. Huazhong Univ. Sci. Technol. 33 (5), 73–75 (Natural Science Edition; in Chinese).
Zeng, Q., 2012. Study on the Thermal Dynamic Characteristics of Combustion System for Coal Fires in Xinjiang Region (Ph.D. Thesis). China University of Mining and Technology, Xuzhou City, China (in Chinese).
Zeng, Q., Nie, J., Pu, Y., 2016. Characteristics of the distribution of typical heavy metals in the soils of underground coal fire. J. China Coal Soc. 41 (8), 1989–1996 (in Chinese).
Zeng, Q., Tiyip, T., Wuttke, M.W., Kurban, B., 2014. Characteristics of the heavy metal distribution in the soils of underground coal fire and the evaluation of its impact on environment. J. China Univ. Min. Technol. 43 (4), 695–700 (in Chinese).

CHAPTER 12

Colloid Technology for Preventing and Extinguishing the Spontaneous Combustion of Coal

Coal fires are global disasters that seriously threaten ecosystems. They cause human-health problems, and contribute to the loss of an important natural resource while causing air, water, and soil pollution. In 2010, spontaneous combustion ignited bituminous coal in the underground Sunjiacha Coal Mine in Shenmu City, Shaanxi Province, China. A road on the premises of the mine, once visible near the center of this picture, was buried in 2018 by fallen rock. Smoke from subsurface combustion is visible at the left and right sides of this picture. The vertical field of view is about 35 m. *Photo by Zhai Xiaowei, 2011.*

CHAPTER CONTENTS

Coal and Peat Fires: A Global Perspective
Edited by Glenn B. Stracher

https://doi.org/10.1016/B978-0-12-849885-9.00012-3

Photo by Zhai Xiaowei, 2011.

12.1 Colloidal Technology for Fighting the Spontaneous Combustion of Coal

Zhai Xiaowei

In 2010, an underground bituminous-coal fire started by spontaneous combustion in the Sunjiacha Coal Mine in Shenmu city. In 2018, fallen rock buried a road that was by the person in the picture.

Introduction

Colloid is defined as a system in which the size of dispersed particles is between 1 and 100 nm. According to the state of the dispersed system, a colloid can be divided into three types: aerosol, lyosol, and solid sols. Under appropriate conditions, dispersed particles of sol or a high polymer solution can interconnect into a network structure. With the aqueous medium filled in the network, the system becomes a semisolid jelly, with fluidity lost. The substance in this state is called gel.

Gel, a special form of colloid, is a special state between solid and liquid. It not only shows the characteristics of some solids (nonfluidity, certain geometry, elasticity, strength and yield values, etc.), but also retains the characteristics of some liquids (the diffusion rate of ions in the gel with water as a medium is like that in the aqueous solution).

Colloidal Mitigation of Spontaneous Combustion

Colloid material, different from solid and liquid, is a special material. The main component of a colloid is water. It has some advantages of fire extinguishing with water, with good heat absorption and cooling performance. It also has some characteristics of a solid, which can block the air leakage passage and cut off the oxygen channel. In addition, most colloids have the ability to prevent oxidation of coal. Therefore, the colloid material has good fire extinguishing performance, and is widely used against the spontaneous combustion of a coal seam. A lot of colloids can be used for mine fire extinguishing. While these colloids have something common, they have their own characteristics, too (Cao et al., 2007; Deng et al., 2007).

There are many colloids used to control mine fires currently, but different mines have different requirements for the material. According to the characteristics of coal mine fire, the selection of gel material for mine-fire control should follow the following principles:

(1) Innocuous. The material has no harm to the health of the underground workers, no corrosion to the equipment, and no pollution to the underground environment.
(2) Inexpensive and simple processing equipment. Fire extinguishing material should be economical and practical. Besides, simple processing equipment is required due to the limitation of the specific working environment at the pit.
(3) Good leaking stoppage performance. The concentration of oxygen in the space will be greatly reduced when the path of air into the coal seam is blocked. Meanwhile, it is difficult for the outside air to enter the goaf.
(4) Good permeability. Colloid can easily get into the loose coal body and then form an organized whole so that the oxygen molecules can hardly enter the coal body.

(5) Good high temperature performance. Because of the high temperature in the area of spontaneous combustion of coal, the material should not decompose easily at high temperature, but should keep its original characteristics to give full play to its fire extinguishing effect.
(6) Good heat absorption performance. The material should have a high specific heat capacity, which can speed up the reduction of the temperature of a high temperature coal seam.

Colloid Composition and Classification

Colloid Composition

Colloid technology uses seriflux, a mixture of a base material, additives, and water. This mixture is used to prevent or extinguish a coal fire and is injected with a pump (grouting system) into the area of the fire. Seriflux reduces the temperature of the area near the injection site and at the same time penetrates cracks in the coal seam and pore spaces while under pump pressure and because of its self-weight. The seriflux will form a gel or colloid at the igniting area to block oxygen diffusion, so as to prevent the coal oxidation, and then reduce the internal temperature of coal to achieve the effect of fire prevention and extinguishment.

For the colloid, which forms by the physical or chemical action of two raw materials in water, the main raw material is called the base material, and the material for promoting gelling is called a coagulant or gelling agent. For example, in preparation for a silica gel with a sodium silicate solution and acid, the sodium silicate solution is referred to as the base material and the acid as a coagulation accelerator (coagulant agent). The gel can then be prepared by adding a raw material into water, such as gelatin (the gelatinizer). Additives such as fly ash, loess, and other filling materials are called aggregates or reinforcers.

Colloid Classification

The colloid for fire extinguishing in a coal mine is different from the colloid system in chemistry, which is a dispersion system of microparticles. While the colloid for mine fire prevention and control mainly refers to viscous fluid, such as water gel, pseudogel, mixed colloid comprised of mud and gel or pseudogel, and thick fluid. The process to form gel automatically by a certain concentration of sol, true solution of macromolecular mixture, or other dispersed system is called gelatination. A large amount of water is mechanically occluded in holes of the gel porous structure, interlayers of clay minerals with hydrophilism (the material has a greater affinity for water, which can attract water molecules or dissolve easily in water), or high polymer materials. The colloid is in a semisolid state and is difficult to flow. The colloid material to prevent and extinguish fire has some characteristics of solids, such as certain strength. But it is different from solid because it consists of both phases of solid and liquid. According to the property and characteristics of a colloid preventing and extinguishing fire, it can be divided into three types: gel, thickened colloid, and composite colloid:

(1) Gel fire extinguishing material: A kind of material is formed through chemical reactions of sol or true solution, without mobility but with geometric shape, or pseudogel of colloidal particles in heaps by absorbing water of pure absorbent resin.
(2) Thickened colloid-fire extinguishing material: A small amount of hydrophilic materials is added to the seriflux of silt and fly ash to increase the consistency of the colloid, making it more uniform macroscopically, more difficult for sedimentation with good pipeline fluidity and good adhesion on coal walls.
(3) Composite colloid-fire extinguishing material: A small amount of polymeric additive is added into the seriflux of mud and sand or fly ash. Owing to the bridging effect of polymer materials, seriflux becomes the composite colloid, which can hardly flow. Some other materials can be added to seriflux like mud and sand and fly ash, and the colloid that has lost its fluidity through chemical reactions is also called composite colloid. The solid granules in this kind of composite colloid only play the role of filling, but do not participate in the formation of the network structure skeleton. Adding more base material into thickened colloid can result in the loss of its fluidity, and can make composite colloid, too.

According to differences in mechanical properties, colloids for extinguishing a fire can be classified as an elastic gel (elastic colloid) and a rigid gel (rigid colloid). In an elastic colloid, the particles can deform without structural damage under an external force. This kind of colloid is also called an elastic colloid. In a rigid colloid, the rigid particle sol (SiO_2, TiO_2, V_2O_5, Fe_2O_3, etc.) is a nonelastic gel. By dewatering, a rigid gel becomes a dry gel. If when placed in water, the dry gel becomes a gel again, the gel is called a reversible gel. If the rigid gel does not absorb water again to form a gel, it is called an irreversible gel.

Colloid Properties for Preventing Combustion

Control of Gelation Time

For a colloid fire extinguishant formed by chemical reaction, the rate of gelation depends on the rate of chemical reaction. The colloid forming rate of the absorbent clay or absorbent resin in mud is mainly determined by the water absorption rate of the base material. The greater the concentration of base material, the more water absorption per unit time. For linear polymer materials, the greater the amount of base material, the greater the amount of solution per unit time, so the gelation time is short. In addition, the larger the size, the longer the gelation time.

Mine fire extinguishing generally chooses gelling materials whose gelling time is from a few seconds to 10 min. The colloid should have a different gelling time according to different conditions. If it is used to seal the leakage and put out the high temperature fire point, the gelling time should be controlled 30 s after the mixed liquid erupts from the spray gun. If it is used to prevent spontaneous combustion of float coal, its gelling time should be controlled 5–10 min after the mixed liquid erupts from the spray gun.

Water Retention Capacity

The colloid has water retention capacity, which can make a certain amount of water fixed in the colloidal network structure skeleton lose fluidity. The mass fraction (90%) of the colloid is water, so as to give full play to the role of water fire extinguishing. A large amount of water is fixed between the colloidal network without fluidity. Colloid fire extinguishing technology makes full use of the water retention of colloids to use the advantages of water in fire extinguishing. It overcomes the disadvantages of water's strong liquidity, which makes fire extinguishment on the top of the tunnel and working face out of reach.

Different colloids have different water retention capacity. For example, silica gel can hold more than 90% of the water in the network structure to make it lose mobility. When the concentration of V_2O_5 in the gel is higher than 0.005%, the colloid will lose fluidity. A colloid can be formed when the gelatin concentration is higher than 1% and when the agar solution concentration is higher than 0.2%.

Pressure Resistance

One of the most important indexes of colloid-fire extinguishing is its plugging property. The pressure resistance of colloids has a great influence on the plugging effect. The greater the intensity of the colloid, the greater the air leakage pressure it can stand, and the better effect of plugging. The pressure resistance of colloids increases with the increase of the concentration of the base material. With the usual concentration of base material, when the colloid thickness reaches 3 cm, it is resistant to air pressure 2942 Pa.

Permeability

In the process of mine fire extinguishing, it is necessary to inject the colloid into the coal body. The burning coal is often a loose coal, which is a porous medium containing a large number of voids and cracks. Permeability is the performance of porous media to transmit fluid, which is not only related to the nature of the framework (particle composition, distribution, size, filling), but also is related to the nature of the fluid. For colloid formed through chemical reactions, the gel is Newton fluid, easy to flow for its very small viscosity. In the process of chemical reaction, the flow resistance increases, and then it is fixed to the coal seam after the colloid is formed. The colloid formed by the linear polymer material in water has obvious thixotropy. The flow velocity in the pipeline is very high. The apparent viscosity of the polymer is infinite. The viscosity is small, and the flow resistance is small. When the colloid enters into the coal seam, the seepage velocity slows down, which results in the increase of the viscosity and eventual retention in the coal seam. An external force greater than its yield value must be added to make the colloid flow. The higher the colloid concentration, the less the colloid passes through the coal. The results show that when the concentration of a colloid is 2%, the yield value of the colloid is greater than that of its own gravity, so the retention among the coal seam is possible.

Heat Energy Absorption

The main component of the colloid is water (water accounts for more than 90% of the total weight). The specific heat of water is very large. As the water temperature rises, it can absorb a large amount of heat energy, so as to reduce the environmental temperature. At a higher temperature, water vaporization in the colloid can absorb part of heat. According to the theoretical calculation, when the concentration of the base material is 6%, the heat absorption by colloid vaporization per cubic meter is more than 4 × 104 KJ. For silicate colloid, if ammonium salt is used as a coagulant, the ammonium salt can absorb a large amount of heat in the process of dissolving. Meanwhile, the water temperature is reduced, and the heat absorbing effect of colloid can be improved.

Temperature Tolerance

Because of the large amount of water solidified in a colloid, the moisture in the colloid at a high temperature evaporates slowly. Therefore, the internal temperature of the colloid will not be very high. So, the colloid will not vaporize rapidly at a high temperature.

Drag Reduction

When the colloid concentration is less than 4%, the flow velocity is greater than 3 m/s. The flow resistance of the colloid is smaller than that of water, showing a certain drag reduction. Considering the improvement of the yield value of the colloid to decrease the colloid flow resistance, the concentration of the colloid is generally 3%–5%.

Chemical Resistant Effect

Colloid-fire extinguishing material has a good chemical resistant effect. The resistant rate of gel material reaches more than 50% at 60°C and more than 70% at 80°C. With the increase of temperature, the resistant rate of gel increases, and particularly, that of fly ash increases obviously. Colloid and coal can also adhere to each other and spread, with an inert effect on the coal surface.

Photo by Zhai Xiaowei, 2011.

12.2 Gel for Extinguishing a Coal Fire

A coal fire in the underground Liujiamao Coal Mine in Shenmu city. The fire started in January of 2016 and was under control by December of 2017. The smoke seen here is coming out of a fracture in a coal seam that is overlain by loess.

Characteristics

Through chemical reactions, sol, or true solution, forms a kind of material that cannot flow and has a certain geometric shape. Pseudogel is formed with a conglomerate of colloidal particles by absorbing water of pure absorbent resin. Gels exhibit certain characteristics of solids, such as geometric shapes, strength, elasticity, and yield values. However, gel is different from solid in the internal structure. It consists of network structure and dispersed medium. In the newly formed hydrogels, not only the network structure, but also the dispersed medium is the continuous phase, which is the main characteristic of gel (Xu et al., 2003, 2004). The pseudogel used in mine fire extinguishing is usually a highly water absorbent material, mainly including water-absorbing clay and resin. High water-absorbing materials are mixed with water at higher concentration. Most of the water molecules are controlled by water-absorbing materials and become a material with high viscosity and solid qualities. The pseudogel is the deposit of small particles of water-absorbent granules, while the network structure of the gel and dispersed medium are continuous. Each water-absorbent particle of pseudogel can also be a small gel molecule.

Both the pseudogel and the gel in mine fire extinguishing materials are called gel materials. FHJ16 composite colloidal additive is a kind of composite material developed by the Xi'an University of Science and Technology. The material is comprised of linear hydrophilic polymers, cross-linked superabsorbent resins, and a small amount of inorganic and organic materials that can improve wettability and inhibition between coal bodies. FHJ16 is a white powder, which forms a colorless transparent gel when dissolved in water. This gel is mainly a linear polymer aqueous solution, so it has a certain degree of viscoelasticity, and it is a composite elastic gel, shown in Figures 12.2.1 and 12.2.2.

Preparation and Injection

The mobile injection technology and injection techniques are developed for the treatment of partial high temperature area of underground coal. The techniques utilize a movable colloidal pressure injection system to mix the various colloidal materials in proportion, and then deliver the mixed solution to the fire area by glue pipe and drill hole. The mixture acts as a gelling agent in the fissures (or pores) of the loose coal seam cracks (or pores), which can block the air leakage channel, absorb the heat accumulated in the coal, reduce the coal temperature, and finally achieve the purpose of fire extinguishing. The techniques are convenient, flexible, and can handle fire in local and high temperature areas quickly and efficiently. However, the injection flow of the mobile injection technology is small. It is difficult for the technology to achieve a large area of high temperature rapid fire extinguishing. Therefore, the main application of mobile injection technology is the mobile colloid injection system.

In this system, the base material, the coagulation accelerator, the intensifier, and water are mixed in proportion and pressed into the coal seam. There are three main types of underground injection systems used in the scene: continuous injection, quantitative proportioning pumping, and fully automatic quantitative injection.

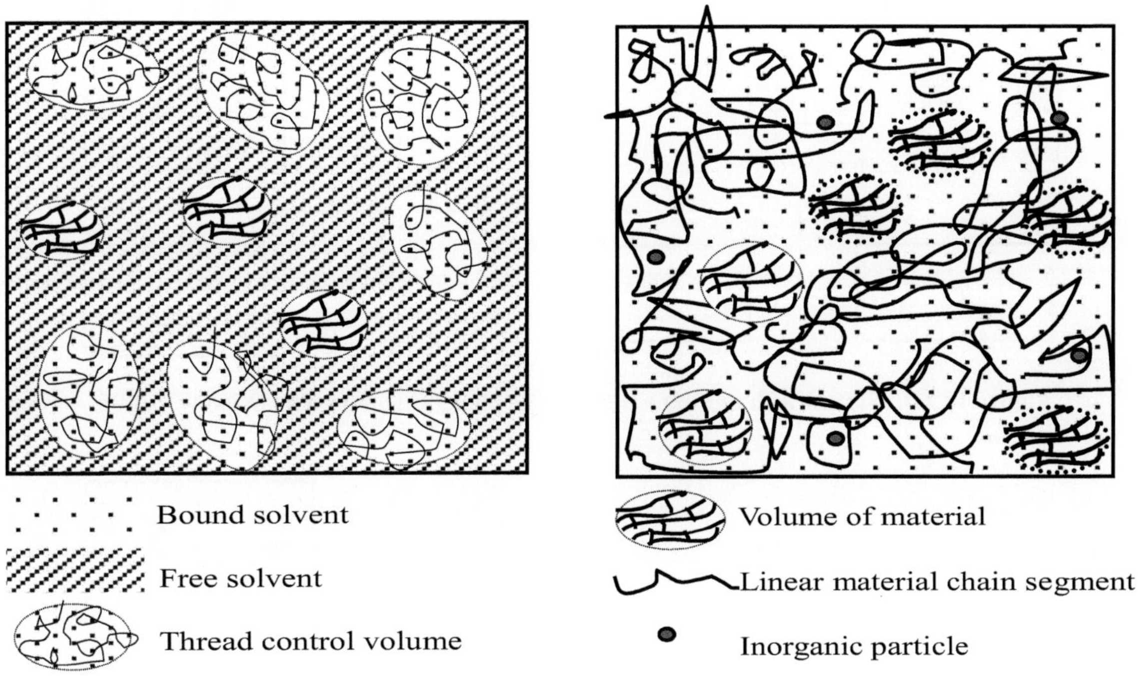

Figure 12.2.1. Process of FHJ16 forming a gel in water. Thinner concentration (left) and gel above the critical concentration (right). Figure by Zhai Xiaowei, 2005.

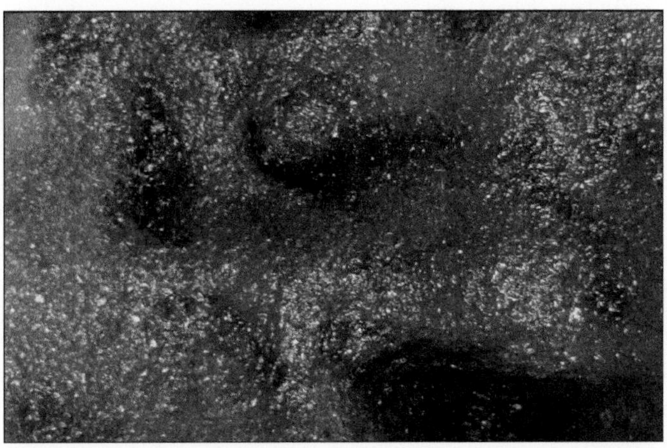

Figure 12.2.2. Gel made with FHJ16 and water. Photo by Deng Jun, 2004.

Continuous Injection System

Four bins are used alternatively to reach continuous injection of glue (one for use, one for dosing). The system can run continuously but needs four boxes with large volume. Since the underground space is relatively narrow, usually it is difficult to operate, and the actual operation can be very complicated.

System of Four Bins and a Single Pump As shown in Figure 12.2.3 for a system of double bins and a single pump, the continuous injection can be achieved by adding two material bins. When the A1 bin and B1 bin valves are open for glue injection, close the A2 and B2 bin valves, so the dosing is under preparation. When the material in the A1 and B1 bins is used up, close the A1 and B1 valves. At the same time, open the A2 and B2 bin valves to ensure continuous operation of the system. After the system stops gelling, the water pump and the pipeline are cleaned.

System of Four Bins and a Double Pump With a system of double bins and double pumps, two bins can be added to form a double-pump system with four bins to achieve continuous injection, as shown in Figure 12.2.4. When A1

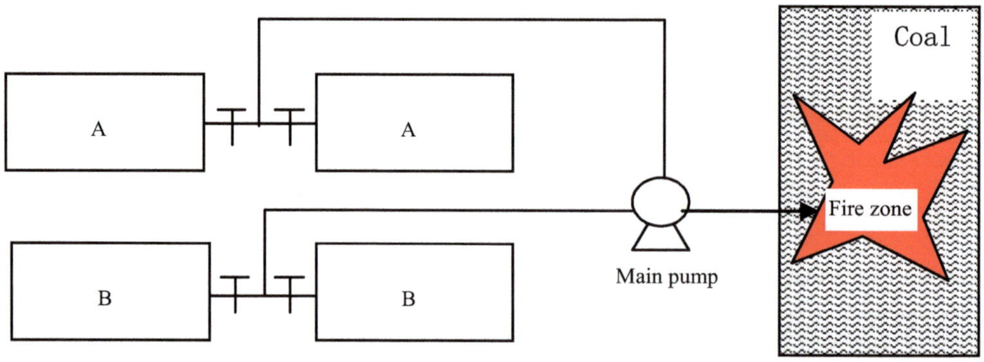

Figure 12.2.3. System with four bins and a single pump. Figure by Zhai Xiaowei, 2005.

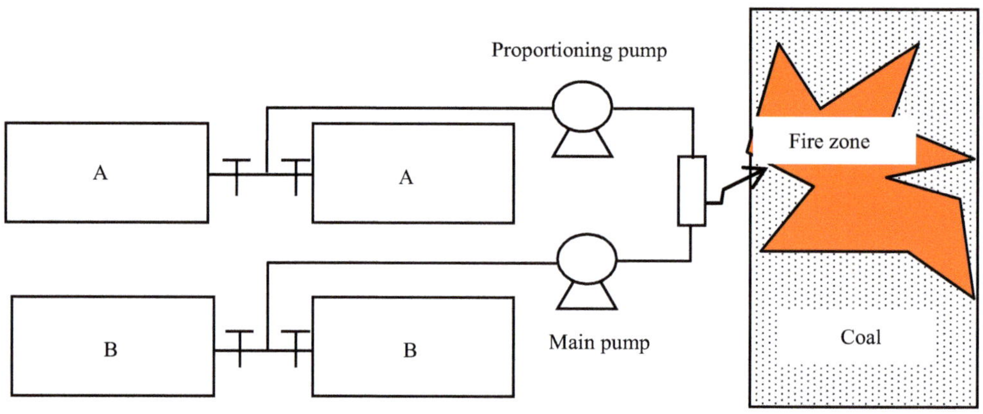

Figure 12.2.4. System of four bins and double pumps. Figure by Zhai Xiaowei, 2005.

and B1 bins run out of material, open the A2 and B2 valve, and then close the A1 and B1 valve. Add water and materials into A1 and B1 bins; continue mixing to get it ready for use, to achieve the effect of continuous operation.

Gel and the Quantitative Proportioning Pumping Injection System The system can be used to realize the quantitative proportioning of the glue material without the manual proportioning by adopting the glue injecting system with the proportioning pump and the main pump. However, since the material ratio is fixed, it cannot be adjusted flexibly for actual needs.

Injection system of quantitative proportioning pumping A single pump or a double-pump system has a clear defect—the dosing must be made one by one. Meanwhile, advantages are that the equipment is simple, the existent pumps of various types in the mine can be used, and the investment is small. In order to reduce human factors to avoid manual underground dosing and to simplify the operation, a quantitative dosing pumping injection system is developed.

The system is mainly comprised of a main piston-type pump, two quantitative proportioning pumps, two material bins, and a water tank. The liquid base material and coagulant accelerator, respectively, are loaded into the A and B bins. Start the main pump and the proportioning pump. After the injection, close the proportioning pump first, and then close the main pump after the pipeline is cleaned (Figure 12.2.5).

Semiautomatic proportioning system Three pumps of the system share one motor, as shown by Figure 12.2.6. When in use, the valve of feed pipe 11, discharge pipe 8, water inlet and outlet pipes 12, 13 are opened, other valves are closed, and the liquid base material and the coagulant, respectively, are poured into A and B bins. Open the water pipe for water to go into the water tank. Start the main pump first, and then start motor 4, making the main pump 3 and proportioning pumps 1 and 2 operate roughly at the same time. The system transports colloid to the required location. During the process, workers do not need to pay attention to the ratio, as long as the base material

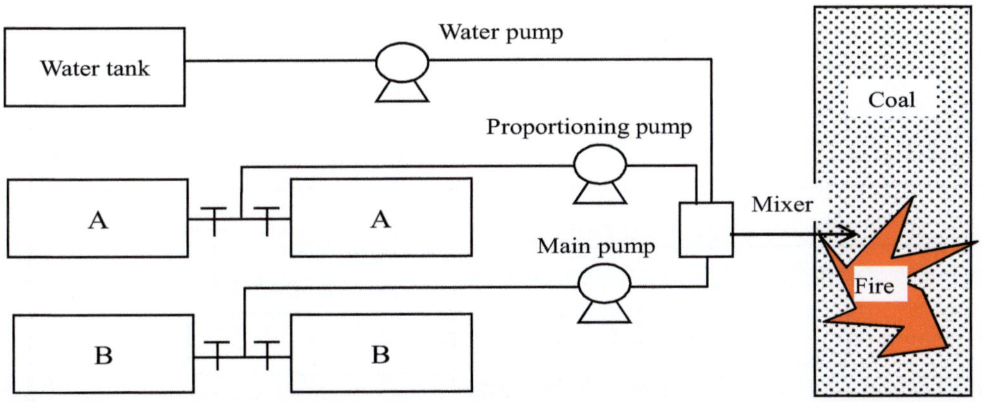

Figure 12.2.5. Schematic diagram of the glue injection system of quantitative proportioning pumping. Figure by Zhai Xiaowei, 2005.

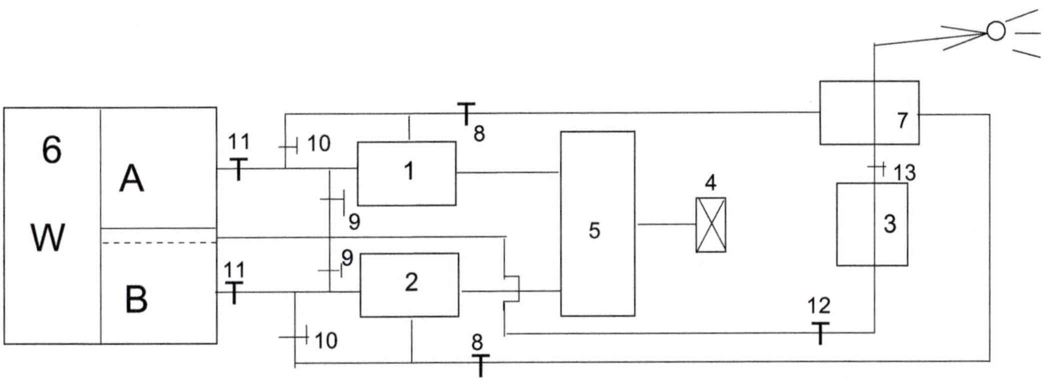

Figure 12.2.6. Semiautomatic proportioning technology. 1, 2 – quantitative proportioning pump, 3 – main pump, 4 – motor, 5 – gearbox, 6 – material bin, 7 – mixer, 8 – discharge valve, 9 – flushing pipe valve, 10 – return material pipe valve, 11 – inlet material pipe valve, 12 – inlet water valve, 13 – outlet water pipe valve. Figure by Zhai Xiaowei, 2005.

and coagulant are constantly poured into A and B bins and water is injected into the tank continuously. When the injection is finished, close the feed pipe valve 11, and open the flushing pipe valve 9 to flush the proportioning pump and the pipeline. Close the proportioning pump first, and then close the main pump.

The system is easy to operate, and it solves the difficult problem of preparing materials in an underground environment to achieve the continuity of the colloid injection so that the superiority of the colloid-fire extinguishing is more prominent.

Gel and the Fully Automatic Quantitative Injection System Fully automatic quantitative glue injection system mainly mixes all kinds of colloidal materials in proportion through the special colloid press machine (Figure 12.2.7) and then transports them to the ignition point. The equipment is mainly designed for FHJ16 colloid fire extinguishing materials. The system is easy to use, and the proportion of the glue material can be adjusted by the equipment (controlling the colloid strength and gelling time).

Water, coagulant, and reinforcing agents are added from their respective material boxes, and the coagulant is sent into the mixer quantitatively through the solid material push device. Water, coagulant, and base material are mixed in the mixer, then output from the main pump, the current diverter to the fire area after being pressed through the pipe and drill. The control of the adding quantity of the base material, the coagulant accelerator, and the reinforcing agent is realized by the stepless speed regulator. The proportion of the materials can be changed by adjusting the speed of the stepless speed variator.

When used, the water pipe is connected with the water inlet of the equipment first, and the outlet of the current diverter is connected with the glue injecting hole through a glue-injecting high-pressure pipe. Then open the inlet

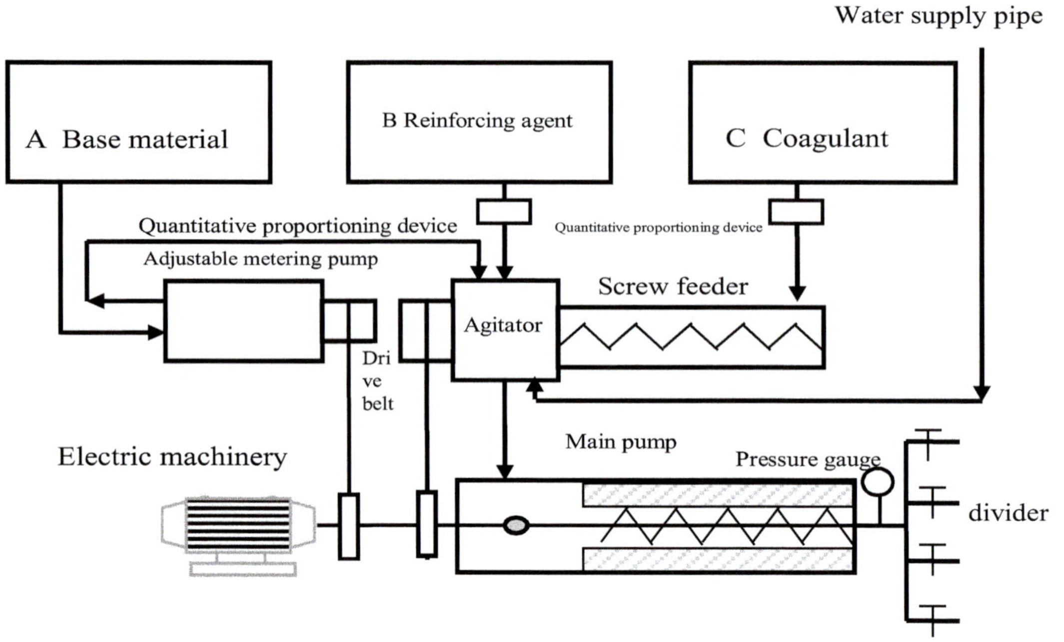

Figure 12.2.7. Diagram of the XKJ (model number) of the colloid injection machine. Figure by Zhai Xiaowei, 2005.

valve to supply water to the equipment. Start the equipment; the base material, the coagulant agent, and the reinforcing agent are added into the corresponding material box when the system is stable, and the amount of various materials is adjusted according to the required proportion. After the equipment is adjusted to be stable, the material is fed into the material box in time, and the equipment is adjusted for the expected proportion to input materials, thus reducing labor intensity and improving work efficiency. When glue injection is over, stop the addition of base material, accelerator, and reinforcing agent, and clean the equipment and pipeline with water, so as not to block the equipment and pipeline. Cut off water and electricity after completion of all these tasks.

Purpose

This equipment is used to control the spontaneous combustion of coal or for emergency management when there is the danger of spontaneous combustion. It is mainly used in a relatively clear fire position like an underground roadway caving zone, coal seam fault, and so on, and the range of use is small. At the same time, it can also be used in the filling of the loose coal in the back of the goaf, and in the prevention of the spontaneous combustion of coal by blocking the air leakage and reducing the temperature of the coal.

Photo by Zhai Xiaowei, 2011.

12.3 Thickened Colloid for Extinguishing a Coal Fire

A goaf fire at the underground bituminous Liujiamao Coal Mine in Shenmu city. The fire started by spontaneous combustion in January of 2016 and was extinguished in December of 2017 by filling this pit with loess. Black coal seams are visible in the cliff. The vertical field of view is about 35 m.

Characteristics

In general, the greater the concentration of slurry, the better the effect of fire control. However, when the slurry concentration is too large, mobility is a problem and the pipeline clogs easily. In addition, slurry requires a large amount of fly ash or loess, but fly ash or loess has transportation cost, so the greater the concentration of slurry, the higher the cost. A small amount of thickening agent and dispersant are added to water or mud to form viscous thickened colloid. In thickened colloid, the concentration of solid substance is not high, but its viscosity is high, and the effect of injecting glue is quite the same as that of pouring viscous mud. The thickened colloid has thixotropy, the flow resistance of the pipeline is small, the dispersion of the colloid is better, and the pipeline cannot be blocked.

There is no big difference between the thickened colloid and the pseudogel in terms of the structure. However, the pseudogel cannot flow naturally, but the thickened colloid can flow, and it is a kind of viscous disperse system. Another characteristic of thickened colloid is that it is not prone to settling for a long time. The base material added in the thickened colloid has both thickening and dispersion properties and can keep the sticky mud from subsiding for a long time. The thickened colloid made with high polymer base and slurry has a good thickening and suspension effect, and it is used widely.

Preparation and Injection

The thickened colloid additive agent is added to a certain proportion of the slurry in a certain proportion, and the mixture is uniformly mixed and then transported directly from the ground filling station to the underground location through the grouting pipe network. The use of a ground-grouting system for the thickening colloid technology is shown in Figure 12.3.1. Hydraulic pressure is used to naturally make slurry, and then a certain amount of thickener is added to the slurry. The slurry is transported to underground grouting site by a grouting line. The main work of the system is completed on the ground, and the underground is mainly for pipeline maintenance. Therefore, the construction of the system is more convenient.

Purpose

Because the preparation process of thickened colloid material is relatively simple, the actual frequency of use is higher. It is mainly used for filling a large area of goaf. By the thickening effect, the mud retention in the goaf is enhanced and improves the effect of the slurry on cooling and plugging.

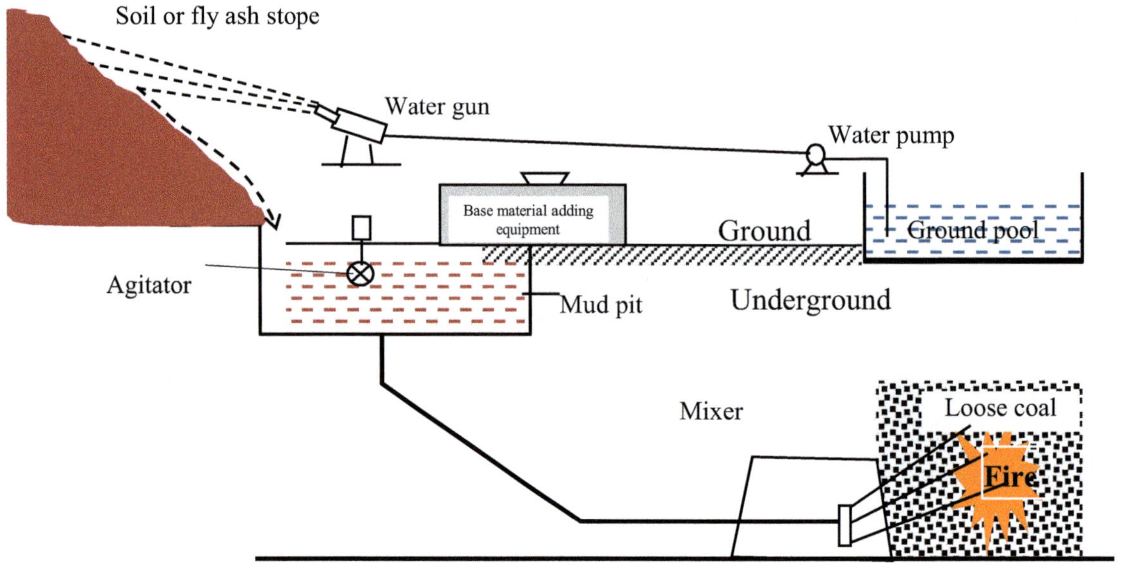

Figure 12.3.1. Injection technology with a ground-grouting system. Figure by Deng Jun, 2005.

12.4 Composite Colloid for Extinguishing a Coal Fire

Photo by Zhai Xiaowei, 2016.

A 3:1 mixture of gel and loess, respectively. This mixture is then reduced with water in the ratio of 1:1000. The diameter of the beaker is 12 cm.

Characteristics

Composite colloid is comprised of water, a base material (product number FCJ12 Gelatinizer), and fly ash (or loess). Its preparation process is as follows: first, the water and fly ash are mixed into a slurry, then a polymer-based material is added in. A mesh skeleton can be formed by the mutual force between the polymer-based materials and solid particles. The particles such as fly ash often become the nodes of the network structure, so the solid particles in the mud would cement together into a whole by the use of the mesh skeleton formed by the polymer-based materials and molecular bridge, forming fire-fighting materials with a gel-like nature (Figures 12.4.1–12.4.3). The high-absorbent water mineral and resin are considered as the base material of the composite gel to enhance its performance and improve the application process. For example, the thick mud can be made into an ideal composite gel by adding FCJ12 Gelatinizer with the quantity of only 0.8‰–1‰ (Zhai et al., 2010, 2011).

Preparation and Injection

As shown in Figure 12.4.4, the XKF-5/18 composite colloid (mud) injection machine is a kind of underground colloid injection and fire extinguishing equipment, which is based on the principle that the absorbent material is mixed

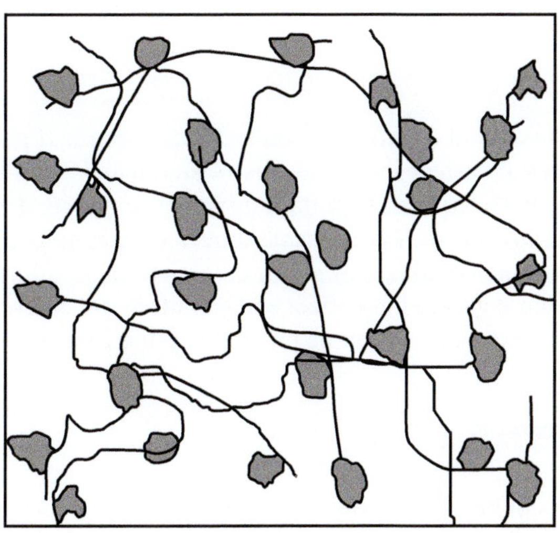

Figure 12.4.1. A structural model of a composite gel. The gray-colored ovals represent a high polymer material and the lines represent chemical bonds. The particles of fly ash or loess are often the nodes in such a reticular structure. The nodes in the high polymer chain are bound by strong forces, so the solid particles in mud can be joined together by the bridge action of the linear polymer material. Figure by Deng Jun, 2005.

Figure 12.4.2. A composite of gel mixed with loess. The proportion of gel to loess is 3:1. This is then mixed with water in a 1:1000 proportion. The vertical and horizontal distances inside the brim of the bowl, are 5 and 35 cm, respectively. Photo by Zhai Xiaowei, 2016.

Figure 12.4.3. A composite of fly ash from a power plant, mixed with fly ash. The proportion of fly ash to gel is 1:3. This is then mixed with water in a 1:1000 proportion. The outside diameter of the bowl is 40 cm. Photo by Zhai Xiaowei, 2013.

with water or mud. The composite colloid (mud) injection machine can mix a powder-base material and water to prepare the colloid and can also mix fly ash or loess with base material and water to prepare the composite colloid. When it is used, water source is connected first, then the equipment is started, and base material is added into the base material box to adjust laying-off speed. If at the same time, a certain amount of loess (or fly ash) is added to the reinforcing material box and the concentration of reinforcing material is adjusted, the composite colloid can be prepared. The composite colloid has a good filling effect and can fill in big cracks at the top, so it has a better performance in fire extinguishing. Then the colloid can be pressed into the coal seam to extinguish fire.

Equipment Features and Applicability

The equipment has the advantages of being lightweight and small. It can be fixed on the tramcar and can be transported to the needed position flexibly where colloid is made. The equipment has the characteristics of less workload and larger injection pressure. Therefore, it is mainly used for partial small area fire, such as roadway faults, roof fall, and so on, or in the mine where the fire prevention and extinguishing has not been set up.

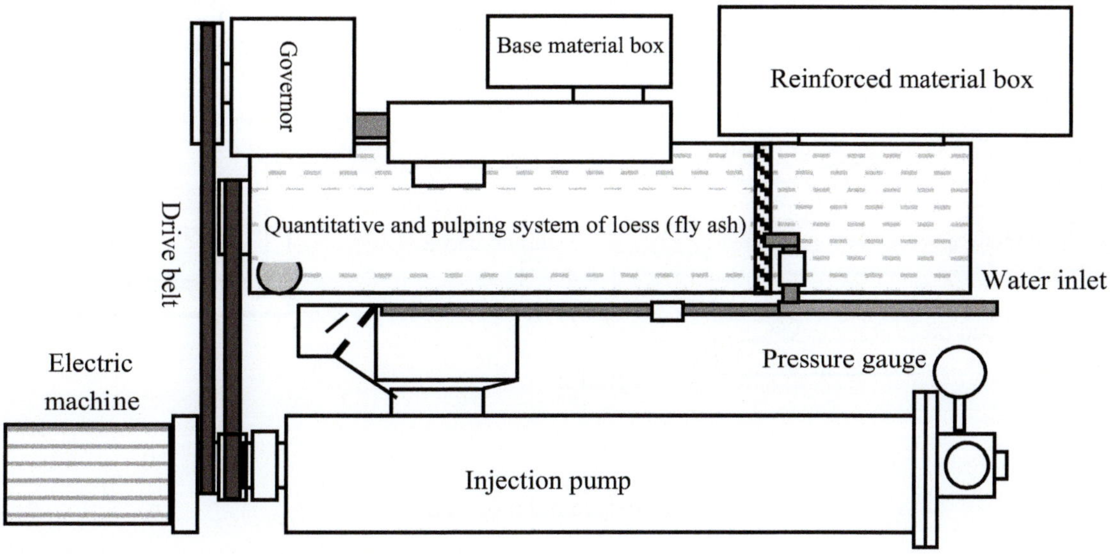

Figure 12.4.4. XKF-5/18 (model number) Schematic diagram of a composite colloid slurry-injection machine (model XKF-5/18). Figure by Deng Jun, 2005.

Ground Compaction and Injection

Depending on the mine fire, the transport of material to the area of combustion may be difficult, especially when the area is large and requires the transport of a large amount of extinguishing material. Mobile injection technology is restricted by the underground environment (underground space is small), and its injection flow is small; the processing fire area needs much time. These problems can be solved by the injection process of the pipe network. This technology can be used in the coal mine ground grouting fire prevention system, sand injection and fire extinguishing system, and dustproof pipeline system.

The preparation of ground dynamic pressure composite colloid-fire extinguishing material results mainly from the pressure of water by which water and soil (fly ash) are made into slurry. To improve the quality of the slurry, a certain amount of suspending agent is added into the slurry, and then it is transported to the underground location by pipeline. Near the place where colloid is needed, special equipment is applied to add the coagulant in proportion to the pipe and mixed with base material to form a mixed composite colloid material. This material is injected through the pipeline into the fire zone, and the intermixture is formed to a composite colloid in the target location. This technology only needs to transport a relatively small quantity of coagulant accelerator to the underground, thus greatly reduces the transportation of fire extinguishing material. The injection flow amount is 30–100 m^3/h, which is quite large. There are no strict restrictions to the solid strengthening agent particles.

The size of the coagulation accelerator can be controlled by a stepless speed regulator to adjust the speed of the push rod. Water and coagulant mixture flow through the main pump into the filling pipeline, and are mixed with a base solution after mixing through the current diverter (according to the situation in the field of design, processing), and drilling pipe into the specified location.

Composite colloid addition technology is shown in Figure 12.4.5. The mixture of the base material is completed in the ground grouting pool, and the addition of the coagulant accelerator is completed by a multifunctional pressure injection device. The outlet of the multifunctional pressure injection equipment is connected to the grouting pipeline. The technological processes are as follows: first, the base material and the slurry are proportionally mixed in the slurry mixing tank of the ground-grouting system (if the pure gel is pressed, the slurry is not required to be made in the filling tank, and only the base material and water need to be fixed). After the completion of all tasks, open the ground grouting pipe gate to release mixed liquid so that the mixed liquid flows along the grouting pipe. When the mixed liquid reaches the connection point of grouting pipeline and coagulant pipe, start the multifunctional injection device and add the coagulant. In order to achieve the objective of gelling in the fire area, according to the mixed solution flow, base material concentration and distance to fire area, coagulant feeding speed is adjusted for appropriate gelling time. After the injection; the injection equipment and pipeline should be cleaned with water to prevent the blockage of the pipeline and equipment.

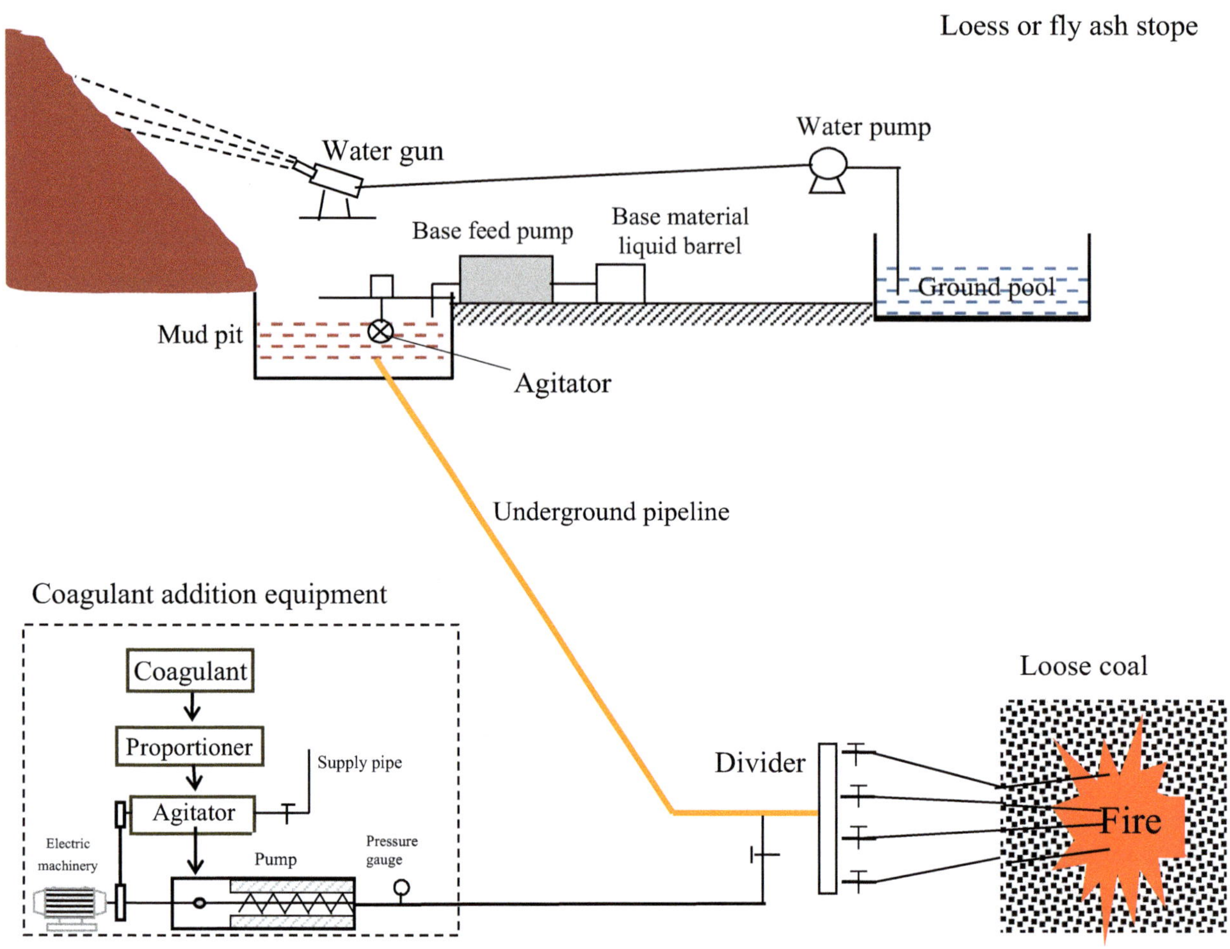

Figure 12.4.5. Technology of an underground pipe network with a large amount of composite colloid injection. Figure by Zhai Xiaowei, 2005.

Equipment Features and Applicability

The preparation and injection system of ground composite material has the advantages of high flow rate, low cost, easy reconstruction, and large flow capacity. However, because natural agitation is adopted for mud preparing, the concentration of slurry is unstable and affects the quality of composite colloid. But it still can play a role in cooling the fire zone and filling loose coal. Therefore, the system is mainly applicable to the work face without considering the influence of water gushing on the related mining seams, such as outcrop fire, goaf fire, and treatment of outcrop fire. Cooling by direct injection colloid in the fire zone is shown in Figure 12.4.6.

Preparation and Injection Technique

The ground mobile colloid-fire extinguishing system (model MYZ-30) is mainly comprised of a slurry storage yard, continuous quantitative pulping machine, slurry pump, coal mine grouting machine, and a pipe network system. The system automatically proportions a slurry concentration and the quantitative addition of gelling agents, and it uses dynamic pressure to transport composite colloids. The system is shown in Figures 12.4.7 and 12.4.8.

System parameters: grouting quantity is 30 m^3/h, input power 45 kW, the output flow rate more than 30 m^3/h, the water pressure greater than 0.4 Mpa, no soluble impurities in water, water–volcano ash ratio: 3–1:1. Pulping material is fly ash, loess, volcano ash supply capacity is 30 m^3/h, suspending agent dosage >0.1% gel, coagulant dosage >0.06% composite colloid.

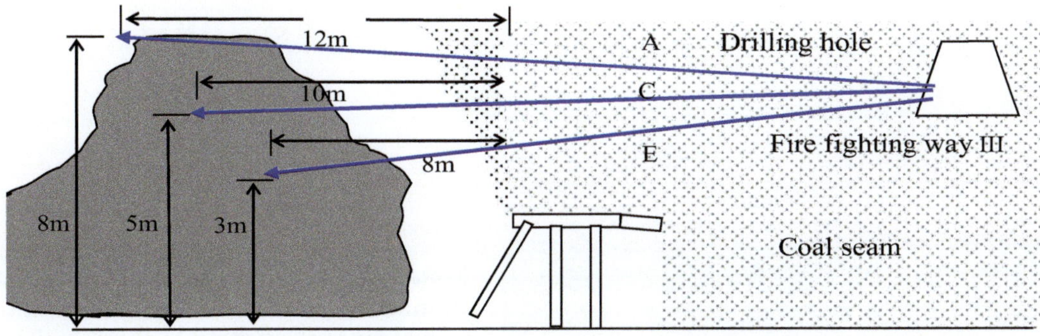

Figure 12.4.6. Cooling by direct injection colloid in fire zone. Figure by Zhai Xiaowei, 2005.

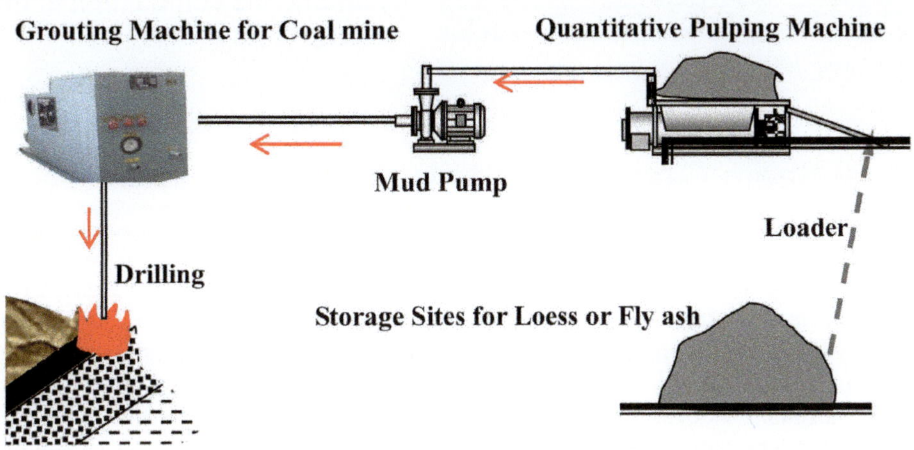

Figure 12.4.7. Flow chart of the MYZ-30 ground mobile colloid-fire extinguishing system. Figure by Zhai Xiaowei, 2005.

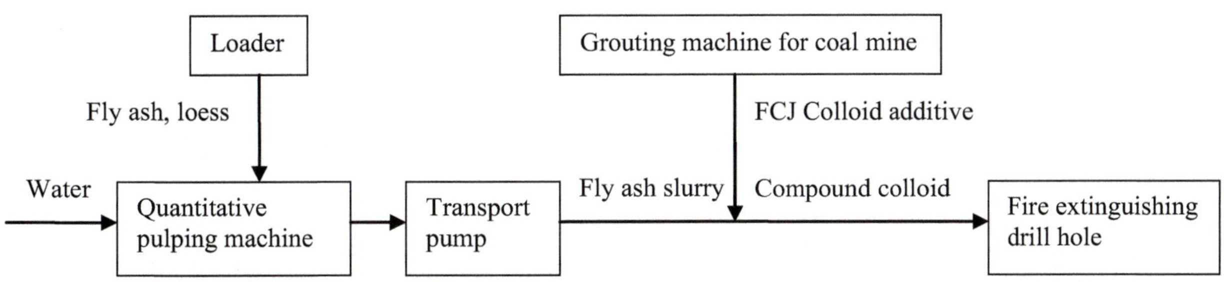

Figure 12.4.8. The colloid injection technological process. Figure by Zhai Xiaowei, 2005.

Equipment Features and Applicability

The preparation and injection system of a composite colloid has the advantage of large flow and good colloid retention. It can have the effect of cooling the fire zones and reducing the oxygen concentration. It is mainly used for the prevention and control of spontaneous combustion of goaf in dip angle working face, underhand stopes working face, and fully mechanized mining working face. It has special use for old goaf fire in abandoned mines and treatment of outcrop fire.

Photo by Zhai Xiaowei, 2006.

12.5 Case Study

Installing instrumentation (man crouched) to measure and record (man standing) gas and temperature data in the Baoding coal field, Panzhihua city, Sichuan Province. In 2003, bituminous coal was ignited by spontaneous combustion in the underground Dabaoding Coal Mine here.

No. 4 Coal Seam Fires in Baoding, Panzhihua

Spontaneous combustion was discovered in 2003 in the No. 4 coal seam in the Baoding coal field, Panzhihua city, Sichuan Province. The seam is 3.5 m in thickness on average and inclination angle is 35 degrees. Based on the measurement, the fire is 685 m long, 25 m wide, and the whole fire area is about 8490 m^2. It was divided into three parts including north fire-field, middle fire-field, and south fire-field for control purpose, as shown in Figure 12.5.1. Some predisaster signs or events are present such as smoke, surface fissures, collapsed crater, and larger-area high-temperature baked metamorphic rock caused by the coal fire. The high temperature of the surface rock exceeds 500°C in some collapsed craters. The sketch map of the fire distribution at Baoding coal field is shown in Figure 12.5.1.

Coal-Fire Detection

Radon

Radon gas exists in most kinds of rock and sediment, including coal and clay. The amount of radon released increases as rock and sediment temperature increases, and the ability of radon to migrate is more intense in a vertical rather than horizontal direction. A schematic of radon migration is illustrated in Figure 12.5.2.

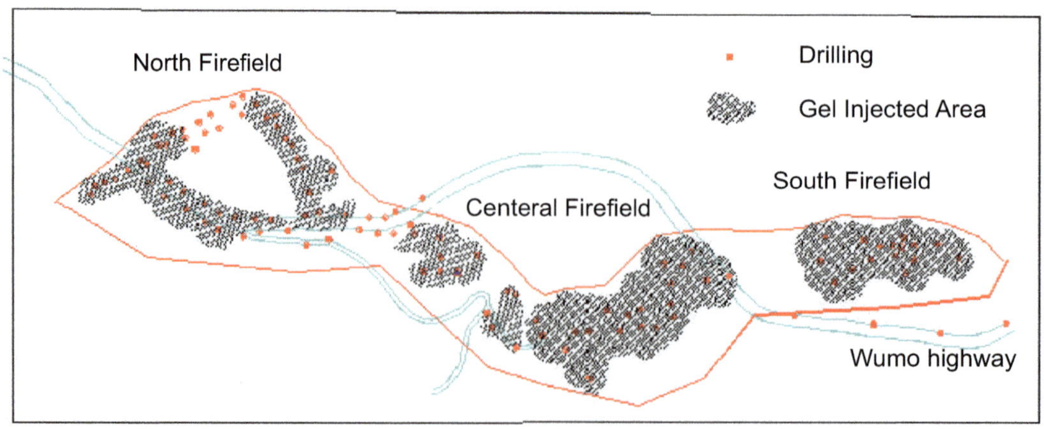

Figure 12.5.1. Sketch map of the distribution of an underground coal fire in the Baoding coal field. The fire started by spontaneous combustion in 2003. The double light-blue lines represent the Wumo Highway System. Figure by Zhai Xiaowei, 2006.

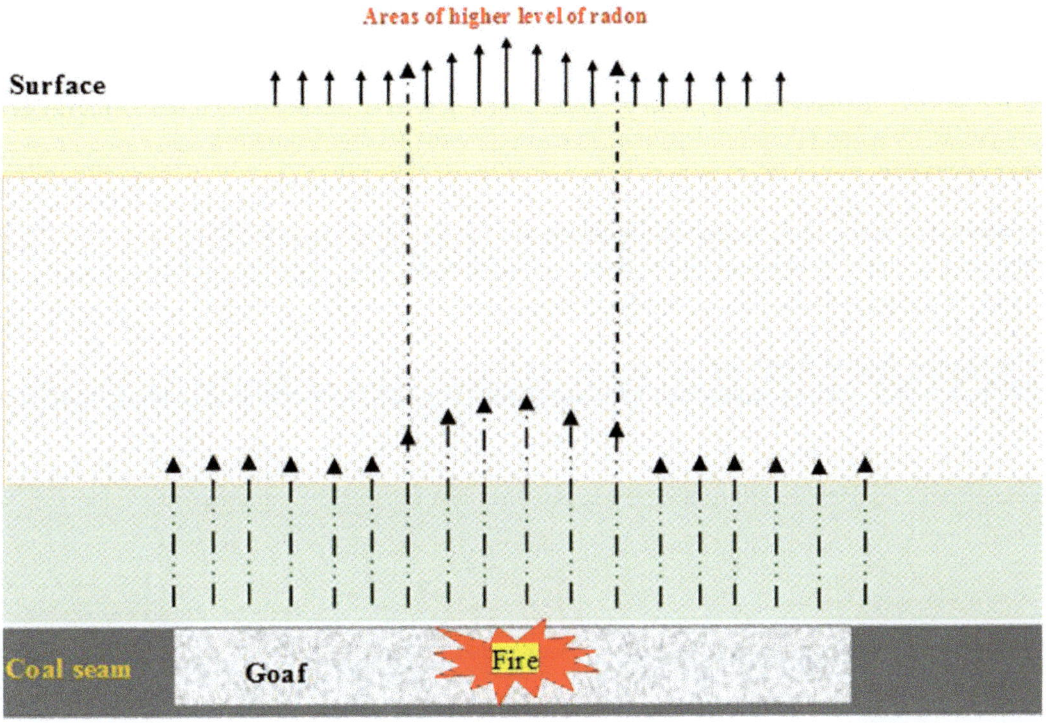

Figure 12.5.2. Detection of an underground coal fire, using radon gas. Figure by Zhai Xiaowei, 2006.

Figure 12.5.3. Marking a detection location for radon gas in the Baoding coal field, Panzhihua city, Sichuan Province. Photo by Zhai Xiaowei, 2006.

Radon Gas Detection of the Location of No. 4 Underground Coal Fires

(1) Determine the surface area of the fire:
Coal combustion emits radon, so the approximate area of a combustion zone can be determined by locating radon emission points on the ground with a Global Positioning System (Figure 12.5.3). The horizontal distance between two radon-gas detection points is 10 m.

(2) Collecting radon gas.
A hole 0.3–0.4 m deep and 0.3 m in diameter was dug to ensure that radon could be adequately collected (Figure 12.5.4). The radon gas was collected with an alpha cup (Figure 12.5.5). The cup was put upside down at the bottom of the hole, the hole covered with a plastic cover, and loess put on top (Figure 12.5.4). The hole was then filled with loess and the radon-detection location and time recorded (Figure 12.5.6).
(3) Radon gas detection.
The cup was dug up after 4 h, ensuring adequate collection of radon. The alpha cup was put into a KZ-D02 (model number) (Figure 12.5.7) alpha testing radon device. The device uses high voltage to quantify the radon adsorbed into the cup. The quantity of the radon is shown on the display screen of the device. The amount of radon was recorded for each collection station. The process of using the radon alpha cup is shown in Figures 12.5.3–12.5.8.
(4) Mapping the distribution of radon.
The distribution of radon was mapped with Surfer software (Golden Software, 2018) and the result is illustrated in Figure 12.5.8.

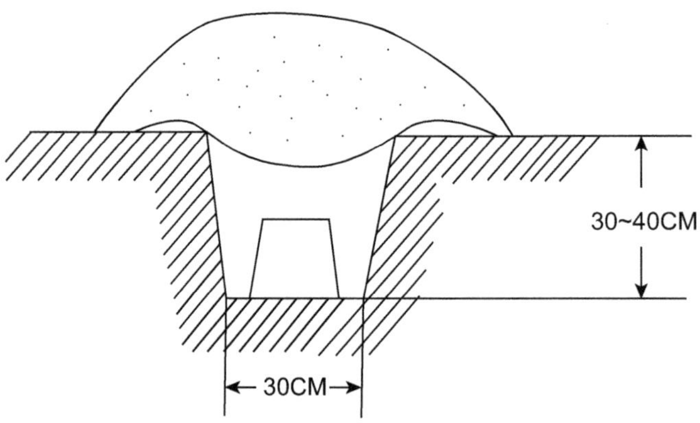

Figure 12.5.4. A radon collection pit with an alpha cup at the bottom. Radon moves upward, so the opening of the cup faces downward to collect the radon. A plastic cover is placed on top of the hole and then loess, shown by the dot pattern between the wavy lines, is placed on top of the plastic cover. Figure by Zhai Xiaowei, 2006.

Figure 12.5.5. An alpha cup. The height and width of the cup are 15 cm and 6 cm, respectively. Photo by Zhai Xiaowei, 2006.

Figure 12.5.6. Burying a radon alpha cup in the Baoding coal field, Panzhihua city, Sichuan Province. Photo by Zhai Xiaowei, 2006.

Figure 12.5.7. Model KZ-D02 radon measuring instrument used at the Baoding coal field, Panzhihua city, Sichuan Province. The dimensions of the instrument are 50 cm by 40 cm by 50 cm. Photo by Deng Jun, 2006.

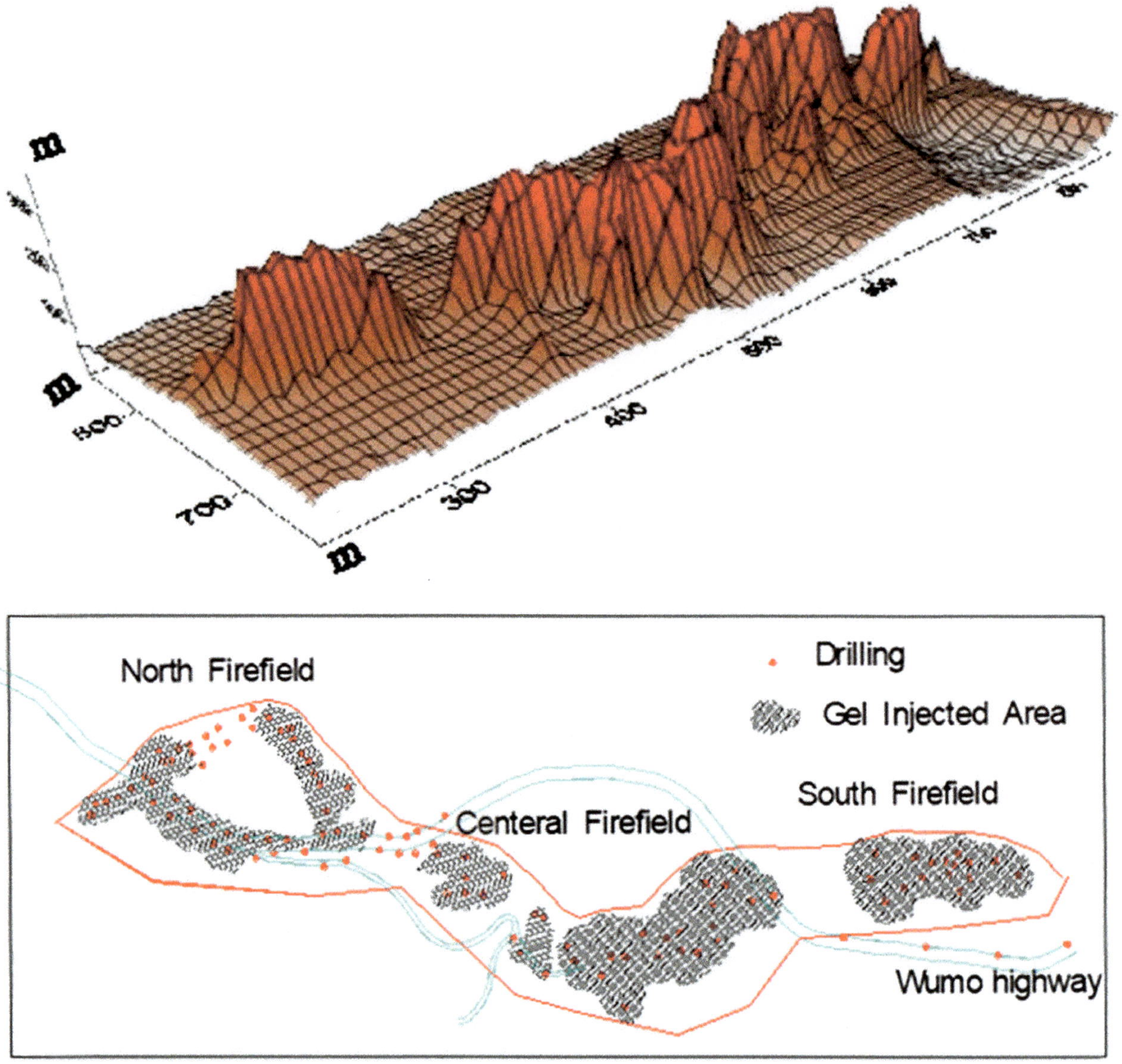

Figure 12.5.8. Radon detection test results for the same coalfield as in Figure 12.5.1. Top: The red peaks designate areas of high radon emissions due to coal combustion. Bottom: The dark-blue line is where radon emissions are abnormally high. The areas outlined with the red-brown colored polygons are where the spontaneous combustion of coal occurred. The double light-blue lines represent the Wumo Highway System. Figure by Zhai Xiaowei, 2006.

Drilling Process

According to the calculation of the actual area of the No. 4 coal seam, 112 boreholes are designed in the whole coal fire area, 59 located in the south fire field, 32 in the middle fire field, and 21 in the south fire field. The temperature and gas in the borehole are monitored after each drilling is completed. Monitoring of gas and temperature during drilling is shown in Figures 12.5.9 and 12.5.10.

Injection Technique

In the process of extinguishing a fire, the manufacture and injection of a composite colloid utilizes the MYZ-30 ground injection colloid-fire extinguishing system. The system consists of a storage site of pulverized coal ash, a continuous quantitative pulping machine, a slurry pump, a coal mine grouting machine, and a pipe network system. The drilling process was discussed in the previous section, and the site construction is shown in Figure 12.5.11.

Figure 12.5.9. Drilling for the collection of gas samples and measurement of gas temperatures at underground coal fire in the Baoding coal field, Panzhihua city, Sichuan Province. Photos by Zhai Xiaowei, 2006.

Figure 12.5.10. The Baoding coal field, Panzhihua city, Sichuan Province. Measuring gas temperatures and collecting gas samples (gas chromatography) after drilling into an underground coal fire. Left: recording the depth of the hole. Right: collecting gas samples. Photos by Deng Jun, 2006.

Treatment Effects

The boreholes were completed in November 2006, and the injection of colloid started. It took 8 months for entire project of the injection of fly ash composite colloid, with the injection volume at 110,494 m^3, which is shown in Table 12.5.1.

After injection, the surface temperature decreased to normal levels and the high-temperature near the collapsed crater decreased from 500°C to 600°C to below 100°C. There are 110 boreholes where the temperature is less than 100°C. The temperature variation of the monitored boreholes is shown in Figures 12.5.12 and 12.5.13. Toxic and harmful gases, including CO, SO_2, and CO_2, disappeared. The effect of fire suppression is remarkable.

Figure 12.5.11. The colloid injection system. (A) Mixing fly ash and water. (B) Adding the gelling agent to the mixture from (A). The diameter of the hose in (C) is 108 mm, and the length and width of the pond in (D) are about 3 and 2.5 m, respectively. Photos by Zhai Xiaowei, 2006.

Table 12.5.1
Statistical table of colloid injection.

Injection date	Injecting sites	Volume of gel (m^3)
8/22/2006–1/25/2007	30 drillings and subsidence pits in the middle fire field	53,124
10/19/2006–4/20/2007	28 drillings and subsidence pits in the north fire field	31,495
10/26/2006–4/10/2007	15 drillings in the south fire field	25,875
Total volume		110,494

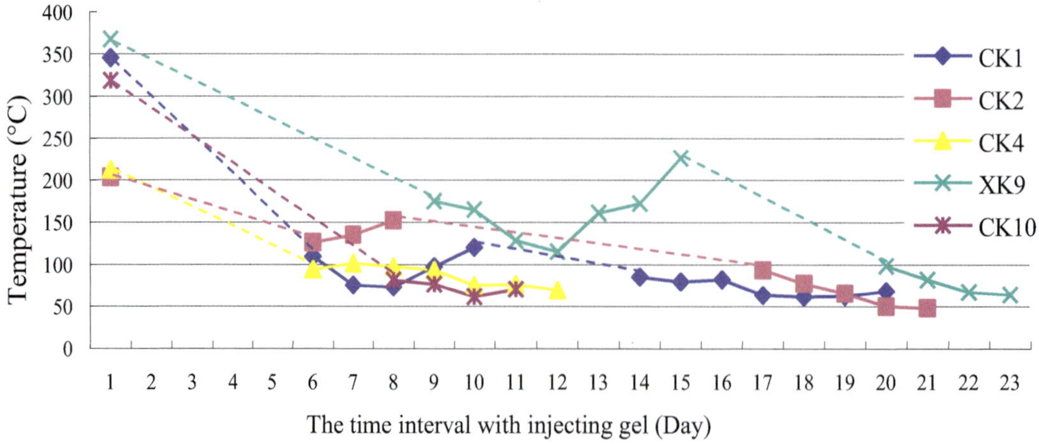

Figure 12.5.12. Temperature changes in holes CK1, CK2, CK4, CK9, and CK10 before and after injecting colloid. Figure by Deng Jun, 2006.

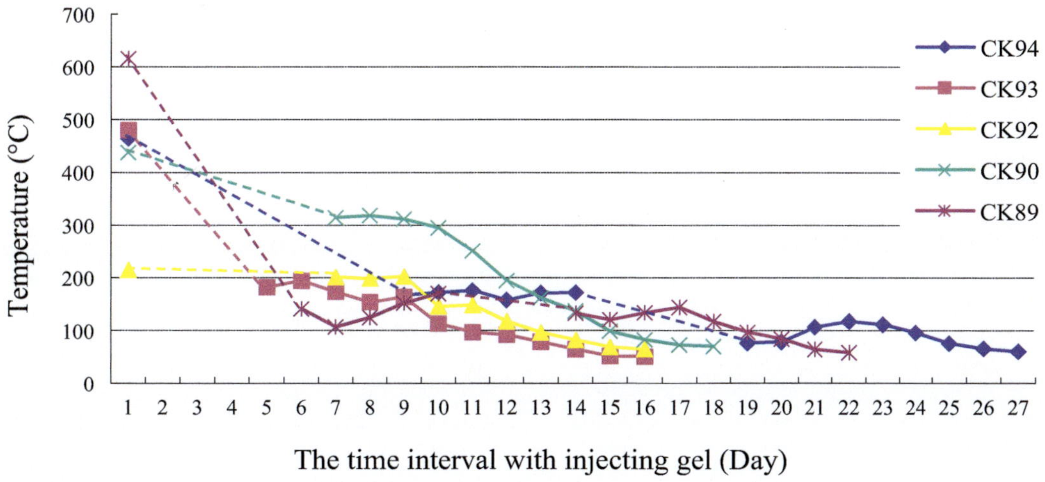

Figure 12.5.13. Temperature changes in holes CK89, CK93, CK92, CK90, and CK89 before and after injecting colloid. Figure by Deng Jun, 2006.

Conclusions

According to the data obtained from field work, the temperature of the coal at each measuring location is generally decreasing after the infusion of the colloid extinguishing agent, and the temperature of all the measuring locations is reduced to below 100°C after 24 days. This shows that colloid is a good material for preventing the development of the spontaneous combustion of coal. At the same time, compared with other materials, colloid-fire extinguishing material has the advantage of easy application over a wide area.

References

Cao, D.Y., Shi, L., Fan, J., Wu, C.C., 2007. Environmental effect analysis of coalfield fire. China Min. Mag. 16, 40–42 (in Chinese).

Deng, J., Sun, B.L., Fei, J.B., Wang, Z.J., Zeng, K.J., Hu, Z.G., 2007. Application of colloidal fire extinguishing technology to fire disaster control of seam outcrop. Coal Sci. Technol. 35, 58–60 (in Chinese).

Golden Software, 2018. http://www.goldensoftware.com (accessed April 2018).

Xu, J.C., Zhang, H., Wen, H., Deng, J., 2003. Theory and Technology of Extinguishing Spontaneous Coal Fire with Gel. China Coal Industry Publishing House, Beijing (in Chinese), Beijing.

Xu, J.C., Ma, L., Zhang, Z.W., Xiao, Y., 2004. Fire extinguishing technology with fly ash complex gel applied to control mine spontaneous combustion fire area. Coal Sci. Technol. 11, 24–26 (in Chinese).

Zhai, W., Deng, J., Zhang, H., 2010. Composite gel as a barrier between mine workings and coal fires. In: The 2nd International Coal Fire Symposium Proceedings, Beijing, China, pp. 345–349.

Zhai, W., Deng, J., Wen, H., Dai, A.P., Wu, J.B., 2011. The research on the thermal destruction of environment and controlling techniques of the no. 4 outcrops fire in haibaoqing. J. Coal Sci. Eng. 17, 345–348.

CHAPTER 13

Crystallochemical Behavior of Slag Minerals and the Occurrence of Potentially New Mineral Species From Lapanouse-de-Sévérac, France

Lapanouse slag sample LdS1 comprised of clinopyroxenes, melilite, and wollastonite. The yellow zones are rich in clinopyroxenes. The vesicular texture of the rock is reminiscent of either parabasalt or buchite, but this rock is devoid of glass, typical of buchite. The scale is in centimeters. *Photo by Łukasz Kruszewski, 2015.*

CHAPTER CONTENTS

Coal and Peat Fires: A Global Perspective
Edited by Glenn B. Stracher

https://doi.org/10.1016/B978-0-12-849885-9.00013-5

13.1 Slag and Potentially New Minerals From Lapanouse-de-Sévérac, France

Łukasz Kruszewski, Pierre Gatel, Vincent Thiéry, Izabela Moszumańska, Danuta Kusy

Photo by Łukasz Kruszewski, 2015.

Vesicular slag, sample LdS2, comprised mainly of clinopyroxenes, with locally yellow zones of melilite. Also present is olivine and minor wollastonite. Some calcite-rich amygdules occur at the top. The scale is in centimeters.

Introduction

Bituminous shales of the Lapanouse-de-Sévérac area are of the Lower Jurassic age and coexist intercalated with limestones and, more rarely, black lignites. These rocks are the protoliths of the slags described in this paper. The slags formed in gobs that once spontaneously combusted. The regional stratigraphic sequence, geological sketch of the oil-shale quarry, exploitation processing of the shales, primary rocks composition, and detailed information about the local gobs and some slag-forming minerals can be found in Gatel et al. (2015). The locations of the two studied heaps, the Western heap and the Eastern heap, are shown in Figure 13.1.1.

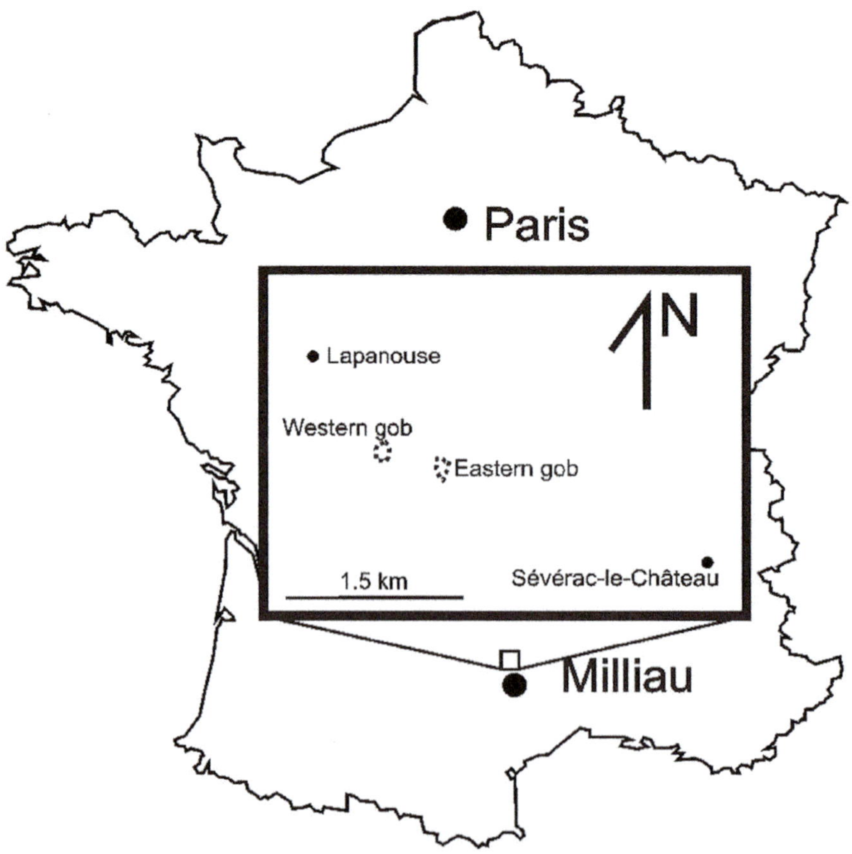

Figure 13.1.1. Localization of the Lapanouse gobs.

Previous Research

The first mineralogical research of the Western and Eastern heaps was done by Eytier et al. (2004). Additional work with the known mineral species and their allocation to prograde, retrograde, secondary, and primary origin classes is given by Gatel et al. (2015). In both papers, we can find information about mineral occurrences and associations, but data about rock and mineral chemistry is lacking. This study was performed to fill this gap; it also focuses on an examination of mineral composition based on rock type and geochemistry. Some of the minerals described were not previously identified from the study site. This is also true for potentially newly discovered mineral species observed and analyzed with the electron microprobe (EPMA). A separate subchapter, "Unnamed Minerals," is devoted to these.

13.2 Analyses and Materials Description

Photo by Łukasz Kruszewski, 2015.

Vesicular slag, sample LdS5. The rounded dark-red specks are hematite-bearing, olivine-group-rich segregations. Wollastonite is a minor component. The scale is in centimeters.

Rock Characteristics

The rock samples collected from two dumps are difficult to classify from the petrological point of view. The term "paralava" (e.g., Sokol et al., 2005; Grapes, 2006), with its better (to our opinion) analogue "parabasalt," fits well in the argument of (1) being vesicular, and (2) being composed of mainly silicates, which is in opposition to the oxide-rich and silicate-low mineralogy of slags. The latter, however, is only correct for the rock portions that formed in expense of carbonate-rich accumulations (such description is not attributable to the so-called industrial slags); typical slag minerals are apatite group, anhydrite, gypsum, and oxide minerals like ye'elimite, spinels, and Ca ferrites (e.g., Sokol et al., 2005). The main argument suggesting the rocks under study being not attributable to parabasalts is the lack of glass phase—it was neither observed in thin sections nor in the diffraction spectra. It is possible, however, that glass inclusions will be found during further, more detailed study.

To be sure, the rocks under study are somewhat comparable to hornfels from the Gurim anticline, a structure belonging to the famous Hatrurim (pyrometamorphic) formation in Israel, which comprises gehlenite-åkermanite, wollastonite, pseudowollastonite (a $CaSiO_3$ polymorph, not approved by the International Mineralogical Association, interestingly, found only in the matrix of the LdS6 sample), titaniferous andradite, fluorapatite, and chromian spinel. Paralavas of this complex may contain wollastonite, melilite, rankinite, Ti-bearing andradite, fluorapatite, kalsilite, nagelschmidtite, magnetite, hematite, perovskite, and tobermorite; a rock made of rankinite, gehlenite, schorlomite, and larnite is also known. However, at least some of the Lapanouse rocks strongly resemble—in mineral composition—melilitolite of Colle Fabri, Italy; this rock comprises titaniferous garnet, Ti–Al–Fe-bearing clinopyroxene, magnetite, perovskite, rankinite, Si-bearing apatite, Fe–Ni sulfides, and carbonate/zeolite minerals (Seryotkin et al., 2012). Due to the common occurrence of carbonate minerals within the Lapanouse rocks, we call them slags. The mentioned melilitolite is of magmatic origin; because the pyrometamorphism is associated with melting, which was also favored at Lapanouse thanks to the presence of flux-acting carbonates, it is suggestible that a process at least partially responsible for the slag formation is similar or identical to anatexis.

The rock samples provenience and mineral composition is given in Table 13.2.1. The common feature of the rocks is the presence of voids that may sometimes be amygdules or "nuts," the latter term taken from Sokol et al. (2005) and understood as remnant carbonate concretions that underwent calcination (decarbonization; i.e., temperature-driven CO_2 removal). Veinlets are also sometimes observed, often filled with gypsum. The rocks are often heterogenic in terms of their lithology, and may contain some clasts, other than amygdules, suggesting primary inhomogeneity of the protolith rather than coalescence of various lithotypes (as no strong borders or contact reaction zones are evident). Some rocks will undergo further study. Within this study, the already obtained mineral compositions will be used for geothermometry.

Analytical Methods

Samples of the Lapanouse slags were first ground in automatic corundum mortar and analyzed for their mineralogical composition using the powder X-ray diffraction (PXRD) method. For analyses, a Bruker axs D8 ADVANCE

Table 13.2.1
Provenience, petrography, and mineral composition of Lapanouse rock samples, France.

Sample	Provenience	Petrography	Mineral composition*
LdS1*	Unknown	With olivine-clinopyroxene-spinel pseudomorphosed lithoclasts	Clinopyroxene group, nepheline, wollastonite, orthoclase, ye'elimite, *hejtmanite*, *tounkite*, *wagnerite*
LdS2*	Unknown	Some separate carbonate segregations (B) present	A: melilite g., clinopyroxene g., kalsilite, perovskite, olivine group, vaterite, *nepheline*, *fluorapofyllite-(Na)*, *alstonite*, *tunisite*, *benstonite*; B: calcite, hydrocalumite (or rapidcreekite), dolomite, *sulfur*
LdSV	Western heap	With gypsum veins	A: leucite, wollastonite-*1A*, clinopyroxene g., anorthite, nepheline, melilite g., apatite g.
LdS5			Clinopyroxene g., nepheline, hematite, forsterite, sanidine, *osumilite or makatite*, *pectolite*, *tridymite*, *analcime*, *α-quartz*
LdS6*	Unknown	With blue-white amygdales/nuts (B)	A: melilite g., clinopyroxene g., wollastonite-*2M* and -*1A*, "pseudowollastonite," leucite, orthoclase, *tinaksite*, sillimanite, andradite, apatite g., perovskite, latiumite; B: calcite, anhydrite, gypsum, ye'elimite, ternesite, melilite g., cuspidine, *rustumite*, wollastonite, *creedite*
LdS7	Western heap	With greenish-gray to white zones (amygdales) (B)	A: melilite g., leucite, wollastonite, orthoclase, *silvialite*, *forsterite*, *clinoenstatite*, *"pseudowollastonite"*; B: portlandite, gypsum, hannebachite, thaumasite, fluorapatite, ye'elimite, *oldhamite*
LdS9*		With minor white zones (amygdales) (B)	A: melilite g., wollastonite-*1A*, leucite, *antigorite*, orthoclase, *vesuvianite*, *wollastonite-2M*, *silicocarnotite*, *liottite*, *forsterite*; B: larnite, melilite g., *liottite*
LdS11*	Eastern heap	(A) With dark "amygdales" (or lithologically different clasts)	A: wollastonite-*2M*, clinopyroxene g., *graftonite*, *calzirtite or burpalite*, perovskite, *offretite*, hematite, *spurrite*; B: leucite, clinopyroxene g., wollastonite-*1A*, *gypsum*, *quartz*
LdS12*	Eastern heap	(A) With internally beige and externally white amygdales (B)	A: clinopyroxene g., leucite, wollastonite, melilite g., anorthite, gypsum, *rhodesite*, *antigorite*, *magnetite*; B: calcite, cuspidine, ettringite or thaumasite, bazhenovite, melilite g., *"meta-sillimanite,"* *magnesite*, *sulfur*, *silvialite*
LdS15	Eastern heap	(A) With elongated blue (B) and white (C) zones/amygdales	A: clinopyroxene g., wollastonite-*2M*, leucite, anorthite, orthoclase (or sanidine), andradite, *frankamenite*, *kaliophilite*, *nepheline*, *hematite*, *xonotlite*; B: anhydrite, gypsum, esseneite, wollastonite, apatite group; C: gypsum, anhydrite, apatite g.

*By PXRD analysis. Mineral species in italics are not confirmed and those not approved by the IMA (International Mineralogical Association) are in quotation marks and listed under Mineral composition on the right side of the table.

diffractometer equipped with superfast linear position-sensitive detector, model VÅNTEC-1, CoKα radiation lamp, and Ni kβ filter (no monochromator) was used. After obtaining the qualitative results, quantitative analysis was conducted using the Rietveld method (Rietveld, 1967). TOPAS software was used for this purpose.

The minerals were chemically examined for their purity using the EPMA method. These analyses were conducted at the Inter-Institute Laboratory of the Microanalysis of Minerals and Synthetic Materials (Faculty of Geology, University of Warsaw), which is equipped in CAMECA SX-100 apparatus. The operating conditions—WDS mode, 15 kV accelerating voltage and 10–20 nA beam current, and 1 μm beam size—were used. The beam current was varying depending on mineral types: clinopyroxenes, olivines, melilites, spinels, hematite and other Fe oxides, sulfides, perovskite, apatite group, and similar (usually silicate) species were analyzed using the 20 nA beam; 10 nA beam was used for less stable, often Na- and S-bearing minerals, like petalite-like species, feldspars, plagioclases, anhydrite, and vaterite. Beam size was enlarged in case of minerals showing destruction and Na-rich minerals like nepheline.

Rock Mineralogy and Mineral Chemistry

Both the mineral composition of the slags (Table 13.2.2) and chemical composition of minerals under scope are complex to very complex, as can be guessed from the following description and figures. The slag textures are also variable and it is suggestible that they contain various metamorphosed lithoclasts (of variable protolith composition), although no fusing (glass formation) was observed. The commonly observed texture variability involves crystal size. Chemical zoning is common and show especially in pyroxenes, with at least two geochemical/crystallochemical types present in a single sample. Some minerals occur as rims on another species (e.g., magnesioferrite-rich spinels around chromite-rich ones). Others contain numerous voids (e.g., wollastonite) or occur as tiny rounded inclusions (e.g., ye'elimite in fluorellestadite, sample LdS15). Figure 13.2.1 shows BSE images of the LdS1 (A) and LdS2 (B) slags, with forsterite-rich aggregates, giuseppettite, depmeierite-tuscanite, "K-petalite," and probably an unnamed amphibole-like phase. Figure 13.2.2 shows corresponding data for the LdS5 slag, with images distinguishing in the presence of titanite and perovskite (A), hyalophane-like phase and pseudobrookite (B). Figure 13.2.3 shows BSE images for the LdS6 (A, B) and LdS9 (C, D) slags. Figure 13.2.4 illustrates the mineralogical character of the LdS7 slag, for example, with V-dominant analogue of fluorapatite. Figure 13.2.5 depicts perovskite- (A) and garnet-rich (B) matrix of the LdS11 sample. Six subfigures shown in Figure 13.2.6 illustrate the very complex nature of the LdS12 slag, with matrix-forming (A, B, C, F) and amygdule-forming (D, E) species. Similar complexity is shown in the four-image set of Figure 13.2.7, describing the LdS15 sample (A, B, D-inset, matrix; C, D, amygdule).

Anhydrite ($CaSO_4$)

Anhydrite is an important constituent of white, usually marginal, zones of amygdules residing within the slag sample LdS15 (Figure 13.2.7C and D). However, it is also found in the blue, apatite-group-bearing internal zones. An analytical report for anhydrite is given in Table 13.2.3. A unique feature of this anhydrite is the molybdenum admixture, which would likely act due to small substitution of the powellite ($CaMoO_4$) component. The anhydrite's empirical formula (n=4) is $Ca_{1.00}[(SO_4)_{0.92}(CO_3)_{0.07}(MoO_4)_{0.02}]_{\Sigma 1.01}$ and the corresponding mean end-member composition is $Anh_{92}Ar_7Pow_2$, where Anh is anhydrite, Ar is the (possibly present) aragonite, and Pow is the powellite end-member. The occurrence of Mo, although confirmed by WDS detector scans, needs to be further confirmed.

Alkali Feldspar ($KAlSi_3O_8$)

Composition of alkali feldspars from chosen slags is shown in Table 13.2.4. The LdS5 material is ferrous, may occasionally bear trace Sr, and is very rich in Ba (up to 11.06 wt.% oxide), as also shown by its empirical formula (n=4): $(K_{0.54}Na_{0.26}Ba_{0.20}Ca_{0.01}Sr_{0.01})_{\Sigma 1.02}(Al_{0.97}Fe_{0.03})_{\Sigma 1.00}(Si_{2.79}Al_{0.19})_{\Sigma 3.01}O_{7.99}$, corresponding to mean end-member $Or_{53}Ab_{25}Cls_{20}(An + Sls)_1$ formula, where Or is orthoclase, Ab is albite, Cls is celsian, An is anorthite, and Sls is slawsonite. The corresponding formula for feldspar from the LdS15 sample (n=10) is $(K_{0.86}Ca_{0.11}Na_{0.02}Ba_{0.01})_{\Sigma 1.00}(Al_{0.95}Fe_{0.06})_{\Sigma 1.01}(Si_{2.89}Al_{0.11})_{\Sigma 3.00}O_{8.02}$, and the related mean end-member formula is $Or_{82}An_{10}Fkf_5Ab_2Cls_1Fan_1$, where Fkf is a hypothetical $KFeSi_2O_6$ end-member and Fan is a hypothetical $CaFe_2Si_2O_8$ end-member. This feldspar is more Fe-enriched but bears no Sr and only trace Ba. It is usually more calcian. The LdS15 feldspar is more complex that the previous one and can actually be divided into three groups using the anorthite content as a discriminant; the first group composition would then be $Or_{91}Fkf_4Ab_2An_2$ (n=4, first four analyses), the second one would be $Or_{85}An_6Ab_2Fkf_6(Cls + Fan)_{\sim 1}$ (n=4, analyses 10–11 and 13–14), and the last one would be $Or_{56}An_{34}Ab_2Fkf_4Fan_3Cls_{<1}$ (n=2, analyses 9 and 12).

Apatite Supergroup ($Ca_5(PO_4,SiO_4,VO_4,SO_4)_3(F,OH,O,Cl)$)

This group represents quite large variability, as shown in Table 13.2.5 (for the samples LdS7, LdS11, and LdS12) and Table 13.2.6 (for the sample LdS15). The apatite group from sample LdS7 (Figure 13.2.4B) is characterized by

Table 13.2.2
Quantitative phase analysis (Rietveld method) of the Lapanouse slags (France), in wt.%. Values below 3% are analytically questionable.

Sample	1	2	5*	5-2	5-3	6	6a	7	7-2	9	11-1	11-2	11-3	12	12o	15	15-b	15-w	V
Cpx[†]	22	10	48	18	82	26		15		6	39		47	22		29		9	30
Ne	25		27			4													
W-*2M*	14	8				10		13		14	32		22	13		15		4	7
W-*1T*-	28	26				6		10		17	17		24	4		1		4	24
Psw						23													
Lc			9		18	8		11		16				26		12			
Kf			1	62				2		2									
Kpe	4																		
Pg											5			20		23			26
Me		33				21	10	41		45	6	85	4	9					12
Q				21															
And																<1			
Nap		7																	
Ol		7	7																
Pe		<1	<1					3		<1						3			
Sp		<1	3					1			1			1		<1			<1
H			6									1							
Ilm								1											
Tt								1											
Ps															5				
Y		<1					3		1										
Cc						2	32	<1	1						29				
Ah							4										30	41	
Gp							32		10						10		35	42	
Ap								2	20								36		
Cs							4								24				
Tn							8												
R							7												
Po																			<1
Bz															12				
Hb									36										
Pt									8										
Th									18										
E															5				
Mg															2				
Bd									4										
Hy		6														11			
Hj	8																		
Dt		2																	
Su												13							
Fr													3			6			
S															10				
Si															4				
R_{wp}[‡]	12.4	8.7	5.7	16.5	16.9	8.5	10	8.9	20.2	9.6	19	17.2	18.1	16.5	12.1	8.5	22.4	18.5	13.6
GOF	2.5	1.8	1.4	1.1	1.1	1.6	1.4	1.7	1.1	1.7	1.2	1.1	1.2	1.1	1.7	1.78	1.2	1.1	2.7
DW	0.4	0.7	1.1	1.7	1.7	0.9	1.2	0.7	1.6	0.7	1.5	1.6	1.6	1.8	0.8	0.7	1.5	1.8	0.4

**LdS5-2*, white needle-like accumulations from a void (test analysis); *LdS5-3*, brown void filling (test analysis); *LdS6-a*, white-black amygdale; *LdS7-2*, an amygdale (test analysis); *LdS11-1*, green matrix (may also contain burpalite and offretite, both not confirmed via the EPMA method); *LdS11-2*, brown-white matrix; *LdS11-3*, dark amygdale (all LdS11 samples: test analyses); *LdS12o*, brown-beige amygdale; *LdS15-b*, blue zone (test analysis); *LdS15-w*, white zone (test analysis).

[†]*Cpx*, clinopyroxene (diopside); *Ne*, nepheline; *W*, wollastonite; *Psw*, pseudowollastonite; *Lc*, leucite; *Kf*, K feldspar; *Kpe*, *K*-petalite; *Pg*, plagioclase; *Me*, melilite group; *Q*, quartz; *And*, andradite; *Nap*, natroapophyllite; *Ol*, olivine group; *Pe*, perovskite; *Tt*, titanite; *Ps*, pseudo-brookite; *Y*, ye'elimite; *Cc*, calcite; *Ah*, anhydrite; *Gp*, gypsum; *Ap*, apatite group; *Cs*, cuspidine; *Tn*, ternesite; *R*, rustumite; *Po*, pyrrhotite; *Bz*, bazhenovite; *Hb*, hannebachite; *Pt*, portlandite; *Th*, thaumasite; *E*, ettringite; *Mg*, magnesite; *Bd*, bredigite; *Sp*, spinel group; *H*, hematite; *Ilm*, ilmenite; *Hy*, hyalophane; *Hj*, hejtmanite; *Dt*, datolite; *Su*, spurrite; *Fr*, frankamenite; *S*, sulfur (alpha form); *Si*, sillimanite.

[‡]R_{wp}, residual weighted-pattern; *GOF*, goodness of fit (χ^2); *DW*, Durbin–Watson statistics.

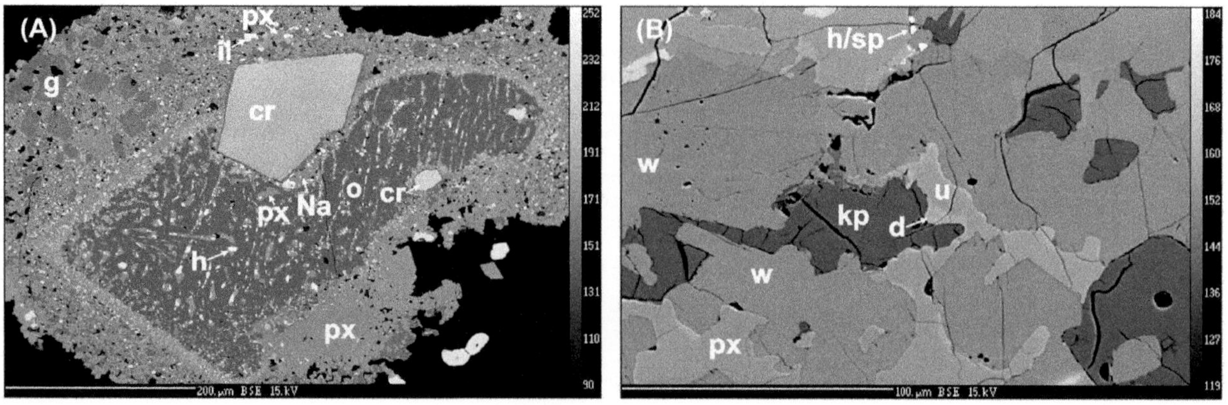

Figure 13.2.1. Backscattered electron (BSE) images of a thin section of the LdS1 (A) and LdS2 (B) slags; *cr*, chromite-rich spinel group (magnesioferrite); *d*, depmeierite-tuscanite; *g*, giuseppettite-like phase; *h*, hematite; *il*, ilmenite; *kp*, "K-petalite"; *o*, olivine group (forsterite); *p*, perovskite (?); *px*, clinopyroxene (*Na*, Na-rich one); *sp*, spinel group (mn-bearing magnesioferrite); *u*, unnamed phase A (?); *w*, wollastonite. Scale bars are 200 and 100 μm, respectively.

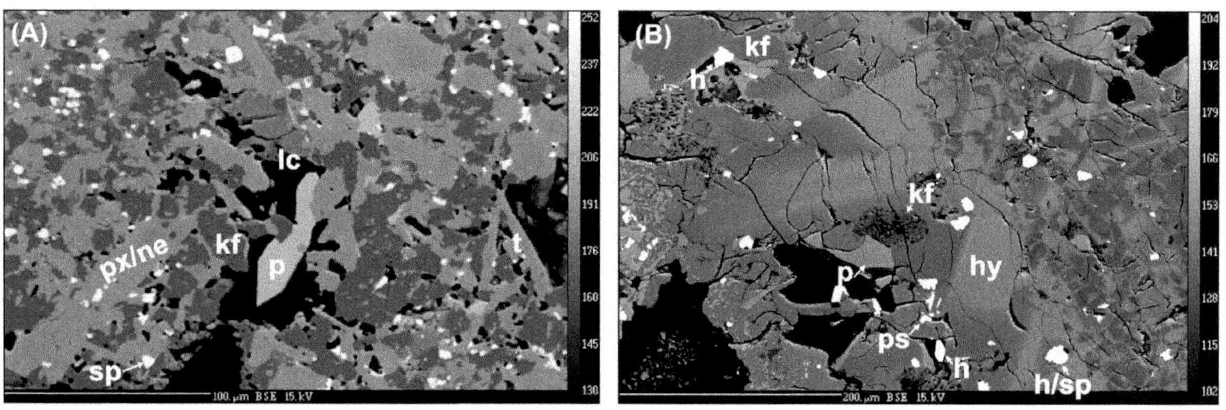

Figure 13.2.2. Back-scatter ..... matrix: (A) titanite-bearing portion, (B) pseudobrookite-bearing portion; *h*, hematite; *hy*, hyalophane-like phase; *kf*, alkaline feldspar; *lc*, leucite; *ne*, nepheline; *p*, perovskite; *ps*, pseudobrookite; *px*, clinopyroxene; *sp*, spinel group. Zoned crystals in the upper-right corner are clinopyroxenes. Scale bars are 100 and 200 μm, respectively.

moderate to extreme enrichment in vanadium, with some members probably containing Cr, which is a new discovery for natural apatites. It is not sure if Ti and Al also enter the structure or exist as a submicron inclusion phase. Slightly V-enriched bearing fluorapatite from the sample LdS7 can be described using the following empirical formula (n=3): $(Ca_{4.97}Sr_{0.03})_{\Sigma 4.99}[(PO_4)_{2.03}(SiO_4)_{0.56}(SO_4)_{0.41}(VO_4)_{0.05}]_{\Sigma 3.05}F_{0.68}$. The corresponding end-member formula is $Fap_{79}Fel_{18}Vfa_2$, with Fap for fluorapatite, Fel for fluorellestadite, and Vfa for a hypothetical $Ca_5(VO_4)_3F$ end-member. The latter is dominant in three remaining analyses, also corresponding to Sr-rich and Pb-bearing representative, mean member formula being $Vfa_{31}Cfe_{13}Fel_{12}Vca_6Svf_6Cce_3Sce_3Cle_2Sfe_2Svc_1$. The latter formula's end-members, most of them of a hypothetical nature, are $Ca_5(SiO_4,CrO_4)_3F$ is Cfe, Cl analogue of Vfa is Vca, Sr analogue of Vfa is Svf, Cl analogue of Cfe is Cce, Sr analogue of Cfe is Sce, Cle is chlorellestadite, Sr analogue of fluorellestadite is Sfe, and Svc is $Sr_5(VO_4)_3Cl$. Formulas of the proposed end-members can always be found under the relevant tables. A two-analysis empirical formula of the V-dominant representative is $(Ca_{4.22}Sr_{0.72}Fe_{0.03}K_{0.02}Ba_{0.01})_{\Sigma 5.00}[(VO_4)_{1.71}(PO_4)_{0.70}(SiO_4)_{0.56}(SO_4)_{0.06}(CrO_4)_{0.13}]_{\Sigma 3.17}[F_{0.65}Cl_{0.07}(OH)_{0.28}]_{\Sigma 1.00}$. Synthetic Pb–Ca–V–F apatite is reported by Dong et al. (2005). Substitution of a small amount of V in natural Si-rich members of the apatite group is known (McConnell, 1938), but a natural V–Ca-rich member was not yet found in nature. However, we discovered a similar V-dominant member in magnesioferrite-gehlenite-(perovskite)-(Ba-mica) slags from a heap in Bytom (Upper Silesian Coal Basin, Poland; see chapter 19 in this book).

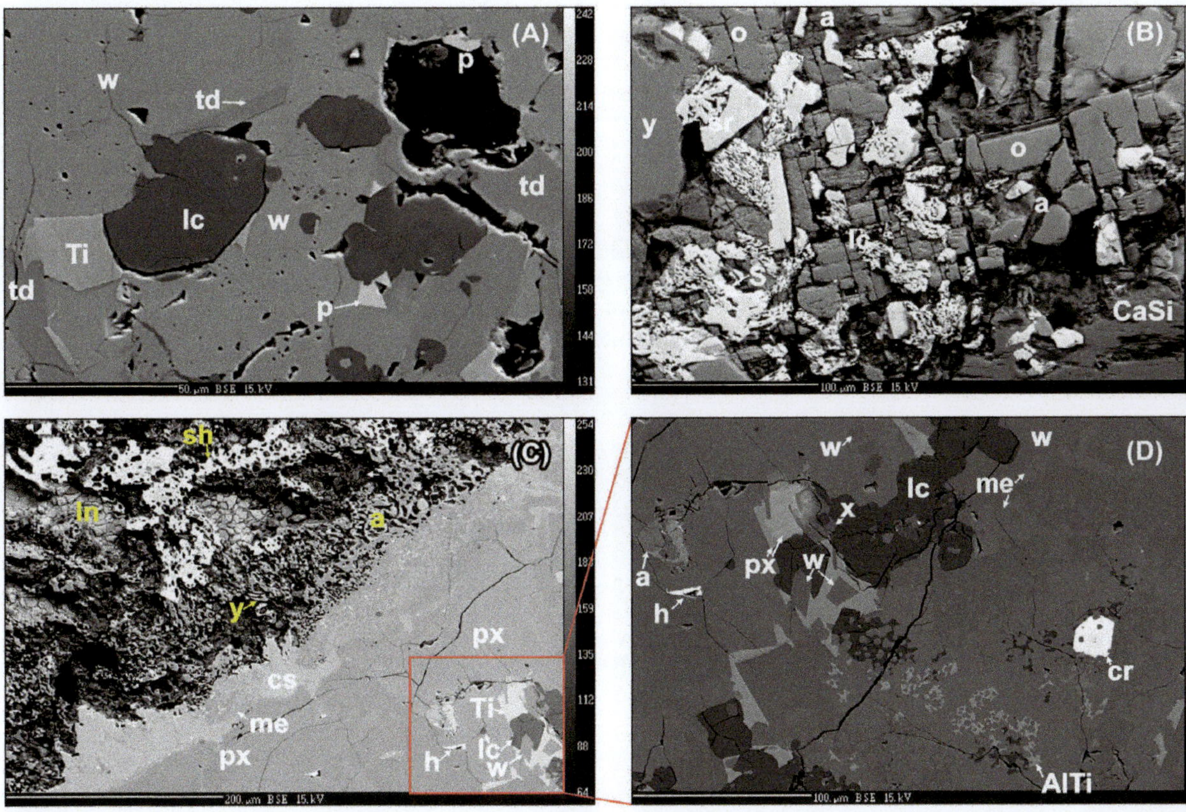

Figure 13.2.3. Backscattered electron (BSE) images of a thin section of the LdS6 (A, B) and LdS9 (C, D) slags. (A) and (C) are the matrix, (B) oldahmite-rich portion, and (C) enlargement of the spinel- and leucite-bearing portion: *a*, fluorapatite (strontian); *AlTi*, Al- and Ti-rich clinopyroxene; *CaSi*, various hydrated Ca silicates; *cr*, chromite-rich spinel group; *lc*, leucite (or K feldspar?); *ln*, larnite; *me*, melilite group (gehlenite); *oh*, oldhamite (strontian); *p*, perovskite (?); *s*, various Ca–Fe(Mn) s-rich (sulfate?) phases; *sh*, complex spinel-hematite aggregates (with various spinels); *sr*, srebrodolskite-brownmillerite (magnesian); *td*, tounkite-depmeierite; *y*, ye'elimite; *x*, unnamed amphibole-like phase (possibly corresponding to phase C). Scale bars are 50, 100, 200, and 100 µm, respectively.

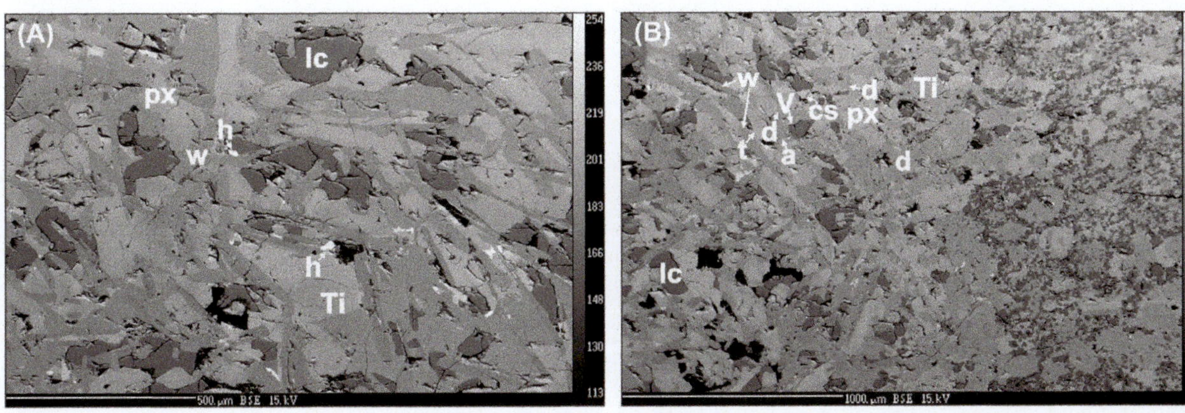

Figure 13.2.4. Backscattered electron (BSE) images of a thin section of the LdS7 slag matrix. (A) The matrix, (B) matrix fragment with V-rich fluorapatite: *a*, vanadiferous apatite; *d*, depmeierite-latiumite; *h*, vanadiferous hematite (with pseudobrookite molecule admixed); *lc*, leucite; *px*, clinopyroxene (ti-low); *t*, titanite (vanadiferous and chromian); *Ti*, Ti-rich clinopyroxene; *V*, Sr(Ba,Pb)-bearing V-dominant analogue of fluorapatite. Scale bars are 500 and 1000 µm, respectively.

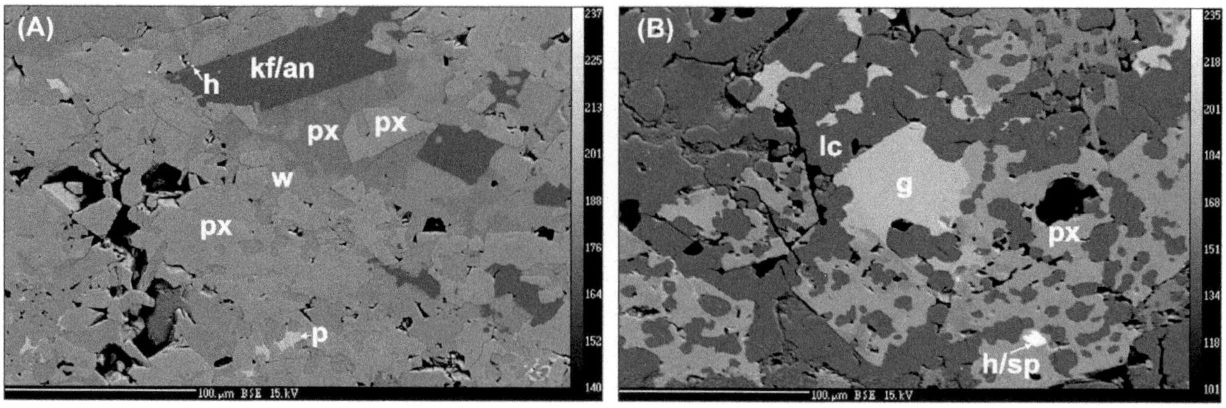

Figure 13.2.5. Backscattered electron (BSE) images of a thin section of the LdS11 slag matrix. (A) Feldspar-bearing portion, (B) garnet-bearing portion: *an*, anorthite; *g*, garnet group; *h*, hematite; *kf*, alkaline feldspar; *lc*, leucite; *p*, perovskite; *sp*, spinel group; *w*, wollastonite.

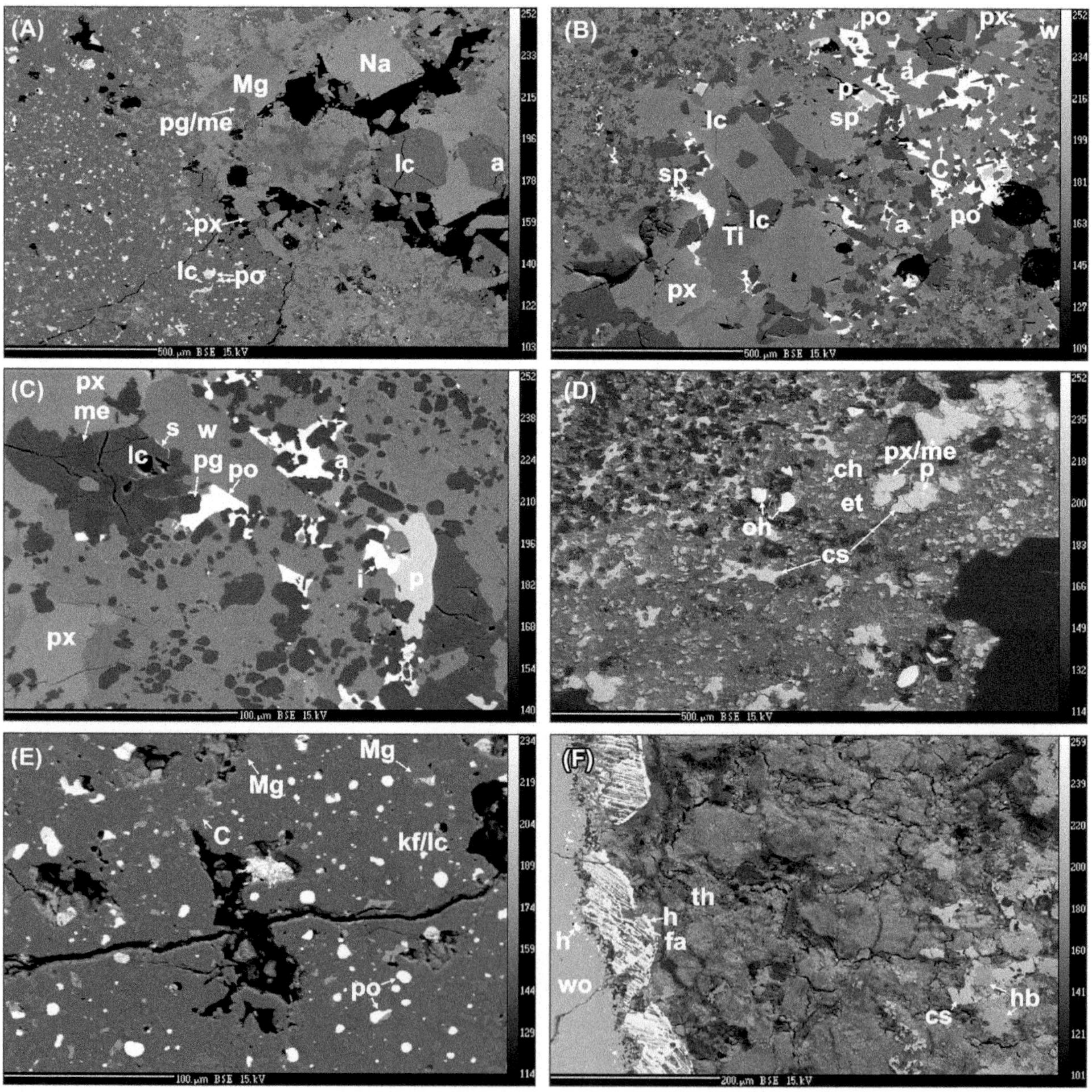

Figure 13.2.6. Backscattered electron (BSE) images of a thin section of the LdS12 slag; (A) two metalithoclasts border zone, (B) a matrix fragment with numerous BSE-bright pyrrhotite, (C) another matrix fragment with perovskite, (D) ca-s-rich amygdule with oldhamite, (E) another metalithoclast matrix, (F) a border of matrix with Ca–Si rich amygdule. Mineral abbreviations: *a*, fluorapatite (vanadiferous); *C*, unnamed phase C; *cs*, cuspidine; *et*, unspecified ettringite-like matrix (possibly composed of hannebachite and/or barzhenovite); *fa*, fayalite-like inclusions; *h*, hematite; *kf*, k feldspar; *lc*, leucite; *me*, melilite group; *Mg*, mg-rich clinopyroxene; *na*, na-bearing clinopyroxene; *oh*, oldhamite; *p*, perovskite; *pg*, plagioclase (anorthite); *po*, pyrrhotite; *px*, clinopyroxene (Mg-low); *th*, thaumasite-like phase; *Ti*, Ti-rich clinopyroxene; *w*, wollastonite.

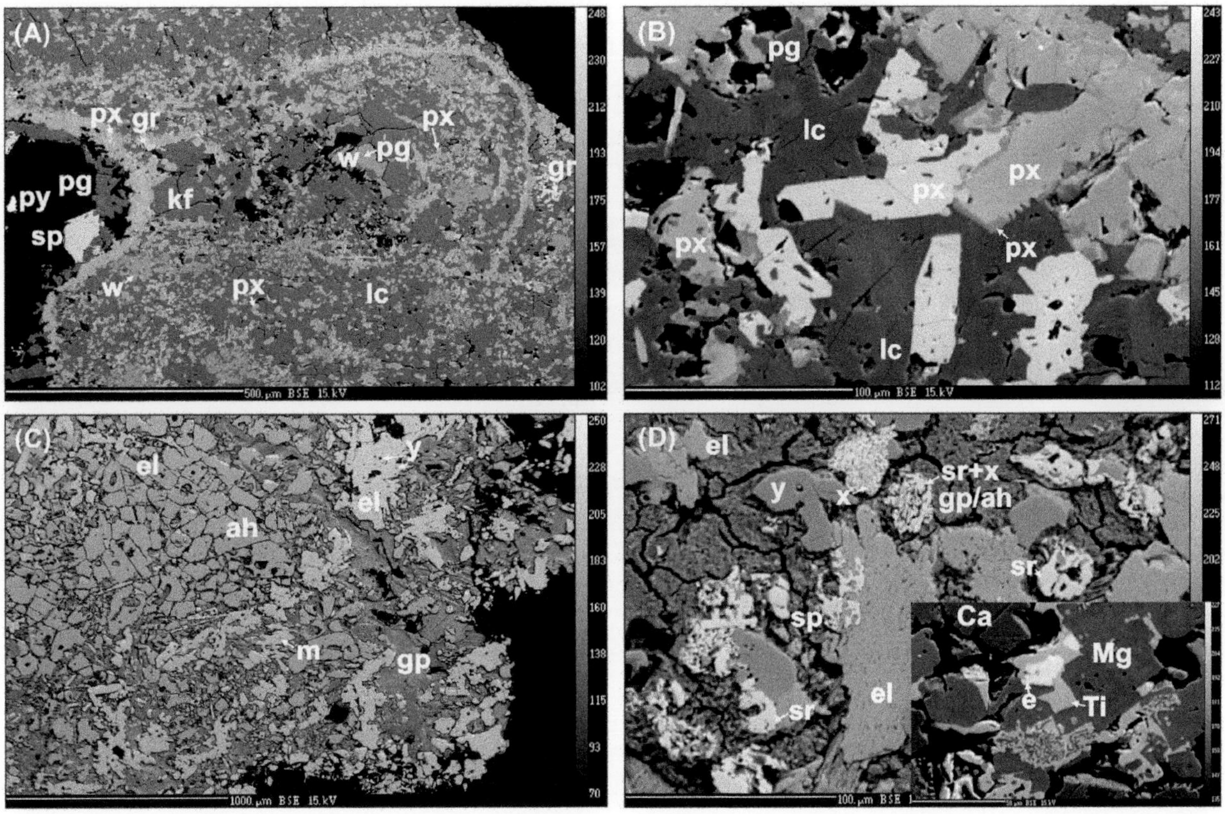

Figure 13.2.7. Backscattered electron (BSE) images of a thin section of the LdS15 slag; (A) general matrix view, (B) zoom of the lowermost leucite-bearing part showing complex clinopyroxene zonation, (C) view of the anhydrite-fluorellestadite-ye'elimite amygdule, (D) central part of the amygdule with Fe oxides (*inset*, the matrix zone with esseneite and other clinopyroxenes). Mineral abbreviations: *ah*, anhydrite; *Ca*, Ca-rich clinopyroxene; *e*, esseneite; *el*, fluorestadite; *gp*, gypsum; *gr*, garnet group; *kf*, K feldspar; *lc*, leucite; *m*, tiny rounded zones (within fluorellestadite) enriched in Mg; *Mg*, Mg-rich clinopyroxene; *pg*, plagioclase (anorthite); *py*, pyrite (?); *px*, clinopyroxene (various compositions); *sp*, spinel group; *sr*, srebrodolskite; *Ti*, Ti-rich clinopyroxene; *w*, wollastonite; *x*, unidentified aggregates/crystals (Al–Ca–Fe–S–Ti–Mn system); *y*, ye'elimite.

Apatites from samples LdS12 and LdS6 are way more similar to known ones, with Sr being the only exotic element occasionally showing. Considering the possible (and commonly observed in pyrometamorphic apatites, e.g.) presence of CO_3^{2-} ions, empirical and end-member formulas corresponding to the LdS12 and LdS6 apatites are $(Ca_{4.96}Sr_{0.02}La_{0.01}Ce_{0.01}K_{0.01})_{\Sigma 5.01}[(PO_4)_{2.68}(SiO_4)_{0.16}(CO_3)_{0.14}]_{\Sigma 2.98}[F_{0.69}(OH)_{0.24}O_{0.06})]_{\Sigma 0.99}$ and $Fap_{62}Oha_{24}Fce_7Oap_5Hce_3Oce_1$ (n=2) and $(Ca_{5.07}Fe_{0.02})_{\Sigma 5.09}[(SiO_4)_{1.51}(SO_4)_{1.34}(PO_4)_{0.06}]_{\Sigma 3.00}[(OH)_{0.49}O_{0.37}F_{0.14}]_{\Sigma 1.00}$ (n=4) corresponding to "oxyellestadite" $Oel_{98}Oap_2$ (n=3) and fluorellestadite $Fel_{57}Oel_{41}Oap_1Fap_1$ (n=1), respectively. Abbreviations used are Oha for hydroxylapatite, Fce for hypothetical "fluorcarboellestadite," Oap for oxyapatite, Hce for "hydroxylcarboellestadite," and Oce for "oxycarboellestadite." A single analysis (no. 7) for the sample LdS12 either corresponds to a nonapatite unnamed $Ca_4Al(PO_4,CO_3)_2(SiO_4)(F,OH)$ phase or is related to an unseen inclusion of an Al-rich phase.

The blue-colored apatite group representative from the LdS15 amygdule (see BSE image in Figure 13.2.7C and D) may be classified as fluorellestadite, as shown by the empirical formula $Ca_{5.00}[(SiO_4)_{1.55}(SO_4)_{0.92}(PO_4)_{0.03}(AlO_4)_{0.02}(CO_3)_{0.51}]_{\Sigma 3.01}[F_{0.55}(OH)_{0.30}]_{\Sigma 0.85}$. The composition of this mineral, again, is complex, especially due to variable but usually observed addition of Al. Although Al-rich species ye'elimite coexists with the fluorellestadite, no Al-rich inclusions were observed within it. Some BSE-dark, rounded, tiny spots inside were shown to be enriched only in Mg. The preceding formula includes a single analysis (no. 10) where hydroxylellestadite is the dominant end-member and a supposed "aluminoapatite" (Aap) hypothetical one is present in the largest-observed amount. A moderately good positive Al/Si correlation (r^2=0.74) is a single evidence of Al entering the fluorellestadite structure. The suggested end-member formula is thus $Fel_{61}Ohe_{35}Aap_2Fap_1Oha_1$, where Ohe designates hydroxylellestadite.

Table 13.2.3
Chemical analyses of anhydrite from the Lds15 sample, France.

	1	2	3	4
wt.%				
SO_3*	55.18	55.78	55.61	55.69
MoO_3†	1.74	1.68	1.57	1.62
CaO	41.87	42.82	42.79	41.88
Σ	98.78	100.27	99.97	99.19
CO_2‡	2.00	2.43	2.54	1.76
apfu				
S	0.92	0.91	0.91	0.93
Mo	0.02	0.02	0.01	0.02
Ca	1.00	1.00	1.00	1.00
C	0.06	0.07	0.08	0.05
end-members§ [%]				
Anh	92	91	91	93
Pow	2	2	1	2
Ar	6	7	8	5

*Na, Mg, Mn, Fe, Cu, Zn, Sr, and Si were below their detection limits.
†Confirmed via a WDS spectrometer scan for Mo.
‡A supposed substituent, from backward (differential) calculation after primary apfu calculation and, in the case of H_2O, after CO_2 calculation.
§*Pow*, powellite ($CaMoO_4$); *Ar*, aragonite.

Table 13.2.4
Chemical analyses of alkali feldspar, Lapanouse-de-Sévérac, France.

	1	2	3	4	5	6	7	8	9	10	11	12	13	14
	LdS5				LdS15									
wt.%														
SiO_2*	57.05	56.17	56.39	56.20	63.62	64.29	64.29	63.62	55.63	61.02	59.51	53.37	62.09	62.36
Al_2O_3	20.66	20.70	20.20	20.09	17.04	17.75	17.75	17.04	22.75	19.21	19.01	24.04	18.05	16.12
FeO	0.48	0.65	0.61	0.55	1.22	1.08	1.09	1.22	2.29	1.55	2.10	2.19	1.44	1.75
CaO	0.23	0.17	0.00	0.00	0.16	0.35	0.35	0.16	6.50	1.00	2.29	8.65	1.14	0.78
SrO	0.56	0.00	0.00	0.00	0.00	0.00	0.00	0.00	0.00	0.00	0.00	0.00	0.00	0.00
BaO	10.30	11.06	10.32	10.44	0.41	0.64	0.64	0.41	0.27	0.62	0.37	0.43	0.00	0.26
Na_2O	2.72	2.68	2.69	2.67	0.22	0.26	0.26	0.22	0.27	0.27	0.28	0.30	0.23	0.22
K_2O	8.60	8.35	8.72	8.53	16.16	14.67	14.67	16.16	11.41	15.88	13.48	9.46	15.24	14.86
Σ	100.58	99.78	98.94	98.48	98.82	99.05	99.05	98.82	99.12	99.55	97.05	98.44	98.20	96.33
Si	2.78	2.77	2.79	2.80	3.00	3.03	3.03	3.00	2.61	2.85	2.86	2.53	2.94	3.02
Al	1.19	1.20	1.18	1.18	0.95	0.99	0.99	0.95	1.26	1.06	1.08	1.34	1.01	0.92
Fe	0.02	0.03	0.03	0.02	0.05	0.04	0.04	0.05	0.09	0.06	0.08	0.09	0.06	0.07
Ca	0.01	0.01			0.01	0.02	0.02	0.01	0.33	0.05	0.12	0.44	0.06	0.04
Sr	0.02													
Ba	0.20	0.21	0.20	0.20	0.01	0.01	0.01	0.01	0.00	0.01	0.01	0.01	0.00	0.00
Na	0.26	0.26	0.26	0.26	0.02	0.02	0.02	0.02	0.03	0.02	0.03	0.03	0.02	0.02
K	0.53	0.52	0.55	0.54	0.97	0.88	0.88	0.97	0.68	0.95	0.83	0.57	0.92	0.92
end-members† [%]														
Or	52	51	54	53	92	90	90	92	61	87	78	51	87	87
Ab	25	25	25	25	2	2	2	2	2	2	2	2	2	2
An	1	1			1	2	2	1	29	5	11	39	5	4
Cls	19	21	20	20	1	1	1	1		1	1	1		
Sls	2													
Fan									2		1	3		
Fkf					5	4	4	5	4	5	6	3	5	7

*Mg, P, Sc, V, Cr, Mn, Ga, Rb, and Cs were below their detection limits, while Ti was either absent or not analyzed.
†*Or*, orthoclase/K feldspar; *Ab*, albite; *An*, anorthite; *Cls*, celsian; *Sls*, slawsonite; *Flc*, ferrilucite; *Fan*, $CaFe_2Si_2O_8$ "ferrianorthite"; *Fkf*, $KFeSi_3O_8$ "ferrifeldspar."

Table 13.2.5
Chemical analyses of apatite-group species, Lapanouse-de-Sévérac, France – part I.

	1	2	3	4	5	6	7*	8	9	10	11	12	13
	LdS7						LdS12			LdS6			
wt.%													
SO_3†	0.63	0.94	6.90	6.44	6.32	0.77	0.00	0.00	0.00	22.98	22.13	22.19	22.63
P_2O_5	7.24	8.34	28.10	28.25	29.49	7.80	25.61	37.11	37.08	1.09	0.90	0.75	0.77
V_2O_5	20.80	22.86	0.99	1.00	0.67	23.67	n.a.	n.a.	n.a.	n.a.	n.a.	n.a.	n.a.
SiO_2	6.34	4.03	7.16	6.27	6.47	4.84	13.02	1.92	1.89	18.29	19.02	19.17	19.85
TiO_2	3.81	0.79	0.00	0.00	0.00	1.08	n.a.	n.a.	n.a.	n.a.	n.a.	n.a.	n.a.
Al_2O_3	0.45	0.28	0.00	0.00	0.00	0.49	10.18	0.00	0.00	0.00	0.00	0.00	0.00
Cr_2O_3	1.50	1.68	0.00	0.00	n.a.	1.60	n.a.	n.a.	n.a.	0.00	0.00	0.00	0.00
Y_2O_3	n.a. ‡	n.a.	n.a.	n.a.	n.a.	n.a.	0.01	0.00	0.00	n.a.	n.a.	n.a.	n.a.
La_2O_3	n.a.	n.a.	n.a.	n.a.	n.a.	n.a.	0.17	0.23	0.20	n.a.	n.a.	n.a.	n.a.
Ce_2O_3	n.a.	n.a.	n.a.	n.a.	n.a.	n.a.	0.27	0.38	0.34	n.a.	n.a.	n.a.	n.a.
FeO	0.60	0.00	0.00	0.00	0.00	0.27	0.62	0.00	0.00	0.34	0.29	0.31	0.00
CaO	35.93	37.78	54.94	55.71	55.79	36.54	43.49	54.35	54.21	59.95	59.62	59.99	59.57
BaO	0.27	0.31	0.00	0.00	0.00	0.43	0.00	0.00	0.00	0.00	0.00	0.00	0.00
PbO	n.a.	n.a.	n.a.	n.a.	n.a.	0.71				n.a.	n.a.	n.a.	n.a.
SrO	11.55	11.80	0.60	0.53	0.42	12.89	0.37	0.34	0.33	0.29	0.00	0.00	0.00
K_2O	0.00	0.31	0.00	0.00	0.00	0.63	0.00	0.00	0.14	0.00	0.00	0.00	0.00
F	1.49	2.35	2.69	2.93	2.12	1.18	1.77	2.46	2.68	0.00	0.00	2.25	0.00
Cl	0.50	0.55	0.00	0.00	0.00	0.61	0.00	0.00	0.00	0.00	0.00	0.00	0.00
Σ	91.12	92.03	101.21	100.95	101.15	92.80	96.37	97.25	97.24	102.94	101.95	104.65	102.81
H_2O§							0.87	0.47	0.37				
apfu													
S	0.05	0.07	0.44	0.40	0.40					2.56	2.48	2.47	2.51
P	0.67	0.74	2.01	1.99	2.08	0.06	1.82	2.68	2.68	0.14	0.11	0.09	0.10
Si	0.69	0.42	0.60	0.52	0.54	0.51	1.09	0.16	0.16	2.71	2.84	2.85	2.94
V	1.50	1.58	0.06	0.06	0.04	1.64							
Ti	0.31	0.06											
Al	0.06	0.03				0.09	1.01						
Cr	0.13	0.14				0.13							
La							0.01	0.01	0.01				
Ce							0.01	0.01	0.01				
Fe	0.05					0.03				0.02	0.02	0.02	
Ca	4.20	4.23	4.97	4.97	4.98	4.10	3.92	4.96	4.95	5.08	5.08	5.09	5.04
Ba	0.01	0.01				0.02							
Pb						0.02							
Sr	0.73	0.72	0.03	0.03	0.02	0.78	0.02	0.02	0.02	0.01			
K		0.04				0.09			0.01				
F	0.52	0.78	0.72	0.77	0.56	0.39	0.47	0.66	0.72			1.06	
Cl	0.09	0.10				0.11							
OH							0.49	0.27	0.21				
end-members‖ [%]													
Aap													
Fap			78	79	81			60	64			1	
Oha								24	20				
Fel	14	11	19	18	17	10						57	
Fce								7	8				
Hce								3	2				
Oce								1	1				
Cle	2	1				3							
Vfa	29	35	2	2	2	30							
Vca	5	5				8							
Cfe	16	12				12							

Continued

Table 13.2.5
Chemical analyses of apatite-group species, Lapanouse-de-Sévérac, France – part I—cont'd.

	1	2	3	4	5	6	7*	8	9	10	11	12	13
	LdS7						LdS12			LdS6			
Oap								5	5	3	2	1	2
Oel										97	98	41	98
Cce	3	2				3							
Svf	5	6				6							
Svc	1	1				2							
Sfe	2	2				2							
Scl						1							
Sce	3	2				2							
Scc						1							

*A possible unnamed nonapatite $Ca_4Al(PO_4,CO_3)_2(SiO_4)$ (F,OH) phase.
†Nd, Mn, Mg, and Na were analyzed but not observed.
‡Not analyzed.
§Calculated by the CAMECA software.
‖*Aap*, "aluminoapatite"; *Fap*, fluorapatite; *Oha*, hydroxylapatite; *Fel*, fluorellestadite; *Cle*, chlorellestadite; hypothetical end-members: *Fce*, $Ca_5(SiO_4,CO_3)_3F$="fluorcarboellestadite"; *Hce*, "hydroxylcarboellestadite"; *Oce*, "oxycarboellestadite"; *Vfa*, $Ca_5(VO_4)_3F$; *Vca*, $Ca_5(VO_4)_3Cl$; *Cfe*, $Ca_5(SiO_4,CrO_4)_3F$; *Oap*, "oxyapatite"=$Ca_5(PO_4)O_{0.5}$; *Oel*, "oxyellestadite"; *Cce*, $Ca_5(SiO_4,CrO_4)_3Cl$; *Svf*, $Sr_5(VO_4)_3F$; *Svc*, $Sr_5(VO_4)_3Cl$; *Sfe*, $Sr_5(SiO_4,SO_4)_3F$; *Scl*, $Sr_5(SiO_4,SO_4)_3Cl$; *Sce*, $Sr_5(SiO_4,CrO_4)_3F$; *Scc*, $Sr_5(SiO_4,CrO_4)_3Cl$ (end-member calculation for the V-rich species assumes Cr entering the *T* site, Ti and Al being impurities, and the *X* site containing vacancies and no OH groups).

Table 13.2.6
Chemical analyses of an apatite-group species from the LDS15 sample, France.

	1	2	3	4	5	6	7	8	9	10	11	12
wt.%												
SO_3*	18.20	20.18	17.79	16.98	16.49	18.15	16.34	13.19	15.39	12.51	17.07	17.05
P_2O_5	0.77	0.65	0.97	3.28	2.82	0.25	0.20	0.17	0.62	0.48	0.36	0.33
V_2O_5	0.00	0.00	0.00	0.00	0.00	0.00	0.00	0.00	0.00	0.00	0.00	0.00
SiO_2	20.04	19.03	19.28	17.56	18.34	19.61	20.17	23.09	19.67	22.99	19.21	18.86
Al_2O_3	0.15	0.00	0.00	0.39	0.53	0.11	0.17	0.58	0.00	1.56	0.09	0.11
FeO	0.00	0.00	0.00	0.43	0.34	0.00	0.00	0.00	0.00	0.00	0.00	0.00
MgO	0.00	0.00	0.00	0.00	0.00	0.00	0.00	0.06	0.00	0.00	0.00	0.00
CaO	60.40	56.48	59.83	58.05	58.94	61.27	60.34	60.45	61.15	59.75	59.44	59.58
F	2.46	2.89	2.54	2.12	2.00	2.30	2.50	1.76	2.29	1.50	2.61	1.89
Σ	102.59	99.54	100.91	99.50	100.22	102.31	100.20	100.13	99.67	99.84	99.19	98.58
H_2O†	0.56	0.32	0.49	0.68	0.75	0.62	0.48	0.84	0.56	0.97	0.41	0.76
apfu												
S	1.01	1.05	1.25	1.01	0.97	1.04	0.95	0.76	0.88	0.73	1.01	1.00
P	0.05	0.05	0.06	0.22	0.19	0.02	0.01	0.01	0.04	0.03	0.02	0.02
Si	1.54	1.57	1.50	1.39	1.43	1.49	1.56	1.78	1.50	1.79	1.51	1.48
Al	0.01			0.04	0.05	0.01	0.02	0.05		0.14	0.01	0.01
Fe				0.03	0.02							
Mg								0.01		0.01		
Ca	4.99	5.00	5.00	4.94	4.93	5.00	5.00	4.99	5.00	4.99	5.00	5.00
F	0.60	0.76	0.63	0.53	0.50	0.55	0.61	0.43	0.55	0.37	0.65	0.47
OH	0.29	0.18	0.25	0.36	0.39	0.31	0.25	0.43	0.29	0.50	0.22	0.40
end-members‡ [%]												
Aap	1			4	5	1	2	5		14	1	1
Fap	3	4	4	13	11	1	1	1	1	1	4	2
Oha	2	1	2	9	8		1	1	1	2	1	1
Fel	63	77	67	44	43	69	47	51	65	35	79	60
Ohe	31	18	27	30	33	28	47	47	32	48	15	36

*Elements analyzed but not observed: Ba, Pb, Sr, K, Na, Y, La, Ce, Nd, Mn, and Cl.
†By difference.
‡*Aap*, aluminoapatite; *Fap*, fluorapatite; *Oha*, hydroxylapatite; *Fel*, fluorellestadite; *Ohe*, hydroxylellestadite.

Calcium Ferrites (Mainly Srebrodolskite, $Ca_2Fe_2O_5$)

Calcium ferrites are usually found as accessory species. As stated earlier, this simple Ca–Fe oxide group is usually represented by srebrodolskite, or actually srebrodolskite-brownmillerite solid solution (Table 13.2.7). The solution from the LdS15 slag is described by the empirical formula of $(Ca_{2.00}Mg_{0.05})_{\Sigma 2.05}(Fe^{3+}{}_{1.06}Al_{0.78}Mn^{3+}{}_{0.05}Ti_{0.03}Si_{0.02}[]_{0.08})_{\Sigma 2.00}O_{5.00}$ (n=3), where [] stands for vacancy. The related mean end-member composition is $Srb_{53}Brm_{39}Mns_3Mgs_1Ctv_1Mbr_1Csv_1$, with Srb for srebrodolskite, Brm for brownmillerite, and Mns for "manganisrebrodolskite," the other hypothetical end-members being only vestigial. Just a single analysis corresponds to grandiferrite (Gdf), ideally $CaFe_4O_7$, and more specifically $(Ca_{0.78}Mg_{0.22})_{\Sigma 1.00}(Fe^{3+}{}_{2.31}Al_{1.32}Mg_{0.16}Mn^{3+}{}_{0.10}Ti_{0.06}Si_{0.05}S_{0.01})_{\Sigma 4.01}O_{7.01}$. The mineral is clearly more complex in terms of diadochy and its end-member composition is $Gdf_{45}Gss_{26}Mgg_{13}Mgs_7Tmg_3Mng_2Smg_2Mmg_1Mvg_1Msv_1Sug_1$, with Gss standing for grossite ($CaAl_4O_7$) equivalent, followed my main

Table 13.2.7
Chemical analyses of calcium ferrites from the LdS15 Sample, France.

	1	2	3	4
wt.%				
SO_3*	0.09	0.09	0.00	0.25
SiO_2	0.33	0.39	0.42	0.90
Al_2O_3	16.06	16.02	15.66	20.21
Fe_2O_3	33.86	33.40	33.84	55.13
MgO	0.77	0.98	0.74	4.55
CaO	44.68	44.80	44.68	13.13
Σ	95.79	95.68	95.34	94.17
apfu				
S	0.00	0.00	0.00	0.01
Si	0.01	0.02	0.02	0.05
Al	0.79	0.79	0.77	1.32
Fe	1.06	1.05	1.07	2.31
Mg	0.05	0.06	0.05	0.38
Ca	2.00	2.00	2.01	0.78
end-members† [%]				
Srb	53	53	54	
Brm	40	39	39	
Mgs	1	2	1	
Mbr	1	1	1	
Mns	3	2	3	
Ctv	1	2	1	
Csv	1	1	1	
Gdf				45
Gss				26
Mng				2
Tmg				3
Smg				2
Mgg				13
Mgs				7
Mmg				1
Mtg				1
Msg				1
Sug				0.5

*P, Ti, V, Cr, Mn, Co, Ni, Cu, Zn, Ga, Sr, Ba, and Pb were measured but not observed.

†*Srb*, srebrodolskite; *Brm*, brownmillerite; *Gdf*, grandiferrite; *Gss*, grossite; hypothetical end-members: *Mgs*, $Mg_2Fe_2O_5$ "magnesiosrebrodolskite"; *Mbr*, Mg–$2Al_2O_5$ "magnesiobrownmillerite"; *Mns*, $Ca_2Mn_2O_5$ "manganisrebrodolskite"; *Ctv*, $Ca_2(Ti_{1.5}[]_{0.5})O_5$; *Csv*, $Ca2(Si_{1.5}[]_{0.5})O_5$; *Mng*, $CaMn_4O_7$ "manganograndiferrite"; *Tmg*, $Ca(Ti_2Mg_2)O_7$; *Smg*, $Ca(Si_2Mg_2)O_7$; *Mgg*, $MgFe_4O_7$ "magnesiograndiferrite"; *Mgs*, $MgAl_4O_7$ "magnesiogrossite"; *Mmg*, $MgMn_4O_7$ "magnesiomanganigrandiferrite"; *Mtg*, $Mg(Ti_2Mg_2)O_7$; *Msg*, $Mg(Si_2Mg_2)O_7$; *Sug*, (Ca, Mg) $(SMg_3)O_7$.

hypothetical end-members "magnesiograndiferrite," "magnesiogrossite," "titanomagnesiograndiferrite," "manganograndiferrite," and "silicomagnesiograndiferrite." The Mgg may be dominant, as is seen in material from sample LdS2 (see the Unnamed Minerals discussion at the end of this section of the chapter). Srebrodolskite is also present in the LdS6 sample, closely associating with oldhamite and ye'elimite (Figure 13.2.3B).

Cancrinite Group (Approximately $(Na,Ca,K)_{7\text{-}8}(Al,Si)_{10\text{-}12}O_{22\text{-}24}(SO_4,PO_4)_{[(1\text{-}x)\text{-}2]}\cdot nH_2O$)

Cancrinite group minerals are sometimes (or locally) rock-forming in the Lapanouse slags, usually as rounded or somewhat elongated crystals (Figure 13.2.1A). It took a while to recognize its diadochy nature and to systematize them correctly. This is mainly due to (1) recurring excess of sulfur, here ascribed to sulfate, sulfhydryl, and polysulfide groups; and (2) the presence of essential iron that is not necessarily in a good correlation with any other elements (Table 13.2.8). Typical dodecaalumino(ferri)silicate group representative, with a good fit to giuseppettite (ideally $(Na,K,Ca)_{7\text{-}8}(Al_6Si_6O_{24})(SO_4,Cl)_{1\text{-}2}$), is only found in the sample LdS1 (Figure 13.2.1A). There is a rather clear S/Cl negative correlation with $r^2=0.85$ suggesting that most, if not all, S is located in the extra framework anion sites. The best correlation of FeO is attained when juxtaposing it with CaO ($r^2=0.75$). As the state of S is unknown, with SO_4^{2-}, S_5^{2-}, and SH^- (the latter two known, e.g., in bystrite and sulhydrylbystrite members, Sapozhnikov et al., 2017) possible, we present two proposals of empirical formulas, which are best recast using the (Al + Si) = 12 basis: (1) $[(Na_{5.49}K_{0.71}Ca_{0.98}Fe^{2+}_{0.13}Mg_{0.02}]_{\Sigma 7.33}[(Al_{5.92}Si_{6.08})_{\Sigma 12.00}O_{24.00}][(SO_4)_{0.94}Cl_{0.47}(S_5)_{0.08}(PO_4)_{0.01}]_{\Sigma 1.50}$; (2) $[(Na_{5.49}K_{0.71}Ca_{0.98}Fe^{2+}_{0.13}Mg_{0.02}]_{\Sigma 7.33}[(Al_{5.92}Si_{6.08})_{\Sigma 12.00}O_{24.00}][(SO_4)_{0.71}(SH)_{0.62}Cl_{0.47}(PO_4)_{0.01}]_{\Sigma 1.81}$. The first formula is more likely due to polysulfide groups being confirmed to exist in bazhenovite (Bindi et al., 2005), which is known from some Lapanouse slags.

Cancrinite-like species of the LdS2 (Figure 13.2.1B) and LdS7 (Figure 13.2.4B) are intermediate between depmeierite—a phosphate-rich cancrinitoid, ideally $Na_8(Al_6Si_6O_{24})(PO_4,CO_3)_{1\text{-}x}\cdot 3H_2O$ ($x<0.5$)—and either tuscanite ($KCa_6(Si,Al)_{10}O_{22}(SO_4)\cdot H_2O$, LdS2 case) or latiumite ($(Ca,K)_8(Si,Al)_{10}O_{22}(SO_4,CO_3)$, LdS7 case). As opposed to the previous example, they are both better recast to decaaluminosilicates. Empirical formulas for the LdS2 and LdS7 are $K_{1.00}(Ca_{5.46}K_{0.54}Fe^{3+}_{0.16}Mg_{0.15}Na_{0.04})_{\Sigma 5.94}[(Al_{3.93}Si_{5.03}S_{1.04})_{\Sigma 10.00}O_{24.80}](PO_4)_{0.49}(SO_4)_{0.17}]_{\Sigma 0.66}\cdot 1.23H_2O$ and $[(K_{1.75}Na_{0.08}Mg_{0.04}Sr_{0.02}Ca_{0.02})_{\Sigma 1.91}(H_2O)_{0.09}])_{\Sigma 2.00}Ca_{6.00}[(Al_{3.74}Fe^{3+}_{0.27})_{\Sigma 4.01}(Si_{5.29}Al_{0.71})_{\Sigma 6.00}]O_{22.23}[(SO_4)_{1.13}(PO_4)_{0.85}]_{\Sigma 1.98}\cdot 0.45H_2O$, respectively. In the last phase, an evident Al/Fe trend can be seen, although the r^2 is just 0.64 (the largest obtained for Fe), possibly suggesting part of Fe entering the alkaline elements sites. Also, the large anion excess can be attributed to the presence of polysulfide groups instead of sulfate groups, a phenomenon that will have to be confirmed, for example, using IR or Raman spectroscopy. The presence of sulfur at the phosphate-occupied sites is confirmed by a negative S/P correlation ($r^2=0.71$). Considering (1) the polysulfide idea, (2) the known mean apfu(S), and (3) the extra-framework anion site occupancy being 1, the latter site could possibly be explained as $[(PO_4)_{0.85}(S_{7.53})_{0.15}]$.

Clinopyroxene Subgroup (Diopside-Esseneite-Hedenbergite ($Ca(Mg,Fe^{3+},Fe^{2+})(Si,Al)_2O_6$)

As the major Lapanouse slag components, these minerals were analyzed thoroughly. The following six tables bear analytical reports from clinopyroxenes from sample LdS1 (Table 13.2.9), LdS2 (Table 13.2.10), LdS5 (Table 13.2.11), LdS7 (Table 13.2.12), LdS12 (Table 13.2.13), and finally for samples LdS15 and LdSV (Table 13.2.14).

The corresponding empirical formulas and mean end-member formulas are:

- LdS1, n = 22, diopside, $(Ca_{0.95}Na_{0.04}Mg_{0.01})_{\Sigma 1.00}(Mg_{0.72}Fe^{3+}_{0.12}Ti_{0.08}Fe^{2+}_{0.06}Al_{0.02}Cr_{0.01})_{\Sigma 1.01}(Si_{1.74}Al_{0.26})_{\Sigma 2.00}O_{6.00}$, $Di_{68}Es_{12}Cta_7Hd_5Aeg_3Ku_2Cre_1$ (Di, diopside; Es, esseneite; Cta, $CaTiAl_2O_6$; Hd, hedenbergite; Aeg, aegirine; Ku, kushiroite; Cre, $CaCrAlSiO_6$).
- LdS2, n = 12, esseneite, $Ca_{1.00}(Fe^{3+}_{0.54}Mg_{0.24}Al_{0.11}Ti_{0.05}Fe^{2+}_{0.04}Ca_{0.03})_{\Sigma 1.01}(Si_{1.26}Al_{0.74})_{\Sigma 2.00}O_{6.00}$, $Es_{54}Di_{24}Ku_{11}Cta_5Hd_4Wo_3$ (Wo, wollastonite).
- LdS5, n = 29, diopside, $(Ca_{0.96}Na_{0.03}Mg_{0.01})_{\Sigma 1.00}(Mg_{0.67}Fe^{3+}_{0.13}Ti_{0.11}Fe^{2+}_{0.08})_{\Sigma 0.99}(Si_{1.68}Al_{0.33})_{\Sigma 2.01}O_{6.00}$, $Di_{65}Es_{13}Hd_8Cta_{11}Hyp_1(Aeg + Nta + Mge)_1(Cre + Mge)_{0.5}$ (Hyp, hypersthene; Nta, $NaTiAlSiO_6$; Mge, $MgFeAlSiO_6$).
- LdS7, n = 23, esseneite, $Ca_{1.00}(Fe^{3+}_{0.64}Mg_{0.23}Ti_{0.05}Fe^{2+}_{0.03}Ca_{0.03}Al_{0.02})_{\Sigma 0.99}(Si_{1.24}Al_{0.76})_{\Sigma 2.00}O_{6.00}$, $Es_{63}Di_{23}Cta_5Ku_3Hd_3Wo_3$.
- LdS12, n = 15, diopside, $Ca_{1.00}(Mg_{0.42}Fe^{2+}_{0.19}Fe^{3+}_{0.12}Al_{0.12}Ti_{0.12}Ca_{0.01}Cr_{0.01})_{\Sigma 0.99}(Si_{1.52}Al_{0.48})_{\Sigma 2.00}O_{6.00}$, $Di_{42}Hd_{19}Es_{12}Ku_{12}Cta_{12}Wo_1Cre_1$.

Table 13.2.8
Chemical analyses of cancrinite-group species: giuseppettite (LdS1), and species of depmeierite-tuscanite (LdS2), and depmeierite-latiumite systems (LdS7), France.

	1	2	3	4	5	6	7	8	9	10	11	14	15	16	17	18	19	20
	LdS1								LdS2									
wt.%																		
SO_3*	10.94	9.28	9.60	10.21	10.74	10.39	8.44	11.71	9.00	9.56	9.24	8.53	8.79	8.90	9.40	9.64	9.59	9.29
P_2O_5	0.00	0.00	0.11	0.06	0.08	0.09	0.15	0.00	3.40	3.39	3.36	3.61	3.75	3.72	2.87	2.84	2.83	3.48
SiO_2	35.06	34.97	35.45	35.41	35.50	34.31	36.04	34.92	28.62	28.03	28.56	29.80	29.04	28.94	27.98	28.87	28.72	29.27
Al_2O_3	29.09	28.60	28.95	29.28	29.74	28.10	29.62	29.52	19.09	19.73	18.94	18.03	18.45	18.77	19.74	18.62	18.77	18.34
FeO	0.90	0.97	1.02	0.96	0.95	0.85	0.93	0.71	1.16	0.97	1.05	1.04	1.31	0.95	0.74	1.16	1.24	1.24
MgO	0.00	0.11	0.09	0.07	0.00	0.12	0.11	0.05	0.39	0.38	0.55	0.61	0.58	0.52	0.66	0.68	0.71	0.71
CaO	5.55	4.91	5.15	5.15	5.24	5.61	4.92	6.00	29.17	29.05	28.95	29.11	29.29	29.34	28.71	29.13	29.20	28.94
K_2O	3.17	4.79	3.97	3.25	2.55	2.90	3.10	2.08	6.83	6.94	7.00	6.66	6.87	6.76	7.21	6.89	6.78	6.62
Na_2O	16.27	16.23	15.97	16.54	16.68	16.65	17.10	15.87	0.11	0.12	0.08	0.12	0.12	0.13	0.12	0.13	0.13	0.09
Cl	1.40	1.72	1.62	1.73	1.19	1.55	2.45	1.08	0.00	0.00	0.00	0.00	0.00	0.00	0.00	0.00	0.00	0.00
Σ	102.38	101.58	101.93	102.66	102.67	100.57	102.86	101.94	97.77	98.17	97.73	97.51	98.20	98.03	97.43	97.96	97.97	97.98
$H_2O^{\dagger}$									2.10	1.72	2.15	2.37	1.65	1.86	2.49	1.91	1.89	1.88
apfu																		
S	1.34	1.14	1.18	1.24	1.31	1.29	1.01	1.45	1.19	1.25	1.22	1.13	1.15	1.17	1.24	1.27	1.26	1.22
P			0.02	0.01	0.01	0.01	0.02		0.51	0.50	0.50	0.54	0.56	0.55	0.43	0.42	0.42	0.52
Si	5.74	5.72	5.80	5.75	5.79	5.68	5.77	5.78	5.03	4.90	5.02	5.25	5.08	5.07	4.91	5.06	5.03	5.14
Al	5.61	5.51	5.59	5.60	5.71	5.48	5.59	5.76	3.95	4.07	3.92	3.74	3.80	3.87	4.08	3.85	3.88	3.79
Fe^{3+}	0.12	0.13	0.14	0.13	0.13	0.12	0.12	0.10	0.17	0.14	0.15	0.15	0.19	0.14	0.11	0.17	0.18	0.18
Mg		0.03	0.02	0.02		0.03	0.03	0.01	0.10	0.10	0.14	0.16	0.15	0.14	0.17	0.18	0.19	0.19
Ca	0.97	0.86	0.90	0.90	0.92	1.00	0.84	1.06	5.49	5.44	5.45	5.49	5.49	5.51	5.40	5.47	5.48	5.44
K	0.66	1.00	0.83	0.67	0.53	0.61	0.63	0.44	1.53	1.55	1.57	1.50	1.53	1.51	1.61	1.54	1.52	1.48
Na	5.16	5.14	5.07	5.21	5.27	5.34	5.31	5.09	0.04	0.04	0.03	0.04	0.04	0.04	0.04	0.04	0.04	0.03
Cl	0.39	0.48	0.45	0.48	0.33	0.43	0.66	0.30										
$H_2O^{\ddagger}$									1.23	1.01	1.26	1.40	0.97	1.09	1.46	1.12	1.11	1.10

Continued

Table 13.2.8
Chemical analyses of cancrinite-group species: giuseppettite (LdS1), and species of depmeierite-tuscanite (LdS2), and depmeierite-latiumite systems (LdS7), France—cont'd.

	21	22	23	24	25	26	27	28	29	30
					LdS7					
wt.%										
SO_3	7.54	7.66	8.52	7.93	7.86	7.83	7.90	7.59	8.19	8.28
P_2O_5	5.35	5.20	4.40	5.69	5.31	5.32	4.98	5.52	5.23	4.72
SiO_2	27.88	27.78	27.46	27.49	27.68	27.66	27.29	27.91	27.54	26.87
TiO_2	0.00	0.00	0.00	0.00	0.00	0.00	0.00	0.00	0.00	0.00
Al_2O_3	19.43	19.44	19.68	19.82	20.06	19.88	20.17	19.12	19.45	20.15
Cr_2O_3	0.00	0.00	0.00	0.02	0.00	0.00	0.00	0.00	0.00	0.00
FeO	1.67	1.67	1.60	1.94	1.62	1.60	1.36	1.72	1.91	1.25
MgO	0.17	0.16	0.19	0.00	0.14	0.15	0.00	0.17	0.18	0.00
CaO	29.11	29.50	29.14	29.11	29.16	29.67	29.36	29.29	29.11	29.60
SrO	0.00	0.00	0.00	0.00	0.00	0.00	0.48	0.43	0.00	0.63
K_2O	7.06	7.18	7.09	7.11	7.14	7.31	7.26	6.88	7.09	7.44
Na_2O	0.23	0.24	0.38	0.23	0.16	0.00	0.15	0.29	0.18	0.19
Cl	0.00	0.00	0.00	0.00	0.00	0.00	0.00	0.00	0.00	0.00
Σ	98.43	98.84	98.50	99.34	99.14	99.42	98.95	98.92	98.88	99.12
$H_2O^{\dagger}$	1.38	0.97	1.32	0.45	0.68	0.41	0.90	0.89	0.91	0.74
S	1.08	1.10	1.23	1.13	1.12	1.12	1.14	1.10	1.18	1.20
P	0.87	0.84	0.72	0.92	0.85	0.86	0.81	0.90	0.85	0.77
Si	5.34	5.33	5.28	5.24	5.26	5.28	5.23	5.38	5.29	5.20
Al	4.39	4.40	4.46	4.45	4.49	4.47	4.55	4.34	4.40	4.60
Fe	0.27	0.27	0.26	0.31	0.26	0.25	0.22	0.28	0.31	0.20
Mg	0.05	0.05	0.05		0.04	0.04		0.05	0.05	
Ca	5.98	6.07	6.00	5.94	5.93	6.06	6.03	6.05	5.99	6.14
Sr							0.05	0.05		0.07
K	1.73	1.76	1.74	1.73	1.73	1.78	1.77	1.69	1.74	1.84
Na	0.09	0.09	0.14	0.08	0.06		0.06	0.11	0.07	0.07
H_2O	0.88	0.62	0.85	0.29	0.43	0.26	0.58	0.57	0.58	0.48

*V, Mn, and occasionally Ba were measured but not observed.
†By difference.
‡Calculated by difference after conversion of FeO to Fe_2O_3; calculated after conversion of FeO to Fe_2O_3.

Table 13.2.9
Chemical analyses of clinopyroxene found in the LdS1 sample, France.

	1	2	3	4	5	6	7	8	9	10
wt.%										
SiO_2*	49.11	49.23	48.51	42.69	46.17	42.89	48.24	53.01	52.42	51.95
TiO_2	1.65	1.88	1.89	3.26	3.01	4.37	2.37	0.14	0.14	0.66
Al_2O_3	5.29	5.04	3.27	6.69	6.18	8.63	4.86	0.67	1.04	1.33
Cr_2O_3	0.00	0.36	0.00	0.00	0.00	0.00	0.23	0.42	0.50	0.18
Fe_2O_3	6.18	5.48	6.48	7.48	8.48	9.48	10.48			
MnO	0.00	0.12	0.31	0.22	0.23	0.00	0.00	0.17	0.00	0.15
MgO	13.64	14.15	12.41	10.11	12.13	11.57	13.81	14.05	14.45	15.18
CaO	23.68	23.69	21.97	21.12	22.86	23.90	23.87	18.06	19.40	22.82
FeO								9.58	8.48	5.24
K_2O	0.00	0.00	0.00	0.08	0.00	0.00	0.00	0.00	0.06	0.00
Na_2O	0.72	0.00	2.00	1.82	1.26	0.85	0.58	3.29	2.79	1.28
Σ	100.27	99.95	96.84	93.47	100.32	101.69	104.44	99.39	99.28	98.79
apfu										
Si	1.81	1.83	1.80	1.63	1.73	1.61	1.79	1.94	1.92	1.92
Ti	0.05	0.05	0.05	0.09	0.08	0.12	0.07			0.02
Al	0.23	0.22	0.14	0.30	0.27	0.38	0.21	0.03	0.04	0.06
Cr		0.01					0.01	0.01	0.01	0.01
Fe	0.17	0.15	0.29	0.38	0.22	0.22	0.17	0.29	0.26	0.16
Mn			0.01	0.01	0.01			0.01		
Mg	0.75	0.78	0.69	0.58	0.68	0.65	0.76	0.77	0.79	0.84
Ca	0.94	0.94	0.87	0.87	0.92	0.96	0.95	0.71	0.76	0.90
Na	0.05		0.14	0.13	0.09	0.06	0.04	0.23	0.20	0.09
end-members† [%]										
Es	10		20	28	18	20	11			
Cta	4	5	5	8	8	12	6			2
Di	69	68	60	50	62	62	72	51	57	75
Hd	6	14			2		6	19	17	13
Aeg	4		14	13	9	6	4	23	20	9
Jd	1									
Nte			1	1	1	1				
Cre		1					1			
Cen	1	4					1	6	4	

*Ni was analyzed but not observed.
†*Es*, esseneite ($CaFeAlSiO_6$); *Di*, diopside; *Hd*, hedenbergite; *Aeg*, aegirine; *Cen*, clinoenstatite; *Jd*, jadeite; hypothetical end-members: *Cta*, $CaTiAl_2O_6$; *Nte*, $NaTiSi_2O_6$; *Cre*, $CaCrAlSiO_6$.

- LdS15, n=18, diopside, $Ca_{1.00}(Fe^{3+}_{0.49}Mg_{0.41}Fe^{2+}_{0.08}Ti_{0.03}Ca_{0.01})_{\Sigma 1.02}(Si_{1.47}Al_{0.43}Fe^{3+}_{0.10})_{\Sigma 2.01}O_{6.00}$, $Di_{41}Es_{38}Fes_{10}Hd_8Cta_2Wo_1Ctf_1$ (Ctf, $CaTiFe_2O_6$).
- LdSV, n=38, diopside, $Ca_{1.00}(Mg_{0.35}Fe^{2+}_{0.25}Fe^{3+}_{0.15}Al_{0.14}Ti_{0.11}Ca_{0.01})_{\Sigma 1.01}(Si_{1.50}Al_{0.50})_{\Sigma 2.00}O_{6.00}$, $Di_{35}Hd_{25}Es_{15}Ku_{15}Cta_{11}Wo_1$.

Although some variations between the various slag samples do exist, the analyzed pyroxenes are usually diopsidic or esseneitic. The pyroxenes usually show very complex zonation, sometimes with >2 varieties existing close to each other (see, e.g., inset in Figure 13.2.7D). Some of the LdS1 sample clinopyroxenes are more sodic, as shown by the mean empirical formula $(Ca_{0.88}Na_{0.12})_{\Sigma 1.00}(Mg_{0.65}Fe^{3+}_{0.25}Ti_{0.08}Fe^{2+}_{0.01}Mn_{0.01}Ca_{0.01})_{\Sigma 1.01}(Si_{1.72}Al_{0.24}Fe^{3+}_{0.04})_{\Sigma 2.00}O_{6.00}$ and end-member formula $Di_{57}Es_{22}Aeg_{12}Cta_7Nte_1Hd_1$ (n=3). The LdS2 pyroxene is closely associated with next unnamed phase, herein referred to phase A (Figure 13.2.1A), of an amphibole-like stoichiometry (see the Unnamed Minerals discussion at the end of this section of the chapter). This phase has a Fe^{3+}-dominant analogue, coded as phase C, found in the sample LdS12 (Figure 13.2.6B and E). Similar to LdS1, the LdS7 clinopyroxene also exists as a second variety, which in this case is more titanian: $Ca_{1.00}(Fe^{3+}_{0.47}Mg_{0.19}Ti_{0.12}Fe^{2+}_{0.10}Al_{0.09}Ca_{0.02}Mn_{0.01})_{\Sigma 1.00}(Si_{1.20}Al_{0.80})_{\Sigma 2.00}O_{6.00}$; $Es_{47}Di_{19}Cta_{12}Hd_{10}Ku_9Wo_2Jh_1$ (n=13). The LdS7 sample also contains a possible unnamed phase (phase B), compositionally related to phase A. These formulas clearly show that only diopside and

Table 13.2.10
Chemical analyses of clinopyroxene from the LdS2 sample, France.

	1	2	3	4	5	6	7	8	9	10
wt.%										
SiO_2*	30.38	30.81	31.59	30.32	31.77	31.47	32.59	32.57	32.31	30.38
TiO_2	2.30	2.00	1.68	1.74	1.72	1.79	1.33	1.34	1.37	2.30
Al_2O_3	19.64	19.50	18.80	16.46	18.81	18.68	17.86	18.23	18.03	19.64
Fe_2O_3	19.82	20.10	19.15	24.51	19.40	19.15	18.85	18.43	18.88	19.82
MgO	3.28	3.51	3.93	2.40	4.09	3.88	4.64	4.50	4.48	3.28
CaO	23.89	24.11	24.50	23.55	24.14	24.28	24.25	24.42	24.35	23.89
K_2O	0.20	0.00	0.00	0.00	0.00	0.00	0.00	0.00	0.00	0.00
Σ	99.51	100.03	99.65	98.98	99.93	99.25	99.52	99.49	99.42	99.51
apfu										
Si	1.20	1.21	1.24	1.23	1.25	1.25	1.28	1.28	1.27	1.27
Ti	0.07	0.06	0.05	0.05	0.05	0.05	0.04	0.04	0.04	0.04
Al	0.92	0.91	0.87	0.79	0.87	0.87	0.83	0.84	0.84	0.84
Fe^{3+}	0.59	0.60	0.57	0.75	0.57	0.57	0.56	0.54	0.56	0.55
Mg	0.19	0.21	0.23	0.15	0.24	0.23	0.27	0.26	0.26	0.27
Ca	1.01	1.02	1.03	1.03	1.02	1.03	1.02	1.03	1.03	1.03
K	0.01									
end-members† [%]										
Es	56	55	53	64	53	53	53	52	54	53
Ku	12	12	12	2	12	12	11	12	11	11
Cta	7	6	5	5	5	5	4	4	4	4
Di	20	21	23	15	24	23	27	26	26	27
Hd	4	5	3	11	4	4	3	3	2	3
Wo	1	2	3	3	2	3	2	3	3	3

*Na, Mn, Cr, and Ni were analyzed but not observed.
†*Es*, esseneite; *Ku*, kushiroite ($CaAlAlSiO_6$); *Di*, diopside; *Hd*, hedenbergite; *Wo*, wollastonite; hypothetical end-members: *Cta*, $CaTiAl_2O_6$.

esseneite are the dominant end-members (single exceptions of kushiroite and hedenbergite from the sample LdSV, analyses 11, 7, and 8 in Table 13.2.14). An important part of kushiroite in the composition of most of the Lapanouse clinopyroxenes links it with extraterrestrial environment, as the kushiroite-dominant member was discovered in a chondrite (Kimura et al., 2009). Similar links concerning minerals from Polish burning coal-mining heaps are reported by Kruszewski et al. (2012). The empirical formula for brown crystals reported by Eytier et al. (2004)—$Ca_{1.01}(Fe^{3+}_{0.75}Mg_{0.19}Fe^{2+}_{0.03}Ti_{0.03})(Si_{1.20}Al_{0.72}Fe^{3+}_{0.08})O_6$—is in a good correspondence to our average observations.

Cuspidine ($Ca_4(Si_2O_7)F_2$)

Cuspidine was identified in three samples (Figures 13.2.3C, 13.2.6D and F). Its chemical analysis is shown in Table 13.2.15. The LdS7 cuspidine is relatively impure, as shown by the empirical formula $(Ca_{3.97}Sr_{0.03})[(Si_{2.00}P_{0.02}Ti_{0.02})O_{7.09}][F_{1.29}(OH)_{0.71}]_{\Sigma 2.00}$. Its composition may be also presented in the end-member form of $Cus_{97}Tcs_1Pcs_1Src_1$ (n=6), with the following minor hypothetical end-members: $Ca_4(Ti_2O_7)(F,OH)_2$, $Ca_4[(P_{1.6}[]_{0.4})O_7](F,OH)_2$, and $Sr_4(Si_2O_7)(F,OH)_2$. The LdS12 and LdS6 cuspidines are way more pure, with chemical formulas being $(Ca_{4.02}Mg_{0.01})_{\Sigma 4.03}(Si_{1.96}O_{6.90})F_{2.44}$ and $(Ca_{4.01}Fe_{0.02})_{\Sigma 4.03}(Si_{1.97}O_{6.90})[F_{0.88}(OH)_{0.71}]_{\Sigma 1.59}$, respectively.

Ettringite Group ($Ca_3[(Si,Al)(OH)_6](SO_4,CO_3)_2 \cdot 12H_2O$)

A representative of this group, which may be approximated as kottenheimite-thaumasite solid solution (or dehydrated ferroan kottenheimite) with minor add of ettringite and other end-members, was located in the sample LdS12. It is undoubtedly of a secondary origin (Figure 13.2.6D). Its composition is shown in Table 13.2.16. Structural complexity of the mineral is shown by its empirical formula $(Ca_{2.69}Fe^{2+}_{0.21}Mg_{0.10}Na_{0.01})_{\Sigma 3.01}[(Si_{0.85}Al_{0.14})_{\Sigma 0.99}(OH)_{5.99}Cl_{0.01}](SO_4)_{1.00}[(CO_3)_{0.51}(SO_4)_{0.49}]_{\Sigma 1.00} \cdot 1.99H_2O$. The CO_2 oxide addition is theoretical and the hydration level is almost

Table 13.2.11
Chemical analyses of clinopyroxene from the LdS5 sample, France.

	1	2	3	4	5	6	7	8	9	10	11	12
wt.%												
SiO_2*	47.34	46.77	41.14	42.74	46.24	46.53	39.23	46.45	42.54	44.96	39.93	48.68
TiO_2	2.49	2.69	5.17	4.34	2.91	2.64	5.52	2.87	4.13	3.61	5.54	2.02
Al_2O_3	4.86	5.20	9.68	8.66	5.89	5.39	10.15	5.42	8.82	6.13	10.53	2.96
Cr_2O_3	0.31	0.23	0.29	0.32	0.00	0.24	0.00	0.13	0.07	0.00	0.32	0.00
Fe_2O_3	6.22	6.75	8.05	7.65	7.19	6.96	9.00	7.15	8.04	9.46	8.12	9.17
MgO	13.75	13.38	10.90	11.55	12.97	13.21	9.85	13.35	11.39	11.06	10.42	12.01
CaO	23.65	23.75	23.38	23.53	23.73	23.92	23.07	23.74	23.74	22.26	23.40	23.09
Na_2O	0.33	0.33	0.42	0.42	0.35	0.31	0.50	0.36	0.39	1.26	0.40	0.70
Σ	98.95	99.10	99.03	99.20	99.27	99.19	97.32	99.48	99.12	98.74	98.66	98.62
apfu												
Si	1.78	1.76	1.57	1.62	1.74	1.75	1.53	1.74	1.62	1.72	1.53	1.86
Ti	0.07	0.08	0.15	0.12	0.08	0.07	0.16	0.08	0.12	0.10	0.16	0.06
Al	0.22	0.23	0.44	0.39	0.26	0.24	0.47	0.24	0.39	0.28	0.48	0.13
Cr	0.01	0.01	0.01	0.01	0.00	0.01					0.01	
Fe^{3+}	0.10	0.11	0.15	0.14	0.11	0.12	0.19	0.13	0.16	0.18	0.16	0.08
Fe^{2+}	0.08	0.08	0.09	0.08	0.09	0.08	0.08	0.07	0.07	0.10	0.07	0.18
Mg	0.77	0.75	0.62	0.65	0.73	0.74	0.57	0.75	0.64	0.63	0.60	0.68
Ca	0.95	0.96	0.96	0.96	0.96	0.97	0.96	0.96	0.97	0.91	0.96	0.95
Na	0.02	0.02	0.03	0.03	0.03	0.02	0.04	0.03	0.03	0.09	0.03	0.05
end-members† [%]												
Es	9	10	14	14	11	11	18	12	16	16	16	7
Cta	7	8	15	12	8	7	16	8	12	10	16	6
Di	71	69	58	62	68	69	55	67	62	57	57	64
Ku		1				1		1		1		
Hd	7	8	8	7	9	7	8	7	7	9	7	17
Hyp	2	1	1	1	1	1		1			1	
Aeg							1			2		
Cre	1	1	1	1		1					1	
Nta							1			1		

*K, Mn, Ni, and Zn were analyzed but not observed.
†*Es*, esseneite; *Di*, diopside; *Ku*, kushiroite; *Hd*, hedenbergite; *Hyp*, hypersthene; *Aeg*, aegirine; hypothetical end-members: *Cre*, $CaCrAlSiO_6$; *Nta*, $NaTiAlSiO_6$.

certainly lowered due to beam-induced decomposition. The related end-member representation is $Kth_{39}Thm_{38}Cte_7Ett_6Fth_3Mth_2Mkt_1Fce1Nah_1$, with Kth for kottenheimite, Thm for thaumasite, Ett for ettringite, and with the following hypothetical end-members: "carboettringite" (Cte), "ferrothaumasite" (Fth), "magnesiothaumasite" (Mth), "magnesiokottenheimite" (Mkt), "ferrocarboettringite" (Fce), and $Na_3[Al(OH)_6]$ phase (Nah). "Magnesiocarboettringite" and "feroettringite" stand for even less frequent hypothetical end-members. The Na/Si and Na/Al correlation is very high, with $r^2=0.95$ and 0.96, respectively, thus approving the Nah end-member part. Fe is strongly correlated with Mg and Ca (if using a polynomial trend line), but not with Al, which may prove its divalent character.

Garnet Group (Andradite, $Ca_3(Fe,Ti)_2(SiO_4)_3$)

Garnets from three samples were analyzed, the results being juxtaposed in Table 13.2.17. Their spatial relation to other matrix-forming minerals can be seen in Figures 13.2.5B and 13.2.7A. Following are empirical formulas and mean end-member compositions:

- LdS2, n=7, $(Ca_{2.99}K_{0.01})_{\Sigma 3.00}(Fe^{3+}{}_{1.33}Ti_{0.41}Al_{0.20}Mg_{0.06})_{\Sigma 2.00}\{[Si_{2.59}Al_{0.41}]_{\Sigma 3.00}O_4\}_{2.99}$, $And_{66}Hut_{20}Grs_{10}Ctm_2(Ktg+Khh+Ctc)_2$, where And is andradite, Hut is hutcheonite, Grs is grossular, Ctm is "calciotitanomajorite," Ktg is "K–Ti-garnet" $(K_2Ca)Ti_2(SiO_4)_3$, Khh is khoharite $(Mg_3Fe_2Si_3O_{12})$, and Ctc is "calciotitanocalciomajorite."

Table 13.2.12
Chemical analyses of Ti-rich (analyses 1–8) and Ti-low (7–12) clinopyroxene from the LdS7 sample, France.

	1	2	3	4	5	6	7	8	9	10	11	12
wt.%												
P_2O_5*	0.00	0.00	0.00	0.00	0.00	0.00	0.00	0.00	0.00	0.00	0.10	0.00
SiO_2	27.03	27.43	29.67	30.89	29.25	28.85	29.91	32.28	30.25	32.12	29.55	29.57
TiO_2	5.98	7.75	3.63	2.82	4.16	5.75	1.45	1.79	1.62	1.48	1.53	1.85
Al_2O_3	20.25	17.08	21.05	19.15	20.39	17.72	16.18	16.15	16.43	15.52	16.47	18.34
Fe_2O_3	18.86	20.43	16.82	18.14	18.73	20.28	23.61	19.82	22.77	21.38	24.94	21.64
MnO	0.63	1.00	0.00	0.00	0.39	0.44	0.00	0.18	0.00	0.00	0.00	0.00
MgO	1.79	1.27	3.04	3.96	2.75	2.15	3.65	4.50	3.62	4.32	3.24	3.23
CaO	23.62	22.80	24.13	23.97	23.89	23.20	23.81	24.12	23.60	24.16	23.68	24.00
Na_2O	0.00	0.20	0.00	0.00	0.01	0.00	0.00	0.00	0.00	0.00	0.00	0.00
Σ	98.15	99.88	98.34	98.94	99.57	98.38	98.62	98.84	98.29	98.98	99.50	98.63
apfu												
Si	1.10	1.13	1.18	1.23	1.16	1.18	1.21	1.29	1.23	1.29	1.19	1.19
Ti	0.18	0.24	0.11	0.08	0.12	0.18	0.04	0.05	0.05	0.04	0.05	0.06
Al	0.97	0.83	0.99	0.90	0.96	0.85	0.77	0.76	0.79	0.73	0.78	0.87
Fe^{3+}	0.46	0.35	0.72	0.55	0.47	0.43	0.72	0.55	0.66	0.61	0.73	0.64
Fe^{2+}	0.12	0.28		0.04	0.09	0.19		0.04	0.03	0.04	0.03	0.01
Mn	0.02	0.04			0.01	0.02		0.01				
Mg	0.11	0.08	0.18	0.23	0.16	0.13	0.22	0.27	0.22	0.26	0.19	0.19
Ca	1.03	1.01	1.03	1.02	1.02	1.02	1.03	1.03	1.03	1.04	1.02	1.03
Na		0.20										
end-members† [%]												
Es	46	31	42	49	47	43	68	55	65	61	69	64
Cta	18	22	11	8	12	18	4	5	5	4	4	6
Di	11	8	18	23	16	13	22	27	22	26	19	19
Ku	8		17	12	12	4		5	2	2		6
Hd	12	27	8	5	9	19		4	3	4	3	1
Jh	2	4			1	2						
Wo	3	1	3	2	2	2		1				
Aeg	1											
Ctf	1											
Fes							2				2	

*F, Cl, S, K, V, Ba, Pb, Sr, and Cr were analyzed but not observed.
†*Es*, esseneite; *Di*, diopside; *Ku*, kushiroite; *Hd*, hedenbergite; *Jh*, johannsenite; *Wo*, wollastonite; *Aeg*, aegirine; hypothetical end-members: *Cta*, $CaTiAl_2O_6$; *Ctf*, $CaTiFe_2O_6$; *Fes*, $CaFeFeSiO_6$.

- LdS15, n=7, $(Ca_{2.90}Mg_{0.09})_{\Sigma 2.99}(Fe^{3+}{}_{1.48}Ti_{0.51}Zr_{0.01})_{\Sigma 2.00}(Si_{2.59}Al_{0.21}Fe^{3+}{}_{0.20})_{\Sigma 3.00}O_{12.05}$, $And_{72}Sch_{12}Hut_{12}Khh_2(Mhu + Msc + Krm)_1$, with Sch for schorlomite ($Ca_3(Ti,Fe)_2[(Si,Fe^{3+})O_4]_3$), Mhu for "magnesiohutcheonite," Msc for "magnesioschorlomite," Krm for kerimasite (Ca–Zr silicoferrate member).
- LdS11, n=1, $(Ca_{2.98}Mn_{0.02})_{\Sigma 3.00}(Fe^{3+}{}_{1.24}Ti_{0.65}Mg_{0.10}Zr_{0.01})_{\Sigma 2.00}(Si_{2.36}Al_{0.36}Fe^{3+}{}_{0.28})_{\Sigma 3.00}O_{11.97}$, $And_{62}Hut_{18}Sch_{14}Ctm_5(Cld + Kmz + Krm)_1$, where Cld is calderite (Mn–Fe^{3+} silicate garnet), and Kmz is kimzeyite (Ca–Zr silicoaluminate).

An evident part of the recently approved end-member hutcheonite, $Ca_3Ti_2(SiAl_2)O_{12}$, is interesting when considering the provenience of the original pure member: Ca–Al inclusions in the Allende chondrite (Ma and Krot, 2014). This is thus next proof of the relation of fossil fuel fires to extraterrestrial environments. The LdS15 garnets may very rarely contain trace amounts of kimzeyite molecule and another hypothetical phosphorus-bearing $Ca_3Fe_2[Al_{1.5}P_{1.5}]O_{12}$ member. Our garnet analyses are much more complex than that reported by Eytier et al. (2004): $Ca_{2.98}(Fe^{3+}{}_{1.89}Al_{0.11})(Si_{2.89}Al_{0.11})O_{12}$.

Table 13.2.13
Chemical analyses of clinopyroxene from the LdS12 sample, France.

	1	2	3	4	5	6	7	8	9	10	11	12
wt.%												
SiO_2*	36.15	38.29	38.21	35.85	35.06	36.45	42.14	41.89	40.97	37.80	38.50	38.98
TiO_2	5.41	3.99	4.57	4.32	5.78	5.73	2.83	3.13	3.02	4.17	4.08	3.71
Al_2O_3	16.09	14.25	14.61	14.86	15.74	14.79	10.81	10.97	11.50	13.25	13.22	12.90
Cr_2O_3	0.54	1.51	0.92	0.00	0.06	0.23	0.97	0.62	0.00	0.00	0.00	0.00
Fe_2O_3	9.29	8.27	8.90	15.81	12.65	10.49	9.04	8.89	10.08	14.28	12.73	12.64
MgO	6.73	7.68	7.42	4.49	5.51	6.54	8.81	8.61	8.25	5.90	6.54	6.70
CaO	24.45	24.49	24.56	24.02	24.13	24.24	24.50	24.45	24.45	23.70	23.97	23.99
K_2O	0.00	0.00	0.00	0.00	0.00	0.00	0.00	0.00	0.19	0.13	0.00	0.00
Σ	98.65	98.47	99.18	99.35	98.93	98.47	99.09	98.56	98.47	99.24	99.04	98.92
apfu												
Si	1.41	1.48	1.47	1.42	1.38	1.43	1.62	1.62	1.59	1.49	1.51	1.53
Ti	0.16	0.12	0.13	0.13	0.17	0.17	0.08	0.09	0.09	0.12	0.12	0.11
Al	0.74	0.65	0.66	0.69	0.73	0.68	0.49	0.50	0.53	0.61	0.61	0.60
Cr	0.02	0.05	0.03			0.01	0.03	0.02				
Fe^{3+}	0.11	0.10	0.10	0.21	0.17	0.12	0.08	0.06	0.12	0.18	0.14	0.13
Fe^{2+}	0.16	0.14	0.16	0.26	0.21	0.19	0.18	0.19	0.17	0.25	0.23	0.24
Mg	0.39	0.44	0.43	0.27	0.32	0.38	0.51	0.50	0.48	0.35	0.38	0.39
Ca	1.02	1.02	1.02	1.02	1.02	1.02	1.01	1.01	1.02	1.00	1.01	1.01
K									0.01	0.01		
end-members† [%]												
Es	11	10	10	21	17	12	8	6	12	18	14	13
Cta	16	12	13	13	17	17	8	9	9	12	12	11
Di	39	44	43	27	32	38	51	50	48	35	38	39
Ku	14	14	14	11	11	11	11	12	11	10	12	12
Hd	16	14	16	26	21	19	18	19	17	25	23	24
Wo	2	2	2	2	2	2	1	1	2		1	1
Cre	2	5	3			1	3	2				

*P, Na, Mn, Ni, and Zn were analyzed but not observed.
†*Es*, esseneite; *Di*, diopside; *Ku*, kushiroite; *Hd*, hedenbergite; *Jh*, johannsenite; *Wo*, wollastonite; *Aeg*, aegirine; hypothetical end-members: *Cre*, $CaCrAlSiO_6$.

Hematite (Fe_2O_3)

Table 13.2.18 (for the samples LdS1 and LdS2) and Table 13.2.19 (for the sample LdS12) show details of the chemical composition of hematite. Usually a simple oxide, the Lapanouse hematite continues the complexity trend, although it is not as complex as the previous minerals. The most important substitution in the Lapanouse hematite is Ti, due to the formation of pseudobrookite (Psb)-hematite (Hm) solid solution, also analyzed, for example, by Lufkin (1976). Aluminum is also present, but likely not as corundum substitution commonly observed for coal fire–related materials (e.g., Ciesielczuk et al., 2015). The following empirical end-member formulas describe the Lapanouse hematites:

- LdS1, n=24, $(Fe_{0.78}Mg_{0.20}Ca_{0.02}Mn_{0.02})_{\Sigma 1.01}(Fe_{0.78}Ti_{0.19}Al_{0.02})_{\Sigma 0.99}O_{3.00}$, $Hm_{77}Psb_{16}Gk_4Afo_2$, where Gk is geikielite ($MgTiO_3$) and Afo is an $AlFeO_3$ hypothetical end-member.
- LdS2, n=4, $(Fe_{0.95}Mg_{0.03}Ca_{0.01}Mn_{0.01})_{\Sigma 1.00}(Fe_{0.90}Al_{0.08}Ti_{0.02})_{\Sigma 1.00}O_{3.00}$, $Hm_{90}Afo_8Psb_2$.
- LdS12, zone 1, n=4, $(Fe_{1.97}Ca_{0.01}Mg_{0.01}Si_{0.01})_{\Sigma 2.00}O_{3.00}$, ~$Hm_{97}Fcs_1Fms_1$, with Fcs=$Fe^{3+}(Ca_{0.5}Si_{0.5})O_3$ and Fms=$Fe^{3+}(Mg_{0.5}Si_{0.5})O_3$ hypothetical end-members.
- LdS12, zone 2, n=2, $Fe_{1.00}(Fe_{0.66}Al_{0.23}Ti_{0.05}Mg_{0.04}Ca_{0.01}Mn_{0.01})_{\Sigma 1.00}O_{3.00}$, $Hm_{66}Afo_{23}Fmt_7Fct_2Fmn_2$, with Fmt for $Fe^{3+}(Mg_{0.5}Ti_{0.5})O_3$, Fct for $Fe^{3+}(Ca_{0.5}Ti_{0.5})O$, and Fmn for $Fe^{3+}(Mn_{0.5}Ti_{0.5})O_3$ hypothetical end-members.

Table 13.2.14
Chemical analyses of clinopyroxene from the LdS15 and LDSV samples, France.*

	1	2	3	4	5	6	7	8	9	10	11	12
	LdS15						LdSV					
wt.%												
SiO_2†	37.49	35.79	34.78	35.48	36.56	34.20	35.43	36.16	39.75	39.78	42.36	40.14
TiO_2	0.83	0.68	0.93	0.95	0.94	0.76	4.97	4.67	2.93	2.71	1.65	2.40
Al_2O_3	8.11	10.37	9.77	9.80	8.85	11.18	13.85	14.10	12.96	12.77	21.52	13.45
Cr_2O_3	0.00	0.00	0.00	0.00	0.00	0.00	0.00	0.00	0.13	0.23	0.12	0.60
Fe_2O_3	20.92	22.30	24.09	22.67	22.23	23.57	17.83	16.56	11.94	11.63	6.53	10.47
MnO	0.00	0.00	0.07	0.00	0.00	0.00	0.00	0.00	0.00	0.00	0.00	0.00
MgO	7.15	6.14	5.95	6.09	6.63	5.58	3.35	4.23	7.01	7.18	2.85	7.90
CaO	22.98	23.50	23.22	23.42	23.11	23.35	23.20	23.46	23.95	24.13	25.12	24.26
K_2O	0.09	0.00	0.00	0.00	0.00	0.00	0.00	0.00	0.00	0.00	0.00	0.00
Na_2O	0.14	0.00	0.00	0.00	0.00	0.00	0.00	0.00	0.00	0.00	0.12	0.00
Σ	98.10	99.06	98.93	98.91	98.78	99.00	99.05	99.35	98.91	98.50	100.32	98.67
apfu												
Si	1.52	1.44	1.41	1.43	1.48	1.38	1.44	1.44	1.55	1.55	1.61	1.54
Ti	0.03	0.02	0.03	0.03	0.03	0.02	0.15	0.14	0.09	0.08	0.05	0.07
Al	0.39	0.49	0.47	0.47	0.42	0.53	0.66	0.66	0.60	0.59	0.96	0.61
Cr										0.01		0.02
Fe^{3+}	0.54	0.61	0.67	0.61	0.57	0.66	0.16	0.17	0.12	0.13		0.14
Fe^{2+}	0.10	0.07	0.07	0.08	0.10	0.06	0.38	0.33	0.23	0.21	0.19	0.16
Mg	0.43	0.37	0.36	0.37	0.40	0.34	0.20	0.25	0.41	0.42	0.16	0.45
Ca	1.00	1.01	1.01	1.01	1.00	1.01	1.01	1.00	1.00	1.01	1.02	1.00
Na	0.01											
end-members‡ [%]												
Es	35	46	42	42	38	49	16	17	12	13		15
Cta	2	2	2	2	2	2	15	14	9	8	5	7
Di	43	37	36	37	40	34	20	25	41	42	16	46
Ku							10	11	15	14	**58**	16
Hd	10	7	7	8	10	6	**38**	**33**	23	21	19	16
Wo	1	1	1	1		1	1			1	3	
Ctf	1		1	1	1							
Fes	9	7	12	9	9	8						

*Values in bold-faced type for the end-members Ku and Hd are for dominant end-members, and they rarely occur in pyroxene.
†Ni and Zn were analyzed but not observed.
‡*Es*, esseneite; *Di*, diopside; *Ku*, kushiroite; *Hd*, hedenbergite; *Hyp*, hypersthene; *Aeg*, aegirine; hypothetical end-members: *Cta*, $CaTiAl_2O_6$; *Ctf*, $CaTiFe_2O_6$; *Fes*, $CaFeFeSiO_6$.

- LdS12, zone 3, subtype A, n=5 (analyses 7–11), $(Fe^{3+}_{0.79}Fe^{2+}_{0.17}Mg_{0.02}Mn_{0.01}Ca_{0.01})_{\Sigma 1.00}(Fe^{3+}_{0.58}Ti_{0.22}Al_{0.19}V_{0.01})_{\Sigma 1.00}O_{3.00}$, $\sim Hm_{54}Afo_{19}Psb_{18}Ilm_4Fps_2Vfo_1$, with Ilm for ilmenite, Fps for ferropseudobrookite ($Fe^{2+}Ti_2O_5$), and Vfo for a $VFeO_3$ hypothetical end-member.
- LdS12, zone 3, subtype B, n=5 (analyses 12–16), $(Fe^{3+}_{0.97}Fe^{2+}_{0.02}Mn_{0.01})_{\Sigma 1.00}(Fe^{3+}_{0.86}Al_{0.11}Ti_{0.02}Si_{0.01})_{\Sigma 1.00}O_{3.00}$, $\sim Hm_{85}Afo_{11}Psb_2Fs_1$, with Fs for ferrosilite ($FeSiO_3$).

Noteworthy are different substitution trends observed for the LdS12 sample. The pseudobrookite-ferropseudobrookite solid solution found there is also mentioned by Lufkin (1976), and it is known to be complete at temperatures above 1150°C. A minor Cr-rich end-member, $CrFeO_3$ (Cfo), and "silicopseudobrookite" (Fe_2SiO_5) may sometimes also go to the hematite composition. Substitution of Al, Cr, and V may also take place via introduction of machiite (the Al analogue of pseudobrookite) and its Cr and V analogues.

Systematic position of Fe-rich oxides in the LdSV sample is very complex. The possible composition of an A_2O_3 is $(Fe^{3+}_{1.29}Al_{0.43}Ti_{0.13}Fe^{2+}_{0.12}Ca_{0.01}Mn_{0.01})_{\Sigma 1.99}O_{3.00}$; the corresponding analyses neither recast to spinel- nor pseudobrookite-like stoichiometry. However, this formula could also be presented as $(Fe^{3+}_{0.87}Fe^{2+}_{0.12}Mn_{0.01})_{\Sigma 1.00}(Al_{0.43}Fe^{3+}_{0.42}Ti_{0.13}Ca_{0.01})_{\Sigma 0.99}O_{3.00}$, with $(Fe^{2+}{}_{+}Mn)$:Ti ratio being 1:1, thus supporting the part of

Table 13.2.15
Chemical analyses of cuspidine, France.

	1	2	3	4	5	6	7	8	9	10	11	12
	LdS7						LdS12			LdS6		
wt.%												
P_2O_5*	0.26	0.22	0.23	0.22	0.29	0.25	n.a.†	n.a.	n.a.	0.00	0.00	0.00
SiO_2	32.08	32.13	32.18	32.21	31.94	32.03	31.30	31.41	31.64	32.18	31.78	32.64
TiO_2	0.40	0.35	0.36	0.44	0.47	0.42	n.a.	n.a.	n.a.	0.00	0.00	0.00
FeO	0.00	0.00	0.00	0.00	0.45	0.00	0.00	0.00	0.00	0.00	0.52	0.42
MgO	0.00	0.00	0.00	0.00	0.00	0.00	0.16	0.16	0.16	0.00	0.00	0.05
CaO	59.44	60.31	59.00	59.30	59.09	59.23	60.24	60.03	60.24	61.15	61.01	61.41
SrO	0.94	0.58	0.70	0.46	0.87	0.82	0.00	0.00	0.00	0.00	0.00	0.00
K_2O	0.00	0.00	0.00	0.15	0.00	0.00	n.a.	n.a.	n.a.	0.00	0.00	0.00
Na_2O	0.00	0.00	0.00	0.21	0.00	0.00	0.00	0.00	0.00	0.00	0.00	0.00
F	7.11	7.49	6.91	6.65	4.91	6.13	12.51	12.04	12.55	4.82	4.68	4.12
Σ	100.23	101.08	99.37	99.63	98.02	98.89	104.22	103.64	104.59	98.15	97.98	98.64
H_2O‡			0.63	0.37	1.98	1.11				1.85	2.02	1.36
apfu												
P	0.01	0.01	0.01	0.01	0.02	0.01						
Si	2.00	1.98	2.02	2.01	1.99	2.00	1.95	1.96	1.97	1.98	1.95	1.98
Ti	0.02	0.02	0.02	0.02	0.02	0.02						
Fe					0.02						0.03	0.02
Mg							0.01	0.01	0.01			
Ca	3.97	3.98	3.97	3.96	3.95	3.97	4.03	4.02	4.02	4.02	4.02	3.99
Sr	0.03	0.02	0.03	0.02	0.03	0.03						
F	1.40	1.46	1.37	1.31	0.97	1.21	2.47	2.38	2.47	0.94	0.91	0.79
OH			0.26	0.16	0.80	0.46				0.76§	0.83	0.55

*Cl, S, Mn, Ba, and Al were measured but not observed.
†Not analyzed.
‡By difference.
§Direct calculation (does not allow for charge balance).

ilmenite-related members. By involving Fe^{2+}-(Al,Fe^{3+}) and (Fe^{3+},Al)–Ca members (e.g., $Fe^{2+}Al_{1.33}O_3$ close to $Fe_{1.33}Al_{1.77}O_4$ close to hercynite, and $Fe_{1.33}CaO_3$ close to $Fe_{1.77}Ca_{1.33}O_4$ close to harmunite), we can finally obtain the $Afo_{38}Hm_{37}Psb_{11}Ilm_2Hc_5Mt_5Hrm_1Kr_1$ end-member formula. If true, this would be the first find of the Afo ($AlFeO_3$) phase in the nature.

It should be noticed that large enrichment in Mg, Ti, and other elements may also be characteristic of maghemite instead of hematite. This is especially possible for the LdS1 material, but specifically for yet another one from the sample LdS5. The latter is compositionally very similar to maghemite reported in the Handbook of Mineralogy (Anthony et al., 1997). The LdS5 oxide may be chemically written as $(Fe^{3+}{}_{0.77}Mg_{0.21}Mn_{0.02})_{\Sigma 1.00}(Fe_{0.76}Ti_{0.21}Al_{0.02})_{\Sigma 0.99}O_{2.99}$ or $Hem_{77}Psb_{16}Gk_4Afo_2Prp_1$, with Prp standing for pyrophanite ($MnTiO_3$). The LdS5 slag has hematite existence confirmed via the PXRD method. To be sure, maghemite could be overlooked due to its low content. On the other hand, maghemite is hardly resolvable via this method in the presence of spinels (reflection coincidence).

Larnite (Ca_2SiO_4)

The best-known natural polymorph of calcium orthosilicate (industrially known as belite), larnite, was exclusively found in white amygdule-like segregations within the LdS9 slag. Its typically intergrown crystals and close association with ye'elimite may be seen in Figure 13.2.3C. Its composition (Table 13.2.20) is not far from the preceding ideal one and is as follows: $Ca_{2.02}(Si_{0.97}Al_{0.01}P_{0.01})_{\Sigma 0.99}O_{3.99}$ (n=4). This formula corresponds to the $Lrn_{98}Cap_2$ mean composition, with Cap used for a Ca aluminate-phosphate hypothetical end-member. The other currently known natural Ca_2SiO_4 polymorph, calcio-olivine, was not found in our material. This is likely due to stabilization of our larnite by minor impurities.

Table 13.2.16
Chemical analyses of kottenheimite from the LdS12 sample, France.

	1	2	3	4	5	6
wt.%						
SO_3*	32.20	29.64	27.29	26.08	29.07	31.92
SiO_2	15.06	13.55	9.98	11.68	11.79	14.01
Al_2O_3	0.89	0.96	1.98	2.96	3.18	1.02
FeO	1.89	1.99	1.14	9.53	7.14	1.52
MgO	0.99	1.07	0.00	1.67	1.52	1.06
CaO	40.06	37.34	34.41	34.78	36.68	39.56
Na_2O	0.00	0.00	0.05	0.13	0.13	0.00
Cl	0.07	0.09	0.11	0.12	0.13	0.07
Σ	91.16	84.63	74.96	86.95	89.64	89.16
CO_2†	5.04	4.93	3.39	8.79	7.22	4.60
H_2O‡	3.80	10.44	21.65	4.26	3.14	6.24
apfu						
S	1.56	1.54	1.63	1.24	1.38	1.58
Si	0.97	0.94	0.79	0.74	0.74	0.93
Al	0.07	0.08	0.19	0.22	0.24	0.08
Fe	0.10	0.12	0.08	0.50	0.38	0.08
Mg	0.09	0.11		0.16	0.14	0.10
Ca	2.77	2.76	2.94	2.36	2.48	2.80
Na			0.01	0.02	0.02	
Cl	0.01	0.01	0.02	0.01	0.01	0.01
C	0.44	0.46	0.37	0.76	0.62	0.42
H_2O	0.82	2.40	5.75	0.90	0.66	1.38
end-members§ [%]						
Kth	49	46	50	14	24	50
Thm	39	40	29	45	39	36
Ett	3	4	12	4	7	4
Mth	1	2		3	2	1
Mkt	2	2		1	1	2
Fth	1	2	1	10	6	1
Fet				1	1	
Fkt	2	2	1	3	4	2
Nah			1	2	2	
Cte	3	3	7	14	12	3
Mce				1	1	
Fce				3	2	

*Mn and K were below their detection limits.
†Calculated backward (by difference) after primary apfu calculation.
‡Calculated by difference after CO_2 calculation.
§*Kth*, kottenheimite; *Thm*, thaumasite; *Ett*, ettringite; hypothetical end-members: *Mth*, "magnesiothaumasite"; *Mkt*, "magnesiokottenheimite"; *Fth*, "ferrothaumasite"; *Fet*, "ferroettringite"; *Fkt*, "ferrokottenheimite"; *Nah*, $Na_3[Al(OH)_6]$; *Cte*, "carboettringite"; *Mce*, "magnesiocarboettr-ingite"; *Fce*, "ferrocarboettringite."

Leucite ($KAlSi_2O_6$)

Composition of important Lapanouse slag constituent, leucites, is reported in Table 13.2.21. The corresponding formulas are as follows:

- LdS5, n=4, $(K_{0.95}Fe^{2+}_{0.02}[]_{0.03})_{\Sigma 1.00}Al_{0.98}Si_{2.04}O_{6.06}$, $Lc_{97}Flc_2$ where Flc is "ferroleucite" $(Fe_{0.5}[]_{0.5})AlSi_2O_6$ hypothetical end-member (r^2 for the Fe/K system is 0.91).
- LdS7, n=13, $(K_{0.96}Na_{0.01}Ba_{0.01})_{\Sigma 0.98}(Al_{0.98}Fe_{0.02})_{\Sigma 1.00}Si_{2.02}O_{6.04}$, $Lc_{96}Flc_2Bts_1Nlc_1$, with Flc for ferrileucite, Bts for "bariotissintite" equivalent $(Ba,[])AlSi_2O_6)$, and Nlc for "natroleucite."

Table 13.2.17
Chemical analyses of garnets, Lapanouse-de-Sévérac, France.

	1	2	3	4	5	6	7	8	9	10	11	12	13	14	15	16	17	18	19	20
	LdS2							LdS15												LdS11
wt.%																				
P_2O_5*	0.00	0.00	0.00	0.00	0.00	0.00	0.09									0.25	0.00	0.00	0.00	
SiO_2	31.59	30.08	28.13	33.57	34.09	29.91	29.41	30.29	30.61	31.01	32.02	30.96	30.69	30.65	30.45	30.01	29.10	28.71	28.62	27.65
TiO_2	4.83	7.94	10.54	3.48	3.27	7.19	7.98	7.35	6.60	6.40	4.84	6.46	6.66	4.45	4.99	7.57	9.08	9.74	9.47	10.17
ZrO_2																0.27	0.37	0.35	0.41	0.25
Al_2O_3	5.72	5.84	6.46	6.51	7.34	5.52	5.71	2.10	1.93	2.06	1.90	2.14	2.14	2.82	2.92	2.08	1.95	1.99	1.99	3.62
Fe_2O_3	22.37	21.48	20.20	21.20	20.15	21.30	21.27	26.24	26.46	26.21	26.53	26.09	25.87	27.94	27.23	25.51	25.50	24.93	24.72	23.62
MgO	0.36	0.47	0.60	0.36	0.41	0.54	0.51	0.67	0.56	0.65	0.63	0.66	0.64	0.31	0.16	0.69	0.78	0.86	0.80	0.78
CaO	32.66	32.71	32.88	33.60	34.61	33.11	33.46	32.07	32.06	32.09	32.20	32.12	31.70	32.24	31.71	31.44	30.53	30.72	30.86	32.57
MnO	0.00	0.00	0.22	0.00	0.00	0.00	0.00	0.24	0.24	0.29	0.29	0.27	0.21	0.52	0.63	0.00	0.29	0.00	0.00	0.25
K_2O	0.14	0.17	0.24	0.00	0.00	0.00	0.00	0.00	0.00	0.00	0.00	0.00	0.00	0.00	0.00					
Σ	97.67	98.69	99.27	98.75	99.87	97.57	98.43	98.95	98.55	98.88	98.48	98.84	98.02	99.09	98.13	98.482	98.103	97.975	97.472	98.91
apfu																				
P							0.01									0.02				
Si	2.67	2.53	2.36	2.79	2.78	2.54	2.48	2.59	2.63	2.65	2.74	2.65	2.65	2.61	2.62	2.60	2.54	2.51	2.51	2.36
Ti	0.31	0.50	0.66	0.22	0.20	0.46	0.51	0.47	0.43	0.41	0.31	0.42	0.43	0.29	0.32	0.49	0.60	0.64	0.63	0.65
Zr																0.01	0.02	0.01	0.02	0.01
Al	0.57	0.58	0.64	0.64	0.71	0.55	0.57	0.21	0.20	0.21	0.19	0.22	0.22	0.28	0.30	0.21	0.20	0.21	0.21	0.36
Fe	1.43	1.36	1.27	1.32	1.24	1.36	1.35	1.69	1.71	1.69	1.71	1.68	1.68	1.79	1.77	1.66	1.67	1.64	1.63	1.52
Mg	0.05	0.06	0.07	0.04	0.05	0.07	0.06	0.08	0.07	0.08	0.08	0.08	0.08	0.04	0.02	0.09	0.10	0.11	0.10	0.10
Ca	2.96	2.95	2.95	2.99	3.03	3.01	3.02	2.94	2.95	2.94	2.95	2.94	2.93	2.95	2.93	2.92	2.85	2.88	2.90	2.98
Mn			0.02					0.02	0.02	0.02	0.02	0.02	0.02	0.04	0.05		0.02			0.02

Continued

Table 13.2.17
Chemical analyses of garnets, Lapanouse-de-Sévérac, France—cont'd.

	1	2	3	4	5	6	7	8	9	10	11	12	13	14	15	16	17	18	19	20
	LdS2							LdS15												LdS11
end-members† [%]																				
And	70	67	63	66	62	68	68	72	75	75	80	75	73	83	82	72	67	65	65	62
Sch								11	10	8	4	8	8	4	3	11	16	18	18	14
Hut	15	25	33	11	10	23	25	12	11	12	11	12	13	10	12	12	12	13	13	18
Ctm	1	1	3	2	2	3	3	3	3	3	3	3	4	1						5
Khh	1	1						1		1	1	1	0.2		1	2	2	3	2	
Grs	12	5		21	24	5	2													
Msc																0.3	0.6	0.7	0.6	
Mhu																0.4	0.4	0.5	0.5	
Cld										1	1		0.4	1	1		0.5			0.4
Skg													1							
Ktg	1	1	1																	
Ftm													0.1							
Krm																0.3	0.4	0.4	0.5	0.2
Paa																2				
Ctc					1	1	1													

*S, Cl, Na, Sc, V, Cr, Ga, Ni, Zn, Sr, Nb, Sn, Sb, Hf, Th, and U were measured but not observed; empty fields denote elements not analyzed.

†*And*, andradite; *Sch*, schorlomite; *Hut*, hutcheonite; *Khh*, khoharite; *Grs*, grossular; *Cld*, calderite; *Skg*, skiagite; *Krm*, kerimasite; hypothetical end-members: *Ctm*, $Ca_3(MgTi)(SiO_4)_3$; *Msc*, $Mg_3Ti_2[(SiFe_2)O_4]_3$; *Mhu*, $Mg_3Ti_2[(SiAl_2)O_4]_3$; *Ktg*, $(K_2Ca)Ti_2(SiO_4)_3$; *Ftm*, $Fe_3(MgTi)(SiO_4)_3$; *Paa*, $Ca_3Fe_2(Al_{1.5}P_{1.5})O_{12}$; *Ctc*, $Ca_3(CaTi)(SiO_4)_3$; 1%–0.3% kimzeyite may occasionally be found.

Table 13.2.18
Chemical analyses of hematite from samples LdS1 and LdS2, France.

	1	2	3	4	5	6	7	8	9	10	11	12	13	14	15	16	17
	LdS1											LdS2					
wt.%																	
SiO_2*	0.06	0.04	0.05	0.04	0.06	0.04	0.03	0.03	0.05	0.01	0.01	0.02	0.01	0.24	0.03	0.00	0.00
TiO_2	7.28	9.06	8.96	9.11	8.90	11.23	11.32	13.11	13.12	13.09	8.02	1.80	0.73	0.67	0.20	0.14	0.14
Al_2O_3	0.89	0.80	0.81	0.81	0.66	0.46	0.41	0.60	0.61	0.57	0.73	2.03	2.66	3.07	3.03	2.55	2.32
Cr_2O_3	0.19	0.24	0.24	0.25	0.38	0.08	0.09	0.11	0.11	0.10	0.59	0.14	0.03	0.03	0.05	0.41	0.35
Fe_2O_3	85.72	83.45	83.90	84.31	84.42	79.82	79.71	77.88	77.75	77.57	84.45	93.37	94.48	93.38	91.35	94.29	95.34
MnO	0.29	0.37	0.25	0.27	0.26	1.40	1.39	1.27	1.19	1.20	0.34	0.00	0.12	0.21	0.59	0.15	0.00
MgO	3.79	4.31	4.25	4.36	4.32	5.47	5.46	6.21	6.09	6.22	4.05	0.15	0.10	0.40	2.10	0.35	0.39
CaO	0.12	0.05	0.07	0.03	0.20	0.10	0.15	0.04	0.04	0.04	0.02	0.23	0.26	0.32	0.21	0.30	0.22
CoO	0.00	0.00	0.01	0.00	0.00	0.01	0.00	0.05	0.02	0.00	0.00	0.00	0.00	0.00	0.01	0.00	0.00
NiO	0.00	0.06	0.02	0.07	0.03	0.07	0.02	0.14	0.03	0.18	0.08	0.00	0.01	0.10	0.17	0.00	0.00
ZnO	0.01	0.00	0.28	0.00	0.12	0.06	0.01	0.24	0.09	0.13	0.00	0.09	0.06	0.04	0.22	0.00	0.00
Σ	98.35	98.38	98.84	99.25	99.35	98.74	98.59	99.68	99.10	99.11	98.27	97.83	98.46	98.46	97.96	98.19	98.76
apfu																	
Si														0.01			
Ti	0.14	0.18	0.17	0.18	0.17	0.21	0.22	0.25	0.25	0.25	0.16	0.04	0.01	0.01			
Al	0.03	0.02	0.02	0.02	0.02	0.01	0.01	0.02	0.02	0.02	0.02	0.06	0.08	0.10	0.09	0.08	0.07
Cr				0.01	0.01						0.01					0.01	0.01
Fe	1.67	1.62	1.62	1.62	1.62	1.53	1.53	1.47	1.47	1.47	1.64	1.88	1.89	1.85	1.79	1.88	1.90
Mn	0.01	0.01	0.01	0.01	0.01	0.03	0.03	0.03	0.03	0.03	0.01				0.01	0.00	
Mg	0.15	0.17	0.16	0.17	0.16	0.21	0.21	0.23	0.23	0.23	0.16	0.01	0.00	0.02	0.08	0.01	0.02
Ca												0.01	0.01	0.01	0.01	0.01	0.01
end-members† [%]																	
Hm	83	79	80	79	80	77	77	73	73	73	81	90	90	88	90	88	90
Afo	3	2	2	2	2	1	1	2	2	2	2	6	8	10	9	8	7
Cfo				1	1						1					1	1
Psb	12	14	14	14	14	16	16	18	19	18	13	4	1	1			
Prf						1	1	1	1	1							
Gk	2	3	3	3	3	4	4	6	6	6	2						

*V was measured but not observed.

†*Hm*, hematites; *Afo*, $AlFeO_3$; *Cfo*, $CrFeO_3$; *Psb*, pseudobrookite (Fe_2TiO_5); *Prf*, pyrophanite ($MnTiO_3$); *Gk*, geikielite ($MgTiO_3$).

Table 13.2.19
Chemical analyses of hematite and hematite-pseudobrookite from the LdS12 sample: zone 1 (1–4), zone 2 (5–6), and zone 3 (7–16), France.

	1	2	3	4	5	6	7	8	9	10	11	12	13	14	15	16
wt.%																
SiO_2*	0.21	1.23	0.14	0.00	0.12	0.16	0.00	0.00	0.00	0.00	0.00	0.29	0.23	0.20	0.27	1.61
TiO_2	0.00	0.00	0.00	0.00	2.71	2.26	13.35	16.37	9.93	10.25	9.09	0.40	1.30	1.21	0.00	1.49
Al_2O_3	0.00	0.00	0.14	0.00	8.30	7.91	6.68	6.35	6.20	6.34	7.21	3.95	4.97	4.89	3.24	1.59
Fe_2O_3	98.32	95.13	98.36	99.53	90.31	89.72	80.74	77.04	85.44	85.21	85.43	96.27	95.69	95.09	98.39	95.11
V_2O_3	0.00	0.00	0.00	0.00	0.00	0.00	0.00	0.27	0.30	0.32	0.31	0.00	0.00	0.00	0.00	0.00
MnO	0.16	0.11	0.12	0.02	0.46	0.50	0.79	0.78	0.64	0.59	0.62	0.27	0.37	0.26	0.00	0.37
MgO	0.04	0.47	0.08	0.01	1.11	1.10	0.46	0.72	0.59	0.68	0.67	0.12	0.12	0.11	0.01	0.03
CaO	0.33	0.47	0.35	0.22	0.32	0.49	0.65	0.48	0.44	0.54	0.31	0.06	0.01	0.04	0.03	0.07
Σ	99.05	97.40	99.18	99.84	103.33	102.12	102.66	102.01	103.54	103.93	103.65	101.37	102.68	101.80	101.94	100.26
apfu																
Si	0.01	0.03										0.01	0.01	0.01	0.01	0.04
Ti					0.05	0.04	0.25	0.31	0.18	0.19	0.17	0.01	0.02	0.02	0.00	0.03
Al					0.24	0.23	0.19	0.19	0.18	0.18	0.21	0.12	0.15	0.15	0.10	0.05
Fe^{3+}	1.98	1.93	1.98	1.99	1.65	1.66	1.31	1.18	1.44	1.43	1.45	1.85	1.79	1.80	1.88	1.81
V								0.01	0.01	0.01	0.01					
Fe^{2+}							0.20	0.26	0.14	0.14	0.13		0.02	0.02	0.02	0.06
Mn					0.01	0.01	0.02	0.02	0.01	0.01	0.01	0.01	0.01	0.01	0.00	0.01
Mg		0.02			0.04	0.04	0.02	0.03	0.02	0.02	0.02					
Ca	0.01	0.01	0.01	0.01	0.01	0.01	0.02	0.01	0.01	0.01	0.01					
end-members† [%]																
Hm	98	93	98	99	65	66	51	45	58	57	57	85	81	81	89	87
Afo					24	23	19	19	18	18	21	12	15	15	10	5
Vfo								1	1	1	1					
Psb							20	23	16	16	15	1	2	2		3
Ilm							5	8	3	3	2					
Fcs	1	3	1													
Gk							1	1								
Fs												1	1	1	1	4
Fms		4														
Fmt					8	7										
Fct					2	2										
Fmn					2	2										
Fps							2	4	1	1	1					
Prp		1														

*Cr, Ni, and Zn were measured but not observed.

†*Hm*, hematite; *Psb*, pseudobrookite; *Ilm*, ilmenite; *Prp*, pyrophanite; *Fs*, ferrosilite; hypothetical end-members: *Afo*, $AlFeO_3$; *Vfo*, $VFeO_3$; *Fcs*, $Fe^{3+}(Ca_{0.5}Si_{0.5})O_3$; *Fms*, $Fe^{3+}(Mg_{0.5}Si_{0.5})O_3$; *Fmt*, $Fe^{3+}(Mg_{0.5}Ti_{0.5})O_3$; *Fct*, $Fe^{3+}(Ca_{0.5}Ti_{0.5})O_3$; *Fmn*, $Fe^{3+}(Mn_{0.5}Ti_{0.5})O_3$; *Fps*, $Fe^{2+}Ti_2O_5$ "ferropseudobrookite."

Table 13.2.20
Chemical analyses of larnite from the LdS9 sample, France.

	1	2	3	4
wt.%				
P_2O_5*	0.39	0.48	0.32	0.49
SiO_2	34.22	33.18	33.78	32.18
Al_2O_3	0.25	0.26	0.18	0.30
CaO	66.03	64.18	64.56	65.01
Σ	100.88	98.08	98.83	97.98
apfu				
P	0.01	0.01	0.01	0.01
Si	0.97	0.97	0.98	0.94
Al	0.01	0.01	0.01	0.01
Ca	2.01	2.01	2.01	2.04
end-members† [%]				
Lrn	98	98	99	98
Cap	2	2	1	2

*F, Na, Mg, S, Cl, K, Ti, Ba, Cr, Fe, Mn, V, Sr, and Zr were measured but not observed.
†*Lrn*, larnite; *Cap*, $Ca(Al_{0.5}P_{0.5})O_4$.

- LdS1, n=1, $(K_{0.69}Ca_{0.31}Mg_{0.07})_{\Sigma 1.07}(Al_{0.73}Fe_{0.18})_{\Sigma 0.91}(Si_{1.99}Ti_{0.01})_{\Sigma 2.00}O_{6.09}$, $Lc_{52}Ts_{23}Flc_{13}Fts_6Mts_5Mft_1$, with Ts for tissintite equivalent, Flc for "ferrileucite," Fts for "ferritissintite," Mts for "magnesiotissintite," and Mft for "magnesioferritissintite."
- LdSV, normal composition, n=9, $(K_{0.69}Na_{0.01})_{\Sigma 0.97}Al_{0.99}Si_{2.05}O_{6.06}$, $Lc_{99}Nlc_1$.
- LdSV, tissintite-rich composition, n=1, $(K_{0.83}Ca_{0.16}Na_{0.01})_{\Sigma 1.00}(Al_{0.86}Ca_{0.10}Fe_{0.01})_{\Sigma 0.97}Si_{2.04}O_{6.05}$, $Lc_{75}\mathbf{Ts_{23}}Wo_2FlcNfl_1$, with Flc for "ferrileucite."

The possible trace addition of the "bariotissintite" equivalent in the LdS7 material is a third relation to an extraterrestrial mineral, in this case to tissintite, the Ca analogue of jadeite known from Tissint Martian meteorite (Ma et al., 2015). Substitution of tissintite sensu stricto may also be an important feature in some leucites analyzed. Although a mineral compositionally similar to leucite, kalsilite ($KAlSiO_4$), is suggested to be entering the composition of some of the slags under scope, no chemical analysis fits to it. A phase of petalite-like stoichiometry is found in the LdS2 sample instead (see the Petalite-like species discussion in this section of the chapter).

Melilite Group ($Ca_2(Al,Mg,Fe^{3+})(Al,Si)(Si,Al)O_7$) and Related Species

Chemical composition of melilites will be illustrated by the example of the samples LdS15, LdS12, and LdS6 (Table 13.2.22). The associated formulas and member representations are:

- LdS15, åkermanite-gehlenite, n=3 (analyses 1, 3, and 4), $(Ca_{2.02}Na_{0.01})_{\Sigma 2.03}(Mg_{0.40}Al_{0.37}Fe^{3+}_{0.19}Mn_{0.01})_{\Sigma 0.97}(Si_{1.47}Al_{0.53})_{\Sigma 2.00}O_{7.01}$, $Ak_{41}Gh_{39}Fgh_9$, with Fgh for ferri-gehlenite ($Ca_2Fe^{3+}AlSiO_7$).
- LdS12, åkermanite-ferri-gehlenite n=23, $(Ca_{1.95}Na_{0.04}K_{0.02})_{\Sigma 2.01}(Mg_{0.49}Fe^{3+}_{0.28}Al_{0.20})_{\Sigma 0.97}(Si_{1.88}Al_{0.12})_{\Sigma 2.00}O_{7.02}$, $Ak_{48}Fgh_{28}Gh_{20}Aak_2Kme_1$, with Aak for alumo-åkermanite (formerly known as soda-melilite), and Kme for "kalimelilite."
- LdS6, n=2, ferri-gehlenite, $Ca_{2.00}(Fe^{3+}_{0.64}Al_{0.15}Mg_{0.15}Ca_{0.06})_{\Sigma 1.00}(Si_{1.20}Al_{0.80})_{\Sigma 2.00}O_{6.99}$, $Fgh_{64}Ak_{16}Gh_{15}Rn_7$, with Rn for rankinite ($Ca_3Si_2O_7$) equivalent.

These formulas differ from that of Eytier et al. (2004): $(Ca_{1.97}K_{0.04})$ $(Al_{0.47}Fe^{3+}_{0.32}Mg_{0.17}Fe^{2+}_{0.02})(Si_{1.25}Al_{0.70})O_7$.

Melilites are quite difficult to locate among the slag matrix. They may be easily mistaken for other species. As an example, during recasting of the LdS15 and LdS6 melilite-ascribed analyzes three of them were shown to better fit to the rhönite (Ro) group stoichiometry, which may be approximated as $Ca_4(Mg,Fe,Al,Ti)_{12}O_4(Fe,Si,Al)_{12}O_{36}$. In both samples, the main end-member is khesinite (Khe), which is a relatively new addition to the group just recently described from the famous fossil-fuel-fire area known as Hatrurim Formation (Galuskina et al., 2017). While the latter area is commonly reported as a natural analogue of coal fires (e.g., Ciesielczuk et al., 2015), our find of

Table 13.2.21
Chemical analyses of leucite, Lapanouse-de-Sévérac, France.

	1	2	3	4	5	6	7	8	9	10	11	12
						wt.%						
	LdS5					LdS7						LdS15
SiO_2*	56.09	56.19	56.93	56.86	56.17	55.62	55.79	56.26	56.85	56.43	56.98	51.13
TiO_2	0.00	0.00	0.00	0.19	0.00	0.00	0.00	0.00	0.00	0.00	0.00	0.23
Al_2O_3	23.05	22.95	23.28	23.83	22.97	22.63	22.33	23.11	23.01	22.90	23.03	16.14
MgO	0.00	0.00	0.00	0.00	0.00	0.00	0.00	0.00	0.00	0.00	0.00	1.29
CaO	0.00	0.00	0.00	0.00	0.00	0.00	0.00	0.00	0.00	0.00	0.00	7.48
FeO	0.90	0.93	0.46	0.51	0.76	0.64	0.73	0.39	0.63	0.75	0.59	5.59
BaO	0.00	0.77	0.89	1.48	0.82	0.91	0.78	1.02	0.71	0.77	0.00	
Na_2O	0.06	0.13	0.00	0.13	0.20	0.23	0.24	0.23	0.22	0.17	0.00	0.10
K_2O	21.33	20.58	20.50	20.62	20.61	20.63	20.37	20.73	20.88	20.78	21.33	13.86
Σ	101.43	101.53	102.06	103.62	101.52	100.66	100.23	101.73	102.31	101.80	101.92	96.05
apfu												
Si	2.02	2.03	2.05	2.02	2.03	2.02	2.04	2.03	2.03	2.03	2.04	1.99
Ti				0.01								0.01
Al	0.98	0.98	0.99	1.00	0.98	0.97	0.96	0.98	0.97	0.97	0.97	0.74
Mg												0.07
Ca												0.31
Fe	0.03	0.03	0.01	0.02	0.02	0.02	0.02	0.01	0.02	0.02	0.02	0.18
Ba		0.01	0.01	0.02	0.01	0.01	0.01	0.01	0.01	0.01		
Na		0.01		0.01	0.01	0.02	0.02	0.02	0.01	0.01		
K	0.98	0.95	0.94	0.93	0.95	0.96	0.95	0.95	0.95	0.95	0.97	0.69
End members [%]												
Lc	97	95	97	95	95	95	95	96	96	95	98	52
Flc	3	3	1	1	2	2	2	1	2	2	2	13
Ts												23
Nlc		1		1	1	2	2	2	2	1		
Mts		1	1	2	1	1	1	1	1	1		5
Fts												6
Mft												1

	13	14	15	16	17	18	19	20	21	22
						wt.%				
						LdSV				
SiO_2*	53.41	58.12	58.44	58.37	57.46	57.87	57.61	57.55	57.72	57.76
Al_2O_3	19.00	23.43	23.96	23.82	23.51	23.66	23.71	23.55	23.90	23.69
CaO	6.23	0.00	0.00	0.00	0.00	0.00	0.00	0.00	0.00	0.00
FeO	0.34	0.00	0.00	0.00	0.00	0.00	0.00	0.00	0.00	0.00
BaO	0.00	0.00	0.00	0.00	0.00	0.00	0.00	0.17	0.00	0.00
Na_2O	0.11	0.09	0.08	0.06	0.07	0.06	0.10	0.12	0.11	0.10
K_2O	16.98	21.20	21.39	21.19	21.38	21.21	21.06	20.93	21.18	21.09
Σ	96.06	102.84	103.88	103.44	102.42	102.80	102.48	102.30	102.90	102.64
apfu										
Si	2.04	2.06	2.05	2.06	2.04	2.05	2.05	2.05	2.04	2.05
Al	0.86	0.98	0.99	0.99	0.98	0.99	0.99	0.99	1.00	0.99
Ca	0.26									
Fe	0.01									
Na	0.01	0.01	0.01				0.01	0.01	0.01	0.01
K	0.83	0.96	0.96	0.95	0.97	0.96	0.95	0.95	0.96	0.95
End members† [%]										
Lc	75	99	99	100	99	100	99	99	99	99
Flc	1									
Nlc	1	1	1		1		1	1	1	1
Ts	23									
Wo	2									

*P, Sc, Cr, V, Ga, Mn, Ni, Zn, Sr, Rb, Cs, and Ti were analyzed but not observed.

†*Lc*, leucite; *Ts*, tissintite (equivalent); *Wo*, wollastonite. Hypothetical end members: *Flc*, "ferroleucite", (Fe,[])$AlSi_2O_6$, "ferrileucite", $KFeSi_2O_6$; *Nlc*, $NaAlSi_2O_6$ "natroleucite"; *Mts*, (Mg,[])$AlSi_2O_6$ "magnesiotissintite"; *Fts*, (Ca,[])$FeSi_2O_6$ "ferritissintite"; *Mft*, (Mg,[])$FeSi_2O_6$ "magnesioferritissintite."

Table 13.2.22
Chemical analyses of melilite (analyses 1, 3, 4, and 7–14) and rhönite groups (analyses 2, 5, 6, and 15–17), France.

	1	2	3	4	5	6	7	8	9	10	11	12	13	14	15	16	17
	LdS15						LdS12						LdS6				
wt.%																	
SiO_2*	33.08	9.70	30.55	29.60	9.05	9.54	39.05	37.24	39.12	38.17	39.95	39.89	24.21	24.07	6.40	6.06	5.92
TiO_2	0.00	0.74	0.05	0.06	0.78	0.75	0.00	0.00	0.00	0.00	0.00	0.00	0.00	0.00	0.95	0.97	0.99
Al_2O_3	17.26	9.75	13.59	17.65	9.18	9.55	6.71	8.91	6.74	8.19	4.47	4.54	15.34	17.05	7.19	7.35	7.20
Cr_2O_3	n.a.†	0.00	0.00	0.00	0.06	0.08	0.00	0.00	0.00	0.00	0.00	0.00	0.00	0.00	0.00	0.00	0.00
MgO	6.58	4.95	5.55	4.74	4.71	5.04	7.96	7.73	8.97	7.88	6.12	6.96	1.84	2.31	2.70	3.04	2.66
CaO	40.69	13.44	38.86	39.44	13.70	13.81	38.73	39.24	40.04	39.40	37.57	37.92	39.06	38.56	14.66	14.33	14.39
MnO	0.00	1.99	0.25	0.24	1.25	1.42	0.00	0.00	0.00	0.00	0.00	0.00	0.00	0.00	0.62	0.74	0.75
FeO	1.09	52.01	8.25	5.06	53.60	52.14	n.a.‡	n.a.	n.a.	n.a.	n.a.	n.a.	n.a.	n.a.	n.a.	n.a.	n.a.
SrO	0.28	n.a.	n.a.	n.a.	n.a.	n.a.	n.a.	n.a.	n.a.	n.a.	n.a.	n.a.	0.00	0.33	0.00	0.00	0.00
Na_2O	0.12	0.00	0.00	0.11	0.00	0.00	0.32	0.18	0.12	0.26	0.54	0.57	0.00	0.00	0.00	0.00	0.00
K_2O	0.00	0.00	0.00	0.00	0.00	0.00	0.36	0.20	0.14	0.29	0.37	0.40	0.00	0.00	0.00	0.00	0.00
Fe_2O_3‡	1.21	57.80	9.16	5.63	59.57	57.94	5.73	4.85	3.85	4.34	10.14	8.63	18.52	15.76	65.27	64.21	65.13
Σ	99.23	98.51	98.40	97.61	98.50	98.32	98.86	98.35	98.96	98.53	99.16	98.92	98.97	98.08	97.80	96.69	97.05
apfu																	
Si	1.53	3.06	1.47	1.42	2.88	3.02	1.85	1.76	1.83	1.80	1.93	1.91	1.20	1.19	1.05	1.01	0.98
Ti		0.18			0.19	0.18									0.12	0.12	0.12
Al	0.94	3.63	0.77	1.00	3.44	3.56	0.37	0.50	0.37	0.46	0.25	0.26	0.90	1.00	1.40	1.44	1.41
Cr																	
Mg	0.45	2.33	0.40	0.34	2.23	2.38	0.56	0.55	0.63	0.55	0.44	0.50	0.14	0.17	0.66	0.75	0.66
Ca	2.02	4.54	2.01	2.03	4.67	4.68	1.96	1.99	2.01	1.99	1.94	1.95	2.08	2.05	2.59	2.55	2.56
Mn		0.53	0.01	0.01	0.34	0.38									0.09	0.10	0.11
Fe^{3+}	0.04	13.73	0.33	0.20	14.25	13.79	0.20	0.17	0.14	0.15	0.37	0.31	0.69	0.59	8.09	8.03	8.15
Sr	0.01													0.01			
Na	0.01			0.01			0.03	0.02	0.01	0.02	0.05	0.05					
K							0.02	0.01	0.01	0.02	0.02	0.02					

Continued

Table 13.2.22
Chemical analyses of melilite (analyses 1, 3, 4, and 7–14) and rhönite groups (analyses 2, 5, 6, and 15–17), France—cont'd.

	1	2	3	4	5	6	7	8	9	10	11	12	13	14	15	16	17
			LdS15							LdS12					LdS6		
end-members[§] [%]																	
Gh	49		24	43			22	26	20	25	17	16	10	19			
Ak	47		41	35			55	55	64	56	43	49	14	17			
Fgh	4		3	21			20	17	14	16	36	31	69	59			
Aak							1	1	1	1	3	3					
Mak			1	1													
Rn													8	5			
Kme							1	1		1	1	1					
Khe		38			41	38									59	59	60
Do		26			25	25									23	23	24
Ro		24			22	24									2	2	2
Cro		6			7	7									10	9	9
Tkh		1			1	1									1	1	1
Tdo		1			1	1									0.5	0.5	0.5
Mro		5			3	4									0.3	0.3	0.3
Tro		0.5			0.4	0.5									0.3	0.3	0.3

*P, S, Cl, V, Ni, Zn, Zr, and Ba were analyzed but not observed.

†Not analyzed.

‡Calculated.

§Melilite group end-members: *Ghl*, gehlenite; *Ak*, åkermanite; *Aak*, alumo-åkermanite; *Rn*, rankinite ($Ca_3Si_2O_7$) equivalent; melilite group hypothetical end-members: *Fgh*, ferri-gehlenite ($Ca_2Fe(Si,Al)_2O_7$); *Mak*, $Ca_2MnSi_2O_7$; *Kme*, $(Ca,K)_2(Al,Mg,Fe)Si_2O_7$ (Ti- and Mn-rich members were omitted in the case of the LdS15 sample melilites). Rhönite group end-members: *Khe*, khesinite (here simplified as $Ca_2(Fe,Mg)_6O_2(Fe,Si)_6O_{18}$, also corresponding to "malakhovite" to show its relationship to dorrite and rhönite); *Dr*, dorrite (simplified as $Ca_2(Fe,Mg)_6O_4(Al,Fe,Si)_6O_{18}$); *Ro*, rhönite (simplified as $Ca_2(Mg,Fe,Ti)_6O_2(Si,Al)_6O_{18}$); hypothetical end-members of the rhönite group: *Cro*, $Ca_2Ca_6O_2Si_6O_{18}$; *Tkh*, $Ca_2Ti_6O_2(Fe_4[]_2)O_{18}$; *Tdo*, $Ca_2Ti_6O_2(Al_4[]_2)O_{18}$; *Mro*, $Ca_2Mn_6O_2Si_6O_{18}$; *Tro*, $Ca_2Ti_6O_2(Si_3[]_3)O_{18}$.

khesinite places Lapanouse as the second world site for this rare species. The empirical formulas and end-member compositions describing the LdS15 and LdS6 khesinites are as follows:

(1) $Ca_{4.00}$ $[(Mg_{2.31}Ca_{0.63}Mn_{0.06})_{\Sigma 3.00}(Fe^{3+}_{8.48}Mn_{0.34}Ti_{0.18})_{\Sigma 9.00}]O_{4.00}[Fe^{3+}_{5.44}Al_{3.54}Si_{2.99}]_{\Sigma 11.97}O_{35.90}$, $Khe_{39}Do_{25}Ro_{23}Cro_{6}Mro_{4}Tdo_{1}Thk_{1}Tro_{0.5}$, where Do is dorrite, Cro is "calciorhönite," Mro is "manganorhönite," Tdo is "titanodorrite," Tkh is "titanokhesinite," and Tro is "titanorhönite."
(2) $Ca_{4.00}(Fe_{9.04}Mg_{1.38}Ca_{1.14}Ti_{0.24}Mn_{0.20})_{\Sigma 6.00}(Fe_{7.14}Al_{2.84}Si_{2.04})_{\Sigma 12.02}O_{39.78}$, ~$Khe_{60}Do_{24}Cro_{9}Ro_{2}Tkh_{1}(Tdo+Mro+Tro)_{1}$.

Nepheline ($(Na,K)AlSiO_4$)

Nepheline (Nph) crystal chemistry is, again, somewhat anomalous (Table 13.2.23). It is exemplified by analyses conducted on the samples LdS1 and LdS5, with the following mean compositions:

- LdS1, n=3, $(Na_{0.77}K_{0.18}Ca_{0.02})_{\Sigma 0.97}(Al_{0.94}Fe_{0.04})_{\Sigma 0.98}Si_{1.05}O_{4.07}$, $Nph_{76}Ks_{18}Fnp_{3}Yo_{2}Fks_{1}$, with Ks for kalsilite $KAlSiO_4$, Yo for yoshiokaite $(Ca,Na)Al(Al,Si)O_4$ (or $Ca_{1-x}(Al,Si)_2O_4$), Fnp for "ferrinepheline," and Fks for "ferrikalsilite."
- LdS5, n=16, $(Na_{0.75}K_{0.12}Ca_{0.05})_{\Sigma 0.92}(Al_{0.97}Fe^{3+}_{0.03})_{\Sigma 1.00}Si_{1.07}O_{4.13}$, $Nf_{78}Ks_{13}Yo_{6}Fnp_{3}$

A very interesting issue is the Ca substitution in the nephelines, related to the yoshiokaite component. Dominant yoshiokaite was originally reported from Fra Mauro Highlands of the Moon (Vaniman and Bish, 1990).

Oldhamite (CaS)

Oldhamite is yet another species typical for meteorites. Although known from the Earth, it is usually related to fossil fuel fires (e.g., Kruszewski et al., 2012). Oldhamite (Old) is present in the LdS9 slag (Figure 13.2.3B), closely associating with srebrodolskite and ye'elimite. It occurs mainly as a slightly Sr-enriched variety, as shown by the empirical formula of $(Ca_{0.99}Sr_{0.01})_{\Sigma 1.00}\{S_{0.70}[(OH)_2]_{0.30}\}_{\Sigma 1.00}\cdot 3.60H_2O$ (n=20), water possibly being adsorbed. It is likely intermixed with or covered by portlandite (Por), which may be shown in the end-member formula of $Old_{69}Por_{31}$. The corresponding Sr-free oldhamite formula is $Or_{66}Por_{34}$, related to $(Ca_{0.98}Fe_{0.02}Mg_{0.01})_{\Sigma 1.01}\{S_{0.66}[(OH)_2]_{0.34}]\}_{\Sigma 1.00}\cdot 8.80H_2O$(ads). Three analyses correspond to portlandite-dominant composition, $(Ca_{0.96}Fe_{0.03}Mg_{0.01})_{\Sigma 1.00}\{[(OH)_2]_{0.84}]_{\Sigma 0.16}\}_{\Sigma 1.00}\cdot 5.09H_2O$(ads), $Por_{84}Old_{16}$, with trace amakinite $Fe(OH)_2$ and brucite $Mg(OH)_2$ end-members involved. Oldhamite also was found as relatively large crystals residing in the beige-brown core of the otherwise white amygdule of the LdS12 slag (Figure 13.2.6D). Table 13.2.24 shows the composition of this mineral. It is relatively pure, with very minor keilite (FeS) substitution of ~1%. The empirical formula is $Ca_{0.99}Fe_{0.01}S_{1.01}$.

Olivine Group ($(Mg,Fe)_2SiO_4$)

Besides fayalite, which possibly occurs as inclusions in hematite from the LdS12 slag, the most commonly observed olivine group member is forsterite. It is exclusively found as rounded aggregates, in which it acts as the matrix phase (Figure 13.2.1A). These aggregates are usually red due to inclusions of hematite; relatively large chromite-rich spinels may also be present inside. The composition of such forsterite, from the sample LdS1, is reported in Table 13.2.25. As opposed to the previous minerals, this forsterite is very pure—$(Mg_{1.97}Fe_{0.02}Mn_{0.01})_{\Sigma 2.00}Si_{1.97}O_{5.94}$, $Fo_{99}Fa_{1}$—possibly due to its secondary origin: it may postdate the chaotic conditions of the local prograde metamorphism.

Perovskite ($CaTiO_3$)

Perovskites (Pe) are one of the most geochemically complex minerals of the Lapanouse slags due to their notorious enrichment in Sr (up to 0.39 wt.% SrO), V (max. 0.89 wt.% V_2O_3), Cr (up to 1.29 wt.% Cr_2O_3), Nb (up to 2.09 wt.% Nb_2O_5), and possible admixture of Zr (max. 0.80 wt.% ZrO_2) and possibly to S (Table 13.2.26). Iron is treated as trivalent, as strongly suggested by the very good negative Ti/Fe correlation ($r^2=0.97$). Following are the corresponding empirical and mean end-member formulas:

- LdS2, composition no. 1 (e.g., analyses 1–6), n=12, (Ca0.98Al0.02Cr0.01)S1.01(Ti0.79Fe0.13Si0.03[]0.05) S1.00O2.93, $Pe_{78}Wo_{6}Afo_{7}Cfo_{5}Vfo_{1}(Tau+Lak)_{0.5}$ with Wo for the wollastonite equivalent, Afo for $AlFeO_3$, Cfo for $CrFeO_3$, Vfo for $VFeO_3$, Tau for tausonite ($SrTiO_3$), and Lak for lakargiite ($CaZrO_3$).

Table 13.2.23
Chemical analyses of nepheline, Lapanouse-de-Sévérac, France.

	1	2	3	4	5	6	7	8	9	10	11	12	13	14	15	16	17	18	19	20
	LdS1										LdS5									
wt.%																				
SiO_2*	44.24	43.38	42.60	42.98	43.48	42.56	43.19	43.26	43.22	42.48	42.31	43.66	42.04	43.02	42.68	42.93	44.66	43.09	45.71	45.86
TiO_2	0.11	0.00	0.13	0.14	0.12	0.13	0.13	0.13	0.12	0.00	0.00	0.00	0.00	0.00	0.20	0.00	0.00	0.00	0.00	0.00
Al_2O_3	33.12	33.24	33.84	33.35	32.49	33.34	32.25	32.25	33.21	33.04	32.86	34.23	32.79	33.57	32.95	34.18	33.00	33.88	33.28	34.40
Fe_2O_3	2.16	2.03	1.62	1.82	2.36	1.49	2.49	3.05	2.03	1.71	1.46	1.25	1.44	1.38	1.47	1.33				
FeO																	1.65	1.39	1.33	1.40
MgO	0.06	0.03	0.02	0.02	0.13	0.04	0.05	0.05	0.05	0.01	0.10	0.00	0.00	0.00	0.09	0.00	0.28	0.00	0.00	0.00
CaO	0.76	0.67	0.67	0.64	0.80	0.67	0.65	0.65	0.75	0.62	1.90	1.70	2.02	1.68	2.18	1.74	2.30	1.93	1.34	2.03
FeO	1.94	1.83	1.46	1.64	2.12	1.34	2.24	2.74	1.83	1.54							1.65	1.39	1.33	1.40
Na_2O	16.42	16.03	16.40	16.27	16.37	16.66	15.99	16.46	16.01	16.05	16.10	16.18	16.01	16.39	15.81	16.40	16.09	15.99	16.24	15.06
K_2O	5.75	5.67	6.16	6.01	5.56	5.96	5.51	5.47	5.65	5.84	4.06	4.13	3.83	3.88	3.93	4.17	3.68	4.12	3.42	4.15
Σ	102.40	100.85	101.28	101.05	101.07	100.70	100.01	101.01	100.84	99.58	98.79	101.15	98.11	99.91	99.31	100.75	101.66	100.41	101.32	102.90
apfu																				
Si	1.06	1.06	1.03	1.04	1.06	1.03	1.06	1.05	1.05	1.04	1.05	1.06	1.05	1.06	1.06	1.05	1.08	1.06	1.11	1.11
Al	0.94	0.95	0.96	0.95	0.93	0.95	0.93	0.92	0.95	0.96	0.96	0.98	0.97	0.97	0.96	0.98	0.94	0.98	0.95	0.98
Fe	0.04	0.04	0.03	0.03	0.04	0.03	0.05	0.06	0.04	0.03	0.03	0.02	0.03	0.03	0.03	0.02	0.03	0.03	0.03	0.03
Ca	0.02	0.02	0.02	0.02	0.02	0.02	0.02	0.02	0.02	0.02	0.05	0.04	0.05	0.04	0.06	0.05	0.06	0.05	0.03	0.05
Na	0.76	0.76	0.77	0.77	0.77	0.78	0.76	0.78	0.76	0.77	0.78	0.76	0.78	0.78	0.76	0.77	0.76	0.76	0.77	0.71
K	0.18	0.18	0.19	0.19	0.17	0.18	0.17	0.17	0.18	0.18	0.13	0.13	0.12	0.12	0.12	0.13	0.11	0.13	0.11	0.13
end-members† [%]																				
Nph	76	77	76	76	76	77	76	76	76	77	79	80	79	80	78	80	78	79	82	77
Ks	18	18	19	19	17	18	17	17	18	18	13	13	12	13	13	13	12	13	11	14
Yo	2	2	2	2	2	2	2	2	2	2	5	5	6	5	6	5	6	5	4	6
Fnp	3	3	2	3	4	2	4	5	3	3	2	2	2	2	2	2	3	2	2	2
Fks	1	1	1	1	1	1	1	1	1	1										

*Mn, V, Cr, Sr, and Ba were below their detection limits.

†*Nph*, nepheline; *Ks*, kalsilite; *Yo*, yoshiokaite $Ca_{(1-x)}(Al,Si)_2O_4$; *Fnp*, $NaFeSiO_4$ "ferrinepheline"; *Fks*, $KFeSiO_4$ "ferrikalsilite."

Table 13.2.24
Chemical analyses of oldhamite from the LdS12 sample, France.

	1	2	3	4	5	6	7	8	9	10
wt.%										
S*	42.78	42.72	43.26	43.33	42.78	43.22	43.58	43.43	43.22	43.03
Fe	1.16	1.10	1.17	1.02	1.10	1.02	1.00	1.12	1.00	1.11
Ca	52.62	52.76	52.96	52.88	53.04	53.02	52.80	52.48	52.72	52.83
Σ	96.56	96.59	97.39	97.23	96.92	97.25	97.39	97.03	96.94	96.98
apfu										
S	1.00	1.00	1.01	1.01	0.99	1.00	1.02	1.02	1.01	1.00
Fe	0.02	0.01	0.02	0.01	0.01	0.01	0.01	0.02	0.01	0.01
Ca	0.98	0.99	0.98	0.99	0.99	0.99	0.99	0.98	0.99	0.99
end-members† [%]										
Old	98	99	98	99	99	99	99	98	99	99
Ke	2	1	2	1	1	1	1	2	1	1
k										

*Co, Ni, Cu, Zn, Mg, As, and Se were below the detection limits.
†*Ke*, keilite.

Table 13.2.25
Chemical Analyses of Olivine Group From the LdS1 Sample, France.

	1	2	3	4	5	6	7	8	9
wt.%									
SiO_2*	42.84	43.00	43.10	42.87	43.27	43.03	43.22	43.16	43.13
MgO	56.32	56.32	56.47	56.07	56.36	56.42	56.52	56.62	56.35
CaO	0.08	0.07	0.17	0.09	0.00	0.00	0.00	0.00	0.00
MnO	0.33	0.30	0.32	0.45	0.29	0.00	0.00	0.26	0.24
FeO	0.65	0.58	0.52	0.86	0.70	0.80	0.81	0.82	0.86
Σ	100.22	100.27	100.58	100.34	100.62	100.25	100.55	100.86	100.58
apfu									
Si	1.01	1.01	1.01	1.01	1.01	1.01	1.01	1.01	1.01
Mg	1.97	1.97	1.97	1.96	1.97	1.97	1.97	1.97	1.97
Fe	0.01	0.01	0.01	0.01	0.02	0.02	0.02	0.02	0.02
Mn	0.01	0.01	0.01	0.01	0.01			0.01	
end-members [%]									
Fo	99	99	99	99	99	99	99	99	99
Fa	1	1	1	1	1	1	1	1	1
k									

*Ni and Al were measured but not observed.

- LdS2, composition no. 2 (e.g., analyses 6–10), n=5, (Ca0.98Al0.02V0.01)S1.01(Ti0.98Fe0.01[]0.01)S1.00O3.00, $Pe_{95}Afo_1(Vfo + Tau)_{\sim 1}$.
- LdS2, Nb-rich composition no. 3, n=1, Ca0.95Sr0.01)S0.96(Ti1.00Fe0.02Nb0.01)S1.03O3.03, $Pe_{96}Cfn_3Tau_1$, with Cfn for $Ca[Fe^{3+}{}_{0.5}Nb^{5+}{}_{0.5}]O_3$ hypothetical end-member.
- LdS7, n=2, $(Ca_{0.99}Al_{0.02}Sr_{0.01}V_{0.01})_{\Sigma 1.03}(Ti_{0.91}Fe_{0.07})_{0.98}O_{2.97}$, $Pe_{99}Tau_1$ (normalized).
- LdS5, n=14, prw, Ca1.05[(Ti0.49Si0.40Fe0.02Al0.02Nb0.01)[]0.06)S1.00O2.93, $Pe_{41}Wo_{41}Cfn_4Can_3$, where Can is the Al analogue of Cfn.
- LdS12, composition no. 1 (related to analyses 19–22), n=10, $(Ca_{0.98}Al_{0.02}V_{0.01})_{\Sigma 1.01}(Ti_{0.97}Fe_{0.03})_{\Sigma 1.00}O_{3.00}$, $Pe_{97}Afo_2Vfo_1$.
- LdS12, composition no. 2 (related to analyses 23 and 24), n=2, $(Ca_{1.00}Mg_{0.01})_{\Sigma 1.01}(Ti_{0.83}Si_{0.07}Al_{0.03}Fe_{0.02}Cr_{0.02}V_{0.01}Zr_{0.01}S_{0.01})_{\Sigma 1.00}O_{2.96}$, $Pe_{74}Shu_{17}Wo_7Gk_1Lak_1$, where Shu is shulamitite ($Ca_3TiFeAlO_8$) and Gk is geikielite.

Table 13.2.26
Chemical analyses of perovskite, Lapanouse-de-Sévérac, France.

	1	2	3	4	5	6	7	8	9	10	11	12
	LdS2										LdS7	
wt.%												
Nb_2O_5*	0.13	0.15	0.12	0.17	0.11	0.16	0.16	0.12	0.13	0.15	n.a.†	n.a.
SiO_2	2.91	1.33	2.32	2.71	2.71	1.60	0.00	0.00	0.00	0.00	0.14	0.00
TiO_2	45.13	48.72	46.98	43.69	46.45	48.51	57.69	57.51	57.66	57.75	51.54	51.95
ZrO_2	0.00	0.39	0.33	0.80	0.00	0.00	0.00	0.00	0.00	0.00	n.a.	n.a.
Al_2O_3	0.46	0.58	0.48	1.42	0.69	0.33	0.73	0.78	0.77	0.69	0.62	0.63
V_2O_3	0.18	0.35	0.24	0.35	0.34	0.00	0.42	0.40	0.32	0.37	0.47	0.50
Cr_2O_3	0.87	0.45	0.66	0.24	0.54	0.68	0.21	0.21	0.19	0.13	0.00	0.00
Fe_2O_3	7.87	6.50	7.01	8.72	6.69	6.75	0.83	0.69	0.90	0.53	3.87	4.14
MgO	0.00	0.00	0.00	0.05	0.06	0.00	0.00	0.00	0.00	0.00	0.00	0.00
CaO	40.37	40.17	40.55	41.11	39.78	40.16	40.72	40.74	40.29	40.34	39.08	39.94
SrO	0.19	0.27	0.30	0.24	0.19	0.18	0.25	0.27	0.26	0.27	0.39	0.37
K_2O	n.a.	n.a.	n.a.	n.a.	n.a.	n.a.	n.a.	n.a.	n.a.	n.a.	0.00	0.14
Σ	98.11	98.91	98.99	99.50	97.56	98.37	101.01	100.72	100.52	100.23	96.10	97.67
apfu												
Si	0.07	0.03	0.05	0.06	0.06	0.04						
Ti	0.78	0.84	0.80	0.74	0.80	0.84	0.97	0.97	0.98	0.98	0.91	0.90
Zr				0.01								
Al	0.01	0.02	0.01	0.04	0.02	0.01	0.02	0.02	0.02	0.02	0.02	0.02
V		0.01		0.01	0.01	0.00	0.01	0.01	0.01	0.01	0.01	0.01
Cr	0.02	0.01	0.01		0.01	0.01						
Fe	0.14	0.11	0.12	0.15	0.12	0.12	0.01	0.01	0.02	0.01	0.07	0.07
Ca	0.99	0.98	0.99	0.99	0.98	0.99	0.98	0.98	0.97	0.98	0.98	0.99
End members‡ [%]												
Pe	77	82	79	73	79	82	95	95	95	96	90	90
Afo	5	6	6	12	6	6	1	1	1	1	0.1	0.1
Vfo		3		3	3						0.1	0.1
Cfo	9	3	6		3	6						
Tau											1	1
Wo	7	3	5	6	6	4	1	1	1	1		
Lak				1								

	13	14	15	16	17	18	19	20	21	22	23	24
	LdS5						LdS12					
wt.%												
SO_3*	n.a.	n.a.	n.a.	n.a.	n.a.	n.a.	0.00	0.00	0.00	0.00	0.55	0.21
Nb_2O_5	2.09	1.82	1.04	1.35	1.24	1.22	0.45	0.00	0.00	0.00	0.00	0.00
SiO_2	29.34	29.49	29.59	29.57	29.62	29.26	0.00	0.00	0.00	0.00	4.22	1.92
TiO_2	36.31	36.92	37.47	36.66	36.93	36.91	55.31	55.83	55.50	55.47	46.36	51.22
ZrO_2	0.38	0.00	0.00	0.52	0.32	0.52	0.00	0.00	0.00	0.00	0.71	0.34
Al_2O_3	0.81	0.68	0.80	0.93	0.89	0.87	0.57	0.63	0.59	0.67	1.26	1.17
V_2O_3	0.00	0.00	0.00	0.00	0.00	0.00	0.47	0.57	0.61	0.55	0.67	0.89
Cr_2O_3	0.00	0.00	0.00	0.00	0.00	0.00	0.00	0.00	0.00	0.00	1.29	1.00
Fe_2O_3	1.54	1.48	1.96	1.64	1.68	1.73	1.66	1.26	1.57	1.50	1.17	1.22
MgO	0.00	0.00	0.00	0.00	0.00	0.00	0.00	0.00	0.00	0.00	0.39	0.13
CaO	27.55	27.30	27.29	27.83	28.00	27.76	39.08	39.36	39.49	39.24	42.07	40.86
SrO	n.a.	n.a.	n.a.	n.a.	n.a.	n.a.	0.19	0.19	0.00	0.23	0.00	0.00
K_2O	n.a.	n.a.	n.a.	n.a.	n.a.	n.a.	n.a.	n.a.	n.a.	n.a.	n.a.	n.a.
Na_2O	0.11	0.10	0.00	0.00	0.00	0.00	n.a.	n.a.	n.a.	n.a.	n.a.	n.a.
Σ	98.14	97.79	98.15	98.51	98.68	98.27	97.72	97.71	97.60	97.51	98.56	98.85

Table 13.2.26
Chemical analyses of perovskite, Lapanouse-de-Sévérac, France—cont'd.

	13	14	15	16	17	18	19	20	21	22	23	24
	LdS5						LdS12					
apfu												
S											0.01	
Nb	0.02	0.01	0.01	0.01	0.01	0.01						
Si	0.39	0.40	0.40	0.39	0.39	0.39					0.09	0.04
Ti	0.49	0.50	0.50	0.49	0.49	0.49	0.97	0.97	0.97	0.97	0.78	0.87
Zr											0.01	
Al	0.02	0.01	0.02	0.02	0.02	0.02	0.02	0.02	0.02	0.02	0.03	0.03
V							0.01	0.01	0.01	0.01	0.01	0.02
Cr											0.02	0.02
Fe	0.02	0.02	0.03	0.02	0.02	0.02	0.03	0.02	0.03	0.03	0.02	0.02
Mg											0.01	
Ca	1.06	1.05	1.05	1.06	1.06	1.06	0.97	0.98	0.98	0.97	1.01	0.99
End members[‡] [%]												
Pe	51	52	52	52	52	52	97	98	97	97	70	78
Afo							2	2	2	2		
Vfo							1	1	1	1		
Gk											1	
Tau											10	4
Wo	41	41	41	42	42	41						
Lak											1	
Shu											17	17
Cfn	4	4	4	3	3	3						
Can	4	3	3	3	3	3						

*F, Cl, S, P, Mn, Ni, Zn, As, Y, Sn, Ba, Ce, Nd, La, and U were analyzed but not observed.
[†]Not analyzed.
[‡]Wollastonite equivalent, *Lak*, lakargiite; *Shu*, shulamitite. Hypothetical end members: *Afo*, $AlFeO_3$; *Vfo*, $VFeO_3$; *Cfo*, $CrFeO_3$; *Cfn*, $Ca(Fe_{0.62}Nb_{0.31}[]_{0.07})O_3$; *Can*, $Ca(Al_{0.62}Nb_{0.31}[]_{0.07})O_3$; an additional minor and rarely observed end member includes $Ca[Fe_{0.5}Nb_{0.5}]O_3$.

The LdS5 perovskite associates with another unnamed phase (phase D), ideally $(Ca,Na)_2(Ti,Nb)O_4$, possibly $(Ca_{1.91}Na_{0.05})_{\Sigma 1.96}(Ti_{0.98}Nb_{0.05}Fe_{0.02})_{1.05}O_{4.03}$. The empirical formula of the perovskite suggests apfu $(Fe^{3+}) = 1/2$ apfu (Nb^{5+}). A relation of the $CaTi^{4+}O_3$ and $Ca(Fe,Nb)O_3$ end-members suggests Nb-rich members may be better interpreted as $Ca[(Fe,Al)_{0.31}Nb_{0.62}[]_{0.07})$, because (1) $(3+)x+(5+)y=(4+)$; x = Fe, y = Nb; (2) $3x+5y=4$; (3) $x=0.5y$; (4) finally: $x=0.31$, $y=0.62$, $[]=1-x-y=0.07$. The calculations argue the $2Ti^{4+} \leftrightarrow Fe^{3+}+Nb^{5+}$ substitution trend ($r^2=0.58$ but a clear negative Ti vs. (Fe+Nb) correlation is graphically evident).

Petalite-Like Species ($XAlSi_4O_{12}$)

Although X in petalite sensu stricto stands for Li, this is not a case of the Lapanouse material, as shown in Table 13.2.27. Here, the alkaline metal site is either dominated by K (LdS2 sample; BSE image in Figure 13.2.1B) or Ba (sample LdS1), the two metals being similar in their ionic radii. "K-petalite" (Kpt, mentioned by Reinsch, 2009) chemistry may be visualized by the $(K_{1.00}Ba_{0.01}Ca_{0.01})_{\Sigma 1.02}(Al_{0.95}Fe_{0.04})_{\Sigma 0.99}(Si_{3.95}Al_{0.04})_{\Sigma 3.99}O_{9.96}$ formula, corresponding to $Kpt_{94}Gil_4Bpt_1Cpt_1$ end-member composition (n = 10), with Gil as gillespite ($BaFeSi_4O_{10}$) equivalent, Bpt as "Ba-petalite" ($BaAl(Si_3AlO_{10})$) and Cpt as "calciopetalite." The analogous data for the "Ba-petalite" is $(Ba_{0.68}K_{0.31}Na_{0.09}Ca_{0.01})_{\Sigma 1.09}(Al_{0.79}Fe_{0.11}Ti_{0.01}[]_{0.09})_{\Sigma 1.00}(Si_{3.74}Al_{0.26})_{\Sigma 4.00}O_{10.08}$, $Bpt_{63}Kpt_{24}Gil_{12}Cpt_1$ (n = 13). The important addition of the gillespite molecule links the Lapanouse slags with the Ba-silicate rocks known mainly from Fresno Co., California, USA. These rocks are mainly composed of sanbornite, which interestingly, was also determined in a pyrometamorphic rock from a heap in Jankowice (Upper Silesian Coal Basin; Kruszewski, another chapter is this book). Another Ba-rich mineral that may exist in Lapanouse (LdS5 sample) is hyalophane.

Plagioclase ($(Na,Ca)Al_{1\text{-}2}Si_{2\text{-}3}O_8$)

Plagioclases stand for minor but not necessarily accessory components of the slags under study. Their composition is shown in Table 13.2.28. The LdS15 anorthite is more Fe-rich than the LdSV one, as can be compared by the two juxtaposed formulas: $(Ca_{0.97}Na_{0.03}K_{0.02})_{\Sigma 1.02}(Al_{1.80}Fe_{0.11})_{\Sigma 1.91}Si_{2.08}O_{8.01}$ (n=8) and $(Ca_{0.99}K_{0.01}Na_{0.01})_{\Sigma 1.01}(Al_{1.92}Fe_{0.02})_{\Sigma 1.94}Si_{2.05}O_{8.01}$ (n=15). These plagioclases can also be described as $An_{90}Fan_5Ab_3Kf_2$ and $An_{97}Fan_1Kf_1Ab_1$. Slawsonite and celsian molecules are seen only rarely and in trace amounts.

Pseudobrookite ($(Fe^{3+},Mg)_2TiO_5$)

Some information about pseudobrookite was already given in this section about Hematite, as the two species may exist in form of the solid solution. Dominant pseudobrookite is rare in the Lapanouse slags and was only noticed in two samples (Table 13.2.29). Single analysis for the LdS5 material (pseudobrookite-armalcolite solid solution; BSE image in Figure 13.2.2B) gives the following formulas: $(Fe^{3+}{}_{1.56}Mg_{0.21}Ti_{0.17}Mn_{0.03}Al_{0.03})_{\Sigma 2.00}(Ti_{0.98}Zr_{0.02})_{\Sigma 1.00}O_{4.97}$, $Psb_{77}Arm_{11}Tio_8Mna_2Aps_2$, with Tio for unnamed $(Ti,Fe,Mg,Mn)_{1-x}Ti_2O_5$ phase (IMA2000-016), Mna "mangano-armalcolite" and Mac for machiite. The mineral is closely associated with hematite and an unnamed Mg–Ti-[] analogue of batiferrite. Its tiny machiite content again relates it with both meteorites and coal-fire-related rocks (see Kruszewski, another chapter in this book). The composition of the LdS1 pseudobrookite is more complex: $(Fe^{3+}{}_{0.73}Mg_{0.24}Al_{0.03}Mn_{0.02}Ca_{0.01})_{\Sigma 1.02}(Ti_{1.26}Fe^{3+}{}_{0.71})_{\Sigma 1.97}O_{5.00}$, $Psb_{44}Hm_{26}Arm_{24}Mna_2Mac_2Afo_1Dks_1$, with Dks standing for "dehydrokassite" (kassite sensu stricto is, ideally, $CaTi_2O_4(OH)_2$). This pseudobrookite is also interesting in occasionally showing elevated levels of Co, Ni, and Zn.

Pyrrhotite ($Fe_{1\text{-}x}S$)

Pyrrhotite is the only IMA-approved Fe sulfide mineral found in the Lapanouse slags. It is especially frequent in the LdSV sample. Pyrrhotite analyses are reported in Table 13.2.30. The only evident difference between the LdS12 and LdSV pyrrhotites is the more frequent Ni enrichment of the latter. Empirical formulas of these two materials are $Fe_{0.89}S_{1.00}$ (x=0.11) and $(Fe_{0.95}Ni_{0.01})_{\Sigma 0.96}S_{1.00}$ (x=0.04), respectively. The first pyrrhotite is more nonstoichiometric, and actually occurs in two varieties described by the formulas $Fe_{0.92}S_{1.00}$ and $(Fe_{0.85}Ni_{0.01})_{\Sigma 0.86}S_{1.00}$. The LdSV pyrrhotite closes by its composition to troilite. It is also associated with an unnamed $Fe_{10}S_9$ sulfide, which occurs very rarely as pyrrhotites rims. Some tiny areas of these rims are enriched in Cu and Ag.

Spinel Group (Magnesioferrite Fe_2MgO_4 and Magnetite Fe_2FeO_4)

Table 13.2.31 (samples LdS1, LdS2) and Table 13.2.32 (samples LdS15 and LdS5) show compositions of the Lapanouse slags' spinels. The mean formulas are:

- LdS1, magnesioferrite (Mfr)-spinel (Sp), n=20, $(Mg_{0.98}Mn_{0.02})_{\Sigma 1.00}(Fe^{3+}{}_{1.00}Al_{0.49}Cr_{0.46}Ti_{0.03}[]_{0.02})_{\Sigma 2.00}O_{4.01}$, $Mfr_{49}Sp_{24}Mc_{r22}Jc_1Ulv_1Mnc_1Gal_1$, with Mcr for magnesiochromite, Jc for jacobsite, Ulv for ulvöspinel, Mnc for manganochromite, and Gal for galaxite.
- LdS1, spinel-magnesiochromite (from a forsterite-rich aggregate), n=1, $(Mg_{0.96}Mn_{0.04}Zn_{0.01})$ $(Al_{0.87}C_{0.73}Fe^{3+}{}_{0.36}Ti_{0.03})O_{4.01}$, $Sp_{42}Mcr_{35}Mfr_{17}Gal_2Jc_1Mnc_1Ulv_1$.
- LdS2, n=14, magnetite-magnesioferrite, $(Fe^{2+}{}_{0.67}Mg_{0.27}Mn_{0.04}Ca_{0.02}Ni_{0.01})_{\Sigma 1.01}(Fe^{3+}{}_{1.91}Al_{0.06}Cr_{0.01})_{\Sigma 1.98}O_{4.00}$, $Mt_{64}Mfr_{25}Jc_4Hrm_2Hc_2Sp_1(Chr + Mnc + Gal + Ulv)_2$, with Hrm for harmunite ($CaFe_2O_4$) equivalent and Chr for chromite.
- LdS15, composition no. 1 (analyses 1–3), magnesioferrite, n=3, $(Mg_{0.90}Mn_{0.10}Ca_{0.05})_{\Sigma 1.05}(Fe_{1.65}Al_{0.29}Cr_{0.01})_{\Sigma 1.95}O_{3.98}$, $Mfr_{73}Sp_{13}Jc_8Hrm_4Gal_1Kr_1$, with Kr for krotite ($CaAl_2O_4$) equivalent.
- LdS15, composition no. 2 (analyses 4 and 5), magnesioferrite-spinel, n=2, $(Mg_{0.91}Mn_{0.07}Ca_{0.05}Cu_{0.01})_{\Sigma 1.04}(Fe_{1.13}Al_{0.82}Ti_{0.02})_{\Sigma 1.97}O_{3.99}$, $Mfr_{51}Sp_{37}Jc_4Gal_3Hrm_3Kr_2Csp_1$, with Csp for cuprospinel.
- LdS5, n=14, spinel-magnesioferrite-magnesiochromite, $(Mg_{0.94}Mn_{0.01}[]_{0.05})_{\Sigma 1.00}(Fe_{0.63}Cr_{0.52}Al_{0.83}Ti_{0.02})_{\Sigma 2.00}O_{4.03}$, $Sp_{42}Mfr_{30}Mch_{25}Qan_1Gal_1Jc_1$.

Some single spinel-related analyses from the LdS1 slag refer to even more chromian compositions: $Mfr_{50}Mch_{24}Sp_{20}Jc_1Qan_1Mt_1Mnc_1Gal_1Chr_1$, with Qan as qandilite (Mg titanate), Chr as chromite, and rarely also Ni entering

Table 13.2.27
Chemical analyses of petalite-like species: "K-petalite" and "Ba-petalite," Lapanouse-de-Sévérac, France.

	1	2	3	4	5	6	7	8	9	10	11	12	13	14	15	16	17	18	19	20
	LdS2										LdS1									
wt.%																				
SiO_2*	53.01	52.79	56.03	54.20	53.82	54.48	52.06	53.00	53.40	53.14	46.88	47.04	46.76	47.00	47.38	47.29	46.60	47.09	46.14	47.22
TiO_2	0.00	0.00	0.00	0.00	0.00	0.00	0.00	0.11	0.19	0.18	0.23	0.10	0.16	0.18	0.14	0.11	0.13	0.15	0.00	0.13
Al_2O_3	23.16	22.73	23.37	22.91	22.67	23.18	22.61	22.80	22.73	22.63	23.00	22.43	22.60	22.51	22.76	22.58	23.01	22.39	22.68	21.41
CaO	0.10	0.09	0.11	0.00	0.00	0.00	0.10	0.09	0.16	0.15	0.10	0.10	0.09	0.00	0.00	0.13	0.00	0.17	0.09	0.13
FeO	0.87	0.70	0.66	0.66	0.83	0.70	0.55	0.58	0.54	0.43	1.46	1.61	1.57	1.74	2.00	1.69	1.54	1.74	1.52	1.80
BaO	0.39	0.30	0.24	0.24	0.29	0.32	0.25	0.26	0.19	0.27	22.66	21.58	22.14	21.93	21.59	21.83	22.83	22.08	22.87	21.10
Na_2O	0.00	0.00	0.00	0.00	0.00	0.00	0.00	0.00	0.00	0.00	1.13	1.25	1.14	1.16	0.99	1.21	1.12	1.29	1.08	1.12
K_2O	21.08	21.20	21.06	21.27	21.02	21.02	21.42	20.97	21.35	21.03	5.96	6.17	5.91	5.88	6.35	6.02	5.80	5.93	5.56	6.26
Σ	98.61	97.81	101.47	99.28	98.63	99.70	96.99	97.81	98.56	97.83	101.42	100.28	100.37	100.40	101.21	100.86	101.03	100.84	99.94	99.17
apfu																				
Si	3.92	3.93	4.00	3.97	3.97	3.97	3.92	3.95	3.94	3.95	3.72	3.74	3.73	3.74	3.74	3.74	3.71	3.73	3.72	3.78
Ti								0.01	0.01	0.01	0.01	0.01	0.01	0.01	0.01	0.01	0.01	0.01		0.01
Al	1.01	1.00	0.98	0.99	0.98	1.00	1.00	1.00	0.99	0.99	1.07	1.05	1.06	1.06	1.06	1.05	1.08	1.05	1.08	1.01
Ca	0.01	0.01	0.01				0.01	0.01	0.01	0.01	0.01	0.01	0.01			0.01		0.01	0.01	0.01
Fe^{2+}	0.05	0.04	0.04	0.04	0.05	0.04	0.03	0.04	0.03	0.03	0.10	0.11	0.10	0.12	0.13	0.11	0.10	0.12	0.10	0.12
Ba	0.01	0.01	0.01	0.01	0.01	0.01	0.01	0.01	0.01	0.01	0.70	0.67	0.69	0.68	0.67	0.68	0.71	0.69	0.72	0.66
Na											0.09	0.10	0.09	0.09	0.08	0.09	0.09	0.10	0.08	0.09
K	0.99	1.01	0.96	0.99	0.99	0.98	1.03	1.00	1.01	1.00	0.30	0.31	0.30	0.30	0.32	0.30	0.29	0.30	0.29	0.32
end-members† [%]																				
Kpt	93	94	95	95	94	95	95	94	94	94	24	25	24	24	24	24	24	24	23	25
Bpt	1	1	1	1	1	1	1	1	1	1	64	62	64	64	62	62	65	62	66	61
Cpt	1	1	1				1	1	1	1	1	1	1			1		1	1	1
Gil	5	4	4	4	5	4	4	4	3	3	11	12	12	13	14	12	11	13	11	13

*Mg, P, Sc, V, Cr, Mn, Ga, Rb, and Cs were below the detection limits.
†*Kpt*, $KAlSi_4O_{10}$ "K-petalite"; *Bpt*, $BaAl(Si_3Al)O_{10}$ "Ba-petalite"; *Cpt*, $CaAl(Si_3Al)O_{10}$ "Ca-petalite"; *Gil*, gillespite ($BaFeSi_4O_{10}$).

Table 13.2.28
Chemical analyses of plagioclase, Lapanouse-de-Sévérac, France.

	1	2	3	4	5	6	7	8	9	10	11	12	13	14	15	16	17	18	19	20
	LdS15									LdSV										
wt.%																				
SiO_2*	43.25	43.89	43.69	43.29	43.67	43.73	43.23	43.49	43.96	43.51	43.57	43.39	43.54	43.37	43.67	43.60	43.51	43.41	43.43	43.82
Al_2O_3	34.94	31.45	31.87	33.50	32.60	31.48	32.27	31.81	31.64	34.86	34.61	34.68	34.72	34.50	34.49	34.24	34.52	34.89	34.98	34.58
Fe_2O_3	0.43	3.35	3.14	1.43	2.74	3.61	2.64	3.43	3.57	0.49	0.48	0.57	0.59	0.55	0.58	0.56	0.62	0.54	0.56	0.48
CaO	19.68	18.88	18.75	19.25	19.12	18.95	18.83	19.02	18.86	19.67	19.85	19.34	19.37	19.39	19.46	19.25	19.50	19.75	19.79	19.48
SrO	0.36	0.00	0.00	0.00	0.00	0.00	0.48	0.00	0.00	0.00	0.00	0.00	0.00	0.33	0.00	0.32	0.00	0.32	0.00	0.00
Na_2O	0.09	0.37	0.39	0.21	0.31	0.32	0.30	0.31	0.29	0.00	0.13	0.00	0.00	0.12	0.18	0.13	0.00	0.00	0.00	0.00
K_2O	0.22	0.36	0.25	0.15	0.28	0.44	0.24	0.34	0.39	0.13	0.18	0.20	0.16	0.24	0.23	0.25	0.20	0.15	0.00	0.21
Σ	98.98	98.45	98.36	98.15	98.82	98.69	97.98	98.50	98.79	98.98	99.12	98.56	98.80	98.50	98.89	98.37	98.73	99.19	99.24	98.92
apfu																				
Si	2.03	2.09	2.08	2.06	2.07	2.08	2.07	2.07	2.09	2.05	2.05	2.05	2.06	2.05	2.06	2.06	2.06	2.04	2.04	2.07
Al	1.93	1.77	1.79	1.88	1.82	1.77	1.82	1.79	1.77	1.93	1.92	1.93	1.93	1.92	1.91	1.91	1.92	1.93	1.94	1.92
Ca	0.99	0.96	0.96	0.98	0.97	0.97	0.96	0.97	0.96	0.99	1.00	0.98	0.98	0.98	0.98	0.97	0.99	0.99	1.00	0.98
Fe	0.02	0.12	0.11	0.05	0.10	0.13	0.09	0.12	0.13	0.02	0.02	0.02	0.02	0.02	0.02	0.02	0.02	0.02	0.02	0.02
Sr	0.01						0.01							0.01		0.01		0.01		
Na	0.01	0.03	0.04	0.02	0.03	0.03	0.03	0.03	0.03		0.01			0.01	0.02	0.01				
K	0.01	0.02	0.01	0.01	0.02	0.03	0.01	0.02	0.02	0.01	0.01	0.01	0.01	0.01	0.01	0.02	0.01	0.01		0.01
end-members† [%]																				
An	96	89	89	92	90	89	90	89	88	98	97	98	98	96	96	95	98	97	99	98
Ab	1	3	3	3	3	3	3	3	3		1			1	2	1				
Kf	1	2	1	2	2	2	1	2	2	1	1	1	1	1	1	1	1	1		1
Fan	1	6	6	3	5	6	5	6	7	1	1	1	1	1	1	1	1	1	1	1
Sls	1						1							1		1		1		

*Mg, Ti, Mn, and Ba were measured but not observed.
†*An*, anorthite; *Ab*, albite; *Kf*, K feldspar; *Fan*, $CaFe_2Si_2O_8$ "ferrianorthite"; *Sls*, slawsonite; some amounts are normalized.

Table 13.2.29
Chemical analyses of pseudobrookite (sample LdS5) and pseudobrookite-hematite solid solution, France.

	1	2	3	4	5	6	7	8
	LdS5	LdS1						
wt.%								
Nb_2O_5*	0.20							
SiO_2	0.00	0.04	0.03	0.02	0.01	0.05	0.10	0.04
TiO_2	37.84	43.26	43.22	43.86	43.64	44.12	43.83	43.78
ZrO_2	0.78							
Al_2O_3	0.65	0.63	0.64	0.64	0.63	0.63	0.66	0.67
Cr_2O_3	0.00	0.04	0.02	0.02	0.02	0.03	0.02	0.00
Fe_2O_3	51.32	49.84	50.04	49.55	50.06	50.07	50.08	50.32
MgO	3.52	4.25	4.13	4.34	4.16	4.17	4.08	4.03
CaO	0.00	0.15	0.18	0.19	0.10	0.10	0.10	0.12
MnO	0.82	0.80	0.70	0.64	0.72	0.80	0.81	0.71
CoO	0.00	0.04	0.01	0.00	0.03	0.04	0.00	0.00
NiO	0.00	0.01	0.00	0.00	0.02	0.00	0.01	0.00
ZnO	0.00	0.00	0.00	0.00	0.00	0.03	0.13	0.00
Σ	95.13	99.06	98.97	99.26	99.39	100.04	99.82	99.67
apfu								
Ti	1.15	1.25	1.25	1.26	1.26	1.26	1.26	1.26
Zr	0.02							
Al	0.03	0.03	0.03	0.03	0.03	0.03	0.03	0.03
Fe^{3+}	1.56	1.44	1.45	1.43	1.44	1.44	1.44	1.45
Mg	0.21	0.24	0.24	0.25	0.24	0.24	0.23	0.23
Ca		0.01	0.01	0.01				
Mn	0.03	0.03	0.02	0.02	0.02	0.03	0.03	0.02
end-members† [%]								
Hm		26	26	25.5	26	26	26	26
Afo		1	1	1	1	1	1	1
Psb	77	43	44	43.5	44	44	44	45
Arm	11	24	24	25	24	24	23	23
Tio	8							
Mna	2	3	2	2	2	3	3	2
Aps	2	2	2	2	2	2	2	2
Dks		1	1	1				

*Cu was analyzed but not observed (Nb and Zr were not analyzed in the LdS1 sample material).
†*Hm*, hematite; *Psb*, pseudobrookite; *Arm*, armalcolite with hypothetical end-members: *Afo*, $AlFeO_3$; *Tio*, $(Ti,Fe,Mg,Mn)_{1-x}Ti_2O_5$; *Mna*, $MnTi_2O_5$ "manganoarmalcolite"; *Aps*, Al_2TiO_5 "aluminopseudobrookite"; *Dks*, $CaTi_2O_5$ "dehydrokassite."

as trevorite. This is true for LdS2 magnetite, which is also noteworthy due to its coexistence with three potentially new mineral species: $(Mg,Mn,Zn)_2(Fe,Al,Mn)_4Cr_2O_{11}$ (phase F), its Fe^{3+} analogue (phase G), and the already mentioned "magnesiograndiferrite." Western-dump material described by Eytier et al. (2004) is magnesioferrite with the $(Mg_{0.97}Ca_{0.03}Fe^{2+}_{0.01})(Fe^{3+}_{1.54}Al_{0.44}Ti_{0.01})O_4$ formula, while magnetite of the eastern dump is described by them as $(Fe^{2+}_{0.74}Mg_{0.24}Ca_{0.02})(Fe^{3+}_{1.90}Al_{0.10})O_4$.

Titanite ($CaTiO(SiO_4)$)

Titanite is a relatively common accessory mineral easily mistaken for perovskite in the slags of Lapanouse (Figure 13.2.2A). Titanite's chemistry is shown in Table 13.2.33. Mean formulas are as follows:

- LdS7 (n=2): $(Ca_{0.99}Mg_{0.01})_{\Sigma 1.00}(Ti_{0.77}Fe^{3+}_{0.10}Al_{0.09}\mathbf{V^{4+}}_{0.04}Cr_{0.01})_{\Sigma 1.01}(Si_{0.97}Al_{0.03})_{\Sigma 1.00}O_{4.00}(O_{0.84}\mathbf{F_{0.14}})_{\Sigma 0.98}$, $Tt_{77}Ffv_{10}Fvu_8Vma_4Cfv_1$, where Ffv is "ferri-fluoro-vuagnatite," ($CaFe(SiO_4)F$), Fvu is "fluorovuagnatite" ($CaAl(SiO_4)F$), Vma is vanadomalayaite (the V-analogue of titanite), and Cfv is "chromo-fluoro-vuagnatite" ($CaCr(SiO_4)F$).

Table 13.2.30
Chemical analyses of pyrrhotite (analyses 1–19) and a potentially new pyrrhotite-like $Fe_{10}S_9$ phase (last two analyses), Lapanouse-de-Sévérac, France.

	1	2	3	4	5	6	7	8	9	10	11	12	13	14	15	16	17	18	19	20	21
	LdS12									LdSV											
wt.%																					
S*	38.16	38.31	38.43	38.42	39.97	39.22	37.70	39.72	40.41	37.25	37.20	37.39	37.38	36.08	37.17	37.10	37.17	36.35	37.04	33.58	33.42
Fe	61.77	61.65	61.05	60.69	59.09	60.13	60.45	59.07	58.74	60.73	60.88	61.12	61.01	59.04	61.16	61.07	60.87	61.12	61.74	38.80	39.80
Co	0.00	0.00	0.00	0.00	0.00	0.00	0.00	0.00	0.00	0.00	0.00	0.00	0.00	0.00	0.00	0.00	0.00	0.00	0.00	0.78	0.58
Ni	0.00	0.00	0.00	0.00	0.00	0.00	0.00	0.48	0.49	0.86	1.10	0.90	0.88	0.98	0.87	0.87	0.81	0.65	0.72	25.29	23.91
Cu	0.00	0.00	0.00	0.00	0.00	0.00	0.00	0.00	0.00	0.00	0.00	0.00	0.00	0.00	0.00	0.00	0.00	0.00	0.00	1.46	2.13
Ag	0.00	0.00	0.00	0.00	0.00	0.00	0.00	0.00	0.00	0.00	0.00	0.00	0.11	0.00	0.00	0.12	0.00	0.11	0.00	0.14	0.13
Σ	99.93	99.96	99.48	99.11	99.06	99.35	98.15	99.27	99.63	98.83	99.18	99.41	99.39	96.10	99.20	99.16	98.85	98.24	99.49	100.05	99.96
apfu																					
S	1.00	1.00	1.00	1.00	1.00	1.00	1.00	1.00	1.00	1.00	1.00	1.00	1.00	1.00	1.00	1.00	1.00	1.00	1.00	9.00	9.00
Fe	0.93	0.92	0.91	0.91	0.85	0.88	0.92	0.85	0.83	0.94	0.94	0.94	0.94	0.94	0.95	0.95	0.94	0.97	0.96	5.97	6.15
Co																				0.11	0.09
Ni								0.01	0.01	0.01	0.02	0.01	0.01	0.02	0.01	0.01	0.01	0.01	0.01	3.70	3.52
Cu																				0.20	0.29
Ag																				0.01	0.01
x†	0.07	0.08	0.09	0.09	0.15	0.12	0.08	0.15	0.17	0.05	0.04	0.05	0.05	0.04	0.04	0.04	0.04	0.02	0.03		

*Zn, As, Se, Pb, Cd, Au, Hg, and Sb were below the detection limits.
†Cation deficiency.

Table 13.2.31
Chemical analyses of spinels, Lapanouse-de-Sévérac, France – part 1.

	1	2	3	4	5	6	7	8	9	10	11	12	13	14	15	16	17	18	19	20
	LdS2			LdS1																
wt.%																				
SiO_2*	0.01	0.01	0.02	0.01	0.12	0.19	0.14	0.00	0.00	0.10	0.19	0.00	0.00	0.09	0.08	0.00	0.11	0.05	0.07	0.02
TiO_2	0.11	0.16	0.10	0.11	1.24	1.37	1.09	1.48	1.81	1.87	1.71	1.28	1.35	1.25	0.98	0.98	1.27	0.78	0.88	0.32
Al_2O_3	1.46	4.53	2.76	1.54	10.42	15.15	16.03	12.47	15.89	16.88	16.09	15.72	15.12	14.27	5.05	5.24	25.55	5.58	4.64	2.80
Cr_2O_3	0.09	0.05	0.04	2.76	15.88	20.39	21.67	13.47	16.05	19.55	17.84	18.32	20.04	20.70	11.52	11.39	31.63	0.37	0.04	13.62
Fe_2O_3	91.59	86.54	90.74	89.33	49.30	38.59	36.70	48.74	41.61	36.53	39.47	40.27	39.52	40.34	60.36	60.10	16.51	71.01	72.39	60.82
MgO	4.46	6.99	5.48	5.02	21.10	21.47	21.45	21.09	21.69	21.64	21.68	21.41	21.41	21.55	19.58	19.60	22.18	19.79	19.31	18.74
CaO	0.71	0.15	0.27	0.22	0.00	0.00	0.00	0.00	0.06	0.14	0.09	0.00	0.00	0.00	0.00	0.00	0.00	0.03	0.08	0.08
MnO	1.11	2.42	1.66	1.50	1.20	0.90	0.69	0.84	0.86	0.87	0.93	0.93	0.98	0.95	1.13	1.00	1.48	0.88	0.97	1.55
CoO	0.00	0.05	0.01	0.00	0.00	0.00	0.00	0.00	0.00	0.00	0.00	0.00	0.00	0.00	0.00	0.00	0.00	0.04	0.01	0.00
NiO	0.36	0.39	0.37	0.46	0.23	0.20	0.00	0.25	0.00	0.00	0.00	0.17	0.00	0.00	0.20	0.26	0.00	0.18	0.26	0.46
ZnO	0.12	0.37	0.47	0.09	0.00	0.00	0.22	0.00	0.00	0.00	0.00	0.00	0.00	0.00	0.00	0.00	0.50	0.15	0.10	0.20
Σ	100.02	101.66	101.92	101.03	99.49	98.26	97.99	98.34	97.97	97.58	98.00	98.10	98.42	99.15	98.90	98.57	99.23	98.86	98.75	98.59
apfu																				
Si						0.01					0.01									
Ti					0.03	0.03	0.03	0.03	0.04	0.04	0.04	0.03	0.03	0.03	0.02	0.02	0.03	0.02	0.02	0.01
Al	0.06	0.19	0.12	0.07	0.39	0.55	0.58	0.46	0.58	0.61	0.58	0.57	0.55	0.52	0.19	0.20	0.87	0.22	0.18	0.11
Cr				0.08	0.39	0.50	0.53	0.33	0.39	0.48	0.43	0.45	0.49	0.50	0.30	0.30	0.73	0.01	0.00	0.36
Fe^{3+}	1.93	1.88	1.87	1.55	1.16	0.90	0.85	1.15	0.97	0.85	0.91	0.94	0.92	0.93	1.49	1.48	0.36	1.74	1.79	1.51
Fe^{2+}	0.67	0.53	0.62	0.65														0.01	0.01	0.02
Mg	0.25	0.37	0.30	0.28	0.99	0.99	0.99	0.99	1.00	0.99	0.99	0.99	0.99	0.99	0.96	0.96	0.96	0.97	0.95	0.93
Ca	0.03	0.01	0.01	0.01																
Mn	0.04	0.07	0.05	0.05	0.03	0.02	0.02	0.02	0.02	0.02	0.02	0.02	0.03	0.02	0.03	0.03	0.04	0.02	0.03	0.04
Ni	0.01	0.01	0.01	0.01	0.01			0.01							0.01	0.01			0.01	0.01
Zn		0.01	0.01														0.01			

Continued

Table 13.2.31
Chemical analyses of spinels, Lapanouse-de-Sévérac, France – part 1—cont'd.

	1	2	3	4	5	6	7	8	9	10	11	12	13	14	15	16	17	18	19	20
end-members† [%]																				
Mfr	24	33	28	26	57	44	42	56	48	42	45	46	45	46	71	71	17	83	83	70
Mch				1	19	24	26	16	19	23	21	22	24	25	14	14	35			17
Sp	1	4	2	1	19	27	29	23	29	30	29	28	27	25	9	10	42	10	9	5
Qan		0.1																1	1	
Jc	3	7	5	4	2	1	1	1	1	1	1	1	1	1	2	2	1	2	2	3
Mnc				0.2	1	1				1	1	1	1	1			1			1
Gal	0.1	1	0.3	0.2	1	1	1	1	1	1	1	1	1	1			2			0.2
Hrm	3	1	1	1																
Kr	0.1	0.1	0.1																	
Mt	65	48	58	60														2	3	
Hc	2	5	4	2																1
Ulv	0.1	0.1	0.1	0.1	1	2	1	2	2	2	2	1	2	1	1	1	1			0.1
Chr	0.1			3																
Trv	1	1	1	1												1			1	0.3
Atv		0.1	0.1																	1
Lb																				0.1
Ncr				0.1																

*Sc, V, Co, Ni, Cu, Zn, Ga, Zr, and Nb were measured but not observed or are below the detection limits.

†*Mfr*, magnesioferrite; *Mch*, magnesiochromite; *Sp*, spinel; *Qan*, qandilite (Mg_2TiO_4); *Jc*, jacobsite; *Mnc*, manganochromite; *Gal*, galaxite; *Hrm*, harmunite equivalent ($CaFe_2O_4$); *Kr*, krotite equivalent ($CaAl_2O_4$); Mt, magnetite; *Hc*, hercynite; *Ulv*, ulvöspinel; *Chr*, chromite; *Lb*, liebenbergite equivalent (Ni_2SiO_4); *Trv*, trevorite ($NiFe_2O_4$); *Ncr*, nichromite ($NiCr_2O_4$) with hypothetical end-member *Atv*, "aluminotrevorite" ($NiAl_2O_4$).

Table 13.2.32
Chemical analyses of spinels, Lapanouse-de-Sévérac, France – part 2.

	1	2	3	4	5	6	7	8	9	10	11	12	13	14	15
	LdS15					LdS5									
wt.%															
SiO_2*	0.00	0.00	0.00	0.00	0.00	0.00	0.00	0.00	0.00	0.11	0.11	0.00	0.00	0.15	0.00
TiO_2	0.00	0.14	0.00	0.37	1.17	0.65	0.70	0.83	0.79	0.14	1.24	0.46	0.99	1.22	0.24
Al_2O_3	9.55	6.51	6.45	23.65	22.38	18.32	18.62	19.06	20.08	27.71	28.58	29.07	25.95	28.81	29.98
Cr_2O_3	0.15	0.33	0.31	0.00	0.00	18.52	18.99	17.36	15.92	21.84	23.94	24.50	27.49	24.71	24.80
Fe_2O_3	64.31	67.19	67.17	50.01	49.72	36.47	35.27	37.27	37.34	26.75	23.01	22.39	22.54	22.10	20.96
MgO	17.90	18.53	18.46	20.83	19.62	21.38	21.22	21.11	21.03	20.38	20.25	19.92	19.92	20.30	20.86
CaO	1.14	1.35	1.34	0.86	2.19	0.00	0.00	0.10	0.10	0.00	0.00	0.00	0.00	0.00	0.00
MnO	3.14	3.70	3.63	3.00	2.62	1.05	1.14	1.17	1.22	0.31	0.32	0.00	0.37	0.33	0.33
NiO	0.00	0.00	0.00	0.00	0.00	0.00	0.00	0.00	0.00	0.38	0.00	0.00	0.00	0.00	0.00
CuO	0.00	0.00	0.40	0.00	0.44	0.00	0.00	0.00	0.00	0.00	0.00	0.00	0.00	0.00	0.00
Σ						93.39	95.94	96.90	96.48	97.62	97.45	96.34	97.25	97.63	97.17
apfu															
Ti				0.01	0.03	0.02	0.02	0.02	0.02		0.03	0.01	0.02	0.03	0.01
Al	0.37	0.25	0.25	0.83	0.80	0.67	0.68	0.69	0.73	0.97	1.00	1.02	0.92	1.00	1.04
Cr		0.01	0.01			0.45	0.47	0.42	0.39	0.51	0.56	0.58	0.65	0.58	0.58
Fe^{3+}	1.61	1.67	1.67	1.13	1.15	0.85	0.82	0.86	0.87	0.60	0.51	0.50	0.51	0.49	0.46
Mg	0.89	0.91	0.91	0.93	0.89	0.99	0.98	0.97	0.97	0.90	0.89	0.89	0.89	0.89	0.91
Ca	0.04	0.05	0.05	0.03	0.07										
Mn	0.09	0.10	0.10	0.08	0.07	0.03	0.03	0.03	0.03	0.01	0.01		0.01	0.01	0.01
Ni										0.01					
Cu			0.01		0.01										
end-members† [%]															
Mfr	71	74	74	51	50	42	40	42	42	28	24	24	24	23	22
Mch						22	23	20	19	24	26	27	31	27	27
Sp	16	11	11	38	35	33	33	33	35	46	47	48	43	47	49
Qan					1	1	1	1	1		1		1	1	
Jc	7	8	8	4	4	1	1	1	1						
Mnc						1	1	1	1						
Gal	2	1	1	3	3	1	1	1	1						
Hrm	3	4	4	2	4										
Kr	1	1	1	1	3										
Csp			1		1										
Tph					0.1										
Ln					0.1										

*Sc, V, Co, Ga, Zn, Zr, and Nb were measured but not observed or are below the detection limits.
†*Mfr*, magnesioferrite; *Mch*, magnesiochromite; *Sp*, spinel; *Qan*, qandilite (Mg_2TiO_4); *Jc*, jacobsite; *Mnc*, manganochromite; *Gal*, galaxite; *Hrm*, harmunite equivalent ($CaFe_2O_4$); *Kr*, krotite equivalent ($CaAl_2O_4$); *Csp*, cuprospinel ($CuFe_2O_4$); *Tph*, tephroite equivalent (Mn_2SiO_4); *Ln*, larnite equivalent (Ca_2SiO_4).

- LdS15 (n=4): $Ca_{1.00}(Ti_{0.88}Fe^{3+}_{0.12}Al_{0.05}Zr_{0.01})_{\Sigma 1.07}Si_{1.02}O_{5.08}$, $Tt_{82}Ffv12Fvu5Zrt1$, Zrt is "zirconotitanite" (a supposed Zr-analogue of titanite), the F-bearing members are introduced assuming F is the cause of lowered totals.
- LdS1 (n=2): $Ca_{0.99}(Ti_{0.94}Fe^{3+}_{0.05}Al_{0.03})_{\Sigma 1.02}Si_{0.99}O_{4.96}$, $Tt_{92}Ffv_{4}Fvu_{3}$ (assuming F is present).

Aluminum enrichment in titanites and the addition of vuagnatite component is known (Oberti et al., 1991).

Vaterite ($CaCO_3$)

Vaterite is a minor constituent of only some of the Lapanouse slags. The material from the sample LdS12 was identified based on (1) the analyzes not recasting to portlandite and (2) its continuous sulfur enrichment, which is known to stabilize this otherwise metastable calcium carbonate polymorph (e.g., Fernández-Díaz et al., 2010).

Table 13.2.33
Chemical analyses of titanite, Lapanouse-de-Sévérac, France.

	1	2	3	4	5	6	7	8	9
	LdS7		LdS15					LdS1	
wt.%									
SiO_2*	28.54	27.82	29.60	29.54	29.66	29.52	29.52	29.75	29.55
TiO_2	29.84	29.81	33.41	33.04	33.70	33.62	33.90	37.56	36.97
ZrO_2	0.00	0.00	n.a.[†]	0.64	0.47	0.46	0.42	n.a.	n.a.
Al_2O_3	2.94	2.75	1.17	1.23	1.19	1.16	1.12	0.69	0.99
V_2O_3	2.79	3.18	n.a.	0.00	0.00	0.32	0.00	n.a.	n.a.
Fe_2O_3	3.81	3.87	5.00	4.90	4.76	4.86	4.59	1.61	2.30
Cr_2O_3	0.22	0.21	0.00	0.00	0.00	0.00	0.00	0.00	0.00
MgO	0.12	0.11	0.00	0.00	0.00	0.00	0.00	0.05	0.04
CaO	27.24	26.81	27.72	26.90	27.01	26.92	26.87	27.33	27.92
Na_2O	0.00	0.00	0.00	0.00	0.00	0.00	0.00	0.00	0.08
K_2O	n.a.	n.a.	0.11	n.a.	n.a.	n.a.	n.a.	0.09	0.00
F	1.21	1.33	n.a.	n.a.	n.a.	n.a.	n.a.	n.a.	n.a.
Σ	96.70	95.89	97.11	96.25	96.79	96.85	96.41	96.92	97.62
apfu									
Si	0.97	0.96	0.99	1.03	1.02	1.02	1.03	1.00	0.98
Ti	0.77	0.78	0.84	0.86	0.88	0.88	0.89	0.95	0.92
Zr				0.01	0.01	0.01	0.01		
V^{4+}	0.04	0.04				0.01			
Al	0.12	0.11	0.05	0.05	0.05	0.05	0.05	0.03	0.04
Fe	0.10	0.10	0.13	0.13	0.12	0.13	0.12	0.04	0.06
Cr	0.01	0.01							
Mg	0.01	0.01							
Ca	1.00	0.99	0.99	1.00	1.00	1.00	1.00	0.98	0.99
Na									0.01
F	0.13	0.15							
end-members[‡] [%]									
Tt	76	77	83	82	83	82	84	93	90
Vma	4	4			1				
Fvu	9	7	5	5	5	4	4	3	4
Ffv	10	10	13	12	12	12	11	4	6
Ccf	1	1							
Zrt				1	1	1	1		

*Nb was measured in LdS7 but was below the detection limit. P, Sc, Zn, Ga, and Sn were measured in LdS15, but not observed.
[†]Not analyzed.
[‡]*Tt*, titanite; *Gk*, geikielite; *Vma*, vanadomalayaite; hypothetical end-members: *Fvu*, $CaAl(SiO_4)F$ "fluorovuagnatite"; *Ffv*, $CaFe(SiO_4)F$ "ferri-fluoro-vuagnatite"; *Cfv*, $CaCr(SiO_4)F$ "chromo-fluoro-vuagnatite" (the " fluoro-vuagnatite" members are proposed, assuming the presence of F, which is possible due to low totals); *Zrt*, $CaZr(SiO_4)O$ "zirconotitanite."

The composition of the vaterite is shown in Table 13.2.34. Although S-bearing, the mineral may rather be considered pure, as shown by its mean formulas: $Ca_{0.99}[(CO_3)_{0.97}(SO_4)_{0.03}]_{\Sigma1.00}$; $Vat_{96}Anh_3(Wo + Mag)_1$, with Anh, Wo, and Mag for anhydrite, wollastonite, and magnesite equivalents, respectively.

Wollastonite ($CaSiO_3$)

As another main component of the slags, wollastonite was analyzed in many samples, shown in Table 13.2.35 (samples LdS2, LdS7 and LdS12) and Table 13.2.36 (samples LdS15 and LdSV). Composition of wollastonite, although quite invariable, is presented in the form of the following mean empirical and end-member formulas:

- LdS2 (n=11): $Ca_{1.00}(Ca_{0.98}Mg_{0.01}Ti_{0.01})_{\Sigma1.00}(Si_{1.96}Fe^{3+}{}_{0.02}Al_{0.01})_{\Sigma1.99}O_{6.00}$, $Wo_{98}Di_1Cta_1$, with Cta for $CaTiAl_2O_6$ hypothetical end-member.
- LdS7 (n=13): $Ca_{1.00}(Ca_{0.98}Mg_{0.02})_{\Sigma1.00}(Si_{1.98}Fe^{3+}{}_{0.01}Al_{0.01})_{\Sigma2.00}O_{5.99}$, $Wo_{98}Di_2$.
- LdS12 (n=17): $Ca_{1.00}(Ca_{0.93}Mg_{0.04}Fe^{3+}{}_{0.03}Ti_{0.01})_{\Sigma1.01}(Si_{1.98}Al_{0.01})_{\Sigma1.99}O_{5.99}$, $Wo_{92}Di_4Hd_3Cta_1$.
- LdS15, (n=7): $Ca_{1.00}(Ca_{0.95}Mg_{0.04}Fe^{2+}{}_{0.01})_{\Sigma1.00}(Si_{1.99}Fe^{3+}{}_{0.01})_{\Sigma2.00}O_{6.00}$, $Wo_{75}Di_{12}Es_8Ctf_3Hd_1$, with Ctf for $CaTiFe_2O_6$ hypothetical end-member.
- LdSV, composition no. 1 (n=24): $Ca_{1.00}(Ca_{0.90}Mg_{0.05}Fe^{2+}{}_{0.04}Ti_{0.01})_{\Sigma1.00}(Si_{1.99}Al_{0.01})_{\Sigma2.00}O_{6.00}$, $Wo_{90}Di_5Hd_4Cta_1$.
- LdSV, composition no. 2 (n=3): $Ca_{1.00}(Ca_{0.71}Al_{0.12}Mg_{0.08}Fe^{2+}{}_{0.07}Cr_{0.01}Ti_{0.01})_{\Sigma1.00}(Si_{1.89}Al_{0.11})_{\Sigma2.00}O_{6.02}$, $Wo_{71}Ku_{13}Di_8Hd_7Cta_1$.

Table 13.2.34
Chemical analyses of vaterite from the LdS12 sample, France.

	1	2	3	4
wt.%				
SO_3	3.16	3.05	1.64	1.82
SiO_2	0.28	0.07	0.51	0.17
Al_2O_3	0.01	0.00	0.14	0.01
MgO	0.10	0.13	0.26	0.06
CaO	55.24	55.54	55.07	56.20
MnO	0.23	0.19	0.11	0.14
FeO	0.06	0.17	0.17	0.00
SrO	0.34	0.34	0.21	0.20
BaO	0.00	0.00	0.02	0.00
Na_2O	0.01	0.00	0.02	0.01
K_2O	0.02	0.07	0.08	0.07
Σ	59.45	59.57	58.22	58.68
CO_2*	41.67	42.14	42.43	42.98
apfu				
S	0.04	0.04	0.02	0.02
Si			0.01	
Mg			0.01	
Ca	0.99	0.99	0.99	1.00
C†	0.96	0.96	0.97	0.97
end-members‡ [%]				
Vat	96§	96§	96	97
Anh	4	4	2	2
Wo			1	
Mag			1	

*Calculated after other *apfu* calculations.
†Calculated as 1 – *apfu*(S) – *apfu*(Si).
‡*Vat*, vaterite; *Anh*, anhydrite equivalent; *Wo*, wollastonite equivalent; *Mag*, magnesite equivalent.
§Normalized.

Table 13.2.35
Chemical analyses of wollastonite, Lapanouse-de-Sévérac, France – part 1.

	1	2	3	4	5	6	7	8	9	10	11	12	13	14	15	16	17	18
	LdS2						LdS7						LdS12					
wt.%																		
SO_3*	0.00	0.00	0.00	0.00	0.40	0.00	n.a.†	n.a.	n.a.	n.a.	n.a.	n.a.	n.a.	n.a.	n.a.	n.a.	n.a.	n.a.
P_2O_5	0.00	0.00	0.00	0.00	0.18	0.00	n.a.	n.a.	n.a.	n.a.	n.a.	n.a.	0.00	0.00	0.00	0.00	0.00	0.00
SiO_2	50.40	50.73	50.59	50.73	49.33	50.36	50.44	51.15	50.59	50.90	50.72	50.68	50.11	50.34	50.72	50.86	50.83	51.05
TiO_2	0.51	0.42	0.41	0.48	0.39	0.49	n.a.	n.a.	n.a.	n.a.	n.a.	n.a.	0.49	0.46	0.37	0.19	0.00	0.22
Al_2O_3	0.19	0.16	0.20	0.19	0.99	0.19	0.18	0.00	0.09	0.08	0.15	0.22	0.19	0.19	0.23	0.10	0.09	0.09
Cr_2O_3	0.00	0.00	0.00	0.00	0.00	0.00	n.a.	n.a.	n.a.	n.a.	n.a.	n.a.	0.00	0.00	0.00	0.00	0.00	0.00
Fe_2O_3	0.47	0.81	0.70	0.69	0.57	0.74	0.40	0.27	0.37	0.36	0.38	0.42	1.36	1.44	1.37	1.26	1.88	1.36
MnO	0.00	0.00	0.00	0.21	0.00	0.00	0.00	0.00	0.00	0.00	0.00	0.00	0.00	0.00	0.00	0.00	0.00	0.00
MgO	0.25	0.28	0.24	0.28	0.27	0.25	0.30	0.32	0.30	0.35	0.29	0.38	1.00	0.84	0.99	0.87	0.20	0.79
CaO	47.41	47.71	47.12	47.46	46.89	47.05	47.43	47.49	47.84	47.71	47.60	47.72	45.75	46.23	45.67	46.10	46.26	46.19
Na_2O	0.00	0.00	0.00	0.00	0.00	0.00	0.00	0.00	0.00	0.00	0.00	0.00	0.00	0.00	0.00	0.00	0.00	0.00
K_2O	0.00	0.00	0.00	0.00	0.35	0.00	0.09	0.00	0.00	0.00	0.04	0.06	0.00	0.00	0.00	0.00	0.00	0.00
Σ	99.23	100.11	99.26	100.04	99.37	99.08	98.83	99.24	99.19	99.39	99.18	99.48	98.90	99.49	99.34	99.38	99.26	99.72
apfu																		
S					0.01													
P					0.01													
Si	1.97	1.96	1.97	1.96	1.92	1.97	1.97	1.99	1.97	1.98	1.98	1.97	1.96	1.96	1.98	1.98	1.99	1.98
Ti	0.01	0.01	0.01	0.01	0.01	0.01							0.01	0.01	0.01	0.01		0.01
Al	0.01	0.01	0.01	0.01	0.05	0.01	0.01				0.01	0.01	0.01	0.01	0.01			
Cr																		
Fe	0.01	0.02	0.02	0.02	0.02	0.02	0.01	0.01	0.01	0.01	0.01	0.01	0.04	0.04	0.04	0.04	0.06	0.04
Mn				0.01														
Mg	0.01	0.02	0.01	0.02	0.02	0.01	0.02	0.02	0.02	0.02	0.02	0.02	0.06	0.05	0.06	0.05	0.01	0.05
Ca	1.98	1.98	1.97	1.97	1.96	1.97	1.99	1.98	2.00	1.99	1.99	1.99	1.92	1.93	1.91	1.92	1.94	1.92
Na																		
K					0.02													
end-members‡ [%]																		
Wo	97	97	97	97	97	97	98	98	98	98	98	98	92	93	91	92	94	92
Di	1	2	1	2	2	1	2	2	2	2	2	2	6	5	6	5	1	5
Hd															2	2	3	2
Es													1	1	1		2	1
Cta	1				1								1	1	1	1		1
Cfa	1	1	1	1		1												

*Cl, Ni, and Zn were analyzed but not observed.
†Not analyzed.
‡*Wo*, wollastonite; *Di*, diopside; *Hd*, hedenbergite; *Es*, esseneite ($CaFeAlSiO_6$); hypothetical end-members: *Cta*, $CaTiAl_2O_6$; *Cfa*, $CaTiFe_2O_6$.

Table 13.2.36
Chemical analyses of wollastonite, Lapanouse-de-Sévérac, France – part 2.

	1	2	3	4	5	6	7	8	9	10	11	12	13	14	15
	LdS15						LdSV								
wt.%															
SiO_2*	50.97	50.85	50.74	50.91	51.00	51.09	50.90	50.77	49.19	49.84	50.39	48.88	45.36	46.63	49.72
TiO_2	0.16	0.16	0.17	0.16	0.16	0.16	0.21	0.18	0.20	0.22	0.19	0.20	0.93	0.00	0.17
Al_2O_3	0.18	0.00	0.00	0.00	0.00	0.00	0.00	0.00	0.23	0.23	0.10	0.00	7.66	2.65	4.70
Cr_2O_3	0.00	0.00	0.00	0.00	0.00	0.00	0.00	0.00	0.00	0.00	0.00	0.00	0.37	0.00	0.00
Fe_2O_3	0.56	0.43	0.54	0.46	0.32	0.37	1.50	1.47	1.51	1.51	1.31	1.34	4.27	1.31	1.57
MgO	0.59	0.57	0.54	0.59	0.61	0.71	0.85	0.85	0.82	0.84	0.89	0.87	2.57	0.82	0.71
CaO	46.43	46.64	46.77	46.67	46.93	46.61	45.36	44.78	43.83	44.30	45.26	43.48	35.67	41.46	42.07
K_2O	0.00	0.00	0.14	0.00	0.00	0.16	0.00	0.00	0.00	0.00	0.00	0.00	0.00	0.00	0.00
Σ	99.04	98.88	99.20	98.97	99.30	99.25	99.00	98.23	96.05	97.18	98.25	94.87	97.13	93.14	99.07
apfu															
Si	1.99	1.99	1.98	1.99	1.99	1.99	1.99	2.00	1.99	1.99	1.99	2.00	1.80	1.94	1.94
Ti							0.01	0.01	0.01	0.01	0.01	0.01	0.03		0.01
Al	0.01								0.01	0.01			0.36	0.13	0.22
Cr													0.01		
Fe	0.02	0.01	0.02	0.01	0.01	0.01	0.04	0.04	0.05	0.04	0.04	0.04	0.13	0.04	0.05
Mn															
Mg	0.03	0.03	0.03	0.03	0.04	0.04	0.05	0.05	0.05	0.05	0.05	0.05	0.15	0.05	0.04
Ca	1.94	1.96	1.96	1.96	1.96	1.95	1.90	1.89	1.90	1.90	1.91	1.90	1.52	1.84	1.76
Na															
K			0.01			0.01									
end-members† [%]															
Wo	98	98	99	98	100	99	90	89	90	90	90	90	53	84	76
Di							5	5	5	5	5	5	15	5	4
Hd	2	1		1			4	4	5	5	4	4	13	4	5
Ku													16	5	16
Cta	1						1	1	1	1	1	1	3		1
Ctf			1												

*Na, Mn, Ni, and Zn were analyzed but not observed.

†*Wo*, wollastonite; *Di*, diopside; *Hd*, hedenbergite; *Ku*, kushiroite ($CaAlAlSiO_6$); hypothetical end-members: *Cta*, $CaTiAl_2O_6$; *Ctf*, $CaTiFe_2O_6$.

The largest calculated amount of the hedenbergite molecule concerns the LdSV wollastonite. The enrichment in ferrous iron and the enrichment of the parent rock in pyrrhotite suggests that the LdSV slag formed in a reducing environment.

Ye'elimite ($Ca_4Al_6O_{12}(SO_4)$)

Ye'elimite is a quite rare mineral species in general. In the Lapanouse slags it is usually found within postcarbonate amygdules (Figures 13.2.3A and 13.2.7D). Table 13.2.37 shows the composition of ye'elimite from amygdules of the LdS15 and LdS9 rocks. The composition of these two ye'elimites is quite similar in (1) very high amount of the ye'elimite end-member and (2) S, Al, and Fe contents, but differs in terms of (1) Si enrichment and (2) addition of alkaline and alkaline earth elements in the latter one. The mean formulas are following:

- LdS15, n=9, $(Ca_{4.07}Mg_{0.01})_{\Sigma 4.08}(Al_{5.78}Fe^{3+}{}_{0.13}[]_{0.09})_{\Sigma 6.00}[(S_{1.03}Si_{0.01})_{\Sigma 1.04}O_{4.15}]O_{11.90}$, $Ye_{98}Fy_2$, with Fy denoting "ferriye'elimite" hypothetical end-member (the Fe^{3+} analogue of ye'elimite).
- LdS15, n=8, $(Ca_{4.03}Sr_{0.03}K_{0.02})_{\Sigma 4.08}(Al_{5.69}Fe^{3+}{}_{0.15}Si_{0.05})_{\Sigma 5.89}(SO_4)_{1.01}O_{12.12}$, $Ye_{95}Fy_2Csy_1Sy_1Ksy_{\sim 1}$, with Csy, Sy, and Ksy denoting hypothetical Si, Sr, and K analogues of ye'elimite.

Ye'elimite—an important cement phase—is known to be unstable in humid conditions (e.g., Cuesta et al., 2014) and it is possible that the LdS9 material is slightly hydrated, with a mean of 0.54 H_2O molecules per formula unit.

Unnamed Minerals

During Lapanouse slags analysis, a rather large number of unnamed compounds that represent potentially new minerals were found. Most of these species are listed in Table 13.2.38. Some of them were already mentioned due to their coexistence with another, approved species. Further unnamed species have their analyses reported in Table 13.2.39. Among many interesting aspects of these unique phases is the find of the phase similar or possibly identical to birunite (ideally possibly $Ca_{18}(SiO_3)_{8.5}(CO_3)_{8.5}(SO_4)\cdot 15H_2O$), which is related, for example, to spontaneous combustion of some bituminous marls (Khoury et al., 1985). New to science substances forming in zones of fossil fuel fires are said by the IMA not to be accepted as new mineral species (Nickel, 1995). On the other hand, this paper says: "if the exposure [of the already accepted species found in environments related to human activity; Ł.K.] was inadvertent (…) then such products can be accepted as minerals." Examples of approved minerals found in burning heaps are known (ammoniomagnesiovoltaite). The origin of the Lapanouse gob slags is largely, if not exclusively, self-ignition. Thus the Lapanouse unnamed species can be treated as potentially new minerals.

Table 13.2.37
Chemical analyses of ye'elimite, Lapanouse-de-Sévérac, France.

	1	2	3	4	5	6	7	8	9	10	11	12	13	14	15	16	17	18	19
	LdS15											LdS9							
wt.%																			
SO_3*	13.21	13.21	13.18	13.41	13.04	12.86	12.86	13.13	12.86	12.94	13.18	13.16	12.80	12.39	12.68	13.12	13.29	13.166	12.63
SiO_2	0.29	0.12	0.11	0.00	0.18	0.17	0.12	0.09	0.19	0.33	0.13	0.24	0.47	0.98	0.80	0.45	0.66	0.46	0.39
Al_2O_3	47.06	46.68	47.52	47.05	47.44	46.76	47.07	46.98	45.96	46.69	47.64	47.47	47.85	45.56	45.32	46.17	45.61	46.39	45.99
Fe_2O_3	1.71	1.62	1.72	1.36	1.40	1.77	1.70	1.78	1.78	1.96	1.52	2.18	2.14	1.70	2.01	1.74	1.90	1.80	1.87
MgO	0.08	0.00	0.06	0.00	0.00	0.06	0.07	0.00	0.10	0.17	0.00	0.00	0.00	0.00	0.00	0.00	0.00	0.00	0.00
CaO	36.68	36.71	36.17	36.41	36.38	36.02	36.19	36.32	36.49	36.38	36.44	36.70	36.04	36.52	35.39	35.78	35.97	36.25	35.55
SrO	n.a.†	n.a.	n.a.	n.a.	n.a.	n.a.	n.a.	n.a.	n.a.	n.a.	n.a.	0.60	0.53	0.55	0.60	0.63	0.57	0.49	0.52
BaO	n.a.	n.a.	n.a.	n.a.	n.a.	n.a.	n.a.	n.a.	n.a.	n.a.	n.a.	0.15	0.00	0.00	0.18	0.20	0.18	0.00	0.00
K_2O												0.13	0.13	0.18	0.20	0.20	0.26	0.21	0.16
Σ	99.03	98.34	98.76	98.22	98.44	97.63	98.01	98.29	97.38	98.48	98.90	100.63	99.96	97.88	97.16	98.28	98.43	98.76	97.10
apfu																			
S	1.03	1.04	1.03	1.05	1.02	1.01	1.01	1.03	1.02	1.01	1.03	1.01	0.99	0.97	1.01	1.03	1.04	1.03	1.00
Si	0.03	0.01	0.01	0.00	0.02	0.02	0.01	0.01	0.02	0.03	0.01	0.02	0.05	0.10	0.08	0.05	0.07	0.05	0.04
Al	5.75	5.75	5.82	5.81	5.82	5.78	5.79	5.79	5.71	5.73	5.82	5.72	5.79	5.63	5.66	5.70	5.63	5.69	5.73
Fe	0.13	0.13	0.13	0.11	0.11	0.14	0.13	0.14	0.14	0.15	0.12	0.17	0.17	0.13	0.16	0.14	0.15	0.14	0.15
Mg	0.01		0.01			0.01	0.01		0.02	0.03									
Ca	4.07	4.11	4.03	4.09	4.06	4.05	4.05	4.07	4.12	4.06	4.05	4.02	3.96	4.10	4.02	4.01	4.03	4.04	4.03
Sr												0.04	0.03	0.03	0.04	0.04	0.03	0.03	0.03
Ba												0.01			0.01	0.01	0.01		
K												0.02	0.02	0.02	0.03	0.03	0.03	0.03	0.02
end-members‡ [%]																			
Ye	97	98	97.8	98	98	97	97	98	97	96	98	95	95	95	94	95	94	95	96
Fy	2	2	2	2	2	2	2	2	2	3	2	3	3	2	3	2	3	2	2
My	0.3		0.2			0.2	0.3		0.4	0.6									
Sy												1	1	1	1	1	1	1	1
Ksy												0.4	0.4	0.6	0.7	0.6	0.8	0.7	0.5
Csy												0.4	1	2	1	1	1	1	1

*P, Na, Cl, Ti, V, Cr, Mn, Co, Ni, Cu, Zn, Ga, Zr, and Pb were below the detection limits.
†Not analyzed.
‡*Ye*, ye'elimite; *Fy*, $Ca_4Fe_6(SO_4)O_{12}$ "ferriye'elimite"; *My*, $Mg_4Al_6(SO_4)O_{12}$ "magnesioye'elimite"; *Sy*, $Sr_4Al_6(SO_4)O_{12}$ "strontioye'elimite"; *Ksy*, $K_4Si_6(SO_4)O_{12}$ "K-silicoye'elimite"; *Csy*, $Ca_4(Si_{4.5}\square_{1.5})(SO_4)O_{12}$.

Table 13.2.38
Potential newly discovered minerals identified from Lapanouse Slags, France.*

Phase Name	Sample	Ideal crystallochemical formula
V analogue of fluorapatite	LdS7	$Ca_5(VO_4)_3F$
Ca–Al phosphate silicate fluoride	LdS12	$Ca_4Al(PO_4,CO_3)_2(SiO_4)$ (F,OH)
"Magnesiograndiferrite"	LdS2	$Mg(Fe,Al,Cr,Mn^{3+})_4O_7$
A	LdS2, LdS7	$Ca_5Mg(Al,Fe^{3+})Si_5(O,OH)_{17}(OH)$
B	LdS7	$Ca_3(Mg,K,Fe)$ $(Al,Fe,Ti)_4Si_4(O,OH)_{18}(OH)$
C	LdS12	$Ca_5Mg(Fe^{3+},Al)Si_5(O,OH)_{17}(OH)$
Ferri-gehlenite	LdS6	$Ca_2Fe^{3+}(Si,Al)O_7$
D	LdS5	$Ca_2(Ti,Nb)O_4$
"Ba-petalite"	LdS1	$BaAl(Si_3,Al)O_{10}$
"K-petalite"	LdS2	$KAlSi_4O_{10}$
Mg–Ti-[] analogue of batiferrite	LdS5	$(Mg,Mn)Ti_2\{(Ti,Fe)[]\}Fe^{3+}{}_8O_{19}$
Pyrrhotite-like phase	LdSV	$Fe_{10}S_9$
E	LdS2	$(Mg,Mn,Zn)_2(Fe,Al,Mn)_4Cr_2O_{11}$
F	LdS2	$(Mg,Mn,Zn)_2(Fe,Al,Mn)_4Fe_2O_{11}$
"Ferro-ferri-shulamitite"	LdSV	$Fe^{2+}{}_3TiFe^{3+}Fe^{3+}O_8$
G	LdS12	$Ca_2(S,Si)_2O_7$

*Arranged in order of their occurrence in the chapter; see the text for a detailed description.

Table 13.2.39

Chemical analyses of the unnamed phases from Lapanouse, France: phase A (analyses 1–2 and 10–15), phase B (analysis 3), phase C (analysis 4), Mg–Ti analogue of batiferrite (analysis 5), phase D (analysis 6), "ferro-ferri-shulamitite" (analyses 7–9), unnamed phases E, F, and "magnesiograndiferrite" (analyses 16, 17, and 18, respectively), phase G (analysis 19), and birunite-like phase (analysis 20).

	1	2	3	4	5	6	7	8	9	10	11	12	13	14	15	16	17	18	19	20
	LdS7			LdS12	LdS5		LdSV			LdS2									LdS12	LdS6
wt.%																				
SO_3	0.00*	0.00	0.00																31.92	2.51
Nb_2O_5					0.00	4.35	0.00	0.00	0.00											
P_2O_5																				0.60
SiO_2	35.26	38.96	33.11	38.36	0.10	0.00	0.00	0.00	0.00	31.12	31.14	30.96	31.17	30.19	29.55	0.00	0.07	0.07	14.01	22.38
TiO_2	0.00	0.00	1.64		23.95	54.08	18.80	20.08	18.63	0.00	0.00	0.00	0.00	0.00	0.00	0.14	0.10	0.00		0.00
Al_2O_3	12.46	7.05	16.34	5.36	0.22	0.07	4.68	4.60	4.46	18.86	18.90	19.49	19.34	20.21	21.65	6.13	5.33	4.23	1.02	0.15
V_2O_3	0.00	0.00	0.00		0.00	0.00	0.00	0.34	0.45							0.00	0.00	0.00		0.00
Cr_2O_3	0.00	0.00	0.00		0.00	0.00	0.00	0.00	0.00	0.00	0.00	0.00	0.00	0.00	0.00	26.56	0.04	4.05		0.00
Fe_2O_3				12.97			80.03	78.92	79.95	5.93	4.91	4.66	5.03	4.95	4.70	48.36	75.76	76.66		0.00
MnO	0.00	0.00	0.00		0.86	0.00	0.28	0.27	0.27	0.00	0.00	0.00	0.00	0.00	0.00	2.89	2.13	2.96		0.00
MgO	5.82	7.99	4.53	3.89	3.41	0.00	0.43	0.47	0.36	3.37	4.02	3.93	3.85	3.52	3.10	13.17	12.80	9.99	1.06	0.35
CaO	38.27	38.65	23.43	36.78	0.00	36.95	0.00	0.00	0.00	38.08	38.56	38.80	38.58	38.65	38.79	0.13	0.12	0.11	39.56	42.80
FeO	4.74	4.68	17.68																1.52	
CoO					0.00		0.00	0.00	0.00							0.18	0.16	0.21		
NiO					0.00	0.00	0.00	0.00	0.00	0.00	0.00	0.00	0.00	0.00	0.00	0.17	0.63	0.50		
ZnO					0.00		0.00	0.00	0.00							0.93	0.57	0.88		
SrO	0.50	0.49	0.00																	0.00
K_2O	0.58	0.49	0.78	0.44						0.52	0.45	0.48	0.49	0.43	0.49				0.00	0.90
Na_2O	0.46	0.31	0.46	0.64		1.06				0.49	0.46	0.42	0.48	0.44	0.40				0.00	0.15
Cl	0.00	0.00	0.00																0.07	0.00
Σ	98.10	98.62	97.50	100.37		97.47	104.23	104.68	104.12	98.37	98.44	98.74	98.94	98.39	98.68	98.66	97.71	99.66	89.26	69.82
$H_2O^{\dagger}$	1.38	0.86	1.78	1.92						1.63	1.56	1.26	1.06	1.61	1.32				10.57	14.00‡
CO_2																				16.17§
apfu																				
S																			1.13	0.71

Continued

Table 13.2.39

Chemical analyses of the unnamed phases from Lapanouse, France: phase A (analyses 1–2 and 10–15), phase B (analysis 3), phase C (analysis 4), Mg–Ti analogue of batiferrite (analysis 5), phase D (analysis 6), "ferro-ferri-shulamitite" (analyses 7–9), unnamed phases E, F, and "magnesiograndiferrite" (analyses 16, 17, and 18, respectively), phase G (analysis 19), and birunite-like phase (analysis 20)—cont'd.

	1	2	3	4	5	6	7	8	9	10	11	12	13	14	15	16	17	18	19	20
Nb						0.05														
P																				0.19
Si	4.01	4.41	3.92	4.98	0.02	0.00				3.55	3.53	3.50	3.52	3.42	3.34		0.01		0.66	8.46
Ti	0.00	0.00	0.15		2.86	0.98	1.05	1.12	1.04							0.01	0.01			
Al	1.67	0.94	2.28	0.41	0.04		0.41	0.40	0.39	2.53	2.52	2.59	2.57	2.70	2.88	0.66	0.59	0.29	0.06	0.06
V								0.02	0.03											
Cr																1.91		0.19		
Mg	0.99	1.35	0.80	0.75	0.81		0.05	0.05	0.04	0.57	0.68	0.66	0.65	0.60	0.52	1.79	1.78	0.88	0.07	0.19
Ca	4.66	4.69	2.98	5.11		1.91				4.65	4.68	4.69	4.66	4.70	4.70	0.01	0.01	0.01	2.01	17.34
Mn					0.12		0.02	0.02	0.02							0.22	0.17	0.15		
Fe	0.45	0.44	1.75	0.63	8.15	0.02	4.47	4.39	4.48	0.51	0.42	0.40	0.43	0.42	0.40	3.31	5.33	3.41	0.06	
Co																0.01	0.01	0.01		
Ni																0.01	0.05	0.02		
Zn																0.06	0.04	0.04		
Sr	0.03	0.03																		
Na	0.10	0.07		0.08		0.05				0.11	0.10	0.09	0.10	0.10	0.09					0.11
K	0.08	0.07	0.12	0.04						0.08	0.07	0.07	0.07	0.06	0.07					0.43
Cl																			0.01	
OH	1.04	0.65	1.40	0.83						1.24	1.18	0.95	0.79	1.22	1.00					
H_2O																			1.67	
C																				

*Empty fields denote that the component was not analyzed; in some cases, it was analyzed but not detected.
†After recalculating FeO to Fe_2O_3.
‡Calculated after recasting the analysis to other elements apfu.
§Calculated by difference after the H_2O calculation.

Photo by Łukasz Kruszewski, 2015.

13.3 Nucleation Temperatures

Lapanouse slag sample LdS15, predominantly clinopyroxene, wollastonite, and leucite. The yellow coating contains anorthite, feldspars, and garnets. The bottom of the sample is an amygdule, with a white gypsum- and anhydrite-rich zone and an apatite-group-rich blue zone. The amygdule is a decarbonized remnant of a carbonate concretion subjected to pyrometamorphism. The ruler scale is given in centimeters.

Discussion

Unique composition of the minerals constituting the Lapanouse slag, outstanding from the norm, suggests a large bias during their formation. This leads to (1) untypical site occupancies with elements entering sites usually not preferred by them (e.g., important Ca admixture in feldspars; Fe^{3+} in the tetrahedral sites of clinopyroxenes); (2) domination or a large part of untypical, rare, or hypothetical end-members (e.g., ferri-gehlenite in the composition of melilites); (3) untypical element ratios (e.g., Si:Al in alkali feldspars being in the 2.4–2.73 range instead of being equal to 3); and (4) formation of potentially new mineral species. Substitution (or dominance) of some end-members (kushiroite, tissintite, yoshiokaite) relates the oil-shale fires to extraterrestrial environments.

The discovery of some minerals, such as depmeierite, kottenheimite, and khesinite, places Lapanouse as the second world occurrence of these rarities and likely the fourth (or third) occurrence of giuseppettite. Formation temperatures of the Lapanouse slags, assuming normal (atmospheric) pressure, may be obtained based on the following information: (1) pseudowollastonite-wollastonite coexistence (LdS6) is possible in ca. 1150–1290°C range (Seryotkin et al., 2012); (2) larnite-gehlenite coexistence may suggest a 1400–1500°C range (Sokol et al., 2010) but in the case of the LdS9 slag gehlenite is a matrix component while larnite resides in voids; (3) pyrrhotite is stable up to ca. 1240°C and so is true for nepheline, both existing in the LdSV sample (Bale et al., 2016); (4) pseudobrookite is stable from ca. 1150°C (however, the LdS5 pseudobrookite exists as a solid solution); and (5) grandiferrite coexists with srebrodolskite in a narrow thermal range around 1200°C, information applicable to the LdS15 slag. Thus the Lapanouse slags may have formed in the mean 1150–1250°C range. This and other petrological issues, however, will be a matter of a further study.

Acknowledgments

This work was financially supported by the Ministry of Science and Higher Education for the statutory objectives of the Institute of Geological Sciences PAS (Polish Academy of Sciences) for the years 2015 and 2016. The corresponding author would like to thank M.Sc. student Lidia Jeżak of the Inter-Institute Laboratory of the Microanalysis of Minerals and Synthetic Materials (Faculty of Geology, University of Warsaw) for assistance with microprobe analyses.

References

Anthony, J.W., Bideaux, R.A., Bladh, K.W., Nichols, M.C., 1997. Handbook of Mineralogy, v. III: Halides, Hydroxides, Oxides. Mineralogical Society of America, Chantilly, VA, USA. 680 p.

Bale, C.W., Bélisle, E., Chartrand, P., Decterov, S.A., Eriksson, G., Gheribi, A.E., Hack, K., Jung, I.-H., Kang, Y.-B., Melançon, J., Pelton, A.D., Petersen, S., Robelin, C., Sangster, J., Spencer, P., Van Ende, M.-A., 2016. FactSage thermochemical software and databases, 2010–2016. In: CALPHAD: Computer Coupling of Phase Diagrams and Thermochemistry, vol. 54, pp. 35–53.

Bindi, L., Bonazzi, P., Dei, L., Zoppi, A., 2005. Does bazhenovite structure really contain hiosulphate group? A structural and spectroscopic study of a sample from type-locality. Am. Mineral. 90, 1556–1562.
Ciesielczuk, J., Kruszewski, Ł., Majka, J., 2015. Comparative mineralogical study of thermally-altered coal-dump waste, natural rocks and the products of laboratory heating experiments. Int. J. Coal Geol. 139, 114–141.
Cuesta, A., Alvarez-Pinazo, G., Sanfélix, S.G., Peral, I., Aranda, M.A.G., De la Torre, A.G., 2014. Hydration mechanisms of two polymorphs of synthetic ye'elimite. Cem. Concr. Res. 63, 127–136.
Dong, Z.L., White, T.J., Sun, K., Wang, L.M., Ewing, R.C., 2005. Electron irradiation induced transformation of $(Pb_5Ca_5)(VO_4)_6F_2$ apatite to $CaVO_3$ perovskite. J. Am. Ceram. Soc. 88 (1), 184–190.
Eytier, J.-R., Eytier, C., Favreau, G., Devouard, B., Vigier, J., 2004. Minéraux de pyrométamorphisme de Lapanouse-de-Sévérac (Aveyron). Le Cahier des Micromonteurs 85 (3), 3–58.
Fernández-Díaz, L., Fernández-González, Á., Prieto, M., 2010. The role of sulfate groups in controlling $CaCO_3$ polymorphism. Geochim. Cosmochim. Acta 74, 6064–6076.
Galuskina, I.O., Galuskin, E.V., Pakhomova, A.S., Widmer, R., Armbruster, T., Krüger, B., Grew, E.S., Vapnik, Y., Dzieżanowski, P., Murashko, M., 2017. Khesinite, $Ca_4Mg_2Fe^{3+}{}_{10}O_4[(Fe^{3+}{}_{10}Si_2)O_{36}]$, a new rhönite-group (sapphirine supergroup) mineral from the Negev Desert, Israel – natural analogue of the SFCA phase. Eur. J. Mineral. 29 (4), 101–116.
Gatel, P., Žáček, V., Kruszewski, Ł., Devouard, B., Thiéry, V., Eytier, C., Eytier, J.-R., Favreau, G., Vigier, J., Stracher, G.B., 2015. Combustion mineralogy and petrology of oil-shale slags in Lapanouse, Sévérac-le-Château, Aveyron, France: snalogies with hydrocarbon fires. In: Stracher, G.B., Prakash, A., Sokol, E.V. (Eds.), Coal and Peat Fires: A Global Perspective, Case Studies – Coal Fires, vol. 3, pp. 681–742.
Khoury, H.N., Salameh, E., Abdul-Jaber, Q., 1985. Characteristics of an unusual highly alkaline water from the Maqarin area, northern Jordan. J. Hydrol. 81 (1–2), 79–91.
Kimura, M., Mikouchi, T., Suzuki, A., Miyahara, M., Ohtani, E., El Goresy, A., 2009. Kushiroite, CaAlAlSiO6: a new mineral of the pyroxene group from the ALH 85085 CH chondrite, and its genetic significance in refractory inclusions. Am. Mineral. 94, 1479–1482.
Kruszewski, Ł., Ciesielczuk, J., Misz-Kennan, M., 2012. What have meteorites to do with coal fires? A case of Upper and Lower Silesian Basins. Mineralogia 40, 28–30 Special Papers.
Lufkin, J.L., 1976. Oxide minerals in miarolitic rhyolite, Black Range, New Mexico. Am. Mineral. 61, 425–430.
Ma, C., Krot, A.N., 2014. Hutcheonite, $Ca_3Ti_2(SiAl_2)O_{12}$, a new garnet mineral from the Allende meteorite: an alteration phase in a Ca-Al-rich inclusion. Am. Mineral. 99, 667–670.
Ma, C., Tschauner, O., Beckett, J.R., Liu, Y., Rossman, G.R., Zhuravlev, K., Prakapenka, V., Dera, P., Taylor, L.A., 2015. Tissintite, $(Ca,Na,\square)AlSi_2O_6$, a highly-defective, shock-induced, high-pressure clinopyroxene in the Tissint martian meteorite. Earth Planet Sci Lett 422, 194–205.
McConnell, D., 1938. A structural investigation of the isomorphism of the apatite group. Am. Mineral. 23 (1), 1–19.
Nickel, E.H., 1995. The definition of mineral. Can. Mineral. 33, 689–690.
Oberti, R., Smith, D.C., Rossi, G., Caucia, F., 1991. The crystal-chemistry of high- aluminium titanites. Eur. J. Mineral. 3, 777–792.
Reinsch, A., 2009. Effects of Al/Si ratio and charge-balancing cation radius on the viscosity of polymerized aluminosilicate melts. In: 2009 Summer Undergraduate Research and Creative Achievements Forum, University of Missouri.
Rietveld, H.M., 1967. Line profiles of neutron powder-diffraction peaks for structure refinements. Acta Crystallogr. 22, 151–152.
Sapozhnikov, A.N., Kaneva, E.P., Suvorova, L.P., Levitsky, V.I., Ivanova, L.A., 2017. Sulfhydrylbystrite, $Na_5K_2Ca(Al_6Si_6O_{24})(S_5)(SH)$, a new mineral with the LOS framework, and re-interpretation of bystrite: cancrinite-group minerals with novel extra-framework anions. Mineral. Mag. 81, 383–402.
Seryotkin, Y.V., Sokol, E.V., Kokh, S.N., 2012. Natural pseudowollastonite: crystal structure, associated minerals, and geological context. Lithos 134–135, 75–90.
Sokol, E.V., Maksimova, N.V., Nigmatulina, E.N., Sharygin, V.V., Kalugin, V.M., 2005. Combustion Metamorphism. Publishing House of the SB RAS, Novosibirsk (in Russian).
Sokol, E., Novikov, I., Zateeva, S., Vapnik, Y., Shagam, R., Kozmenko, O., 2010. Combustion metamorphism in the Nabi Musa dome: new implications for a mud volcanic origin of the Mottled Zone, Dead Sea area. Basin Res. 22, 414–438.
Vaniman, D.T., Bish, D.L., 1990. Yoshiokaite, a new Ca,Al-silicate mineral from the Moon. Am. Mineral. 75, 676–686.

CHAPTER 14

The Burning Coal Heap at La Ricamarie, Loire Coal Basin, France

The La Ricamarie coal heap overlooking the commune for which it is named. The heap is the red-colored and grassy area surrounded by trees, near the center of the photo. The heap began burning in the 1940s and is still active today. According to the present quarry owners, some attempts were made to extinguish the fire using fly ash. The waste rock in the heap came from a nearby shaft named the *Puits Pigeot*. It closed in 1983 and is located to the left of the picture. *Photo by Vincent Thiéry, 2018.*

CHAPTER CONTENTS

Coal and Peat Fires: A Global Perspective
Edited by Glenn B. Stracher

https://doi.org/10.1016/B978-0-12-849885-9.00014-7

Photo by Vincent Thiéry, 2014.

14.1 The Burning Heap at La Ricamarie

Vincent Thiéry
Bernard Guy
Łukasz Kruszewski
Jean-François Carpentier

Bernard Guy examining paralava and clinker inside the quarry. The quarry opened in the 1980s in order to exploit burnt colliery spoil and various wastes as aggregate. The quarry is currently still exploited.

Introduction

The burning coal heap at La Ricamarie, so-called *Terril Saint-Pierre* (Saint-Pierre heap) or *Crassier Saint-Pierre* (Saint-Pierre dump), is located within the Loire coal basin near Saint-Étienne, toward the center/southeastern part of France (Figure 14.1.1). The Loire coal basin is the type locality for the Stephanian age layers (now discarded). Its geology is well known and has been described since the pioneering works of the French geologist Emmanuel-Louis Grüner (1809–83) who published in 1882 a very detailed synthesis of the basin (Grüner, 1882a, 1882b).

Located in the southern flank of a hillside, at an altitude of around 600 m above sea level, this huge, impressive heap, with its red color and its fumaroles, is an integral part of the landscape (see the picture on the chapter's front page).

The rocks are altered sandstones, conglomerates, and mudstones/shales (locally improperly called schists). In places where coal is present, it is found more often in close association with shales. The thickness of the beds may vary from centimeter to meter scale (Figure 14.1.2). Sandstones are poorly sorted and, in addition to quartz grains, still contain feldspar and mica; all minerals originate from the eroded granites cropping out in the mountains surrounding the coal basin. The grain size may reach centimeter scale and some remnants of granite rocks may also be found, and more rarely pieces of rocks of volcanic origin. The sandstones show channels typical of a fluvial origin. The sandstone has been quarried in multiple localities near Saint-Etienne as a raw material for building houses. The shales are composed of clay minerals, together with quartz and micas (postulated of same origin as in the sandstone beds). The argillaceous rocks may be black, depending on the abundance of organics, including coal matter. More rarely, some cinerite beds of volcanic origin may be found in place of the shales in the shale/sandstone alternations.

Stephanian coal basins are present in various places in France; it has been demonstrated that their content has been deposited at high altitude (at least 4500 m above sea level), mountainous conditions (Becq-Giraudon et al., 1996). Presently, the Loire coal basin in the La Ricamarie vicinity crops out in a valley at 500–550 m above sea level and is surrounded by small mountains reaching 800–900 m.

Historical Coal Fires in the Stephanian Coal Seams

Worth mentioning are old fires known to occur in the coal seams of the Loire coal basin. They have strongly impressed the inhabitants, and they have been known for at least 200 years as being active for "at least 300 years" (de Bournon, 1785). It is of interest to mention that some coal seams have been named *brûlante* (burning) (Grüner, 1882a,b). There is still a place in the Ricamarie city called *le brûlé* (the burnt). A description by the French Geologist Amédée Burat (1851) mentions that around 1800, the coal seams from La Ricamarie were known only because they were burning, forming locally phenomena like solfatares.

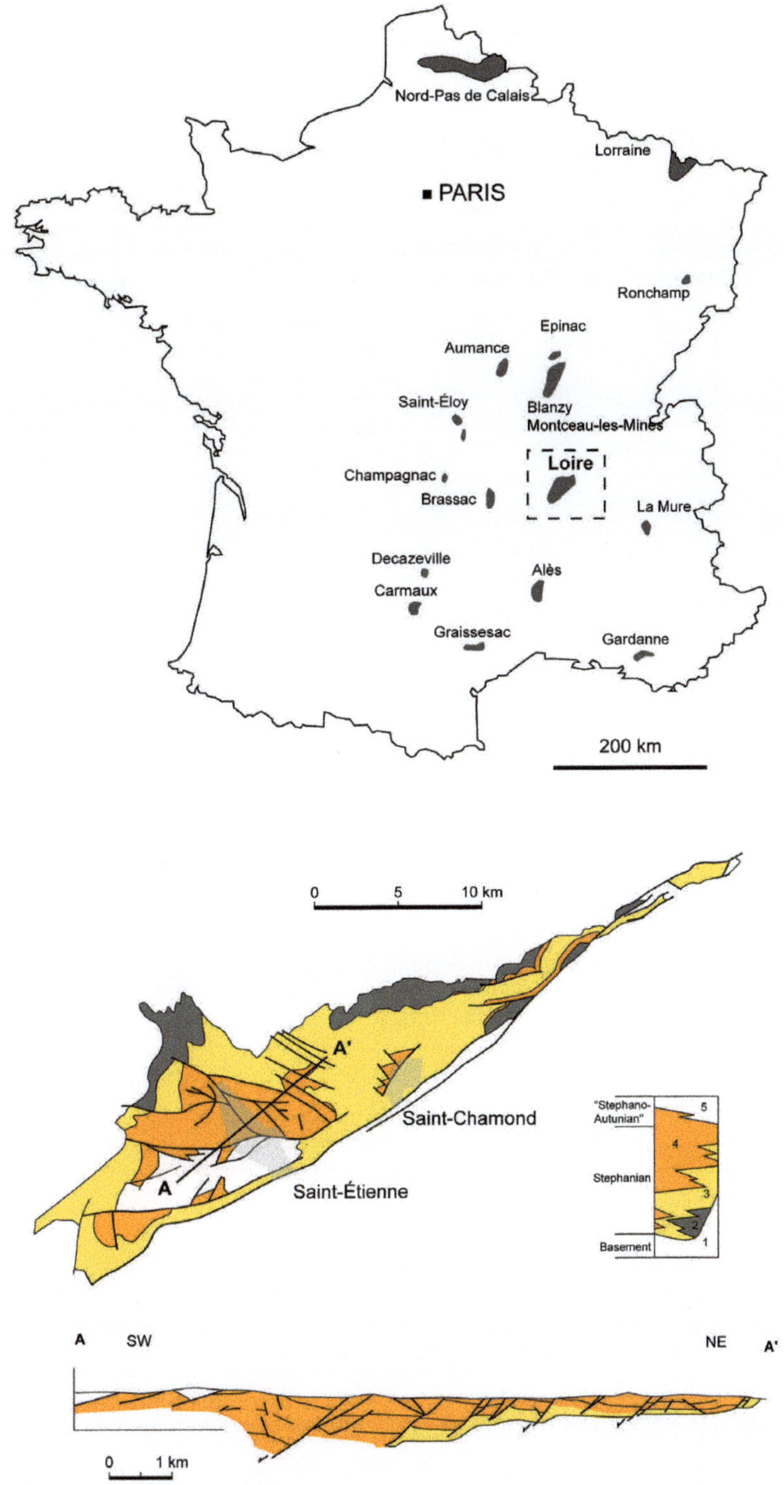

Figure 14.1.1. The Saint-Étienne coal basin. 1: metamorphic basement; 2: proximal breccias; 3: alluvial fan conglomerates and sandstones; 4: fluvio-palustrine sandstones, pelites, and coals; 5: alluvial plain conglomerates and sandstones. Coarse lines: faults. From Becq-Giraudon et al. (1995) with modifications by Vincent Thiéry, 2017.

The famous paralavas described worldwide associated with various coal fires (Grapes et al., 2009; Sharygin et al., 2014) have been termed likewise because of their similarities with natural lavas and are an analogue. In the Loire coal basin, such analogies have been proposed since the first half of the 19th century (Girardin, 1831, p. 89) in the following terms (direct translation from French): "Surrounding rocks undergo modifications because of the continuous action of heat: some of them, mainly clays, undergo half-vitrification, become hard to the point they sparkle when struck by steel, take the aspect of enamel, with various colors; in one word, they are turned as porcelain-jasper

Figure 14.1.2. Argillites and sandstones enclosing coal beds in the Stephanian sediments at Saint-Etienne (Firminy, Loire). Photo by Bernard Guy, 2012.

Table 14.1.1
Historical descriptions of minerals associated with coal seams in the Loire basin.

Mineral	Stoichiometric formula	Reference
Native bismuth	Bi	Daubrée (1887), §40, p. 134
Epsomite	$MgSO_4 \cdot 7H_2O$	Landrin (1856) p. XLI
Salamonniac	NH_4Cl	Drian (1849) p. 73 as ammonium chlorhydrate; Damour (1884) also mentions the presence of I and Br in the analysis
Vivianite	$Fe_3^{2+}(PO_4)_2 \cdot 8H_2O$	Drian (1849) as phosphated iron

or porcellanites; others are totally vitrified, even the most refractory ones such as sandstones, some of them very fusible, schistous clays being mainly in this case, take the aspect of scoria almost similar to volcanic ones; in most cases, terrigenous rocks associated with coal seams are only slightly calcined and look like man-made bricks and tiles."

"Schistous clays ….. take the aspect of scoria almost similar to volcanic ones."

Porcellanites were also described early on in the study of this basin (Jannettaz, 1874), but the most remarkable description of the effects of coal fires on associated rocks is due to the famous French mineralogist Jacques Louis de Bournon (1751–1825, in honor of whom the mineral bournonite was named). In a seminal book (de Bournon, 1785) he describes thoroughly, although with an ancient vocabulary, a huge variety of baked shales, sandstones, and argillites.

Various minerals have also been described historically in association with the burning of coal seams, either by their grandfathered name according to the international mineralogical association terminology, or by their chemical composition (Table 14.1.1).

Photo by Vincent Thiéry, 2015.

14.2 The La Ricamarie Coal Heap

Fairy chimney (hoodoo) formed during the exploitation of burnt-colliery spoil in the La Ricamarie coal heap. The chimney is about 50 cm high.

From the Mine to the Heap

As coal crops out in the Loire basin, its presence was probably noticed since time immemorial. Its exploitation is attested since at least the 13th century (Perrin, 1930). The mining activity ceased in 1983, but some remnants are still present such as coal heaps and mine shafts (one of them being a museum).

The Ricamarie heap is well documented thanks to touristic postcards of the 1980s available from the area, showing how it looked prior to its exploitation (Figures 14.2.1 to 14.2.3).

However, the present study documents a phenomenon that can be seen as recent in the history of the mining of the basin. It concerns a coal heap whose edification had begun in the 1940s. The waste rocks coming along with coal consisted mostly of sandstone, with silica-clastic sediments, carbonaceous shales, and conglomerates (see Section 14.1 concerning rocks accompanying coal seams).

Figure 14.2.1. Postcard from the mid-1980s showing the La Ricamarie coal heap as seen from the southeast. Personal collection of Vincent Thiéry.

Figure 14.2.2. Postcard from the mid-1980s showing part of the La Ricamarie coal heap as seen from the south. Personal collection of Vincent Thiéry.

Figure 14.2.3. Photographs from the final period of the exploitation (1983) of the Pigeot shaft, the Saint-Pierre heap at La Ricamarie being in the background. Documents stored at the Archives Nationales du Monde du Travail, Roubaix, France, under the reference 2007 009 0291.

A survey based on aerial pictures from the French geographical institute (Institut Geographies National) allows a close monitoring of the successive steps of edification of the heap (Figure 14.2.4). In 1942, small (no more than 80 m in diameter) heaps were present very close to buildings in a wide, open-space area corresponding to a former quarry. From 1942 to 67, screes marking the cutting face of the former quarry can be seen at the northern boundary between the fields and the future place of the heap. From 1967 to 89, although dumping on the heap continued, its shape (Figure 14.2.1) took form. It looked from the above like an elongated peer of ca. 600 m long by 300 m wide.

During the 1990s, the quarrying operations have little by little revealed the interior of the heap. The exploitation has begun on the southwestern side. Nowadays, the entire heap is affected by quarrying operations and it is difficult to imagine its original shape (compare Figure 14.2.1 and the picture on the opening page of this chapter).

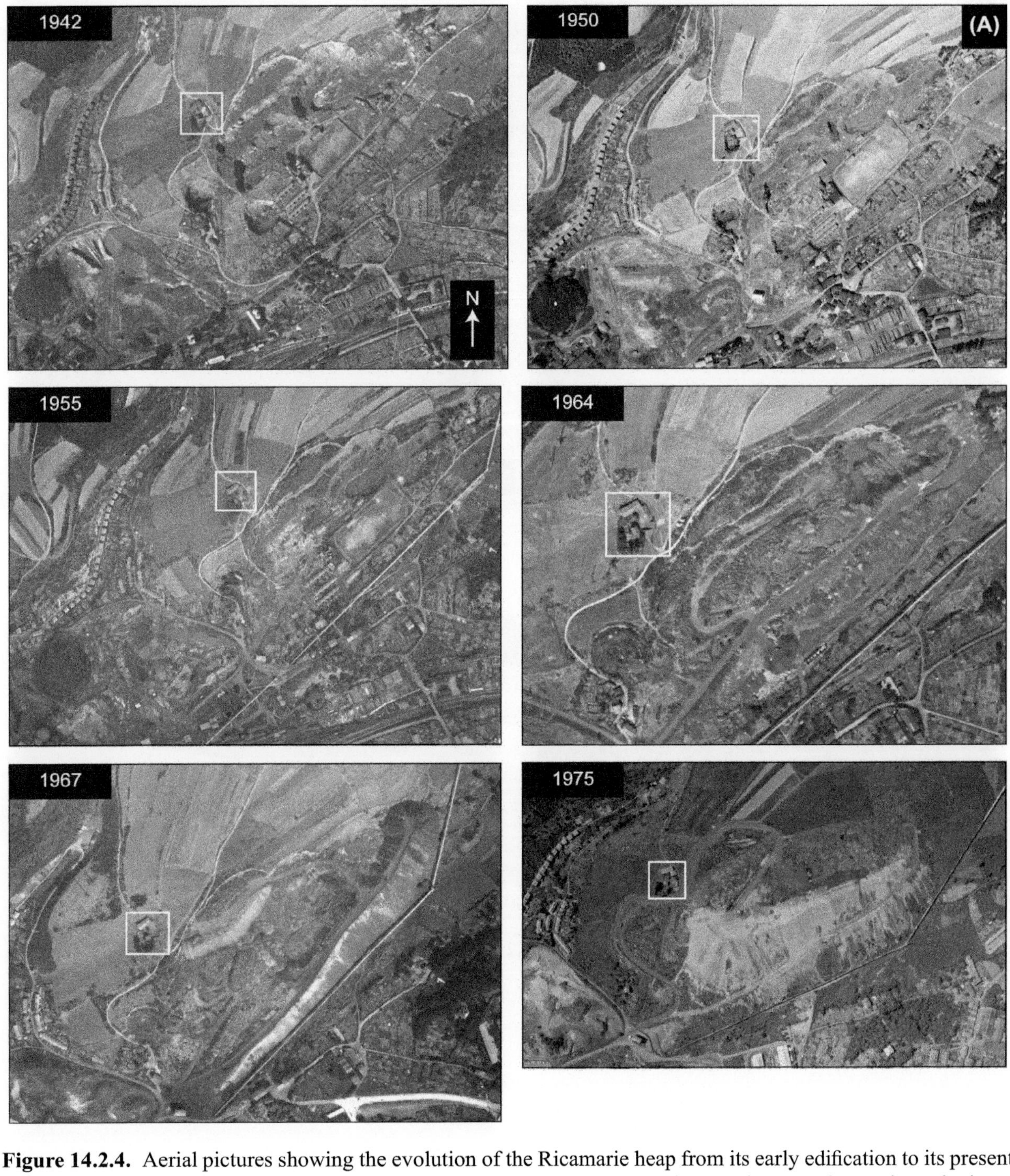

Figure 14.2.4. Aerial pictures showing the evolution of the Ricamarie heap from its early edification to its present state. Since scale varies among the photos, the white square (ca. 80 × 80 m) locates the same feature in each photo. Illustrations by Vincent Thiéry, 2018, based on pictures available from the French geographical service (IGN).

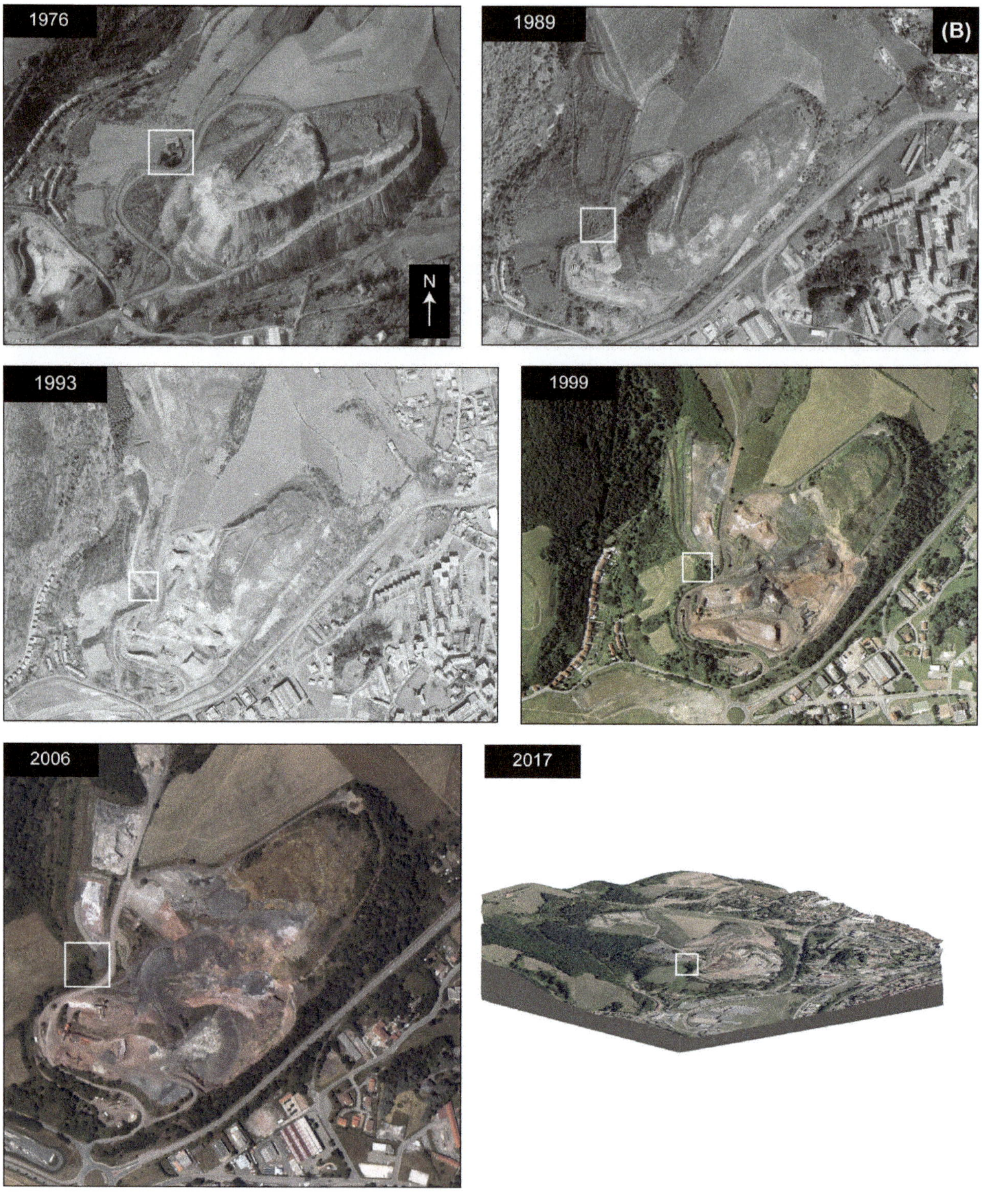

Figure 14.2.4, cont'd.

Photo by Vincent Thiéry, 2014.

14.3 The Inner Coal Heap: Thermal Transformations and Geothermometry

Layering from successive deposits of coal waste, visible in a quarry cut in the La Ricamarie heap. The quarry is still active in most parts of the heap. The horizontal field of view in the foreground is 15 m.

Petrography of Clinker and Paralava

The quarrying operations allow for a direct observation of the variety of rocks contained inside the heap. Initially, various rocks coming along with coal (sandstone, limestone, shales, etc.) were dumped, as usually is done in such contexts (Skarżyńska, 1995). Various wastes, including demolition and metallic wastes, and fine particles resulting from the cleaning of coal as well as fly ashes are also present.

The successive layers resulting from the dumping of waste materials are revealed by the quarrying operations (see the picture at the beginning of this section). They display a wide range of colors due both to the initial diversity of materials as stated earlier but also because of the subsequent evolution of the heap. The complex high temperature mineralogical evolution of the materials results in their annealing or melting, and thus they display strong mechanical properties: cemented within each other, the blocks can hold vertically when quarried.

Clinker

Clinker from the Ricamarie heap are versicolored products, having kept a layered texture, displaying white, grey, orange, red colors (Figure 14.3.1). Under the optical microscope, it is difficult to identify minerals in the layered parts, with the exception of quartz and mica flakes (Figure 14.3.2). Elongated pores, 50–100 μm long, are parallel to the bedding. Scanning electron microscope-back scattered electron (SEM-BSE) images show poor contrast except for dense phases. At higher magnifications (×2000), the complexity of the microtexture can be deciphered; small cordierite crystals (less than 10 μm long, 3–4 μm wide) are present in a complex matrix revealing the abundance of needles (Figure 14.3.3).

Locally, such clinker contains dark porous areas (Figures 14.3.1 to 14.3.10) that are very similar to the gas channels described in Sokol (2005) from spoil heaps of the Chelyabinsk basin. Under the optical microscope, an increase in grain size can be easily seen, and the texture is typical of a crystallized melt. The mineralogy of such zones is indeed very close to the descriptions made by Sokol (2005): euhedral cordierite crystals, 10×25 μm, coexist with Fe–Mg anhedral olivine and pyrrhotite. Optical determinations were confirmed by energy dispersive spectroscopy (EDS) analysis.

A SEM-BSE image (Figure 14.3.6), coupled with EDS maps of a gas channel in contact with clinker (Figure 14.3.7), shows the changes in the shape of porosities (rounded in gas channels but elongated in the clinker; see earlier). The chemical changes highlight the differences of mineralogy between the two zones: Mg, Fe, and Na to a lesser extent gradually increase toward the gas channel whereas Al is much more present in the clinker zone. It is worth mentioning the existence of an intermediate zone between the clinker and the completely melted gas channel, characterized by a smaller grain size and a different chemistry (Figure 14.3.7).

Paralava

Glassy, slag-like paralavas were described by Thiéry and Guy (2015). They are characterized by a ropy aspect and are highly vacuolar (Figures 14.3.11 to 14.3.13); pores up to 1 cm across, but generally in the 200–500 μm range,

Figure 14.3.1. Versicolored clinker sample (polished surface). Dark, porous areas are melted zones, probably associated with gaseous emissions. Photo by Vincent Thiéry, 2017.

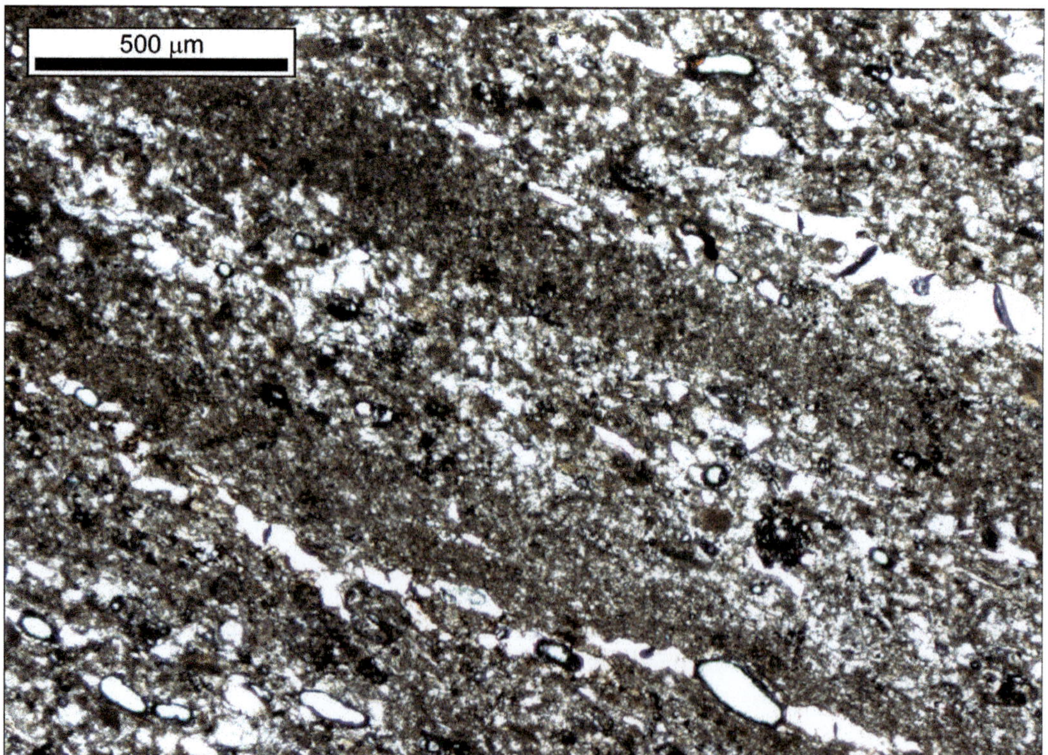

Figure 14.3.2. Thin section (plane polarized light) through the sample illustrated in Figure 14.3.1 showing layering that consists mainly of quartz, mica flakes, and fired clays. Photo by Vincent Thiéry, 2014.

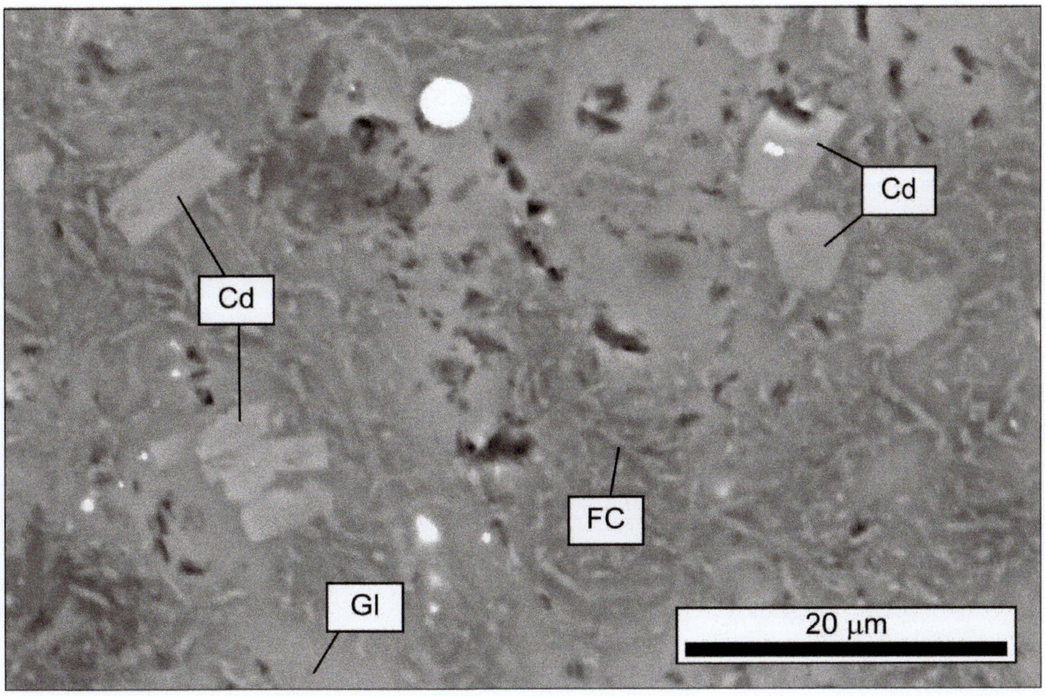

Figure 14.3.3. SEM-BSE image of a polished thin section (sample in Figure 14.3.1) showing clinker microstructure at high magnification. Idiomorphic cordierite (Cd) crystals can be seen in a complex matrix of glass (Gl) and fired clays (FC). Photo by Vincent Thiéry, 2014.

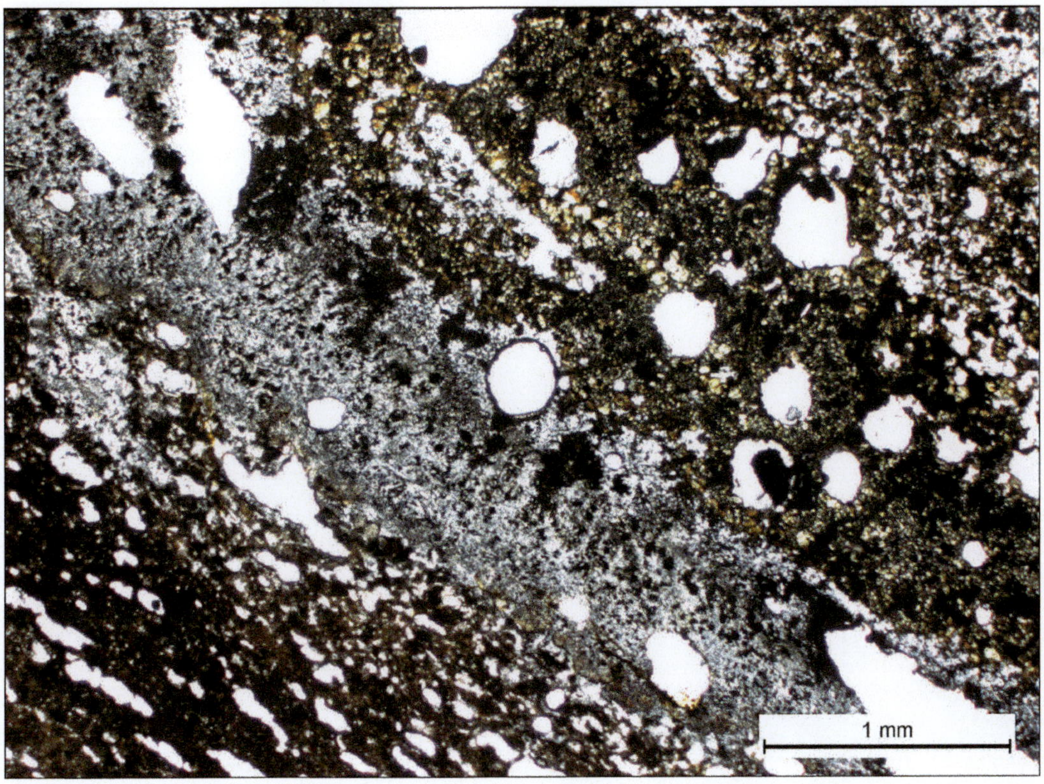

Figure 14.3.4. Thin section of the sample in Figure 14.3.1, showing the transition between clinker (lower left) to porous gaseous emission zone (dark porous zone in Figure 14.3.1). Plane polarized light. Photo by Vincent Thiéry, 2018.

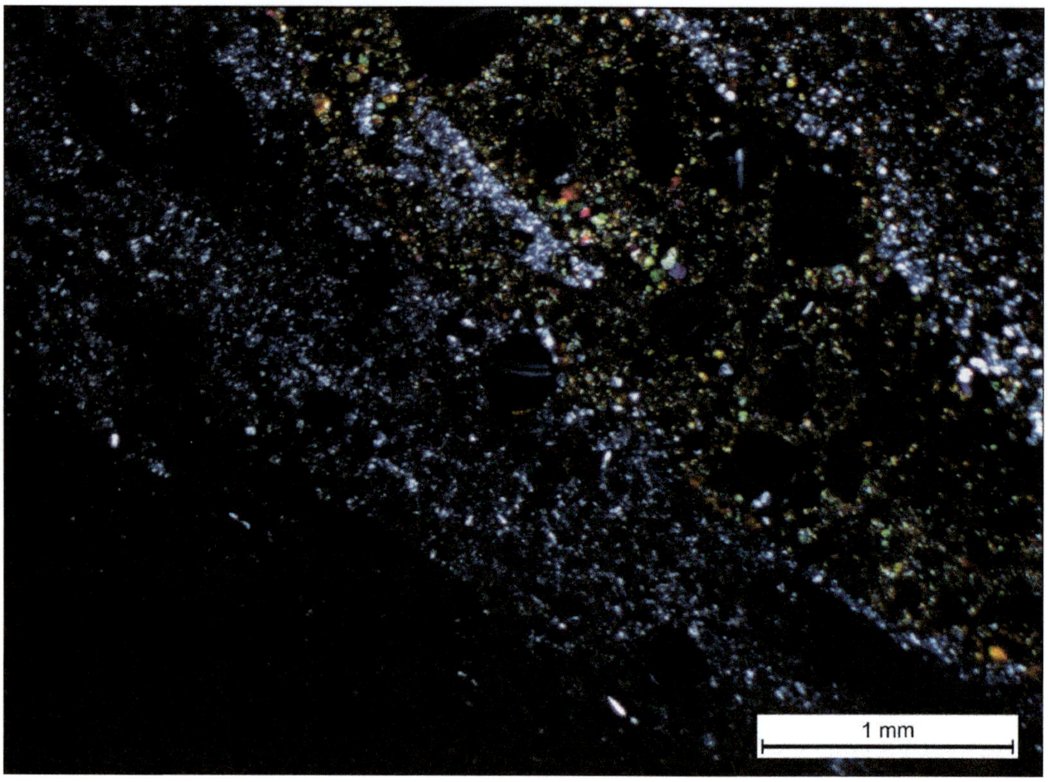

Figure 14.3.5. Thin section of the sample in Figure 14.3.1 (same field of view as Figure 14.3.4 but with crossed polars) showing the transition between clinker (lower left) to the porous gaseous emission zone. Photo by Vincent Thiéry, 2018.

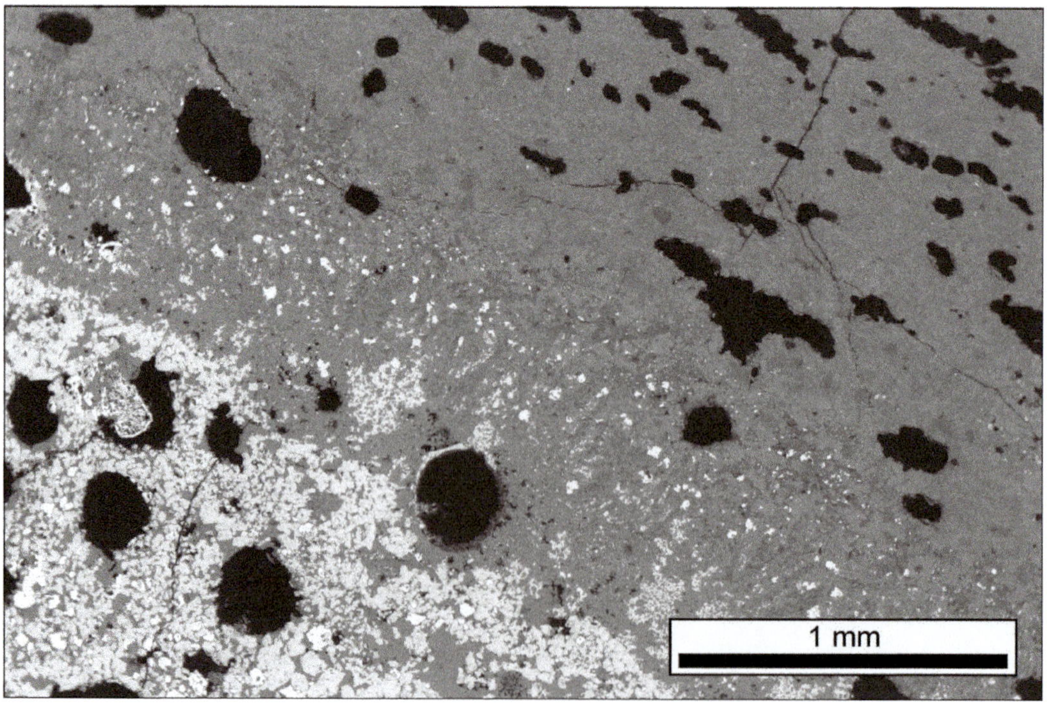

Figure 14.3.6. SEM-BSE image of the sample in Figures 14.3.4 and 14.3.5 but rotated 180°. Note the change in grain size between clinker (upper right) and crystallized paralava (lower left). Photo by Vincent Thiéry, 2018.

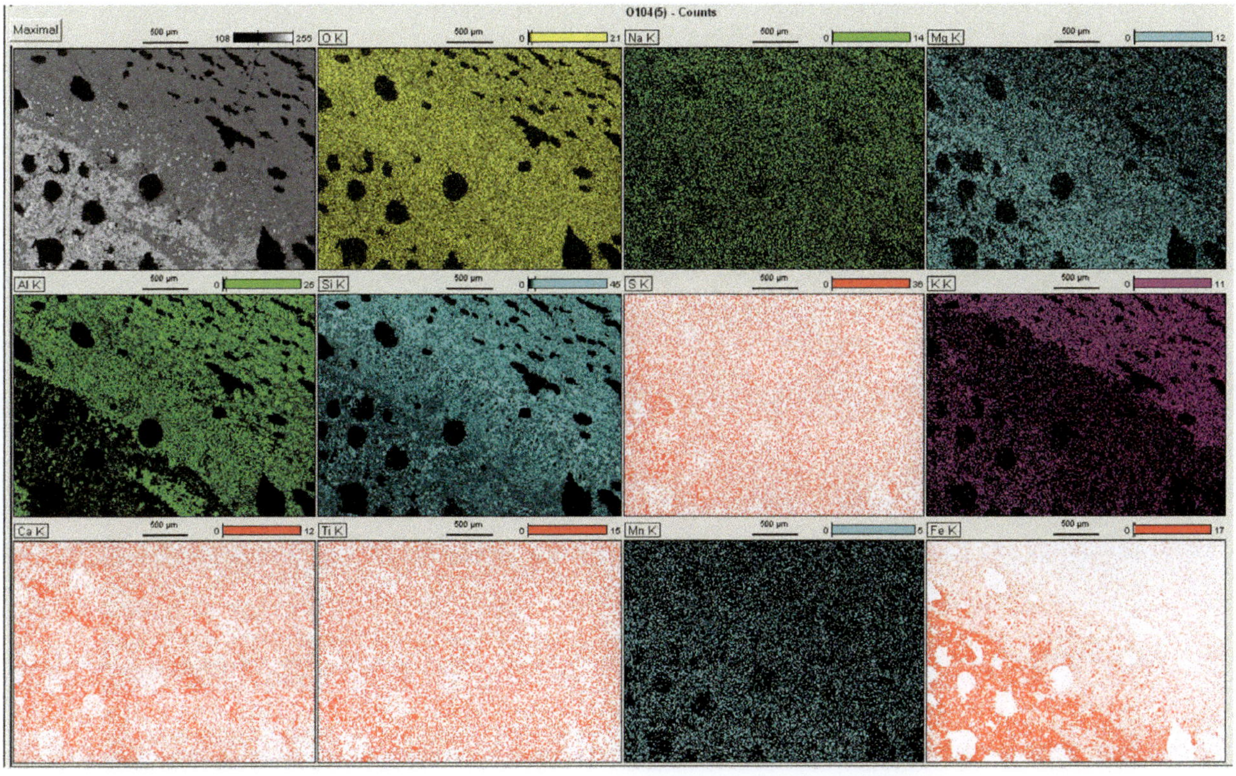

Figure 14.3.7. X-ray elemental map corresponding to the clinker–paralava transition zone shown in the upper-left corner of this figure and illustrated in Figure 14.3.6. Photo by Vincent Thiéry, 2018.

Figure 14.3.8. Thin section through the interface between clinker (extreme lower part) and paralava containing prismatic cordierite (Cd). Plane polarized light with a gypsum plate. Photo by Vincent Thiéry, 2018.

Figure 14.3.9. Thin section through clinker (dark porous area in Figure 14.3.1), showing the coexistence of cordierite (Cd) and olivine (Ol). Both phases were confirmed by EDS analysis. Photo by Vincent Thiéry, 2017.

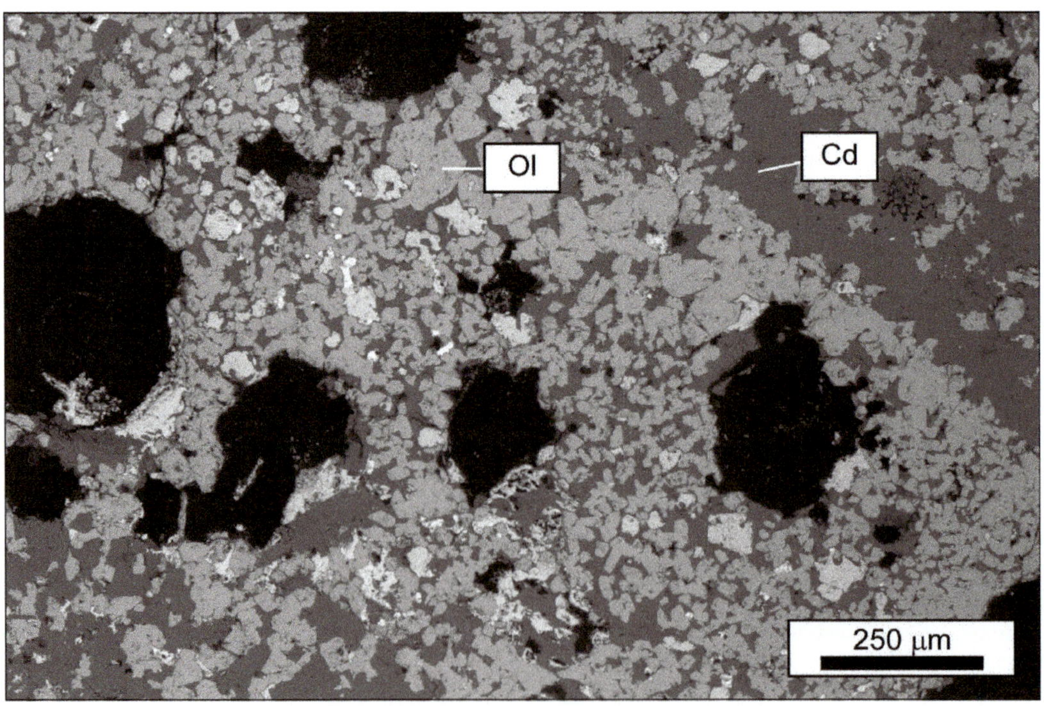

Figure 14.3.10. SEM-BSE image of a porous dark zone illustrated in Figure 14.3.1. Ol is olivine and Cd is cordierite. Photo by Vincent Thiéry, 2017.

Figure 14.3.11. Highly vesicular paralava. The sample was collected in 2015. Photo by Vincent Thiéry, 2017.

Figure 14.3.12. Porous paralava. The sample was collected in 2015. Photo by Vincent Thiéry, 2017.

Figure 14.3.13. Polished slab of paralava, showing numerous vesicles. The sample was collected in 2000. Photo by Vincent Thiéry, 2017.

constitute almost half of the paralavas. Their color is generally red or black and can coexist on a single sample. Under the optical microscope, their high content in glass (ca. half the samples) can be appreciated. The only recognizable minerals are elongated prisms of mullite up to 100 μm long (Figure 14.3.14) and a few micrometers across, as well as Al–Mg–Fe spinel, which forms dendrites, 50–100 μm across (Figures 14.3.15 and 14.3.16).

Geothermometry

Geothermometry consists of an estimation of temperature, based on mineral equilibrium for geological phenomena, generally in metamorphism but also in magmatism. The case of coal heaps can be considered a peculiar kind of metamorphism—anthropogenic pyrometamorphism (Grapes, 2011)—that occurs at almost ambient pressure in coal heaps. The various materials present in the dump are likely to be studied by geothermometry or by comparison with laboratory experiments (Ciesielczuk et al., 2015).

Melted Chain

In the specific case of the La Ricamarie burning heap, geothermometry was proposed on the basis of the presence, inside the heap, of a melted iron chain (Guy et al., 2001). The sample (Figure 14.3.17) is characterized by a mixture between metallic parts that surround various pieces of clinker. The microstructure of the chain shows the presence of native Fe as exsolutions at the grain boundaries of hematite (Figure 14.3.18). A Fe–O as a function of temperature phase diagram (Figure 14.3.19) estimates that such native iron has formed at temperatures of at least 1371°C.

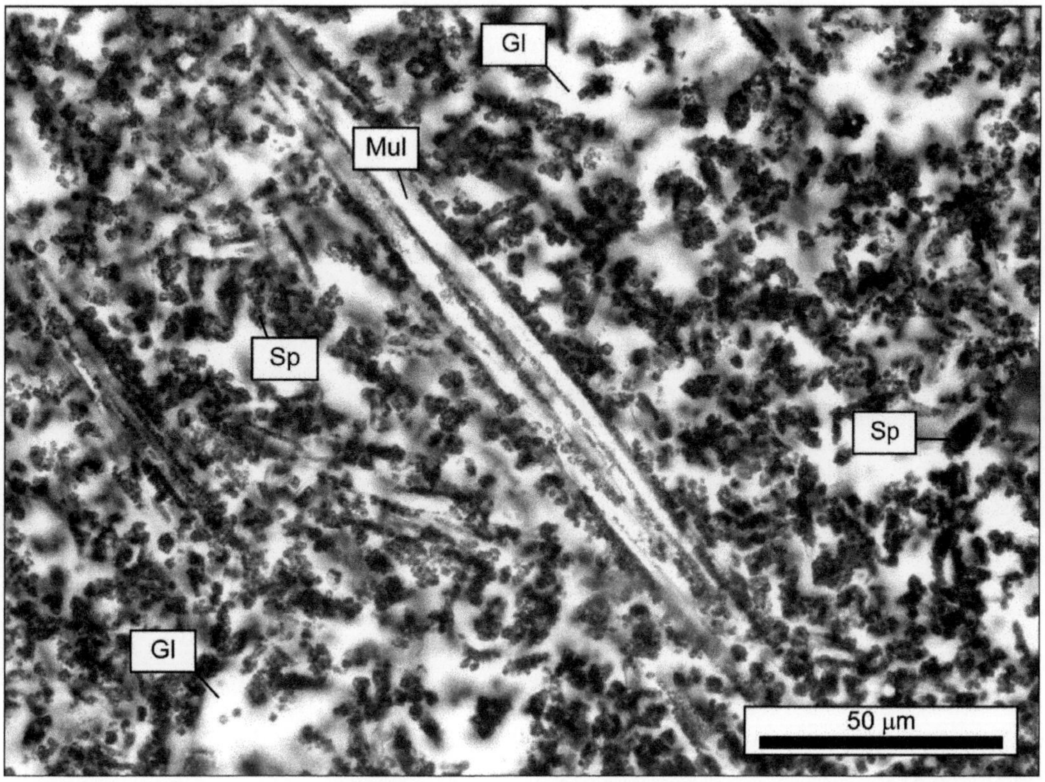

Figure 14.3.14. Thin section of the sample illustrated in Figure 14.3.13, showing the glassy groundmass (Gl), spinel (Sp), and some long needles of mullite (Mul), confirmed by EDS analysis. Plane polarized light. Photo by Vincent Thiéry, 2015.

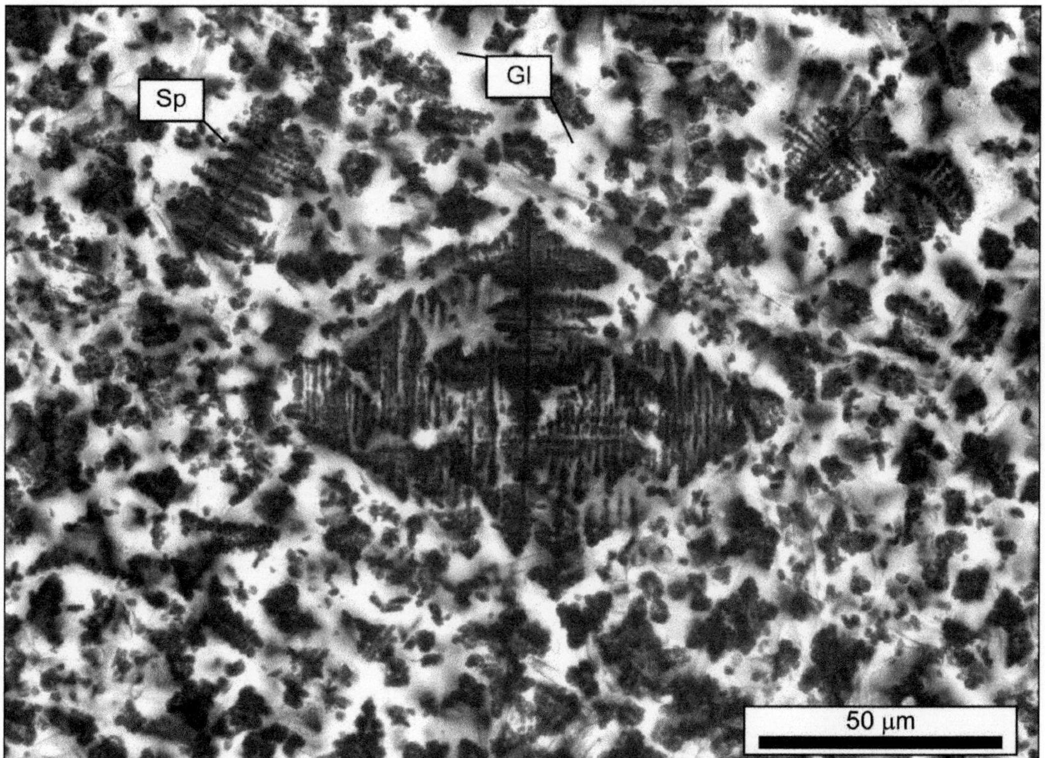

Figure 14.3.15. Thin section through a glassy paralava (Figure 14.3.13) showing dendritic spinel (Sp) in a glass (Gl) matrix. Photo by Vincent Thiéry, 2015.

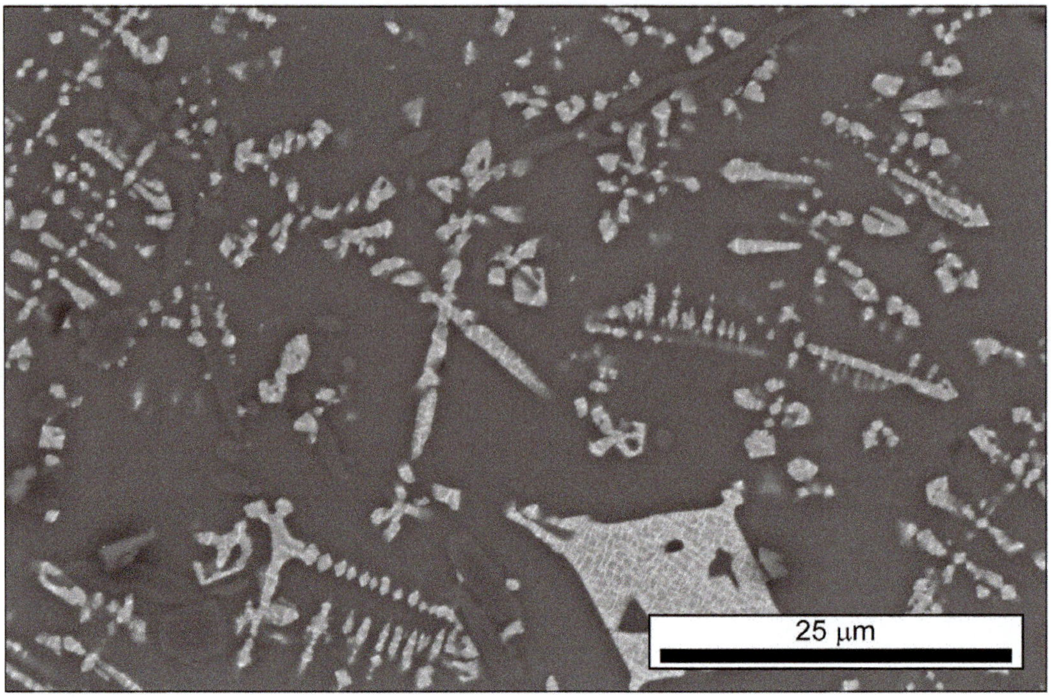

Figure 14.3.16. SEM-BSE image of the sample in Figure 14.3.5, showing dendritic spinel in a glass matrix. Photo by Vincent Thiéry, 2013.

Figure 14.3.17. Melted metallic chain, surrounded by pieces of clinker and all glued together during crystallization of molten metal. The initial form of the chain, seen in cross section, can be approximated. Photo by Vincent Thiéry, 2017.

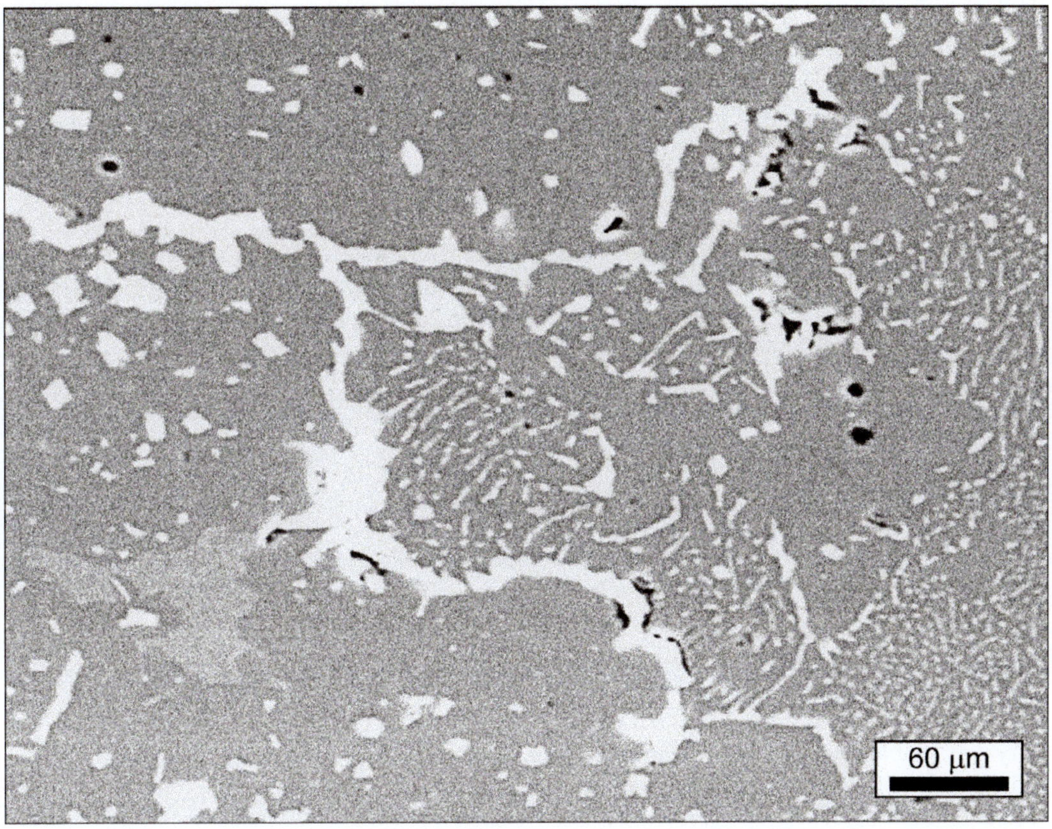

Figure 14.3.18. SEM-BSE image of native iron exsolution (white) present at hematite grain boundaries in the molten chain in Figure 14.3.17. From Guy et al. (2001).

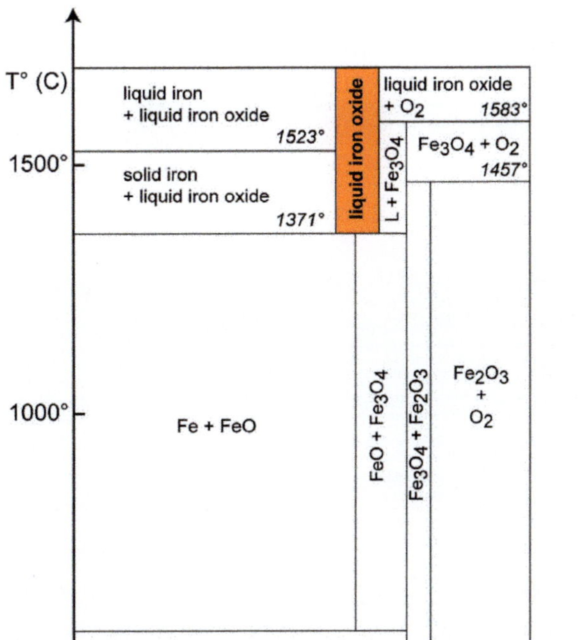

Figure 14.3.19. Fe–O_2 phase diagram illustrating the occurrence of native iron at temperatures higher than 1371°C. Redrawn from Guy et al. (2001).

Photo by Vincent Thiéry, 2014.

14.4 The Outer Heap: Fumaroles and Efflorescence

Fumaroles in an active part of the burning heap at La Ricamarie. The horizontal field of view in the foreground is 20 m.

Minerals

Rare minerals have been described in various places worldwide in association with the combustion of coal heaps including the Chelyabinsk coal basin in Russia (Chesnokov and Tsherbakova, 1991), the Anna Mine in Germany (Witzke et al., 2015), various places in the Czech Republic (Žáček and Skála, 2015), the coal basin in northern France (Masalehdani et al., 2009), and the Silesia basin in Poland (Kruszewski, 2013; Sejkora et al., 2009).

The Ricamarie heap is famous among mineral collectors for magnificent crystals (Laurent, 1995). This section illustrates crystals (Table 14.4.1) that can be found at gas vents at the surface of the heap (Figures 14.4.1 and 14.4.2). Samples are not easy to collect and store because of their extreme friability (Figures 14.4.3 to 14.4.5). Compared to other burning coal heaps, mineral diversity seems less wide in the present state of knowledge of the heap. However, this locality is one of the few worldwide occurrences of bonazziite (β-As_4S_4), discovered here before its formal description; discussed below (Table 14.4.1).

Amorphous As–S Compounds

These compounds, visually identified thanks to their resemblance with the ones described in Žáček and Skála (2015), were confirmed by EDS analyses. They can be yellow or red and form aggregated balls (Figures 14.4.6 and 14.4.7).

Figure 14.4.1. Sulfur and various mineral efflorescence along cracks on an active burning part of the La Ricamarie coal heap. The left side illustrates a truck path from the quarry, indicating the approximate field of view at the foreground. The fracture openings are a few centimeters across. The heap consists of both burnt and unburnt colliery spoil, various demolition waste, clinker, and paralava. Photo by Vincent Thiéry, 2014.

Figure 14.4.2. Close-up view of sulfur, among other unidentified phases, nucleated on the La Ricamarie coal heap. Photo by Vincent Thiéry, 2014.

Figure 14.4.3. Hand sample from the surface of the La Ricamarie coal heap, showing chips of clinker and paralava cemented in a matrix of sulfur and As–S compounds. Photo by Vincent Thiéry, 2018.

Figure 14.4.4. Hand sample from the surface of the La Ricamarie coal heap, encrusted with sulfur and As–S compounds. Photo by Vincent Thiéry, 2018.

Figure 14.4.5. Hand sample from the surface of the La Ricamarie coal heap, partially encrusted with As–S phases. Photo by Vincent Thiéry, 2018.

Table 14.4.1
Minerals (efflorescence) from the La Ricamarie burning heap.

Mineral	Structural formula
Arsenic*	As
Arsenolite	As_2O_3
Bismuthinite*	Bi_2S_3
Bonazziite	β-As_4S_4
Claudetite	As_2O_3
Copiapite*	$Fe^{2+}Fe^{3+}_4(SO_4)_6(OH)_2\cdot20H_2O$
Galena*	PbS
Melanterite*	$FeSO_4\cdot7H_2O$
Orpiment*	As_2S_3
Pharmacosiderite*	$KFe^{3+}_4(AsO_4)_3(OH)_4\cdot6–7H_2O$
Realgar	As_4S_4
Salamonniac	NH_4Cl
Sulfur	S
Tellurite*	TeO_2
Vivianite*	$Fe_3^{2+}(PO_4)_2\cdot8H_2O$

*Recorded in Mindat (https://www.mindat.org/loc-213684.html, accessed 2/25/2018), but not observed or illustrated in the present chapter.

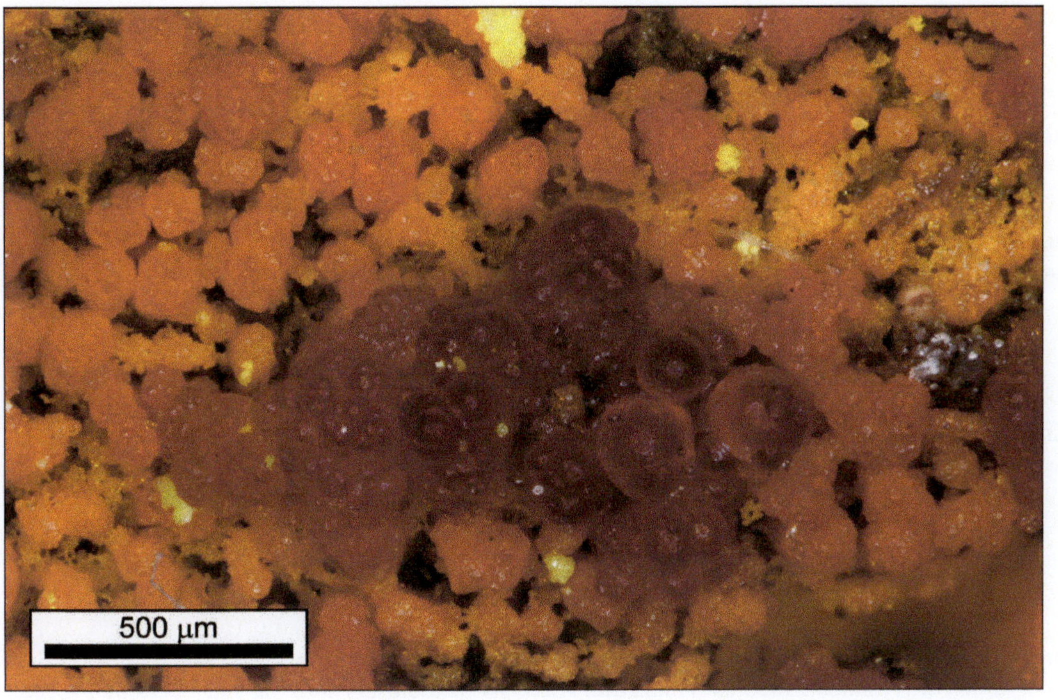

Figure 14.4.6. Red spherules of an amorphous As–S compound nucleated on a sample collected in 2014 at the La Ricamarie coal heap. Photo by Vincent Thiéry, 2018.

Figure 14.4.7. Yellow spherules of an amorphous As–S compound nucleated on a sample collected in 2014 at the La Ricamarie coal heap. Photo by Vincent Thiéry, 2018.

Under the SEM (Figure 14.4.8), minute crystals a few micrometers across appear with a botryoidal texture. They are too tiny to be analyzed by EDS.

Arsenolite and Claudetite: As_2O_3

These compounds are both As_2O_3 polymorphs. Arsenolite (isometric) forms euhedral octahedral hoppers of around 150–200 μm across (Figure 14.4.9). Claudetite (monoclinic) forms platy crystals up to 1 cm long (Figures 14.4.10 and 14.4.11).

Bonazziite (β-As_4S_4 Polymorph)

This mineral, formally described in 2014 (Bindi et al., 2014, 2015), was identified as early as 2001 at Lihir Island, New Ireland Province, Papua New Guinea (Burns and Percival, 2001; Bonazzi et al., 2003) and at La Ricamarie in 2009 (Carpentier and Favreau, 2009). A unique specimen collected on Saint-Pierre burning heap, La Ricamarie, initially eye-identified as possible realgar (As_4S_4, monoclinic $P2_1/n$, $\beta=106.88°$, $V=799.54\,Å^3$), featured an unusual, intriguing yellow-orange color, reminiscent of alacranite (As_8S_9, monoclinic P2/c, $\beta=101.94°$, $V=857.1\,Å^3$), another typical species encountered in burning coal dumps. X-ray diffraction studies performed on several single-crystals of this La Ricamarie specimen revealed all bonazziite (monoclinic C2/c, $\beta=102.03°$, $V=802.3\,Å^3$), the natural analogue of synthetic β-As_4S_4. At this stage, only a single specimen from La Ricamarie (cf. photographs) was authenticated as bonazziite. Analysis of a closely looking like specimen revealed realgar. The actual abundance of bonazziite at La Ricamarie heap is therefore unknown, yet likely quite low. La Ricamarie is one of the only five or six reported occurrences of this rare species, according to the Mindat online database (www.mindat.org). On the other hand, alacranite, dimorphite (As_4S_3, orthorhombic, $V=729.97\,Å^3$), and uzonite (As_4S_5, monoclinic, $\beta=100.1°$,$V=448.44\,Å^3$), other nonstoichiometric arsenic sulfide close members of bonazziite, have not been yet formally identified at La Ricamarie (Figure 14.4.12).

Figure 14.4.8. SEM-BSE image of a yellow, amorphous As–S compound with a botryoidal texture, collected in 2014 at the La Ricamarie coal heap. Bright dots might represent another phase with a heavier molecular weight but these locations could not be analyzed because of their small size. Photo by Vincent Thiéry, 2017.

Figure 14.4.9. Arsenolite crystals collected at the La Ricamarie coal dump. The horizontal field of view is 12.5 mm. Photo by Pascal Chollet, 2017.

Figure 14.4.10. Tabular crystals of claudetite, associated with arsenolite octahedron, collected at the La Ricamarie coal heap. The horizontal field of view is 41 mm. Photo by Pascal Chollet, 2017.

Figure 14.4.11. Tabular crystals of claudetite in the vesicles of a paralava sample collected at the La Ricamarie coal dump. Photo by Vincent Thiéry, 2017.

Figure 14.4.12. Bonazziite from the La Ricamarie coal heap, on a complex aggregate of various burnt and annealed rocks, including paralava and clinker. The picture was taken prior to the published description of this mineral by Bindi et al. (2014, 2015). Photo from Carpentier and Favreau (2009).

Realgar (AsS)

This mineral occurs as beautiful, translucent, and shiny crystals (Figures 14.4.13 and 14.4.14) up to 300–400 μm long.

Salammoniac (NH_4Cl)

Described in the middle of the 19th century (Drian, 1849) in burning coal seams from the Loire coal basin, it is also abundant within efflorescence from the Ricamarie coal heap. It forms dendritic aggregates up to 500 μm long (Figure 14.4.15), and is usually white, but some EDS analyses of yellow crystals, at first inferred to be sulfur, actually revealed salammoniac.

Sulfur (S)

Sulfur is very common around gas vents and fissures on the La Ricamarie coal heap (Figures 14.4.1 and 14.4.2); it forms elongated crystals, 50–200 μm long (Figure 14.4.16).

Figure 14.4.13. SEM image of bonazziite crystals on a complex aggregate of various burnt and annealed rocks, including paralava and clinker, from the La Ricamarie coal heap. Photo by Jean-François Carpentier, 2009.

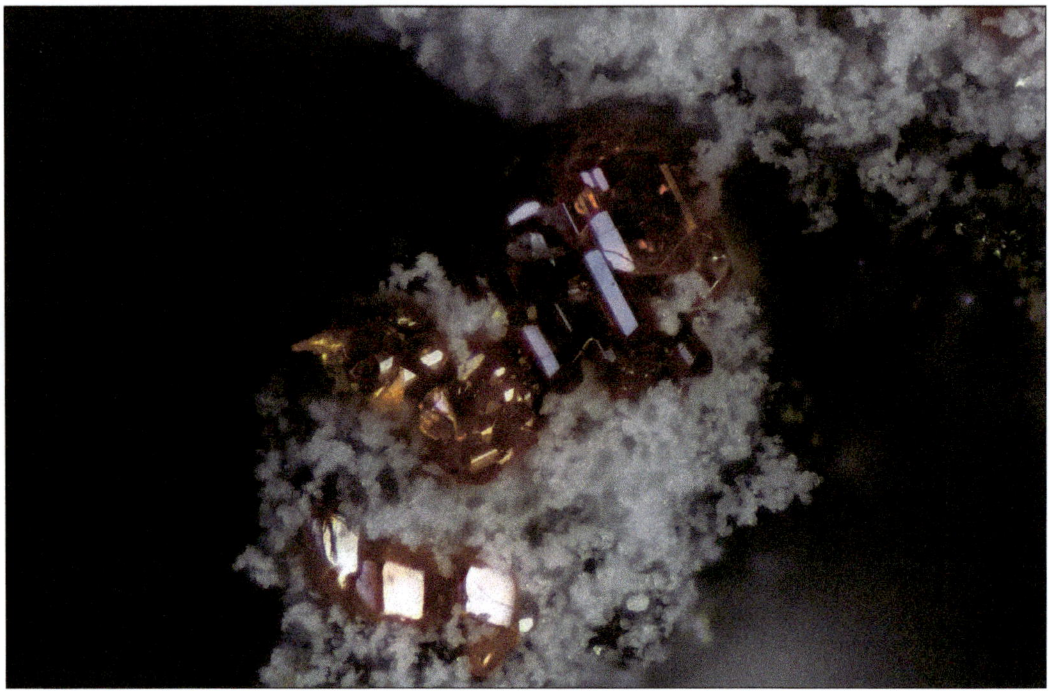

Figure 14.4.14. Realgar crystals from the La Ricamarie coal heap, on a complex aggregate of various burnt and annealed rocks, including paralava and clinker. The horizontal field of view is 0.8 mm. Photo by Jean-Pierre Barral, 2018.

Figure 14.4.15. Salammoniac crystals nucleated on a complex aggregate of various burnt and annealed rocks, including paralava and clinker. Photo by Vincent Thiéry, 2018.

Figure 14.4.16. Sulfur crystals nucleated on a complex aggregate of various burnt and annealed rocks, including paralava and clinker. The horizontal field of view is 1 mm. Photo by Jean-Pierre Barral, 2018.

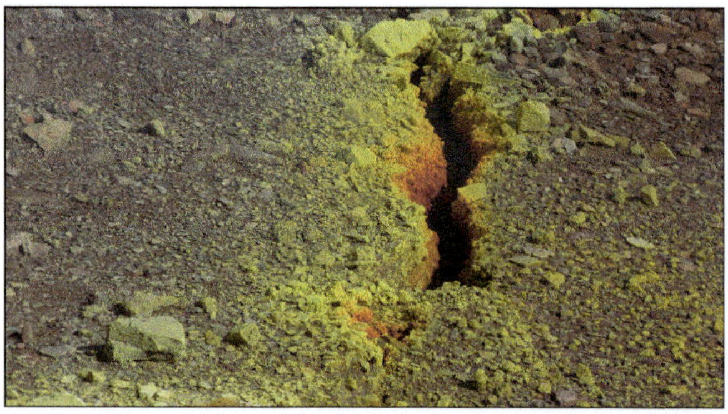

Photo by Vincent Thiéry, 2014.

14.5 Conclusions

The inferno of the La Ricamarie burning coal heap. The vent opening is about 10 cm wide and 5 m long. The vent is encrusted with sulfur crystals, realgar, and other As–S phases. The temperature at the vent was not measured but you could put your bare hand here without getting hurt.

From the coal seam fires known since historical times to the present burning heap, which is one of the last remnants of mining activity now long gone in the Loire coal basin, the Stephanian Loire coal basin has been marked by coal fires. During the 18th and 19th centuries in France, the Loire coal basin around La Ricamarie has been worthy of investigation to understand the effect of heat energy on rocks that we now commonly term porcellanite, paralava, and clinker.

Mineral collectors are attracted by the beauty and the scarcity of minerals that can be found at the surface of the heap. This will help preserve the memory of the heap once it is completely quarried as a source of aggregate. The discovery of rare phases that may become newly discovered minerals such as the case with bonazziite, a mineral and phase of interest in materials science (Bindi et al., 2015) is direct evidence that such places are worth investigating.

Acknowledgments

The authors are indebted to Jean-Pierre Barral, Pascal Chollet, and Georges Favreau, members of the French micro-mineralogical association, for outstanding pictures and good advice. The successive owners of the Ricamarie heap quarry are acknowledged for providing easy access to the quarry for sampling. We thank personnel at the Archives Nationales du Monde du Travail in Roubaix, France, for having granted access to old pictures.

References

Becq-Giraudon, J.F., Mercier, D., Jacquemin, H., 1995. La flore fossile du bassin houiller de Saint-Etienne, in Doubinger Jeanne, Vetter Pierre, Langiaux Jean, Galtier Jean et Broutin Jean, Mémoires du muséum national d'histoire naturelle tome 164 Paléobotanique, pp. 349–351 (in French).

Becq-Giraudon, J.-F., Montenat, C., Van Den Driessche, J., 1996. Hercynian high-altitude phenomena in the French Massif Central: tectonic implications, Palaeogeography, Palaeoclimatology. Palaeoecology 122, 227.

Bindi, L., Pratesi, G., Muniz-Miranda, M., Zoppi, M., Chelazzi, L., Lepore, G.O., Menchetti, S., 2014. Bonazziite, IMA 2013-141. CNMNC Newsletter No. 20, June 2014, page 554. Mineral. Mag. 78, 549–558.

Bindi, L., Pratesi, G., Muniz-Miranda, M., Zoppi, M., Chelazzi, L., Lepore, G.O., Menchetti, S., 2015. From ancient pigments to modern optoelectronic applications of arsenic sulfides: bonazziite, the natural analogue of β-As_4S_4 from Khaidarkan deposit. Kyrgyzstan. Mineral. Mag. 79, 121–131.

Bonazzi, P., Bindi, L., Popova, V., Pratesi, G., Menchetti, S., 2003. Alacranite, As8S9: structural study of the holotype and re-assignment of the original chemical formula. Am. Mineral. 88, 1796–1800.

Burat, A., 1851. De la houille - traité théorique et pratique des combustibles minéraux (houille, anthracite, lignite, etc.). In: Delacroix-Comon. Paris, 482 p.

Burns, P.C., Percival, J.B., 2001. Alacranite, As_4S_4: a new occurrence, new formula, and determination of the crystal structure. Can. Mineral. 39, 809–818.

Carpentier, J.F., Favreau, G., 2009. β-As_4S_4 de La Ricamarie, Loire, 2009. Le cahier des micromonteurs 103, 12 (in French).

Chesnokov, B.V., Tsherbakova, E.P., 1991. The Mineralogy of Burned Heaps in the Chelyabinsk Coal Basin. 152 pp. H. Nauka, Moscow, Russia.
Ciesielczuk, J., Kruszewski, Ł., Majka, J., 2015. Comparative mineralogical study of thermally-altered coal-dump waste, natural rocks and the products of laboratory heating experiments. Int. J. Coal Geol. 139, 114–141.
Damour, A., 1884. Note sur un sel ammoniac iodifère. Bulletin de la Société Minéralogique de France 7, 347–348 (in French).
Daubrée, A., 1887. Les eaux souterraines à l'époque actuelle: leur régime, leur température, leur composition, au point de vue du rôle qui leur revient dans l'économie de l'écorce terrestre (tome 2). 302 pp. (in French).
de Bournon, J.L., 1785. Essai sur la lithologie des environs de Saint-Etienne-En-Forez, et sur l'origine de ses charbons de pierre: Avec des observations sur les silex, pétro-silex, jaspes et granits. 134 pp. (in French).
Drian, A., 1849. Minéralogie et pétralogie des environs de Lyon disposées suivant l'ordre alphabétique. 539 pp. (in French).
Girardin, J., 1831. Considérations générales sur les volcans, et examen critique des diverses théories qui ont été successivement proposées pour expliquer les phénomènes volcaniques. 248 pp. (in French).
Grapes, R., 2011. Anthropogenic pyrometamorphism. In: Pyrometamorphism. Springer, pp. 235–288.
Grapes, R., Zhang, K., Peng, Z.L., 2009. Paralava and clinker products of coal combustion, yellow river, Shanxi Province, China. Lithos 113, 831–843.
Grüner, E.L., 1882a. Bassin houiller de la Loire. Deuxième partie, Description détaillée des districts houillers. 506 pp. (in French).
Grüner, E.L., 1882b. Bassin houiller de la Loire. Première partie: Description générale du bassin. 235 pp. (in French).
Guy, B., Garcia, D., Le Coze, J., Theisse, V., 2001. Estimation de la température atteinte dans le crassier Saint-Pierre de la Ricamarie (Loire) à partir de l'observation de textures de solidification/oxydation dans un élément de chaîne métallique prélevé sur le site. Les cahiers de la Rotonde – Publications de l'Université de Saint-Étienne 61–69 (In French).
Jannettaz, E., 1874. Les roches: description de leurs éléments: méthode de détermination: guide pratique à l'usage des ingénieurs, géologues, minéralogistes, agronomes, des elèves des écoles du gouvernement. 285 pp. (in French).
Kruszewski, Ł., 2013. Supergene sulphate minerals from the burning coal mining dumps in the Upper Silesian Coal Basin, South Poland. Int. J. Coal Geol. 105, 91–109.
Landrin, H., 1856. Dictionnaire de minéralogie, de géologie et de métallurgie. 506 pp. (in French).
Laurent, H., 1995. Minéralisations des houillères embrasées de Saint-Étienne (Loire). Le règne minéral 2, 41–45.
Masalehdani, M.N.N., Mees, F., Dubois, M., Coquinot, Y., Potdevin, J.-L., Fialin, M., Blanc-Valleron, M.-M., 2009. Condensate minerals from a burning coal-waste heap in Avion, Northern France. Can. Mineral. 47, 573–591.
Perrin, M., 1930. Le bassin houiller de la Loire. Ann. Geograph. 39, 359–375 (in French).
Sejkora, J., Škovíra, J., Čejka, J., Plášil, J., 2009. Cu-rich members of the beudantite–segnitite series from the Krupka ore district, the Krušné hory Mountains, Czech Republic. J. Geosci. 54, 354–371.
Sharygin, V.V., Sokol, E.V., Belakovsky, D.I., 2014. Mineralogy and origin of Fayalite-sekaninaite paralava: ravat coal fire, Central Tajikistan. In: Coal and Peat Fires: A Global Perspective, Vol. 3, pp. 582–607.
Skarżyńska, K.M., 1995. Reuse of coal mining wastes in civil engineering – Part 1: properties of minestone. Waste Manag. 15, 3–42.
Sokol, E., 2005. Combustion metamorphism. 250 pp. Publishing house of the Siberian Branch of the Russian Academy of Science, Novosibirsk (in Russian).
Thiéry, V., Guy, B., 2015. Some aspects of petrography of burnt colliery spoil. In: Çopuroğlu, O. (Ed.), Proceedings of the 15th Euroseminar on Microscopy Applied to Building Materials. Delft University of Technology, pp. 71–76.
Witzke, T., de Wit, F., Kolitsch, U., Blaß, G., 2015. Mineralogy of the burning Anna 1 coal mine dump, Alsdorf, Germany. In: Stracher, G.B., Prakash, A., Sokol, E.V. (Eds.), Coal and Peat Fires: A Global Perspective, Volume 3: Case Studies – Coal Fires. Elsevier, pp. 203–240.
Žáček, V., Skála, R., 2015. Mineralogy of burning-coal waste piles in collieries of the Czech Republic. In: Stracher, G.B., Prakash, A., Sokol, E.V. (Eds.), Coal and Peat Fires: A Global Perspective, Volume 3: Case Studies – Coal Fires. Elsevier, pp. 109–159.

CHAPTER 15

Burning Coal-Mine Collieries in the Jharia Coalfield of India

A coal fire in an open-pit coal mine located near Dhanbad, in the Jharia coalfield, India. As a consequence of open-pit mining in Jharia, most coal fires start by spontaneous combustion. The middle flame is about 7 m (23 ft) high. *Source: Daniel B. Sanger, 1994; Courtesy of Stanley R. Michalski, GAI Consultants, Inc., Homestead, Pennsylvania, USA.*

CHAPTER CONTENTS

Coal and Peat Fires: A Global Perspective
Edited by Glenn B. Stracher

https://doi.org/10.1016/B978-0-12-849885-9.00015-9

Photo by Ashwani Raju, 2011; from Prakash et al. (2013).

15.1 Burning Coal-Mine Collieries in the Jharia Coalfield

Hartwig Gielisch
Glenn B. Stracher[a]

Coal is being extracted in the presence of a fire in an open-cast colliery in the Sijua area of the Jharia coalfield. As the unburnt coal is excavated, the burning part of the seam is doused with water to keep the combustion under control, a common mining technique used in Jharia.

Introduction

The **Jharia coalfield** (JCF) is located east and north of the Damodar River in eastern India; it is named after the principal mining city of Jharia (Figure 15.1.1). With an area of about 456 km^2, it is India's largest coalfield and produces low, medium, and high volatile **bituminous coal** (Ranawat and Veena, 2016).

In 1774, commercial coal mining began in the coalfields of eastern India. Initially slow, production increased in 1883 with the arrival of steam-powered locomotives. Mining began in 1894 in the JCF, now the country's primary source of **coking coal** (Gupta and Prakash, 1998; Prakash and Gupta, 1998; Prakash et al., 2013).

Between 1971 and 1973, coal mining was nationalized with the government takeover of private coal mines. In 1975, **Coal India Limited** (CIL), a government holding company headquartered in Kolcata (formerly Calcutta), took control of coal mining in India. CIL was founded to improve safety standards and augment the development of JCF resources. It is the largest coal producing company in the world, with numerous coal-mining subsidiaries abroad (CIL, 2016). JCF production in 2014 was 668 Mt (Coal Information, 2015; World Coal Association, 2016). In India, CIL production is about 15% from **underground mines** and 85% from **opencast mines** (open pit mines) (Prakash et al., 2013).

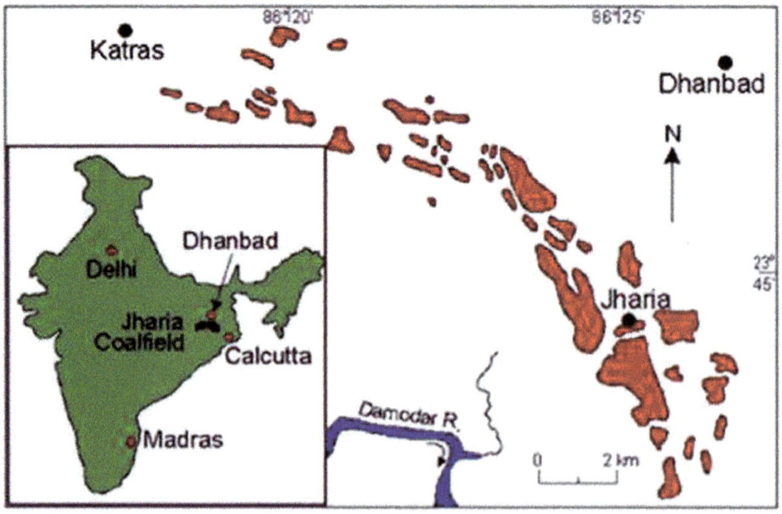

Figure 15.1.1. Location of the Jharia coalfield, India. The Jharia deposits are colored red. From Stracher and Taylor (2011).

---

[a]From Stracher and Taylor (2004), with modifications.

Today, India is the third largest coal-producing nation in the world. Two-thirds of its energy requirements are supplied by coal, 66% of which is used to generate electrical power (CIL, 2016), and the remainder is used to produce steel and as boiler fuel in factories.

Jharia Coalfield Fires

Coal seams in the JCF occur in perm-carboniferous and tertiary shales and sandstones that strike east-west, with shallow dips to the south (Verma et al., 1979, 1989). With 18 major and over 40 minor coal seams, a long history of opencast and underground mining, naturally occurring events like lightning strikes and **spontaneous combustion**, and anthropogenic activities like mining accidents and the illegal distillation of alcohol in some abandoned underground mines, the JCF is highly susceptible to **coal fires** (Stracher et al., 2002; Gupta, 2003; Gangopadhyay et al., 2005; Saxena et al., 2005). Today, many fires in the JCF are ignited by improper mining practices and by the spontaneous combustion of coal after opencast and deep mining.

Coal fires in the JCF have occurred since the earliest days of mining there and the first coal fire broke out at Bhowrah in 1916. Today, the coalfield contains the densest complex of such fires in the world, with over 65 burning in a 450 km^2 area encompassing both opencast and underground fires (Sinha, 1986; Gupta and Prakash, 1998; Michalski, 2004; Prakash et al., 2013). Since the first recorded fire in 1916, over 37 million tons of coal has been lost to JCF fires with temperatures reaching as high as 731°C and as much as 1453 million tons locked up under fire. At this burn rate, combustion of the remaining coal could last another 3500 years (Stracher, 2010).

Coal fires in the JCF don't just consume a valuable energy resource. The particulate matter and toxic gases they release pollute and destroy ecosystems and plague human communities with sometimes fatal diseases like chronic bronchitis and asthma. People have disappeared into the abyss because of dangerous land subsidence, caused by the volume reduction of coal burning in the subsurface. The fires have displaced communities of people as they consume buildings, roads, and render railroads inoperable (Stracher and Taylor, 2004).

Opencast and underground coal fires in the JCF have been extensively mapped and tracked for over 25 years by satellites and with ground-based field work. This data is then integrated into a **Geographic Information System** (Stracher and Taylor, 2004; Prakash et al., 2013) used to keep a record of fire location, areas of ground subsidence, and the locations where borehole temperatures were made.

Temporal and multispectral data acquired with the **Landsat Thematic Mapper** (TM) demonstrated in the 1990s that underground fires are more widespread in the JCF than surface fires (Prakash et al., 1997). Surface fires with temperatures as high as 342°C to 731°C were detected with data corrections for TM-5 and TM-7 short wave–infrared radiation; these temperatures were confirmed with ground-based thermal detectors (Prakash and Gupta, 1999). In addition, underground fires at depths of 45–55 m were identified utilizing TM-6 thermal infrared data that included surface thermal anomalies detected at temperatures of 25.6–31.6°C (Saraf et al., 1995; Prakash et al., 1997).

According to a study by the National Center for Atmospheric Research in Boulder, Colorado (Collins, 2000; Perkins, 2001), sulfate aerosols from JCF fires combine with those from industrial emissions. These aerosols decrease the intensity of sunlight that reaches Earth's surface because they scatter or absorb solar radiation; in the case of India, the reduction is as high as 15%. A TM satellite study by Mishra et al. (2012) revealed that air pollution from the gas components SO_2, CH_4, and NO_x and from particulate matter is higher over the eastern part of the JCF than the western. This is because of more intense mining activity in the eastern JCF.

Because of the population density in the JCF, large communities of people have been and will be relocated from endangered regions to newly constructed settlements in non-coal–producing areas. A master plan approved by the Indian government in 2009 includes the current resettlement of people from land on the lease hold of **Bharat Coking Coal Limited**, a subsidiary of CIL. Relocation is expected to be completed by 2018 (BCCL, 2016).

Various methods are used to contain or extinguish coal fires in the JCF by attempting to cut off the oxygen supply to these fires. For underground fires, sand-bentonite slurry flushing, underground injection with fly ash, surface sealing with incombustible materials such as soil, and inert gas injection are used (Michalski and Gray, 2001; Stracher and Taylor, 2004; Donnelly and Bell, 2011). Fires at the surface are dealt with by firebreak trenching, surface sealing, dousing with water and spraying with liquid-bitumen compounds, and constructing fire-retaining walls (Stracher and Taylor, 2004; Donnelly and Bell, 2011; Prakash et al., 2013). In some cases, such methods have been successful. However, coal fires in the JCF continue to burn in precarious areas where inaccessible underground workings and ground subsidence present a gridlock to fire fighters. In addition, using water for fire control may contaminate surface and ground water, affect the stability of underground pillars, or produce steam and ash that obstruct opencast mining and affect nearby communities (Stracher and Taylor, 2004; Donnelly and Bell, 2011).

Burning Collieries

Figure 15.1.2 illustrates the locations of 41 collieries currently on fire in the Jharia coalfield; the remaining figures (Figures 15.1.3–15.1.9) and accompanying captions illustrate and briefly describe select colliery fires.

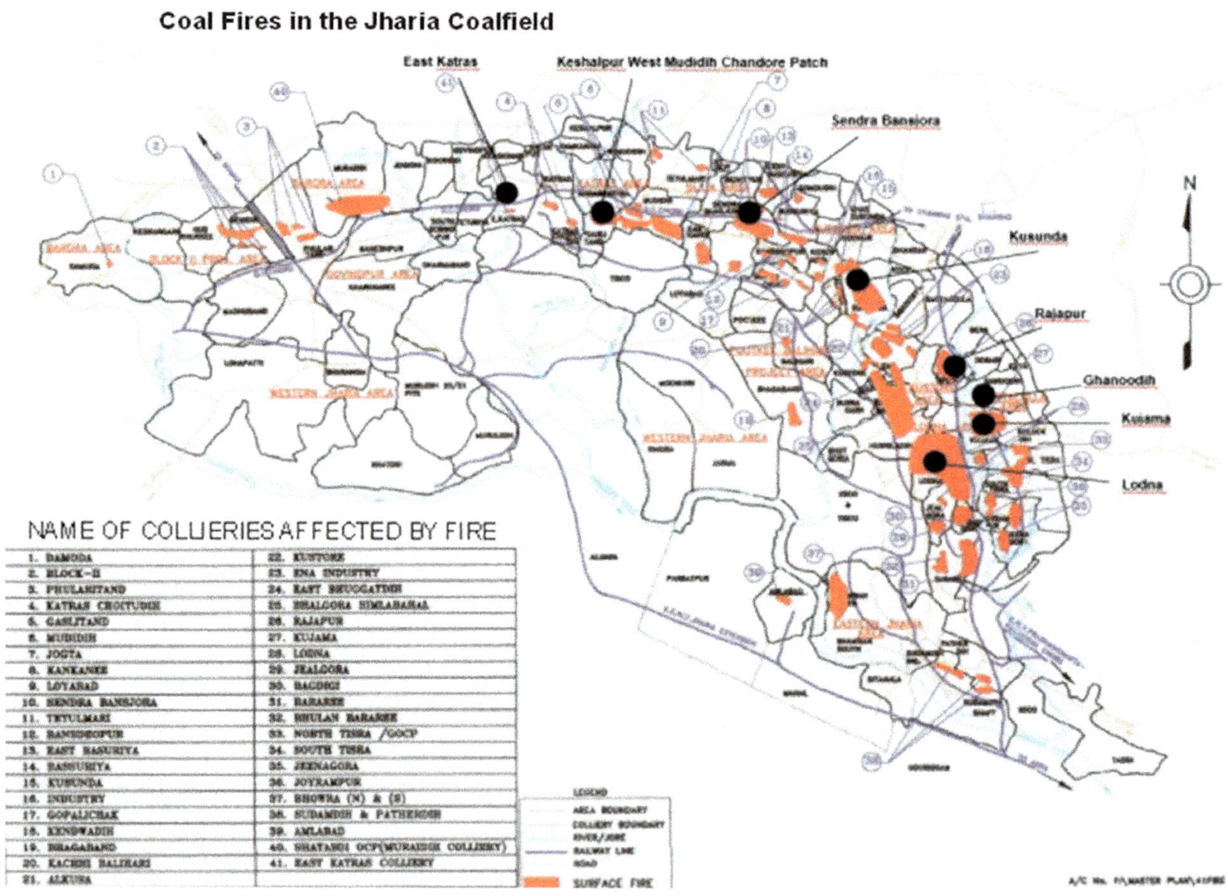

Figure 15.1.2. The names of 41 collieries currently burning in the Jharia coalfield. The Ghanoodih Colliery fire is relatively new and plotted on the map but not recorded in the table. See Figure 15.1.1 as a reference for scale. Courtesy of Hartwig Gielisch, DMT GmbH & Co. KG, Geo Engineering & Exploration, Essen, Germany, 2015.

Examples of Actively Burning Collieries

Ghanoodih

Figure 15.1.3. A burning-coal seam located at the Ghanoodih Colliery, directly beneath the village of Jharia. The fire travelled here through underground mining tunnels from Lodna, in the eastern part of the coalfield (Figure 15.1.2). It was detected at Ghanoodih in 2005. Location and elevation: 45 Q 441285, 2625990, 202 m. Photo by Hartwig Gielisch, May 2015.

Kusunda

Figure 15.1.4. (A) A coal seam burning at the opencast Kusunda Colliery. The seam was ignited by unsafe mining practices. (B) A 17 m burning-coal seam in the pit shown in (A). Location and elevation: 45 Q 437759, 2629138, 238 m. The background fields of view in (A) and (B) are about 500 and 1000 m, respectively. Photo by Hartwig Gielisch, May 2015.

Rajapur

Figure 15.1.5. (A) A coal fire in the opencast Rajapur Colliery. Location and elevation: 45 Q 440978, 2626898, 202 m. (B) A burning-coal seam at Rajapur shown in (A). The field of view here is about 150 m. (C) Raja Shiv Prasad College, about 300 m southwest of the fire illustrated in (A) and (B). Water pumps here are used to mitigate the fire. Photo by Hartwig Gielisch, May 2015.

Sendra Bansjora

Figure 15.1.6. The Dhanbad-Chandrapura railroad in Gopal Gaderia is endangered by an active coal fire about 15 m from the tracks, in an area where mining is prohibited. The fire was detected in 2002 in the nearby Sendra Bansjora Colliery. It was ignited there by unsafe mining practices. Location and elevation: 45 Q 434684, 2631395, 227 m. The field of view in the foreground is about 100 m. Photo by Hartwig Gielisch, May 2015.

Mudidih

Figure 15.1.7. The burning Amalgamated Keshalpur West Mudidih Chandore Colliery. The fire started by unsafe mining practices. A completely burned coal seam (not illustrated) occurs here. Location and elevation: 45 Q 431232, 2632481, 215 m. The field of view in the foreground is about 300 m. Photo by Hartwig Gielisch, May 2015.

Kujama

Figure 15.1.8. The burning opencast Kujama Colliery, ignited about 10 years ago by unsafe mining practices. A 13 m thick coal seam is burning in the foreground, where white ash is visible. Hartwig Gielisch is sanding at the base of the burning seam. Location and elevation: 45 Q 441696, 2624791, 142 m. Photo by Hartwig Gielisch, May 2015.

Figure 15.1.9. A burnt area (inclined surface) in the Kujama Colliery. Photo by Hartwig Gielisch, May 2015.

Acknowledgments

The authors wish to thank Elsevier's Editorial Project Manager, Tasha Frank, and the staff at Elsevier for their review and assistance with the publication of this work.

Important Terms

Bharat Coking Coal Limited
bituminous coal
coal fires
Coal India Limited
coking coal
Geographic Information System
Jharia Coalfield
Landsat Thematic Mapper
opencast mines
spontaneous combustion
underground mines

References

Bharat Coking Coal Limited, 2016. Jharia Master Plan. http://www.bccl.gov.in. (accessed February 2016).

Coal India Limited, 2016. A Government of India Undertaking: About Us. Newtown, Rajarhat, Kolkata https://www.coalindia.in. (accessed February 2016).

Coal Information, 2015. International Energy Agency. 674 p.

Collins, W., 2000. Human-Induced Climate Change: Improved Knowledge and Continuing Uncertainties. In: American Geophysical Union Fall 2000 Meeting, AGU Press Conference Release No. 00–29A, December 17, San Francisco.

Donnelly, L.J., Bell, F.G., 2011. Geotechnical and environmental problems: coal and spontaneous combustion. In: Stracher, G.B., Prakash, A., Sokol, E.V. (Eds.), Coal and Peat Fires, A Global Perspective, Volume 1, Coal – Geology and Combustion. Elsevier, pp. 83–100.

Gangopadhyay, P., Lahiri-Dutt, K., Saha, K., 2005. Application of remote sensing to identify coal fires in the Raniganj Coalbelt, India. Int. J. Appl. Earth Observation Geoinformation 8, 188–195.

Gupta, R.P., Prakash, A., 1998. Reflectance aureoles associated with thermal anomalies due to subsurface mine fires in the Jharia coalfield, India. Int. J. Remote Sens. 19 (14), 2619–2622.

Gupta, R.P., 2003. Remote Sensing Geology, second ed. Springer. 656 p.

Michalski, S.R., Gray, R.E., 2001. Ash disposal – mine fires – environment: an Indian dilemma. Bull. Eng. Geol. Environ. 60, 23–29.

Michalski, S.R., 2004. The Jharia mine fire control technical assistance project: an analysis. In: Stracher, G.B. (Ed.), Coal Fires Burning Around the World: A Global Catastrophe: International Journal of Coal Geology, 59, pp. 83–90 Issues 1–2.

Mishra, R.K., Pandey, J., Chaudhary, S.K., Khalkho, A., Singh, V.K., 2012. Estimation of air pollution concentration over Jharia coalfield based on satellite imagery of atmospheric aerosol. Int. J. Geomat. Geosci. 2 (3), 723–729.

Perkins, S., 2001. Pollution in India may effect climate. Sci. News 159 (1), 15.

Prakash, A., Gupta, R.P., 1998. Land-use mapping and change detection in a coal mining area–a case study in the Jharia coalfield, India. Int. J. Remote Sens. 19 (3), 391–410.

Prakash, A., Gupta, R.P., 1999. Surface fires in Jharia coalfield, India-their distribution and estimation of area and temperature from TM data. Int. J. Remote Sens. 20 (10), 1935–1946.

Prakash, A., Gupta, R.P., Saraf, A.K., 1997. A Landsat TM based comparative study of surface and subsurface fires in the Jharia coalfield, India. Int. J. Remote Sens. 18 (11), 2463–2469.

Prakash, A., Gens, R., Prasad, S., Raju, A., Gupta, R.P., 2013. Coal fires in the Jharia coalfield, India. In: Stracher, G.B., Prakash, A., Sokol, E.V. (Eds.), Coal and Peat Fires, A Global Perspective, Volume 2, Photographs and Multimedia Tours. Elsevier, pp. 153–177.

Ranawat, P.S., Veena, S., 2016. Coal Resources of India. http://www.geologydata.info/index.htm. (accessed February 2016).

Saraf, A.K., Prakash, A., Sengupta, S., Gupta, R.P., 1995. Landsat-TM data for estimating ground temperature and depth of subsurface coal fire in the Jharia coalfield, India. Int. J. Remote Sens. 16 (12), 2111–2124.

Saxena, N.C., Singh, G., Singh, K.N., Jodhpur, B.N., 2005. Reinventing Jharia Coalfield: Scientific. ISBN: 81-7233-398-6. 246 p.

Sinha, P.R., 1986. Mine fires in Indian coalfields. Energy 11, 1147–1154.

Stracher, G.B., Taylor, T.P., 2004. In: Stracher, G.B. (Ed.), Coal Fires Burning Out of Control Around the World: Thermodynamic Recipe for Environmental Catastrophe, pp. 7–17. International Journal of Coal Geology, v. 59, issues 1–2.

Stracher, G.B., Taylor, T.P., 2011. The Effects of Global Coal Fires. In: Stracher, G.B., Prakash, A., Sokol, E.V. (Eds.), Coal and Peat Fires: A Global Catastrophe, Volume 1, Coal - Geology and Combustion. Elsevier, pp. 101–114.

Stracher, G.B., Taylor, T.P., Prakash, A., 2002. Coal fires: a synopsis of their origin, remote sensing detection, and thermodynamics of sublimation. In: Shannon, S. (Ed.), Case Histories of Mine Reclamation and Regulation, Environmental Technology for Mining. Robertson GeoConsultants Inc., Reno, NV; Vancouver, BC, pp. 1–7. http://technology.infomine.com/enviromine/case_hist/coal%20fires/stracher_et_al.html. (accessed January 2016).

Stracher, G.B., 2010. Coal fires: the rising global interest. Earth 55 (9), 46–55.

Verma, R.K., Bhuin, N.C., Mukhopadhyay, M., 1979. Geology, structure and tectonics of the Jharia coalfield, India — a three-dimensional model. Geoexploration 17 (4), 305–324.

Verma, R.P., Jaipuriar, A.M., Paul, P.R., 1989. Compendium on updated and revised geology of Jharia coalfield. Central Mine Planning and Designing Institute Ltd., Ranchi, Jharkland, India. 282 p.

World Coal Association, 2016. https://www.worldcoal.org. (accessed January 2016).

CHAPTER 16

Evidence of Human Health Impacts From Uncontrolled Coal Fires in Jharia, India

Burning coal ignited by spontaneous combustion at the Rajpur opencast coal mine in the Balrampur district, state of Madhya Pradesh, Jharia coalfield, India. Miner's housing is at the top of the hill, where children play and people are constantly exposed to toxic coal-fire gas. *Photo from Stracher and Taylor (2011).*

CHAPTER CONTENTS

Coal and Peat Fires: A Global Perspective
Edited by Glenn B. Stracher

https://doi.org/10.1016/B978-0-12-849885-9.00016-0

Photo by Prasun Gangopadhyay, 2006.

16.1 Human Health Impacts From Coal Fires in Jharia, India

Aarinola Halimah Balogun
Upasana Dhar
Robert B. Finkelman
Sukalyan Chakraborty
Olalekan Olanipekun
Wasim Akram Shaikh
Glenn B. Stracher

Dwellings in Jharia, India, destroyed by a coal fire that was ignited by spontaneous combustion.

Introduction

Burning coal waste piles and **uncontrolled coal fires** have been reported from numerous coal-bearing countries. *Geology of Coal Fires: Case Studies from Around the World* (Stracher, 2007) contain chapters describing these coal fires from New Zealand, Israel, the United States, China, and India. In Volume 2 of *Coal and Peat Fires: A Global Perspective* (Stracher et al., 2013) there are discussions of coal fires in Australia, Azerbaijan, Canada, China, Colombia, Czech Republic, England, France, Germany, India, Borneo, Israel, Italy, Kazakhstan, Poland, Portugal, Romania, Russia, Scotland, South Africa, Spain, the United States, and Venezuela. Volume 3 (Stracher et al., 2015) adds Brazil, Malawi, and Tajikistan. In addition, there are or were coal fires in Egypt, Indonesia, and Norway.

The fires can be caused by spontaneous combustion, sparks from machinery, lightning strikes, grass or forest fires, and intention. Underground fires are notoriously difficult to extinguish and can persist for decades (Guptal, 2016) or millennia (Heffern and Coates, 2004). Combustion of coal waste piles can be quite vigorous (Figure 16.1.1) and both surface and underground fires can be found near human habitation (Figure 16.1.2).

Both underground and surface coal fires mobilize potentially toxic elements such as sulfur, arsenic, selenium, fluorine, lead (Lapham et al., 1980), and mercury (Pone et al., 2007), as well as dangerous organic compounds such as benzene, toluene, xylene, ethyl-benzene and deadly gases such as CO_2 and CO (Pone et al., 2007; Kolker et al., 2009; O'Keefe et al., 2010). Numerous studies have demonstrated that exposure to arsenic, selenium, fluorine, lead, and the like from coal fires cause widespread and severe health problems. Finkelman et al. (1999) describe hyperpigmentation, hyperkeratosis, and squamous cell carcinoma caused by exposure to arsenic mobilized by burning high-As coals in a residential environment in southwest China. In the same region, fluorine released from coal briquettes has caused extensive dental and skeletal fluorosis. Also, in southwest China, burning selenium-rich carbonaceous shales (called stone coal) has caused selenosis, resulting in the loss of hair and finger- and toenails. Burning coal in a residential setting in Xuan Wei County, Yunnan Province, China released fine quartz particles and polycyclic aromatic hydrocarbons that likely contributed to the highest incidence of women's lung cancer in the world (Large et al., 2009). Liang et al. (2014) found substantial amounts of mercury being released from coal fires in northern China. Finkelman and Tian (2017) suggested that mercury released from residential coal fires is the cause of blindness in a village in Guizhou Province, China. Burton (2011) concluded that indoor coal fires in the Czech Republic curbed children's growth.

Considering all these serious health problems caused by the proximity to coal fires, the potential exists for equally serious health problems to be caused by uncontrolled coal fires. It is therefore rather surprising and disappointing that so little has been written about their health impacts. Finkelman (2004) and Finkelman and Stracher (2010) have provided general discussions on the range of potential health problems that can be caused by uncontrolled coal fires. No detailed discussions could be found of the health impacts of the widespread coal fires in the Jharia region of India.

To determine if uncontrolled coal fires have caused health problems, we developed a brief **questionnaire** for villagers and administered them to residents of the Jharia, India region where uncontrolled coal fires have covered an extensive residential area for decades (Figures 16.1.3 and 16.1.4).

Figure 16.1.1. Entrance to the Mulga Gob Fire (top), site of the former underground-bituminous Mulga Coal Mine in the Black Warrior Basin, Jefferson County, Alabama, USA. The height of the burning gob pile (bottom) where the bulldozer is moving waste to create an earthen fire break is about 30 m from its base to the top. Located about 15 miles northwest of Birmingham, Alabama, the fire started around 1991, possibly by spontaneous combustion, and was extinguished in March of 2010. During combustion, residents of the adjacent town of Mulga suffered from respiratory problems. The air for at least a couple of miles from this location was filled with the smell of burning coal and during the gob fire, there were anecdotal reports of an increase in the number of early morning traffic accidents in an adjacent valley because smoke from the fire combined with morning fog, reducing visibility. From Stracher and Carroll (2013).

The Jharia Coalfield

The Jharia coalfield is situated in the Dhanbad district, Jharkhand, India, in the heart of the Damodar River Valley (Figure 16.1.3). It is located about 260 km NW of Kolkata city and about 1150 km SE of Delhi. The coalfield is confined between lat. 23°52′00″ N and long. 86°08′00″–86°30′00″, extending 38 km from east to west and 19 km from north to south, covering an area of 450 km^2, with a coal-bearing area of about 380 km^2 (Pandey et al., 2016).

Figure 16.1.2. An underground coal fire burning since May of 1962 beneath Centralia, Pennsylvania, USA. From Stracher and Nolter (2013) with modifications.

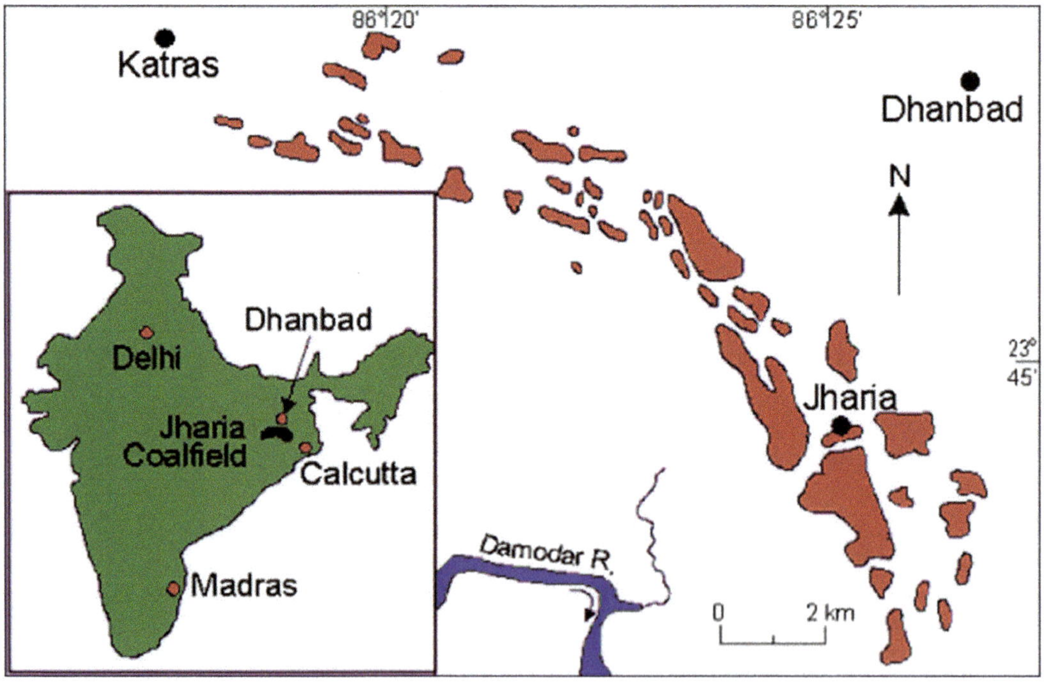

Figure 16.1.3. The Jharia Coalfield, India. The coal deposits are colored red. From Stracher and Taylor (2011).

The Jharia coalfield is the third largest coking coal producing area in the world. It has a rapidly changing landscape and topography due to the mining activities during the past two decades, resulting in extensive barren land and piles of overburden (Gupta et al., 2013).

The present population of the Jharia coalfield region is around 1.3 million, 83% of whom directly or indirectly depend on mining activities, 15% are traders, and the remaining 2% are farmers. There are 16 urban and 172 rural settlements with 68% of the total population living in urban areas. Several residential colonies and miner's dwellings are in areas impacted by fire-induced subsidence; people are often left homeless and ultimately migrate to safer locations. Recently, the Jharia Rehabilitation and Development Agency conducted a demographic survey under a

Figure 16.1.4. A coal fire in an opencast coal mine operated by Bharat Coking Coal Limited in the Jharia coalfield. The mine fire is near Ghanudih village, about 32 km west of Jharia, in the Dhanbad district. The flames on the vertical face are a few meters high. Photo by Sheochandra Prasad, 2008.

master plan to identify unstable houses. It found 98,314 houses, including private homes, encroached by coal fires and the Bharat Coking Coal Limited quarters to be under threat (Pandey et al., 2016).

Depending upon the land use pattern, the entire area has been broadly divided into two zones. One is the area covered mainly by quarries and overburden dumps, which together cover about 25% of the area included within the coalfield. The other zone is the undisturbed surfaces comprising the nonmining area and the underground mining blocks, occupying about 75% area of the field (Ghosh and Ghosh, 1990).

Geology

The Permian Barakar Formation is the main coal-bearing horizon in the Jharia coalfield. The Barren Measures overlay the Barakar Formation followed by the coal-bearing Raniganj Formation. The Barakar Formation does not show any evidence of high-intensity tectonic deformation except for two normal gravity faults (Ghosh and Mukhopadhyay, 1985; Sengupta, 1980). The Barakar Formation consists of buff-colored coarse- to medium-grained feldspathic sandstones, grits, carbonaceous shales, and coal seams. Igneous intrusive rocks (mica peridotities) are also found in wells in the Jharia coalfield. The rock strata contain mainly sandstones and shales. Several minor geological structures such as folds, faults, and fractures are present in the area (Pandey et al., 2016).

Coal Fires

Coal mining in the Jharia coalfield started in 1894 and the first fire was reported in 1916. Mine fires gained widespread attention in India after 1930, with the occurrence of several major fires in the Jharia coalfield and in the Raniganj coalfield. The mining activities intensified in Jharia in the 1920s and thereafter they have been growing extensively and exponentially. At present, there are around 35 large underground and opencast mines in the area (Figure 16.1.4).

The coal fire observed in 1916 was in the 12/14 coal seam at the Bhowrah colliery, followed by a fire in 1917 also located in the 12/14 seam at the Sendra colliery. However, up to 1928, the basin was deeply forested and had good environmental conditions. With increasing mining activities, the coal fires gradually spread over large areas thus endangering the lives of hundreds of thousands of people (Pandey et al., 2016). The spatial coverage of surface and subsurface coal fires was found to change from 0.42 to 2.06 km^2, respectively, in 1992, to 0.33 and 1.36 km^2, respectively, in 1996 and to 0.08 and 1.60 km^2, respectively, in 2001. The propagation of coal fire was found to be more erratic than regular in nature (Chatterjeel et al., 2007).

More than 65 fires were burning in the coalfield in 1996 (Michalski et al., 1997). This likely was the largest complex of above- and underground coal fires in an actively mined area in the world. From the initial fires in 1916 through the present, the fires have spread and grown to affect coal production, the environment, and the health, safety, and wellbeing of over one million people living in the region (Schori et al., 1997). Most of the fires are caused by spontaneous combustion resulting from poor mining practices and the nature of the coal deposits (Munshi, 1995). New fires appear regularly, and many existing fires continue to expand (although a few fires have been extinguished and others reduced in size). Michalski et al. (1997) reported that many fires have coalesced into much larger fires, engulfing adjacent coal seams and collieries. The coalfield contains as many as 46 coal seams averaging 4.5 m thick, but locally up to 55 m in thickness (Munshi and Sen, 1993). It has one of the highest concentrations of thick coal seams in the world (Fox, 1930). About 25 seams account for most of the coal mined. In ascending order, seams 1 through 8 produce boiler fuel whereas seams 9 and higher are best suited for coking purposes. Most of the fires are contained in the uppermost or coking coal seams in the eastern half of the coalfield. The great majority of fires (about 85%) involve a series of burning underground coal seams partially exposed in open pits (quarries). Of these, at least five include burning overburden strata and three burning overburden debris piles. Taking all fires together, active burning conditions extend over 9.4 km^2 (Figure 16.1.5). This is the sum total ground surface area over which fire has been observed (Munshi, 1995). Individual fires extend to maximum depths of 15–140 m below ground surface and average a maximum depth of 50 m. Over time, most of the fires have migrated and changed in size. Where the seam area of the fire increased in size, it expanded, on average, by about half (Michalski, 2004).

In 1994, the Central Mining Research Institute (CMRI) (presently, the Central Institute of Mining and Fuel Research: CIMFR), Dhanbad collected data on 196 fires reported to exist in different coalfields of India. In the Jharia coalfield, among approximately 1864 million tonnes of coal there are about 65 fires spread over an area of 17.32 km^2 affecting the prime coking coal reserve in seams XVII to IV. This constitutes a staggering 12% of the total coal reserves. In 1996, international consultants (M/s GAI, Pittsburgh, USA and M/s MET-CHEM, Montreal, Canada under Joint Venture) studied the problem of the Jharia coalfield under a World Bank–supported project and reported that the area under fire is 9 km^2 (estimated from the surface). The fires not only induced considerable economic losses to the mining companies, but also endangered the lives of the mine workers and their families. About 75% of the coal fires appear to be due to spontaneous combustion (Singh, 2013).

The collieries affected by coal-mine fires are Lodna, Bagdigi, North Tisra, Bhulanbarari, Kujama, Jiyalgarha, Bansjora, Kusunda, and Kendudih in the eastern part of the Jharia coalfield and Dumra, Nadkhurkee, Shatabdi OCP, and Ghanoodih (Sections 16.3 and 16.4 introductory photos) in the western part of the Jharia coalfield. Landsat-7 ETM+ data were used to derive a coal-mine fire map (Figure 16.1.5) and classified the fires into surface and subsurface fires. The map reveals that the coal fires are distributed mostly in the eastern part of the Jharia coalfield. Two major coal fire zones are present in the eastern part of the field: one is the Lodna–Tisra–Bhulanbarari–Kujama–Jellagora area and other is the Kusunda–Kendudih area. The western fire zone, on the other hand, is restricted to the Dumra–Nadkhurkee–Jayramdih area. In addition to these fire zones, another significant fire zone is found west of Ghanoodih. The coal-mine fire map was divided into three categories: (1) temperature >52°C was of high intensity, representing surface fires; (2) temperature >32°C and <52°C was of medium intensity, indicating subsurface fires; and (3) temperature <32°C was of low intensity, indicating the background or water body temperature. Residents near fire-affected areas having opencast mines and dumps are still doing their daily chores despite the intensity of nearby fires (179°C). Children seem to enjoy the heat during winter from the nearby fire area where thermal images show temperatures about 518°C (Figures 16.1.6). The fire having a temperature in the order of 470°C or more was advancing in abandoned opencast mine benches (Mishra et al., 2011).

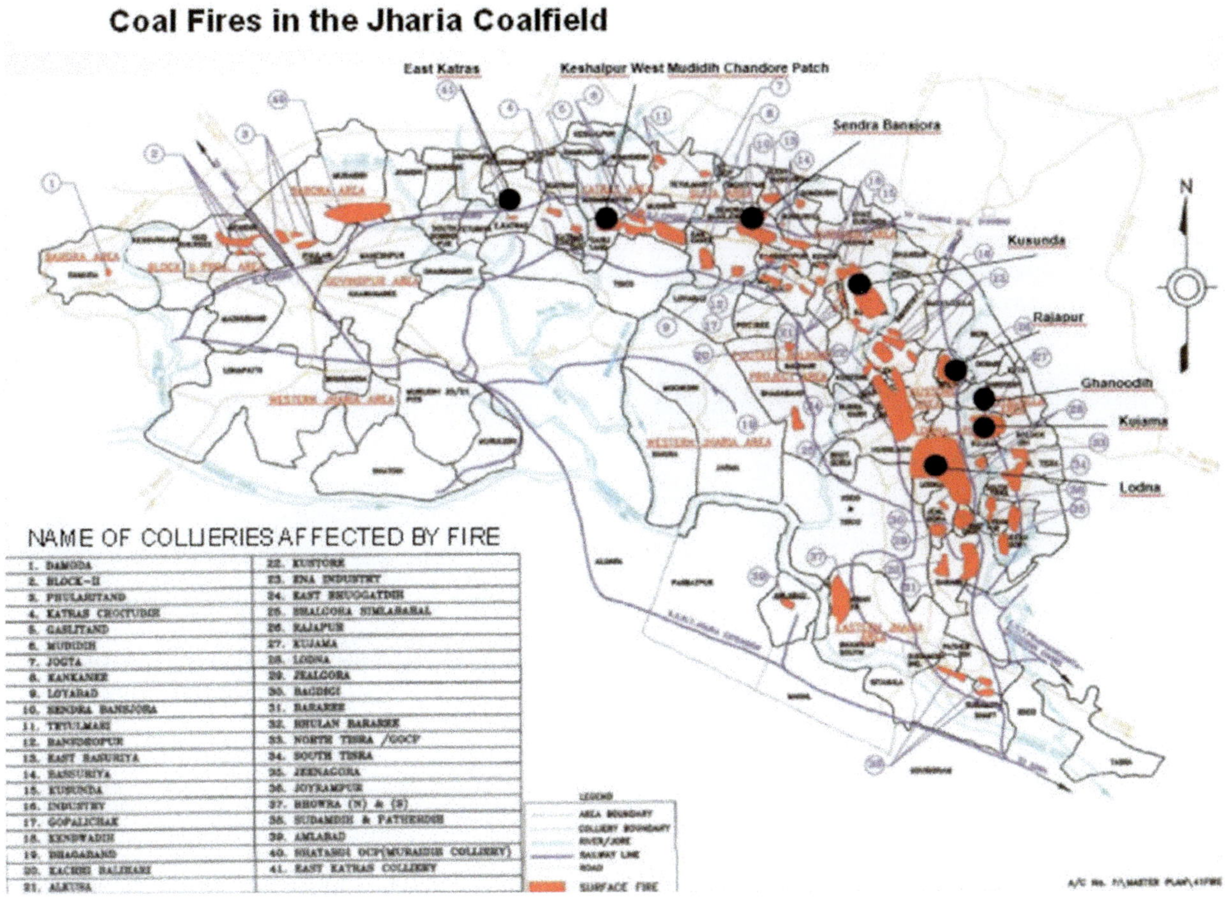

Figure 16.1.5. Map of 41 coal mines (collieries) burning in the Jharia coalfield, India. The location of the relatively new fire at the Ghanoodih mine (Sections 16.3 and 16.4 introductory photos) is plotted on the map, but it is not recorded in the table. From Gielisch and Stracher (2018). See chapter 15 in this book.

Figure 16.1.6. Children exposed to toxic coal-fire gas at a miner's slum in the Jharia coalfield, India. Only the very poorest mining families working in a nearby opencast mine were residing here at the time this photo was taken. Most buildings have cracks in them because of ground subsidence. From Stracher (2007, p. xi); with permission of the Geological Society of America.

16.2 Jharia Coal

From Michalski (2008).

The entrance to a drift-shaft underground coal mine in the Jharia coalfield, state of Jharkhand, India. The coal in underground mines is extracted by the room and pillar (also called board and pillar) or longwall methods of mining.

Coal Analysis

The Jharia coalfield contains subbituminous to high-volatile bituminous rank coking coals. High rank Permian coals (Rv max 0.93%–1.27%) occur in the Jharia basin from the Karharbari Formation (Early Permian), the Barakar Formation (Early Permian), and the Raniganj Formation (Late Permian) (Bhowmik and Dutta, 2013). The coalfield has been categorized as a nonacidic coal field since the coal deposits here are not associated with pyrite bands and the sulfur content in the coal is low (Saini et al., 2016).

Data pertaining to organic petrology and thermal maturity of the Barakar and Raniganj Formations are of importance in determining the coal bed methane potential in the Jharia coalfield. The vitrinite macerals are dominant in the shallower coals of the Barakar Formation. The vitrinite content ranges from 40% to 80%, exinite content varies in the range of 1% to 4%, and the vitrinite reflectance value ranges from 0.83% to 1.69% in the Barakar coal seams of Jharia (Hazra et al., 2003). The volatile matter on a dry ash free-basis indicate that the Barakar coals are high volatile "A" bituminous to low volatile bituminous in rank (Rudra and Hajra, 2009). As the depth increases, in general, it has been observed that ash content increases with consequent decrease in vitrinite content and increase in inertinite content. The range of inertinite content for deeper coals in the Jharia coalfield is from 35% to 80% and vitrinite content ranges from 20% to 62% (Hazra et al., 2003). No information was found on the content of potentially toxic elements in the coals.

16.3 Coal-Fire Health Issues in Jharia

Photo by Sheochandra Prasad, 2009.

A coal fire at the opencast Ghanoodih coal mine operated by Bharat Coking Coal Limited in the Jharia coalfield, India. The miner's dwellings behind the trees near the top of the photo are engulfed in smoke.

Coal Fires and Health

Polluted air may lead to various diseases such as asthma and bronchitis. During rainfall, oxides of nitrogen (NO_x) and sulfur dioxide (SO_2) from the coal fires combine with water to form acidic droplets that contribute to respiratory problems. Coal-mine fires also cast a serious impact on the environment. Oxides of N and S react with volatile organic compounds (released by the coal fires) in the presence of sunlight to form smog, which in turn leads to a number of harmful effects on humans, plants, and animals. Occurrence of acid rain has also been reported in Dhanbad city due to these toxic gases released from the Jharia coalfield (Singh et al., 2007).

In a coal region such as in the Jharia coalfield, which has long been associated with coal fires, large amounts of greenhouse gases like CO_2, CH_4, and oxides of nitrogen are released. Andrup-Henriksen et al. (2007) reported a conservative estimate of three potent greenhouse gases, CO_2, CO and CH_4, at about 108 kg/year for the Jharia coalfield. Saini et al. (2016) reported that in the fire zone, the concentration of respirable particulate matter, SO_2 and NO_x, was higher than the permissible limits, thus posing a danger to the inhabitants of the area. As mentioned earlier, heavy metals along with polycyclic aromatic hydrocarbons can cause a range of health problems such as cancer, neurotoxicity, cardiotoxicity, immunotoxicity, arsenosis, fluorosis, CO poisoning, and so on.

Recent studies in the Jharia coalfield have indicated severe pollution of water bodies in the area. Huge volumes of polluted water from underground mines are channeled into the Damodar River. Coal mining operations in the Jharia coalfield have resulted in the deterioration in the quality of the groundwater (Singh, 1987). Tripathy et al. (2009) studied pollution from heavy metals in soil samples collected from three areas—nonfire, intermediate fire, and fire zones—and reported a high concentration of heavy metals such as Cu, Fe, Mn, Zn, Cd, Cr, and Ni in fire areas as compared to the nonfire and intermediate areas. Rai et al. (2010, 2011) studied the impact of physicochemical properties (bulk density, grain size distribution, pH, electrical conductivity, etc.) on overburden dump materials in the Jharia coalfield and reported the following results: (1) high bulk density resulting in decreased growth of plants, (2) large grain size leading to quick infiltration of water, (3) acidic pH, (4) high to moderate electrical conductivity, and (5) poor organic carbon, available nitrogen, and available phosphorus due to lower amounts of microbial activities in the overburden samples. Thus it is observed that dump material has poor potential for plant growth (Saini et al., 2016).

Masto et al. (2011) collected soil samples from areas near an opencast coal mine and a coal fire affected area located in the Jharia coalfield. The control site was abandoned land approximately 15 km away from the sources of contamination. These samples were analyzed for trace elements including Cr, V, Co, Ni, Cu, Zn, Ga, Rb, Zr, Ba, Th, and U. The soils at the opencast mine were enriched with Cr and Ni; this is attributed to the mining activities in view of the absence of other sources of pollutants. In the case of the coal fire affected area, the soils were enriched with Cr, V, Ni, and Zn. However, the concentrations of Cr, Ni, and Zn in both the soils were well below the US EPA soil screening levels for human health risk assessment. The levels of Co and V exceeded the soil screening limits. Human exposure risks were evaluated for Co and V. The total intake of V concentration exceeded the EPA's reference dose, which may pose adverse health risks.

To what extent the fires affect the health of the people residing there, and what the specific **health issues** are, is still unknown.

Photo by Sheochandra Prasad, 2009.

16.4 Jharia Coal Fires Questionnaire

A coal fire near miner's dwellings at the opencast Ghanoodih coal mine operated by Bharat Coking Coal Limited in the Jharia coalfield, India.

Results of the Jharia Questionnaire

A total of 60 questionnaires (Appendix) were administered in English to residents of two villages in the coal fire region. The distance in miles from the homes of the survey respondents and the location of the coal fires was categorized into two groups: those who live less than 1 mile from the coal fires and those who live further than 5 miles. Thirty-two of the participants reside less than 1 mile from the burning fires (exposed group), and 28 of the interviewees live more than 5 miles from the coal fires (control group).

The health status of the participants was assessed and classified into two groups. The villagers who responded Yes to any respiratory and/or cutaneous symptoms were in one group, and those without any symptoms were categorized in another group. The descriptive information of the respondents such as age, gender, occupation, and the number of people in the household were obtained.

Probit logistic regression was used for the analysis of the data. The health status of the participants is the outcome variable, and the predictors are the villagers' ages along with the distance from their homes to the coal fires. The permissible type 1 error (alpha) level of significance was set at 0.05. The final model was selected with a stepwise procedure with an entry alpha of 0.05 and removal alpha of 0.1. SAS 9.4 was utilized for the statistical analysis.

The distance from the homes of the villagers to the location of the coal fires was a significant predictor of disease status (p-value <0.0001), however, age was not a significant predictor in the model ($p = 0.2203$). The adjusted odds ratio of the distance from the homes of the villagers to the coal fires on disease status having controlled for age is 0.017 (95% Confidence Interval 0.003, 0.084; p-value < 0.0001).

Compared to those who live less than 1 mile from the coal fires and after controlling for the age of respondents, villagers who live more than 5 miles from the burning fires are 98.3% less likely to report health problems, such as respiratory/cutaneous disease symptoms. Alternatively, we can interpret our results by stating that the odds of having health problems (respiratory/cutaneous disease symptoms) are increased by about 58.8 times in the villagers who reside less than 1 mile away from the coal fires, compared to the villagers who live 5 miles or more from the fire.

Figure 16.4.1 shows the results of the data compiled from the responses to the health questionnaire (Appendix). The scatter plot, depicts the total percentage of health problems faced by each individual in the survey versus the

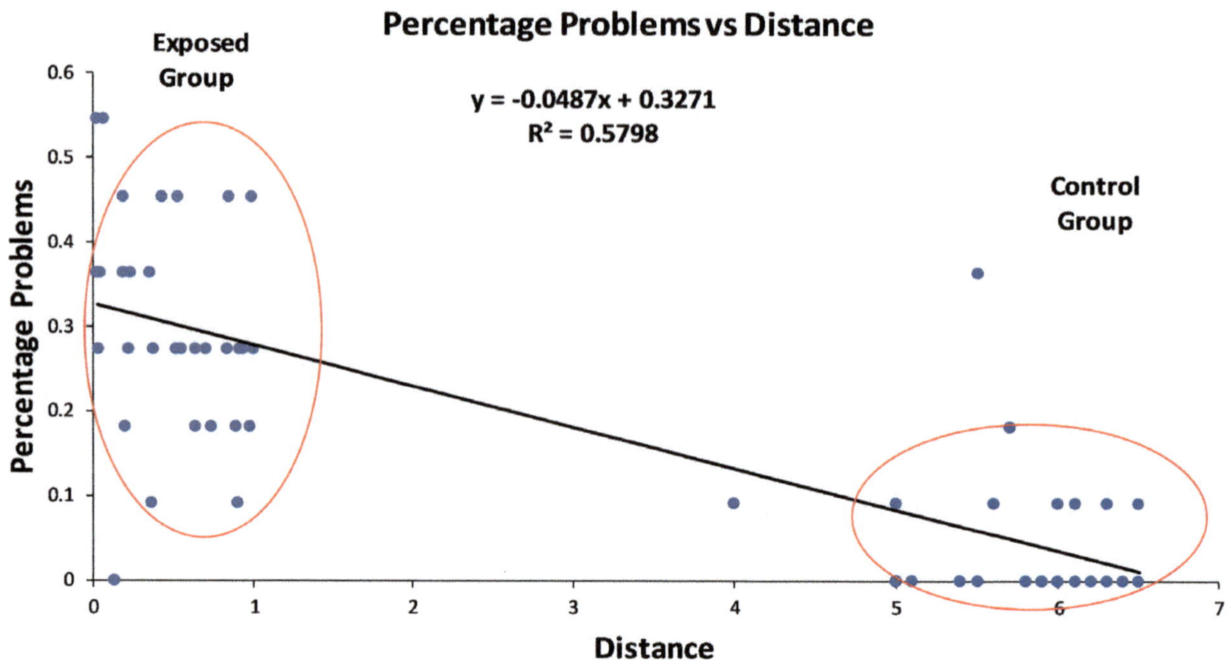

Figure 16.4.1. Correlation between percentage of health problems and distance from the fires.

distance from the uncontrolled coal fires. For the scatter plot, we used the distance as random numbers for each of the two groups—random numbers between 0 and 1 for the exposed group and random numbers between 5 and 6.5 for the control group, staying over 5 miles.

Discussion

It should not be surprising that the responses to the questionnaire reveal a statistically significant greater incidence of health problems in those people living near the uncontrolled coal fires than in those people living some distance away from the fires. Daily exposure to the particulates, acid gases, organic compounds, and potentially toxic elements are taking a toll on the health of the people living closer to these fires.

It is appropriate to ask, What can be done, what should be done? Ideally extinguishing the fires would be the best solution but this approach is often impractical. Millions of dollars were spent in several unsuccessful attempts to extinguish the fires underneath Centralia, PA (Quigley, 2009) but they continue to burn to this day. Moreover, unless extreme precautions are taken the exposed coal and coal waste will often reignite. A more practical approach would be to remove residences from the vicinity of the coal fires and provide the miners with adequate protective gear. This does not mean to imply that the fires should be left unchecked. All reasonable efforts should be made to extinguish the fires, prevent fires from spreading, and prevent new fires from starting.

Our questionnaire contained questions on the general health of the residents including questions about symptoms such as skin lesions, respiratory issues, joint pain, white spots, or lines on their teeth. We made no attempt to determine the cause of the health problems or of the various symptoms. It is essential that a study be conducted to determine the cause or causes of the health problems of those people exposed to the **effluents** of the coal fires. Are the health issues caused by inhalation of particulates, hazardous gases, toxic elements, or a combination of these health hazards? Knowing the cause or causes of the health problems would allow the public health officials to design effective remedies to cure the health problems or prevent them from getting worse.

Important Terms

effluents
health issues
questionnaire
uncontrolled coal fires

References

Andrup-Henriksen, G., Prakash, A., Blake, D.R., 2007. Estimation of gas emissions from shallow subsurface coal fires in Jharia coalfield, India, using FLIR data and coal fire gas analysis. In: Geological Society of America, Abstract with Program, vol. 39, p. 298.

Bhowmik, S., Dutta, P., 2013. Adsorption rate characteristics of methane and CO_2 in coal samples from Raniganj and Jharia coalfields of India. Int. J. Coal Geol. 113 (1), 50–58.

Burton, A., 2011. Coal fire emissions curb children's growth. Environ. Health Perspect. 119, a246.

Chatterjeel, R.S., Wahiduzzaman, M., Shah, A., Raju, E.V.R., Lakhera, R.C., Dadhwal, V.K., 2007. Dynamics of coal fire in Jharia Coalfield, Jharkhand, India during the 1990s as observed from space. Curr. Sci. 92 (1), 10.

Finkelman, R., 2004. Potential human health impacts of burning coal beds and waste piles. Int. J. Coal Geol. 59 (1–2), 19–24.

Finkelman, R.B., Belkin, H.E., Zheng, B., 1999. Health impacts of domestic coal use in China. Proc. Natl. Acad. Sci. U.S.A. 96, 3427–3431.

Finkelman, R.B., Stracher, G.B., 2010. Environmental and health impacts of coal fires. In: Stracher, G.B., Prakash, A., Sokol, E.V. (Eds.), Coal and Peat Fires: A Global Perspective, Volume 1, Coal – Geology and Combustion. Elseviere, The Netherlands, pp. 115–125 (Chapter 7).

Finkelman, R.B., Tian, L., 2017. The Impacts of Coal Mining and Coal Combustion on Health in China: An Overview. International Geology Review. https://doi.org/10.1080/00206814. 2017.1335624 (accessed March 2018).

Fox, F.S., 1930. The Jharia coal field. In: Memoirs of the Geological Survey of India, LVI, 255 p.

Ghosh, R., Ghosh, D.N., 1990. Landuse map of Jharia coalfield, eastern India, aided by remote sensing. J. Indian Soc. Remote Sens. 18 (1 & 2), 23–28.

Ghosh, S.K., Mukhopadhyay, A., 1985. Tectonic history of the Jharia basin — an intracratonic Gondwana basin of eastern India. Q. J. Geol. Mineral. Metall. Soc. India 57, 33–58.

Gielisch, H., Stracher, G.B., 2018. Burning coal-mine collieries in the Jharia coalfield of India. In: Stracher, G.B. (Ed.), Coal and Peat Fires: A Global Perspective, Volume 5: Case Studies – Advances in Field and Laboratory Research. Elsevier, pp. 334–341.

Gupta, N., Anwar, M.S., Soarabh, S., 2013. Mine reclamation planning of Jharia area (Bastacola) through multi temporal remote sensing and GIS. In: 14th ESRI India User Conference, December 10–12, Delhi, India, pp. 1–8. https://www.researchgate.net/profile/Neel_Gupta/publication/282331924_Mine_reclamation_planning_of_Jharia_area_Bastacola_through_multi_temporal_remote_sensing_and_GIS/links/560e8d5508ae4833751 60cc3.pdf (accessed April 2018).

Guptal, J., February 10, 2016. India Creates a Record, Allows Jharia Coalmine Fires to Burn for a Century. The Times of India. https://timesofindia.indiatimes.com/city/kolkata/India-creates-a-record-allows-Jharia-coalmine-fires-to-burn-for-a-century/articleshow/50928024.cms (accessed March 2018).

Hazra, P.N., Rudra, M., Guha, S., Kar, M.K., Basumatary, J.K., Kumar, A., 2003. Geochemical characterization of coalbed gas of Jharia & Raniganj basins and its implications. In: Proceedings of First Association of Petroleum Geologists (APG) Conference & Exhibition, Mussoorie, India, September 28–29.

Heffern, E.L., Coates, D.A., 2004. Geologic history of natural coal-bed fires, Powder River basin, USA. In: Stracher, G.B. (Ed.), Coal Fires Burning Around the World, pp. 25–47 Int. J. Coal Geol., 59, 1–2.

Kolker, A., Engle, M., Stracher, G., Hower, J., Prakash, A., Radke, L., ter Schure, A., Heffern, E., 2009. Emissions from Coal Fires and Their Impact on the Environment. U.S. Geological Survey Open File Report 3084. https://pubs.usgs.gov/fs/2009/3084/pdf/fs2009-3084.pdf (accessed March 2018).

Lapham, D.M., Downey Jr., W.F., Finkelman, R.B., 1980. Mineralogy Associated with Burning Anthracite Deposits of Eastern Pennsylvania: Pennsylvania Geological Survey. Mineral Resource Report 79, 82 p.

Large, D.J., Kelly, S., Spiro, B., Tian, L., Longyi, S., Finkelman, R., Zhang, M., Somerfield, C., Plint, S., Ali, Y., Zhou, Y., 2009. Silica-volatile interaction and the geological cause of the Xuan Wei lung cancer epidemic. Environ. Sci. Technol. 43 (23), 9016–9021.

Liang, Y., Liang, H., Zhu, S., 2014. Mercury emission from coal seam fires at Wuda, Inner Mongolia of China. Atmos. Environ. 83, 176–184.

Masto, R.E., Ram, L.C., George, J., Selvi, V.A., Sinha, A.K., Verma, S.K., Rout, T.K., Priyadarshini, Prabal, P., 2011. Impacts of opencast coal mine and mine fire on the trace elements' content of the surrounding soil vis-à-vis human health risk. Toxicol. Environ. Chem. 93 (2), 223–237.

Michalski, S.R., 2004. The Jharia mine fire control technical assistance project: an analysis. In: Stracher, G.B. (Ed.), Coal Fires Burning Around the World, pp. 83–90 Int. J. Coal Geol., 59, 1–2.

Michalski, S.R., 2008. Coal-mining techniques and coal fires. In: Stracher, G.B., Prakash, A., Sokol, E.V. (Eds.), Coal and Peat Fires: A Global Perspective, Volume 1: Coal – Geology and Combustion. Elsevier, pp. 29–46.

Michalski, S.R., Custer Jr., E.S., Munshi, P.L., 1997. Investigation of the Jharia coalfield mine fires – India. Proc. Am. Soc. Min. Reclam. 211–223.

Mishra, R.K., Bahuguna, P.P., Singh, V.K., 2011. Detection of coal mine fire in Jharia coal field using Landsat-7 ETM+ data. Int. J. Coal Geol. 86 (1), 73–78.

Munshi, P.L., 1995. Experience of dealing with fires in Jharia coalfield – introduction to diagnostic technique. In: National Seminar on Mine Fires. (Varanasi. Feb. 24-25, 1995). Dept. of Mining Engineering, Institute of Technology, Banaras Hindu University, pp. 82–90.

Munshi, P.L., Sen, S.K., 1993. Status of fires in BCCL mines and abatement Measures. In: Seminar on Damodar Basin – Environmental Problem and Issues. (Dec. 1993). Dhanbad, India.

O'Keefe, J.M.K., Henke, K.R., Hower, J.C., Engle, M.A., Stracher, G.B., Stucker, J.D., Drew, J.W., Staggs, W.D., Murray, T.M., Hammond III, M.L., Adkins, K.D., Mullins, B.J., Lemley, E.W., 2010. CO_2, CO, and Hg emissions from the Truman Shepherd and Ruth Mullins coal fires, eastern Kentucky, USA. Sci. Total Environ. 408 (7), 1628–1633.

Pandey, J., Kumar, D., Singh, V.K., Mohalik, N.K., 2016. Environmental and socio-economic impacts of fire in Jharia coalfield, Jharkhand, India: an appraisal. Curr. Sci. 110 (9), 1639–1650.

Pone, J.D., Hein, K.A.A., Stracher, G.B., Annegarn, H.J., Finkelman, R.B., Blake, D.R., McCormack, J.K., Schroeder, P., 2007. The spontaneous combustion of coal and its by-products in the Witbank and Sasolburg coalfields of South Africa. Int. J. Coal Geol. 72, 124–140.

Quigley, J., 2009. The Day the Earth Caved in. Random House, NY. 237 p.

Rai, A.K., Paul, B., Singh, G.A., 2010. Study on the bulk density and its effect on the growth of selected grasses in coal mine overburden dumps, Jharkhand, India. Int. J. Environ. Sci. 1, 677–684.

Rai, A.K., Paul, B., Singh, G.A., 2011. Study on physico chemical properties of over-burden dump materials from selected coal mining areas of Jharia coalfields, Jharkhand, India. Int. J. Environ. Sci. 1, 1350–1360.

Rudra, M., Hajra, P.N., 2009. Isotopic Composition of Coalbed Methane Desorbed from Barakar Coals of Damodar Valley Gondwana Coalfields and Its Implication. Petrotech 09–496 Jan. 11–15, New Delhi, India, ID No.: P496 , pp. 1–6. https://www.researchgate.net/profile/Malay_Rudra/publication/237510627_ISOTOPIC_COMPOSITION_OF_COALBED_METHANE_DESORBED_FROM_BARAKAR_COALS_OF_DAMODAR_VALLEY_GONDWANA_COALFIELDS_AND_ITS_IMPLICATION/links/595d0d2b0f7e9b3aefade6d1/ISOTOPIC-COMPOSITION-OF-COALBED-METHANE-DESORBED-FROM-BARAKAR-COALS-OF-DAMODAR-VALLEY-GONDWANA-COALFIELDS-AND-ITS- IMPLICATION.pdf (accessed April 2018).

Saini, V., Gupta, R.P., Arora, M.K., 2016. Environmental impact studies in coalfields in India: a case study from Jharia coal-field. Renew. Sustain. Energy Rev. 53, 1222–1239.

Schori, A., Scrymgeour, A.H., Munshi, P.L., 1997. Environment management plan for the Jharia Coalfield. In: Proceedings 14th Annual Meeting of the American Society for Surface Mining and Reclamation, (CD-ROM), Austin, TX, May 10–15, pp. 224–228.

Sengupta, N., 1980. A Revision of the Geology of the Jharia Coalfield with Particular Reference to Distribution of Coal Seams (Ph.D. Thesis). Indian School of Mines, Dhanbad, India.

Singh, G., 1987. Mine water quality deterioration due to acid mine drainage. Int. J. Mine Water 6 (1), 49–61.

Singh, A.K., Mondal, G.C., Kumar, S., Singh, K.K., Kamal, K.P., Sinha, A., 2007. Precipitation chemistry and occurrence of acid rain over Dhanbad, coal city of India. Environ. Monit. Assess. 125 (1–3), 99–110.

Singh, R.V.K., 2013. Spontaneous heating and fire in coal mines. Procedia Eng. 62, 78–90.

Stracher, G.B., 2007. Geology of coal fires: case studies from around the world. In: Reviews in Engineering Geology XVIII. The Geological Society of America, Boulder, Co 283 p.

Stracher, G.B., Carroll, R.E., 2013. United States coal fires: Alabama. In: Stracher, G.B., Prakash, A., Sokol, E.V. (Eds.), Coal and Peat Fires: A Global Perspective, Volume 2: Photographs and Multimedia Tours. Elsevier, pp. 446–450.

Stracher, G.B., Nolter, M.A., 2013. Pennsylvania: the centralia mine fire. In: Stracher, G.B., Prakash, A., Sokol, E.V. (Eds.), Coal and Peat Fires: A Global Perspective, Volume 2: Photographs and Multimedia Tours. Elsevier, pp. 498–509.

Stracher, G.B., Prakash, A., Sokol, E.V. (Eds.), 2013. Coal and Peat Fires: A Global Perspective. Volume 2: Photographs and Multimedia Tours. Elsevier. 554 p.

Stracher, G.B., Sokol, E.V., Prakash, A. (Eds.), 2015. Coal and Peat Fires: A Global Perspective, Volume 3: Case Studies - Coal Fires. Elsevier. 786 p.

Stracher, G.B., Taylor, T.P., 2011. The effects of global coal fires. In: Stracher, G.B., Prakash, A., Sokol, E.V. (Eds.), Coal and Peat Fires: A Global Perspective, Volume 1: Coal – Geology and Combustion. Elsevier, pp. 101–114.

Tripathy, D.P., Panigrahi, D.C., Singh, G., 2009. Determination of soil pollution index and soil infiltration rates in some non-fire and fire areas of Jharia coal field. In: Proceedings of the 9th International Mine Ventilation Congress.

Additional Reading

Finkelman, R.B., 2010. Potential human health impacts from coal fire emissions. In: Drebenstedt, C., Fischer, C., Meyer, U., Jianjun, W., Bing, K. (Eds.), Latest Developments in Coal Fire Research, Second International Conference on Coal Fire Research (ICCFR2), May 19-21, Berlin, Germany, pp. 299–300.

Gurgaon Workers News, 2015. Gurgaon Workers News no.9/40, Gurgaon, State of Haryana, India. https://gurgaonworkersnews.wordpress.com (accessed April 2018).

Appendix

Questionnaire

Respondent's Information

Name:

Date of birth:

Gender:

Number of people in the household:

Occupation:

(Instructions: Please select the answers below that apply.)

Exposure Survey

1. Do you see coal fires in your village?	Yes	No
2. How long has the fire been burning: 1–5 years, 5–10 years, Over 10 years		
3. Do you live on the coal mines?	Yes	No
4. If no, how close do you live to the burning fires? <1 mile, 1–3 miles, >3 miles		
5. How long have you lived in the house? 1–5 years, 5–10 years, Over 10 years		
6. Is coal used for cooking in your house?	Yes	No
7. Is coal used for drying in your house?	Yes	No
8. Is coal for used for staying warm in your house?	Yes	No
9. Do you stay away from the coal when using it in the house?	Yes	No

Human Health Survey

10. How would you describe your health?		
11. How often do you visit the hospital?		
12. When was the last time you visited the hospital/clinic?		
13. What was the reason for visit? Routine check-up or Ill-health		
14. If ill-health, what was the diagnosis?		
15. Do you have white lines or spots on your teeth? If yes, for how long?	Yes	No
16. Do you have joint pain? If yes, for how long?	Yes	No
17. Do you have dark spots on your skin? If yes, for how long?	Yes	No
18. Do you have red/brown spots on your body? If yes, for how long?	Yes	No
19. Do you have any wounds on your body? If yes, for how long?	Yes	No
20. Do you find it hard to breath? If yes, how often?	Yes	No

21. Does your nose bleed? Yes No
    If yes, for how long?
22. Do you cough out blood? Yes No
    If yes, for how long?
23. Do you cough out colored mucus? Yes No
    If yes, for how long?
24. Do you have thick and sticky saliva? Yes No
    If yes, for how long?

Illustrations

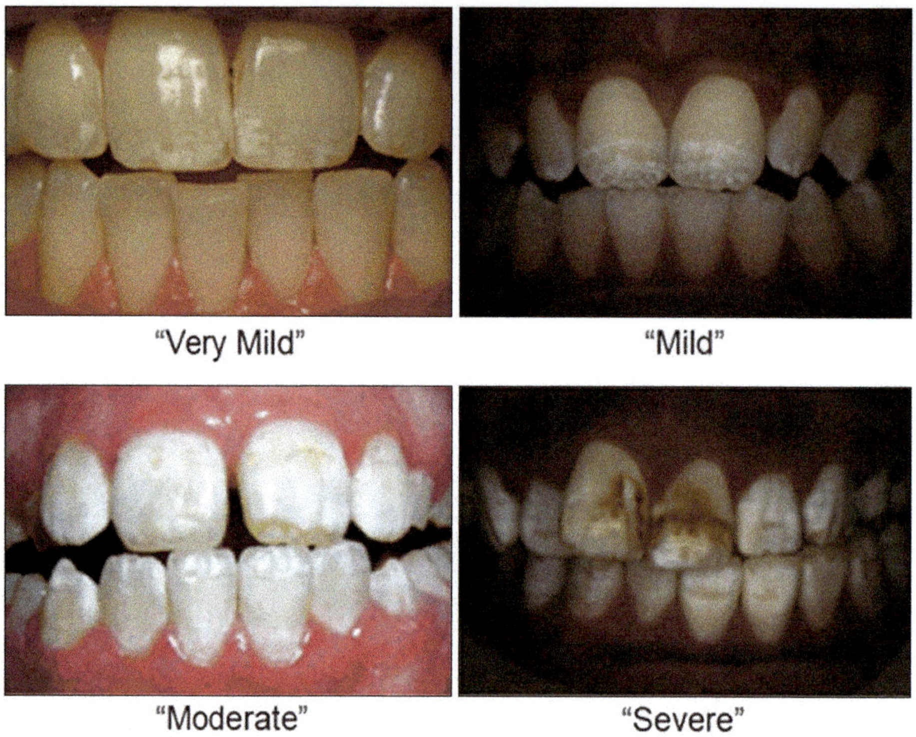

Appendix Figure 1. *White lines* on the teeth (see question 15).

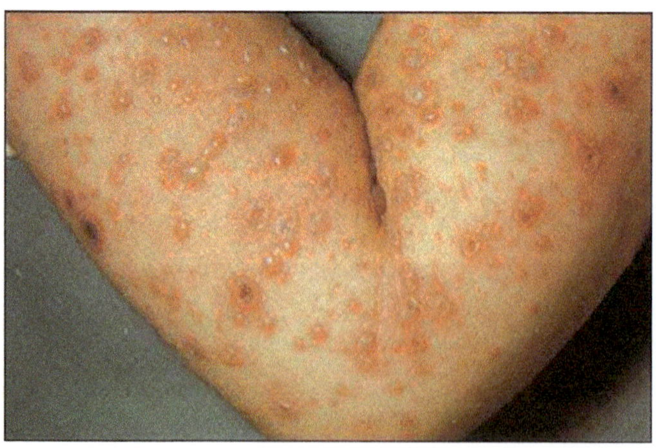

Appendix Figure 2. *Red spots* on the skin (see questions 17 and 18).

CHAPTER 17

Environmental Monitoring in the Jharia Coalfield, India: Vegetation Indices and Surface Temperature Measurements

Coal fire in an opencast mine in the Bokapahari area of the Jharia coalfield, India. The photographs show the mine in the (A) morning; (B) evening, and (C) night. The entire coal seam is on fire and flames are visible with the naked eye in the evening. The distance across the landscape is about 130 m. *Photo by Varinder Saini, 2014.*

CHAPTER CONTENTS

Coal and Peat Fires: A Global Perspective
Edited by Glenn B. Stracher

https://doi.org/10.1016/B978-0-12-849885-9.00017-2

17.1 Environmental Monitoring in the Jharia Coalfield

Varinder Saini
Ravi P. Gupta
Manoj K. Arora

Photo by Varinder Saini, 2014.

Flames emitted from a surface crack in a bituminous coking-coal seam, Rajapur opencast mine, Jharia coalfield, India. The flames are about 1 m high.

Introduction

Coal is one of the most abundantly available fossil fuels around the globe and meets a major part of the energy needs for human consumption. It is used in various industries like power, steel, cement, alumina refineries, and for domestic purposes. Coal accounts for 55% of India's energy needs (MCGI, 2015) and is the major source for power generation. India is the third largest coal producer in the world after China and the United States (CDI, 2014). Geological Survey of India has estimated a cumulative total of 301.5 billion tons of coal resources in the country as of April 1, 2014. The State of Jharkhand tops the list with maximum coal reserves at 80 billion tons (MCGI, 2015). In Jharia, mining started toward the end of the 19th century (Gee, 1940). Mining activities intensified in 1925. The Jharia coalfield (JCF) is important in the Indian context as it is the most important storehouse of prime coking coal that feeds the major percentage of industrial demand.

However, **coal mining** adversely affects the environment of the area. A typical coal mining area has many negative **environmental impacts** such as deforestation; changes in vegetation pattern; damage to the ecological system; loss of agricultural land; loss of biodiversity; degradation of the air, water, and soil quality; and changes in topography and land use/land cover (LULC). Apart from these, surface and subsurface fire is also a recurring cause of concern in some coalfields, as these also impact environment and pose operational difficulty in mining by increasing the cost of production (Saini et al., 2015a,b). The areas affected by surface mining have a bearing on the climatic parameters and eco-hydrological characteristics. Further, deterioration of the environment limits growth of vegetation, which is a key indicator of a healthy ecological system (Van Wijngaarden, 1991; Xia et al., 2007). However, this domain has received only marginal attention (Wechsung, 2000; Pokornỳ, 2001; Pecharova et al., 2006; Pokornỳ et al., 2007; Moreno-de las Heras et al., 2009; He and Yin, 2010).

In JCF alone, a large area is currently affected by surface and subsurface mining, out of which a significant portion is without vegetation cover. Also, the vegetation pattern changes. For example, the diverse vegetation with full grown trees and undercover is changed to herbaceous vegetation growing on the overburden (OB) coal dumps and on areas immediately around the mines. JCF also frequently encounters **coal fires**. This aspect cannot be ignored in any study conducted in this coalfield. A comprehensive review of different environmental impact assessment studies carried out in JCF have been reported in Saini et al. (2016).

The focus of this chapter is on the specific aspects of geoecology functioning at different spatial and temporal scales with reference to JCF. It covers two broad topics, changing pattern of vegetation density and health, and impact of high surface temperature areas on the vegetation, as fires lead to reduction in potential of soil to support plant growth. Further, an analysis showing the spatio-temporal variation of fire areas along with a **correlation** between high temperature areas and vegetation has also been included.

Location

The JCF is located in the Dhanbad district of the State of Jharkhand, and is named after the main mining center of Jharia (Figure 17.1.1). It is situated in the heart of the Damodar river valley and is about 250 km northwest of Kolkata. JCF covers an area about 450 km^2 confined between latitude 23° 38′ N - 23° 50′ N and longitude 86° 07′ E - 86° 30′ E, with an average elevation of about 220 m above mean sea level. JCF is well connected by road, rail, and air. Two national highways pass through the area. Eastern railways and southeastern railways connect the coalfield to other parts of the country. The coalfield is sickle-shaped with the basin axis running roughly NW–SE. The maximum length is about 38 km from east to west and 19 km from north to south.

The sickle-shaped JCF is shown in Landsat TM FCC (Figure 17.1.1). On the standard FCC, densely vegetated areas appear in shades of red and sparsely vegetated areas in faint red. Water bodies, coal bands, and coal dumps appear black and are indistinguishable from each other. Barren wastelands and OB dumps appear in shades of dirty white and gray. However, OB dumps appear in the vicinity of mining areas. For processing, analysis, and interpretation,

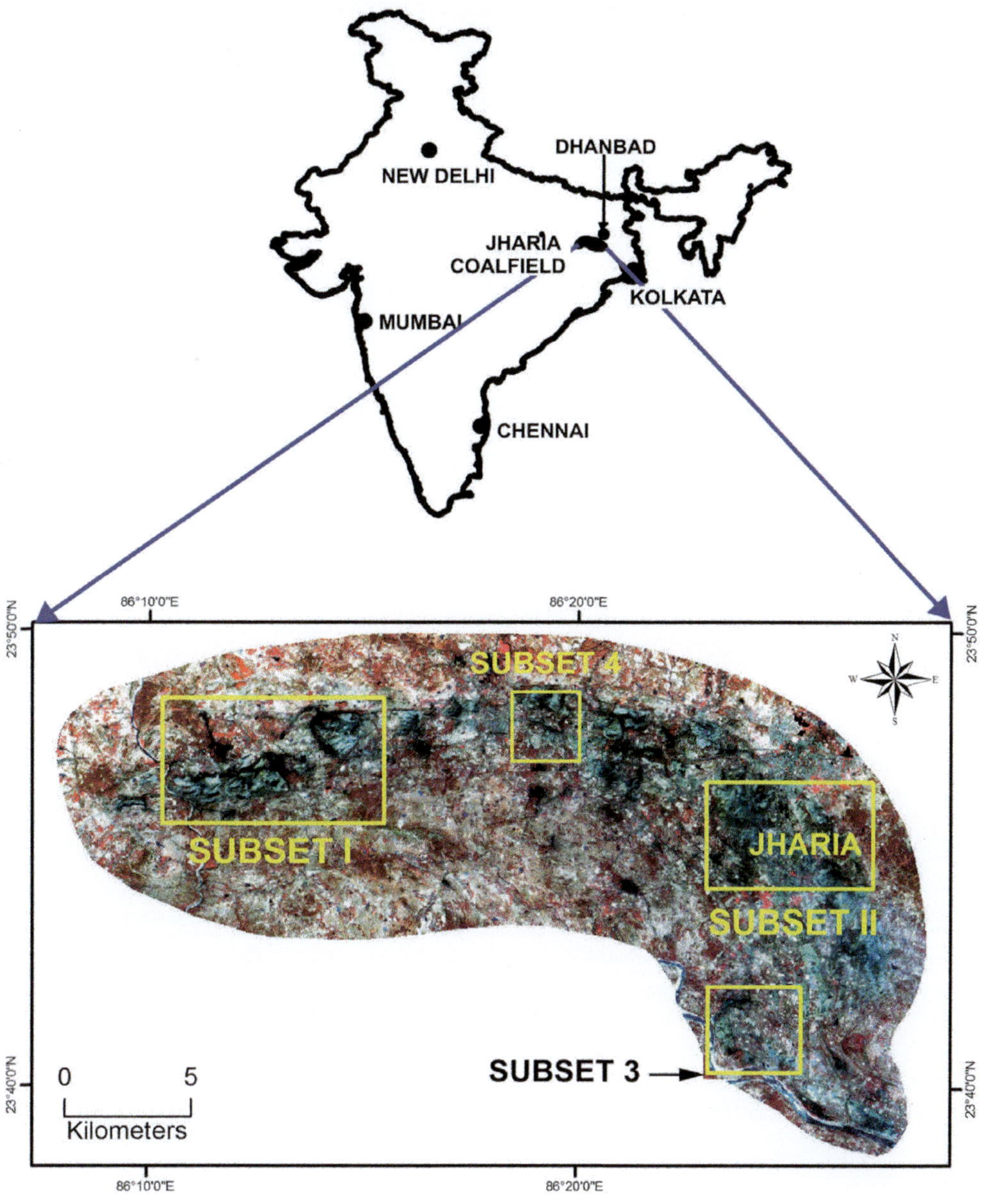

Figure 17.1.1. Location map of the Jharia coalfield. Landsat image (FCC: R=NIR, G=Red, B=Green, January 2010) of the study area showing subsets 1–4.

two peculiar subsets from the image have been created, and have been named Shatabdi opencast mine area (called subset 1, area: 4410 ha) and mining area in and around the main Jharia town (called subset 2, area: 2963 ha) (yellow rectangles shown in Figure 17.1.1), for ease in understanding.

Additionally, for studying changes in vegetation cover using the normalized difference vegetation index (NDVI) over a short span of time, Resourcesat-2 LISS IV images have been used. Two new subsets of LISS IV images—Bhowrah colliery and surrounding areas (called as subset 3, area: 1891 ha) and Angarpathar area (called as subset 4, area: 988 ha)—have been used for the analysis (discussed in detail in Section 17.3.3). The reason for using new subsets for this analysis is that the LISS IV images have a high resolution (5.8 m) as compared to Landsat images (30 m), therefore the changes in smaller subsets could be seen with greater clarity using LISS IV data.

Photo by Varinder Saini, 2012.

17.2 Surface Temperatures

A fire in a barren area near an overburden dump in Lodna, India. The fire migrated to here from a nearby bituminous coking coal-mine on fire in Lodna.
The temperature here, measured by Varinder Saini using an infrared thermometer, was 73°C. The crack is about 10–12 m long and 50 cm wide.

Introduction

Coal fire is a ubiquitous phenomenon prevalent in a number of coalfields all over the world. Coal fires are marked by a sudden rise in ground temperature relative to the background. This rise is very conspicuous when the fire is exposed on the surface, but it is less pronounced in the case of subsurface fires. As the coal fire makes the area inaccessible, collection of data in regions of coal fire through field surveys becomes extremely dangerous and difficult. Remote sensing images acquired in **thermal infrared** regions of electromagnetic spectrum may therefore serve as a key dataset to study fires.

Land surface temperature (LST) has been defined as the temperature of the interface between the earth's surface and its closest atmosphere, and thus it is a crucial variable to understand the land–atmospheric interactions. Deriving LST from satellite sensor-based thermal infrared (TIR) measurements require radiometric calibration, **emissivity**, and atmospheric corrections (Vidal, 1991; Li and Becker, 1993). A number of different approaches have been proposed to derive LST from TIR bands, using a variety of methods, which account for emissivity and atmospheric effects (Price, 1983, 1984; Susskind et al., 1984; Becker and Li, 1990; Hook et al., 1992; Kerr et al., 1992; Kealy and Hook, 1993; Wan and Dozier, 1996; Pozo Vazquez et al., 1997; Wan and Li, 1997; Gillespie et al., 1998; Qin et al., 2001; Tonooka, 2001; Jiménez-Muñoz and Sobrino, 2003) as inputs to the computation model of LST.

In the present work, the NDVI-based emissivity method has been used to derive LST (Valor and Caselles, 1996; Sobrino and Raissouni, 2000). It is based on a statistical relationship between the NDVI derived from the visible- and near-infrared bands and the land surface emissivity (LSE) in the TIR bands (Li et al., 2013). The amount of vegetation determines the LST by latent heat flux from the surface to atmosphere via evapotranspiration. Areas with high NDVI are usually found to have lower LSTs and vice versa (Yuan and Bauer, 2007). In this work, LST has been determined to identify the fire areas, investigate its spatio-temporal pattern, and establish its correlation with vegetation (discussed in Section 17.3.4).

The spectral data in the form of digital number (DN) values of pixels of the TIR image have been converted into radiant temperature and subsequently to LST. The LST values have been suitably adjusted using emissivity values obtained from regression relationships with NDVI (Eqs. 17.2.5 and 17.2.6). The field measured temperatures using an infrared thermometer have been used to validate the LST derived from TIR images.

LST Determination

Landsat TM and ETM+ sensors have only one thermal band. Therefore, the well-known and accepted split-window algorithm (Sobrino et al., 2001) and temperature/emissivity separation method (Gillespie et al., 1998) may not be applied. Therefore, an NDVI-based single-window algorithm for deriving LSE has been used here. Out of the different approaches given in the literature (Van de Griend and Owe, 1993; Valor and Caselles, 1996; Sobrino and Raissouni, 2000), a modification of the NDVI Thresholds Method ($NDVI^{THM}$) (Van de Griend and Owe, 1993) has been used here. It involves following steps to retrieve LST from TIR images.

Calculation of Radiance Images

The raw DN values are converted to radiance using Eq. (17.2.1) (Chander et al., 2009):

$$L^* = (L_{max} - L_{min})\ DN + L_{min} / DN_{max} \quad (17.2.1)$$

where L* is the spectral radiance received at the sensor; L_{min} and L_{max} are the minimum and the maximum spectral radiance for the sensor, respectively; DN_{max} is the maximum DN. For TIR, Eq. (17.2.2) is used (USGS, 2013):

$$L^* = M_L \cdot DN + A_L \quad (17.2.2)$$

where M_L is the band-specific multiplicative rescaling factor and A_L is the band-specific additive rescaling factor. The values of these factors are obtained from the metadata.

Calculation of Radiant Temperature

The radiance images derived from thermal bands are used to calculate radiant temperature using Eq. (17.2.3) (Chander et al., 2009):

$$Tr = K2 / \ln (K1/L^* + 1) \quad (17.2.3)$$

where T_r is the radiant temperature (in Kelvin); K1 and K2 are the prelaunched calibration constants 1 and 2, respectively ($W/(m^2 \cdot sr \cdot \mu m)$).

Creation of NDVI Images

NDVI is based on observed reflectance of vegetation in red and **near-infrared (NIR)** bands of the Landsat TM or ETM+ sensors. Atmospherically corrected bands have been used as the NDVI values derived from these may be more representative of the natural surfaces than derived from the raw DNs. NDVI values for each pixel have been calculated using Eq. (17.2.4) (Rouse et al., 1973):

$$NDVI = NIR - Red / NIR + Red \quad (17.2.4)$$

where red and NIR are the spectral reflectance of vegetation in the red band and the NIR band, respectively.

Calculation of Emissivity

Sobrino et al. (2004) considered 49 soil spectra included in the ASTER spectral library and obtained the following expression for deriving emissivity from pixels having proportion of both soil and vegetation:

$$\varepsilon = 0.004 \cdot p_v + 0.986 \quad (17.2.5)$$

where p_v is the proportion of vegetation, derived from the NDVI image using Eq. (17.2.6) (Carlson and Ripley, 1997):

$$p_v = [NDVI - NDVI_{min} / NDVI_{max} - NDVI_{min}]^2 \quad (17.2.6)$$

where $NDVI_{max}$ and $NDVI_{min}$ are assumed to be 0.5 and 0.2, respectively, for scenes, which have the presence of bare soil and vegetation, respectively. It has been reported that this equation is most suitable if bare soil and vegetation appear in a pixel (Sobrino et al., 2001), which is the case in the JCF.

Calculation of LST

Eqs. (17.2.3) and (17.2.5) are used to determine LST using Eq. (17.2.7) (Weng et al., 2004):

$$LST = Tr / 1 + (\lambda \cdot Tr/\rho) \ln\varepsilon \tag{17.2.7}$$

where λ is the central wavelength (in μm) of the Landsat TM or ETM+ TIR band; ϱ is assumed to be 1.438×10^{-2} m K.

LST was computed from the TIR bands of all the Landsat TM, ETM+ and OLI-TIRS images procured for the period 1991 to 2013 for the two subsets. These images were used to study the spatio-temporal pattern of temperature in the JCF.

Field-Based Temperatures

A handheld infrared thermometer was used to estimate the ground temperature at different surfaces and in proximity of cracks/fire areas. This instrument operates in a wide range of temperatures (−20 to 560°C) over a spectral coverage from 8 to 14 μm. A laser pointer makes it easy to point to an object from a distance. Two field expeditions—first during December 12–19, 2012 and second during January 10–19, 2014—were conducted in JCF. The temperature was recorded at a total of 92 sample locations during the two field surveys. The sample locations correspond to areas having different surface conditions such as open pit mines, OB dumps, coal dumps (both under fire and non-fire conditions); vegetated areas; barren areas; and water bodies among others. For the majority of sample locations, the distance from the object was kept as 1 m. However, for areas with surface fires visible, the observations were recorded from a distance of 3–5 m. The GPS coordinates of each sample location were also measured using a Garmin handheld GPS device and are given in Table 17.2.1. Figure 17.2.1 shows the field photographs of high temperature measurements at some of the locations in JCF.

Remote Sensing LST

From numerous available Landsat OLI-TIRS images, the image acquired on December 13, 2013 was used to derive the LST image. The field data acquired during the second field expedition were used to validate the LST, as it was the closest one to the date of acquisition of the satellite image. Of the 92 sample locations, field data of only 9 sample locations, representing each type of land cover, have been taken for validation. The field observed temperature values and the corresponding LST values at these locations are shown in the form of bar graphs (Figure 17.2.2).

Table 17.2.1
Sampling data for the Jharia coalfield, India.

Sampling Number	Land Cover	Location	Latitude (N) (DMS)*	Longitude (E) (DMS)	Field Temp. (°K)	LST† (°K)
1	Vegetated area	Sijua	23°47′ 01″	86°19′ 49″	298.5	296.5
2	Barren area	Sijua	23°46′ 59″	86°19′ 46″	301.2	298.1
3	Water body	Shamdih	23°49′ 16″	86°16′ 50″	297.8	293.8
4	OB‡ dump with fire	Kankanee	23°46′ 60″	86°20′ 47″	705.4	313.5
5	OB dump without fire	Lodna	23°42′ 55″	86°25′ 20″	299.4	298.4
6	Coal dump with fire	Shatabdi OCM§	23°47′ 30″	86°14′ 07″	741.2	299.3
7	Coal dump without fire	Sudamdih	23°39′ 25″	86°25′ 54″	297.2	295.5
8	Mine with fire	Rajapur	23°45′ 03″	86°25′ 07″	777.2	313.2
9	Mine without fire	Bhowrah	23°40′ 42″	86°24′ 02″	297.8	297.6

*DMS, degree, minute, second.
†LST, land surface temperature.
‡OB, overburden dump.
§OCM, opencast mine.

Figure 17.2.1. Field photographs showing temperature measurements at some of the sample locations in JCF. These temperature data correspond to the LST derived from Landsat TIR image. (A) Surface temperature (50°C) in a mining area, location: Ena colliery; (B) temperature (396°C) of a surface fire, location: Chasnalla; (C) temperature (73°C) of an overburden dump, location: Kankanee; (D) temperature (100°C) of a freshly developing crack, location: Rajapur mine; (E) temperature (520°C) of a subsurface fire, location: Ghanudih; (F) temperature (468°C) of fire in a coal dump, location: Shatabdi opencast mine. Photos by Varinder Saini, 2012 and 2014.

The temperature from field and LST values are comparable for the majority of the locations except those under fire. The reason for any mismatch may be attributed to the coarse spatial resolution of the TIR band of the remote sensing sensor (i.e., 100 m) vis-à-vis field of view (FOV) of the field thermometer, as it directly focuses on fires only due to close-range and its small FOV. As a result, small or deep coal fires often would not be detected. But for low temperature areas, the field temperature corroborates well with LST values.

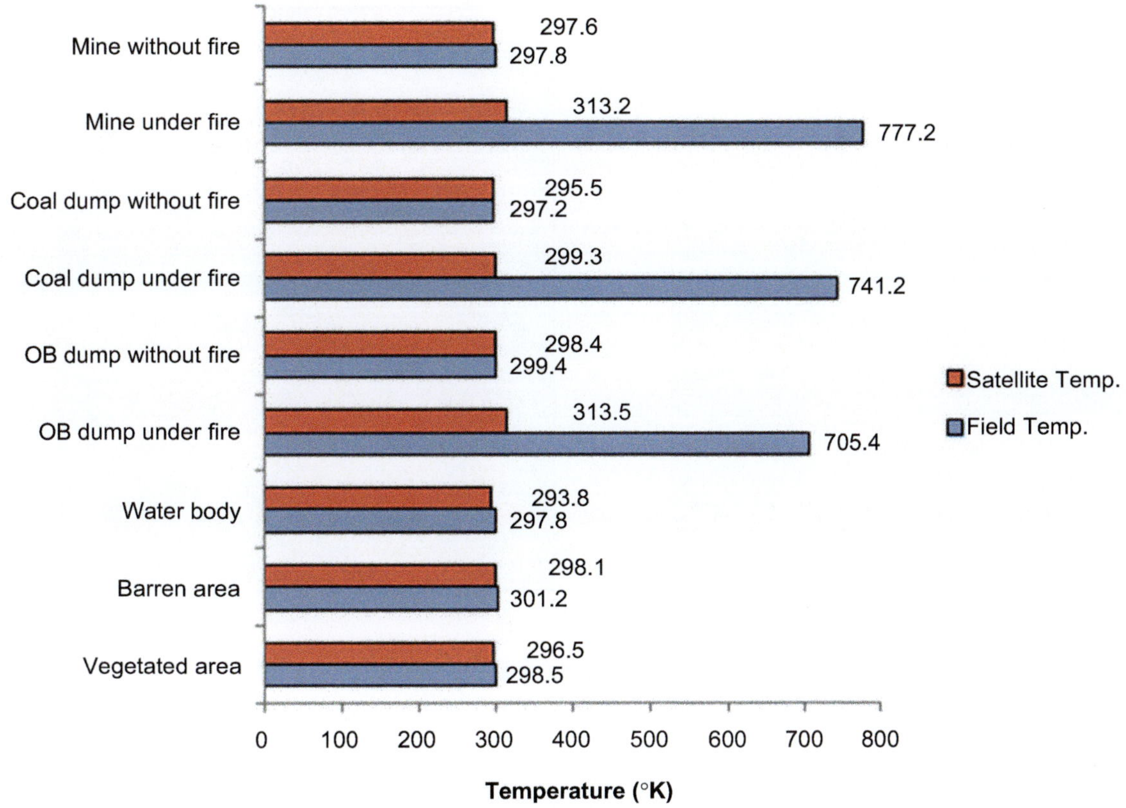

Figure 17.2.2. Bar graphs showing comparison of field measured temperature and LST derived from TIR image over different sample locations. Note that except the areas engulfed with fire (including mines, coal dumps, and OB dumps under fire), there is a good correlation between field temperature and LST values.

Spatio-Temporal Analysis of LST Images

Figures 17.2.3 and 17.2.4 show the spatial variation of LST over a period of 23 years from 1991 to 2013 for the two subsets 1 and 2, respectively. The high temperature areas have been shown in shades of yellow and red and correspond to the barren surfaces, OB dumps, or subsurface fires, whereas the low temperature areas are depicted by shades of blue and correspond to water bodies and vegetation. It can be observed that the minimum and maximum temperature in JCF has been increasing since 1991. For the area covered in subset 1 (Figure 17.2.3), it is clearly visible how the high temperature areas have increased over time. The fires have moved toward the western side from 1991 to 2001. Similarly, for the area in subset 2, high temperatures have increased from 1991 to 2001. This may happen due to the propagation (lateral and/or vertical) of subsurface fire to the surface or starting of the new fires. But by years 2011 and 2013, the high temperature areas seem to consolidate in some patches. The probable reasons for this phenomenon may be that:

(1) There must be a small amount of coal that might have been exhausted due to continuous burning over time.
(2) The coal present there may have been extracted so that there is no fuel left for the fire.
(3) The fire has been controlled due to various reclamation and fire control measures adopted by Bharat Coking Coal Limited.

High temperature areas also generally tend to have lower NDVI values. This aspect has also been corroborated from the analysis of the relationship between LST and NDVI, discussed in detail in the following section.

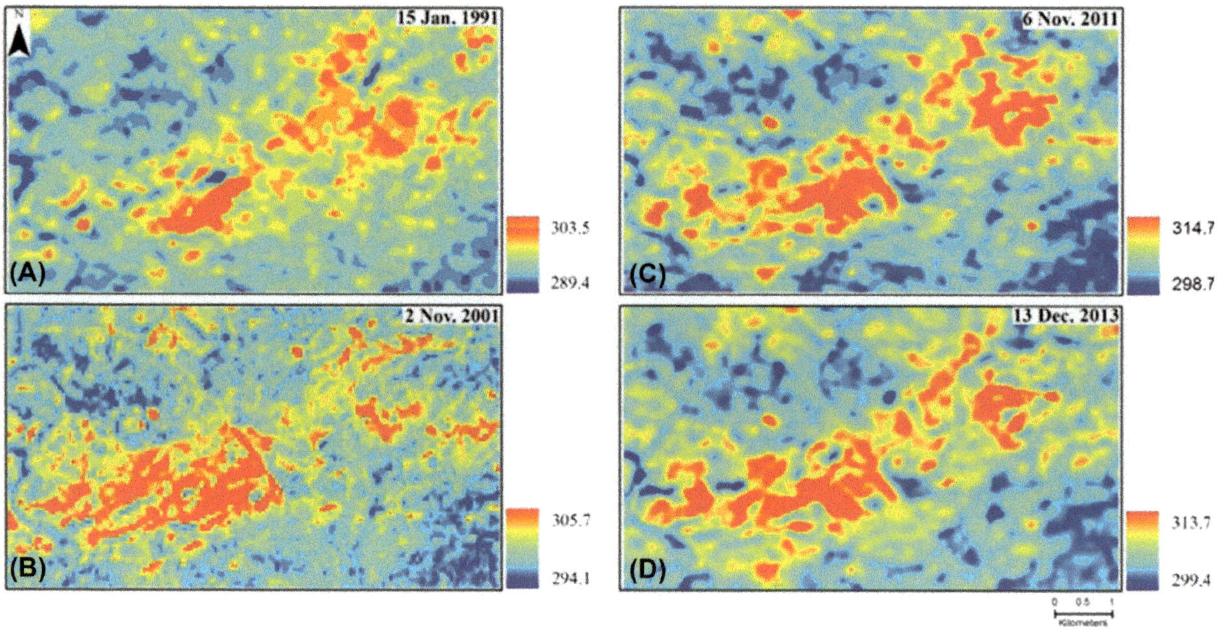

Figure 17.2.3. Subset 1: Spatial variation of LST for the years (A) 1991, (B) 2001, (C) 2011, and (D) 2013. All values are in Kelvin.

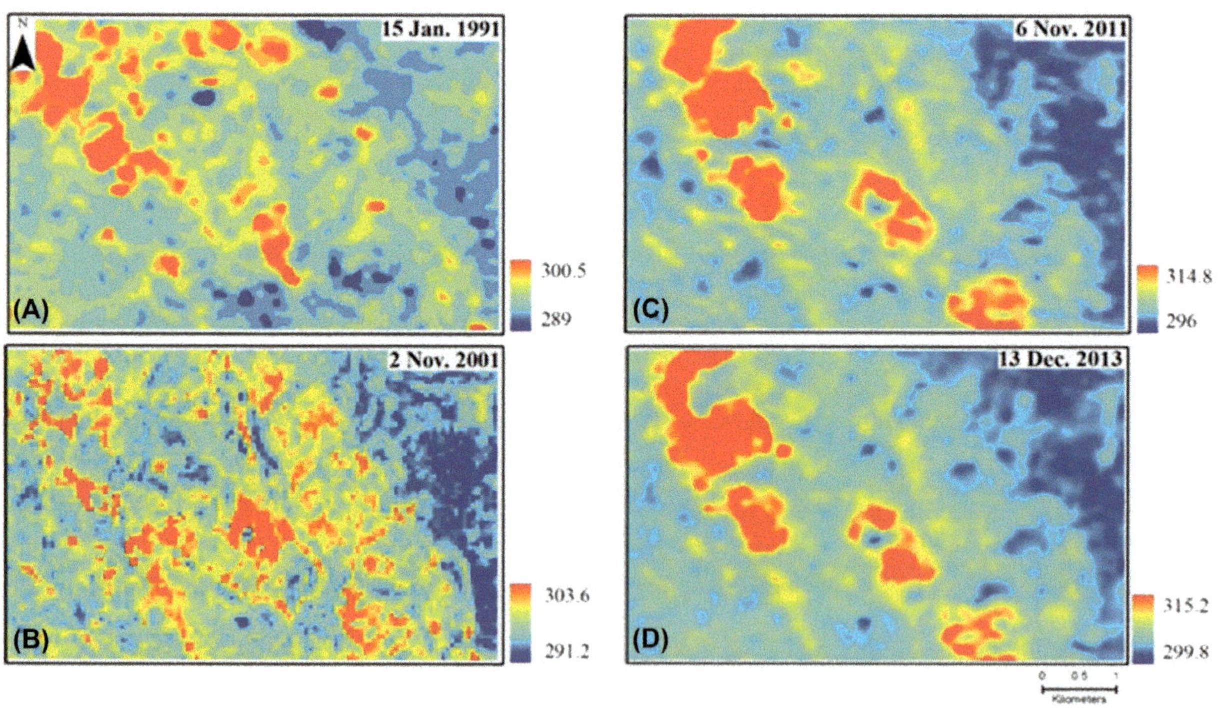

Figure 17.2.4. Subset 2: Spatial variation of LST for the years (A) 1991, (B) 2001, (C) 2011, and (D) 2013. All values are in Kelvin.

17.3 NDVI Index

Photo by Varinder Saini, 2014.

A fire at an open patch of land in Sijua. The fire migrated here from a nearby underground bituminous coking coal-mine in Sijua. The black area behind the dead tree, near the center of the photo, is an exposed coal seam on fire. The black area in front of the tree is two uprooted tree roots. The horizontal field of view at the center of the photo is about 10 m.

Introduction

Vegetation indices derived from multispectral remote sensing data are one of the primary sources of information for operational monitoring of earth's land cover (Gilabert et al., 2002; Baig et al., 2014). The applications of these indices have ranged from a simple leaf level study to global vegetation cover assessment (Viña et al., 2011). Vegetation indices have been aimed at enhancing the vegetation signature while minimizing the soil background effects (Jackson and Huete, 1991). The principle behind this is the high absorption of green healthy vegetation in the red wavelength due to the leaf pigments such as chlorophyll and high reflectance of vegetation due to internal leaf structure in NIR wavelength of the electromagnetic spectrum. On the other hand, unhealthy or sparse vegetation reflects more in red wavelength than healthy vegetation. Barren land and soil reflect moderately in both the red and NIR wavelengths (Holme et al., 1987). Each index has its own advantages and limitations.

NDVI is a numerical indicator that uses red and NIR bands of remote sensing data to assess whether the land cover includes live and green vegetation or not. The index, first conceived by Rouse et al. (1973), is based on the absorption of light by vegetation in the red band and its reflectance in the NIR band. The NDVI is derived using Eq. (17.2.4).

Its values range from −1 to +1, where −1 corresponds to water and barren land and +1 to very dense forests. It is especially suitable for monitoring low to moderate density vegetation (Justice et al., 1985; Townshend and Justice, 1986; Toren and Ünal, 2001; Lee et al., 2002; Anyamba and Tucker, 2005).

The applicability of NDVI has been reported to be either the same (Huete et al., 1997; Mróz and Sobieraj, 2004) or better (Purevdorj et al., 1998; Kariyawasam et al., 2014) than some of the other indices. The main advantage in studying NDVI is that it can be derived for data from most of the remote sensing satellite sensors since the launch of the first earth resource mapping satellite (i.e., Landsat MSS) in 1972. The majority of these satellite sensors have at least one red and one NIR band. Moreover, since NDVI is a ratio of two bands, it helps compensate for differences both in illumination within an image due to slope and its aspect, and differences between images due to factors like time of day or season of image acquisition (The Landscape Toolbox, 2017). Thus NDVI may be very useful to compare images over different time periods to assess ecologically significant changes. This is the primary reason for using the NDVI in this study. Later, the NDVI analyses have been supplemented by another popular index, **tasseled cap transformation** (TCT).

Long Duration Medium Spatial Resolution NDVI

In the present study, six Landsat images belonging to MSS (1972), TM (1992, 2005, 2009), ETM+ (1999), and OLI (2013) sensors have been used. All the images have been atmospherically corrected and then converted to NDVI images using Eq. (17.2.4). The images belong to nearly the same month to avoid any seasonal variation in vegetation cover. For ease in analysis of the temporal variations of NDVI over the years, NDVI images have been density sliced into three classes as reported by Gangopadhyay et al. (2006):

(1) Nonvegetated area: Pixels having NDVI values below 0.2. It includes water bodies, barren lands, open pits, exposed OB dumps, built-up areas, coal processing plants, and other such areas having no vegetation.
(2) Sparse vegetation: Pixels having NDVI values between 0.2 and 0.5. It includes areas covered with shrubs, herbs, small trees, and so on.
(3) Moderate to dense vegetation: Pixels having NDVI values above 0.5. It includes natural forests/dense vegetation, reclaimed and similar areas.

Figures 17.3.1 and 17.3.3 show the density-sliced NDVI images for the areas covered in subsets 1 and 2, respectively. The bar diagrams corresponding to the distribution of the NDVI-derived vegetation classes, as defined earlier, have been shown in Figures 17.3.2 and 17.3.4. In subset 1 (Figure 17.3.2), moderate to dense vegetation has drastically decreased from 1972 to 2013. There has been a corresponding increase in sparse vegetation and non-vegetated areas. In subset 2 (Figure 17.3.4), the non-vegetated areas have grown to a large extent over the years, thereby reducing the moderate to dense vegetation areas to very low levels.

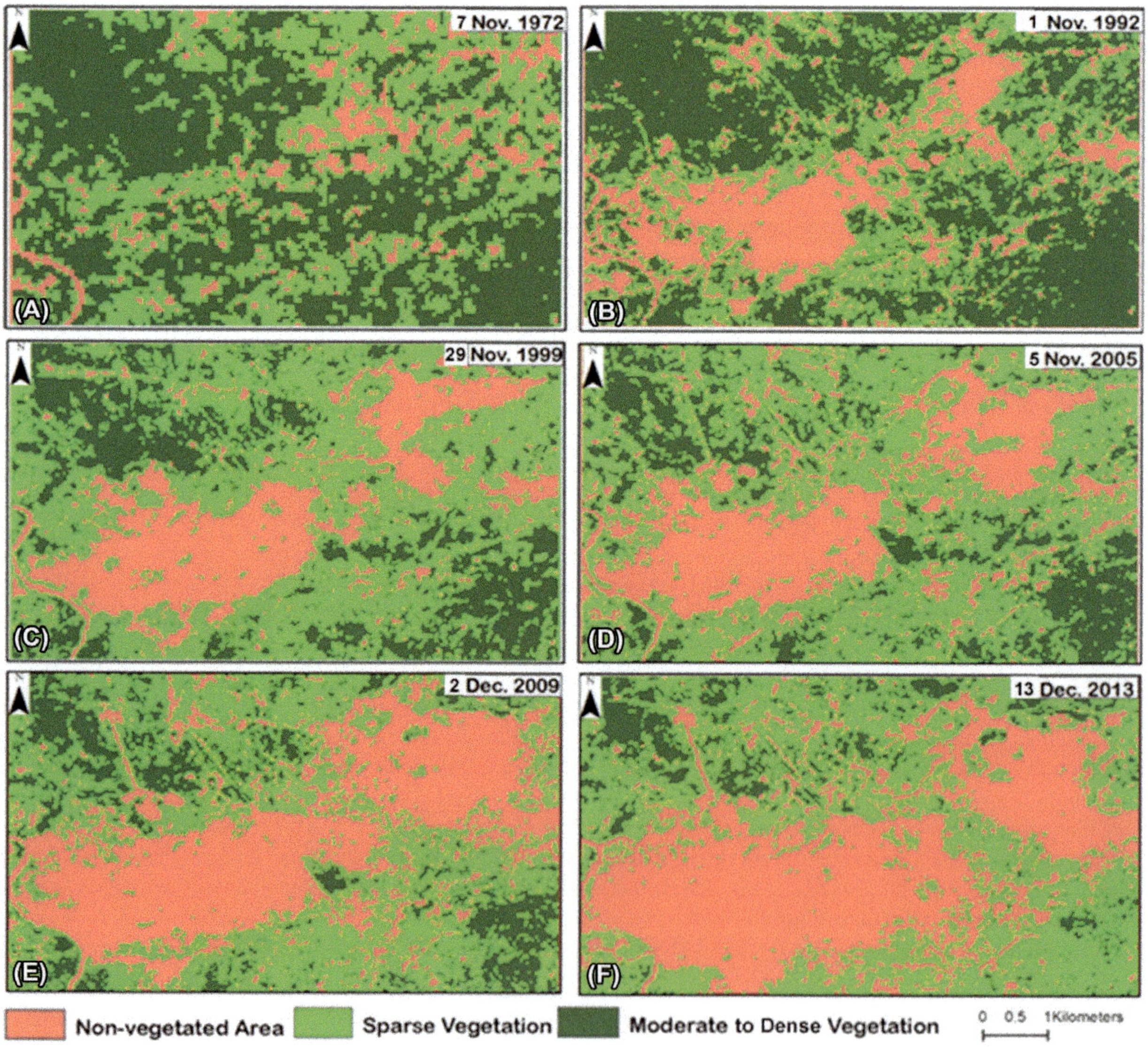

Figure 17.3.1. Subset 1: Density-sliced NDVI images generated for the years (A) 1972, (B) 1992, (C) 1999, (D) 2005, (E) 2009, and (F) 2013. Note how the spatial distribution of vegetation has changed over the period.

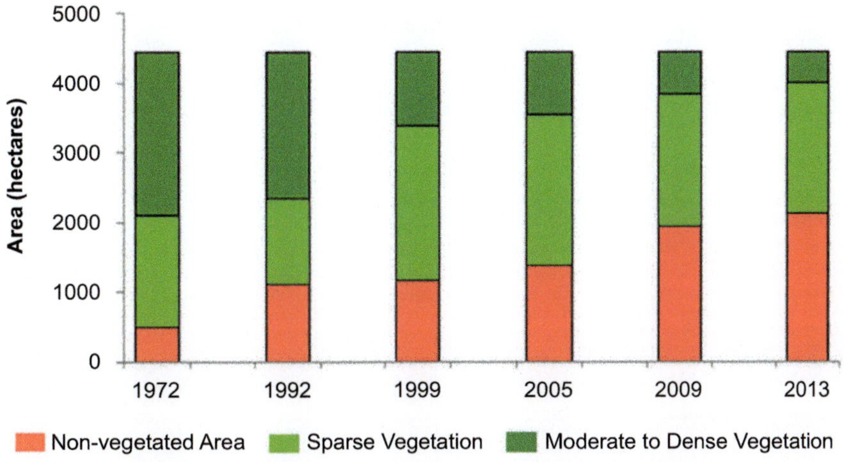

Figure 17.3.2. Subset 1: Bar graph showing areas corresponding to each NDVI class shown in Figure 17.3.1. Note the significant decrease in the area under moderate to dense vegetation and increase in sparse vegetation and non-vegetated areas.

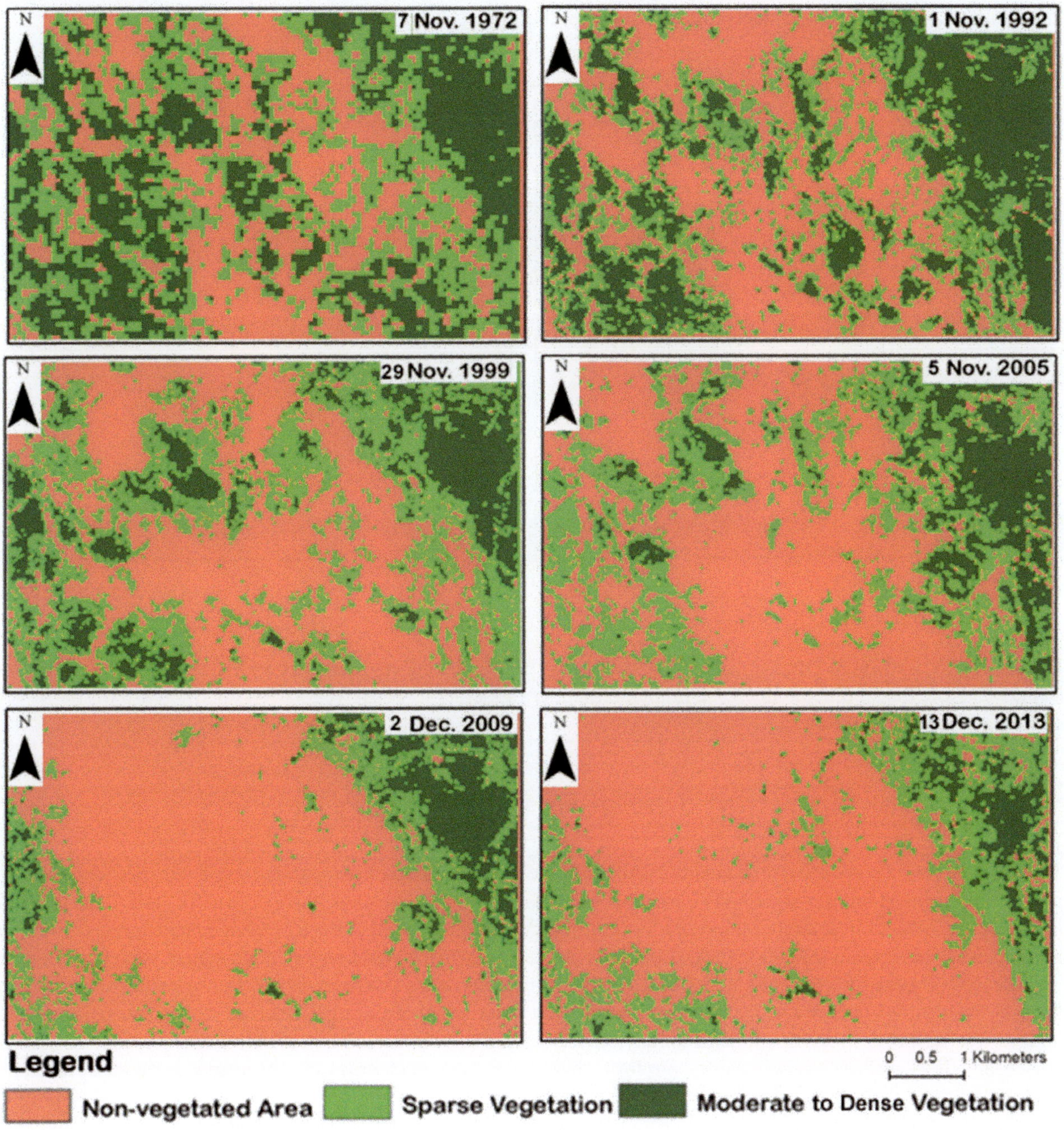

Figure 17.3.3. Subset 2: Density-sliced NDVI images generated for the period 1972–2013. Note the significant decrease in the area under moderate to dense vegetation and increase in sparse vegetation and non-vegetated areas.

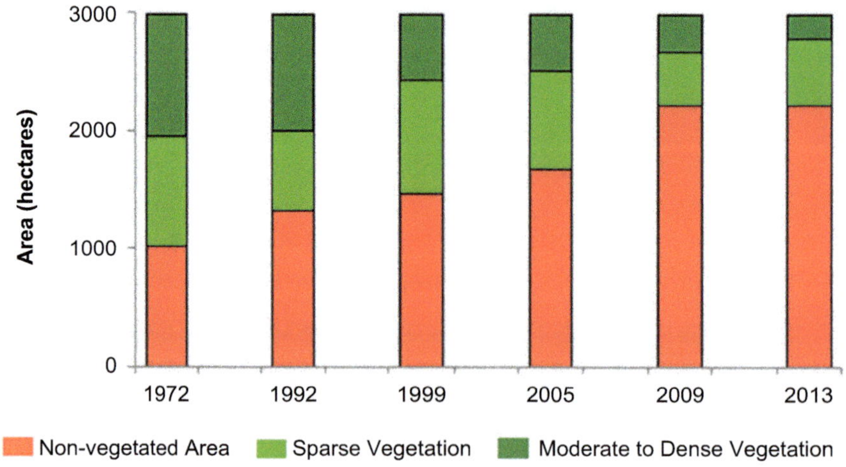

Figure 17.3.4. Subset 2: Bar graph showing areas corresponding to each NDVI class shown in Figure 17.3.3. Note the significant increase in non-vegetated areas during the forty-year period.

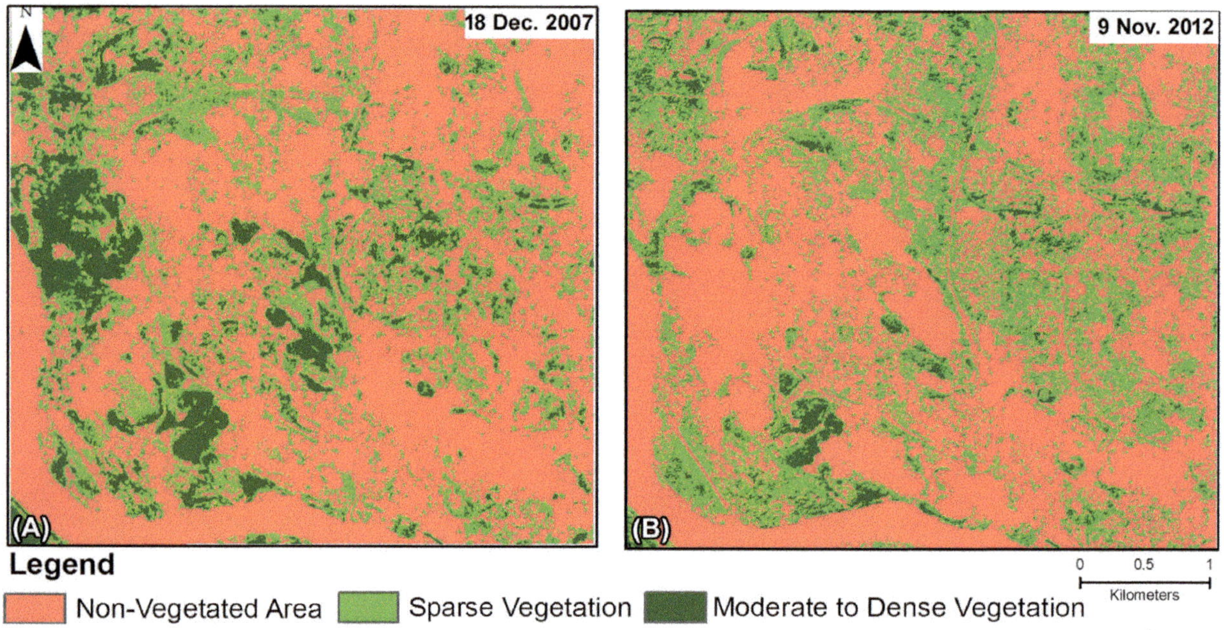

Figure 17.3.5. Subset 3: Density sliced NDVI images (Bhowrah colliery and surrounding areas) of (A) 2007, and (B) 2012. Note the significant decrease in the area under moderate to dense vegetation and increase in sparse vegetation and non-vegetated areas.

Short Duration High Spatial Resolution NDVI

The LISS-IV sensor aboard the Resourcesat-2 satellite is a relatively high resolution multispectral sensor having immense potential for generation of quality NDVI composites. These data possess excellent capability for LULC management including applications in agriculture, forestry, water resources, urban planning, disaster management, and so on. Landsat images, being of medium resolution data, do not permit the production of land cover classifications with enough detail and accuracy required for the operational LULC management activities (Wenger, 1984). Further, some of the mining features such as OB dumps and reclaimed areas are very small, and may not get detected in 30 m spatial resolution images. The number of such small features may be very large. Therefore, their cumulative effect may not be ignored. Hence, an NDVI analysis has been carried out for high spatial resolution (5.8 m) LISS IV images in the study area.

The data belong to years 2007 and 2012. Two new subsets of LISS IV images described in Section 17.1.2 (Figure 17.1.1) have been used for the NDVI analysis. The NDVI images have been density sliced as described in Section 17.3.2. The density-sliced images for both subsets are shown in Figures 17.3.5 and 17.3.7. The bar graphs

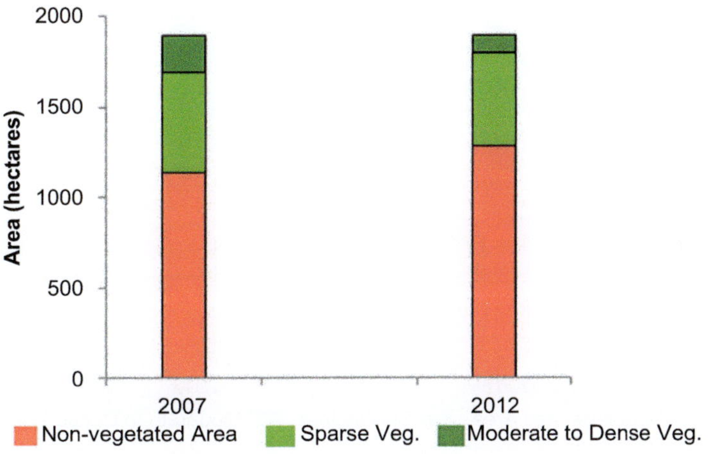

Figure 17.3.6. Subset 3: Bar graph showing areas corresponding to each NDVI class shown in Figure 17.3.5.

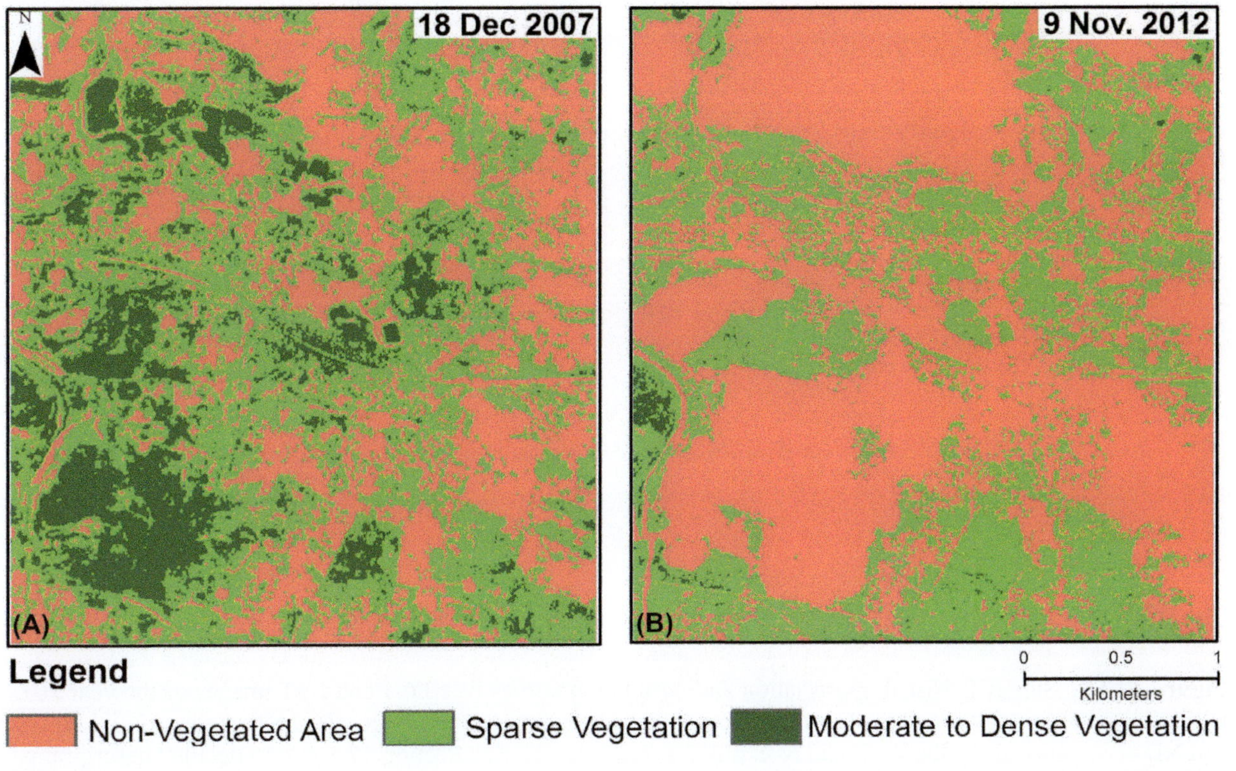

Figure 17.3.7. Subset 4: Density-sliced NDVI images of subset 4 (Angarpathar area) of (A) 2007 and (B) 2012. Note the significant decrease in the area under moderate to dense vegetation and the increase in sparse vegetation and non-vegetated areas.

corresponding to the **areal distributions** are shown in Figures 17.3.6 and 17.3.8. In subset 3, a thick patch of vegetation in the western side has changed to a non-vegetated area just within a span of 5 years. The proportion of non-vegetated area has increased by 150 ha from 2007 to 2012. The changes have occurred because of the mining-related activities such as clearing the surface of existing vegetation, excavation for mining, dumping the OB in nearby areas, dumping mined out coal on the surface, and other such related activities. Similarly, in subset 4, it can be clearly seen that within a span of just 5 years, a densely vegetated area has changed to non-vegetated area due to mining activities. This is a typical example of clearing the surface cover for initiating mining. This can be corroborated from the areal distribution, which shows that the moderate to dense vegetation has decreased significantly from 162 ha in 2007 to 8.3 ha in 2012 (Figure 17.3.8).

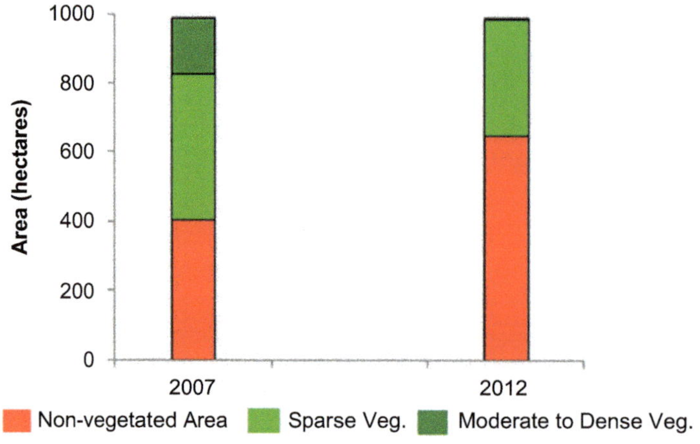

Figure 17.3.8. Subset 4: Bar graph showing areas corresponding to each NDVI class shown in Figure 17.3.7.

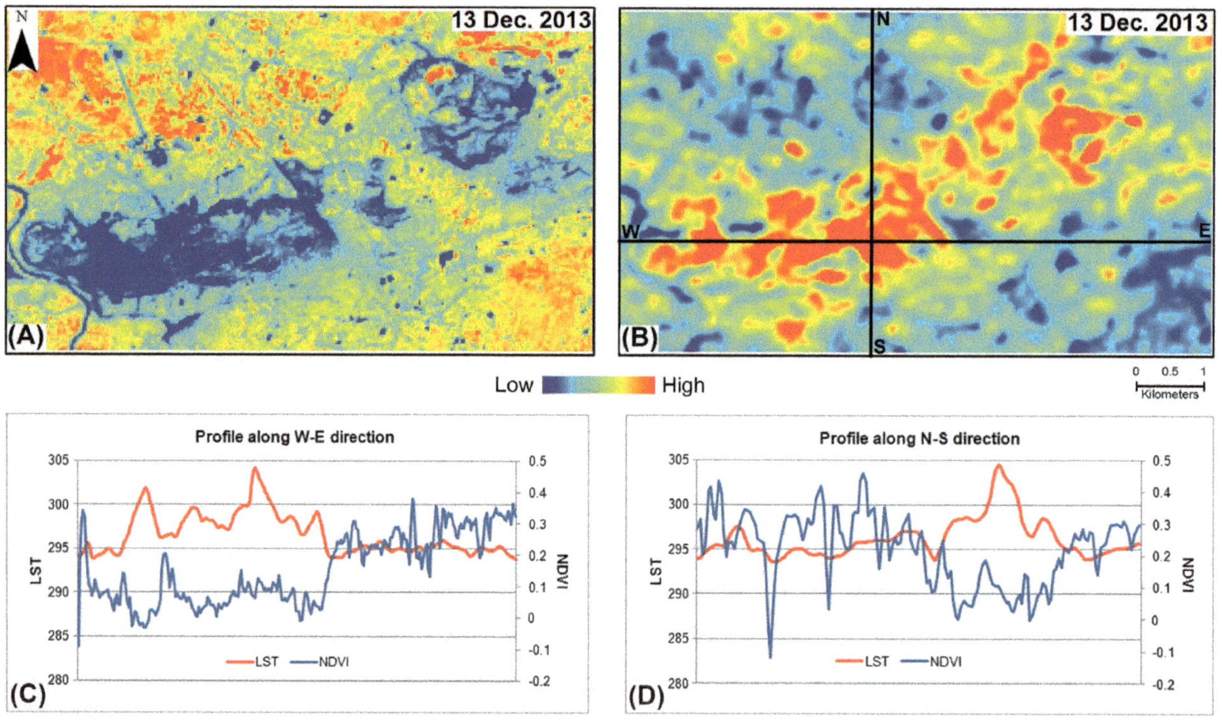

Figure 17.3.9. Subset 1: Visual appreciation and correlation curves of NDVI and LST images of the year 2013. Black lines represent N–S and W–E profiles across the subset. (A) NDVI image, (B) LST image, (C and D) LST and NDVI variance along the W–E and N–S profiles, respectively. See Section 17.3.4 for a detailed description.

Correlation Between NDVI and LST

The NDVI and LST images of 2013 for subset 1 and subset 2 are interpreted together to improve understanding of the NDVI and LST patterns. The contrast between NDVI and LST indicates an apparent correlation (Figures 17.3.9 and 17.3.10). In order to study the variance of NDVI and LST in different directions, two profiles were taken. The pixel values of NDVI and LST images, one across W–E direction (Figures 17.3.9C and 17.3.10C) and the other across N–S (Figures 17.3.9D and 17.3.10D) direction, were plotted against each other. The lines (black in color) along which pixel values were taken are shown in Figures 17.3.9B and 17.3.10B. The profiles show that the peaks of the LST are usually the areas where there are open pits/mines under fire, while the troughs are the areas where the water bodies and vegetation are found. The peaks of NDVI appear at vegetated areas including agricultural lands. Thus, NDVI and LST show an obvious negative correlation. In other words, the NDVI values are low (or even negative) where LST is high and vice versa.

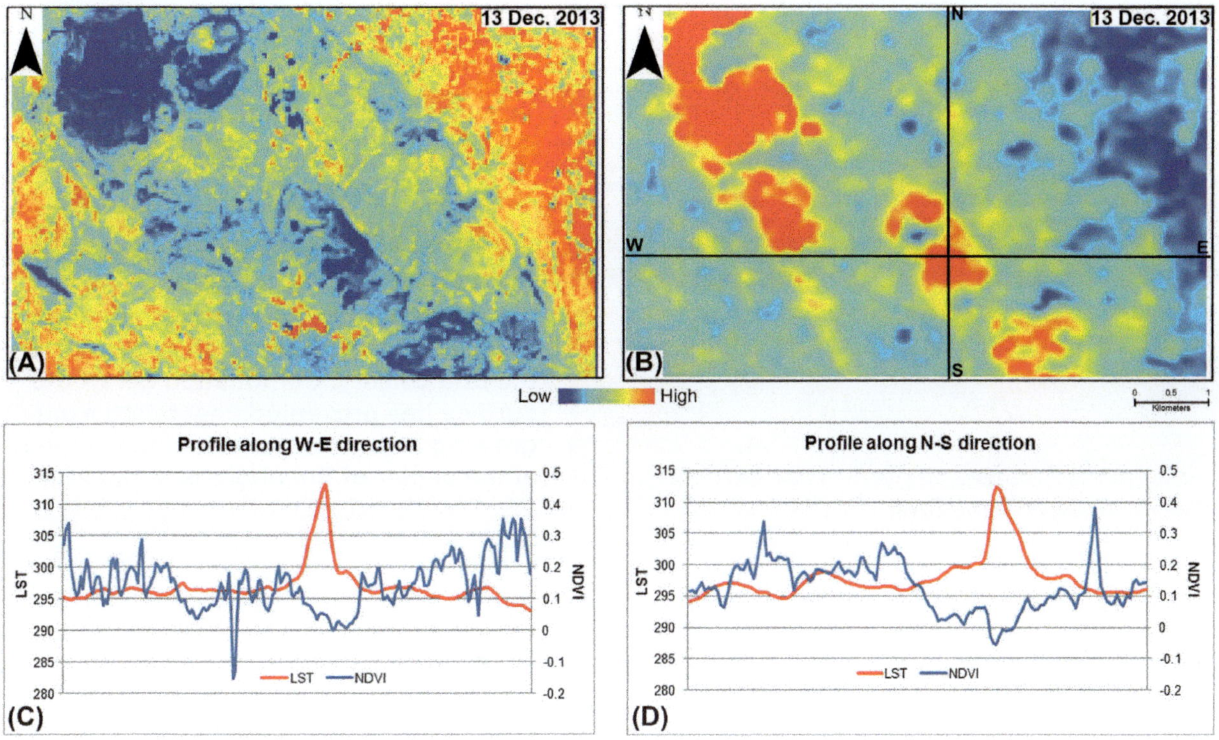

Figure 17.3.10. Visual appreciation and correlation curves of NDVI and LST images of the year 2013 for subset 2. Black lines represent N–S and W–E profiles across the subset. (A) NDVI image, (B) LST image, (C and D) LST and NDVI variance along the W–E and N–S profiles, respectively. See Section 17.3.4 for a detailed description.

17.4 Tasseled Cap Transformation

Photo by Varinder Saini, 2014.

A subsurface fire migrated to this patch of land from an abandoned coal mine at the Ena colliery. Combustion below the surface has prevented vegetation from growing here. The horizontal field of view at the center of the photo is about 20 m.

Introduction

The concept of TCT was given by Kauth and Thomas in 1976. It was originally formulated for Landsat MSS data to understand the important phenomenon of crop development in the spectral space. A new coordinate system to represent soil, vegetation, and their interrelationship was defined. The TCT orthogonally transforms the data of the original image into three new components, which enhances the spectral information of the data. The brightness and greenness components of TCT are equivalent to the soil brightness index and the green vegetation index (GVI), respectively. Together brightness and greenness components contain almost 95%–98% of the scene information (Jensen, 1996).

Here, TCT has been used to derive brightness and greenness images from the Landsat MSS, TM, ETM+, and OLI sensors for the period 1972–2013 to understand the development and health of vegetation in JCF area. The study has been conducted for the years 1972 (MSS), 1992 (TM), 2001 (ETM+) and 2013 (OLI). Small areas from subset 1 and subset 2 have been taken for studying the interrelationship between brightness and greenness (highlighted in yellow in Figures 17.4.1 and 17.4.3). The scatter plots between brightness and greenness components are shown in Figures 17.4.2 and 17.4.4. The values in these plots have been normalized. Finally, a correlation was established between NDVI and TCT to study their interrelationships. The methodology adopted to transform the multisensor Landsat data to TCT brightness and greenness images is described later.

For the MSS sensor, the TC components are derived as:

$$\text{TCi} = (\text{Coeff}_1 * \text{band1}) + (\text{Coeff}_2 * \text{band2}) + (\text{Coeff}_3 * \text{band3}) + (\text{Coeff}_4 * \text{band4}) \tag{17.4.1}$$

where TCi is the calculated tasseled cap index for brightness or greenness, depending on the coefficients (Coeff_i) used.

For the TM/ETM+ sensor, the TC components are derived as:

$$\text{TCi} = (\text{Coeff}_1 * \text{band1}) + (\text{Coeff}_2 * \text{band2}) + (\text{Coeff}_3 * \text{band3}) + (\text{Coeff}_4 * \text{band4}) + (\text{Coeff}_5 * \text{band5}) + (\text{Coeff}_6 * \text{band7}) \tag{17.4.2}$$

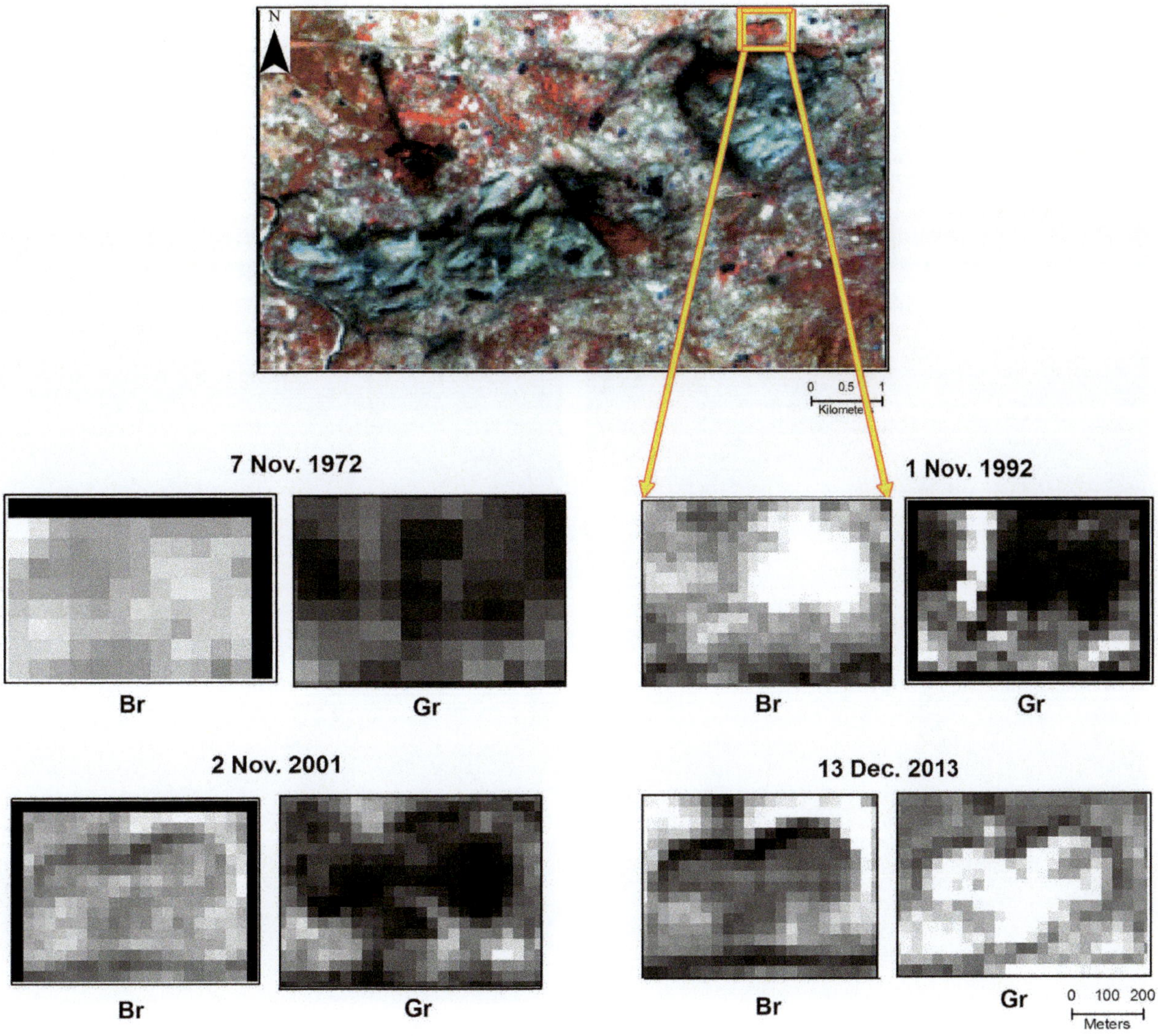

Figure 17.4.1. Subset 1: A small area (highlighted in yellow) showing variation in brightness (Br) and greenness (Gr) components of TCT in an OB dump area from 1972 to 2013. See Section 17.4.2 for a detailed description.

For the OLI sensor, the TC components are derived as:

$$TCi = (Coeff_1 * band2) + (Coeff_2 * band3) + (Coeff_3 * band4) + (Coeff_4 * band5) + (Coeff_5 * band6) + (Coeff_6 * band7) \quad (17.4.3)$$

The coefficients in Eqs. (17.4.1), (17.4.2), and (17.4.3) have been taken from Kauth and Thomas (1976), Crist and Cicone (1984), Huang et al. (2002), and Baig et al. (2014) for MSS, TM, ETM+, and OLI sensors, respectively.

Spatio-Temporal Analysis of TCT

Figure 17.4.1 shows an area that was later turned into an OB dump. In 1972, the area had very sparse vegetation but by 1992, the OB dump started taking shape. From the TCT analysis of the 1992 image, it is clearly visible that the brightness is high and the corresponding areas in **greenness component** are dark. This situation continues until

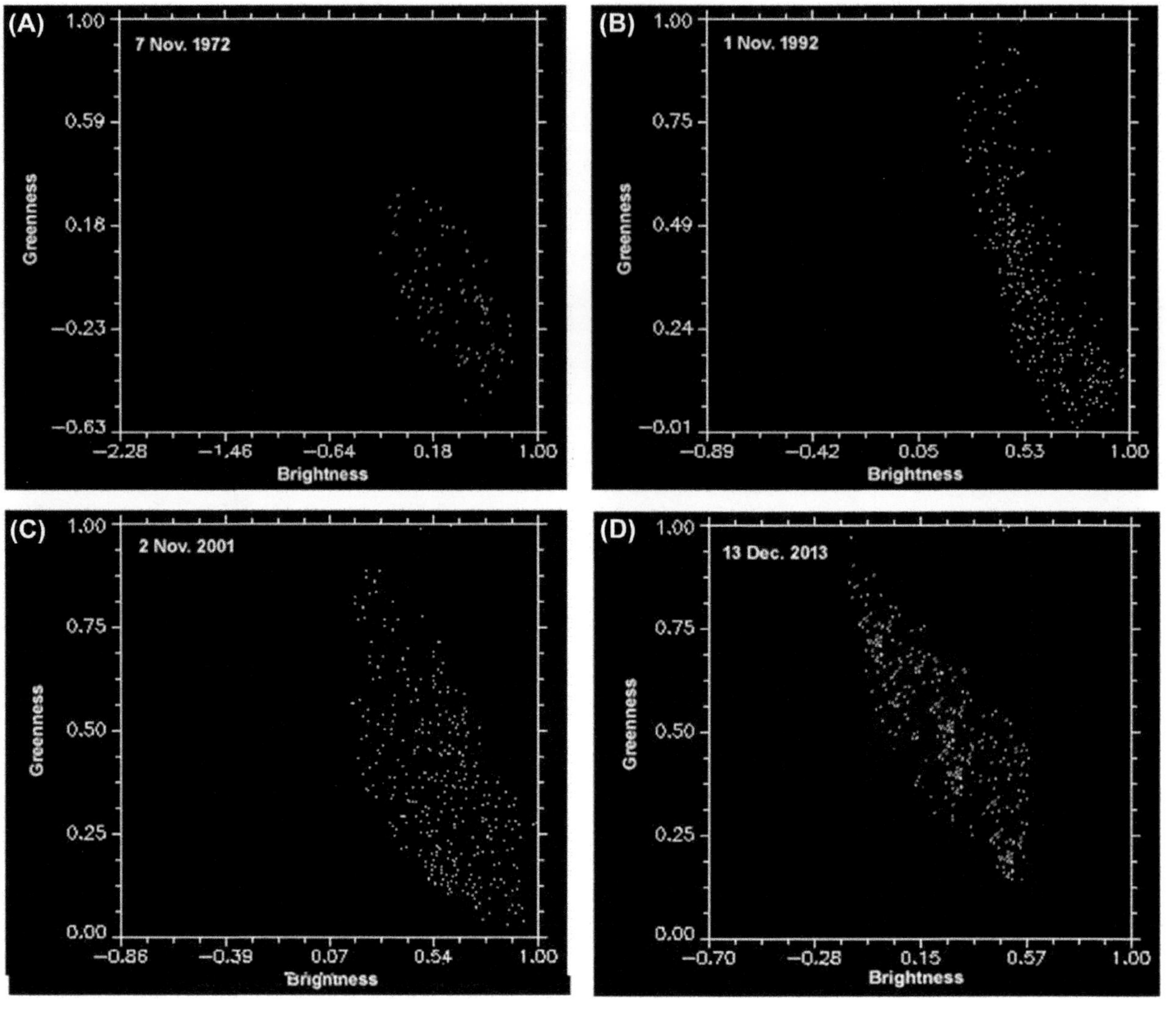

Figure 17.4.2. Scatter plots of greenness vs. brightness corresponding to areas shown in Figure 17.4.1 for the years (A) 1972, (B) 1992, (C) 2001, and (D) 2013. Note the negative correlation between the two components in all the years.

2001, as can be observed from the corresponding TCT components of the 2001 image. After that, the vegetation started growing on the OB dump and by 2013, a high reflectance in the greenness component and corresponding decrease in **brightness component** may be seen.

Similarly, in Figure 17.4.3, an area having vegetation has been chosen from subset 2. In the whole subset, this is the only area that appears to have retained vegetation over the years. In 1972, the whole area showed high greenness, which continued until 2001, but by 2013 the brightness had increased.

Typically in TCT images, the brightness component is highly correlated with bare soil while the greenness component relates to the vegetation biomass. The greater is the amount of bare soil, the brighter the response. The bare soil lies parallel to the brightness component in the brightness–greenness spectral space plot (Kauth and Thomas, 1976) while the greenness component progresses perpendicular to the plane of soils during a growing season. As the amount of green vegetation increases, it moves farther away from the plane of the soil (Jensen, 1996). This phenomenon may be seen in the cluster plots shown in Figures 17.4.2 and 17.4.4. The reflectance of green vegetation is located to the left, at the tip of the plots. The reflectance of soil lies close to brightness line. Figure 17.4.2(A–D) shows the scatter plots of greenness versus brightness for all four years. For each year, all plots illustrate a negative correlation of greenness and brightness. Therefore, it

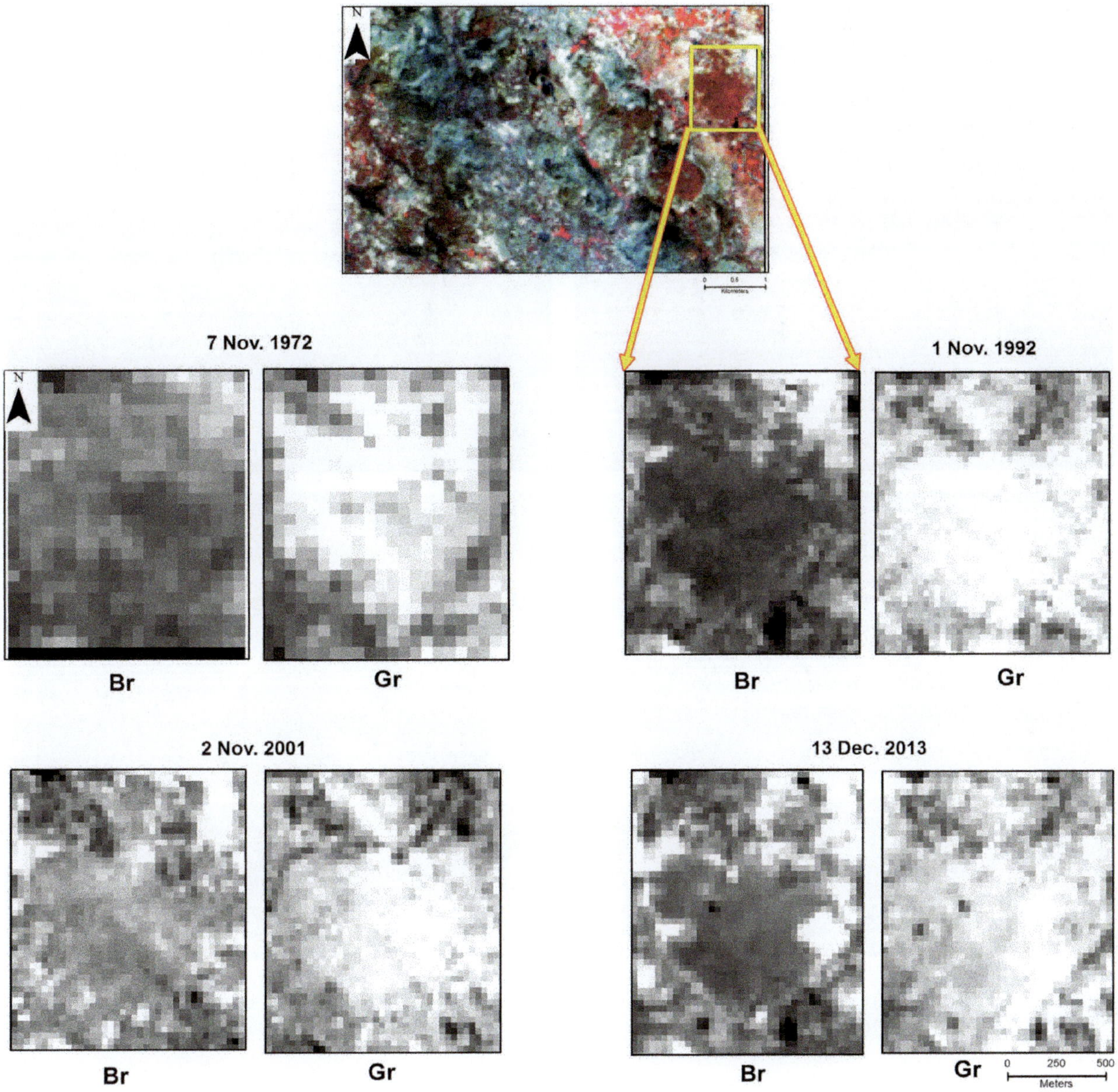

Figure 17.4.3. Subset 2: A small area (highlighted in yellow) showing variation in brightness (Br) and greenness (Gr) component of TCT in a vegetated area from 1972 to 2013. See Section 17.4.2 for a detailed description.

may be concluded that the reflectance of soil is higher during 1972–2001 as compared to 2013. This means that the area has shown an increase in vegetation since 2001. Such analysis may therefore help in identifying the success of reforestation/reclamation measures being carried out in mining areas. Tasseled cap components along with field data may also have the potential to map the success of such measures taken up by mine planners.

In the area shown in subset 2, it is expected that with time, such vegetated areas may get destroyed completely because of the expansion of mining. This can be substantiated from the corresponding scatter plots for all four years as shown in Figure 17.4.4. In the scatter plot, for the year 2013 (Figure 17.4.4D) it may be observed that the distribution of pixels has moved close to the soil line (x-axis), indicating decrease in vegetation biomass since 2001. The areas undergoing deforestation may therefore be identified using TCT components and further damage may be prevented.

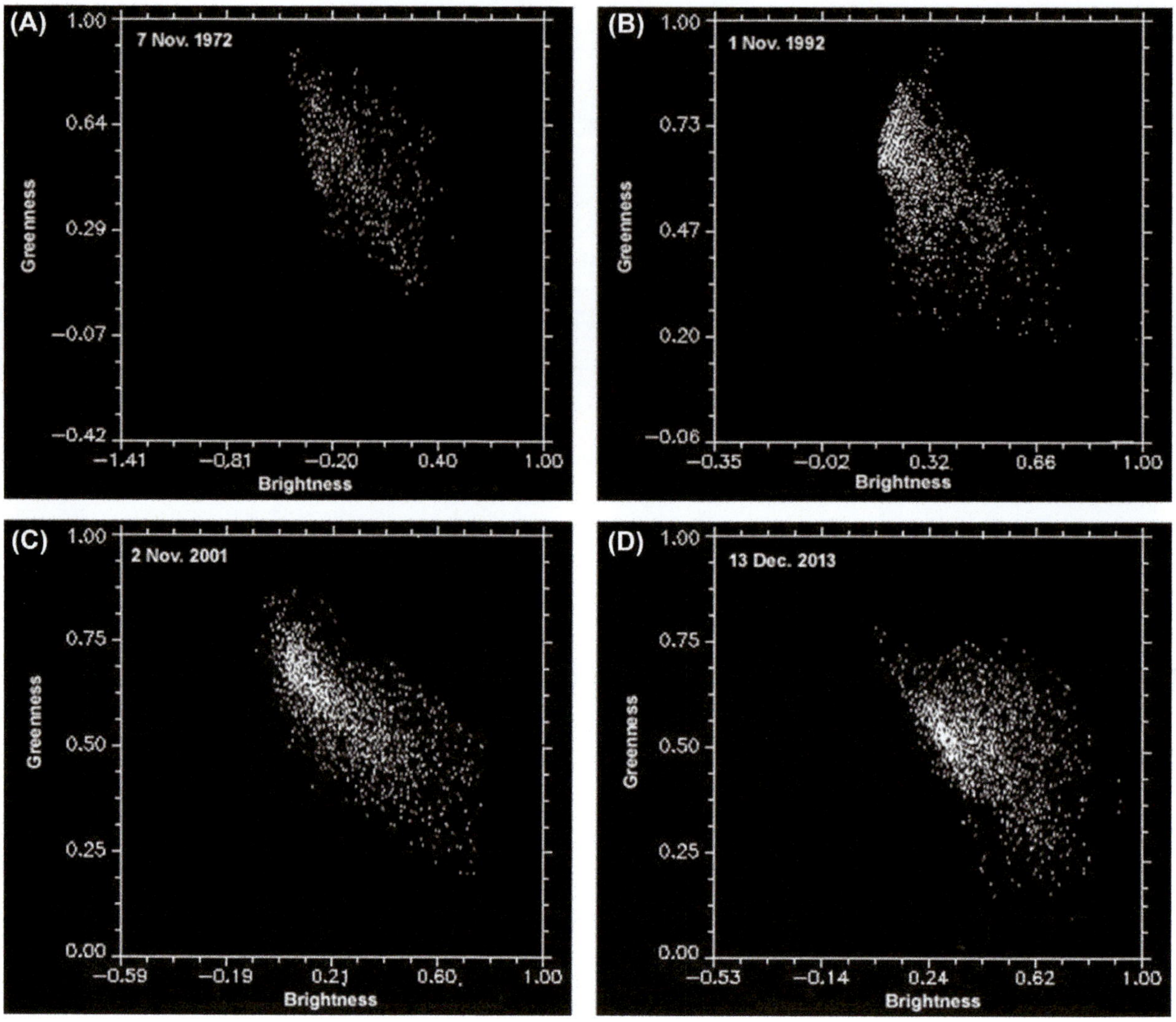

Figure 17.4.4. Scatter plots of greenness *vs.* brightness corresponding to areas shown in Figure 17.4.3 for the years (A) 1972, (B) 1992, (C) 2001, and (D) 2013.

Correlation Between NDVI and TCT

In the present study, two vegetation indices, NDVI and TCT, have been studied to assess the impact of coal mining on vegetation. As already discussed, the greenness component of TCT is equivalent to GVI, hence it is intuited to be positively correlated to NDVI. In order to further analyze the correlation between greenness and NDVI, a scatter plot has been drawn with greenness component of TCT on the x-axis and NDVI on the y-axis. These scatter plots have been drawn for the years 1972, 1992, 2001, and 2013 for both subset 1 and subset 2 (Figures 17.4.5 and 17.4.6). As is evident from the scatter plots, NDVI has strong positive correlation with greenness. Additionally, the correlation coefficients and regression equations have been computed for the data of all mentioned years.

In summary, NDVI and TCT analyses reveal a strong positive correlation indicating that one strongly affects the other and in direct proportion. More specifically, the coefficient of correlation (R) ranges from 0.88 in 1972 to 0.98 in 2013 for subset 1 while it ranges from 0.92 in 1972 to 0.97 in 2013 for subset 2. The high correlation between the two suggests that the greenness component may be used to study changes in vegetation like NDVI. For all the cases, there is negligible probability that the regression output is random (significance of F of the regression $<< 0$), and Y-intercept and coefficients are reliable with 95% confidence level (P-value $<< 0$). Since TCT segregates the brightness component from the greenness component, the results may be better in areas covered with sparse vegetation since NDVI measurements in such cases may be hampered due to high soil reflectance.

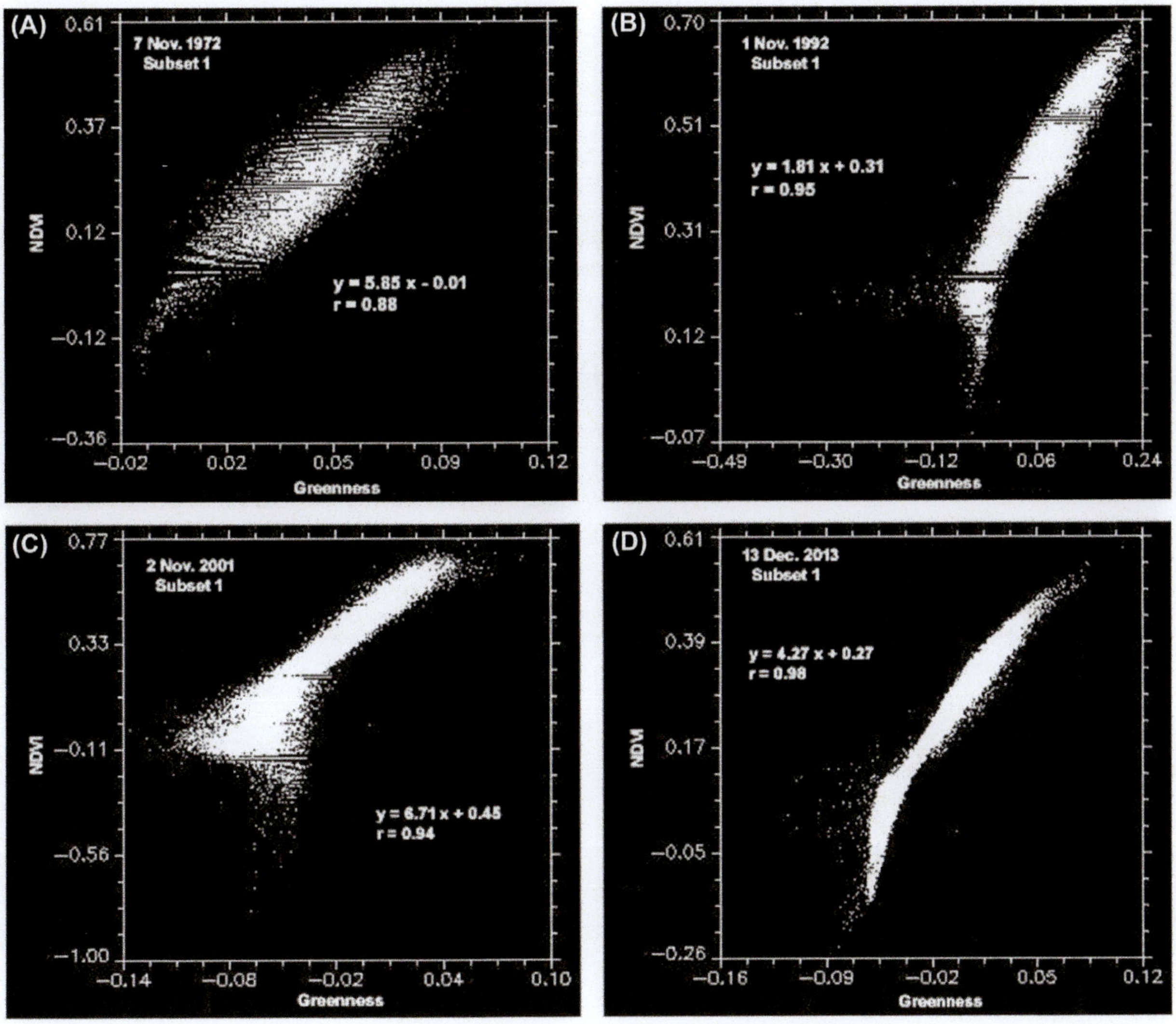

Figure 17.4.5. Subset 1: Scatter plots showing strong positive correlation between greenness component *(x-axis)* of TCT and NDVI *(y-axis)* for the years (A) 1972, (B) 1992, (C) 2001, and (D) 2013.

Discussion and Conclusions

Coal mining is a dynamic process, resulting in continual change of one land cover to the other. These changes may be easily quantified using temporal change detection from remote sensing satellite data. Haphazard mining with no regard to environmental considerations has led to mine fires in JCF, which have been burning for more than a century. The analysis of LST images in this study shows that the minimum and maximum LST has increased from 1991 to 2013. The fires may be seen moving to newer areas. Thus repetitive satellite data may be effectively used for studying the changes in fire affecting coal mines, while such studies may be difficult to conduct through field-based methods.

The NDVI analysis highlights the fact that the trend in LULC is the same whether the change is evaluated over a four-decade period or a five-year period. Further, the effect of high temperature areas (basically associated with mine fires) on vegetation was studied. A negative correlation between NDVI and LST has been found for all the years, which should be the case. The implication of temperature increase is that the areal extent of vegetation and agricultural lands may decline further. This happens because the soil loses moisture and gets baked due to the fire making it less potent for plant growth. Loss of vegetation has negative impacts on the ecology of the area. The accumulation of dust over plants leads to reduction in photosynthetic capability of the plants. Moreover, the vegetation-free areas become prone to wind and water erosion as well as habitat deterioration and degradation of living conditions for humans and animal species.

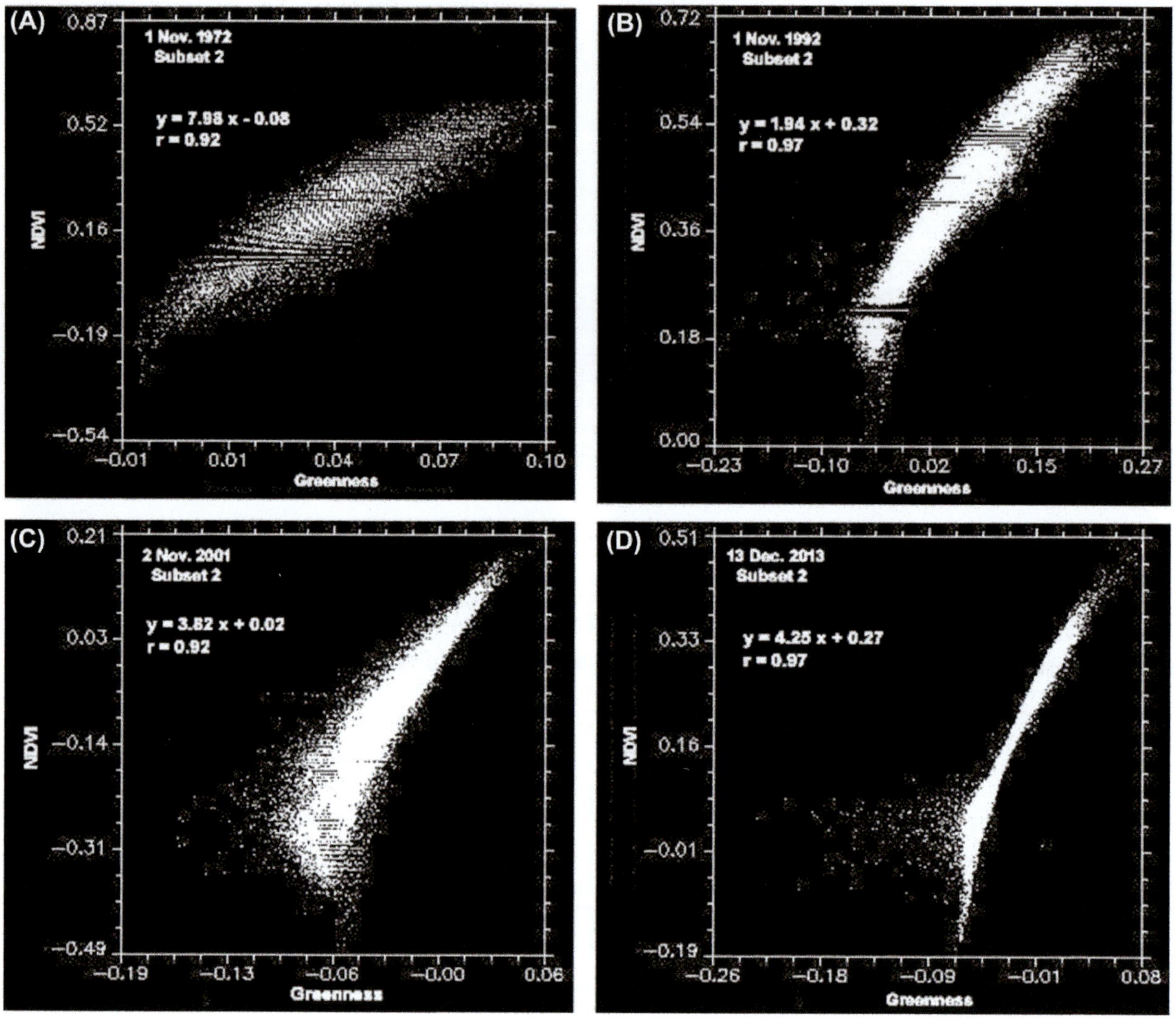

Figure 17.4.6. Subset 2: Scatter plots showing strong positive correlation between greenness component *(x-axis)* of TCT and NDVI *(y-axis)* for the years (A) 1972, (B) 1992, (C) 2001, and (D) 2013.

TCT analysis has been used to evaluate the condition of vegetation coverage from 1972 to 2013. The vegetation increase has been associated with an increase in the greenness index, and a decrease in brightness of soil. This study also shows that the analysis of difference images derived from TCT components are an effective method for delineating the areas, where changes in brightness and greenness components have occurred over the years due to mining activities. The greenness derived from TCT has the potential to link with the NDVI. Linear regression between greenness and NDVI for different years shows a high positive correlation (>85% for all cases) at 95% confidence level. Thus the greenness component of TCT may be more useful in measuring green biomass in sparsely vegetated areas.

The results presented in this chapter show that remote sensing data analysis may help in monitoring coal fires and vegetation patterns, identifying degraded areas, and planning reclamation and remedial measures in such areas.

Acknowledgments

The first author is grateful to the University Grants Commission (UGC), New Delhi, for providing financial assistance in the form of a Research Fellowship under Grant number 6405-14-061. The authors thank the USGS (http://glovis.usgs.gov/) for providing the Landsat data used in this work free of cost through the Internet.

Important Terms

areal distribution
brightness component
coal fires
coal mining
correlation
emissivity
environmental impacts
greenness component
land surface temperature (LST)
near-infrared (NIR)
normalized difference vegetation index (NDVI)
tasseled cap transformation (TCT)
thermal infrared

References

Anyamba, A., Tucker, C.J., 2005. Analysis of Sahelian vegetation dynamics using NOAA- AVHRR NDVI data from 1981–2003. J. Arid Environ. 63, 596–614.

Baig, M.H.A., Zhang, L., Shuai, T., Tong, Q., 2014. Derivation of a tasselled cap transformation based on Landsat 8 at-satellite reflectance. Remote Sens. Lett. 5, 423–431.

Becker, F., Li, Z.-L., 1990. Towards a local split window method over land surfaces. Int. J. Remote Sens. 11, 369–393.

Carlson, T.N., Ripley, D.A., 1997. On the relation between NDVI, fractional vegetation cover, and leaf area index. Remote Sens. Environ. 62, 241–252.

Chander, G., Markham, B.L., Helder, D.L., 2009. Summary of current radiometric calibration coefficients for Landsat MSS, TM, ETM+, and EO-1 ALI sensors. Remote Sens. Environ. 113, 893–903.

Coal Directory of India 2012-2013, 2014. Coal Statistics. Ministry of Coal Government of India. http://www.coal.nic.in/sites/upload_files/coal/files/coalupload/coaldir12-13.pdf (accessed May 2015).

Crist, E.P., Cicone, R.C., 1984. A physically-based transformation of thematic mapper data-the TM tasseled cap. IEEE Trans. Geosci. Remote Sens. GE–22, 256–263.

Gangopadhyay, P.K., Lahiri-Dutt, K., Saha, K., 2006. Application of remote sensing to identify coalfires in the Raniganj coalbelt, India. Int. J. Appl. Earth Obs. Geoinform. 8, 188–195.

Gee, R.E., 1940. History of coal mining in India. Geological Surv. India 6, 313–318.

Gilabert, M., González-Piqueras, J., García-Haro, F.J., Meliá, J., 2002. A generalized soil-adjusted vegetation index. Remote Sens. Environ. 82, 303–310.

Gillespie, A.R., Rokugawa, S., Matsunaga, T., Cothern, J.S., Hook, S., Kahle, A.B., 1998. A temperature and emissivity separation algorithm for advanced spaceborne thermal emission and reflection radiometer (ASTER) images. IEEE Trans. Geosci. Remote Sens. 36, 1113–1126.

He, F., Yin, J., 2010. Environmental effect of land cover change: a case study of coal mining areas in Beijing, China. In: Proceedings of 18th International Conference on Geoinformatics, 18–20 June, Beijing, China, pp. 1–5.

Holme, A.M.R., Burnside, D.G., Mitchell, A.A., 1987. The development of a system for monitoring trend in range condition in the arid shrublands of Western Australia. Aust. Rangel.J. 9, 14–20.

Hook, S.J., Gabell, A.R., Green, A.A., Kealy, P.S., 1992. A comparison of techniques for extracting emissivity information from thermal infrared data for geologic studies. Remote Sens. Environ. 42, 123–135.

Huang, C., Wylie, B., Homer, C., Yang, L., Zylstra, G., 2002. Derivation of a tasseled cap transformation based on Landsat 7 at-satellite reflectance. Int. J. Remote Sens. 23, 1741–1748.

Huete, A.R., Liu, H.Q., Batchily, K., van Leeuwen, W.J.D., 1997. A comparison of vegetation indices over a global set of TM images for EOS-MODIS. Remote Sens. Environ. 59, 440–451.

Jackson, R.D., Huete, A.R., 1991. Interpreting vegetation indices. Prev. Vet. Med. 11, 185–200.

Jensen, J.R., 1996. Introductory Digital Image Processing: A Remote Sensing Perspective, second ed. Prentice Hall, Inc., Upper Saddle River, New Jersey, USA. 316 p.

Jiménez-Muñoz, J.C., Sobrino, J.A., 2003. A generalized single-channel method for retrieving land surface temperature from remote sensing data. J. Geophys. Res. 108, 4688–4695.

Justice, C.O., Townshend, J.R.G., Holben, B.N., Tucker, C.J., 1985. Analysis of the phenology of global vegetation using meteorological satellite data. Int. J. Remote Sens. 8, 1271–1318.

Kariyawasam, N., Raval, S., Shamsoddini, A., 2014. Incorporating remote sensing as a tool to assist rehabilitation monitoring in a dolomite mining operation in south Australia. In: Proceedings of FIG Congress, Engaging the Challenges – Enhancing the Relevance, 16–21 June, Kuala Lampur, Malaysia, Paper 6966, pp. 1–13.

Kauth, R.J., Thomas, G.S., 1976. The tasseled cap-A graphic description of the spectral- temporal development of agricultural crops as seen by Landsat. In: Proceedings of the Symposium on Machine Processing of Remotely Sensed Data. Purdue University, West Lafayette, Indiana. p. 4B-41-4B-50.

Kealy, P.S., Hook, S.J., 1993. Separating temperature and emissivity in thermal infrared multispectral scanner data: implications for recovering land surface temperatures. IEEE Trans. Geosci. Remote Sens. 31, 1155–1164.

Kerr, Y.H., Lagouarde, J.P., Imbernon, J., 1992. Accurate land surface temperature retrieval from AVHRR data with use of an improved split window algorithm. Remote Sens. Environ. 41, 197–209.

Lee, R., Yu, F., Price, P., Ellis, J., Shi, P., 2002. Evaluating vegetation phenological patterns in Inner Mongolia using NDVI time-series analysis. Int. J. Remote Sens. 12, 2505–2512.

Li, Z.-L., Becker, F., 1993. Feasibility of land surface temperature and emissivity determination from AVHRR data. Remote Sens. Environ. 43, 67–85.

Li, Z.-L., Wu, H., Wang, N., Qiu, S., Sobrino, J.A., Wan, Z., Tang, B.-H., Yan, G., 2013. Land surface emissivity retrieval from satellite data. Int. J. Remote Sens. 34, 3084–3127.

MCGI, 2015. Ministry of Coal Government of India. Coal Reserves http://www.coal.nic.in/content/coal-reserves (accessed May 2015).

Moreno-de las Heras, M., Merino-Martin, L., Nicolau, J., 2009. Effect of vegetation cover on the hydrology of reclaimed mining soils under Mediterranean- Continental climate. Catena 77, 39–47.

Mróz, M., Sobieraj, A., 2004. Comparison of several vegetation indices calculated on the basis of a seasonal SPOT XS time series, and their suitability for land cover and agricultural crop identification. Tech. Sci. 7, 39–66.

Pecharova, E., Hais, M., Svoboda, I., 2006. Changes in landscape energy balance as a result of different land use during three time periods. Ekologia Bratislava 25, 205–214.

Pokorný, J., 2001. Dissipation of solar energy in landscape—controlled by management of water and vegetation. Renew. Energy 24, 641–645.

Pokorný, J., Sima, M., Rejskova, A., Brom, J., 2007. The role of vegetation in water cycling and energy dissipation. In: Proceedings of the 1st Natural Sequence Farming Workshop 'Natural Sequence Farming: Defining the Science and the Practice'. 31st October- 1st November, 2006, Bungendore, NSW, Australia.

Pozo Vazquez, D., Olmo Reyes, F.J., Alados Arboledas, L., 1997. A comparative study of algorithms for estimating land surface temperature from AVHRR data. Remote Sens. Environ. 62, 215–222.

Price, J.C., 1983. Estimating surface temperatures from satellite thermal infrared data – a simple formulation for the atmospheric effect. Remote Sens. Environ. 13, 353–361.

Price, J.C., 1984. Land surface temperature measurements from the split window channels of the NOAA 7 AVHRR. J. Geophys. Res. 89, 7231–7237.

Purevdorj, T., Tateishi, R., Ishiyama, T., Honda, Y., 1998. Relationships between percent vegetation cover and vegetation indices. Int. J. Remote Sens. 19, 3519–3535.

Qin, Z., Karnieli, A., Berliner, P., 2001. A mono-window algorithm for retrieving land surface temperature from Landsat TM data and its application to the Israel-Egypt border region. Int. J. Remote Sens. 22, 3719–3746.

Rouse, J.W., Haas, R.H., Schell, J.A., Deering, D.W., 1973. Monitoring vegetation systems in the great plains with ERTS. In: Proceedings of 3rd ERTS Symposium, NASA SP-351 I, pp. 309–317.

Saini, V., Gupta, R.P., Arora, M.K., 2015a. Environmental issues of coal mining – a case study of Jharia coal-field India: special publication of Journal of Engineering Geology. In: Proceedings of International Conference on Engineering Geology in New Millennium, 27–29 Oct. 2015, New Delhi, pp. 989–998.

Saini, V., Gupta, R.P., Arora, M.K., 2015b. Spatio-temporal pattern of eco-environmental parameters in Jharia coalfield. In: India: Proceedings of SPIE 9644, Earth Resources and Environmental Remote Sensing/GIS Applications VI, 96441H, SPIE Remote Sensing-2015, 21–24 Sept., Toulouse, France.

Saini, V., Gupta, R.P., Arora, M.K., 2016. Environmental impact studies in coalfields in India- A case study from Jharia Coal-field. Renew. Sustain. Energy Rev. 53, 1222–1239.

Sobrino, J.A., Jimenez-Munoz, J.C., Paolini, L., 2004. Land surface temperature retrieval from Landsat TM 5. Remote Sens. Environ. 90, 434–440.

Sobrino, J.A., Raissouni, N., 2000. Toward remote sensing methods for land cover dynamic monitoring: application to Morocco. Int. J. Remote Sens. 21, 353–366.

Sobrino, J.A., Raissouni, N., Li, Z.-L., 2001. A comparative study of land surface emissivity retrieval from NOAA data. Remote Sens. Environ. 75, 256–266.

Susskind, J., Rosenfield, J., Reuter, D., Chahine, M.T., 1984. Remote sensing of weather and climate parameters from HIRS2/MSU on TIROS-N. J. Geophys. Res. 89, 4677–4697.

The Landscape Toolbox, 2017. Remote Sensing Methods: Normalized Difference Vegetation Index. http://wiki.landscapetoolbox.org/doku.php/remote_sensing_methods:normalized_difference_vegetation_index?s[]=ndvi (accessed March 2016).

Tonooka, H., 2001. An atmospheric correction algorithm for thermal infrared multispectral data over land – a water-vapor scaling method. IEEE Trans. Geosci. Remote Sens. 39, 682–692.
Toren, T., Ünal, E., 2001. Assessment of open pit coal mining impacts using remote sensing: a case study from Turkey. In: Proceedings of 17th International Mining Congress and Exhibition of Turkey, 19–22 June, Ankara, Turkey, pp. 461–466.
Townshend, J.R.G., Justice, C.O., 1986. Analysis of the dynamics of African vegetation using the normalized difference vegetation index. Int. J. Remote Sens. 11, 1435–1445.
USGS, 2013. Using the USGS Landsat 8 Product. https://landsat.usgs.gov/using-usgs-landsat-8-product (accessed September 2014).
Valor, E., Caselles, V., 1996. Mapping land surface emissivity from NDVI: application to European, African, and South American areas. Remote Sens. Environ. 57, 167–184.
Van de Griend, A.A., Owe, M., 1993. On the relationship between thermal emissivity and the normalized difference vegetation index for natural surfaces. Int. J. Remote Sens. 14, 1119–1131.
Van Wijngaarden, W., 1991. The green cover of the Earth: a dynamic resource in a changing environment. ITC J. 3, 113–121.
Vidal, A., 1991. Atmospheric and emissivity correction of land surface temperature measured from satellite using ground measurements or satellite data. Int. J. Remote Sens. 12, 2449–2460.
Viña, A., Gitelson, A.A., Nguy-Robertson, A.L., Peng, Y., 2011. Comparison of different vegetation indices for the remote assessment of green leaf area index of crops. Remote Sens. Environ. 115, 3468–3478.
Wan, Z., Dozier, J., 1996. A generalized split-window algorithm for retrieving land-surface temperature from space. IEEE Trans. Geosci. Remote Sens. 34, 892–905.
Wan, Z., Li, Z.-L., 1997. A physics-based algorithm for retrieving land-surface emissivity and temperature from EOS/MODIS data. IEEE Trans. Geosci. Remote Sens. 35, 980–996.
Wechsung, F., 2000. May land use change reduce the water deficiency problem caused by reduced brown coal mining in the state of Brandenburg? Landsc. Urban Plan. 51, 177–189.
Weng, Q., Lu, D., Schubring, J., 2004. Estimation of land surface temperature–vegetation abundance relationship for urban heat island studies. Remote Sens. Environ. 89, 467–483.
Wenger, K.F., 1984. Forestry Handbook, second ed. John Wiley & Sons, New York, p. 1159.
Xia, L., Zhen-qi, H., Wei-jie, L., Xiao-yan, H., 2007. Vegetation growth monitoring under coal exploitation stress by remote sensing in the Bulianta coal mining area. J. China Univ. Min. Technol. 17, 479–483.
Yuan, F., Bauer, M.E., 2007. Comparison of impervious surface area and normalized difference vegetation index as indicators of surface urban heat island effects in Landsat imagery. Remote Sens. Environ. 106, 375–386.

WWW Addresses: Additional Reading

(1) Bharat Coking Coal Limited (BCCL)
http://www.bccl.gov.in
http://www.bccl.gov.in/PDFs/MPLANBCCL-2008.pdf

(2) Coal India Limited (CIL)
http://www.coalindia.in

(3) Government of India, Ministry of Coal
http://www.coal.nic.in
http://www.coal.nic.in/content/coal-reserves

(4) Jharia coalfires
http://jharia.jharkhand.org.in
http://www.gi.alaska.edu/~prakash/coalfires/faq_india.html

(5) The Landscape Toolbox
http://www.wiki.landscapetoolbox.org/doku.php/remote_sensing_methods:normalized_difference_vegetation_index

CHAPTER 18

Remote Sensing Techniques for Detecting Self-Heated Hot Spots on Coal Waste Dumps in Upper Silesia, Poland

The cone-shaped Anna coal-mine dump ignited by self-heating in the Ruch Anna Coal Mine, Poland. Burning today, some gas vents in the dump, like the one shown here, are surrounded by sulfur at the surface. The temperatures at a depth of 30 cm inside the vent, measured in September and November of 2016, varied from 82 to 84°C. The diameter of the vent is 30–40 cm. *Photo by Ádám Nádudvari, 2016.*

CHAPTER CONTENTS

Coal and Peat Fires: A Global Perspective
Edited by Glenn B. Stracher

https://doi.org/10.1016/B978-0-12-849885-9.00018-4

Photo by Ádám Nádudvari, 2016.

18.1 Remote Sensing Detection of Hot Spots on Coal Dumps in Upper Silesia, Poland

Ádám Nádudvari
Justyna Ciesielczuk

Like today, the Anna dump exhibited thermal activity more than 20 years ago, as revealed by satellite thermal maps. The top of the cone was flattened between 1.5 and 2 years ago in an unsuccessful attempt to extinguish hot spots. Self-heating and combustion may intensify in the future.

Introduction

This study summarizes the possibilities of remote self-heating detection on six chosen coal-waste dumps in Upper Silesia (Poland) using Landsat 4–5 TM, ETM+, and ASTER images with validation by in situ temperature measurements. On these dumps, self-heating of varying intensity has taken place since they were first formed. Coal-waste dumps are characteristic elements of the Upper Silesian mining and industrial region where intensive coal mining from greater depths started at the turn of the 19th and 20th centuries (Dulias, 2003). These **coal-waste dumps** typically contain appreciable amounts of coal from uneconomic seams and a variety of host rocks like pyrite-bearing carbonaceous shales and sandstones. The organic matter in coal waste usually comprises 20%–30% carbon. This starts to weather immediately after deposition and leads to **exothermic oxidation of the carbon** (Zhang and Kuenzer, 2007; Carras et al., 2009; Masalehdani et al., 2009; Misz-Kennan and Fabiańska, 2011). When large amounts of these materials occur in humid surroundings, condensation and adsorption of moisture lead to generation of additional heat that is transported into the bulk material and may lead to ignition (Clemens and Matheson, 1996; Lohrer et al., 2005a). **Susceptibility to self-heating** is increased by air temperature; wind; organic-matter rank; ash content; surface area exposed; particle size; moisture and oxygen content; and the shape, layering, and compaction of a dumped pile (Skarżyńska, 1995; Pone et al., 2007; Lohrer et al., 2005b). When wet and dry coals are combined, the interface between them becomes a heat exchanger (Lu et al., 2004; Misra and Singh, 1994; Misz-Kennan and Fabiańska, 2010). Self-heating is responsible for changes in the physical and chemical properties of coal waste (Carras et al., 2009). During self-heating, high concentrations of toxic acidic gases and compounds such as polycyclic aromatic hydrocarbons are released (Ribeiro et al., 2010; Pone et al., 2007; Lighty et al., 2000; Fabiańska et al., 2013).

Self-heating of coal-waste dumps (**hot spots**) may appear as high-temperature surface anomalies. Temperature variations of background rocks due to solar influences (e.g., different albedo, slope, aspect or vegetation cover) can significantly influence thermal anomalies (Zhang and Kuenzer, 2007). However, extended observation permits persistent heat sources due to self-heating to be recognized, as well as their migration, intensification, and disappearance, when they are hot enough for detection with satellite sensors (Nádudvari, 2014). Remote-sensing techniques have been previously applied to recognize and analyze coal-seam fires using nighttime satellite images (Guha and Kumar, 2012), night and winter images (Tetzlaff, 2004; Voigt et al., 2004), Landsat 7 ETM+ with day and night images (Zhang, 2004; Chatterjee, 2006; Huo et al., 2014; Prakash et al., 2011), and, using winter-time Landsat images, hot spots on snow-covered coal waste (Nádudvari, 2014).

The aim of the research was to improve the detection of the growth and migration of hot spots using Landsat satellite, and Sentinel 2 and ASTER images, with applied atmospheric correction. Nighttime summer and snow-cover images as well as temperature measurements were used for the research.

The Study Area

Intense thermal activity has characterized the selected coal-waste dumps over the entire period involved in the collection of data (1993–2015; Figure 18.1.1). After the early 1990s, many Upper Silesian Coal Basin (USCB) coal mines were closed to reduce operating costs (Cabała and Ćmiel, 1999; Cabała et al., 2004). However, their long-time operations had resulted in more than 220 dumps and landfills, now covering >4000 ha. The greatest number of dumps occur in the central and southwestern part of the USCB, typically in areas with high populations (Konopko, 2010; Cabała et al., 2004). At present, the burnt-out waste material is used for road construction (Misz-Kennan et al., 2013). Until the late 1980s, little was done to prevent dump self-heating and negative impacts on the environment (Nádudvari and Fabiańska, 2016).

Makoszowy Dump

The Makoszowy mine, together with an adjacent coking plant, opened in 1906 (The History of the Makoszowy Coal Mine, 2016). Though there is no detailed information about the history of the dump or the amount of deposited material, it was common practice to dump waste close to the source mine (Figure 18.1.1). The full extent of the dump on a map complied between 1958 and 1962 and vegetation cover on orthophotos from 1996 (Otwarty Regional NY System, 2017) are illustrated in Figure 18.1.2. In the early 2000s, excavation started on the western side and, by about 2013, all waste there had been removed. Though probably visible on the surface beforehand, excavation served to intensify thermal activity on the dump. Though there are no available photographs of self-heating on the dump, burnt reddish material was evident in 2009.

Czerwionka-Leszczyny Dump

This dump, mostly forested, consists of three cones, the highest of which is ~100 m high (Figure 18.1.1). On the top of the highest steep cone, intense **surface pseudo-fumarolic activity** is associated with surface gas vents surrounded by sulfate crusts (Parafiniuk and Kruszewski, 2010). On the top of this cone, there were also several puddles of **expelled bitumen** (Figure 18.1.3). Heating under pyrolytic conditions in the dump caused thermal cracking of the coal-waste organic-matter macromolecule. The resulting bitumen migrated from the heating zone and accumulated on colder coal-waste surfaces (Nádudvari and Fabiańska, 2016). The scale of this thermal activity, extant for more than 30–40 years, is decreasing today and burnt-out material is visible in the forested parts of the dump (Nádudvari, 2014).

Rymer Dump

The Rymer dump is unused at present (Figure 18.1.1). The original waste material was essentially burnt out. From 1994 to 1999, the cones were redeveloped and spread with additional coal-waste material. However, as the original burnt-out material was hot, self-heating restarted in this added coal waste (Tabor, 2002). Around at the beginning of the 21st century, the surface of the dump was covered with concrete panels and ash to block air access. Despite this, self-heating intensified. Finally, the panels were removed from the western part (Misz-Kennan et al., 2013). Today, heating occurs mainly on the still-covered eastern slope. Bitumen with an elevated amount of phenols and naphthalenes spills through cracks in the panels (illustrated in Nádudvari and Fabiańska, 2016). The concrete panels and the adsorbent ash covering the waste inhibited leaching and evaporation of these compounds.

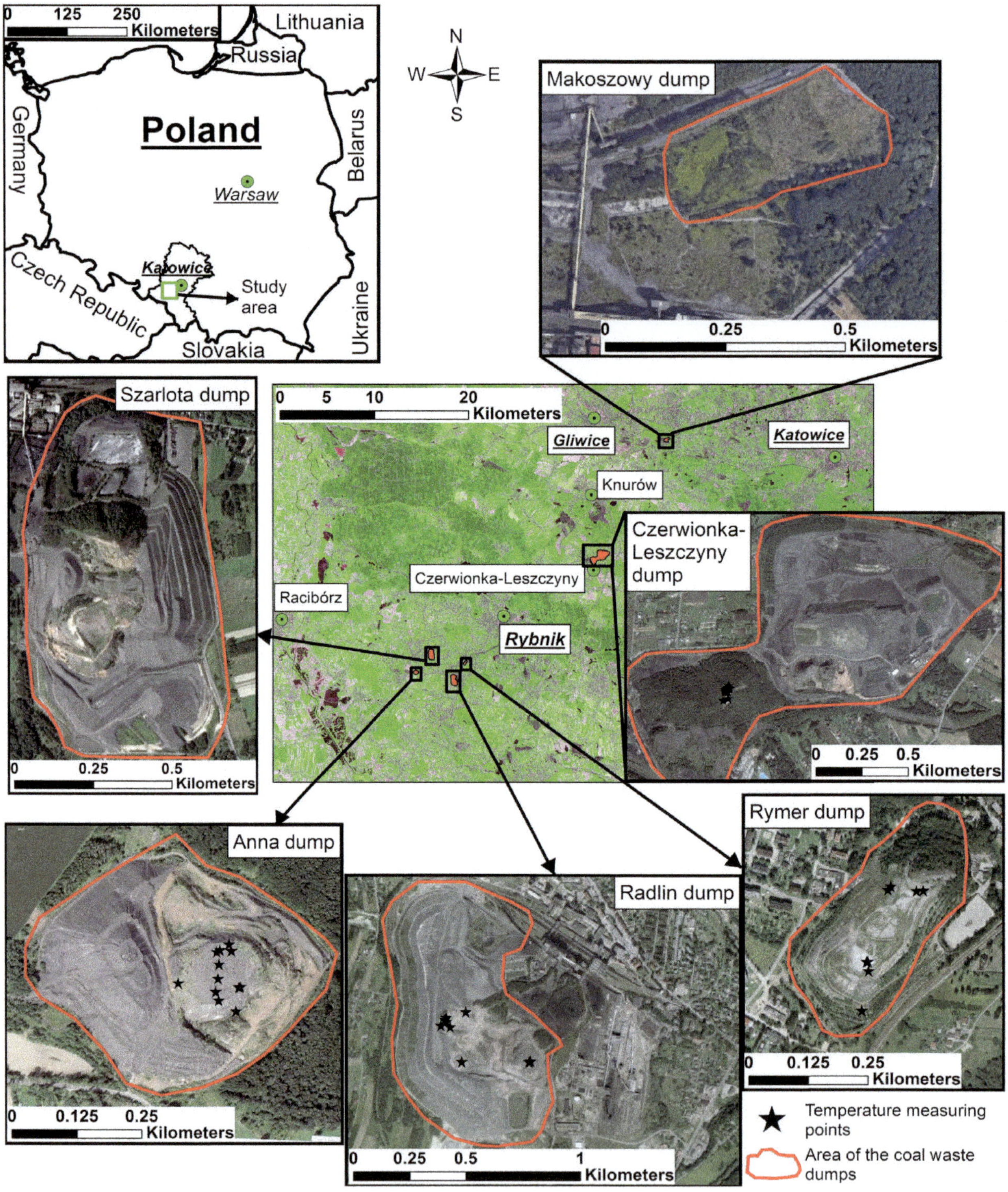

Figure 18.1.1. The locations of the analyzed coal-waste dumps showing points where surface temperatures were measured. Self-heating on these dumps (except Makoszowy) was previously reported by Misz-Kennan and Tabor (2011), Misz-Kennan et al. (2013), and Parafiniuk and Kruszewski (2010). Therefore, these dumps served as good study sites. Figure by Ádám Nádudvari, 2016, with georeferenced Google Maps (2017) and Sentinel 2 satellite images (EarthExplorer, 2017).

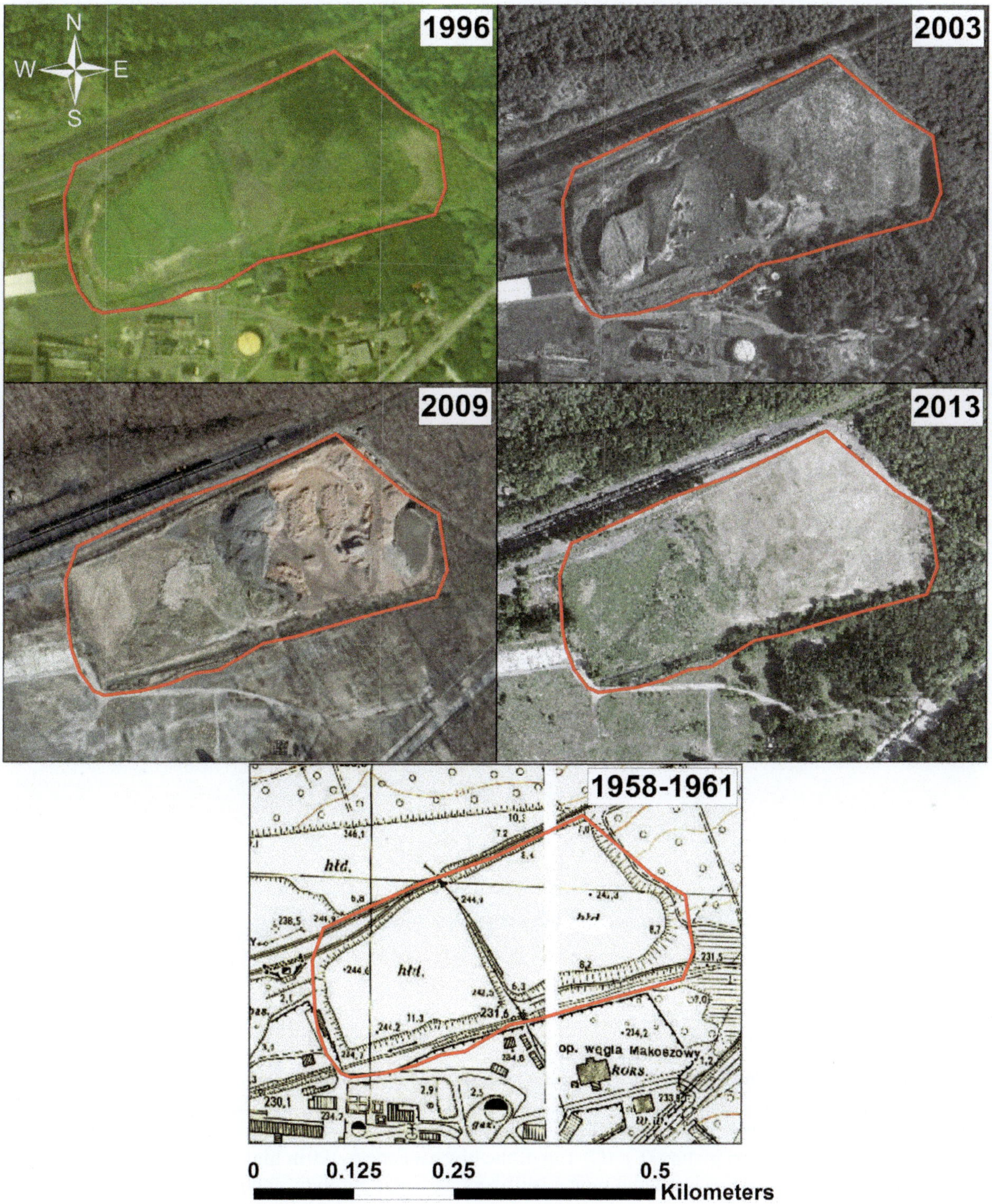

Figure 18.1.2. Excavation of the Makoszowy dump as revealed by an archival map and orthophotos. From the Otwarty Regional NY System (2017); edited and georeferenced by Ádám Nádudvari, 2017.

Radlin Dump

The Radlin dump located close to the Marcel Coal Mine is a combination of four old dumps formed at different times (Figure 18.1.1). The first part (the oldest cone) dates from the end of the 19th century. The capacity of two other cones covering an area of ~0.13 km^2 was estimated at 12 × 10^6 m^3 in 1986 (Barosz, 2003; Misz-Kennan and Tabor, 2011). Self-heating of varying intensity occurred throughout both. In 1982, at the base of the old cones,

Figure 18.1.3. The Czerwionka-Leszczyny dump. (A) Thermally active southern slope of the highest cone. When the surface temperature was 30°C, it was 50°C at a depth of 30 cm. (B) View to the east; from the highest cone in the dump, showing reclaimed forest growing on burnt-out waste rock. In the background, the former place of the reclaimed inactive cone; see on picture (E). (C) Top of the highest cone (~100 m high from foot of the dump, data from contour line maps) with precipitated bitumen (self-heating under pyrolytic conditions with low oxygen content or even its absence has caused thermal cracking of organic matter and high bitumen production; later that generated bitumen evaporated from the self-heating zone, migrated and accumulated on colder coal-waste surfaces) and white salammoniac (NH_4Cl) crust on surface of condensation zone nucleated from gas vents as it cooled. The surface temperatures when measured here varied from 30–40°C in zero wind blow and from 24–70°C on a windy day even in cold weather. (D) Three forested cones seen at the middle and toward the west, the highest of which is thermally active at the top. On the eastern side of the dump, there is an inactive cone. (E) The inactive burnt-out cone now being reclaimed (approx. 40–50 m high, picture was taken in 2015). The height of the trees on the top are ~10 m. Photos by Ádám Nádudvari, 2016.

Figure 18.1.4. (A) Reusing burnt-out coal waste for road constructions. (B) An active hot spot near the top of the photo, from which smoke is exhaled; below this are two self-heated and burnt layers. (C) and (D) Smoke emitted from hot spots. Photos by Justyna Ciesielczuk, 2011.

a new fourth dump was created on a flat surface covering an area of 0.33 km^2 and with a capacity of $3.7 \times 10^6 m^3$. A newer dump surrounding the two older cones covers an area of 0.56 km^2 and has a $24.9 \times 10^6 m^3$ capacity (Barosz, 2003). Though material from current mining replacing the exploited wastes is being compacted with vibrating rollers (Figure 18.1.4), self-heating occurs at several sites (Misz-Kennan et al., 2013).

Anna Dump

This dump stores wastes from the Ruch Anna Coal Mine that opened in 1954 (Figure 18.1.1). There is one cone covering an area of 0.43 km^2. The oldest part is ~0.20 km^2 and ~50 m high. The planned capacity was $>3 \times 10^6 m^3$ (Barosz, 2003). Exploitation of the waste, by enabling oxygen access, intensified the self-heating that later obstructed further exploitation of the waste. Toxic fumes and odorous gases are a problem in the nearby town of Pszów (Misz-Kennan et al., 2013). Self-heating that is still occurring is reflected in puddles of bitumen and salammoniac crusts (Figure 18.1.5).

Szarlota Dump

The Szarlota dump (Figure 18.1.1) is located beside the active Rydułtowy Coal Mine where mining started in 1900 (Barosz, 2003; Misz-Kennan et al., 2013). This dump covers an area of 37 ha, has a capacity of 13.3 million m^3, and is 134 m high. The cone is surrounded by waste heaps derived from former cones that, due to waste removal (disposal), were cut to the height (<90 m) of an adjacent flat dump (Figure 18.1.5). Since its beginning, self-heating has occurred with varying intensity. The waste is thermally altered to different degrees. Self-heating still happening in some parts of the dump has caused landslides (Pieczyrak, 2010; Barosz, 2003; Misz-Kennan et al., 2013). Today, the dump is being extended to the south. Entrance to the entire area of this dangerous dump is only with permission.

Figure 18.1.5. (A) The thermally active slope on the Anna dump, locally covered with white salammoniac (NH_4Cl); where rain water has eroded a large gully into the dump. At the foot of the gully, a temperature of ~400°C was recorded at a depth of 30 cm. (B) The top of the Anna dump with a large crack encrusted with salammoniac. The temperature at a depth of 30 cm was 90–93°C. (C) A crack encrusted with sulfur, at the top of the self-heated slope of the Anna dump. The temperature at a depth of 30 cm was 82–84°C. (D) Reddish burnt-out coal waste in the Szarlota dump. The highest point reaches 134 m from foot of the dump (Pieczyrak, 2010). The picture was taken from Anna dump, the field of view up to Szarlota dump is ~3 km. (E) Reddish thermally altered carboniferous rocks on the active side of the Szarlota dump. The top of the flat shaped side reaches ~90 m from foot of the dump (Pieczyrak, 2010). The length of the pipes at the left-middle side of photo is about 3 m. Photos by Ádám Nádudvari, 2016.

Applied Remote Sensing Methodology

Remote sensing using the thermal band offers a cost-effective and time-saving method for mapping thermal anomalies such as coal-seam fires, self-heating on coal-waste dumps, and forest fires. This technology has almost replaced older methods such as in situ measurement of surface temperature (Gangopadhyay et al., 2005; Guha and Kumar, 2012). **Thermal infrared remote sensing** relies on the fact that everything above absolute 0K (−273.8°C) emits thermal radiation. A perfect black body emits the maximum amount of radiation at each wavelength (Chatterjee, 2006; Gangopadhyay et al., 2005; Tetzlaff, 2004). The wavelength of thermal infrared radiation from an object is controlled mainly by the characteristics of its surface emissivity and temperature, whereas the entire electromagnetic spectrum is considered as thermal infrared; wavelengths 3–5 and 8–12 μm are specifically used in thermal remote sensing (Gangopadhyay et al., 2005). The temperature of earth materials is estimated from the thermal radiation using Planck's radiation equation (Rees, 2001).

Self-heating in coal-waste dumps can be recognized as high-temperature surface anomalies. However, from the viewpoint of thermal satellite sensors, these anomalies can be substantially influenced by many factors such as temperature variations of background rocks due to solar radiation, albedo differences, slope, aspect, and/or vegetation cover. These factors can commonly generate false-positive signals for potential self-heating sites on a coal-waste dump, especially if only a few images are analyzed (van Genderen et al., 1996; Zhang and Kuenzer, 2007). By continuous addition of detected hot spots to a stack, anomalies can be seen to reflect persistent heat sources in coal-waste dumps that can also be seen to migrate, intensify, or disappear when they are long-lasting and hot enough for detection with a satellite sensor (Nádudvari, 2014). Application of the thermal bands of Landsat images can cause several problems when the surface temperatures are extremely high. **Coal-seam fires** with extended surface temperatures of 150–250°C, or even higher, serve as a good example; these emit more energy than a simple self-heating site on a coal-waste dump (Cracknell and Mansor, 1992; Chatterjee, 2006; Engle et al., 2012; Huo et al., 2014). Thus, the short-wavelength infrared (SWIR) Landsat bands—bands 5–7—can detect 160°C –277°C hot spots, <420°C in the case of coal-seam fires, using nighttime images to avoid solar radiation (Rothery et al., 1992; Huo et al., 2014). However, for the **thermal band** (thermal infrared (TIR) sensor) sensitivity of Landsat TM, ETM+, the maximum is 70°C (Chatterjee, 2006; Huo et al., 2014).

In some of the pixels of thermal-band data, where surface coal fire occupies the major part of the pixel area, the pixel-integrated temperature attains the sensor sensitivity ~70°C (Chatterjee, 2006). In addition, high-temperature zones with mean temperatures of ≥50°C are generally unaffected by normal air temperature or by the background temperature of the pixel (Chatterjee, 2006). Surface temperatures on coal wastes generally vary from 7°C to 85°C, but in open fires can be >200°C (Misz-Kennan, 2010). Thus, the applied Landsat thermal bands are suitable for detecting hot spots on coal wastes. In general, at a heated spot on a dump, the difference between the surface and air temperatures is >3°C (Barosz, 2002; Tabor, 2002). The hot spot is commonly located about 1.5–2.5 m below the dump surface (Urbański, 1983; Walker, 1999). Heating inside a coal-waste dump is highly variable with temperatures ranging from ambient to >500°C (Carras et al., 1994, 1999). However, in the subsurface, self-combustion temperatures can reach up to 1000°C –1300°C (Ribeiro et al., 2010; Querol et al., 2008). In some cases, there may be little or no surface indication of the presence of intensely hot zones even within a few meters of the dump surface (Day et al., 2015).

Landsat TM, ETM+ snow-cover and nighttime images, one ASTER snow-cover image, and one ASTER nighttime image were used to assess the thermal histories of self-heating sites on the chosen dumps. Landsat TM, ETM+, ASTER, and Sentinel 2 summertime images were chosen to develop vegetation-cover maps (**Normalized Difference Vegetation Index**; NDVI) on the Czerwionka-Leszczyny dump, where the vegetation cover is increasing as a hot spot is waning. These images are freely available to download from the US Geological Survey satellite data source. Summer images showing the most developed vegetation cover were chosen. The NDVI does not measure the amount of vegetation, and its values can be influenced by many factors such as viewing angle, soil background, and by differences in row direction and spacing in the case of agricultural crops (Weng et al., 2004). The Landsat TM thermal infrared band operates in the 10.40–12.50 μm band with 120 m resolution (resampled to 30 m). The Landsat ETM+ thermal infrared band (TIR) (10.40–12.50 μm) operates in two gain types—low gain (band 6/1) and high gain (band 6/2) with 60 m resolution (resampled to 30 m). As the visible- and near-infrared (VNIR) and SWIR bands have 30 m resolution, the NDVI, **Normalized Difference Snow Index** (NDSI), and **Normalized Difference Snow and Ice Index** (NDSII) indices can provide better resolution (US Geological Survey Landsat Missions, 2016). The resolutions of ASTER images are for VNIR (15 m), SWIR (30 m), and TIR bands (90 m, resampled to 30 m by the Authors) (ASTER Users Handbook, 2011).

The image processing was carried out with ENVI 5.3 and the final map created with ArcGis 10.2 software. For the VNIR through the SWIR bands of the Landsat images, a relatively fast and effective atmospheric correction—the quick atmospheric correction (details in Harris Geospatial Solutions, 2016)—was applied to calculate the surface temperatures. To calculate surface temperatures, the *At-satellite brightness temperature (TS)* first had to be counted from radiance (US Geological Survey Landsat Missions, 2016). The thermal emissivity map also had to be generated based on the spectral absorption properties of the surface materials. Thermal *emissivity (e)* was calculated using *Proportion of vegetation (Pv)* values, as surface temperatures are directly related to surface physical properties (Table 18.1.1). The *Land Surface Temperature (LST)* was then calculated using the thermal *e* values and *TS* (Chatterjee, 2006; Weng et al., 2004; Jiménez-Muñoz et al., 2009; Suresh et al., 2016). For ASTER images, *brightness temperatures* were calculated using *radiometric calibration* then *emissivity normalization* tools in ENVI 5.3. In addition, a more robust atmospheric correction module—fast line-of-sight atmospheric analysis of hypercubes (FLAASH) (details in Harris Geospatial Solutions, 2016)—was applied to give more accurate results for calculating NDVI (for summertime images), NDSI, and NDSII values.

To obtain more precise localizations for the self-heated areas, the NDSI and NDSII indices were used. The high and low reflectance of the snow is in the visible (green) and SWIR region and the NDSI values are defined for pixels

Table 18.1.1
Indices calculated from formulas and parameters.

Calculated indices	Formula	Parameters	Additional information
NDVI	(NIR – VIS)/(NIR + VIS)	Near-infrared (NIR) and visible red (VIS) bands	Varies between −1.0 and +1.0. Values close to 1.0 indicate abundant vegetation.
NDSI	(Green – SWIR)/ (Green + SWIR)	Short-wave infrared and green bands	Varies between −1.0 and +1.0. Values close to 1.0 indicate abundant snow cover.
NDSII	(Green –NIR)/ (Green + NIR)	Green and near-infrared bands	Varies between −1.0 and +1.0. Values close to 1.0 indicate abundant snow cover.
Proportion of vegetation (Pv)	$((NDVI - NDVI_{min})/(NDVI_{max} - NDVI_{min}))^2$	$NDVI_{min,max}$ = minimum and maximum values of NDVI image	
Emissivity (e)	0.004Pv + 0.986	Pv = Proportion of vegetation	
Land Surface Temperature (LST)	$TS/1 + DN_{TIR}\,(TS/p)\ln(e)$	TS = Temperature at satellite; DN_{TIR}* = DN values of original thermal band of Landsat TM†, ETM+‡; p = hC/S where h = Planck's Constant (6.626×10^{-34} Js); C = 2.998×10^{8} m/s, i.e., velocity of light; S = 1.38×10^{-23} J/K, i.e., Boltzmann Constant; $P = 1.438 \times 10^{-2}$ mK	

*DN_{TIR} = Digital Number (DN) with a range between 0 and 255. The acquired temperature data stored in this information.
†TM = Landsat Thematic Mapper.
‡ETM+ = Landsat Enhanced Thematic Mapper Plus.

>0.4 approximately; that is, 50% or more of the given area is covered by snow (Dozier, 1989; Hall et al., 1995, 1998; Negi et al., 2009). The results of a second index (NDSII), which uses reflectance values of red and mid-infrared spectra (Xiao et al., 2001) and which may give more accurate results, were compared with the NDSI values. The degree of snow cover on a coal-waste dump is influenced by several factors like sun angle, surface geomorphology, wind direction, vegetation cover, air temperature, depth of hot spots, and snow thickness. Nevertheless, together with thermal mapping, snow-cover mapping can aid precise definition of self-heating areas on coal-waste dumps (Nádudvari, 2014).

Results and Discussion

Typically, hot spots appeared only on certain parts of the dumps. Surrounding surface temperatures close to the detected heat anomalies are lower than the hot-spot temperatures by 6–14°C on average, and even more in the Makoszowy dump where active self-combustion was intense. These results compare with those on shallow coal-seam fires reported by Prakash et al. (2011) where Landsat thermal band pixels (summertime images) corresponded to surface temperatures persistently higher by about 5–14°C than those of the background. The coal-waste dumps may be good sites for comparing NDVI, NDSI, and NDSII indexes and thermal band data. As reported by Nádudvari (2014), active hot spots on the Radlin, Anna, Szarlota, Rymer, and Czerwionka–Leszczyny dumps were a constant presence, although with passing time, the self-heated areas decreased. In the Makoszowy dump, some developing hot spots with elevated temperatures are obvious near the KWK Makoszowy coal mine (Figure 18.1.6). From the initial stage, through development to final extinction, the self-heating processes are well seen using the different image types. All thermally active sites in this dump were extinguished by 2007; at present, no more have appeared. The dump has been exploited and does not exist now. Moreover, the ASTER images (with high resolution of 90 m) were sensitive enough to detect high temperature hot spots on the dump.

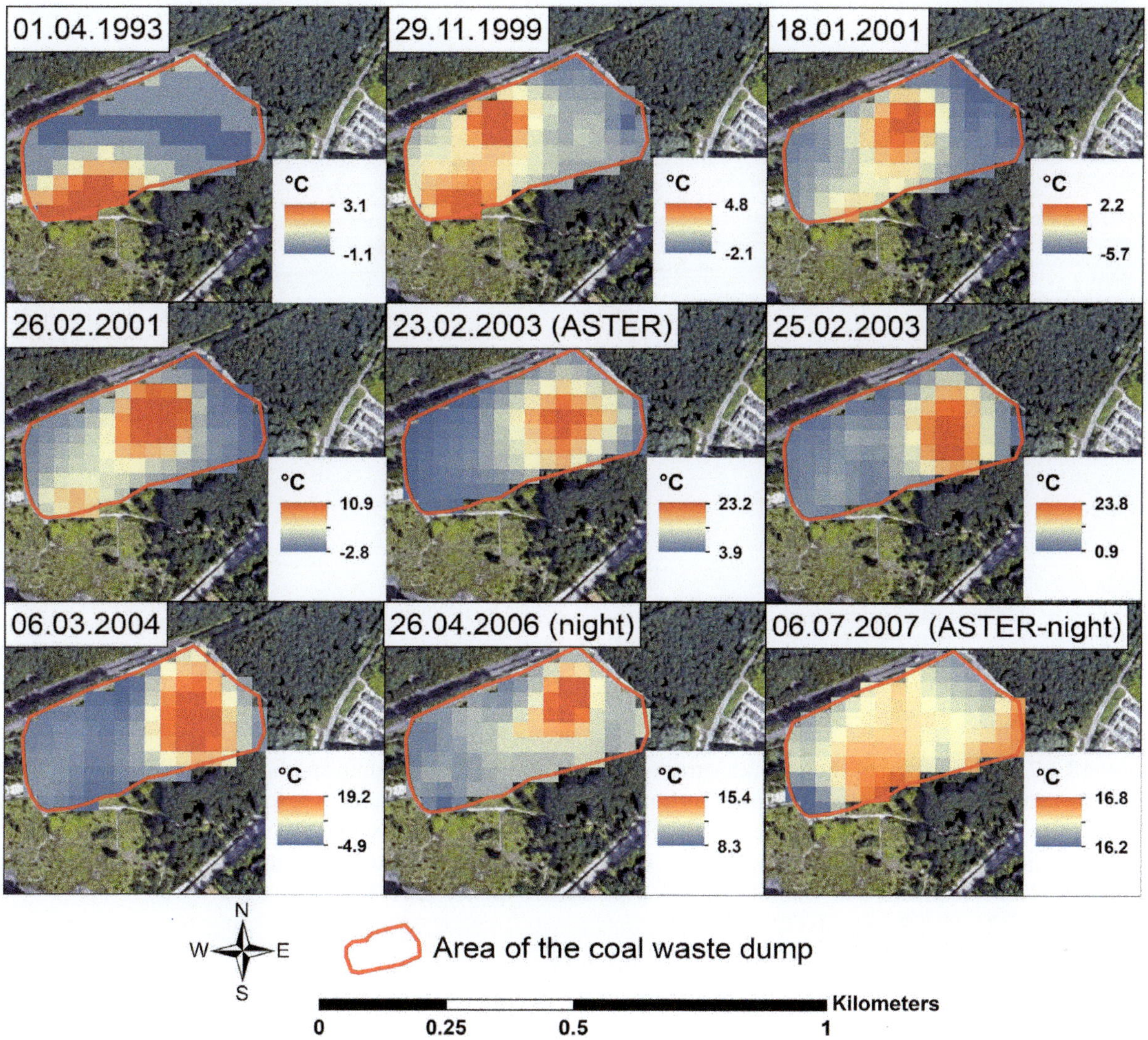

Figure 18.1.6. A series of thermal maps (surface temperatures, except ASTER) of the Makoszowy dump on which a hot spot (red colors) is evident. The day, month, and year is recorded in the upper left-hand corner of each image. Figure by Ádám Nádudvari, with satellite images from EarthExplorer (2017).

The four coal-waste dumps have been chosen to illustrate the scope of thermal maps and of the snow-cover indexes NDSI and NDSII. Abundant snow cover was indicated by >0.4 values that indicate where snow has melted (<0.4) on the hot spots, and in their surroundings, can outline the extended hot spots (Figure 18.1.7). Thus these bare surfaces indicate that the anomalies are not related to solar heating. NDSI values are useful in specifying the exact locations of self-heating areas. Moreover, the index helps reduce false anomalies caused by frozen lakes that appear to have relatively high temperatures but, on the snow-cover maps, are seen to be frozen. Snow-cover maps (Figure 18.1.7) also show the impact of cold weather on surface anomalies; where background surface temperatures are close to −10°C (Figure 18.1.7D), the surface temperatures of hot spots proportionally decrease. However, where background surface and air temperatures are similarly low, (i.e., −15°C to −11°C), hot spot surface temperatures show positive values with low (<0.4) NDSI values that may indicate self-heating (Nádudvari, 2014). In addition, it merits noting that if hot spots occur at shallow depths or have low thermal activity with small extension, cold air temperatures can overcome the self-heating, and any thermal response by the satellite sensor is diminished.

In Table 18.1.2, reference air temperatures for validating surface temperatures are shown. As reference points, the nearest cities of Racibórz and Katowice were chosen. Obviously, geomorphological factors (e.g., valleys) can influence measured surface and air temperatures.

In comparing the thermal maps and NDVI, it is proposed to use the data for the highest cone of the Czerwionka-Leszczyny dump (Figures 18.1.3D and 18.1.8). As it is a spontaneously forested dump, the extinguishing of a

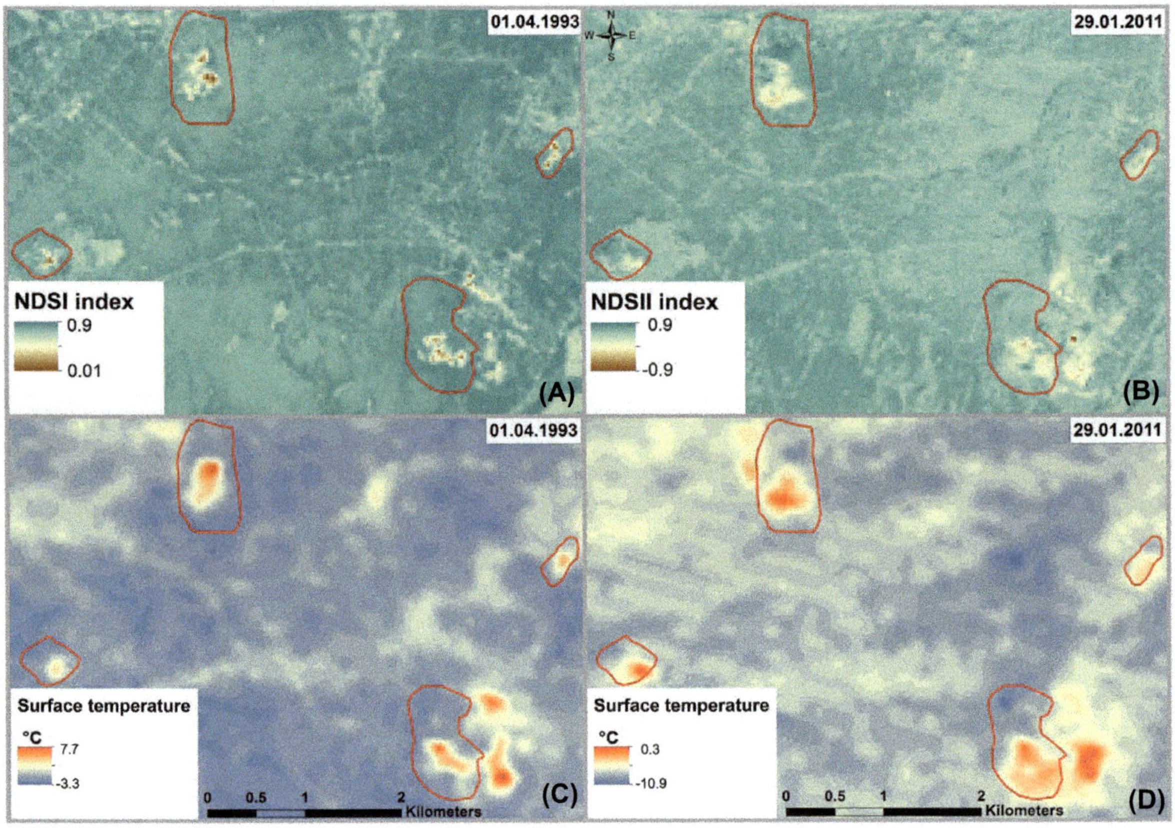

Figure 18.1.7. Comparison of snow-cover terrain (A and B) on the Radlin dump and the respective thermal images (C and D) that show hot spots (red colors). The hot spots that are not encircled in the lower right-side of (C and D) are not on the coal-waste dump. They are due to industrial heat-energy emissions from a coking plant. The scales for all four photos are the same. The day, month, and year is recorded in the upper right-hand corner of each image. Figure by Ádám Nádudvari, with satellite images from EarthExplorer (2017).

Table 18.1.2
Air temperatures (°C) from the Racibórz and Katowice meteorological stations in Poland.*

	Racibórz			Katowice		
Date of satellite image	Max.	Min.	Aver.	Max.	Min.	Aver.
April 1, 1993	7.1	−4.0	1.7	–	–	–
November 29, 1999	2.5	−1.3	−0.4	5.4	−4.1	1.1
February 24, 2000	4.9	−2.4	0.0	2.1	−5.6	−1.7
January 18, 2001	−2.4	−7.1	−4.6	−2.5	−9.3	−6.4
February 26, 2001	1.9	−18	−7.8	1.0	−14.2	−6.5
February 23, 2003	4.4	−7.6	−2.5	3.1	−11.1	−3.8
February 25, 2003	7.4	−6.5	−0.5	7.6	−12.0	−1.1
March 6, 2004	0.2	−18.2	−7.4	−1.0	−17.4	−9.2
April 26, 2006	21.2	9.0	14.6	21.2	6.4	13.7
July 6, 2007	18.6	11.9	14.8	17.1	12.0	13.5
January 29, 2011	−0.1	−9.6	−5.1	0.2	−11.1	−5.1
February 17, 2015	3.6	−4.7	−1.9	2.5	−6.2	−2.7
Date temp. measured	Max.	Min.	Aver.	Max.	Min.	Aver.
October 22, 2010	8.0	0.9	4.2	8.8	0.9	4.3
April 11, 2011	13.0	0.7	7.6	12.3	−1.5	6.4
March 15, 2015	9.8	1.7	3.9	9.8	1.5	3.8
September 22, 2015	20.0	4.4	12.0	18.5	3.2	10.2
September 23, 2015	22.5	6.1	16.0	20.0	8.1	14.6
May 16, 2016	13.3	2.9	8.0	12.8	1.7	7.6
September 30, 2016	24.4	13.9	19.6	22.9	13.6	18.4
November 12, 2016	4.9	0.6	1.8	5.4	0.3	1.7

*Weather Information Service (2017).

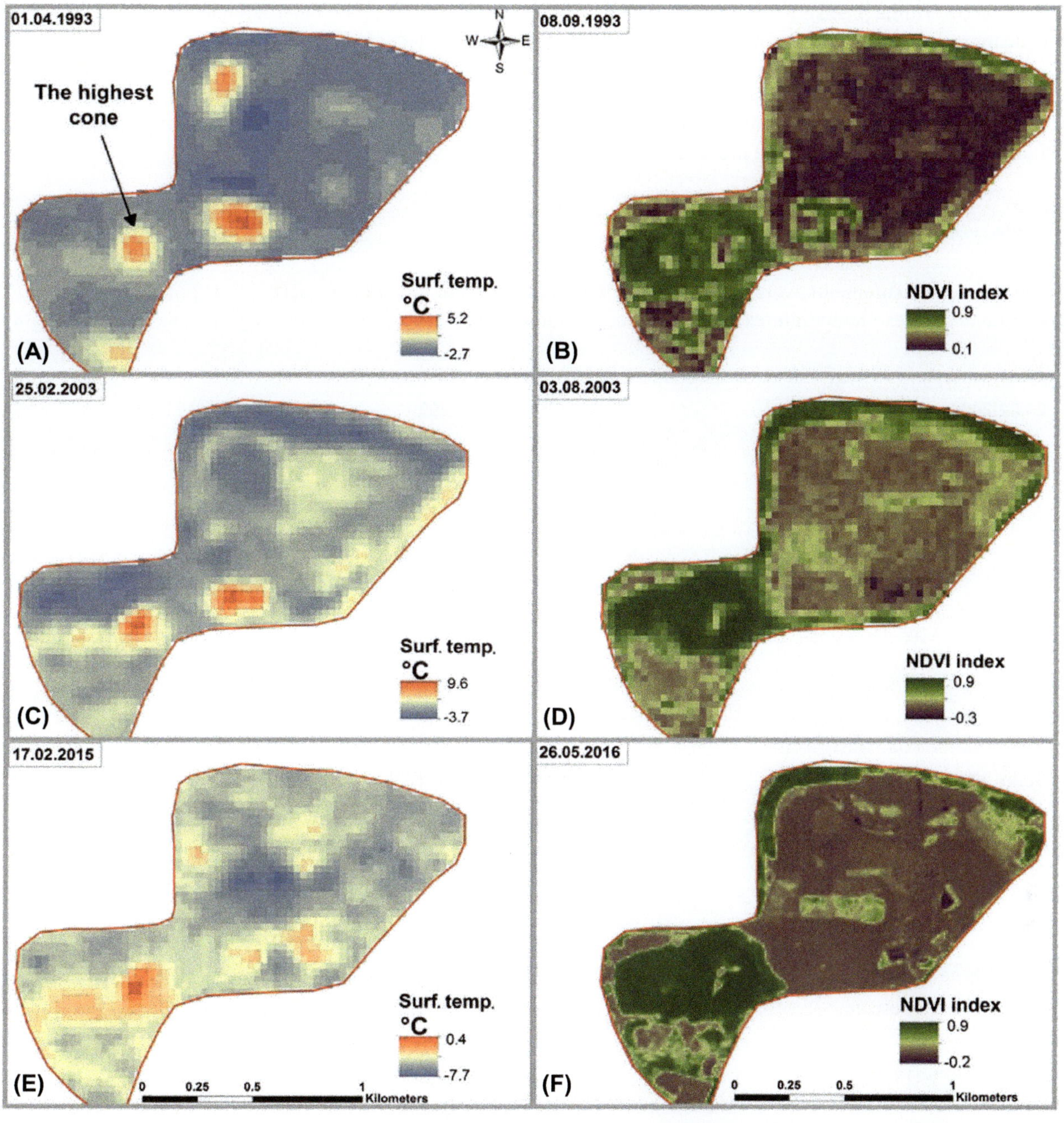

Figure 18.1.8. Comparison of surface temperature and vegetation cover (NDVI - Normalized Difference Vegetation Index) on the Czerwionka-Leszczyny dump. The negative NDVI values indicate the presence of a body of water at the surface. The scales for all six photos are the same. The day, month, and year is recorded in the upper left-hand corner of each image. Figure by Ádám Nádudvari, with satellite images from EarthExplorer (2017).

self-heating site gives a new place for reforestation even as coal waste material is continuously burning out and the hot spot reaches the top of the cone. Ahmed et al. (2015) has described a similar situation of low vegetation covering hot spots on coal-seam fires. In 1993, the spread of vegetation on the highest cone was greater than in the following 10 years. In 1993, a lack of covering vegetation was also evident on other cones located a few hundred meters to the east of the highest cone where there is another self-heating site (Figures 18.1.3E and 18.1.8B). However, during 2003–2016, these burnt-out cones (Figure 18.1.3E) were reshaped and their removal was initiated in 2016. It is worth noting that the resolution (10 m) of the Sentinel 2 image (Figure 18.1.8F) is sufficient; combined with thermal maps, it serves as a new image source for monitoring snow cover in the future.

The images on Figure 18.1.9 illustrate several errors that can significantly influence hot-spot identifications. The reference image is Figure 18.1.9A and B, which shows abundant snow cover and well-defined self-heating sites. In Figure 18.1.9C, the presence of snow cover is not consistent and is influenced by the terrain and the direction of snow fall. As the southern part of the dumps where self-heating occurs is not covered by snow, this image cannot be used to determine precisely the extension of hot spot. However, on the related thermal map (Figure 18.1.9D), the self-heating sites are visible. Another problem with snow cover is represented by an ASTER image (Figure 18.1.9E; resolution 15 m) where low NDSI values cause difficulties in detection of the hot spots. The nighttime image was taken (Figure 18.1.9F) in the evening when the surface was strongly influenced by remaining solar radiation. Except in the case of the Makoszowy dump (Figure 18.1.6), the hot spot nighttime images cannot be identified with certainty. These images would be more effective if taken on a snow-covered landscape.

Surface Temperature Measurements

Temperature measurements were carried out on the dumps on the locations of active hot spots (Figure 18.1.10). Unfortunately, these temperatures could not be measured at the time of satellite-image acquisition. In addition,

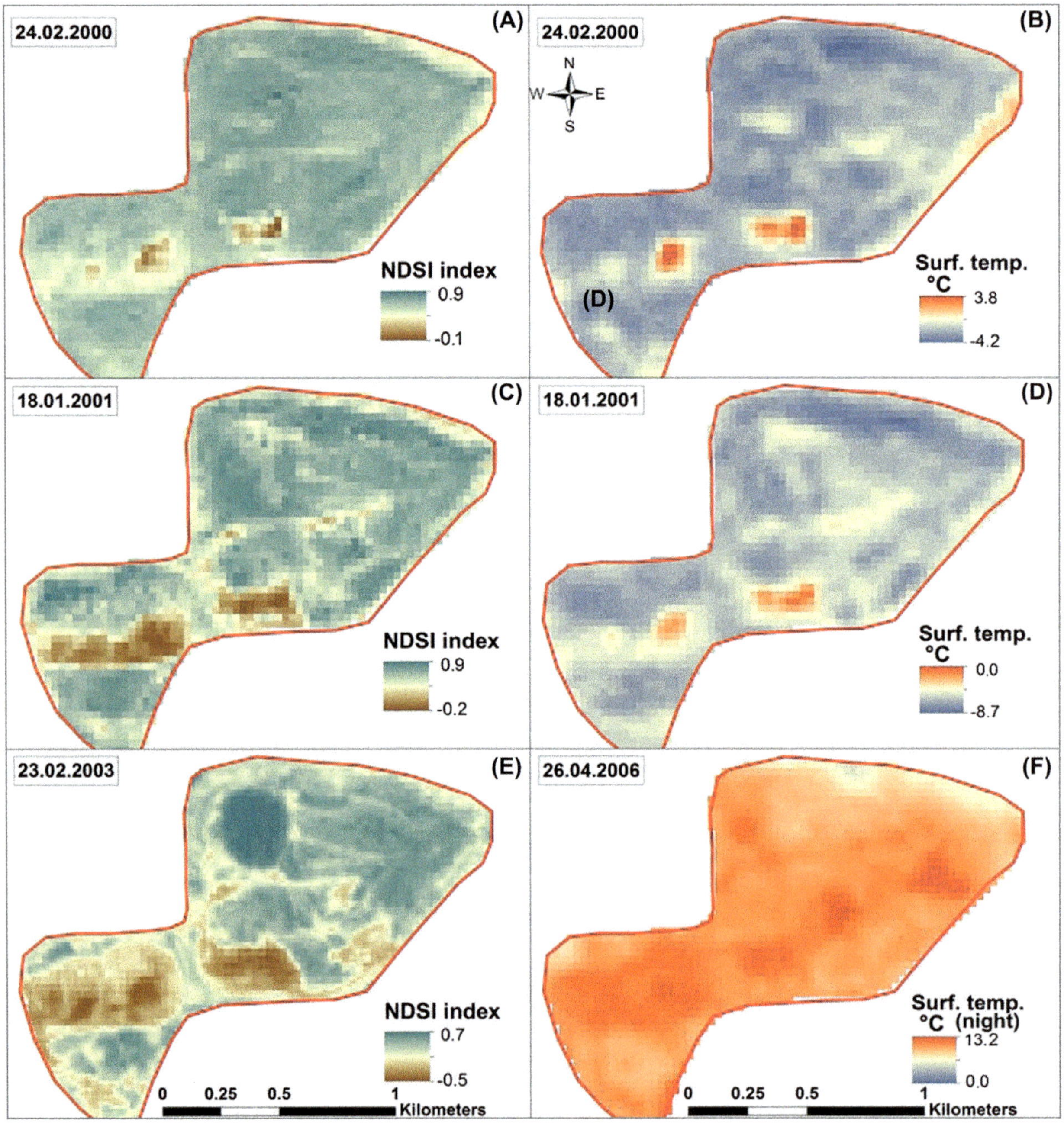

Figure 18.1.9. Illustrations of errors that can influence the calculation of NDSI or surface temperatures; for example, when snow cover is not equally distributed or when there is lack of abundant snow cover. The scales for all six photos are the same. The day, month, and year is recorded in the upper left-hand corner of each image. Figure by Ádám Nádudvari, with satellite images from EarthExplorer (2017).

measurements can differ significantly even with the space of a few square meters, for example, around cracks, holes, gas vents, or bitumen crusts. However, self-heating temperatures can change rapidly too because of excavating of smoldering material or efforts made to extinguish a hot spot, for example. The in situ temperature measurements were carried out on the surface of the self-heating site and at ~0.3 m depth.

In the Czerwionka-Leszczyny dump, the measured surface temperature ranges from ~24°C to 70°C and, at ~0.3 m depth, from 50°C to 90°C. At measuring point 9 where there was no evident heating, the temperatures vary from 27°C to 29°C. On the Anna dump, the surface temperatures range from 36°C to 154°C though, at measuring points 2 and 7, values of 24–32°C indicate no self-heating, only solar heating. At ~0.3 m depth, the measured temperatures vary from 66°C to 412°C with measurements at point 1 (150–300°C) and 9 (154–412°C) having been measured at the same place on different dates. On the Rymer dump surface, temperatures vary from 5°C to 73°C; the hot spots are seen to be waning. The extremely high temperatures on the surface at the Radlin dump are related to gas vents close to measuring points 6 and 8–9 (220–330°C; Table 18.1.3). Notably, measurements taken from almost the same places on the Czerwionka-Leszczyny dump revealed higher temperatures (24–70°C) on November 11, 2016, a day of cold air with strong and cold wind, than on September 30, 2016 (27–43°C), a day of sunshine and no wind. In most cases, the reverse was true; surface temperatures increased on windy days. Misz-Kennan and Fabiańska (2010)

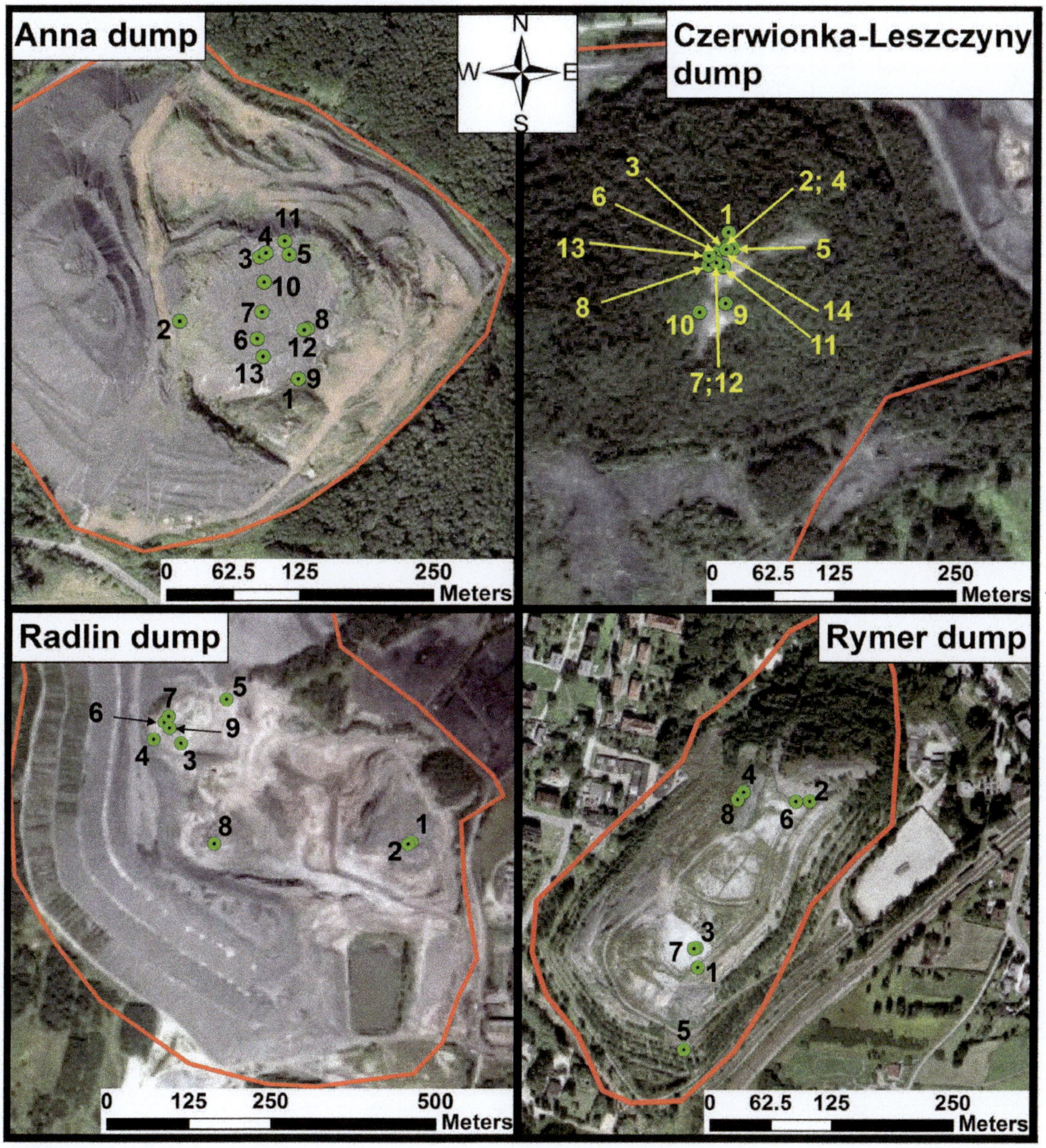

Figure 18.1.10. Locations of surface and subsurface temperature measurements. These are recorded in Table 18.1.3. Photos by Ádám Nádudvari, with satellite images from EarthExplorer.

Table 18.1.3
Surface temperature measurements at four coal waste dumps in upper Silesia, Poland.*

	Czerwionka-Leszczyny Coal Dump			Anna Coal Dump			Rymer Coal Dump			Radlin Coal Dump		
Dump Location[†]	Surface temperatures (°C)	Temperatures (°C) at ~30 cm depth	Date	Surface temperatures (°C)	Temperatures (°C) at ~30 cm depth	Date	Surface temperatures (°C)	Temperatures (°C) at ~30 cm depth	Date	Surface temperatures (°C)	Temperatures (°C) at ~30 cm depth	Date
1.	49	57	30/09/2016	154	412	30/09/2016	73	80	23/09/2015	60–66	69	22/09/2015
2.	34	59	30/09/2016	29	24	30/09/2016	50	75	23/09/2015	60	77	22/09/2015
3.	37	89	30/09/2016	36	68	30/09/2016	50	77	23/09/2015	66	77	22/09/2015
4.	43	90	30/09/2016	42	66	30/09/2016	75	–	23/09/2015	117	127	22/09/2015
5.	35	68	30/09/2016	43	83	30/09/2016	33	60	23/09/2015	210	–	11/04/2011
6.	30	80	30/09/2016	76	93	30/09/2016	5	60	15/03/2015	250–300	–	22/10/2010
7.	28	78	30/09/2016	32	24	30/09/2016	12	50	15/03/2015	30	163	22/10/2010
8.	27	58	30/09/2016	76	84	30/09/2016	16	28	15/03/2015	220–330	–	22/10/2010
9.	29	27	30/09/2016	150	300	12/11/2016	40–45	–	16/05/2016	220	–	22/10/2010
10.	30	50	12/11/2016	54	72	12/11/2016						
11.	60	80	12/11/2016	70	85	12/11/2016						
12.	65	85	12/11/2016	70	82	12/11/2016						
13.	70	90	12/11/2016	75	90	12/11/2016						
14.	24	55	12/11/2016									

*See Table 18.1.2 for the dates that air temperatures were measured at the Racibórz and Katowice meteorological stations.
[†]See Figure 18.1.10.

reported that, on the Rymer dump, the highest temperatures occurred on the western and southern slopes and the lowest on the eastern slopes. This phenomenon reflects the dominant direction of winds driving air into the dump. Surface temperatures can also be influenced by wind.

Conclusions

The use of the NDVI, NDSI, and NDSII indices and thermal maps, combined with Landsat TM, ETM+, ASTER, and Sentinel 2 images, can help to identify and localize self-heating sites in coal-waste dumps. Nighttime images can show hot spots with high intensity as in the Makoszowy dump, for example. The advantages of the methods described here are that the satellite images used are freely available, and that they localize the hot spots by drawing information from large areas. Though the method does not require it, field temperature measurements provide a useful validation. However, several disadvantages are apparent; for example, snow-cover or nighttime images are not always available due to atmospheric disturbances and technical problems with the Landsat ETM+. The distribution and amount of snow can significantly influence the perceived location of hot spots. Furthermore, self-heating sites of small spatial extent are often difficult to detect as they are too small in relation to the pixel size of the thermal data. Finally, the detection of hot spots with low thermal intensity in shallow locations, and those on steep slopes, can be problematic, due to the limitations of the thermal sensor. As these types of hot spots commonly provide smooth values, attempts to distinguish these anomalies from the surrounding pixel values, and solar effects, can give misleading results.

The method has potential for monitoring coal-waste dumps in USCB, and wherever similar self-heating problems exist. In the future, drones with thermal infrared cameras will likely replace the more expensive aircraft and alleviate detector sensitivity problems.

Acknowledgments

The authors would like to thank to Dr. Monika J. Fabiańska for her valuable suggestions concerning surface temperature measurements; and Dr. Pádhraig Kennan, University College, Dublin, Ireland, for his review of the manuscript.

Important Terms

coal-seam fires
coal-waste dumps
exothermic oxidation of the carbon
expelled bitumen
hot spots
Normalized Difference Snow and Ice Index
Normalized Difference Snow Index
Normalized Difference Vegetation Index
thermal band
thermal infrared remote sensing
surface pseudo-fumarolic activity
susceptibility to self-heating

References

Ahmed, T., Singh, D., Gupta, S., Raman, B., 2015. An efficient application of fusion approach for hot spot detection with MODIS and PALSAR-1 data. Geocarto Int. 31, 715–738.

Barosz, S., 2002. Monitoring of the dismantling and reclamation of the coal waste dumps. In: Proceedings – VII Conference "Long Term Proecological Undertakings in the Rybnik Coal Area", October 2002. Rybnik, pp. 149–156 (in Polish).

Barosz, S., 2003. Technical, Economical and Environmental Conditions of Management of Coal Waste Dumps Using the Mines from the Rybnik Coal District as Examples (Unpublished PhD dissertation). Academy of Mining and Metallurgy, Cracow, Poland (in Polish).

Cabała, J.M., Cmiel, S.R., 1999. Dynamics of ground surface deformation caused by mining in the period 1973–97 in USCB on the example of Kazimierz trough. In: Documenta Geonica of Akademy of Sciences of Czech Republic, Proc. 2 Czech-Polish Geomechanical Symposium. Prague, pp. 243–252.
Cabała, J.M., Cmiel, S.R., Idziak, A.F., 2004. Environmental impact of mining activity in the upper Silesian Coal Basin (Poland). Geol. Belg. 7 (3–4), 225–229.
Carras, J.N., Bainbridge, N.W., Saghafi, A., Szemes, F., Roberts, O.C., Haneman, D., 1994. The Self-heating of Spoil Piles from Open Cut Coal Mines, Vol. 1 Characterisation, Field Measurements and Modelling. NERRDC 1609 Final Report. Australian Coal Association, Brisbane, p. 128.
Carras, J.N., Bus, J., Roberts, O.C., Szemes, F., 1999. Monitoring of Temperature and Oxygen Profiles in Selfheating Spoil Piles. ACARP Project C6003 Final Report. Australian Coal Association, Brisbane, p. 19.
Carras, J.N., Day, S.J., Saghafi, A., Williams, D.J., 2009. Greenhouse gas emissions from low temperature oxidation and spontaneous combustion at open-cut coal mines in Australia. Int. J. Coal Geol. 78, 161–168.
Chatterjee, R.S., 2006. Coal fire mapping from satellite thermal IR data – a case example in Jharia Coalfield, Jharkhand, India. ISPRS J. Photogramm. Remote Sens. 60, 113–128.
Clemens, A.H., Matheson, T.W., 1996. The role of moisture in the self-heating of low-rank coals. Fuel 75 (7), 891–895.
Cracknell, A.P., Mansor, S.B., 1992. Detection off sub-surface coal fires using landsat thematic mapper data. The international archives of the photogrammetry. Remote Sens. Spatial Inf. Sci. (ISPRS Arch.) 29 (b7), 750–753.
Day, S., Bainbridge, N., Carras, J., Lilley, W., Roberts, C., Saghafi, A., Williams, D., 2015. Spontaneous combustion in open-cut coal mines: Australian experience and research (Chapter 1). In: Stracher, G.B., Prakash, A., Sokol, E.V. (Eds.), Coal and Peat Fires: A Global Perspective, Volume 3: Case Studies – Coal Fires. Elsevier, Amsterdam, pp. 1–36. ISBN: 978-0-444-59509-6.
Dozier, J., 1989. Spectral signature of alpine snow cover from the Landsat Thematic Mapper. Remote Sens. Environ. 28, 9–22.
Dulias, R., 2003. Subsidence Depressions in Upper Silesian Coal Basin. Geomorfologický sborník 2. ČAG, ZČU v Plzni., pp. 11–16.
EarthExplorer, 2017. US Geological Survey. https://earthexplorer.usgs.gov (accessed February 2017).
Engle, M.A., Radke, L.F., Heffern, E.L., O'Keefe, J.M.K., Hower, J.C., Smeltzer, C.D., Hower, J.M., Olea, R.A., Eatwell, R.J., Blake, D.R., Emsbo-Mattingly, S.D., Stout, S.A., Queen, G., Aggen, K.L., Kolker, A., Prakash, A., Henke, K.R., Stracher, G.B., Schroeder, P.A., Román-Colón, Y., ter Schure, A., 2012. Gas emissions, minerals, and tars associated with three coal fires, Powder River Basin, USA. Sci. Total Environ. 420, 146–159.
Fabiańska, M.J., Ciesielczuk, J., Kruszewski, Ł., Misz-Kennan, M., Blake, D.R., Stracher, G., Moszumańska, I., 2013. Gaseous compounds and efflorescences generated in self-heating coal-waste dumps — a case study from the Upper and Lower Silesian Coal Basins (Poland). Int. J. Coal Geol. 116–117, 247–261.
Gangopadhyay, P.K., Malthuis, B., van Dink, 2005. ASTER derived emissivity and coal-fire related surface temperature anomaly a case study in Wuda, North China. Int. J. Remote Sens. 26, 5555–5571.
Guha, A., Kumar, K.V., 2012. Structural controls on coal fire distributions - remote sensing based investigation in the Raniganj Coalfield, West Bengal. J. Geol. Soc. India 79, 467–475.
Hall, D.K., Riggs, G.A., Salomonson, V.V., 1995. Development of methods for mapping global snow cover using moderate resolution imaging spectroradiometer data. Remote Sens. Environ. 54, 27–140.
Hall, D.K., Foster, J.L., Verbyla, D.L., Klein, A.G., 1998. Assessment of snow cover mapping accuracy in a variety of vegetation-cover densities in central Alaska. Remote Sens. Environ. 66, 129–137.
Huo, H., Jiang, X., Song, X., Li, Z.-L., Ni, Z., Gao, C., 2014. Detection of coal fire dynamics and propagation direction from multi-temporal nighttime landsat SWIR and TIR data: a case study on the Rujigou Coalfield, Northwest (NW) China. Remote Sens. 6, 1234–1259.
Jiménez-Muñoz, J.C., Sobrino, J.A., Plaza, A., Guanter, L., Moren, J., Martínez, P., 2009. Comparison between fractional vegetation cover retrievals from vegetation indices and spectral mixture analysis: case study of PROBA/CHRIS data over an agricultural area. Sensors 9, 768–793.
Konopko, W., 2010. Wydobycie węgla i destrukcja górotworu w Górnośląskim Zagłębiu Węglowym. Przeglad Gorn. 66, 1–10.
Lighty, J.S., Veranth, J.M., Sarofim, A.F., 2000. Combustion aerosols: factors governing their size and composition and implications to human health. J. Air Waste Manag. Assoc. 50, 565–618.
Lohrer, C., Schmidt, M., Krause, U., 2005b. Influence of environmental parameters on the self- ignition behavior of coal. In: Proceedings of the International Conference on Coal Fire Research, November-December 2005, Beijing, pp. 210–232.
Lohrer, C., Schmidt, M., Krause, U., 2005a. A study on the influence of liquid water and water vapour on the self-ignition of lignite coal-experiments and numerical simulations. J. Loss Prev. Process. Ind. 18, 167–177.

Lu, P., Liao, G.X., Sun, J.H., Li, P.D., 2004. Experimental research on index gas of the coal spontaneous at low-temperature stage. J. Loss Prev. Process. Ind. 17, 243–247.

Google Maps, 2017. Poland coal-Waste Dumps. https://www.google.com/maps/@50.1393357,18.6972167,12z/data=!3m1!1e3?force=lite (accessed January 2017).

Masalehdani, N.N., Mees, F., Dubois, M., Coquinot, Y., Potdevin, J.-J., Fialin, M., Blanc- alleron, M.-M., 2009. Condensate minerals from a burning coal-waste heap in Northern France. Can. Miner. 47, 867–881.

Misra, B.K., Singh, B.D., 1994. Susceptibility to spontaneous combustion of Indian coals and lignites: an organic petrographic autopsy. Int. J. Coal Geol. 25, 265–286.

Misz-Kennan, M., 2010. Thermal alterations of organic matter in coal wastes from Upper Silesia, Poland. Mineralogia 41 (3–4), 105–236.

Misz-Kennan, M., Fabiańska, M.J., 2010. Thermal transformation of organic matter in coal waste from Rymer cones (upper Silesian Coal Basin, Poland). Int. J. Coal Geol. 81, 343–358.

Misz-Kennan, M., Fabiańska, M.J., 2011. Application of organic petrology and geochemistry to coal waste studies. Int. J. Coal Geol. 88, 1–23.

Misz-Kennan, M., Tabor, A., 2011. The thermal history of selected coal waste dumps in the Upper Silesian Coal Basin (Poland) (Chapter 15). In: Stracher, G.B., Sokol, E.V., Prakash, A. (Eds.), Coal and Peat Fires: A Global Perspective, Volume 3-Case Studies. Elsevier, Amsterdam, pp. 431–462. ISBN: 978-0-444-59509-6.

Misz-Kennan, M., Ciesielczuk, J., Tabor, A., 2013. Coal-waste dump fires of Poland (Chapter 15). In: Stracher, G.B., Prakash, A., Sokol, E.V. (Eds.), Coal and Peat Fires: A Global Perspective Volume 2: Photographs and Multimedia Tours. Elsevier, Amsterdam, pp. 233–311. ISBN: 0978-0-444-59412-9.

Nádudvari, Á., 2014. Thermal mapping of self-heating zones on coal waste dumps in Upper Silesia (Poland) — a case study. Int. J. Coal Geol. 128–129, 47–54.

Nádudvari, Á., Fabiańska, M.J., 2016. Use of geochemical analysis and vitrinite reflectance to assess different self-heating processes in coal-waste dumps (Upper Silesia, Poland). Fuel 181, 102–119.

Negi, H.S., Kulkarni, A.V., Semwal, B.S., 2009. Estimation of snow cover distribution in Beas basin, Indian Himalaya using satellite data and ground measurements. J. Earth Syst. Sci. 118 (5), 525–538.

Otwarty Regional NY System, 2017. Archival Map of the Anna Coal-Mine Dump. (in Polish) http://mapy.orsip.pl/imap/?locale=pl&gui=new&sessionID=68582 (accessed February 2017).

Parafiniuk, J., Kruszewski, Ł., 2010. Minerals of the ammonioalunite_ammoniojarosite series formed on a burning coal dump at Czerwionka, Upper Silesian Coal Basin, Poland. Miner. Mag. 74 (4), 731–745.

Pieczyrak, J., 2010. Characteristics and engineering usage of waste from coal mining. Archit. Civil Eng. Environ. 1, 77–84.

Pone, J.D.N., Hein, K.A.A., Stracher, G.B., Annegarn, H.J., Finkleman, R.B., Blake, D.R., 2007. The spontaneous combustion of coal and its by-products in the Witbank and Sasolburg coalfields of South Africa. Int. J. Coal Geol. 72, 124–140.

Prakash, A., Schaefer, K., Witte, W.K., Collins, K., Gens, R., Goyette, M.P., 2011. A remote sensing and GIS based investigation of a boreal forest coal fire. Int. J. Coal Geol. 86, 79–86.

Querol, X., Izquierdo, M., Monfort, E., Alvarez, E., Font, O., Moreno, T., Alastuey, A., Zhuang, X., Lud, W., Wangd, Y., 2008. Environmental characterization of burnt coal gangue banks at Yangquan, Shanxi Province, China. Int. J. Coal Geol. 75, 93–104.

Rees, W.G., 2001. Physical Principles of Remote Sensing (second ed.). Cambridge University Press, Cambridge. In: Chatterjee, R.S., 2006. Coal fire mapping from satellite thermal IR data — a case example in Jharia Coalfield, Jharkhand, India. ISPRS J. Photogramm. Remote Sens. 60, 113–128.

Ribeiro, J., Ferreira da Silva, E., Flores, D., 2010. Burning of coal waste piles from Douro Coalfield (Portugal): petrological, geochemical and mineralogical characterization. Int. J. Coal Geol. 81, 359–372.

Rothery, D.A., Borgia, A., Carlton, R.W., Oppenheimer, C., 1992. The 1992 Etna lava flow imaged by Landsat TM. Int. J. Remote Sens. 13, 2759–2763.

Skarżyńska, K.M., 1995. Reuse of coal mining wastes in civil engineering – Part 1: properties of minestone. Waste Manag. 13, 3–42.

Suresh, S., Suresh, A.V., Mani, K., 2016. Estimation of land surface temperature of high range mountain landscape of Devikulam Taluk using Landsat 8 data. Int. J. Res. Eng. Technol. 5, 92–96.

Tabor, A., 2002. Monitoring of coal waste dumps, re-cultivated dumps and other collection sites of Carboniferous waste rocks in the light of many years experience. In: Proceedings - VII Conference "Long Term Proecological Undertakings in the Rybnik Coal Area", October 2002 Rybnik, pp. 131–141 (in Polish).

Tetzlaff, A., 2004. Coal Fire Quantification Using ASTER, ETM and BIRD Satellite Instrument Data (PhD thesis) Ludwig Maximilians University, Munich, Germany, pp. 2–148. http://edoc.ub.uni-muenchen.de/4398/1/tetzlaff_anke.pdf (accessed June 2016).

Urbański, J., 1983. Technical re-cultivation of mine waste dumps with particular reference to fire protection. In: The Association of Mining Engineers and Technics Training Materials, Katowice, p. 62 (in Polish).
van Genderen, J.L., Cassells, C.J.S., Xiangmin, Z., 1996. The synergistic use of remotely sensed data for the detection of underground coal fires. In: The International Archives of the Photogrammetry, Remote Sensing and Spatial Information Sciences (ISPRS Archives) XXXI, (Part B7. Vienna), pp. 722–726.
Voigt, S., Tetzlaff, A., Zhang, J., Künzer, C., Zhukov, B., Strunz, G., Oertel, D., Roth, A., van Dijk, P., Mehl, H., 2004. Integrating satellite remote sensing techniques for detection and analysis of uncontrolled coal seam fires in North China. Int. J. Coal Geol. 59, 121–136.
Walker, S., 1999. Uncontrolled Fires in Coal and Coal Wastes. IEA Coal Research, London (CCC/16). ISBN 92-9029-3247-1.
Weather Information Service, 2017. http://www.ogimet.com (accessed January 2017).
Weng, Q., Lu, D., Schubring, J., 2004. Estimation of land surface temperature–vegetation abundance relationship for urban heat island studies. Remote Sens. Environ. 89, 467–483.
Xiao, X., Shen, Z., Qin, X., 2001. Assessing the potential of VEGETATION sensor data for mapping snow and ice cover: a Normalized Difference Snow and Ice Index. Int. J. Remote Sens. 22 (13), 2479–2487.
Zhang, J., 2004. Spatial and Statistical Analysis of Thermal Satellite Imagery for Extraction of Coal Fire Related Anomalies (PhD thesis). Technical University of Vienna, Austria, pp. 2–145.
Zhang, J., Kuenzer, C., 2007. Thermal surface characteristics of coal fires 1: results of in-situ measurements. J. Appl. Geophys. 63, 117–134.

WWW Addresses: Additional Reading

(1) Archival Temperatures Retrieved from the Weather Information Service
http://www.ogimet.com
(2) Technical Information about ASTER Images
http://web.gps.caltech.edu/gislab/HowTo/aster_landsat_bands_v2.pdf
(3) Technical Information about Landsat Bands (US Geological Survey)
http://landsat.usgs.gov/band_designations_landsat_satellites.php
(4) Technical Information about Landsat Images and Processing Methods (US Geological Survey)
https://landsat.usgs.gov/using-usgs-landsat-8-product
(5) Technical Information and Details about FLAASH Atmospheric Correction Module
http://www.harrisgeospatial.com/docs/FLAASH.html
(6) Technical Information and Details about Quick Atmospheric Correction (QUAC)
https://www.harrisgeospatial.com/docs/quac.html
(7) The History of the Makoszowy Coal Mine (available only in Polish)
http://zzgmakoszowy-zabrze.pl/?page_id=1327
(8) US Geological Survey Satellite Data
http://earthexplorer.usgs.gov

CHAPTER 19

Geochemical Behavior of Trace Elements in the Upper and Lower Silesian Basin Coal-Fire Gob Piles of Poland

Pyrometamorphic complex in Radlin, near Wodzisław Śląski, Upper Silesian Coal Basin. The protolith material comes from the local Marcel mine. The gob is constantly being reworked and its shape changes, which has allowed the fire to constantly burn for at least 11 years. Each visit to this heap brings samples of new, often unique, mineral composition. The cliff here is mainly built of clinker. *Photo by Łukasz Kruszewski (2012).*

CHAPTER CONTENTS

Coal and Peat Fires: A Global Perspective
Edited by Glenn B. Stracher

https://doi.org/10.1016/B978-0-12-849885-9.00019-6

Photo by Łukasz Kruszewski (2013).

19.1 Geochemical Behavior of Trace Elements in Silesian Coal-Fire Gob Piles

Łukasz Kruszewski

Flow-texture of a parabasalt from a heap in Świętochłowice (Upper Silesian Coal Basin), winter of 2013. The white material is snow. The texture and color of the rock distinguish it from more commonly observed buchites, which are usually rounded and less dark (often orange). The parabasalts of the basin are usually composed of cordierite/indialite, anorthite, hematite, and minor amounts of various representatives of clinopyroxenes, spinels, and olivines. Accessory minerals include mullite, while quartz is present as a relic phase.

Introduction

Coal is a rock known to accumulate many elements, including rare and precious metals and semimetals such as beryllium, gallium, germanium, vanadium (Lapham et al., 1980; Bonnett, 1996; Wagner and Hlatshwayo, 2005), and many more (Table 19.1.1). Affinition of trace elements to particular minerals in coals is reported by Chatziapostolou et al. (2006) and Parzentny and Lewińska-Preis (2006). Sulfide minerals present in coal deposits may concentrate arsenic, copper, cadmium, lead, cobalt, nickel, antimony, selenium, and zinc (Lapham et al., 1980; Wagner and Hlatshwayo, 2005; Parzentny and Lewińska-Preis, 2006). As shown by the latter authors cited, at least some of these elements may also be enriched in coal-associated carbonate (veins), and coal organic matter.

There are numerous papers concerning coal element geochemistry. Coal Clarke values for many elements are reported by Ketris and Yudovich (2009). Probably the most well known is germanium; there are many papers dealing with this matter, including Hower et al. (2002), Zhuang et al. (2006), Höll et al. (2007), and Yudovich (2003). Mercury concentrations and enrichment processes in coals are reported by Goodarzi and Goodarzi (2004) and Yudovich and Ketris (2005a); Liang et al. (2014) reports data on Hg determination in the Wuda coal fire gas. Related information on arsenic can be found in Yudovich and Ketris (2005b) and Wang et al. (2006), while that on selenium include Yudovich and Ketris (2005c) and Woo et al. (1987). The last paper is also unique in that it also reports data on tellurium. Enrichment in platinum-group elements (PGEs) and rhenium in some coals is also known (Yang, 2006; Baioumy et al., 2011); Rhenium enrichment in U-rich coals is described by Maksimova and Shmariovich (1983). Data on lithium and beryllium is rather scarce, with exceptions being papers by Sun et al. (2010) and Eskenazy (2006), respectively. Xiao et al. (2016) give us some clue on concentrations on some more rarely determined elements in some Chinese coals, namely Hf, Ga, Gd, Li, Sr, Th, U, Zn, and Zr. Information on boron can be found in papers by Goodarzi and Swaine (1994) and Hower et al. (2002). Rare earth elements in some Polish coals are in turn analyzed by Adamczyk et al. (2015). Table 19.1.1 presents juxtaposed data on the concentrations of some trace elements in coals and coal ashes from Poland.

Coal firing ash is known to catch even more elements that tend to be immobilized in it. The ash, in terms of worldwide analyses, may contain even 1500 ppm As, 5000 ppm B, 1308 ppm Be, 41 ppm Cd, 780 ppm Cr, 84,000 ppm Ge, 97 Hg, 7400 ppm Ni, 2300 ppm Pb, 15,000 ppm Sb, 77 ppm Se, 35 ppm Tl, and 2100 ppm V (Hower et al., 2002; Seredin, 2003; Yudovich and Ketris, 2005a,b; Eskenazy, 2006). Experiments with coal pyrolysis show that some trace elements like mercury do not show a tendency to condense from coal fire gases (this is, however, in opposition to the following data), while volatile Se and As compounds may be immobilized in mineral condensates. For example, native sulfur from pseudofumaroles of Ravat, Tajikistan, may contain as much as 1300 ppm of Se

(Srebrodolskiy, 1989). According to some research, cobalt, vanadium, chromium, lead, and nickel are rather not entering the volatile products of coal pyrolysis (Liu et al., 2006; Konieczyński et al., 2012).

Extreme geochemical enrichment in coal is a well-described phenomenon. Especially high amounts of trace elements are released during fires of secondary-mineralized coal due to hydrothermal ore-forming processes. The latter lead to the formation of geochemical and hydrochemical anomalies, such as the ones at Donbass, Ukraine (mercury: Panov et al., 1999) or China (mercury, arsenic, selenium, and fluorine: Finkelman, 2004). A review on some element concentrations in worldwide metalliferous coals is provided by Seredin and Finkelman (2008). Based on this paper, the maximum Ag concentration in such coals reaches 100 ppm (epigenetic coals of Russia); however, ashes of the Zeravshan Depression coals (Tajikistan) have only 663 ppm. As much as 32,000 ppm of arsenic may be found in such coals; much less is associated with the Wulatunga coal (China)—1321 ppm—and in coals of the Primorskiy Kray (Russia), with a still-important 790 ppm. Gold concentration in worldwide coals may reach 100 ppb, but 14,500 ppb is reported for uraniferous Kol'dzhatsk coals of Central Asia; up to 9800 ppb is found in Dnieper brown coals of Ukraine, and 9000 in siderite from Sinegorsk coal. Maximum Au content in Wyoming coals is 2800 ppb. Also, some peats may capture large amounts of this precious metal: this is especially true for Mannar, Sri Lanka, with 880 ppb of Au (plus up to 7900 ppb Pt). Coals may catch up to 9000 ppm of B, 22,000 of Ba, 170 ppm of Cd, 932 ppm of Co, 3200 ppm of Cr, 45 ppm of Hf, 1000 ppm of Hg, 747 ppm of Li, 3200 ppm of Mo, 300 ppm of Nb, 3490 ppm of Ni, 1900 ppm of Pb, 46 ppm Re, 5800 ppm Sb, 408 ppm Rb, 2700 ppm Se, 7 ppm Sn, 2800 ppm Sr, 220 ppm Ta, 500 ppm Th, 420 ppm Tl, 14,100 ppm U, 1582 ppm Zr, and 19,000 ppm Zn. The largest quantities of the first element in samples from Primorskiy Kray, 474 ppm, are known from Rakovsk. Some coals may also concentrate as much as 330 ppm Be; large amounts, ca. 85 ppm, concern the Primorskiy Kray again. Some coals may contain 700 ppm Ce, but 899 ppm are found in the Primorskiy Kray samples and 1185 ppm in some Guizhou coals (China). Also, lanthanum may be coal-concentrated, with some samples having 370 ppm, or 627 (Primorskiy Kray) and even 2790 ppm (Nizhneilinsk, Kazakhstan) in extreme cases. Yttrium-rich coals may capture as much as 1160 ppm of the element, though maximum values for epigenetic infiltration-type uraniferous coals of Russia are around 1400 ppm. Palladium levels in enriched coals may reach 4240 ppb, with enormous 6100 ppb in the Tulamin deposit, British Columbia, Canada. The latter rocks are also extremely rich in Pt, containing >10,000 ppb, and also enriched in Ru (80 ppb), Rh (61 ppb), Os (24 ppb), and Ir (22 ppb). Concentrations of Ir are still far greater in coals of the Zapadno-Kapital'naya mine, with 3800 ppb, which is ~38 times the Ir abundance at the Cretaceous/Paleogene border rocks. Meanwhile, higher rhodium concentrations, reaching 300 ppb, concern the Guizhou coals. Scandium levels in coal may be as high as 230 ppm, but even 700 ppm is found in ashes of the Minusa basin coal. Vanadium content in enriched coals may be as high as 1292 ppm; or 1400 ppm at the Dong mine, Heshan, Guangxi, China; 2000 ppm at El'ginsk, Southern Yakutia, Russia; or enormous 10,600 ppm in Western Kentucky coal bed no. 9. One of the largest quantities of W in coal is 1500 ppm, exceeded by 2800 ppm in Mugla coals (Bulgaria). Antimony levels in worldwide coal ashes may rise up to 15,000 ppm (Seredin, 2003). Maximum content of tellurium in Australian Great Grata coals is only 0.08 ppm. Some methods for extraction for Ge and Ga from coals are reported by Fang and Gesser (1996), Welz and Sperling (1999), Arroyo and Fernández-Pereira, (2008), and Xu et al., (2013). Not much is known about bismuth and indium concentrations in worldwide coals.

Coal is prone to undergo natural, spontaneous combustion (Sokol et al., 2005). In Poland and worldwide the process is known in both natural (Kwiecińska et al., 1992) and anthropogenic (Kruszewski, 2006, 2012, 2013; Kruszewski et al., 2012) geological environments, the latter known as burning coal-mining heaps (BCMHs), bingsteads, or dumps. This process triggers further mineralogenic processes, like high-temperature pyrometamorphism, mid-temperature pneumatolytic- and hydrothermal-like exhalative processes, and low-temperature supergene (surface weathering) processes (Srebrodolskiy, 1989, and references therein; Parafiniuk and Kruszewski, 2009).

Contrary to the preceding geochemical data concerning coal itself, the one on BCMH materials is scarce. Probably the most extensive data in this matter is given by Sokol et al. (2005) for the Chelyabinsk Coal Basin (CCB) objects. Fused metapelites from those heaps contain, on average, 600 ppm Ba, 2.1 ppm Be, 10 ppm Cs, 96 ppm Cu, 3 ppm Ge, 26 ppm Pb, 1.2 ppm Se, 170 ppm V, 102 ppm Zn, but <1 ppm Mo; they are also devoid of (higher concentrations of) Hg. The mean concentrations of elements in the CCB clinkers (in ppm) are 200 Cl, 570 Ba, 35 Co, 300 F, 60 Li, 1140 Mn, 15 Pb, and 160 V. The "black-block" metapelites (soot-rich, pyrometamorphosed under oxygen-low conditions, often lens-shaped zones, bearing reduced mineral species; Sokol et al., 2005) of the CCB are of special interest, since they contain, on average (in ppm), 53 As, 500 Ba, 105 Br, 750 Cl, 32 Co, 140 Cr, 850 F, 80 Ga, 985 Mn, 230 Ni, 1200 Rb, 1.3 Se, and 600 Sr. Slag-like Ca- and Mg-rich postcarbonate rocks are also interesting, with mean 1800 ppm Cl and 435 ppm Mn, but otherwise do not seem to accumulate large quantities of other elements. The thermally changed Fe-rich carbonates may be a source of Ba (310 ppm), Cr (130 ppm), Cu (73 ppm), Mn (1900 ppm), and Zn (170 ppm). Thermoanthracites do not pose an important carrier, with 100 ppm Br and 49 ppm

Table 19.1.1

Concentrations of chosen trace elements (in ppm) in coals, coal ashes, and coal-associated rocks and minerals of the Upper Silesian Coal Basin (USCB, Poland), in ppm; 1, 2, 3, 5, 7, 11, 13: coals; 4, 6, 8, 12, 14, 15: ASHES; 16, 17—rocks; 9, 10, 18—minerals. Note that no data are given for bismuth, caesium, iodine, hafnium, tin, tantalum, tellurium, rhenium, uranium, tungsten, and platinum-group elements.

samples*:	1	2	3	4	5	6	7	8	9	10	11	12	13	14	15	16	17	18
Ag															2.0		1.3	33†
As		53		378									5.1	41	82		38	
Au																		448
B																	12	
Ba													249		3620		148	
Be					34	20									26		26	
Cd		0.1					1.1	11	8.1	6.8			0.64	9.0	18		2.0	
Ce	223															81		
Co					4	46	45	598	83	187			17	104	167		15	
Cr		27	50		47	145							19		232		62	
Cu					26		24	308	124	282			24	305	509		59	1180
Ga														35			58	
Ge						7.6											24	
Hg		0.09	0.76	0.008									0.02		0.23			210†
La	104															46		
Li					17												8.0	
Mn		170	508		45								124		1190		170	
Mo					87										7.0			
Ni		23	47		38		42	993	128	202			20	441	359		42	210
Pb		19	54	196	42		40	475	412	289	372	183	39	153	248		62	2230
Rb													21					
Sb													1.0					714
Sc	54															20		
Se														8.0				
Sr													159			1.4		
Tl														3.0				
V					130								49	462	675		170	
Y	139															28		
Zn		57	238	360	117		62	1094	501	272	1243	1507	107	431	714	110		29

*1 – coal, Orzeskie beds, mean, Adamczyk et al. (2015); 2 – Polish hard coals of the USCB, mean, Juda-Rezler and Kowalczyk (2013); 3 – as before, maximum; 4 – as "2" but on the ash basis, mean; 5 – crude coal, mean (from 2 fractions), Lewińska-Preis et al. (2008); 6 – USCB coal fly ash, mean (3 samples), Makowska and Wierońska (2016); 7 – USCB coals, mean from 17 samples of 2 mines, Parzentny and Lewińska-Preis (2006); 8 – as before but on the ash basis; 9 – as before but in coal-associated carbonate minerals, mean; 10 – as before but in coal-associated sulfide minerals, mean; 11 – USCB coals, maximum, Parzentny et al. (1999); 12 – as before but for coal shales; 13 – USCB coals, mean (of 44 samples), Smoliński et al. (2014); 14 – Wesoła colliery coal fly ash, mean, Wyszomirski and Brylska (1996) and Zajusz-Zubek and Konieczyński (2003); 15 – USCB coal fly ash (Hg: in coal), mean, Zajusz-Zubek & Konieczyński (these authors also report 61 ppm Hg in coal, which is a suspiciously high value); 16 – claystones of the Mudstone and Sandstone Series of the USCB, mean, Kokowska-Pawłowska (2016); 17 – coal-associated rocks of the Poruba beds, maximum, Kokowska-Pawłowska (2015); 18 – USCB coal-disseminated iron sulfide minerals, mean (calculated from wt.% of WDS analyses), Bielowicz and Misiak (2016).

†Content is usually below detection limits.

V. Chloride-rich exhalative crusts capture mean 910 ppm Br, 400,000 ppm Cl, 100 ppm Cr, 900 ppm Mn, and 50 ppm Se. The sal-ammoniac-bearing crusts have even more Cl: 600,000 ppm on average; they are also enriched in Br (782 ppm). Sulfate crusts of a similar origin bear the largest amounts of As (880 ppm, mean) and are also enriched in Se (11 ppm on average) and Mn (655 ppm on average). Bitumens are the most Hg-enriched products of the local coal fires, with mean 21 ppm. In comparison, the maximum mercury concentration in some mineral samples from the Anna mine dump in Alsdorf, Germany, reaches 290 ppm, and concentration of selenium may exceed 2300 ppm (Sindern et al., 2005). The high amount of Cl (and Br) in coals is connected, *i.a.*, with formation brines penetrating the coal deposits. While the maximum Cl content in UK coals is 11,000 ppm, and that in lignites of Karlovo (Bulgaria)—15,900 ppm—the amount in UK coal-deposit brines reaches 100,000 ppm (Yudovich and Ketris, 2006).

The data on the concentrations of trace and rare elements in Polish BCMH is only preliminary (Kruszewski, 2009; Porowski et al., 2015; Kruszewski et al., 2012). Thus the purpose of this paper is to present the preliminary geochemical results of analyses of materials coming from those objects. However, possibly the more important task here is to point to minerals (including possible own, i.e., element-dominated phases) that may act as carriers of elements that are interesting or important to both science and industry.

Photo by Łukasz Kruszewski (2013).

19.2 Mineralization Processes in Coal-Fire Gob Piles

Salammoniac on black coal-bearing shale from the Marcel mine heap in Radlin, near Wodzisław Śląski, Upper Silesian Coal Basin. The sample is enriched in Br, I, Tl, Pb, Sn, and Bi. The orange color is likely due to Tl bromide/iodide inclusions. Tiny aggregates of hydrokenoralstonite, $Na_{0.5}(Al,Mg)_2(F,OH)_6 \cdot H_2O$ and minor amounts of sulfur occur with the salammoniac.

Gob Pile Fires

Self heating and fire-related processes are observed in many dumps in the Upper Silesian Coal Basin (Wagner, 1980; Kruszewski, 2013; Fabiańska et al., 2013; Ciesielczuk et al., 2015). Spontaneous coal combustion is caused by oxidation of the coal matter, with a catalytic role of coal-associated sulfide minerals (Srebrodolskiy, 1989). Important factors influencing the initiation of coal combustion include coal petrography (maceral composition) and coal structure (Wagner, 1980; Cebulak et al., 2005). More data on the phenomena associated with coal fires in the heaps may be found in these papers.

Mineral Formation

The formation of minerals in this environment takes place in three stages: (1) high-temperature pyrometamorphism, that is, the burnout stage (temperatures usually up to ca. 1200°C); (2) the exhalative (pseudofumarolic) stage (typically 50–450°C); and (3) a low-temperature ($T < 50°C$) supergene stage (Srebrodolskiy, 1989; Nasdala and Pekov, 1993; Sokol et al., 2005; Stracher, 2007; Kruszewski, 2013). Large quantities of gas are formed during the combustion processes, the gas resulting from coal and barren rock degassing. Gaseous jets on the BCMH contain a wide range of compounds, with H_2O, H_2S, HCl, NH_3, and SO_2 being mineralogically the most important (Panov et al., 1999; Sokol et al., 2005; Fabiańska et al., 2013). Many exhalative minerals originate from these compounds (Kruszewski, 2012).

Analytical Methods

The BCMH studied are in the Upper and Lower Silesian Coal Basins (USCB and LSCB, respectively). Descriptions of most of these heaps and the lithology of their waste rocks may be found in the following papers: Kruszewski (2013), Ciesielczuk et al. (2015), and Fabiańska et al. (2013). The heaps location is shown in Figure 19.1.1. Large heap in Czerwionka (USCB, ca. 100 ha area) is known for its intense hydrothermal-like and exhalative (fumarolic) activity. Locally, coal-bearing clayey sediments partially metamorphosed into red clinker and black soot-rich lenses known as a "black block." An even larger heap in Rydułtowy (said to be the highest one in Europe) consists, *i.a.*, of a large cone totally made of pyrometamorphic rocks; some aspects of their mineralogy may be found in Kruszewski (2008). These rocks occur in the form of breccia, which contains clinkers and minor slags, parabasalts, porcellanites, and trace weakly metamorphosed black shales, all fused together. A similar description is applicable for a heap in Bytom (USCB), which however, characterizes in local coverage of large crystals of gehlenite and mullite; many minerals here are associated with slags. Minor pyrometamorphic samples are derived from (1) chimney-like structures once present in the middle part of a large heap in Radlin (Rybnik Coal Area, USCB), and (2) a large heap in Przygórze (near Nowa Ruda, LSCB). The last heap contains numerous

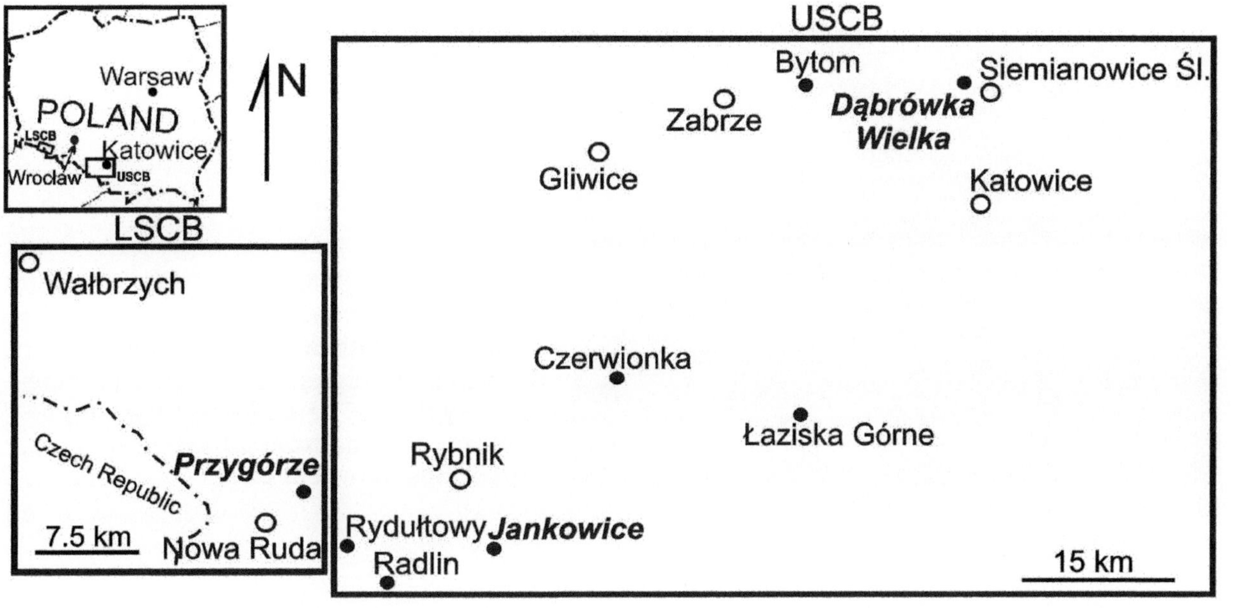

Figure 19.1.1. Location of the gob study areas in the Lower (LSCB) and Upper Silesian (USCB) Coal Basins of Poland.

pyrometamorphic rock types, mainly clinkers and porcellanites, but also slags, red and yellow metapelites, light-toned metapelites, and light-toned buchites; parabasalts are very scarce here. Locally, white hydrothermal-like veins cut the burnt rock complex. Mineralogy of this heap is preliminary reported in Kruszewski et al. (2014a,b). A single slag sample comes from a heap located at the border of Piekary and Siemianowice Śląskie, described in detail, in Kruszewski (2006). Some results concerning similar materials from Jankowice (near Rybnik) are also reported.

Bulk samples of pyrometamorphic rocks were analyzed using the inductively coupled plasma–mass spectroscopy (ICP-MS) method (Acme Labs); pulverized samples were fused with $LiBO_2/Li_2B_4O_7$; the obtained samples were divided. For the purpose of analysis one aliquot was digested in HNO_3, while the other was digested in aqua regia. Exhalative mineral mixtures (EMMs) were also analyzed in ACME Labs using ultratrace ICP-MS after aqua regia digestion. Single supergene sulfate samples were analyzed, after digestion in HCl, using the optical emission spectroscopy variant of the ICP (ICP-OES) method at the Faculty of Geology, University of Warsaw (Poland). More detailed information on geochemistry of the EMM will be published elsewhere.

Pyrometamorphic minerals were chemically examined using the electron microprobe (EPMA) method. These analyses were conducted at the Inter-Institute Laboratory of the Microanalysis of Minerals and Synthetic Materials (Faculty of Geology, University of Warsaw), which is equipped in CAMECA SX-100 apparatus. The typical operating conditions—wavelength dispersion spectroscopy (WDS) mode, 15 kV accelerating voltage and 10–20 nA beam current, and 1 μm beam size—were used for the analysis of main and minor elements. Trace elements were analyzed by elevated beam currents reported in the corresponding tables. Prior to this step the presence of each element suspected to be present in a mineral species, due to geochemical and chemical considerations like its location in the periodic table, was checked via WDS scans of the detector crystals. This procedure is also important for a correct set of the background around the line of interest.

Photo by Łukasz Kruszewski, 2011.

19.3 Geochemistry of Fire-Related Rocks and Minerals

Golden crystals of acicular mullite, 2.3 mm in diameter, nucleated on phaneritic crystals of hematite in a pyrometamorphic breccia from a heap in Bytom, Upper Silesian Coal Basin.

Rock Geochemistry

The results of geochemical analyses of bulk materials, namely pyrometamorphic rocks, are reported in Table 19.3.1. For comparison, data for some coals and shales is also given.

Geochemistry of Elements

The results given below concern geochemistry of trace elements measured in various minerals. As such, elements like titanium, manganese, chlorine, and fluorine are omitted because they are commonly found in many BCMH minerals as minor or main constituents. Titanium is common especially in spinels and pyroxenes, but it also enters glasses and some sulfate minerals. Manganese occurs in all these compounds; interesting sink of Mn are Ca ferrites (Srebrodolskite, $Ca_2Fe_2O_5$; "grandiferrite," $CaFe_4O_7$; and harmunite, $CaFe_2O_4$), in which Mn most likely occurs in a trivalent state. Chlorine is present in various chlorides, the most important being salammoniac, sometimes found in relatively large quantities. Apatite group minerals of the BCMH are practically devoid of this halogen; some crystals from a single Rydułtowy slag sample contain up to 124 ppm Cl (Table 19.3.2), and a single analysis of the Przygórze slag material has 0.04 wt.% Cl. A prominent exception is fluorellestadite from a powdery hydrothermalite (composed of calcite, melilite group, andradite, tobermorite, and possibly langbeinite, kilchoanite, haüyne, troilite, scapolite group, rostite, apophyllite group, and moganite) of Przygórze (sample BK5), which always bears Cl in the range of 0.10–0.16 wt.%. A unique case of Cl enrichment is the mayenite supergroup. A phase like chlormayenite and chlorkyuygenite occurs in the Przygórze hydrothermal veins; this phase concentrates 7.10–7.84 wt.% Cl. Both F and Cl are present in the structure of ferrian fluormayenite-fluorkyuygenite-chlormayenite from a slag from a heap in Rybnik-Niewiadom; the composition of this mineral is $Ca_{11.94}(Al_{13.55}Fe^{3+}_{0.51})_{\Sigma 14.06}[([]_{2.38}(H_2O)_{1.10}F_{0.52})_{\Sigma 4.00}(F_{1.82}Cl_{0.18})_{\Sigma 1.95}]$, corresponding to wt.% F and Cl, respectively (with [] denoting vacancy). Notable examples of fluorine include (1) an unnamed Mg–Ca(Fe) silicate sulfate (phosphate) from the Bytom slag, which contains 1.60 to 3.75 wt.% F; and (2) a phase like aquacreptite (of Harada) and talc, filling interstices in the Bytom slag, which contains 1.22–1.64 wt.% F. However, the most important minerals concentrating F in the BCMH are fluorapatite-fluorellestadite solid solution, less common cuspidine, and more rarely encountered fluorite. Cuspidine is found in the Piekary slag; the mineral has mean 10.19 wt.% F and 0.19 wt.% Cl, with maximum Cl content being 0.48 wt.%. Chlorine was not detected in any other cuspidines (from the Bytom and Przygórze slags); its presence in Piekary reflects unique coal-fire gas composition, which is also responsible for deposition of minor single crystals of a Ca chloride.

The sample from Piekary also contains very rare sellaite (MgF_2). Whitlockite of the Czerwionka parabasalt is also a carrier of trace halogens, with maximum 0.18 wt.% F and 0.04 wt.% Cl suggesting very minor substitution of bobdownsite and chlorophosphate end-members of the whitlockite group. A single analysis of fluorite from the Piekary heap shows 0.13 wt.% Cl. Two of four trace-mode analyses have shown Cl, with two values being 158 and 819 ppm. Fluorite of the hydrothermal veins from Przygórze is always enriched in trace Cl, with the range being 0.07–0.89 wt.%.

Table 19.3.1
Whole-rock geochemistry of chosen Upper Silesian Coal Basin pyrometamorphics by means of ICP-MS. Notable enrichment and maximum values are given in bold.

Sample:	BBR27a*	CDbb	Jn-b	Jn-p	LZ-kn	RDT-b	RDT16	RDT-p
ppm								
Ag	<0.10	**0.20**	**0.20**	<0.10	<0.10	<0.10	**0.20**	<0.10
As	**61**	14	5.3	1.0	0.5	1.4	**39**	9.2
Ba	289	**4320**	**652**	**754**	584	**611**	45	475
Be	3.0	1.0	**5.0**	4.0	2.0	4.0	1.0	**9.0**
Bi	<0.10	**0.80**	<0.10	<0.10	<0.10	<0.10	<0.10	<0.10
Cd	<0.10	**0.90**	<0.10	<0.10	<0.10	<0.10	<0.10	<0.10
Ce	51	55	73	**81**	**79**	**83**	13	76
Co	6.9	16	**35**	19	12	23	9.6	15
Cs	5.6	13	4.4	11	**23**	**27**	0.30	20
Cu	22	**48**	**56**	13	16	16	27	31
Dy	3.5	6.2	6.5	**6.8**	4.8	5.5	4.6	**7.1**
Er	1.7	2.9	3.2	**3.9**	2.9	3.3	2.7	**3.8**
Eu	0.87	1.5	**2.0**	1.6	1.3	1.6	1.0	**2.0**
Ga	16	20	**61**	**49**	**38**	23	4.3	26
Gd	3.9	6.9	**8.1**	7.2	5.1	6.4	4.5	**7.6**
Hf	2.9	2.4	2.8	**5.9**	**4.7**	**4.9**	0.10	2.6
Hg	<0.01	**0.24**	<0.01	0.01	<0.01	<0.01	0.08	0.03
Ho	0.57	1.1	1.2	1.3	0.84	1.1	0.91	**1.4**
La	24	30	35	**41**	**49**	39	4.6	38
Lu	0.23	0.35	0.37	**0.52**	0.45	**0.49**	0.28	0.50
Mo	1.6	0.80	1.0	0.70	<0.10	0.20	**1.8**	0.90
Nb	8.6	8.6	9.4	15	**21**	**17**	0.50	12
Nd	23	26	**36**	**39**	35	**38**	10	**39**
Ni	12	30	**42**	22	17	16	34	26
Pb	1.0	**39**	0.60	0.80	2.5	0.70	0.40	3.3
Pr	5.3	7.0	9.4	**9.8**	**10**	9.5	2.1	9.5
Rb	53	100	50	125	**182**	**187**	1.6	118
Sb	1.9	1.8	0.70	<0.10	0.10	0.30	**5.5**	1.1
Se	<0.50	**1.10**	<0.50	<0.50	<0.50	<0.50	**1.00**	<0.50
Sm	4.1	6.3	7.5	**7.9**	7.1	7.5	3.5	**8.3**
Sn	4.0	4.0	<1.0	<1.0	**6.0**	4.0	4.0	4.0
Sr	**673**	260	128	216	141	153	485	159
Ta	0.50	0.70	0.60	1.2	**1.7**	1.2	<0.10	1.1
Tb	0.59	**1.1**	**1.1**	**1.2**	0.84	0.97	0.76	**1.1**
Th	8.6	11	9.9	**16**	**21**	**16**	0.90	**17**
Tl	<0.10	**0.20**	<0.10	<0.10	<0.10	<0.10	<0.10	<0.10
Tm	0.25	0.42	0.4	**0.54**	0.43	**0.50**	0.35	**0.56**
U	3.9	5.1	4.1	6.9	6.4	6.6	5.9	**10**
V	96	86	74	136	**177**	**149**	21	**146**
W	22	1.5	**3.6**	**3.4**	3.0	**3.4**	1.4	**4.2**
Zn	61	237	14	19	2.0	3.0	12	14
Y	17	28	**36**	**36**	24	32	28	34
Yb	1.6	2.6	2.8	**3.6**	2.9	**3.7**	2.3	**3.9**
Zr	100	88	100	**204**	**181**	165	5.1	101
ppb								
Au	<0.50	<0.50	**4.9**	0.7	0.5	0.8	<0.50	0.6

*BBR27a – slag, Bytom; CDbb – Czerwionka, black block metapelite; Jn – Jankowice (b – buchite, p – parabasalt); LZ-kn – Łaziska, clinker; RDT – Rydułtowy (b – buchite, 16 – slag, p – parabasalt).

Table 19.3.2
Results of microchemical trace-element analyses of fluorapatite-fluorellestadite solid solution from the slag sample no. 1 (analyses 1–10 and 14–30) and no. 2 (analyses 11–13) from Rydułtowy, Poland.*

	1	2	3	4	5	6	7	8	9	10
ppm										
Cl		102			124					94
Sr	1592	1544	1625	1518	1627	1554	1452	1419	859	918
Ce					313			352		
	11	12	13[†]	14	15	16	17	18	19	20
ppm										
Sr[‡]			4945	1500	1300	1300	1700	1700	1600	1900
Y	142									
La		440	315							
Ce	522		351							
	21	22	23	24	25	26	27	28	29	30
ppm										
Sr	2400	1200	1600	1500	1300	1300	1700	1600	1800	1900

*Beam current was 50 (analyses 1–13) to 70 nA (analyses 14–30).
[†]13 of total 20 analyses with beam current of 50 nA showed the presence of given trace elements.
[‡]Y, La, and Ce were below the detection limits when beam currents of 70 and 100 nA were used.

Barium and strontium, which are not as common in nature as Ti, Ca, Mg, and Al, are included in this chapter, although their contents (especially in case of barium) suggest they are rather minor and not trace elements.

Lithium

Lithium was not measured during microprobe sessions due to the lack of adequate (windowless) detector. Geochemical analyses of the EMM show especially enlarged Li concentration in (Na)–Mg sulfates (mainly hexahydrite, kieserite, with minor löweite and vanthoffite) and to a lesser extent in tschermigite, all of exhalative origin (Radlin); the amount may exceed 350 ppm Li and the element is supposed to follow the known geochemical trend of Mg–Li substitution. A minor possible NH_4–Li substitution not only concerns tschermigite, but also mascagnite (and maybe also letovicite), as shown in case of the Rydułtowy material. Some secondary (and/or late-stage) sulfates associated with sulfate crust from the same site, containing trace alum-(K) and voltaite, may contain >100 ppm, with the supposed K–Li substitution. Much lesser amounts are found in gypsum-rich accumulations (Pszów). Some clay minerals (Rydułtowy) may stand for an important source of Li, too.

Beryllium

Beryllium was not measured during microprobe sessions for the same reason as lithium. Both beryllium and lithium (and possibly cesium) are suspected to be present in elevated amounts in indialite-ferroindialite and cordierite-sekaninaite solid solution crystals; such enriched varieties of those minerals are known from nature (Gadas et al., 2016). Beryllium concentration in the EEM samples is low, always being below 10 ppm. The maximum amount was found in supergene Mg sulfates from Radlin. Tschermigite from the same heap may also contain slightly elevated amounts of Be.

Boron

Boron was not measured during microprobe sessions due to analytical difficulties (light element) and expected low concentrations. Almost all the EMM samples show B concentration below 20 ppm. The only, and thus prominent, example is alunite-silica shale—a product of hydrothermal alteration of clayey shales collected from the surroundings of a sulfate crust in the apical part of the Radlin heap; this material contains 33 ppm of B, presumably localized in the alunite structure.

Scandium

The first transition metal, scandium, has not been positively determined in most of the minerals analyzed. An exception is a single analysis of perovskite from the Przygórze slag (clinopyroxene-rich), which contains Sc but just 0.03 wt.% Sc_2O_3. In the EMM samples, the largest amounts of the element (mean of 11 ppm) concern Radlin samples, Mg sulfates of both exhalative and supergene origin. A similar amount was found in godovikovite-rich core of the sulfate crust in Rydułtowy; the corresponding object from Radlin is less enriched.

Vanadium

The most prominent vanadium enrichment in the BCMH minerals is undoubtedly the apatite group. A most likely V-rich to V-dominant representative of the group from the Bytom slag (Table 19.3.3) contains 18.42 wt.% V_2O_5. Considering the high (close to 100%) totals, this mineral does not seem to contain any CO_2. However, the proposed empirical formula (n=5), $(Ca_{5.08}Fe^{2+}{}_{0.03}Mg_{0.02})_{\Sigma 5.12}[(PO_4)_{1.33}(SiO_4)_{0.59}(VO_4)_{0.56}(SO_4)_{0.34}]_{\Sigma 2.82}$ $[F_{0.66}O_{0.34}]_{\Sigma 1.00}$, has the T position slightly unoccupied, and the whole formula is slightly unbalanced; this is suggestive of the presence of some $CO_3{}^{2-}$. Interestingly, the last reported analysis corresponds to potentially new mineral species, which is the V-dominant analogue of fluorapatite; considering possible admixture of (unanalyzed) CO_2, its empirical formula may take the form of $(Ca_{5.16}Fe_{0.03}Mg_{0.02})_{\Sigma\ 5.21}[(VO_4)_{1.06}(PO_4)_{0.84}(SiO_4)_{0.53}(SO_4)_{0.30}$ $(CO_3)_{0.20}$ $(AsO_4)_{\Sigma 0.05}]_{\Sigma 2.99}[F_{0.54}O_{0.46}]$. Vanadium was clearly shown in the WDS scan conducted on this phase. However, it was not observed in similar scans conducted on another group representative from hematite-clinopyroxene-rich slag from Rydułtowy.

Another silicate mineral that may occasionally concentrate vanadium is leucite. However, this is only Przygórze slag material; 3 of 8 analyses have shown V_2O_3 content from 0.11 to 0.17 wt.% V_2O_3. The clinopyroxene (rich in the esseneite molecule) does not concentrate V, as shown by a WDS scan. Interestingly, a supposed

Table 19.3.3
Results of microchemical analyses (wt.%) of a supposed V-rich member of the apatite group from the Bytom slag.

	1	2	3	4	5
SO_3	5.27	5.09	5.67	5.68	4.62
P_2O_5	23.36	14.50	21.08	22.51	11.46
V_2O_5	5.29	15.55	5.03	4.97	18.42
As_2O_5	0.92	0.90	0.90	1.01	1.05
SiO_2	6.83	6.75	7.70	7.20	6.06
TiO_2	0.05	0.08	0.00	0.13	0.07
CaO	56.27	53.20	57.15	57.47	55.29
FeO	0.41	0.70	0.48	0.44	0.46
MgO	0.21	0.11	0.20	0.18	0.16
F	2.91	2.20	2.75	2.47	1.97
Σ	101.51	99.09	100.96	102.05	99.55
apfu* (basis: 8 cations)					
S	0.33	0.33	0.36	0.35	0.30
P	1.66	1.07	1.49	1.58	0.84
V	0.29	0.90	0.28	0.27	1.06
As	0.04	0.04	0.04	0.04	0.05
Si	0.57	0.59	0.64	0.60	0.53
Ti	0.00	0.01	0.00	0.01	0.00
Ca	5.05	4.99	5.13	5.10	5.16
Fe	0.03	0.05	0.03	0.03	0.03
Mg	0.03	0.01	0.03	0.02	0.02
F	0.77	0.61	0.73	0.65	0.54

*Atoms per formula unit.

ilmenite-perovskite-ferrosilite-wollastonite solid solution, gehlenite-åkermanite-alumoåkermanite, albite, and $K_2NaAl_5Si_7O_{25}$, (K, Na)$_2Al_4Si_4O_{15}$, $Ca_3MgAl_2Si_2O_{11}$ and an amphibole-like phase of a sample of Przygórze slag gave negative results in terms of vanadium and other trace elements measured. Vanadium-enriched mullite occurs in multicolor buchite from Ludwikowice Kłodzkie; 4 of 5 analyses (Table 19.3.4) contain V_2O_3, up to 0.23 wt.% amount. The corresponding empirical formula (n=5) is $(Al_{5.89}Ti_{0.04}Fe_{0.02}K_{0.02}Ca_{0.02}Na_{0.01}V_{0.01})_{\Sigma 5.99}$ $(Si_{1.71}Al_{0.29})_{\Sigma 2.00}O_{12.83}$ (x=0.17).

Important and rich V carriers are the TiO_2 polymorphs, rutile and anatase (Table 19.3.5). The mean V_2O_3 content in TiO_2 phase (both rutile and anatase possible) from the Przygórze light-toned metapelite is 1.86 wt.% (n=2), which in anatase of the Przygórze porcellanite is 0.37 wt.% (n=4), and which in anatase from the red-violet metapelite is 0.54 wt.% (n=8), with single analysis of silicon- and aluminum-rich representative containing 0.62 wt.% V_2O_3. Rutile from the Przygórze slag (Table 19.3.6) is evidently vanadiferous, as illustrated by its mean empirical formula (n=15): $(Ti_{0.98}Fe^{3+}_{0.01}V_{0.01})_{\Sigma 1.00}O_{1.99}$; the formula corresponds to mean 0.46 wt.% V_2O_3.

Detailed inspection of the last-mentioned rutile analyses pinpointed two other unique and stoichiometrically related Ti oxides, which are also enriched in vanadium: pseudorutile (from the Przygórze slag), $(Fe^{3+}_{1.43}Mg_{0.21}Ti_{0.17}Ca_{0.05}Al_{0.05}Mn_{0.04}K_{0.02}V_{0.01}Cr_{0.01})_{\Sigma 1.99}(Ti_{2.89}Si_{0.11})_{\Sigma 3.00}O_{8.92}$, corresponding to 0.27 wt.% V_2O_3; and machiite, a mineral recently described from Murchison meteorite (Knot, 2016), here found among rutile from a light-toned metapelite of the Przygórze breccias. Its formula is $(Al_{0.73}Fe_{0.69}Mg_{0.32}V_{0.12}Ti_{0.11}Cr_{0.03})_{\Sigma 2.00}(Ti_{2.99}Si_{0.01})_{\Sigma 3.00}O_{8.90}$, corresponding to even larger 2.57 wt.% V_2O_3. The concentration is most likely due to the role of the schreyerite end member in machiite's composition. Another oxide with elevated V (0.27 wt.% V_2O_3) is pseudobrookite. A single analysis concerning the hematite-rich zones of the Bytom slag gave the empirical formula as $(Fe^{3+}_{1.62}Al_{0.32}Mg_{0.04}Ti_{0.02})_{\Sigma 2.00}Ti_{1.00}O_{4.99}$, suggesting the importance of the hypothetical aluminopseudobrookite end member in the composition. Not only is rutile an important carrier of V, but so is perovskite (Table 19.3.7). The Przygórze slag material may have up to 0.40 wt.% V_2O_3, while the mean for material from powdery accumulations

Table 19.3.4
Results of representative analyses (wt.%) of mullites from white Przygórze (1–8) and white-gray-black Ludwikowice Kłodzkie (9–13) buchites, LSCB, Poland.

	1	2	3	4	5	6	7	8	9	10	11	12	13
SiO_2	23.71	23.97	25.37	24.89	24.48	24.87	25.16	23.70	24.76	24.25	24.21	23.07	24.10
TiO_2	0.81	0.84	0.73	0.79	0.63	0.67	0.55	0.77	0.80	0.76	0.90	0.92	0.63
Al_2O_3	74.05	73.05	70.95	71.83	71.69	70.14	70.48	72.62	72.48	73.19	74.16	74.61	74.37
Fe_2O_3	0.75	0.52	0.52	1.18	1.89	3.24	2.80	0.44	0.29	0.25	0.34	0.24	0.40
Cr_2O_3	0.13	0.14	0.11	0.16	0.57	0.47	0.32	0.17	0.00	0.00	0.00	0.00	0.00
V_2O_3*									0.15	0.00	0.23	0.20	0.20
MgO	0.00	0.06	0.08	0.10	0.17	0.18	0.20	0.06	0.06	0.04	0.05	0.04	0.08
CaO	0.00	0.00	0.00	0.00	0.00	0.00	0.00	0.00	0.18	0.14	0.27	0.20	0.24
Na_2O	0.11	0.05	0.14	0.21	0.12	0.09	0.15	0.14	0.00	0.00	0.10	0.00	0.16
K_2O	0.00	0.07	0.25	0.16	0.11	0.08	0.11	0.05	0.28	0.16	0.17	0.06	0.17
Σ	99.56	98.79	98.14	99.33	99.66	99.74	99.77	97.95	99.00	98.78	100.42	99.34	100.35
apfu (base: 8 cations)													
Si	1.69	1.72	1.83	1.78	1.75	1.79	1.80	1.71	1.77	1.74	1.71	1.64	1.70
Ti	0.04	0.05	0.04	0.04	0.03	0.04	0.03	0.04	0.04	0.04	0.05	0.05	0.03
Al	6.21	6.17	6.04	6.05	6.04	5.94	5.95	6.18	6.11	6.18	6.16	6.26	6.17
Fe	0.04	0.03	0.03	0.06	0.10	0.18	0.15	0.02	0.02	0.01	0.02	0.01	0.02
Cr	0.01	0.01	0.01	0.01	0.03	0.03	0.02	0.01					
V									0.01		0.01	0.01	0.01
Mg		0.01	0.01	0.01	0.02	0.02	0.02	0.01	0.01				0.01
Ca									0.01	0.01	0.02	0.02	0.02
Na	0.01	0.01	0.02	0.03	0.02	0.01	0.02	0.02			0.01		0.02
K		0.01	0.02	0.01	0.01	0.01	0.01		0.03	0.01	0.02	0.01	0.02
$x^{\dagger}$	0.15	0.14	0.11	0.14	0.14	0.12	0.13	0.15	0.14	0.13	0.18	0.19	0.20

*V was not analyzed in the Przygórze material; Mn was analyzed but not detected.
†Oxygen deficiency.

Table 19.3.5

Results of representative microchemical analyses (wt.%) of rutile/anatase from various pyrometamorphics from Przygórze (LSCB, Poland): light-colored metapelite within breccia (1–2), porcellanite within breccia (3–4), separate porcellanite (5–8), and red-violet metapelite (10–16).

	1	2	3	4	5	6	7	8	9	10	11	12	13	14	15	16
Nb_2O_5	0.00	0.00	0.29	0.21	0.00	0.00	0.00	0.00	0.13	0.14	0.18	0.00	0.19	0.23	0.00	0.22
SiO_2	1.90	0.23	5.97	13.29	8.61	3.79	9.24	6.71	1.38	0.13	0.56	1.89	0.41	1.85	0.44	0.14
TiO_2	92.54	97.17	95.68	78.26	77.52	85.45	72.06	77.60	94.81	97.32	97.13	95.39	97.36	94.01	97.30	99.61
ZrO_2	0.00	0.14	0.52	3.17	0.00	0.00	0.00	0.00	0.00	0.00	0.00	0.00	0.00	0.00	0.00	0.00
Fe_2O_3	0.51	0.25	0.90	1.48	0.63	0.45	0.56	0.63	0.54	0.36	0.32	0.30	0.42	0.62	0.38	0.34
Al_2O_3	3.09	1.44	0.78	2.53	7.27	2.86	7.17	5.26	0.89	0.13	0.33	0.20	0.26	1.24	0.23	0.12
Cr_2O_3	0.55	0.42	0.00	0.08	0.00	0.00	0.00	0.00	0.00	0.00	0.00	0.00	0.00	0.00	0.00	0.00
V_2O_3	2.01	1.72	0.22	0.00	0.39	0.40	0.35	0.36	0.57	0.59	0.41	0.59	0.58	0.52	0.50	0.55
MnO	0.00	0.00	0.00	0.10	0.00	0.00	0.00	0.00	0.00	0.00	0.00	0.00	0.00	0.00	0.09	0.00
MgO	0.03	0.00	0.00	0.16	0.06	0.06	0.10	0.07	0.00	0.00	0.00	0.00	0.00	0.00	0.00	0.00
CaO	0.03	0.00	0.00	0.07	0.26	0.29	0.40	0.45	0.14	0.00	0.10	0.03	0.05	0.22	0.04	0.03
Σ	100.65	101.37	104.30	99.29	94.76	93.29	89.86	91.08	98.46	98.68	99.03	98.40	99.26	98.70	98.99	101.01
apfu (base: cation sum of 1)																
Si	0.02		0.07	0.17	0.11	0.05	0.13	0.09	0.02		0.01	0.03	0.01	0.02	0.01	
Ti	0.90	0.95	0.90	0.75	0.76	0.89	0.74	0.80	0.95	0.99	0.98	0.96	0.98	0.94	0.98	0.99
Zr				0.02												
Fe			0.01	0.01	0.01		0.01	0.01	0.01					0.01		
Al	0.05	0.02	0.01	0.04	0.11	0.05	0.12	0.09	0.01		0.01			0.02		
Cr	0.01															
V	0.02	0.02							0.01	0.01		0.01	0.01	0.01	0.01	0.01
Ca							0.01	0.01								

Table 19.3.6

Results of representative microchemical analyses (wt.%) of rutile from the Przygórze (LSCB, Poland) slag.

	1	2	3	4	5	6	7	8	9	10
SiO_2*	0.12	1.77	0.23	0.33	0.05	0.23	0.00	0.00	0.00	0.00
TiO_2	100.68	96.75	99.07	99.98	100.85	99.90	97.95	100.45	99.91	100.35
Fe_2O_3	0.53	1.60	1.45	0.67	0.48	0.44	1.44	0.52	0.67	0.52
Al_2O_3	0.00	0.59	0.00	0.00	0.00	0.00	0.00	0.00	0.00	0.00
Cr_2O_3	0.07	0.11	0.08	0.09	0.00	0.08	0.00	0.00	0.06	0.00
V_2O_3	0.48	0.51	0.48	0.56	0.48	0.51	0.43	0.54	0.46	0.50
MnO	0.00	0.00	0.00	0.00	0.00	0.00	0.00	0.00	0.11	0.00
MgO	0.00	0.15	0.08	0.00	0.00	0.00	0.23	0.00	0.00	0.00
CaO	0.16	0.16	0.27	0.52	0.37	0.11	0.25	0.43	0.68	0.24
K_2O	0.06	0.38	0.05	0.06	0.00	0.10	0.04	0.00	0.00	0.04
Na_2O	0.00	0.04	0.00	0.00	0.00	0.00	0.00	0.00	0.00	0.00
Σ	102.08	102.05	101.70	102.20	102.23	101.39	100.34	101.94	101.88	101.66
apfu (1 cation basis)										
Si		0.02								
Ti	0.98	0.93	0.97	0.97	0.98	0.98	0.97	0.98	0.98	0.99
Fe	0.01	0.02	0.01	0.01	0.00	0.00	0.01	0.01	0.01	0.01
Al		0.01								
V	0.01	0.01	0.01	0.01	0.01	0.01		0.01		0.01
Ca				0.01	0.01			0.01	0.01	

*Nb, Zr, and Y were not measured, while Ba, Sr, F, and Cl were not detected.

(containing magnesioferrite, calcite, aragonite, diopside, oldhamite, periclase, anorthite, ettringite, nepheline, brucite, magnesite, maghemite, and possibly scawtite, babingtonite and scapolite group), named BK8 and likely derived from local hydrothermal veins, is 0.69 wt.%.

Fe oxides may also carry vanadium. However, only 2 of 12 analyses of hematite occurring in a red fine-vesicular buchite from Radlin gave positive V_2O_3 determination (maximum of 0.16 wt.%). For coexisting magnesioferrite-spinel only a single analysis gave a positive result, with 0.12 wt.%. Buchite surrounding the light-colored Przygórze metapelite contains spinels ($Hc_{55}Sp_{42}$) with mean 0.01 apfu (atoms per formula unit). V corresponds to a very minor coulsonite component role. Also a single determination concerning Fe-rich corundum from Jankowice parabasalt—$(Al_{1.01}Fe_{0.84}Mg_{0.11}Mn_{0.01}Ti_{0.01}[]_{0.02})_{\Sigma 2.00}O_{2.95}$ (n = 7)—shows quite low V_2O_3 amount of 0.10 wt.%. Crystals from the dark (brown-violet) buchite from Rydułtowy gave only a few positive determinations, always ca. 0.13 wt.%. Results of WDS scans on both hematite and magnesioferrite forming exsolution intergrowths in the Fe-rich clinopyroxene-bearing slag from Rydułtowy were negative. Table 19.3.8 shows results of trace-mode analyses of hematite from the Rydułtowy parabasalt, with 3 per 10 positive determinations; the maximum obtained concentration is 250 ppm. Some vanadium may also be found in other hematites and some magnesioferrites (Tables 19.3.8 and 19.3.9, respectively), and also a melilite group (Tables 19.3.10 and 19.3.11, respectively). The content in melilites is usually lower than that found in other minerals.

Vanadium is only rarely found in glasses. A notable exception is the multicolor buchite-like rock from Przygórze. Analyzes of white, black, and mullite-free zones have shown single points with positive V_2O_3 determination, the values being 0.13 (similar to that found in a single analysis of glass of a breccia-included porcellanite named WH), 0.14 (the same as a single analysis from a white buchite), and 0.16 wt.%, respectively. However, the yellow and brown-yellow glasses are enriched in the element in a more persistent way, with the corresponding means being 0.27 and 0.21 wt.%.

The largest V amounts in the EMM samples always denote sulfate crusts, with >140 ppm in Radlin and 90 ppm in Rydułtowy case. It should be noted that the values may vary depending on the sampling site and probably the ICP method and apparatus used: the initial concentration obtained for two steklite-millosevichite crusts of Radlin was 33 ppm V. Smaller amounts concern late-stage sulfates associated with the first object, with the potential concentrator being alunite. Abundant ammonium-rich alunite-group accumulations from Czerwionka have mean 64 ppm V. However, a similar level of enrichment concerns some local coals. Much lower concentration (17 ppm) was found in butlerite-dominated supergene veins from the latter heap. Kremersite-rich condensates contain only 3 ppm.

Table 19.3.7
Results of representative microchemical analyses (wt.%) of perovskite from various pyrometamorphites: a Przygórze slag (1–4), hydrothermal powdery sample BK8 (5–6), Przygórze slag (other geochemical type; 7–8), Bytom slag (9–10), and Bytom breccia (11–14). All samples come from Poland.

	1	2	3	4*	5	6	7†	8	9‡	10	11§	12	13	14
Nb_2O_5	0.17	0.14	0.40	0.15	0.21	0.28					0.76	0.35	0.29	0.36
SiO_2	0.89	0.17	1.78	0.08	0.51	1.06	0.98	0.13	1.14	0.27	0.08	0.00	0.00	0.00
TiO_2	56.80	56.88	53.53	58.75	54.83	54.55	57.29	50.79	47.73	49.55	46.81	52.43	51.91	51.81
ZrO_2	0.00	0.18	0.00	0.00	0.30	0.28								
Al_2O_3	0.15	0.14	0.81	0.48	0.43	0.58	0.15	0.05	0.44	0.42	2.36	0.77	0.78	0.79
Cr_2O_3‖					0.33	0.21								
V_2O_3	0.00	0.40	0.23	0.32	0.32	0.69								
Y_2O_3					0.06	0.00					0.59	0.23	0.25	0.33
Fe_2O_3									9.88	7.00	7.84	4.71	4.50	4.44
FeO	1.06	1.12	0.85	17.96	1.26	0.77	2.25	21.34						
MnO							0.15	0.41						
MgO	0.21	0.25	0.89	2.67	0.00	0.17	1.00	2.63	0.02	0.03	1.03	0.00	0.06	0.00
CaO	40.38	40.83	39.30	16.41	41.12	40.19	37.72	22.50	40.52	39.95	33.64	39.85	39.84	39.80
Σ							99.54	97.83						
apfu (total 2 cations basis)														
Nb											0.01			
Si	0.02		0.04		0.01	0.02	0.02		0.03	0.01				
Ti	0.97	0.97	0.92	1.08	0.94	0.94	0.98	0.90	0.81	0.86	0.86	0.91	0.90	0.90
Al			0.02	0.01	0.01	0.02			0.01	0.01	0.07	0.02	0.02	0.02
Cr					0.01									
V		0.01		0.01	0.01	0.01								
Y											0.01			
Fe^{3+}								0.10	0.17	0.12	0.14	0.08	0.08	0.08
Fe^{2+}	0.02	0.02	0.02	0.37	0.02	0.01		0.32						
Mn								0.01						
Mg	0.01	0.01	0.03	0.10		0.01	0.03	0.09			0.04			
Ca	0.98	0.98	0.96	0.43	1.00	0.98	0.92	0.57	0.98	0.99	0.88	0.98	0.99	0.99

*A supposed member of a perovskite-ilmenite-geikielite series.
†Nb, Zr, Cr, V, and Y were not analyzed, while Ni and F were not detected in analyses 7 and 8.
‡U, Nb, Ni, V, and Cr were below their detection limits, and Zr, Mn, and F were not analyzed (analyses 9 and 10).
§Sc, U, W, Ta, Hf, Ni, and V were below their detection limits, and Zr, Mn, and F were not analyzed (analyses 11–14).
‖Cr and Y were not measured in the 1–4 analytical range.

Table 19.3.8
Results of microchemical trace-element analyses of hematite from the Rydułtowy parabasalt, Poland.*

	1	2	3	4	5	6	7	8	9	10
ppm										
Co†	620	520	500	570	470	630	630	510	530	560
Ga			150	200		180	180			210
V					140				110	250

*Beam current was 100 nA.
†Zn, Cu, Ni, Cr, In, and Tl were below the detection limits.

Chromium

Chromium is quite often found as a trace substitute in the analyzed materials, sometimes in association with vanadium enrichment. It is rather commonly found in various oxide minerals. Three of 11 analyses of the Bytom slag hematite have shown the presence of chromium, up to 0.56 wt.% oxide. This hematite is associated with minor magnesiograndiferrite, (Mg, Mn) (Fe, Al)$_4$O$_7$ phase; a single analysis of it has shown only ca. 0.10 wt.% Cr_2O_3. Hematite from the Przygórze porcellanite is less geochemically engaged, with 1 per 9 analyses showing 0.33 wt.% Cr_2O_3. Chromium was not observed

Table 19.3.9
Results of microchemical trace-element analyses of magnesioferrite from the Bytom slag, Poland.*

	1	2	3	4	5	6	7	8†
ppm								
V	230							
Cr	844	149		257				180
Co	826	379					364	
Ni	1636	915	649	455	801	369		1091
Ga			222					
In						225‡		
	9	10	11	12	13	14	15	16
ppm								
V					268			
Cr		1243						
Co	243	204	345	335	384			194
Ni						301		
Ga								
In						430	241	

*Beam current was 60 nA.
†16 of total 20 analyses showed the presence of given trace elements.
‡Concentration close to detection limit.

Table 19.3.10
Results of microchemical trace-element analyses of melilite from the "olive rock" sample no. 1, Przygórze (LSCB, Poland).*

	1	2	3	4	5	6	7	8	9	10
ppm										
V			106		111		97†	81†	147	95†
Cr	93†		91†		187	153		84†		
Ga	168	98		197	123	186	115	249	221	365
Sb	480	395	497	585	425	384	467	476	808	592

*Beam current was 100 nA.
†Concentrations close to the detection limits.

Table 19.3.11
Results of microchemical trace-element analyses of gehlenite from the Bytom slag, Poland.*

	1	2	3	4	5	6	7	8	9	10†
ppm										
V			201‡							
Ga	382	507	325		572		247	282	273	429
Ge				205‡		275				
	11	12	13	14	15	16	17	18	19	
ppm										
Cr								215		
Ga	468	230		403	256		412	330	243	
Ge									210‡	
Zn			488			354				

*Beam current was 50 nA.
†19 of total 27 analyses showed the presence of given trace elements.
‡Concentration close to the detection limit.

Table 19.3.12
Results of representative microchemical analyses (wt.%) of Ti-bearing maghemite from the contacting buchite and slag from Bytom, Poland.

	1	2	3	4	5	6	7	8	9	10	11	12
SiO_2*	0.05	0.03	0.01	0.04	0.06	0.11	0.06	0.02	0.05	0.08	0.04	0.06
TiO_2	4.79	4.51	2.95	3.20	3.33	4.13	4.23	3.68	3.56	3.41	3.69	3.36
Cr_2O_3	0.14	0.07	0.00	0.05	0.07	0.09	0.10	0.08	0.08	0.06	0.05	0.09
Al_2O_3	3.94	3.90	5.43	5.50	5.38	6.01	5.95	6.32	6.39	3.00	6.19	3.03
Fe_2O_3	84.30	86.14	87.16	87.32	86.93	85.54	86.45	85.54	86.33	88.61	86.33	88.52
CoO	0.00	0.09	0.00	0.11	0.19	0.10	0.09	0.00	0.11	0.33	0.27	0.51
MnO	0.37	0.30	0.13	0.00	0.14	0.20	0.23	0.12	0.17	0.33	0.27	0.51
MgO	2.30	2.35	1.35	1.40	1.56	1.72	1.76	1.65	1.56	1.72	1.62	1.73
CaO	0.00	0.00	0.00	0.00	0.04	0.00	0.00	0.00	0.00	0.00	0.00	0.00
Σ	95.89	97.39	97.04	97.63	97.69	97.90	98.86	97.39	98.24	97.54	98.63	97.81
apfu (base: 3 oxygen atoms)												
Ti	0.10	0.09	0.06	0.06	0.07	0.08	0.08	0.07	0.07	0.07	0.07	0.07
Al	0.12	0.12	0.17	0.17	0.16	0.18	0.18	0.19	0.19	0.09	0.19	0.09
Fe	1.68	1.69	1.72	1.71	1.70	1.66	1.66	1.67	1.67	1.76	1.66	1.75
Co	0.00	0.00	0.00	0.00	0.00	0.00	0.00	0.00	0.00	0.01	0.01	0.01
Mn	0.01	0.01	0.00	0.00	0.00	0.00	0.00	0.00	0.00	0.01	0.01	0.01
Mg	0.09	0.09	0.05	0.05	0.06	0.07	0.07	0.06	0.06	0.07	0.06	0.07

*V, Cu, Sn, and Sb were measured but not detected.

in hematite from siderite-rich aggregates (most likely initial slags) occurring in plenty among pyrometamorphic and supergene mineralization zones in Czerwionka. Only a single analysis of hematite from the Radlin buchite has determined Cr_2O_3 at the 0.30 wt.% level. The largest concentration in $(Al, Fe)_2O_3$ concerns the $AlFeO_3$ phase (single analysis): 1.07 wt.% oxide (0.02 apfu Cr). A single analysis of hematite from a melanocractic zone of the Rydułtowy buchite characterizes in the concentration of 0.21 wt.% Cr_2O_3. The content in crystals found in the parabasalt-buchite border zone is less interesting, with 2 analyses having 0.07 wt.% on average. Maghemite from buchite surrounding the Bytom slag almost always bears some Cr (Table 19.3.12), but the amount is low, averaging 0.07 wt.% Cr_2O_3, with maximum concentration equal to 0.14 wt.%. Formation of this mineral in the slag-buchite border zone is likely not accidental, as this mineral is not usually found in buchites; such an occurrence is suggestive of element diffusion processes and their exchange between the various lithologies. The Fe^{3+}-bearing corundum from Jankowice breccia is also an important carrier of chromium: the element of interest is proven in most of the analyses and its amount reaches 0.13 wt.% oxide. However, when doing trace-mode analyses on a separate series, a mean of 0.05 wt.% oxide was measured (n=23). These analyses also have shown lack of V, Co, Ni, Cu, Zn, Ga, In, and Sb. Chromium is also found (0.25–2.75 wt.% on average, n=16) in an associated (in a parabasalt) hercynite-spinel, with chromite molecule content of 1%. Magnesioferrite from the gehlenite-rich slag of Rydułtowy may have as much as 0.12 wt.% oxide, but Cr is not seen in most of the analyses.

Chromium is rarely determined in Ca ferrites. Only 1 of 20 analyses of srebrodolskite from one of the Rydułtowy slags (the gehlenite-rich one) has 0.06 wt.% oxide. A single determination for the same sample in the trace mode gave 156 ppm Cr (Table 19.3.13). The element is only slightly more frequently seen in the trace-mode analyses of the corresponding Bytom slag crystals (Table 19.3.14), with an amount peaking at 265 ppm Cr.

Mean chromium content in anatase of the Przygórze light-toned metapelite is 0.48 Cr_2O_3 wt.% (see Table 19.3.5). Hercynite-spinel form the surrounding buchite characterizes in mean 0.5% of the chromite end-member content, corresponding to 0.02 apfu Cr; similar numbers denote hercynitic spinels from attached parabasalts. Other TiO_2 polymorphs analyzed are usually very low in Cr or devoid of it. There are more Ti-bearing oxides that may act as potential Cr carriers in the BCMH studied. The porcellanite fragment WH from the same sample contains a Ti oxide corresponding to $(Ti_{0.75}Si_{0.17}Al_{0.04}Zr_{0.02}Fe^{3+}_{0.01})_{\Sigma 0.99}O_{1.95}$ or possibly $(Ti,Zr,Fe)_{15}AlSi_3O_{37}$, which contains 0.08 wt.% Cr_2O_3. Pseudorutile of the Przygórze slag has 0.19 wt.%, which is higher than the corresponding values found in the associated vanadian rutile (up to 0.11 wt.%); only 6 of 15 analyses of the latter point to Cr presence. As compared to those above Przygórze mineral, machiite from the light-toned metapelite characterizes in larger enrichment, with 0.56 wt.% Cr_2O_3. Most of analyzed perovskites do not contain measurable chromium; the BK8 one has 0.33 wt.% Cr_2O_3 on average. The maximum content of Cr (>1200 ppm) was found in magnesioferrite (see Table 19.3.9). Some rutiles may contain up to 0.55 wt.% Cr_2O_3 (see Table 19.3.5).

Table 19.3.13
Results of microchemical trace-element analyses of srebrodolskite from the slag sample no. 1 (analyses 1–9) and no. 2 (analyses 10–19) from Rydułtowy, Poland.*

	1	2	3	4	5	6	7	8	9	10
ppm										
Co	318		233			286	270	249	199	334
Ga	477	260	319	295	772	283		428	418	472
	11	12	13	14	15	16	17	18	19[†]	
ppm										
Cr							156			
Co	257			401	323		200	215	283	
Ga	669	474	441		441	533		399	274	

*Beam current was 50 nA.
[†]19 of total 20 analyses showed the presence of given trace elements.

Table 19.3.14
Results of microchemical trace-element analyses of srebrodolskite from the Bytom slag, Poland.*

	1	2	3	4	5	6	7	8	9	10[†]
ppm										
Cr							162			
Co		498	712	480	266	309	539		466	406
Ga	343		610		703			386		
	11	12	13	14	15	16	17	18	19	
ppm										
V			266	428					223	
Cr	265	145				201				
Co	549	376	439	636				787		
Ga			454		305[‡]		691	305		

*Beam current was 50 nA.
[†]19 of total 21 analyses showed the presence of given trace elements.
[‡]Concentration close to detection limit; Ni, Cu, and Zn were below the detection limits.

A single case, namely of a white buchite from Przygórze (of unconfirmed natural genesis), concerns chromium entering the mullite structure (see Table 19.3.4). The mean concentration is 0.20 wt.% Cr_2O_3, and the empirical formula of the mineral is $Al_{6.00}(Si_{1.76}Al_{0.00}Fe_{0.07}Ti_{0.04}Na_{0.01}Cr_{0.01}K_{0.01})_{\Sigma 2.00}O_{12.86}$ (x=0.14). Concentration of chromium was also measured in multicolor surrounding glasses, but the element was only found in the brown-yellow V-enriched glass; 2 analyses have shown 0.10 wt.% Cr_2O_3. Large-crystalline mullite found in voids of the Bytom breccia (and coexisting supposed $(Mg, K)_2Al_4(Si, Al)_5O_{18}$ phase), and mullites from Ludwikowice Kłodzkie buchite(s), do not share such geochemistry. Some mullites may also bear vanadium.

Chromium is rarely found in other silicates. A diopside-rich clinopyroxene ($Di_{68}Hd_{19}Cta_7Hip_4Cre_2$, n=3, where Cta=$CaTiAl_2O_6$ molecule, and Cre stands for a hypothetical chromoesseneite end-member) is found in a pinkish buchite from the Jankowice breccia. Two of 3 analyses have as much as 0.81 wt.% Cr_2O_3 on average (0.02 apfu Cr). As in case of vanadium, some chromium may also enter the melilite-type structures (see Tables 19.3.10 and 19.3.11 in the Gallium section); the amount is usually low in this rather rare case.

A single of 30 trace-mode analyses of glass from the Czerwionka black-block metapelite has 186 ppm Cr. Mean concentration in glasses forming porcellanite fragments (WH type) in breccia from Przygórze is 0.13 wt.% oxide. An interesting behavior of Cr is noticed in the border zone of the light-toned metapelite and light-toned buchite from Przygórze. The mean Cr_2O_3 concentration in the first lithological type is 0.24 wt.%, which is very close to 0.25 found in the core of the latter and 0.26 in a quartz-spinel rich zone located near the contact zone. Interestingly, no Cr was detected at the exact contact. Table 19.3.15 shows results of microprobe analyses of glasses of various other lithological types from Przygórze. The mean values (wt.% Cr_2O_3) are 0.26 for a microbreccia, and white metapelite

Table 19.3.15

Results of representative analyses (wt.%) of glasses from brecciated rocks from Przygórze: buchite-like microbreccia (analyses 1–5), white metapelite no. 1 (analyses 6–9), and no. 2 (analyses 10–13) yellow-red metapelite (analyses 14–17), red metapelite (analyses 18–26), and parabasalt (analyses 27–34).

	1	2	3	4	5	6	7(1)	8(2)	9	10	11	12	13	14(3)	15(4)	16	17
SiO_2(5)	52.25	48.84	48.05	44.08	47.73	45.14	50.10	50.10	49.81	54.46	53.47	54.13	53.30	39.95	54.25	49.20	50.13
TiO_2	0.76	0.64	1.20	0.60	0.55	0.90	0.49	0.62	0.92	0.44	0.00	0.00	0.00	1.12	1.19	0.91	0.86
Cr_2O_3	0.25	0.17	0.29	0.31	0.29	0.27	0.27	0.16	0.34	0.13	0.00	0.00	0.13	0.51	0.20	0.34	0.15
Al_2O_3	38.76	42.55	43.03	50.15	44.72	47.14	40.56	40.50	42.64	38.12	42.60	42.51	41.27	49.99	33.52	39.73	38.38
Fe_2O_3	3.65	3.31	4.05	2.47	2.94	3.20	4.34	4.11	3.41	2.55	1.19	0.79	1.05	3.66	3.26	2.87	2.92
MgO	0.00	2.14	2.65	1.84	2.02	2.01	2.48	2.27	1.86	1.43	0.82	0.60	0.90	1.13	1.77	1.28	1.43
CaO	0.11	0.10	0.08	0.00	0.12	0.15	0.17	0.11	0.11	0.13	0.12	0.10	0.12	1.53	3.08	2.57	2.58
Na_2O	0.20	0.33	0.61	0.19	0.16	0.09	0.32	0.34	0.48	0.61	0.58	0.47	0.67	0.23	0.34	0.52	0.58
K_2O	0.78	0.76	0.82	0.45	0.60	0.33	0.57	0.67	0.73	0.84	0.80	0.74	0.83	1.33	2.05	1.77	1.94
Σ	96.77	99.00	100.78	100.09	99.12	99.22	99.30	98.88	100.30	98.71	99.57	99.32	98.27	100.23	99.66	99.28	98.98
	18	19	20(6)	21	22	23	24	25	26(7)	27	28	29	30	31	32	33	34
SO_3	0.00	0.00	0.00	0.00	0.00	0.00	0.00	0.00	0.00	0.45	0.56	0.46	0.33	0.00	0.00	0.24	0.26
SiO_2	51.04	49.33	45.43	46.50	65.94	67.53	65.21	48.43	49.91	55.15	56.05	55.99	55.06	55.06	55.09	56.12	53.90
TiO_2	1.15	1.02	0.98	1.01	1.12	0.92	1.24	0.42	0.00	0.00	0.00	0.00	0.00	0.00	0.00	0.00	0.00
Cr_2O_3	0.12	0.18	0.19	0.20	0.00	0.00	0.00	0.00	0.00	0.00	0.00	0.00	0.00	0.00	0.00	0.00	0.00
Al_2O_3	36.94	37.81	45.61	44.92	18.45	25.35	22.36	32.38	32.72	13.69	13.84	13.55	13.64	13.59	13.32	13.39	13.46
Fe_2O_3	5.26	6.10	3.81	3.52	3.25	1.45	2.95	11.55	10.64	21.70	22.36	21.22	21.67	23.20	21.63	21.47	23.91
ZnO	0.00	0.00	0.00	0.00	0.00	0.00	0.00	0.12	0.22	0.00	0.00	0.00	0.00	0.00	0.00	0.00	0.00
MgO	2.55	2.23	1.56	1.63	1.39	0.34	0.91	7.30	7.69	0.38	0.40	0.34	0.34	0.52	0.63	0.57	0.46
CaO	0.22	0.20	0.11	0.21	0.55	0.13	0.61	0.00	0.00	5.63	5.62	5.82	6.03	5.89	6.69	5.84	6.29
Na_2O	0.39	0.48	0.38	0.37	0.97	0.70	1.13	0.00	0.05	0.39	0.34	0.00	0.30	0.19	0.25	0.37	0.43
K_2O	1.62	1.54	1.07	1.12	6.61	4.68	6.26	0.29	0.25	1.40	1.56	1.56	1.34	1.39	1.27	1.36	1.16
Σ	99.30	98.89	99.12	98.54	98.27	101.10	100.67	100.49	101.49	98.80	100.72	98.93	98.71	99.83	98.88	99.34	99.86

(1)This point has 0.11 wt.% Co in addition.
(2)This point has 0.11 wt.% Ni in addition.
(3)This point has 0.14 wt.% V in addition.
(4)This point has 0.15 wt.% NiO in addition.
(5)P, V, and Mn were measured but not detected.
(6)This point has 0.12 wt.% CuO in addition.
(7)This point has 0.12 wt.% V_2O_3 in addition.

Table 19.3.16
Results of microchemical analyses (wt.%) of native iron of green buchite from Nowa Ruda (LSCB, Poland).

	1	2	3	4	5	6	7	8	9	10
Si*	0.00	0.10	0.00	0.46	0.70	0.78	0.04	1.45	0.05	0.53
P	0.29	0.10	0.10	0.21	0.33	0.24	0.31	0.20	0.22	0.24
Cr	0.00	0.13	0.12	0.24	0.09	0.00	0.09	0.09	0.23	0.26
Fe	97.16	97.84	98.86	96.72	97.62	97.86	97.98	97.68	98.35	97.91
Co	0.22	0.19	0.19	0.14	0.00	0.00	0.22	0.00	0.00	0.00
Ni	0.58	0.28	0.29	0.48	0.53	0.54	0.56	0.29	0.44	0.35
Cu	1.29	1.21	0.96	1.25	1.34	1.48	0.84	1.35	1.19	0.75
S	0.16	0.00	0.00	0.00	0.00	0.00	0.00	0.00	0.00	0.15
Σ	99.70	99.75	100.52	99.50	100.61	100.90	100.04	101.24	100.48	100.19
apfu (1 atom basis)										
Si	0.00	0.00	0.00	0.01	0.01	0.02	0.00	0.03	0.00	0.01
P	0.01	0.00	0.00	0.00	0.01	0.00	0.01	0.00	0.00	0.00
Fe	0.97	0.98	0.98	0.97	0.96	0.96	0.98	0.95	0.98	0.97
Ni	0.01	0.00	0.00	0.00	0.01	0.01	0.01	0.00	0.00	0.00
Cu	0.01	0.01	0.01	0.01	0.01	0.01	0.01	0.01	0.01	0.01

*Mn was not measured, and Al and As were not detected.

of the first type; 0.07 for the second type of white metapelite; 0.30 for yellow-red metapelite; and 0.08 for red metapelite. Parabasalt glass shows no Cr concentration.

Native iron from the green Nowa Ruda buchite contains 0.16 wt.% Cr on average ($n=17$), with maximum at 0.28, but the element is not detected in some analyses (Table 19.3.16). A similar fragment from Ludwikowice white-gray-black buchite, of a possible ($Fe_{0.89}Al_{0.04}Cr_{0.03}Ni_{0.02}Mn_{0.01}$) composition, has a much higher Cr_2O_3 amount: 3.10 wt.%. The gupeiite/suessite of the Ludwikowice multicolor buchite (Table 19.3.17) does not bear Cr admixtures, while the similar Fe_3Si_{1-x} reduced phase has mean 0.09 wt.% ($n=5$). There seems to be a positive correlation between Cr and P contents in similar phases of this rock: the supposed gupeiite/suessite-schreibersite solid solution always contains Cr (mean 0.29 wt.%, $n=6$), and even more Cr (~1 wt.%) is in the Fe_7P_3 and/or $Fe_5(P,Si)_2$ species.

For vanadium, the largest amounts of chromium (also >100 ppm) in the EMM samples were found in association with sulfate crust of Radlin, with alunite as the main concentrator. The mean abundance in alunite-group aggregates from Czerwionka is 42 ppm Cr. A similar value is the maximum found initially in the steklite-rich crust of Radlin. From Czerwionka, 28 ppm is found in tamarugite-dominated supergene accumulations. The Rydułtowy sulfate crust is also enriched, but with lesser amounts observed. As expected, the local coal samples are way less Cr-enriched than in case of V. Butlerite and metavoltine-rich accumulations (Czerwionka) bear very low V amounts (up to 3 ppm).

Cobalt

Trace amounts of cobalt are ubiquitous in Fe-rich minerals of the BCMH under study, as sometimes shown even by the main/minor element EPMA analyses. The most important sinks for cobalt are Srebrodolskite and hematite, with lesser importance of Ca ferrites other than the first phase mentioned. Direct measurements on srebrodolskite from the Rydułtowy slags gave only a single positive result, with 0.10 wt.% Co; similar content (0.11) denotes a single analysis of associated harmunite. However, another series of analyses ($n=20$) of crystal associated with fluorellestadite gave seven positive determinations, the maximum being 0.13 wt.% oxide. Trace-mode analyses (see Table 19.3.13) show cobalt content, in 13 of 19 analyses, reaching 401 ppm Co; the mean is 274 ppm. For Srebrodolskite from the Bytom slag (see Table 19.3.14), cobalt was also seen in 13 of the total 19 analyses, with maximum at 787 and mean at 497 ppm.

Red iron-rich buchite from Rydułtowy contains abundant hematite, with only 2 analyses having cobalt (ca. 0.13 wt.% CoO). This also is true for the dark brown-violet buchite, which gave only a single positive determination (0.14 wt.%). On the other hand, all the trace-mode analyses of hematite from parabasalt-buchite contact

Table 19.3.17

Results of microchemical analyses (wt.%) of reduced iron minerals from the white-gray-black (1–7) and multicolor buchite (8–16) from Ludwikowice, Poland.

	1*	2	3	4	5	6	7	8	9	10	11	12	13	14	15	16	17
Si	7.29	0.00	0.00	0.08	0.00	0.00	0.08	13.52	13.27	12.29	13.12	13.12	12.91	9.08	10.25	3.95	5.86
P	0.30	0.10	0.00	0.10	0.10	0.02	0.10	0.17	0.13	1.85	0.45	0.30	0.41	5.84	4.72	14.67	12.03
As	0.00	0.00	0.00	0.00	0.00	0.00	0.00	0.00	0.00	0.00	0.00	0.00	0.00	0.00	0.00	0.13	0.00
Al	0.00	0.00	0.00	0.04	0.00	0.00	0.04	0.00	0.00	0.00	0.00	0.00	0.00	0.00	0.00	0.00	0.00
Cr	0.00	0.00	0.00	0.00	0.00	0.00	0.00	0.00	0.11	0.17	0.09	0.13	0.00	0.52	0.30	1.02	1.07
Fe	92.00	67.27	64.17	63.07	67.27	64.17	63.07	85.11	85.06	84.79	84.89	85.14	83.56	82.57	82.23	77.09	78.19
Mn								0.00	0.00	0.00	0.00	0.00	0.00	0.00	0.00	0.25	0.00
Co	0.34	0.17	0.14	0.00	0.17	0.14	0.00	0.33	0.58	0.77	0.53	0.55	0.66	0.60	0.70	0.78	0.65
Ni	0.90	0.00	0.00	0.00	0.00	0.00	0.00	0.58	1.42	1.51	1.54	1.38	1.37	1.49	1.49	1.54	1.36
Cu	0.13	0.00	0.63	0.00	0.03	0.67	0.00	0.00	0.00	0.00	0.00	0.00	1.82	0.00	0.00	0.00	0.00
S	0.00	32.90	36.19	36.52	32.90	36.19	36.52	0.00	0.00	0.00	0.00	0.00	0.00	0.00	0.00	0.00	0.00
Σ	100.97	100.43	101.14	99.81	100.45	101.20	99.81	99.70	100.56	101.38	100.63	100.61	100.74	100.10	99.69	99.42	99.15
apfu																	
Si	0.93							0.95	0.93	0.85	0.92	0.92	0.92	0.90	0.64	0.68	0.71
P	0.04	0.02						0.01	0.01	0.12	0.03	0.01	0.02	0.03	0.37	2.30	1.33
As																0.01	
Cr										0.01					0.02	0.09	0.07
Fe	5.92	7.02	1.02	0.99	1.08	1.00	0.99	3.01	2.99	2.95	2.98	3.00	2.99	2.95	2.90	6.70	4.78
Mn																0.02	
Co	0.02	0.02						0.01	0.02	0.03	0.02	0.02	0.02	0.02	0.02	0.06	0.04
Ni	0.06							0.02	0.05	0.05	0.05	0.04	0.05	0.05	0.05	0.13	0.08
Cu	0.01	0.00	0.01			0.01								0.06			
S		5.98	1.00	0.99	0.92	0.99	1.00							0.00			

*analysis 1 – Fe_6Si phase; analysis 2 – Fe_7S_6 phase; analysis 3–4 and 6–7 – troilite; analysis 5 – pyrrhotite; analysis 8–9 – gupeiite or suessite; analysis 11–13 – Fe_3Si_{1-x} phase (or gupeiite/suessite); analysis 10 and 14–15 – $Fe_3(Si,P)$ phase (possibly gupeiite-schreibersite or suessite-schreibersite solid solution); analysis 16 – Fe_7P_3 phase; analysis 17 – $Fe_5(P,Si)_2$ or $Fe_7(P,Si)_3$ phase.

Table 19.3.18
Results of microchemical trace-element analyses of siderite from the hematite-siderite rock, Czerwionka (USCB, Poland).*

	1	2	3	4	5	6	7	8†
ppm								
Ni	259	182‡		203		241		
Co†	112	124	112					
Ga			290		160		133‡	
Pb			267‡		362	484		284‡

*Beam current was 100 nA.
†8 of total 13 analyses showed the presence of given trace elements.
‡Concentrations close to detection limits.

(see Table 19.3.8) contain cobalt, with maximum and mean being 630 and 554 ppm Co, respectively. As opposed to Srebrodolskite, the Co enrichment level in this hematite is rather constant. Two-thirds of the analyses done on crystals of the Łaziska mullite-rich clinker show ca. 0.11 wt.%. Many more analyses (8 of 11) of the mineral from the Bytom slag gave positive results (up to 0.17 wt.% CoO); an additional session gave 9 analyses with 4 positive determinations and mean of 0.14 wt.%. The associated magnesiograndiferrite (hypothetical $MgFe_4O_7$ end-member) also has elevated Co (up to 0.18 wt.% oxide). The true grandiferrite ($Gdf_{64}Mgd_{15}Mng_6Gr_6Cso_3$, where Mgd is magnesiograndiferrite, Mng is manganograndiferrite, $MnFe_4O_7$, and Gr is grossite $CaAl_4O_7$ equivalent, and Cso is a hypothetical $Ca(Si_3[])O_7$ end-member) analysis shows 0.11 wt.% CoO in a single analysis only. An intensive search in another coexisting Fe-rich oxide, magnesioferrite-spinel, gave positive results only in 6 of 42 analyses, with mean 0.15 wt.% CoO. Trace-mode analyses of this material (see Table 19.3.9) gave 9 positive results among 16 analyses done, with Co content up to 826 ppm, but usually below 400 ppm. This spinel occasionally may also contain very large amounts of Cr (>1200 ppm, single spot) and rarely also bears V. Only a single positive determination of cobalt in magnesioferrite from a slag of Rydułtowy exists; the amount is 0.13 wt.% oxide.

The Bytom slag also contains numerous potentially new species, like $Mg_2(Fe, Al, Mn)_5O_9$ and $Mg_3(Fe, Al, Mn)_7O_{13}$, suggestible for some disequilibrium formation conditions of the whole assemblage. The mentioned compounds usually do not concentrate either cobalt or nickel, as is the case of ferrograndiferrite, $(Fe,Mn)Fe_4O_7$. Some other species like $(Mg,Fe, Mn)_4(Fe, Al)_6O_{13}$ and $(Ca,Mn)Mg(Fe, Al, Si)_6O_{11}$ may incidentally contain Co and Ni oxides in the amount up to 0.17 and 0.20 wt.%, respectively. The corresponding values for $(Mg, Fe, Mn)_3(Fe, Al)_4O_9$ phase are ca. 0.12 wt.%. Another similar phase, $Mg_4(Fe,Mn)_5(Fe, Al)_6O_{14}$, may occasionally encompass up to 0.15 wt.% CoO. Of 18 analyses of magnesioferrite-rich spinel from this slag, 5 have CoO (up to 0.13 wt.%) and 6 have NiO (up to 0.40 wt.%) reported. On the contrary to earlier data, cobalt is only rarely found in hematite from the Przygórze porcellanite, with ca. 0.17 wt.% CoO in a pinch. This porcellanite also contains a similar phase, possibly maghemite, with the formula $(Fe_{1.85}Al_{0.06}Mn_{0.03}Ti_{0.03}Mg_{0.01}Si_{0.01})_{\Sigma 1.99}O_{3.00}$ ($n = 11$), with cobalt (up to 0.23 wt.% CoO) often seen.

Another rock of Przygórze, parabasalt, contains low- and high-aluminum hematite; cobalt is present in the first variety but in very low amount (0.07 wt.% CoO on average). The variation in CoO concentration in hematite from white-gray-black Ludwikowice buchite is in the 0.00–0.33 wt.% range ($n = 3$). Most of 13 analyses of maghemite from buchite surrounding the Bytom slag contains cobalt, with much larger concentration reaching 0.74 wt.% CoO (0.01 apfu Co; Table 19.3.12). Just a few of the analyses of hematite-bearing siderite from initial slag (Czerwionka) show Co enrichment (Table 19.3.18). Less of this metal oxide, and only in two cases, concerns hematite found in the Radlin buchite; the amount is 0.14 wt.%, which is the same as in the case of the hematite in this rock (1 of 3 analyses with positive determination). A single analysis of associated magnesioferrite-spinel has shown 0.11 wt.% CoO. A very small amount of Co is found—as in the case of V—in Fe-rich corundum from Jankowice, the number being 0.07 wt.% CoO.

An important cobalt and nickel carrier in the black-block metapelites in Czerwionka is murashkoite (Ciesielczuk et al., 2015) FeP, a mineral recently approved by the International Mineralogical Association. The mineral contains up to 0.31 and 0.42 wt.% Co and Ni, respectively. The graftonite/sarcopside surrounding murashkoite, a supposed alteration product, is practically devoid of them: a single of 5 analyses has shown 0.15 and 0.12 wt.% CoO and NiO, respectively. Ni-bearing pyrrhotite is a quite common accessory phase found in a metasandstone from Czerwionka; its cobalt content is, however, incidental, with one-third of analyses showing a positive result, maximum of 0.35 wt.% Co. A similar incidental case concerns ye'elimite from the Bytom slag.

Table 19.3.19
Results of microchemical trace-element analyses of fayalite from the Rydułtowy parabasalt, Poland.*

	1	2	3	4	5	6	7	8	9	10	11
ppm											
Co†	480		500		460	480		420		550	

*Beam current was 50 nA.
†Zn, Cu, Ni, and Ge were below the detection limits.

Cobalt usually seems to not enter the silicate species. A single exception is fayalitic olivine from the Rydułtowy parabasalt, which is enriched in Co (3 analyses with ~0.11 wt.% oxide) and does not accumulate Ni; this suggests a siderophile behavior of Co here. This behavior is confirmed via the trace-mode analyses (Table 19.3.19), with cobalt found in 6 of 11 analyses; the content is relatively invariable, being in the 420–550 ppm range. A single of 30 trace-mode glass analyses, Czerwionka black-block metapelite, has 165 ppm elemental Co. In comparison, 6 standard analyses on the Przygórze slag have mean 0.12 wt.% CoO.

Native iron from the green Nowa Ruda buchite usually lacks Co (Table 19.3.16); the maximum observed content is 0.22 wt.%. Analyses of associated siderite (with 12% calcite and 10% magnesite components) show a single case (1 of 11) of 0.16 wt.% CoO amount. A similar metallic fragment from Ludwikowice white-grey-black buchite has just 0.08 wt.% Co. The gupeiite/suessite of multicolor buchite of the same site (Table 19.3.17) has mean 0.56–0.60 wt.%. While the corresponding value for the associated P-rich silicide phases is similar in the case of the $Fe_3(Si, P)$ phase (0.61 on average), the phosphides bear more (as in case of Cr), that is mean 0.71 wt.%. Mean concentration in the Ludwikowice pyrrhotite is 0.16 wt.%, a similar one concerns the Fe_7S_6 phase, while troilite contains 0.00–0.14 wt.% of the metal. Of the latter association, the Fe_6Si phase has the largest 0.34 wt.% amount (0.02 apfu Co).

The enrichment of Co in the EMM samples seems to follow at least some of the preceding trends, namely that of the Co-Mg substitution. The largest Co amount (ca. 90 ppm) denotes exhalative Mg sulfates from Radlin. Neither local coal and pyrite nor sulfate crusts of other heaps contain more than a quarter of this amount (maximum of 13 ppm for steklite-rich ones, Radlin). Ammonioalunite-ammoniojarosite from Czerwionka is low in cobalt, with mean being 2.3 ppm. Tamarugite-hexahydrite-pickeringite efflorescences from this heap are more geochemically engaged, with 110 ppm Co abundance. Also metavoltine-boussingaultite occurring in the vicinity have much Co (74 ppm). Concentration of cobalt in kremersite is only 6 ppm.

Nickel

Nickel is not that commonly met as a trace substitution in minerals of the BCMH. However, as shown in Figure 19.3.1, it may sometimes be present in the form of its own mineral species. A single analysis of the red iron-rich buchite hematite (Rydułtowy) has shown 0.12 wt.% NiO. The element stands, however, for a constant admixture in hematite from melanocractic zones of more typical buchites from this site, with very similar mean being 0.13 wt.%. Most of the analyses conducted on the Łaziska clinker material have Ni, ranging from 0.05 wt.% to 0.20 wt.% oxide. Of a total of 11 analyses of the Bytom slag hematite, 4 have nickel above its detection limit, with maximum 0.22 wt.% oxide. More nickel is associated with magnesiograndiferrite—up to 0.36 wt.% NiO and presence in all 3 analyses—with Ni possibly following the known trend of coassociation with Mg. The 42-analysis series of the associated magnesioferrite-spinel gave just a single positive determination (0.12 wt.% oxide). The only other example of Ni-enriched spinel is hercynite-spinel from the Jankowice breccia (parabasalt zone); although only 6 analyses gave positive determination, the NiO content is often close to 0.20 wt.%. The Przygórze maghemite (from the porcellanite) has 4 of 11 analyses showing nickel (up to 0.44 wt.% NiO). The corresponding Bytom buchite material rarely contains the element, with one of two positive determinations being 0.31 wt.% NiO. Only 1 of 10 analyses of hematite of the Czerwionka siderite-rich initial slag has shown nickel, although in relatively high amounts equal to 0.68 wt.% oxide. The siderite itself may sometimes also bear Ni, with maximum at 259 ppm Ni (trace-mode; Table 19.3.18). The maximum NiO content of the corresponding material from the white-gray-black Ludwikowice buchite is somewhere in between, at 0.36 wt.%. The Al-rich hematite from the Przygórze parabasalt is relatively Ni-poor (maximum of 0.14 wt.%).

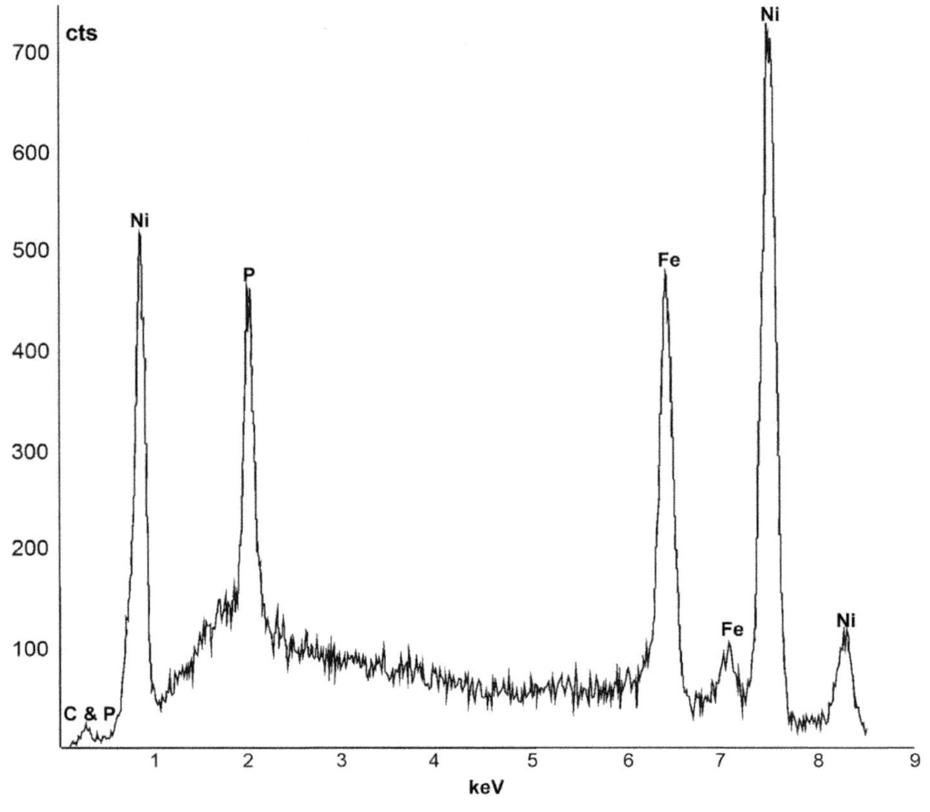

Figure 19.3.1. EDS spectrum of an Ni–Fe–P phase, likely a phosphide, within a matrix of a hematite-siderite proto-slag from Czerwionka. cts = counts per second and kev = 1 000 electron volts.

As suspected, native iron from the green Nowa Ruda buchite is enriched in Ni (see Table 19.3.16), with mean 0.40 and maximum 0.58 wt.%. This also is true for the supposed iron from the Ludwikowice white-to-black buchite, which has more than thrice of that: 2.04 wt.% Ni. The silicide and phosphide phases of the Ludwikowice multicolor buchite (see Table 19.3.17) have more Ni, that is 1.12, 1.39, 1.44, and 1.45 mean wt.% concentration in Fe_3Si, Fe_3Si_{1-x}, $Fe_3(Si, P)$ and the phosphides. The Czerwionka metasandstone pyrrhotite enrichment is evident, with mean 0.53 wt.% Ni (0.18–0.82 wt.% range). Nickel arsenide was observed as a single ~1 μm crystal within the gehlenite-rich slag from Rydułtowy. In case of the Ludwikowice buchite, nickel, as opposite to cobalt, is only found in the Fe_6Si phase (0.90 wt.%).

Enrichment in Ni among the EMM samples is quite common. For Co, relatively large amounts (up to ca. 130 ppm) are associated with exhalative Mg-rich sulfates of Radlin. Like Cr and V, Ni also enters sulfate crust, preferentially locating in the cores, with lesser amounts found in late-stage Al sulfates. The K–Al-rich crust of Radlin has only 22 ppm. The mean content in alunite-group species of Czerwionka is 7.2 ppm. As in the cobalt case, nickel is found relatively abundant (210 ppm) in the tamarugite-hexahydrite-pickeringite zone. Metavoltine-boussingaultite aggregates contain 120 ppm, the local butlerite-rich veins have 78 ppm, but much less is found in kremersite-rich aggregates (18 ppm).

Copper

Only a single analysis of spinel (ca. 50/50 magnesioferrite-spinel with minor harmunite, krotite, and ringwoodite equivalents; and even less of qandilite, galaxite, and jacobsite end members) from the Bytom slag has a positive determination of Cu, with 0.49 wt.% CuO (i.e., 0.01 apfu Cu). Much less is occasionally found in magnesioferrite of some Rydułtowy slags (maximum of 0.18 wt.% CuO). Quite exceptional Cu enrichment in spinels denote also magnesioferrite, from the gehlenite-rich slag of Rydułtowy; half of the analytical set has positive determination, CuO ranging from 0.13 to 0.27 wt.%. One of 9 analyses of the Piekary slag pyrrhotite shows 0.23 wt.% Cu. Two of 30 trace-mode glass analyses, Czerwionka black-block metapelite, have copper (287 and 693 ppm Cu).

Relatively high amounts of copper are found in native iron from Nowa Ruda (see Table 19.3.16), with mean at 0.98 and maximum at 1.48 wt.%. Much less (as opposed to other elements described) is found in the corresponding material from the Ludwikowice white-to-black buchite: 0.31 wt.%. Pyrrhotite from nearby Ludwikowice buchite has 0.35 wt.% on average, with lesser amount (0.13) present in the Fe_6Si phase (see Table 19.3.17). Some analyses of iron silicides, namely the Fe_3Si_{1-x} phase, show copper presence (mean of 1.82 wt.%, $n = 2$).

The largest Cu concentrations among the EMM samples were found in the case of Radlin K–Al-rich sulfate crust (slightly above 100 ppm, although alternative value from another laboratory shows 67 ppm as the maximum). Associated anhydrite-rostite-alum accumulations are less enriched. A value close to the previous one also denotes the Mg-rich sulfate crust of Radlin. However, the largest amount (>200 ppm) was found in coal from Pszów. Interestingly, slightly elevated Cu content, comparable to that in the Rydułtowy sulfate crust, was found in both native sulfur and surrounding sulfur-bearing shale from Katowice-Wełnowiec. Kremersite from Radlin has only 33 ppm Cu. The mean content in ammonioalunite-ammoniojarosite of Czerwionka is even lower, at 6.5 ppm. As opposite to nickel and cobalt, the largest Cu concentrations in the Czerwionka supergene zone concern butlerite-rich veins, which have 340 ppm Cu. Three times less is found in metavoltine-boussingaultite assemblage.

Copper may also exist as its own species. We know this thanks to a recent find of Wojciech Sierny made within a once hot zone in Radlin. Samples collected were shown to be composed of chalcanthite and the very rare dolerophanite (first find in Poland), with possible occurrence of chalcocyanite and tenorite.

Zinc

As expected from the literature data on the USCB coals geochemistry, zinc is quite rich in some of the BCMH waste materials. However, this element is usually not encountered when analyzing the pyrometamorphic products. There are, however, three exceptions. The first is periclase-wüstite solid solution from the Piekary-area slag, described by Kruszewski (2006) and mentioned in Ciesielczuk et al. (2015); this phase is constantly enriched in zinc, with mean 1.36 wt.% ZnO (i.e., 0.01 apfu Zn). Coexisting are Zn-bearing sulfides, forming tiny hexagonal plates, up to ~10 μm in diameter. The mean empirical formula ($n = 3$) is $(Zn_{0.60}Fe_{0.33}Ca_{0.04}Mn_{0.01})_{\Sigma 0.98}S_{0.99}$; some crystals may likely bear trace Cl. The mineral almost certainly represents the wurtzite-buseckite solid solution. The third example may be hercynitic spinel (powder X-ray diffraction fit) from the Rydułtowy buchite; a preliminary inductively coupled plasma (ICP) analysis of this rock has shown a large amount of Zn; although not necessarily confirmed by the further ICP analysis, the element is supposed to be concentrated in the mentioned spinel phase as a minor gahnite substitution. The last example denotes glasses; a single analysis of the red Rydułtowy buchite gave 0.12 wt.% ZnO. Some analyses of the red metapelite of Przygórze may have ZnO peaking at 0.22 wt.% (see Table 19.3.15). A separate case is the Przygórze slag gehlenite, with just 2 analyses having Zn above the detection limit (see Table 19.3.10).

Zinc enrichment in EMM samples is, in general, larger than that in previous elements. Almost 800 ppm were found in recrystallized samples coming from the Mg-sulfate-rich crust of Radlin, originally containing up to ca. 230 ppm Zn. A very similar amount was found in local steklite-rich sulfate crust (although initially this sample was shown to contain maximum of 69 ppm Zn), virtually devoid of Mg-bearing sulfates. In this zone, an interesting unnamed $(NH_4)_3ZnCl_5$ phase, also rich in manganese, copper, and cobalt, was found (Kruszewski, 2012; IMA proposal in preparation). The phase associates with kremersite (which has 53 ppm Zn), scacchiite ($MnCl_2$), and most likely belloite, Cu(OH)Cl.

The Rydułtowy sulfate crust contains less Zn. Like copper, zinc was also found clearly enriched in native sulfur of Katowice-Wełnowiec (twice as much as in the surrounding sulfur-bearing shale). Even larger amounts of Zn concern native sulfur associated with the Rydułtowy sulfate crust. An important Zn concentrator seemed to be mascagnite, as confirmed in both Rydułtowy and Rybnik-Rymer exhalative accumulations. The latter case also concerns koktaite-rich flues. The mean content in ammonioalunite-ammoniojarosite of Czerwionka is 13 ppm. Much more is found in the tamarugite-rich efflorescences, the amount being 220 ppm, and the element most likely entering hexahydrite. The amount in the local metavoltine-boussingaultite aggregates is 54 ppm.

Gallium

This rare and precious element is a ubiquitous admixture in many pyrometamorphic minerals analyzed, like cobalt. Probably the most unique and rich source of this element was first mentioned by Kruszewski (2006) and concerns

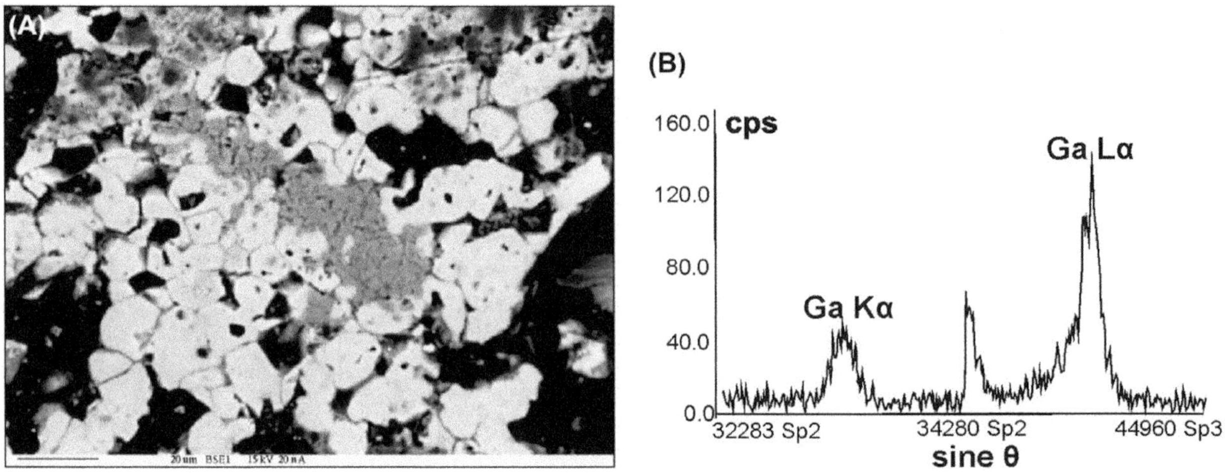

Figure 19.3.2. Secondary-electron (SEM) image of BSE (back-scattered-electron)-darker Ga-rich (mayenite group member) among BSE-bright crystals of oldhamite (Piekary area); cps = counts per second. The scale bar is 20 μm.

oldhamite-rich zones of the Piekary slag, which also contains unique **germaniferous cuspidine** described later. Gallium in this sample is present in an oxyhalide phase, likely of the mayenite supergroup (Figure 19.3.2). Analysis of the gallium-richest point (element wt.% basis: 3.17 wt.% Ga, corresponding to 4.26 wt.% Ga_2O_3) recast based on 26 total cations gives an atypical formula of $Ca_{12.00}(Al_{11.14}Ca_{1.30}Ga_{0.71}Si_{0.47}Fe_{0.16}Mg_{0.08})_{\Sigma 13.86}(O_{30.46}Cl_{0.71}F_{0.70}S_{0.13})_{\Sigma 32.00}$ $[\{x\}_{3.91}\ (Cl_{1.56}F_{0.53})_{\Sigma 2.09}]$, with x being either [] (vacancy) or H_2O. Recalculation of the obtained element wt.% to oxide wt.% gives very good totals (around 100 wt.%) for all the analyses (Table 19.3.20), which suggests that two of three points of the Si-low phase do not bear water. The final empirical formula (n = 3) is $Ca_{12.00}(Al_{11.97}Ca_{0.65}Ga_{0.58}Si_{0.44}Mg_{0.19}Fe_{0.15}Ti_{0.01})_{\Sigma 13.99}(O_{31.22}F_{0.46}Cl_{0.21}S_{0.11})_{\Sigma\ 32.00}\{([]_{3.10}(H_2O)_{0.75})Cl_{2.15}\}$. These formulas are correct assuming (1) Ca, Si, Fe, and Mg may enter the Al sites; (2) some of the excess Cl, F, and S may enter the O sites; (3) S is present as S^{2-} anions (as in oldhamite); and (4) S does not come from submicron inclusions of oldhamite, which seems to be the parent phase. The analysis (4), corresponding to a point of the same back scattered electron (BSE) brightness, recast to mayenite-type stoichiometry gives the formula $Ca_{12.00}(Al_{12.36}Ca_{0.30}Ga_{0.51}Si_{0.42}Fe_{0.15}Mg_{0.25}Ti_{0.01\ \Sigma\ 14.00}(O_{31.57}Cl_{0.26}F_{0.07}S_{0.10})_{\Sigma\ 32.00}\{[]_{3.72}Cl_{2.28}]$; we should be careful due to the high total of ~104%. Although further studies are needed for confirmation, the given formulas suggest possible complexity of the mayenite-type structures.

Gallium may be found in hematite, as in the case of some crystals from the Rydułtowy parabasalt (see Table 19.3.8); the enrichment level is quite constant, as the range is 150–210 ppm Ga. The element can also enter srebrodolskite; as shown in Tables 19.3.13 and 19.3.14, both the Rydułtowy and Bytom crystals contain Ga. The corresponding mean values for crystals of two slag types from Rydułtowy are 407 and 462 ppm, while that for the Bytom material is 475 ppm (n = 8). Gallium is a common trace element in this mineral as cobalt is, at least in the Rydułtowy case. The only known example of Ga entering the spinel structure is magnesioferrite-spinel from the Bytom slag (see Table 19.3.9). Unfortunately, this is only a single-analysis case, as shown in Table 19.3.9. Also, only a few siderite analyses concerning the Czerwionka proto-slag (see Table 19.3.18) have Ga over the detection limit; the amount is lower than the previously reported ones.

An important gallium and antimony enrichment concerns the Przygórze slag gehlenite-åkermanite-alumoåkermanite (see Table 19.3.10). Gallium occurs in a rather wide range, 98–365 ppm, but is usually well above the detection limit. Another melilite-group sample containing gallium comes from Bytom (see Table 19.3.11). In this case, gallium is also usually well above the detection limit, but the maximum amount is even larger, reaching 572 ppm. This Bytom melilite gallium anomaly coincides with small germanium anomaly. Another silicate that may sometimes immobilize gallium is indialite. This is, however, a single case of crystals coming from the Rydułtowy parabasalt (Table 19.3.21). Five out of 8 analyses have Ga, and the amount is relatively low but not very variable, being in the 100–220 ppm Ga range. Gallium may also sometimes be observed enriched in plagioclase. A single case, again, has to do with anorthite from the Bytom slag (Table 19.3.22); the maximum is 301 ppm metal. An essential Ga carrier is also the oxide-sulfate mineral ye'elimite, $Ca_4Al_6(SO_4)O_{12}$. Crystals from the Bytom slag contain up to 580 ppm Ga (Table 19.3.23). Small amounts of Ga were also found, thanks to 6 of 30 trace-mode analyses, in the Czerwionka black-block metapelite glass; the peak accumulation is 275 ppm, while the mean is 182 ppm Ga.

Table 19.3.20
Results of microchemical analyses (wt.%) of a potential gallium-rich member of the mayenite supergroup from the Piekary slag (USCB, Poland).

	1	2	3	4
SiO_2*	1.56	1.82	15.66	1.82
TiO_2	0.07	0.03	0.38	0.00
Fe_2O_3	0.62	0.82	0.46	0.94
Al_2O_3	42.89	36.07	21.60	40.95
Ga_2O_3	3.04	4.26	2.76	3.35
MgO	0.05	0.20	0.07	1.31
CaO	45.66	47.38	52.19	46.08
S	0.19	0.26	0.40	0.24
F	0.19	1.48	7.30	0.00
Cl	5.80	5.12	3.81	6.20
Σ	100.05	97.44	104.62	100.88
H_2O†	0.00	2.25	0.00	0.00
apfu (26 metallic cations basis)				
Si	0.39	0.48	4.69	0.45
Ti	0.01	0.01	0.09	
Fe^{3+}	0.12	0.16	0.10	0.18
Al	12.69	11.19	3.81	12.03
Ga	0.49	0.72	0.53	0.54
Mg	0.02	0.08	0.03	0.49
Ca	12.28	13.36	16.75	12.31
S	0.09	0.13	0.22	1.93
F	0.15	1.23	6.92	0.00
Cl	2.47	2.28	1.93	2.62
H_2O		2.25		
[]‡	3.64	1.47	0.59	3.86

*K, Mn, Na, and Pb were below the detection limit.
†By difference.
‡Vacancy.

Table 19.3.21
Results of microchemical trace-element analyses of indialite-ferroindialite† from the Rydułtowy parabasalt, Poland.*

	1	2	3	4	5	6	7	8
ppm								
V‡		120						
Cr			130					
Ga	220		120		130		130	100

†Some crystals analyzed may be a cordierite-sekaninaite solid solution.
*Beam current was 100 nA.
‡Zn, Cu, Ni, Co, Ge, and Pb were below the detection limits.

Table 19.3.22
Results of microchemical trace-element analyses of anorthite from the Bytom slag, Poland.*

	1	2	3	4	5	6	7	8	9	10	11	12†
ppm												
Ga		261		210				189				301
Ba	1059	847	658	1004	1642	739	394	380	333	359	859	456

*Beam current was 50 nA.
†All 12 analyses showed the presence of these elements. Cs, Ge, Sr, and Zn were below the detection limits.

Table 19.3.23
Results of microchemical trace-element analyses of ye'elimite from the Bytom slag, Poland.*

	1	2	3	4	5	6†
ppm						
Co					246	
Ga	273	385	580	272		463

*Beam current was 60 nA.
†6 of total 9 analyses showed the presence of these trace elements.

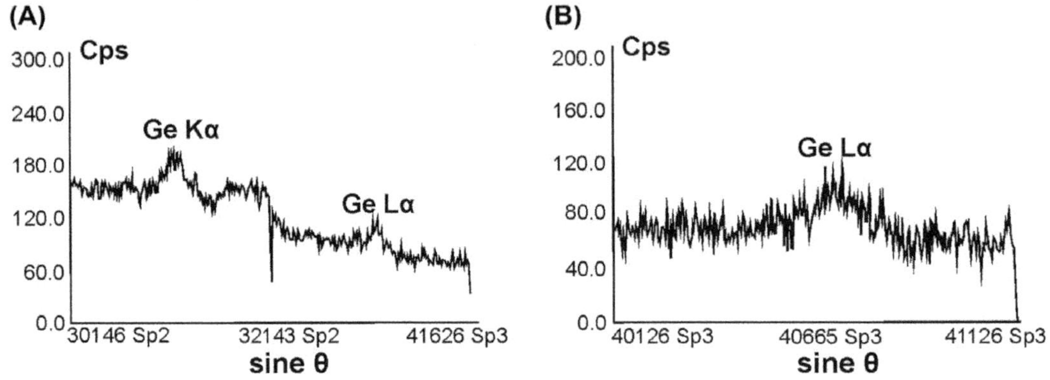

Figure 19.3.3. WDS detectors (A=TAP and B=LIF detector crystals) scans confirming the presence of germanium in cuspidine from the Piekary area. cps = counts per second.

The single metallic iron-like fragment from the Ludwikowice white-to-black buchite has 0.11 wt.% Ga. Gallium was not found in any of the analyzed hematites or spinels, but 3 of 8 analyses of siderite from Czerwionka (a siderite-hematite proto-slag) show Ga analyzed in trace mode, with the maximum at 290 ppm metal (see Table 19.3.18).

For Cr and V, an important Ga concentrator is alunite (Radlin material, exhalative origin, ca. 70 ppm). A slightly elevated amount, although ~6 times less, denotes both flue-associated clays (Rydułtowy) and Hg-rich calcite-anhydrite-gypsum-mogánite accumulations (Rybnik-Rymer). The earlier alunite enrichment is > 100 times the mean amount found for local coals and 25 times the maximum found in local shales.

Germanium

The search for germanium enrichment in the pyrometamorphic minerals was somewhat disappointing. However, an important find, **germaniferous cuspidine** (likely the first natural occurrence of such species), was made when analyzing the Piekary slag. This sample was a choice due to four facts: (1) the preliminary ICP analysis has shown slightly elevated Ge, (2) the F-rich mineral cuspidine is present in this sample, (3) fluorine is an important element in coal-fire gas jets and thus most likely acting in the (semi)metal gas transport (Sokol et al., 2005), and finally (4) Ge compounds like fluoride(s) are commonly known to be very volatile. Indeed, the K and L lines of germanium were found in the WDS scans, as shown in Figure 19.3.3. Trace element analyses (Table 19.3.26) determined very large concentrations of this precious semimetal, peaking at 941 ppm (0.14 wt.% GeO_2). The mean Ge content is 895 ppm (crystal no. 5) and 761 ppm (whole range). The cuspidine crystals differ in Ge enrichment, with 18% difference between the lowest and highest values. Unfortunately, the Ge presence was not observed in any other cuspidine-bearing samples. Small germanium anomaly concerns Ga-rich gehlenite from the Przygórze slag (see Table 19.3.10); although only 3 analyses have Ge above the detection limit, the amount is relatively similar, that is, not crossing 275 ppm Ge (Table 19.3.24).

As opposed to these finds, germanium enrichment does not concern any of the analyzed EMM mixtures. The larger amounts (mean of 0.60 ppm Ge) were found in gypsum- and koktaite-rich, sulfur-bearing flue in Rybnik-Rymer. A similar amount concerns the mentioned Hg-rich calcite- and mogánite-bearing accumulations from the same site. Sulfur arising near the Rydułtowy sulfate crust is also slightly enriched in Ge, but the value obtained is essentially identical to that found in pyrite from Pszów.

Table 19.3.24
Results of microchemical trace analyses of germanium in cuspidine from the Piekary Śląskie slag, USCB, Poland.

	1	2	3	4	5	6	7	8
Ge (ppm)	465*	929	174	945	793	926	918	941†
Beam current (nA)	40	100	100	100	100	100	100	100
Crystal No.	1	2	3	4	5	5	5	5

*Corresponds to 0.068 wt.% GeO_2.
†Corresponds to 0.136 wt.% GeO_2.

Table 19.3.25
Results of microchemical trace-element analyses of fluorapatite from the Przygórze slag, Poland.*

	1	2	3	4	5	6	7	8
ppm								
Y†	1320	640	1930	450	210	230	190	610
Sr	2460	3470	2950	4010		310		280
As					1550	910	300	

*Beam current was 50 nA.
†There was only a single positive determination of V, Ce, and Cl.

Arsenic

Some apatite-group species analyzed contain trace but measurable arsenic. Trace-mode EPMA of the Przygórze slag fluorapatite have positive results in 38% of the analyses, with mean being 0.28 wt.% element. An exception is the vanadiferous member from the Bytom slag (see Table 19.3.3), which is constantly enriched in As, with maximum 1.05 wt.% As_2O_5 (mean is also high, at 0.96 wt.%). A single analysis made on the Rydułtowy slag material shows 0.23 wt.% As_2O_5. Arsenic was neither found in the BK8 sample fluorellestadite form Przygórze nor the Czerwionka whitlockite. Relatively large but occasional and very variable enrichment concerns fluorapatite from the Przygórze slag, with maximum of 1550 ppm As determined (Table 19.3.25). As an arsenic- and zinc-rich feldspar (filatovite) was described quite recently, the elements were added to analytical trace-mode set of celsian from the Czerwionka black-block metapelite; they were always below their detection limits.

Arsenic content in murashkoite from the Czerwionka black-block metapelite is constant, being 0.01 apfu (Ciesielczuk et al., 2015), corresponding to maximum 1.12 wt.% As. On the contrary, only a single and trace-mode analysis of closely associated pyrite has positive determination with 575 ppm As. The coexisting graftonite/sarcopside does not readily catch the element; only a single analysis has shown 0.14 wt.% As_2O_5. Two of 9 analyses of pyrrhotite from the metapelite of the same locality show arsenic (maximum is 0.08 wt.% As). Another reduced As phase is a supposed arsenopyrite, found as a single ca. 8 μm crystal in the Łaziska mullite-rich clinker. A single-case enrichment denotes an iron phosphide of the reduced assemblage from Ludwikowice buchites (Table 19.3.17); the mineral catches 0.13 wt.% of the element. The Nowa Ruda native iron has no elevated As.

A totally different level of As enrichment concerns the unique exhalative zone found in Radlin (Kruszewski, 2012; Fabiańska et al., 2013); a single though rather large (few cm) aggregate of skeletal As-rich sulfur (possibly the so-called arsensulfurite) was found there, in association with As-devoid sulfur and ammonium fluorosilicates.

The largest As anomaly among the EMM samples, with 200 ppm of the element, concerns the steklite-bearing crusts of Radlin. Kremersite, once existing in their vicinity, has only 40 ppm. About 27 ppm As was determined in the gypsum-rich outer part of the otherwise Mg-rich sulfate crust from the same heap. This low value is very similar to that found in the Pszów pyrite. The other known sulfate crust contains less As. Even less As (ca. 15 ppm) is found in gypsum-bearing exhalative EMM (both the koktaite-bearing and the Hg-rich mogánite-bearing) of

Rybnik-Rymer. All the analyzed native sulfur samples are low in As, with amounts not exceeding 4 ppm. No arsenic was found in ammonioalunite-ammoniojarosite from Czerwionka. On the contrary, the tamarugite-rich assemblage of this heap has 140 ppm, with much less (21 ppm) in metavoltine-boussingaultite.

Selenium

Although selenium analyses were done in oldhamite and periclase from the Piekary heap Se was not found. Thus the only known Se behavior concerns the coal-fire gas flues. Native selenium was not found in any of the BCMH under scope. However, quite large Se enrichment, > 100 ppm Se, was found for sulfur associated with the Rydułtowy sulfate crust. Interestingly, a native sulfur sample from the same heap contains ca. 50 times less Se, and the same is for the Pszów samples, while the Wełnowiec one has only 13 ppm. Twice this value was found in supergene thénardite-löweite accumulations once existing in the Rybnik-Rymer heap.

Bromine and Iodine

The two heavy halogens are known to be exclusively concentrated in some halide minerals of the BCMH studied. As already shown by Kruszewski (2012), this concentration usually concerns salammoniac. Semiquantitative analyses of abundant, multicolor (orange to colorless), skeletal material from a single site in Radlin (association: sulfur, hydrokenoralstonite) show maximum 4.68 wt.% Br, corresponding to 0.08 apfu Br. The corresponding values for iodine are 2.01 wt.% and 0.02 apfu. These values concern the orange-colored part, which was later shown to contain relatively large amounts of Tl, Br, and Sn. It is to be confirmed, if the Tl and Br(I) concentrations may be referred to inclusions of a Tl mineral, possibly similar to natalyamalikite; the high-temperature form of thallium iodide, which is stable at room temperature if substituted, is red in color (Saima Sultana, 2009). Also, tiny crystals of most likely thallium chloride (possibly lafossaite) were found in Radlin (Figure 19.3.4) Bromine- and most likely indium-enriched Sn chloride, possibly panichiite, is known from the same BCMH (Kruszewski, 2012; Figure 19.3.5). Kremersite from a separate flue of the same heap may also contain some small amounts of Br (single analysis, 0.64 wt.% Br). The associated salammoniac may contain as much as 1.28 wt.%. Analyses of both species gave negative results for iodine. Single crystals of salammoniac from Rydułtowy (sulfate crust area) may bear even more Br.

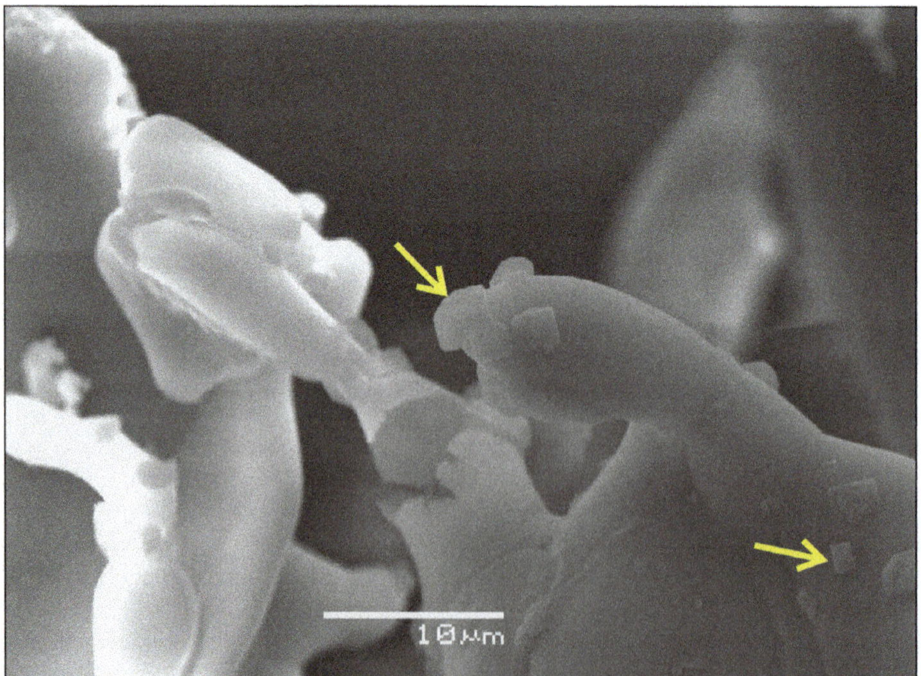

Figure 19.3.4. Salammoniac crystals overlain by minute isometric crystals (arrows) of some variety of thallium chloride, possibly lafossaite (SEM image, Radlin).

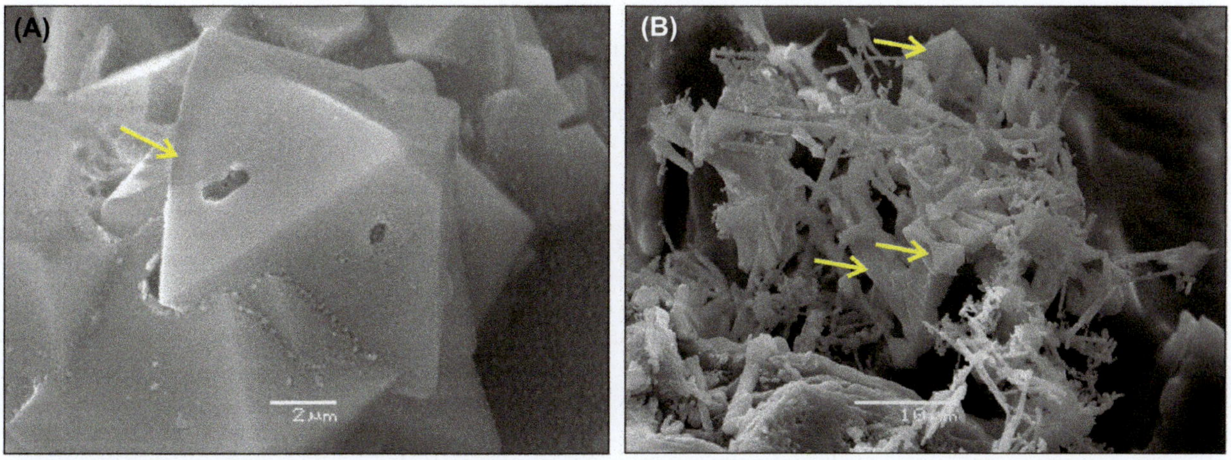

Figure 19.3.5. Bromine- and iodine-bearing Sn (In) chloride (panichiite?) crystals (arrows) overlying salammoniac. (A) Single isometric crystal. (B) Isometric crystals nucleated with other crystals that are likely of the same chemical composition. (SEM image, Radlin).

Rubidium

Rubidium was only measured in feldspars. Three of 10 conducted analyses of K-feldspar ($Kfs_{89}Ab_{11}$) of the red-violet metapelite from Przygórze have Rb_2O, with a concentration in the 0.11–0.15 wt.% range. Sanidine ($Kfs_{69}Ab_{30}An_1$) found as a separate ~2 cm aggregate on a clinker from Rydułtowy has Rb_2O positively determined in 6 of the entire 17 analyses, with maximum of 0.14 wt.%, while the associated plagioclase ($Ab_{73}An_{21}Kfs_6$) positively determined 7 analyses of the entire 26-analyses range, with mean 0.11 wt.% oxide.

Gas-related alunite accumulations from Radlin (see Cr, V, and Ga) again stand for the most Rb-rich EMM sample, with 283 ppm. Over twice less is found in voltaite- and alum-(K)-bearing secondary accumulations of the same area and the Mg-rich sulfate crust of this heap. The Radlin tschermigite is also slightly enriched in this alkaline element. Slight lower enrichment degree denotes the Rydułtowy sulfate crust.

Strontium

Apatite-group minerals are the most important sink of strontium in the heaps waste material (see Tables 19.3.2 and 19.3.25). SrO content in the Przygórze slag fluorapatite crystals varies from 0.12 to 0.20 (series 1) or 0.14 to 0.29 (series 2); third series shows 0.08–0.62 variation (mean, $n = 19$, is 0.32 wt.%). Much lower concentrations and rarer enrichment case denote the BK8 hydrothermalite: 0.00–0.14 wt.% SrO. Trace EPMA analyses of the fluorapatite-fluorellestadite solution conducted on the Rydułtowy slags material (see Table 19.3.2) show continuous Sr enrichment at least in one of two rock types, with mean 0.14 wt.% Sr (beam current 50 nA) or 0.15 (beam current 70 nA) in the trace-mode analyses. The largest amount in the Rydułtowy case concerns 0.49 wt.% Sr. Mean content in hydroxylapatite-rich Si- and S-low member of the clinopyroxene-bearing slag (Rydułtowy) has a maximum of 0.25 wt.% SrO (mean is 0.17). Similar trace-mode research on the Przygórze slag (Table 19.3.25) gave much higher mean of 1.69 wt.% metal, with 75% of the analyses giving positive determination.

Quite unique Sr–Ba geochemistry concerns a yellow metapelite from the Jankowice breccia (see the Barium section for more details). It contains a celsian-hyalophane solid solution from 1.61 wt.% (0.05 apfu) SrO, on average. Interestingly, it coexists with an unnamed $(Ba,Sr,Ca,Fe,Mg)_2(Si,Al)_3O_8$ phase, which is much more Sr-enriched, with corresponding means being 8.05 wt.% and 0.34 apfu. More detailed study has shown three additional phases in this association: (1) celsian-slawsonite solid solution, with a mean 8.87 SrO (or 0.36 apfu Sr); (2) also celsian-slawsonite solid solution, with mean 3.31 wt.% SrO (up to 8.49 wt.% SrO, corresponding to 0.32 apfu Sr); and (3) strontian sanbornite, with mean 6.71 wt.% SrO. As compared to the Jankowice material, the large-crystalline celsian from the Czerwionka black-block metapelite, when traced for Sr (4 analyses), has shown 0.08 to 0.14 wt.% Sr.

Strontium may also enter the melilite-group structures. Such behavior is observed in case of both the Przygórze and Bytom slags. In the first case, one-third of the gehlenite analyses has some Sr, the average oxide content clustering around 0.22 wt.%. The associated anhydrite is devoid of Sr. The second-case gehlenite-åkermanite-ferrigehlenite-alumoåkermanite characterizes in mean strontium concentration of 0.22 wt.% SrO only in 6 of the whole 26 analyses. The only more Sr-rich phases noticed in some slags are strontian baryte (see the Barium section), and a single, 5 μm crystal of celestite, found in the Rydułtowy melilitic slag.

As expected, some strontium is also present in carbonate minerals standing most likely for retrograde-metamorphic components of most slags. In case of the Srebrodolskite- and fluorapatite-rich slag of Rydułtowy, the element was positively determined in 5 of a total 8 analyses of a phase corresponding to bassanite, with the content varying from 0.00 to 0.67 wt.% SrO. The element was not found in magnesite of this sample. One of 4 analyses of fluorite from the fluorellestadite-bearing slag of Rydułtowy has 0.33 wt.% Sr attributable to trace substitution of the strontiofluorite component. This phenomenon is not observed in fluorite of the Przygórze hydrothermal veins. No barium or lead was found in any of the fluorites analyzed.

Following these concentration trends (sulfates and carbonates), the largest amounts of Sr found in bulk EMM samples concern gypsum-dominated accumulations. A single sample from Pszów has >940 ppm, but much lower (even 8 times) in other samples from this heap. The most Sr-rich gypsum of Radlin has almost 580 ppm, while in Rybnik it is 364 ppm. Calcium-rich accumulations of the same heap have 481 ppm. Alunite and alunite-rich shales are also slightly enriched in strontium (but<200 ppm). Alunite-group accumulations from Czerwionka contain 45 ppm Sr on average. The coal content is very variable, with maximum 235 ppm in the Rydułtowy rock. Very large concentrations of Sr, 2000 ppm, were found during initial ICP session (a different laboratory than the EMM batch) in the butlerite-rich sample from Czerwionka, which also contains some gypsum. Approximately half of this value is reported for metavoltine-boussingaultite assemblage.

Yttrium

Yttrium is not a common trace element in the BCMH. As in case of strontium, it is usually expected to be found in the apatite-group minerals. Indeed, the trace-mode EPMA on the Przygórze slag crystals always show Y (Table 19.3.25), but the amounts vary from 0.02 to 0.19 wt.% metal (0.07 wt.% on average). On the contrary, the Rydułtowy material usually does not show Y enrichment. Fluorite from Piekary may contain 0.12–0.16 wt.% Y, but on the other hand Y is not observed in 50% of the analyses conducted. No yttrium was detected in the hydrothermal-vein Przygórze material. Some yttrium may be found in perovskite; the BK8 mineral has 0.00–0.06 wt.% Y_2O_3. A much higher amount concerns the Bytom breccia mineral, with mean 0.27 wt.% (but possibly as much as 0.59). The associated $(Mg,Ca)_3Fe(Al,Ti)_2TiO_9$ phase contains 0.13 wt.% Y.

For EMM, the largest amount of Y (>70 ppm) was found in supergene Mg-rich sulfates from Radlin. The second concentrator may be gypsum: values around 30 ppm are noticed in a few samples from Pszów and Radlin. Anhydrite-rich secondary accumulations associated with Radlin steklite- and millosevichite-rich sulfate crust bear similar amounts.

Zirconium

Some unidentified Ca–Zr phases are observed in two types of slags. In the case of gehlenite-rich slag from Rydułtowy, the BSE-bright phase associated with melilite, anorthite, and aragonite (Figure 19.3.6A) seem to fit to either baghdadite or dovyrenite. The first phase would possibly fit, as it belongs to the cuspidine group, and cuspidine is a relatively common constituent of slags in general. Dovyrenite-like phase (also known from a contact-metamorphic site) was noticed in a slag from Przygórze (Figure 19.3.6B). On the other hand, the only confirmed Zr concentrations in pyrometamorphic minerals of the BCMH under scope concern perovskite, which is suggestible for the existence of perovskite-lakargiite solid solution, the latter end-member also described from the contact-metamorphic environment. Returning to perovskite, the Przygórze one may incidentally show as much as 0.18 wt.% ZrO_2, while the mean for the one of the BK8 hydrothermal sample is 0.30 wt.% (see Table 19.3.7). More elevated concentration of 3.17 wt.% was found in the rutile-like Ti–Zr–Fe–Al–Si oxide from the WH breccia-included porcellanite from Przygórze.

The Rydułtowy and Radlin sulfate crust (steklite-millosevichite in the latter case) bear the largest amount of Zr, being ca. 50 and 20 ppm, respectively. Gypsum-bearing samples from Pszów (a single case) and Rybnik-Rymer (the calcite-rich one) contain ca. 16 ppm Zr. Similar amount concerns the Radlin Mg-rich sulfate crust.

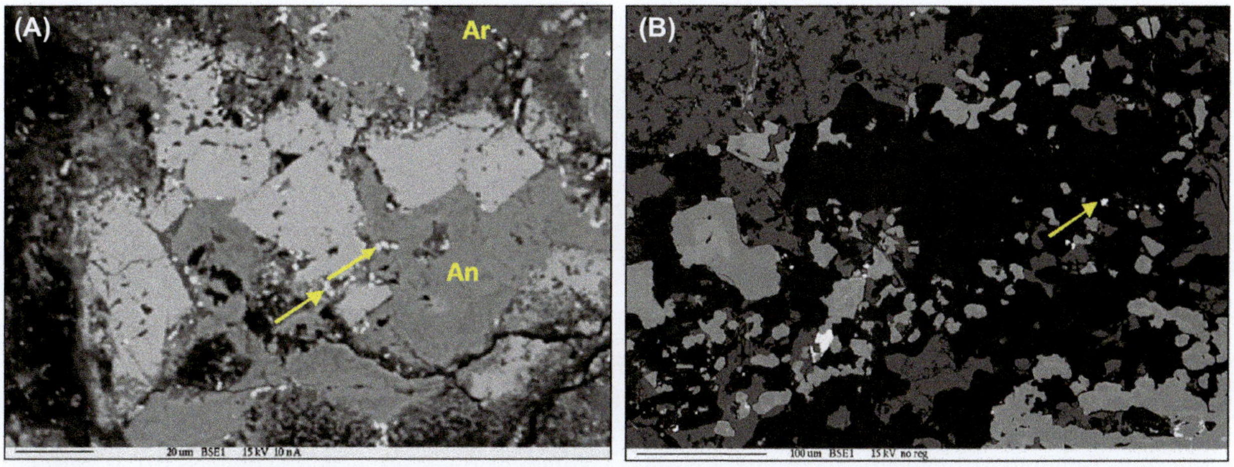

Figure 19.3.6. BSE-bright Zr-rich phases (possibly baghdadite and/or dovyrenite) in slags. (A) Rydułtowy (gehlenite-rich slag). (B) Bytom (gehlenite-rich slag). The scale bars in (A) and (B) are 50 and 100 μm, respectively.

The obvious primary Zr mineral, zircon, is quite often found in some metapelites, a possible black-block organic-rich metapelite of Przygórze; mean composition of this mineral is $(Zr_{1.00}Hf_{0.01})_{\Sigma 1.01}Si_{0.98}O_{4.00}$; 2 of 5 analyses show Y_2O_3 (0.08 wt.% on average), and 1 analysis shows 0.13 wt.% Yb_2O_3.

Niobium and Tantalum

As expected, the main carriers of Nb in the BCMH are titanium oxides like perovskite and rutile. Crystals from the Przygórze slag were shown to be constantly enriched in Nb_2O_5, with maximum concentration being 0.40 wt.% (see Table 19.3.7). The corresponding value of the BK8 sample is 0.28 wt.%, and that of the Bytom breccia is 0.34 wt.% on average (and possibly as high as 0.76 wt.%). The unnamed $(Mg,Ca)_3Fe(Al,Ti)_2TiO_9$ phase contains 0.21 wt.% Nb_2O_5. Anatase of the Przygórze red-violet metapelite contains 0.14 wt.% on average (see Table 19.3.5). More niobium (0.21 wt.% oxide) enters the structure of a rutile-like Ti–Zr–Fe–Al–Si oxide from the WH porcellanite. An interesting unnamed phase was found associated with a more-or-less Nb-enriched TiO_2 polymorph within the Przygórze slag; the ideal formula is $(Ca,Mg)Fe_2Ti_2O_7$, while the empirical formula ($n = 1$) is $(Ca_{0.52}Mg_{0.47}Mn_{0.02})_{\Sigma 1.01}(Fe_{1.96}Mn_{0.04})_{\Sigma 2.00}(Ti_{1.95}Al_{0.02}Nb_{0.01}Si_{0.01})_{\Sigma 1.99}O_{6.98}$, which corresponds to 0.29 wt.% Nb_2O_5.

Some Nb anomalies are observed also within the EMM, but with maximum 4 ppm Nb (anhydrite-rostite-alum-voltaite assemblage). Three ppm Nb are found in the Rydułtowy sulfate crust, which is still nine times more than the maximum concentration found in primary waste rocks. The gypsum-rich inner flue deposits in Rybnik have mean 1.6 ppm niobium. As opposed to this element, tantalum enrichment was not observed either during EPMA sessions or among the ICP analyses results, the latter always showing concentrations below 10 ppm. Although measured, tantalum was not detected in any perovskite crystal.

Molybdenum

Although molybdenum species like ilsemannite ($Mo_3O_8 \cdot nH_2O$) are known from some of the world BCMH (Jírasek, 2001), these were not observed in Poland yet. Molybdenum was analyzed only within the EMM sample set. Values larger than those found in primary waste constituents, that is 0.98 ppm in pyrite and shale-maximum of 0.62 ppm (Pszów), and a mean 0.69 in coals, are scarce. The pyrite-like values denote (1) koktaite-rich outer zone of otherwise gypsum-dominant flue in Rybnik, and (2) the Radlin K–Al-rich sulfate crust. Slightly over twice as much (maximum), is for Ca- and Hg-rich mogánite-bearing efflorescences from Rybnik.

Platinum Group Elements

The PGEs were not measured directly, that is, in separate mineral species. However, some single anomalies, although far from those found in some special coals worldwide, were noticed in case of the EMM. Palladium and platinum were the only PGEs measured. Palladium is found usually in amounts below 10 ppb. Thirteen ppb is determined for the Ca- and Hg-rich gypsum-bearing accumulations from Rybnik, while 15 ppb is found in gypsum-rich outer zone of the Radlin Mg-rich sulfate crust. Platinum amount is usually below 2 ppb. The palladium anomaly of the last-reported gypsum zone coincides with >40 ppb Pt content, the largest observed. The first palladium anomaly (Rybnik) is also connected with Pt enrichment, but the content is much lower at 9 ppb. A similar amount is found in a single gypsum-rich sulfur-bearing sample from Pszów. Fifteen ppb Pt was determined for alunite from Rybnik. Slightly less (12 ppb) is found for gypsum- and halite-rich accumulations associating with hot salammoniac zone in the inner part of the Rybnik heap.

Silver

Silver, cadmium, and selenium were searched in pyrite from the Czerwionka black-block metapelite using WDS scans but were not observed. The largest amount of Ag in the primary waste rocks concerns 0.96 ppm found in shale from Pszów. Interestingly, the associated pyrite contains a half, which is a value smaller than that found in native sulfur from Wełnowiec (0.79 ppm). Interestingly, a value 3.5 times the shale one denotes colorless Br-, I-, Sn-, Tl-, and Pb-rich salammoniac from Radlin, but the orange zone is practically devoid of silver. It is not yet sure whether silver substitutes for the NH_4^+ cation or forms discrete inclusions of chlorargyrite, bromargyrite, and/or iodargyrite. This find seem to be negligible when compared with an enormous 10 ppm value obtained for the gypsum-rich inner part of accumulations from Rybnik, where koktaite was also found in the outer zone. Association of Ag with gypsum was also observed in a single sample from Pszów, but only after recrystallization; the recrystallized material, which contained alunogen and some NH_4-rich sulfates, has 6.42 ppm Ag, which is over 49 times the original amount in gypsum. This suggests the presence of pore-waters, which are relatively enriched in Ag.

Cadmium

As reported earlier, cadmium was not found in any of the microprobed minerals. The current description thus concerns only the bulk ICP analyses of the EMM samples. Not abundant in coal, shales, and pyrite, cadmium may undergo enrichment during coal-fire processes. The Cd amount, 84 ppm, found in anhydrite-rostite-alum-voltaite assemblage of the apical part of Radlin K–Al-rich sulfate crust is over 300 times the maximum observed coal content (0.27 ppm, Pszów); the element is supposed to be entering mainly anhydrite. Such large amounts were not observed in any other sample. The Czerwionka tamarugite-rich supergene assemblage has 6 ppm. Slightly over 4 ppm of Cd were found in gypsum-bearing samples from the Mg-rich sulfate crust of Radlin, but only after recrystallization (of the pore water). The same value concerns steklite-millosevichite crust (maximum) and the once-coexisting kremersite. A value close to 2 ppm concerns weathering Mg-rich sulfates of the same heap and the mentioned "raw" crust. Values over 1 ppm still occur in case of koktaite-rich part of a flue in Rybnik, recrystallized gypsum-rich sample from Pszów, and alunite-rich shale from Radlin. As in the case of arsenic, cadmium was not found enriched in alunite-group crusts from Czerwionka.

Indium

A single but interesting case of In enrichment in pyrometamorphic minerals concerns magnesioferrite from the Bytom slag (see Table 19.3.9). Although found in only 3 analyses, with maximum of 430 ppm, In was not expected to be found in this particular oxide. A search for indium in hematites did not bring similar results.

A rather large amount of 220 ppm In was found in tamarugite-hexahydrite-pickeringite-alum-bearing supergene assemblage found in Czerwionka; however, this value has limited credibility. On the other hand, a single BSE bright spot was once found in a sample of closely associated vein-forming butlerite-rich accumulations, showing trace indium in the energy dispersion spectroscopy (EDS) spectrum obtained. Such large values as given earlier were never found in any of the EMM samples analyzed using later ICP system. The maximum of 9.9 ppm concerns anhydrite-rostite-alum-voltaite assemblage of Radlin, with metal most likely entering the Al sulfates. Six ppm is reported for the thénardite-rich supergene sample from Rybnik-Rymer. The amount found for the godovikovite-rich core of the Rydułtowy sulfate crust is 0.13 ppm, which is still much larger (over two times) than the maximum found in local coals (0.05 ppm, Pszów). Even larger amount (ca. 0.30 ppm) denotes gypsum from the Mg-rich crust of Radlin. A similar amount characterizes the local alunite-rich shale. As reported in the Bromine section, indium may enter some Sn minerals (possibly panichiite in this case), if present.

Tin

A single Sn determination during EPMA sessions concerns quite pure magnesioferrite from the BK8 hydrothermal sample; the analysis shows 0.10 wt.% SnO_2. Tin amounts crossing the maximum shale (29 ppm) and pyrite-coal (mean 1 ppm) values are very rarely observed among the EMM samples. One exceptional case is the Br(I)-rich bismuth- and thallium-bearing salammoniac from Radlin and anhydrite-rich assemblage from the same heap, both containing >100 ppm Sn. The local alunite bears Sn in amount close to the shale maximum. Kremersite from the same zone has 34 ppm, but the local steklite-rich crust has only 18 ppm. Tin is somewhat enriched in the Wełnowiec sulfur, and when compared to the shale matrix, has a behavior like that of tungsten and gold. Other anomalies seem to be negligible.

Antimony

A single but interesting case of antimony immobilization in the neomineralization of the BCMH concerns gehlenite from the Przygórze slag (see Table 19.3.10). This gallium-bearing phase is also constantly enriched in Sb, with maximum and mean at 808 and 511 ppm, respectively. As compared to those values, the ones found during ICP session in both primary waste rocks and EMM samples of the USCB are extremely low. Still, the 6.7 ppm value of the anhydrite-rostite-alum-voltaite assemblage of Radlin is quite impressive in relation to the coal and shale maximum, namely 1.03 and 1.68 ppm (Pszów). However, as expected, a larger value is found in local pyrite (8.6 ppm). Antimony enrichment in relation to coal and shale also concerns the Ca- and Hg-rich EMM from Rybnik (over 3 times the coal value). Quite a large amount is also noticed in a second gypsum-rich flue deposit of this heap, with almost 3 ppm in the outer koktaite-rich zone.

Tellurium

Tellurium was not analyzed in any separate mineral species in terms of the EPMA method. An interesting, though single enrichment find concerns the Ca- and Hg-rich calcite-anhydrite-gypsum-mogánite assemblage from Rybnik; its bulk analysis has shown ca. 11 ppm Te, which is 135 times the maximum found in local barren rocks (coal from Pszów). Possibly nonaccidentally, a single other over-unit value is found in anhydrite-rich sulfate-crust-related accumulations from Radlin. The Radlin bromian salammoniac contains quite a large 0.55 ppm amount concentrated in the colorless zone. A gypsum-rich sample (not related to any of the sulfate crusts) and the alunite shale have a similar value with mean at 0.34 ppm Te. Native sulfur contains much less Te, with a maximum of 0.10 ppm found in Rydułtowy (sulfate crust zone).

Cesium

A sample cesium search was done on celsian from the Czerwionka black-block metapelite, showing the element being under the detection limits. The reference value for the following EMM samples analysis description is 8.2 ppm, which is the maximum value found in the primary waste rocks (coal, Pszów). Most of the EMM values are below this level. Quite an exception is twice that value found in tschermigite-gypsum flue deposits in Radlin. In this case, following Rb and likely K, Cs most likely substitutes the NH_4^+ cation of tschermigite. Very similar Cs enrichment level is found for anhydrite-rostite-alum-voltaite assemblage of Radlin. In this case cesium most likely enters the structures of both the alum-(K) and voltaite. It is also found, in a slightly lesser amount of 14 ppm, in the coexisting alunite.

Barium

The most Ba-rich phase known from the BCMH under scope is a phase giving EDS spectra resembling strontian baryte; this phase is a very minor constituent of some slags. A ca. 10-μm-sized aggregate found between fluorapatite-fluorellestadite and Srebrodolskite-brownmillerite in the Rydułtowy slag was analyzed, with a very low total and SO_3 amount obtained; strong instability of the mineral and the presence of trace sulfur is suggestible for a carbonate mineral. Analysis recast attempt gave a possible formula as $(Ba_{0.82}Ca_{0.16}Sr_{0.02})_{\Sigma 1.00}[(CO_3)_{0.90}(SO_4)_{0.10}]\cdot 3.60H_2O$. A similar phase, though Sr-free, was found in the Piekary slag in the form of a single, triangular crystal, almost 1 μm in diameter. A few baryte-like crystals were encountered among glass cenospheres within the Łaziska mullite-bearing clinker.

More often barium is found in silicate phases. Extreme Ba enrichment is reflected in the formation of celsian, which is typical for black-block metapelites and some parabasalts. The first case concerns both Ludwikowice and Czerwionka, with the mean BaO content being 38.58 wt.% ($n = 2$; corresponds to 90% of the celsian end-member with 3% of a hypothetical $BaFe_2Si_2O_8$ end-member) and 29.57 wt.% (n = 11; corresponds to mean 0.69 apfu Ba and 70% of the celsian end-member), respectively. Celsian crystals are shown in Figure 19.3.7. Atoll-like aggregates of celsian of the Czerwionka parabasalt (A) in Figure 19.3.7) contain 32.69 wt.% BaO (i.e., 0.77 apfu Ba) on average (Ciesielczuk et al., 2015). The mean BaO content in the associated hyalophane at Ludwikowice is 6.47 wt.%, while in the surrounding glass it is 0.33 wt.%. Glasses of other pyrometamorphic rocks are enriched in Ba: The content in glass of a multicolor buchite is 0.20 (mean) and 0.61 (average) wt.%. The mean content of Czerwionka rocks is 0.98 wt.% BaO, while the Rydułtowy buchite material may contain up to 0.22 wt.%. Extremely Ba-enriched glasses exist in the celsian-bearing zones of the black-block metapelite of Czerwionka; two analyses cluster around 8.5 and two others around 2.5 wt.% BaO. A glass or an unnamed phase of the possible $(Ba,Na,K)(Mg,Al)_2Al_4Si_8O_{27}$ composition is connected with up to 10.57 wt.% BaO enrichment. Four of 14 analyses of a K–Mg–Fe silicate or a glass phase from the red-violet Przygórze metapelite show Ba, with the maximum at 3.53 wt.%. The yellow clinker-like metapelite and a porcellanite from Przygórze may concentrate 0.73–0.76 wt.% BaO, but such results are based on single positive determinations.

The unique Ba–Sr mineralization from Jankowice was already mentioned in the Strontium section; this includes the first documented occurrence of the rare Ba mineral sanbornite in the BCMH environment; the empirical formula is $(Ba_{0.38}Ca_{0.21}Sr_{0.14}Fe_{0.09}Mg_{0.08}K_{0.01}Na_{0.01})_{\Sigma 0.92}(Si_{1.72}Al_{0.29}Ti_{0.09})_{\Sigma 2.07}O_{4.92}$, corresponding to mean 26.23 wt.% BaO. The celsian-slawsonite of this sample may be described as two mean formulas: (1) $(Ba_{1.11}Sr_{0.36}Ca_{0.22}Fe_{0.12}Mg_{0.09}Na_{0.02})_{\Sigma 1.92}(Si_{2.69}Al_{0.44}Ti_{0.05})_{\Sigma 2.09}O_{7.92}$, corresponding to mean 40.35 wt.% BaO; and (2) $(Ba_{0.87}Sr_{0.12}Ca_{0.10}Mg_{0.06}K_{0.01}Na_{0.01})_{\Sigma 1.20}(Si_{2.19}Al_{1.49}Fe_{0.11}Ti_{0.02})_{\Sigma 3.81}O_{7.95}$, corresponding to mean 35.60 wt.% BaO. The first-type phase may alternatively be attributed to an unnamed $Ba_3Sr(Ca,Mg,Fe)(Al,Fe)(Si,Al,Ti)_8O_{22}\cdot H_2O$ species.

Not only celsian-dominant feldspars contain Ba. It was also detected in 4 of 10 analyses conducted on K-feldspar of the red-violet metapelite from Przygórze; mean BaO content of these analyses is 0.24 wt.% (i.e., 0.004% of the celsian end-member), while the maximum is 0.82. Sanidine from Rydułtowy is usually devoid of Ba: only 1 of the whole 17 analyses has shown 0.21 wt.% BaO. The associated plagioclase has shown up to 0.12 wt.% in 7 of the whole 23 analyses conducted. Barium was also detected in all 12 trace-mode analyses of anorthite from the Bytom slag (see Table 19.3.22); the content of Ba is very variable, being in the 333–1642 ppm range.

Another silicate concentrating essential amounts of Ba is mica-group species from an organic-rich metapelite from Przygórze, possibly related to a black-block zone. The empirical formula of this mica is (n = 17) $(K_{0.84}Ba_{0.12}Na_{0.02}Ca_{0.02})_{\Sigma 1.00}(Mg_{2.81}Al_{0.11}Fe_{0.07})_{\Sigma 2.99}(Si_{2.81}Al_{1.19})_{\Sigma 4.00}O_{10.02}[F_{1.41}(OH)_{0.60}]_{\Sigma 2.01}$; it is thus fluorphlogopite-phlogopite-fluorokinoshitalite with minor kinoshitalite, fluoromuscovite (hypothetical end member), muscovite, annite, and fluorannite members: $Fph_{55}Phl_{23}Fks_8Ks_3Fmu_2Mu_1Fan_1An_1$. This composition corresponds to mean 4.10 wt.% BaO.

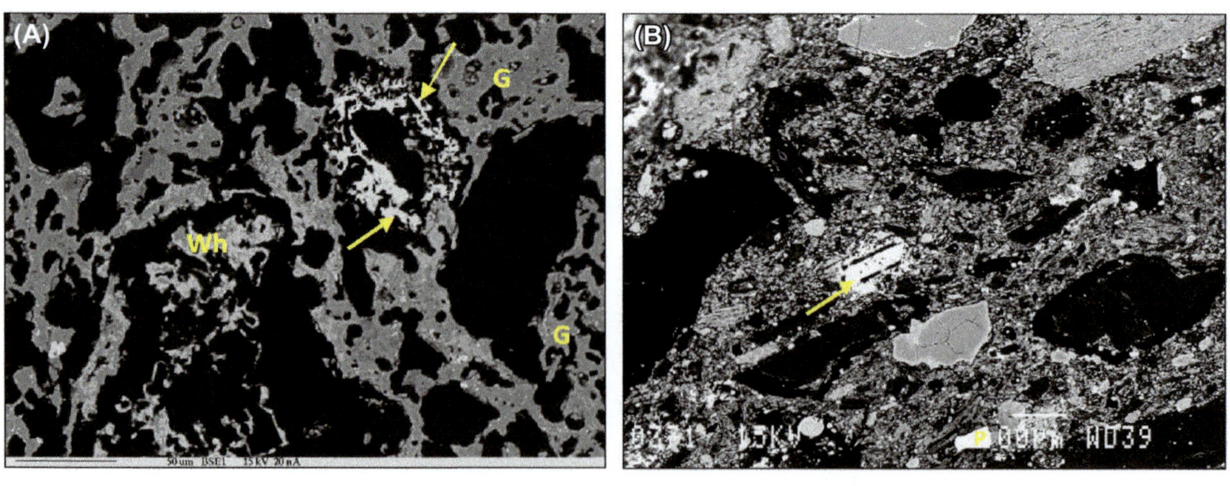

Figure 19.3.7. Celsian (arrows) from Czerwionka. (A) Atoll-shaped accumulations of Celsian with less bright BSE-imaged whitlockite (Wh) within glass (G) from a parabasalt. An is anorthite and Ar is aragonite. (B) A crystal of Celsian (arrow), about 200 µm long, with pyrite (P, lower left of center) in a "black block" metapelite. The metapelite was fused in a low Eh (electron activity) environment. The scale bars in (A) and (B) are 20 and 50 µm, respectively.

Barium may also be found in leucite and similar species, exemplified by the Przygórze slag, where leucite (Kruszewski et al., 2014a) is associated with unnamed $(Na,K,Ca,Ba)_3Al_3Si_4O_{14}$ phase. While only half of the leucites show BaO, up to 0.55 wt.%, the unnamed species is nearly constant with a BaO content of 0.20–0.39 wt.%. The third associate, K-petalite ($KAlSi_4O_{10}$) does not seem to contain barium. A single analysis of K-feldspar of this sample shows 0.29 wt.% BaO. No barium concentrations higher than 206 ppm (the Rydułtowy coal) were found in any of the EMM samples. The metal was found slightly enriched in alunite-rich shale (Radlin), Ca- and Hg-rich carbonate-sulfate segregations (Rybnik), and clay sample from Rydułtowy. Ammonium-rich alunite-group accumulations from Czerwionka contain 4 ppm Ba on average. The maximum content in sulfates from Czerwionka is 61 ppm (butlerite-rich sample).

Lanthanides (Ln)

Like strontium, the most important carriers of the Ln series elements are the apatite-group minerals. However, almost all these minerals contain only trace amounts of the Ln and positive Ce and La determinations only concern the Rydułtowy slags materials. Only 4 of the total 13 analyses (see Table 19.3.2) have positive cerium determination, with the amount always being above 300 ppm. Lanthanum is observed only occasionally.

A single La–Mg–Zn phase was detected in the periclase-rich zone of the Piekary slag; the phase seems to be present as two intergrown 5-μm-sized hexagonal-shaped crystals; its chemical formula, probably (Mg,Zn,Na) (La,Ce,Pr,Nd) $O_3(OH,F,[])_2$, does not fit to a known synthetic compound and seems to follow the Mg–Zn substitution trend observed in the surrounding periclase. The main mineral source of the Ln is undoubtedly monazite-(Ce), which is found as an accessory phase in some metapelites (from Przygórze).

Cerium and lanthanum were the only elements analyzed in bulk EMM samples via the ICP method. The largest Ce concentration in the primary waste material concerns coal from Rydułtowy, which has 42 ppm. Twice as much is found in a single gypsum sample from Pszów, with a second sample still having more than the coal value. By this behavior, Ce follows Y. Indeed, the largest Ce concentration, 100 ppm, was found in gypsum from Radlin. The anhydrite-rich assemblage from this heap also has slightly more Ce than the previous coal value. Lanthanum follows Y and Ce, and is usually found concentrated in relation to coal-maximum (14 ppm) in most gypsum- and many anhydrite-bearing assemblages. The maximum La amount of 33 ppm is associated with a gypsum sample from Pszów.

Hafnium

Being a very rare element, hafnium was not expected to be found at levels measurable within the EPMA constraints, and thus it was not measured by this technique. A single exception is perovskite, in which Hf was always below the detection limits. Very low concentrations of Hf were determined in the EMM samples. Hafnium behavior is similar to that of Zr, as expected. The largest amount found, 1.2 ppm, concerns the Rydułtowy sulfate crust; the value exceeds the largest coal concentration almost 4.5 times. A bit less than the coal content is the maximum for gypsum samples of Pszów. Twice the coal value is found in the Radlin K–Al-rich sulfate crust, and a little more in the coexisting anhydrite-dominated segregations.

Tungsten

Tungsten was not detected in any of the analyzed minerals, including perovskite, in which it was occasionally measured. Interesting finds concern W behavior in the EMM samples. Many of them show W enrichment in comparison to the largest observed primary waste value (0.2 ppm, coal, Rydułtowy). Although the largest amount of 1.4 ppm is found in koktaite-gypsum aggregates from Radlin, more interesting seem to be a similar value (1.2 ppm) in native sulfur from Katowice-Wełnowiec. In this case one can reject the matrix influence since the surrounding shale content is below 0.2 ppm. The same enrichment has to do with calcite-anhydrite-gypsum-mogánite accumulations from Rybnik; it is yet to be confirmed if W is somehow codeposited salammoniac from a single spot from this heap. Interestingly, quite similar amounts of W still four times larger than the coal maximum observed denote sulfate crusts. The values are 0.8 and 0.7 ppm for Radlin K–Al-rich one (equal to the one in anhydrite-rich assemblage) and Rydułtowy one, respectively.

Rhenium

Microprobe-measurable enrichment in Re was not suspected in the minerals analyzed. However, as in the PGE elements, some small Re anomalies were found in a few EMM samples. The maximum primary waste (coal) content is 5 ppb (Pszów). Almost 9.5 times more (47 ppb) is found in mascagnite-letovicite from a zone surrounding the Rydułtowy sulfate crust; this value stands for the dataset record. The crust itself contains only 2 ppb Re. About 11 ppb is found in gypsum of the outer part of Mg-rich Radlin sulfate crust. The core of the local K–Mg-rich crust has only 3 ppb of this exceedingly rare metal.

Gold

Gold was only measured in terms of ICP analyses conducted on the EMM samples. As expected, the primary carrier of gold is pyrite; the Pszów sample has 0.6 ppb. Coal contents are below 0.2 ppb. An interesting behavior of gold was found, as this precious metal is often found enriched in the EMM. As much as 6.3 ppb is found cocondensed in mascagnite-letovicite from Rydułtowy, the value being the largest observed. Three ppb is found in gypsum from Radlin not related to sulfate crusts; only slightly less occurs in alunite-rich shale of supposed hydrothermal origin. Interestingly, the same amount concerns gypsum-halite aggregates from Rybnik. Almost 2 ppb Au exist in the apical, hydrated part of the Rydułtowy sulfate crust, while its core contains about 1 ppb of the metal. More than 1 ppb occurs in (1) a single gypsum-dominant sample from Pszów, (2) koktaite-gypsum aggregates (Rybnik), (3) Mg-rich sulfate crust (Radlin), (4) mascagnite from Rybnik, and (5) a clay sample from Rydułtowy. Concentration over four times the pyrite value is also found in orange part of the Br(I)-rich salammoniac of Radlin, the enrichment following that of bismuth. Also very interesting is over 3 ppb enrichment in native sulfur from Rydułtowy not related to sulfate crust. The corresponding crystals from Katowice-Wełnowiec are still 3.5 times more abundant in gold than the mentioned pyrite.

Mercury

A unique discovery concerns tiny gypsum- and calcite-rich accumulations localized nearby a single vent in the interior part of a heap in Rybnik-Rymer. As shown in Figure 19.3.8, some zones within gypsum-dominant areas contain large amounts of mercury. The bulk sample contains over 50 ppm of Hg. Some grains analyzed seem to contain Hg, Ca, and Fe. It is not yet known if a single phase (Fe-enriched cinnabar or a Hg–Fe sulfide), or gypsum alone, or both are responsible for this geochemical anomaly. Neither mercury nor thallium lines were found in the WDS scans conducted on pyrite from the Czerwionka black-block metapelite.

Beside the above enormous enrichment, other mercury anomalies are quite common in EMM samples, especially when compared to the constant 0.03 ppm coal content, mean 0.51 ppm shale content, and 1.24 ppm pyrite content. The second-largest anomaly concerns gypsum-sulfur condensates from Pszów, with Hg content reaching 35 ppm. Third, much lower anomaly with 11 ppm occurs within mascagnite-letovicite condensates from Rydułtowy. Quite a large amount (8.6 ppm) is also coprecipitated with the local native sulfur, while the other sulfur sample outside the sulfate crust is over three times richer in Hg. About 1.5 ppm Hg is the mean value for the sulfate crust itself. The Wełnowiec sulfur contains 5 ppm of the metal, still quite a large amount. Radlin samples contain less Hg, with the peak of 3.9 ppm found in alunite. The local sulfate crust (core) has 3 ppm. Gypsum samples from this heap contain slightly more than 1 ppm. Three ppm of Hg are found in the inner, gypsum-encrusted part of the otherwise koktaite-bearing flue from Rybnik. Salammoniac from this heap has much lower concentration, but still larger than the coal value.

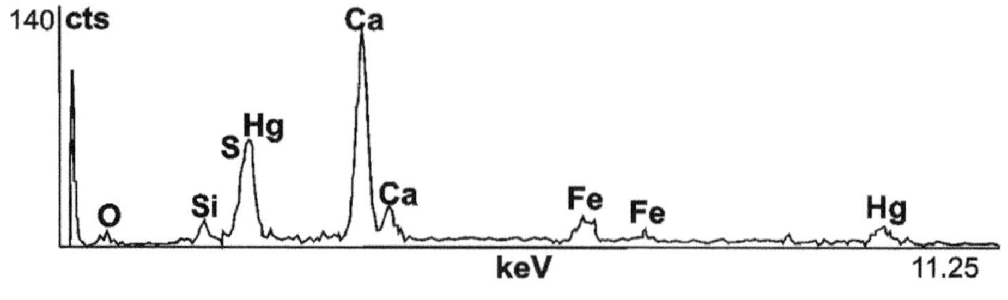

Figure 19.3.8. EDS spectrum of Hg- and Fe-bearing zones in gypsum from Rybnik-Rymer. cts = counts per second and kev = 1 000 electron volts.

Thallium

Thallium was analyzed via microprobe in a few heap minerals, but with no positive results. As opposed to EPMA, the ICP results are quite interesting. Exceptionally high concentrations of thallium, close to or equal to 40 ppm, were found for both orange and colorless intergrown bromine-rich and iodine-bearing salammoniac crystals from Radlin. This mean value is six times larger than the maximum found in primary materials (pyrite, Pszów). However, the seven times larger anomaly (282 ppm) is found in anhydrite-rostite-alum-voltaite secondary aggregates of the apical part of the local K–Al-rich sulfate crust. The crust itself does not show Tl enrichment, the metal being most likely incorporated into the structure of the later minerals, probably both alum-(K) and voltaite. If true, the alum incorporation would suggest a very minor substitution of lanmuchangite component. In addition to a given mineral, Tl also enters alunite, but the amount is lower than the pyrite one. A rather large amount of 44 ppm Tl concerns the Czerwionka butlerite-rich veins. The metavoltine-boussingaultite aggregates occurring in the vicinity have 15 ppm. Ten ppm is found the Radlin kremersite, with Tl most likely fitting the $K(NH_4)$ site. In Rybnik, some Tl (less than the pyrite value but much more than the coal maximum, that is 0.03 ppm, and shale maximum equal to 0.29 ppm) is found in both mascagnite- and possibly meixnerite-rich accumulations from a conical part of the heap, and in Ca- and Hg-rich flue deposits. Also, mascagnite-boussingaultite aggregates from Radlin characterize in a similar extent of Tl enrichment. Some gypsum samples from Pszów show 1.5 times enrichment compared to the shale. Twice the shale value is found in the Rydułtowy sulfate crust.

Lead

Like zinc, lead was suspected to be present in elevated amounts in the studied waste materials due to local coal geochemistry considerations. An interesting Pb-bearing species was found within masses of mainly godovikovite, composing the so-called sulfate crust in Czerwionka. The 20-μm-sized crystals somewhat resemble cerussite in shape; EPMA analyses show small amounts of sulfur, which is suggestible for an S-substituted carbonate mineral (S stabilization is known, in vaterite, and is also observed for the BCMH vaterite). Back-calculation of the CO_2 wt.% suggests the presence of water, but considering the instability of material and lack of suitable light-element detectors the mineral may be a Pb-substituted calcite or rather the tarnowitzite variety of aragonite; nevertheless, mean of 2 Pb-rich analyses may lead to the following empirical formula: $(Ca_{0.86}Mg_{0.07}Pb_{0.06}Fe_{0.01})_{\Sigma 1.00}[(CO_3)_{0.98}(SO_4)_{0.02}]_{\Sigma 1.00}\cdot 0.72H_2O$. Following the carbonate trend, lead was positively determined in half of the trace-mode analyses siderite from Czerwionka hematite-bearing proto-slag (see Table 19.3.18). The maximum amount is 484 and the mean (n=4) is 349 ppm.

Due to the known Pb–Ba coexistence in the nature, lead was also traced in the Czerwionka black-block metapelite celsian; it was way below the detection limits. Another silicate very rich in Pb and K, and containing Ba and Na, observed in form of automorphic crystals in few samples from Rydułtowy, is most likely an artifact. No Pb lines were found in the WDS scans of the Czerwionka black-block metapelite pyrite and the Bytom slag V-rich apatite group.

With relatively large abundance in primary wastes—from ca. 30 ppm in coal and pyrite (Pszów) and as much as 80 ppm in shales—lead was suspected to be enriched within the EMM neomineralization. This is usually not true. However, although values larger than 80 ppm were not found in Pszów, Rybnik, and Rydułtowy, single Pb anomalies were detected in Radlin. The largest such anomaly, with 1219 ppm, is associated with anhydrite-rich assemblage. Over twice the shale value is also found in the vicinity, within alunite. Much less concerns the local steklite-rich crust (35 ppm), and the kremersite (30 ppm). The mean of 40 ppm is also found in alunite-group representatives from Czerwionka. The value obtained for the butlerite-rich sample from this heap is 51 ppm, but the metavoltine-boussingaultite aggregates have ca. 4 times less Pb.

Bismuth

Extreme bismuth enrichment, following Sn and Tl enrichment, is found in the orange parts of the Br(I)-rich salammoniac of Radlin. The determined amount of 593 ppm is almost 250 times larger than the coal maximum found. An amount over two times less, equal to the given difference, is found in both the colorless salammoniac crystals and anhydrite-dominant accumulations found in another part of the heap. Bismuth also enters the alunite structure, and is found in kremersite (8 ppm). In Rydułtowy, the largest amounts of Bi are found in clay sample and in mascagnite-letovicite aggregates. The latter case is 14-fold enrichment as compared to local coal content. Like tungsten, tin, and gold, bismuth is found strongly enriched in native sulfur from Wełnowiec compared to the shale matrix. Some gypsum samples from Pszów and Rybnik may also be slightly enriched.

Thorium and Uranium

Both the radioactive elements were not detected in any of the analyzed minerals, including perovskite, in which it was measured. In the case of thorium and uranium, the reference value is the coal maximum, being 12 and 8.4 ppm, respectively (Pszów samples in both cases). The amount close to the Th coal maximum is only reported for alunite from Radlin and a clay sample from Rydułtowy. In the latter heap, thorium is found enriched, especially in the sulfate crust; the 9.1 ppm value is 18 times larger than in the local coal. Three times less than the crust content is found in spatially related mascagnite-letovicite aggregates. Uranium behavior is to some extent like that of yttrium: the largest amount, about two times larger than the coal maximum, was found in supergene Mg-rich sulfates of Radlin. The local Mg-rich sulfate crust is somewhat enriched, too. In Rydułtowy U anomalies coincide with the Th ones, and the sulfate crust content is 11 times larger than the local coal content.

Conclusions

The bulk rock ICP analyses make it possible to group the elements found by the rock types they preferentially enter. These groups are:

- Elements somewhat enriched in parabasalts (group 1)
- Elements somewhat enriched in buchites (group 2)
- Elements somewhat enriched in slags (group 3)
- Elements somewhat enriched in the black-block metapelite (group 4)
- Elements somewhat enriched in the clinker (group 5)

Group 1 elements are Se, Th, Nd, Er, Hf, Ho, Lu, Tm, Zn, Yb, and Zr. Elements belonging to the second group are Co, Ga, Gd, Ni, and Au. The third group contains only arsenic, molybdenum, antimony, and tin. The group 4 contains many environmentally important elements; group members are barium, bismuth, cadmium, mercury, lead, scandium, thorium, and yttrium. Strontium, samarium, terbium, and to some extent also lanthanum, niobium, and uranium fall to the last, fifth group. The remaining elements are found enriched in two or more lithological types. This division should be treated cautiously, as (1) Ni is found enriched also in the black-block metapelite, where it even forms its own phase; (2) the above systematics is based on relatively few rock examples.

Elements measured in minerals and EMM samples may be grouped into different categories:

- Elements with common anomalous heap-content compared to largest known contents in Polish coal, coal ash, and coal-associated sulfide minerals and various rocks (group 1A)
- Elements with common anomalous heap-content compared to Polish coals alone (group 2A)
- Elements occasionally concentrated in heap materials (single or small anomalies known) (group 3A)
- Elements not preferentially concentrated in the BCMH (group 4A)
- Elements with possible anomalies but hardly comparable to primary materials (known from Poland) due to data lacking (group 5A)
- Elements forming their own mineral species (i.e., dominant) (group 6A)

The group 1A elements are arsenic, barium, cobalt, gallium, nickel, strontium, thallium, vanadium, and zinc. Elements falling to the second category are Cr, Ge, Li, Hg, Pb. Single heap anomalies (group 3A membership) concern Cd, Ce, Hg, La, Pb, Rb, Sb, Sc, and Y. Molybdenum is a representative of the group 4A. The fifth (5A) group contains Bi, Cs, Hf, In, Nb, Se, Sn, Te, Th, U, W, and Zr. Barium, nickel, zinc, copper, zirconium, vanadium, and to some extent also strontium, arsenic, tin, thallium, and possibly mercury and lead enter the last group; this group may interfere with the group 1A.

The enrichment of Ge in cuspidine and of Sb enrichment in gehlenite offer new insights into the crystal chemistry of these minerals. The detection of Ge in $GeCl_4$ and in the BCMH jets are the subject of a future study that will be submitted for publication in the International Journal of Coal Geology.

Acknowledgments

This work was financially supported by the Ministry of Science and Higher Education for the statutory objectives of the Institute of Geological Sciences, Polish Academy of Sciences (PAS) for the years 2013–2015, and in part by the grant (of the Narodowe Centrum Nauki, NCN) no. 2013/11/B/ST10/04960. The author would like to thank M.Sc. Lidia Jeżak of the Inter-Institute Laboratory of the Microanalysis of Minerals and Synthetic Materials for assistance and help with microprobe analyses. The late Piotr Dzierżanowski, Ph.D., is acknowledged for his help in confirming the presence of some elements.

References

Adamczyk, Z., Białecka, B., Całusz Moszko, J., Komorek, J., Lewandowska, M., 2015. Rare earth elements of Orzeskie beds of south-west part Upper Silesian Coal Basin (Poland). Arch. Min. Sci. 60 (1), 157–172.

Arroyo, F., Fernández-Pereira, 2008. Hydrometallurgical recovery of germanium from coal gasification fly ash. Solvent extraction method. Ind. Eng. Chem. Res. 47 (9), 3186–3191.

Baioumy, H.M., Eglinton, L.B., Peucker-Ehrenbrink, B., 2011. Rhenium–osmium isotope and platinum group element systematics of marine vs. non-marine organic-rich sediments and coals from Egypt. Chem. Geol. 285, 70–81.

Bielowicz, B., Misiak, J., 2016. Siarczki w pokładach węgla kamiennego warstw orzeskich s.s. serii mułowcowej (westfal B) we wschodniej części GZW. Gospodarka Surowcami Mineralnymi Min. Resour. Manag. 32 (3), 23–38 (in Polish).

Bonnett, R., 1996. Porphyrins in coal. Int. J. Coal Geol. 32, 137–149.

Cebulak, S., Smieja-Król, B., Tabor, A., Misz, M., Jelonek, I., Jelonek, Z., 2005. Oksyreaktywna Analiza Termiczna (OTA) – dobra i tania metoda oceny samozapalności węgli na składowiskach – wstępne wyniki badań. Wydawnictwo Geologiczne, Warszawa 135–138 (in Polish).

Chatziapostolou, A., Kalaitzidis, S., Papazisimou, S., Christanis, K., Vagias, D., 2006. Mode of occurrence of trace elements in the Pellana lignite (SE Peloponnese, Greece). Int. J. Coal Geol. 65, 3–16.

Ciesielczuk, J., Kruszewski, Ł., Majka, J., 2015. Comparative mineralogical study of thermally-altered coal-dump waste, natural rocks and products of laboratory heating experiments. Int. J. Coal Geol. 139, 114–141.

Eskenazy, G.M., 2006. Geochemistry of beryllium in Bulgarian coals. Int. J. Coal Geol. 66, 305–315.

Fabiańska, M., Ciesielczuk, J., Kruszewski, Ł., Misz-Kennan, M., Blake, D.R., Stracher, G., Moszumańska, I., 2013. Gaseous compounds and efflorescences generated in self-heating coal-waste dumps – a case study from the Upper- and Lower Silesian Coal Basins (Poland). Int. J. Coal Geol. 116–117, 247–261.

Fang, Z., Gesser, H.D., 1996. Recovery of gallium from coal fly ash. Hydrometallurgy 41, 187–200.

Finkelman, R.B., 2004. Potential health impact of burning coal beds and waste banks. In: Stracher, G.B. (Ed.), Coal Fires Burning Around the World: A Global Catastrophe, pp. 9–24 International Journal of Coal Geology, v. 59, Issues 1-2.

Gadas, P., Novák, M., Szuszkiewicz, A., Szełęg, E., Vašinová Galiová, M., Haifler, J., 2016. Manganoan Na,Be,Li-rich sekaninaite from miarolitic pegmatite at Zimnik, Strzegom-Sobótka massif, Sudetes, Poland. Can. Mineral. 54 (4), 971–987 (in press).

Goodarzi, F., Goodarzi, N.N., 2004. Mercury in Western Canadian subbituminous coal – a weighted average study to evaluate potential mercury reduction by selective mining. Int. J. Coal Geol. 58, 251–259.

Goodarzi, F., Swaine, D.J., 1994. The influence of geological factors on the concentration of boron in Australian and Canadian coals. Chem. Geol. 118, 301–318.

Höll, R., Kling, M., Schroll, E., 2007. Metallogenesis of germanium – a review. Ore Geol. Rev. 30, 145–180.

Hower, J.C., Ruppert, L.F., Williams, D.A., 2002. Controls on boron and germanium distribution in the low-sulfur Amos coal bed, Western Kentucky coalfield, USA. Int. J. Coal Geol. 53, 27–42.

Jírasek, J., 2001. Thermal Changes of the Rocks in the Dump Pile of the Kateřina Colliery in Radvanice (Eastern Bohemia), vol. 541. VSB – Technical University of Ostrava, Institute of Geological Engineering, Ostrava, p. 69.

Juda-Rezler, K., Kowalczyk, D., 2013. Size distribution and trace element contents of coal fly ash from pulverized boilers. Pol. J. Environ. Stud. 22 (1), 25–40.

Ketris, M.P., Yudovich, Y.E., 2009. Estimations of Clarkes for Carbonaceous biolithes: world averages for trace element contents in black shales and coals. Int. J. Coal Geol. 78, 135–148.

Knot, A.N., 2016. Machiite, IMA 2016-067, CNMNC Newsletter No. 34, December 2016, p. 1317. Min. Mag. 80, 1315–1321.

Kokowska-Pawłowska, M., 2015. Petrographic and mineral variability of the rocks accompanying selected coal seams of the Poruba beds and their influence of the trace elements content. Gospodarka Surowcami Mineralnymi Min. Resour. Manag. 31 (2), 73–92 (in Polish).
Kokowska-Pawłowska, M., 2016. Pierwiastki ziem rzadkich (REE) w iłowcach z wybranych pokładów węgla kamiennego serii mułowcowej i piaskowcowej Górnośląskiego Zagłębia Węglowego. Gospodarka Surowcami Mineralnymi Min. Resour. Manag. 32 (3), 39–66 (in Polish).
Konieczyński, J., Zajusz-Zubek, E., Jabłońska, M., 2012. The release of trace elements in the process of coal coking. Sci. World J. 2012 294927 (PMC3361233).
Kruszewski, Ł., 2006. Oldhamite-periclase-portlandite-fluorite assemblage and coexisting minerals of burnt dump in Siemianowice Śląskie – Dąbrówka Wielka area (Upper Silesia, Poland) – preliminary report. Mineralogia Polonica 28, 118–120.
Kruszewski, Ł., 2008. Apatite-ellestadite solid solution and associated minerals of metacarbonate slags from burning coal dump in Rydułtowy (Upper Silesia). Miner. Spec. Papers 32, 100.
Kruszewski, Ł., 2012. Unique chloride assemblage of exhalative origin from burning coal mining dump in Radlin (Rybnik Coal Area, S Poland). Miner. Spec. Papers 40, 90–92.
Kruszewski, Ł., 2013. Supergene sulphate minerals from the burning coal mining dumps in the Upper Silesian Coal Basin, South Poland. Int. J. Coal Geol. 105, 91–109.
Kruszewski, Ł., Ciesielczuk, J., Misz-Kennan, M., 2012. Geochemia pierwiastków śladowych w środowisku płonących hałd górnictwa węglowego – przykłady z Górnośląskiego Zagłębia Węglowego: II Polski Kongres Geologiczny. Warszawa (in Polish; poster, unpublished).
Kruszewski, Ł., Ciesielczuk, J., Misz-Kennan, M., 2014a. Mineralogy of Some Metacarbonate Rocks from Burned Coal-Mining Dump in Przygórze (Lower Silesian Coal Basin) and its Analogy to "olive" Rocks from the Hatrurim Formation: 4th Central-European.
Kruszewski, Ł., Ciesielczuk, J., Misz-Kennan, M., Fabiańska, M., 2014b. Chemical composition of glasses and associating mineral species in various pyrometamorphic rocks from coal-mining dumps of the Lower Silesia. Miner. Spec. Papers 42, 70–71 Mineralogical Conference - CEMC 2014 (Proceedings of the international symposium CEMC 2014; ISBN 978-80-210-6832-2), p. 77.
Kwiecińska, B.K., Hamburg, G., Vleeskens, J.M., 1992. Formation temperatures of natural coke in the Lower Silesian coal basin, Poland: evidence from pyrite and clays by SEM-EDX. Int. J. Coal Geol. 21, 217–235.
Lapham, D.M., Barnes, J.H., Downey Jr., W.F., Finkelman, R.B., 1980. Mineralogy associated with burning anthracite deposits of Eastern Pennsylvania: mineral resource report 78. In: Pennsylvania Geological Survey. Commonwealth of Pennsylvania, Harrisburg.
Lewińska-Preis, L., Fabiańska, M.J., Parzentny, H., Kita, A., 2008. Geochemical characteristics of the macromolecular part of crude and biodesulphurised flame coal density fractions. Chemie der Erde 68, 279–293.
Liang, Y., Liang, H., Zhu, S., 2014. Mercury emission from coal seam fire at Wuda, Inner Mongolia, China. Atmos. Environ. 83, 176–184.
Liu, S., Wang, Y., Yu, L., Oakey, J., 2006. Volatilization of mercury, arsenic and selenium during underground coal gasification. Fuel 85, 1550–1558.
Makowska, D., Wierońska, F., 2016. Analysis of content of selected critical elements in fly ash. Geosci. Eng. LXII (1), 31–36.
Maksimova, M.F., Shmariovich, Y.M., 1983. Rhenium in infiltration uranium-coal deposits. Int. Geol. Rev. 25 (2), 189–195.
Nasdala, L., Pekov, I.V., 1993. Ravatite, $C_{14}H_{10}$, a new organic mineral species from Ravat, Tadzhikistan. Eur. J. Mineral 5, 699–705.
Panov, B.S., Dudik, A.M., Shevchenko, O.A., Matlak, E.S., 1999. On pollution of the biosphere in industrial areas: the example of the Donets coal Basin. Int. J. Coal Geol. 40, 199–210.
Parafiniuk, J., Kruszewski, Ł., 2009. Minerals of the ammonioalunite-ammoniojarosite series formed on a burning coal dump at Czerwionka, Upper Silesian Coal Basin, Poland. Mineralogical Magazine 74 (4), 731–745.
Parzentny, H.R., Lewińska-Preis, L., 2006. The role of sulphide and carbonate minerals in the concentration of chalcophile elements in the bituminous coal seams of a paralic series (Upper Carboniferous) in the Upper Silesian Coal Basin (USCB), Poland. Chemie der Erde 66, 227–247.
Parzentny, H.R., Różkowska, A., Róg, L., 1999. Relationship between bed thickness, average ash content, and Zn and Pb content in coal in the Upper Silesian Coal Basin. Geol. Q. 43 (3), 365–374.
Porowski, A., Kruszewski, Ł., Lewandowski, M., 2015. Development of Geochemical Studies of Water and Wastes with Commercial Potential: The ING PAN Experience [Poster]: International Congress on Water, Waste and Energy Management Porto (Portugal), 16–18 July 2014, Abstracts Book. Instituto Politécnico de Portalegre, PortugalInstituto Politécnico de Portalegre, Portugal.

Saima Sultana, R., 2009. Electrical conductivity in TlI–TiO_2 composite solid electrolyte. Phys. B Condens. Matter 404 (1), 36–40.
Seredin, V.V., 2003. Anomalous trace elements contents in the Spetsugli Germanium deposit (Pavlovka Brown coal deposit) Southern Primorye: communication 1. Antimony. Lithol. Miner. Resour. 38 (2), 154–161.
Seredin, V.V., Finkelman, R.B., 2008. Metalliferous coals: a review of the main genetic and geochemical types. Int. J. Coal Geol. 76, 253–289.
Sindern, S., Warnsloh, J., Witzke, T., Havenith, V., Neef, R., Etoundi, Y., 2005. Mineralogy and geochemistry of vents formed on the burning coal mining waste dump Anna I, Alsdorf, Germany. Eur. J. Mineral 17, 130.
Sokol, E.V., Maksimova, N.V., Nigmatulina, E.N., Sharygin, V.V., Kalugin, V.M., 2005. Combustion Metamorphism: Publishing House of the SB RAS. Novosibirsk (in Russian).
Smoliński, A., Rompalski, P., Cybulski, K., Chećko, J., Howaniec, N., 2014. Chemometric study of trace elements in hard coals of the Upper Silesian Coal Basin, Poland. Sci. World J. 2014 ID 234204, 12 pp.
Srebrodolskiy, B.I., 1989. Tainy Sezonnykh Mineralov. Nauka, Moscow (in Russian).
Stracher, G.B., 2007. The origin of gas-vent minerals: isochemical and mass-transfer processes. In: Stracher, G.B. (Ed.), Geology of Coal Fires: Case Studies from Around the World. The Geological Society of America, pp. 91–96. Reviews in Engineering Geology XVIII.
Sun, Y., Li, Y., Zhao, C., Lin, M., Wang, J., Qin, S., 2010. Concentrations of lithium in Chinese coals. Energy Explor. Exploit. 28 (2), 97–104.
Wagner, M., 1980. Przemiany termiczne węgla kamiennego w strefach pożarów hałd kopalnianych. Zeszyty Naukowe Akademii Górniczo-Hutniczej – Geologia 6 (2), 5–14 (in Polish).
Wagner, N.J., Hlatshwayo, B., 2005. The occurrence of potentially hazardous trace elements in five Highveld coals, South Africa. Int. J. Coal Geol. 63, 228–246.
Wang, M., Zheng, B., Wang, B., Li, S., Wu, D., Hu, J., 2006. Arsenic concentrations in Chinese coals. Sci. Total Environ. 357, 96–102.
Welz, B., Sperling, M., 1999. Atomic Absorption Spectrometry, third ed. Wiley-VCH. 941 pp.
Woo, I.H., Watanabe, K., Hashimoto, Y., Lee, Y.K., 1987. Determination of selenium and tellurium in coal by graphite furnace atomic absorption spectrometry after coprecipitation with arsenic. Anal. Sci. 3, 49–52.
Wyszomirski, P., Brylska, E., 1996. Fly ash in Polish building ceramics – threat or proecology? Appl. Geochem. 11, 351–353.
Xiao, L., Zhao, B., Duan, P., Shi, Z., Ma, J., Lin, M., 2016. Geochemical characteristics of trace elements in the No. 6 coal seam from the Chuancaogedan mine, Jungar coalfield, inner Mongolia, China. Minerals 6, 28–42.
Xu, D., Chen, Y.W., Guo, H., Liu, H.Q., Xue, Y.D., Zhang, W.F., Sun, Y.C., 2013. Review of germanium recovery technologies from coal. Appl. Mech. Mater. 423–426, 565–573.
Yang, 2006. Modes of occurrence and origins of noble metals in the Late Permian coals from the Puan Coalfield, Guizhou, southwest China. Fuel 85, 1679–1684.
Yudovich, E.A., 2003. Notes on the marginal enrichment of Germanium in coal beds. Int. J. Coal Geol. 56, 223–232.
Yudovich, E.A., Ketris, M.P., 2005a. Mercury in coal: a review – Part 1. Geochemistry. Int. J. Coal Geol. 62, 107–134.
Yudovich, E.A., Ketris, M.P., 2005b. Arsenic in coal: a review. Int. J. Coal Geol. 61, 141–196.
Yudovich, E.A., Ketris, M.P., 2005c. Selenium in coal: a review. Int. J. Coal Geol. 67, 112–126.
Yudovich, E.A., Ketris, M.P., 2006. Chlorine in coal: a review. Int. J. Coal Geol. 67, 127–144.
Zajusz-Zubek, E., Konieczyński, J., 2003. Dynamics of trace elements release in coal pyrolysis process. Fuel 82, 1281–1290.
Zhuang, X., Querol, X., Alastuey, A., Juan, R., Plana, F., López-Soler, A., Du, G., Martynov, V.V., 2006. Geochemistry and mineralogy of the Cretaceous Wulantuga high-germanium coal deposit in Shengli coal field, Inner Mongolia, Northeastern China. Int. J. Coal Geol. 66, 119–136.

CHAPTER 20

Peat Fires in Ireland

Fire is common in the Irish landscape, despite Ireland being well known for its wet weather. Peatlands occupy 17% of the surface, and there is evidence from scientific records that indicate that fire has been a natural phenomenon since early times. Nowadays, fire in Irish peatlands occurs mainly as an anthropic phenomenon. Peat fires happen under extreme drought conditions, generally as flames from vegetation fires heat the surface of the peat, generating spontaneous combustion in peat that is hundreds of years old. The authors have compiled scientific and technical information about a range of fires in Irish Peatlands and include a discussion about the history of peatland fires, an overview of the current peatland fire regimes, and some examples of Irish peat fires and the ecological effects of these fires. *Photo by Nuria Prat-Guitart, 2015.*

CHAPTER CONTENTS

Coal and Peat Fires: A Global Perspective
Edited by Glenn B. Stracher

https://doi.org/10.1016/B978-0-12-849885-9.00020-2

Photo by Donna Hawthorne, 2012.

20.1 History of Irish Peatland Fires

Nuria Prat-Guitart
Ciaran Nugent
Enda Mullen
Fraser J.G. Mitchell
Donna Hawthorne
Claire M. Belcher
Jonathan M. Yearsley

Lake sediment core sampling at Cuckoo Lough, Killarney.

Evolution of Irish Peatlands and Fires

Peatlands are one of Ireland's most characteristic natural features, making up about 17% of the Irish landscape. Their formation began at the end of the last glaciation (about 11,700 years ago) and was initially confined to shallow lakes and wet hollows. However, over time there was a transition from open water to fen and then acid ombrotrophic raised bogs. Blanket bog also spread over extensive areas of poorly drained land in the uplands and western seaboard (Figure 20.1.1). This spread may have taken place in areas as early as 7000 years ago but was more

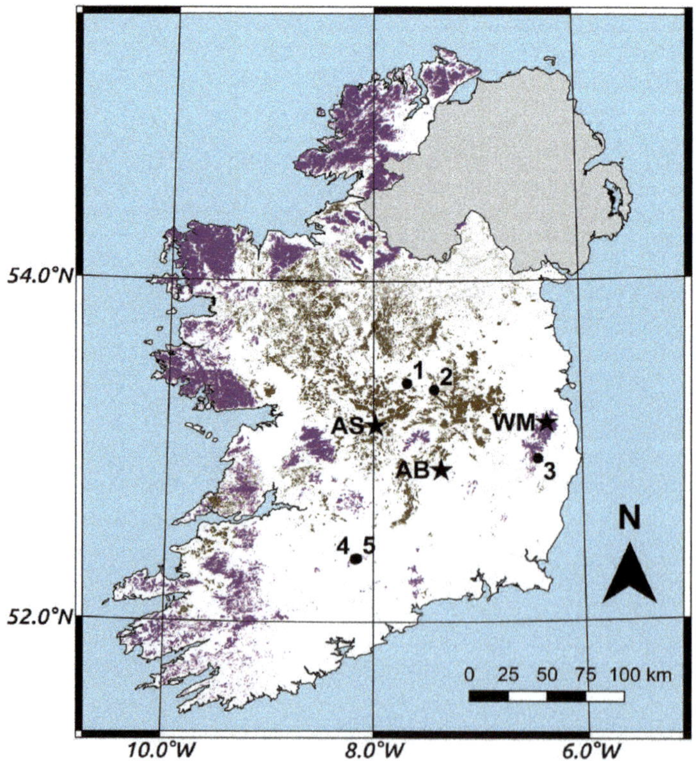

Figure 20.1.1. Distribution of blanket peat (purple) and other peat sources (brown). Sampling locations used to study fire record: (1) Ballinderry Lough, (2) Cornaher Lough, (3) Kelly's Lough, (4) Diheen Lough, and (5) Borheen Lough. Distribution of three examples of peatland fires described in Section 20.3 (WM = Wicklow Mountains, AS=All Saints Bog, and AB=Abbeyleix Bog). Source: https://data.gov.ie/dataset/national-soils-map (accessed February 2018). Figure by Jonathan M. Yearsley, 2018.

widespread by 4000 years ago when the climate may have become wetter. Ireland's peatlands have been exploited for fuel over the last 400 years, reducing the original extent of the peatland by almost 47%.

Nowadays major steps are being taken to conserve the remaining areas of Irish peat. Peatlands are largely characterized as wetland areas and tend to occur where there is high rainfall. Paradoxically this wetland habitat provides perfect anaerobic and acidic conditions for the preservation of charcoal particles providing an important archive of past fire history.

Fires generated from natural causes have been occurring in Ireland throughout the Holocene but changes in vegetation, land use, agriculture, and policy in recent years have also promoted an increase in fires in the Irish landscape. A number of studies have examined global trends in wildfire and identified that globally fire was low in the Late Glacial and Early Holocene, subsequently increasing throughout the Holocene (Marlon et al., 2013; Power et al., 2008; Daniau et al., 2012), with fluctuations dictated by factors such as climate, CO_2 concentration, and human activity. Disentangling natural and anthropogenic fires in Ireland has become a challenge because during the past two centuries anthropogenic activities have dominated the global biomass burning signal (Marlon et al., 2012). Fire is a complex ecological process driven by a variety of factors at a range of spatial scales.

Record of Peat Fires

Here we briefly review the long-term fire history in Ireland to provide context for recent and contemporary data. Despite Ireland's wet oceanic climate, the Holocene era, since the end of the last glaciation, is punctuated by repeated fires that vary in intensity as well as in temporal and spatial resolution. Records of fire history in Ireland are derived from three principal sources, charcoal in archaeological settings and microscopic and macroscopic charcoal fragments from within sedimentary sequences. There has been a considerable volume of charcoal analysis completed from archaeological sites, which uses taxonomic resolution to explore wood selection and use with some associated quantitative analyses (O'Donnell, 2016; OCarroll and Mitchell, 2016). Microscopic charcoal analysis has tended to be associated with pollen analysis, where the charcoal on pollen slides is quantified under the microscope. During pollen preparation, samples are passed through a sieve of 100–130 μm, the charcoal fragments are thus sufficiently small to be carried significant distances prior to deposition. A limited number of discrete samples tend to be prepared down a sequence and so the charcoal record lacks contiguity. The larger size of macroscopic charcoal (≥125 microns) ensures that it is of local origin such as a lake catchment or peatland. It is washed from sediment samples and can be quantified using digital image analysis. This allows significant throughput of samples, which enables contiguous samples from entire sequences; this facilitates the reconstruction of fire dynamics including specific fire events, fire frequency, and fire return times (Hawthorne and Mitchell, 2016).

The most comprehensive background context for Irish Holocene fire history is derived from contiguous macroscopic charcoal analysis of lake sediments (Hawthorne and Mitchell, 2018). This revealed a variety of fire return times (ranging from 65.8 $yr\,fire^{-1}$ at Diheen Lough to 995 $yr\,fire^{-1}$ at Kelly's Lough; Figure 20.1.1) and intensities that were related to geographical setting but overall the Irish composite record fits reasonably well with European and global tends (Figure 20.1.2). This suggests that fire activity was generally lower in the early Holocene and increased toward the present day.

Macroscopic charcoal is routinely noted in descriptions of peat stratigraphy (e.g., Molloy and O'Connell, 1995) but to date, there has been no attempt to collate this information on a national scale to explore spatial and temporal patterns. Dodson (1990a, b) published a series of microscopic charcoal records from blanket peat on the Dingle Peninsula, S.W. Ireland, which highlighted a continuous record of charcoal but fire did not appear to impact on vegetation development in the wider landscape. The same is true from further north in Galway and Mayo (O'Connell et al., 1988; Molloy and O'Connell, 1995). It is thus important to separate the microscopic charcoal records, which represent charcoal that could have been blown on to the peatland surface from fire on the adjacent mineral soil, from macroscopic charcoal records that indicate the burning of the original peatland surface. These records serve to illustrate that Irish peatlands provide a valuable archive of past burning on both minerals soils (i.e., via microscopic charcoal records) and the peatlands themselves (i.e., via macroscopic charcoal records). It should also be noted that charcoal layers are often found at the base of Irish blanket bogs—at the bog/soil interface—suggesting that fire has played a role in their establishment (O'Connell, 1990).

Calculating historic fire return intervals (i.e., number of years between fire events) helps us characterize past fire regimes (Iglesias et al., 2014) and can be useful for predicting future fire patterns (Riley and Loehman, 2016). Fire

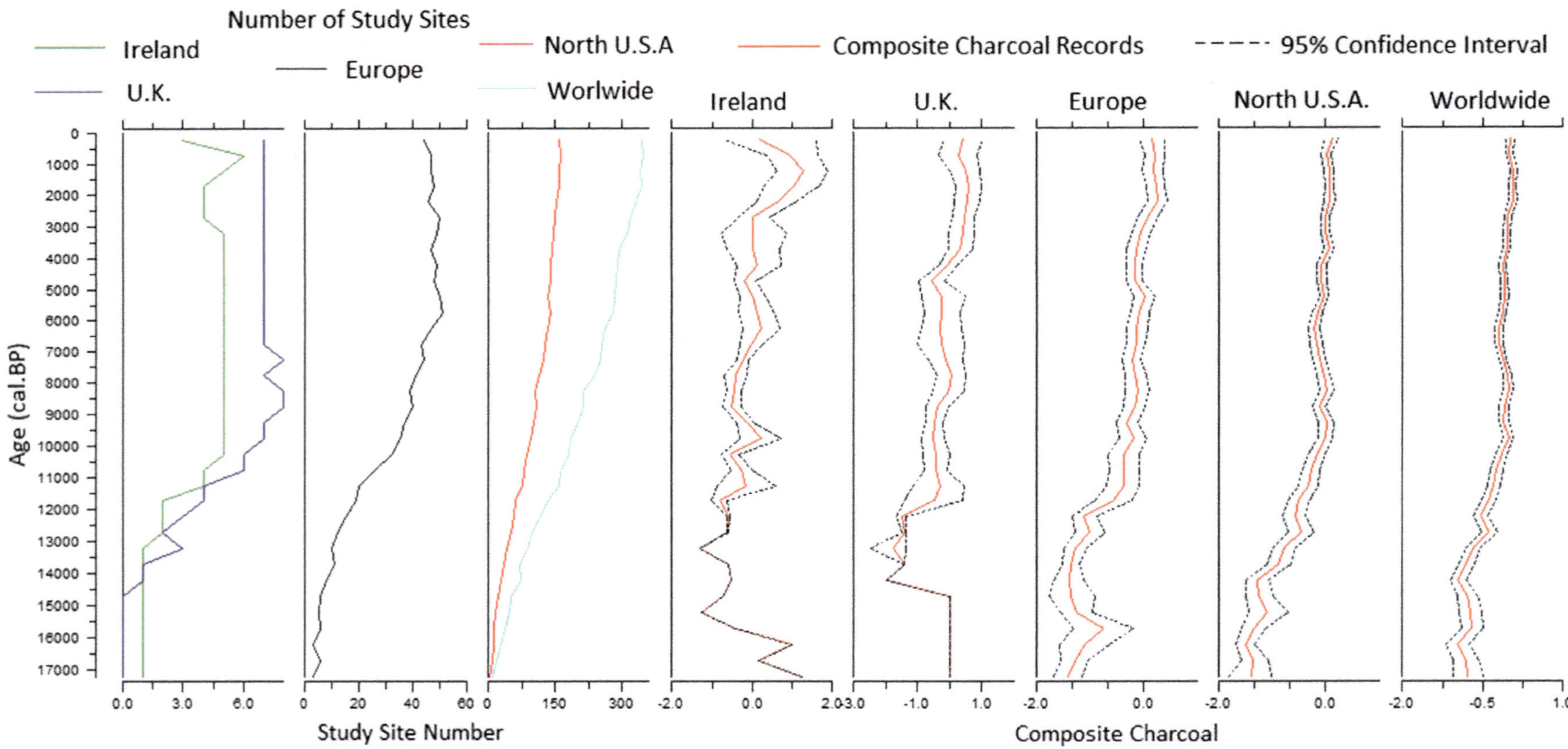

Figure 20.1.2. Synthesized records of fire activity based on composite charcoal data from multiple sites (included in the Global Charcoal Database), from Ireland, Britain, Europe, North America, and the world (right), including 95% confidence interval and number of study sites included (left). From Hawthorne and Mitchell (2018).

parameters such as return intervals are dictated by local site conditions, such as vegetation, altitude, and climate, and also factors that affect fire ignition such as fuel load, moisture index, and prevailing wind (Bistinas et al., 2014: Kloster et al., 2015).

Historic fire regimes for Irish peatlands have not yet been estimated; however, a range of average fire return intervals have been calculated at five study sites in Ireland, representing a range of upland, lowland, and midland locations (Figure 20.1.1), and represent fires in the surrounding catchment, which could be flaming or peat fires. Ballinderry Lough in the Irish midlands produced an average fire return interval of 157.5 $yr\,fire^{-1}$, while Cornaher Lough also located nearby produced an interval of 936 $yr\,fire^{-1}$. Borheen and Diheen Lough in the Galty Mountains produced fire return intervals of 404 $yr\,fire^{-1}$ and 65.8 $yr\,fire^{-1}$, respectively, and finally at Kelly's Lough in the Wicklow Mountains a return interval of 995 $yr\,fire^{-1}$ was calculated. The latter is relatively close to the peat fire event analyzed in Section 20.3, Wicklow Mountains Fire. These figures reflect the variability of the fire regime at each of the study sites, and no definitive pattern between average fire return intervals can be prescribed based on topography or vegetation type.

Photo by Nuria Prat-Guitart, 2015.

20.2 Irish Peatland Fire Regimes

An ongoing vegetation and smoldering fire in the Wicklow Mountains. The photo was taken 24 h after ignition next to a road, on April 22, 2015.

Fire Regimes in Peatlands

The temperate, moist, oceanic climate of Ireland is not usually associated with wildfire occurrence, and yet Ireland is subject to a fire regime with an annual cycle of wildfire events, with varying levels of extent and severity including smoldering combustion of organic soils (Nugent and Casey, 2014). This cycle is interspersed with episodic major wildfire spate events in years where sufficiently dry weather conditions permit.

Recent estimates suggest up to 75,000 ha of land, comprised mainly of peatlands and associated land types, have been burned over in Ireland between 2010 and 2015 (Schmuck et al., 2013). As seen in recent years, major fire events (Figure 20.2.1) have the capacity to severely challenge available emergency service resources, cause considerable impacts to key habitat areas, and inflict high levels of disruption to affected human communities living in

Figure 20.2.1. Killarney National Park Fire, April 2015 (1500 ha). The horizontal field of view in the background is 15 km. Photo: Ciaran Nugent, Forest Service, Department of Agriculture, Food and the Marine (DAFM).

affected areas. Such events increasingly precipitate adverse public reactions, media commentary, and political responses (Fogarty, 2017).

Despite the records of fire history in deposits of charcoal and carbon arising from previous fires (Section 20.1), little is known about historical fire regimes on open peatlands in Ireland. This in turn presents a limiting factor on understanding the fire patterns and for the attempts at managing current fire problems (Nugent and Casey, 2014). Remote sensing and aerial methods may provide additional data regarding more recent fire history; however, these methods have yet to be applied in Ireland (Millin-Chalabi et al., 2011; Allen et al., 2016).

Spatial Factors

Spatial analysis of satellite-derived fire detection data since 2002 (e.g., MODIS and VIIRS satellite outputs, Figure 20.2.2) suggests that Irish peatlands, particularly upland heaths, blanket bogs, and lowland blanket bogs have the strongest association with wildfire (vegetation and peat fires) of any land type in Ireland. This is due in part to the flammable potential of vegetation types arising on peatlands and presence of seasonal ignition risk arising from human activities on these land types during the peak fire risk season (Section 20.2).

Temporal Patterns

Most wildfire incidents in Ireland occur in spring, most notably, March, April, and May. Fire activity in uplands and peatlands rapidly subsides during late May and June, as fresh, live, and nonflammable vegetation states dominate.

Analysis of 1760 MODIS fire points detected between 2002 and 2017 shows that 50% of these detections occurred during April (873), with 24% occurring during May (421), and 12% during March.

A further 5.9% of detections occurred during June. The spatial distribution of the MODIS fire points (Figure 20.2.3) suggests that these may involve permitted burning of agricultural residues as opposed to wildfire incidents.

A confluence of flammable, mature, fine fuel vegetation and dry weather makes fires on Irish peatlands possible. In spring, most upland vegetation is in a mature state and capable of rapid drying when weather conditions permit. However, under normal early spring conditions, only surface vegetation is capable of sustaining combustion and litter and soil layers remain moist. Should prolonged drought conditions exist during or immediately following the mature phase of vegetation, it is possible for fire to transfer the heat from surface vegetation to dry litter and organic soil of subsurface layers below, deriving to peat fires. The smoldering combustion in organic soils can last for weeks (Rein, 2016) and ignite new vegetation fires if drought conditions prevail.

An indication of multiannual cyclical fire incidence is provided by Irish Fire and Rescue Service data between 2000 and 2015 (Figure 20.2.4). Peaks and troughs are in line with prevailing spring weather for given years (Figure 20.2.3).

The nature of Irish peatland vegetation means that Irish wildfires occur several months earlier than fires elsewhere in Europe, and can take place at far lower fire weather index levels (Figure 20.2.5) compared to southern Mediterranean countries rising above an index of 30 during the summer season (Schmuck et al., 2013).

Fuels

The most critical factor is the small fuel diameter profile and low live moisture content of shrub species such as *Calluna vulgaris*, *Erica tetralix*, *Ulex Galii*, *Erica cinerea*, and *Ulex Europaeous*; and mature winter and spring phases of Graminoid vegetation such as *Molinia caerulea*, *Deschampsia*, and *Eriophorum* species that dominate peatland and upland site types (Nugent and Casey, 2014). Most of these species conform to the standard for one-hour fine fuels under NWCG standards (National Wildfire Coordinating Group, 1996).

The dominance of these types of fuels in peatland landscapes (Figure 20.2.6) means that when fires do occur in suitable conditions, they are capable of building and spreading to cover large areas extremely quickly. These fuel types having small diameter fuel particles (<6 mm) have large surface area-to-volume ratios and their moisture

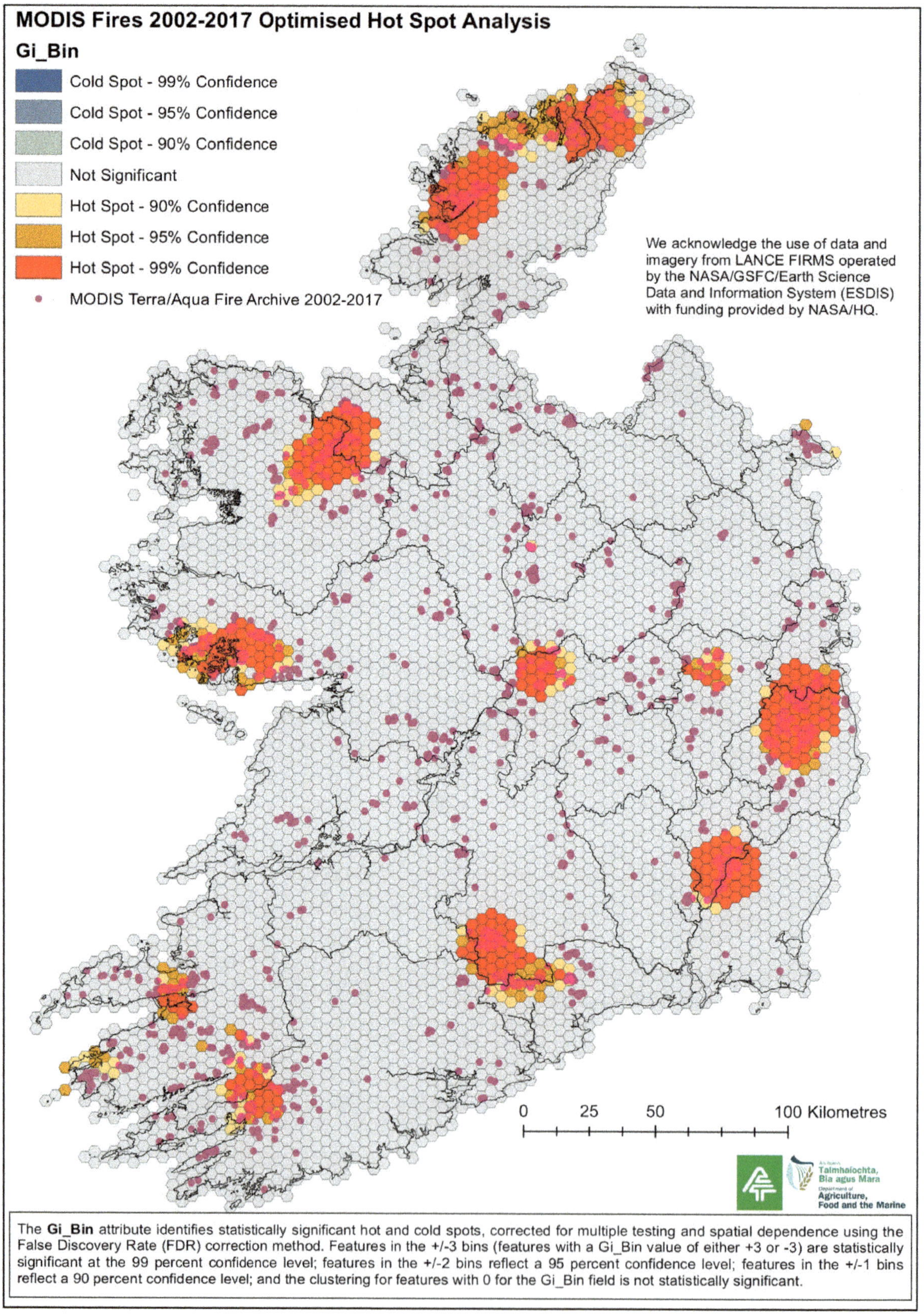

Figure 20.2.2. Optimized analysis for fire detections in Ireland 2002–2017. The Hot Spots location indicates occurrence of fires. Source: Ciaran Nugent, Forest Service, Department of Agriculture, Food and the Marine (DAFM).

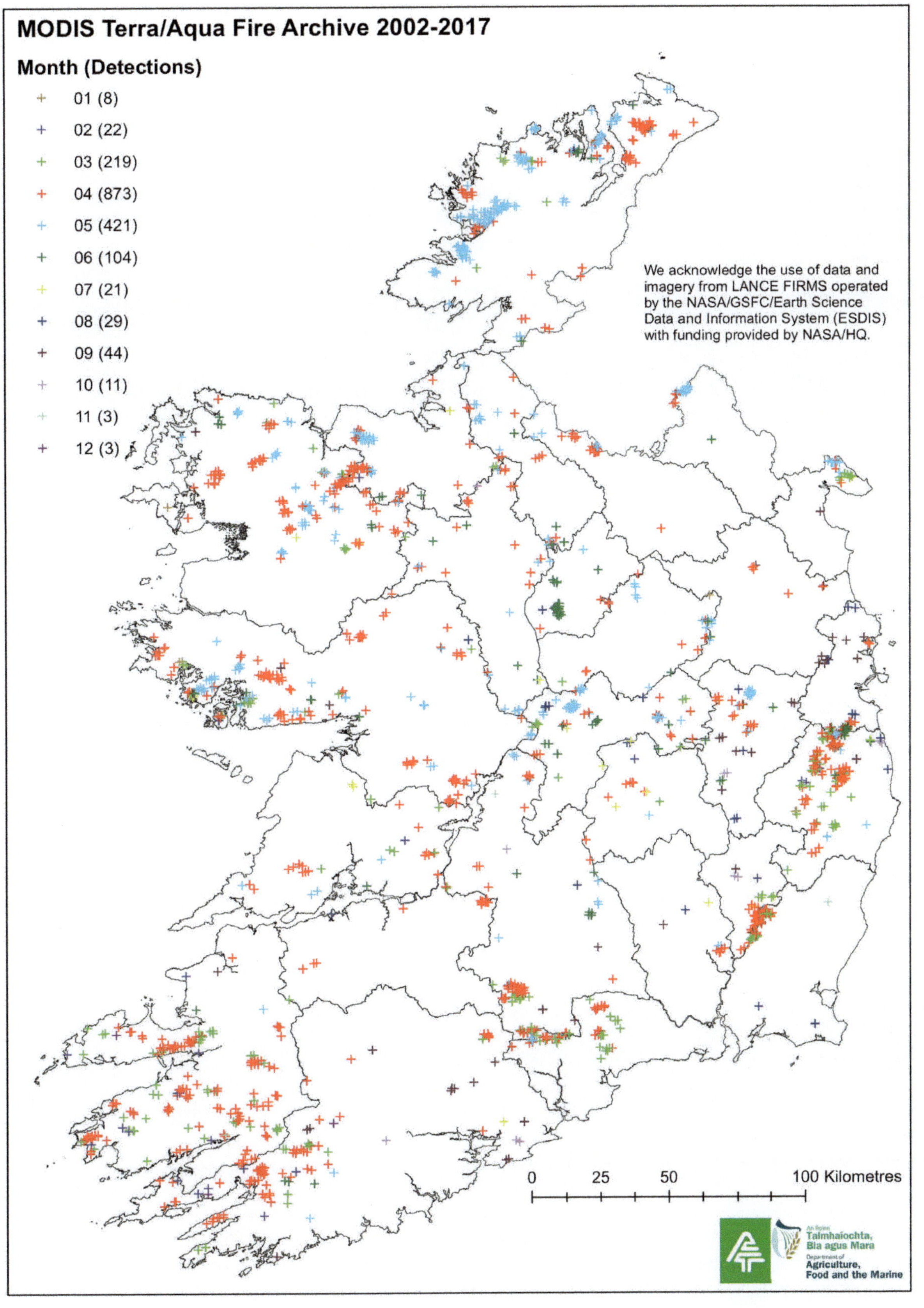

Figure 20.2.3. Spatial representation of fire incidence 2002–2017 by month. Source: Ciaran Nugent, Forest Service, Department of Agriculture, Food and the Marine (DAFM).

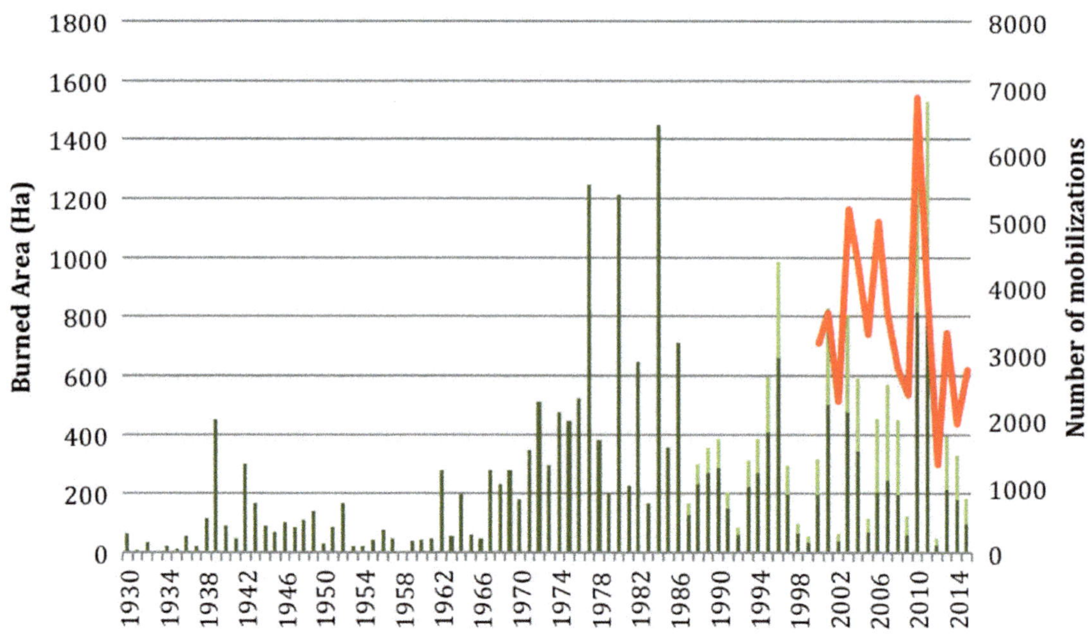

Figure 20.2.4. Combined area burned and annual Fire Service mobilizations to bog, grass, and forest fires. Bars: area burned, dark green public area, light green private area. Red line: number of mobilizations. Source: Department of Housing, Planning, Community and Local Government and Ciaran Nugent, Forest Service, Department of Agriculture, Food and the Marine (DAFM).

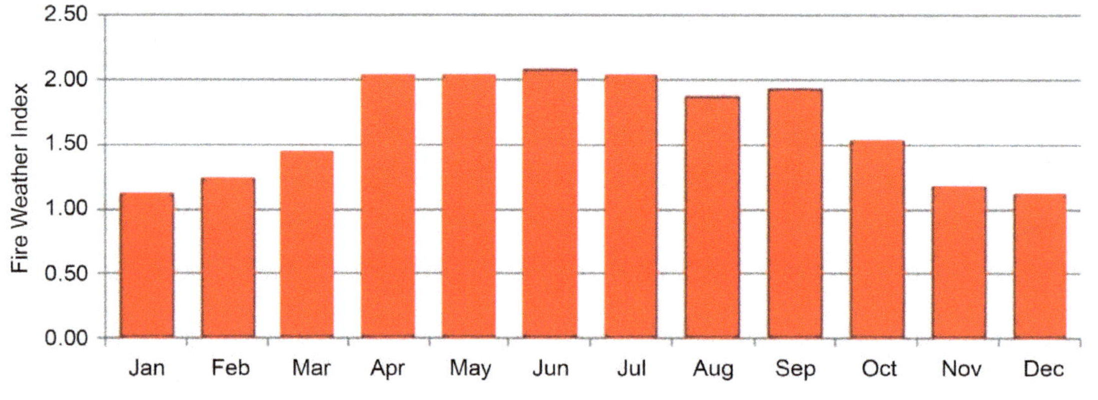

Figure 20.2.5. Average fire weather index (Van Wagner, 1987) for Ireland. 1971–2013. Source: Met Eireann and Ciaran Nugent, Forest Service, Department of Agriculture, Food and the Marine (DAFM).

contents can change rapidly with changes in temperature and relative humidity (National Wildfire Coordinating Group, 1996).

Due to their structural arrangements, grasses and dwarf shrub species make ready fine fuels, capable of shedding moisture rapidly as ambient weather dries, and capable of ignition and fire build-up even at quite high ambient humidity levels (Scottish Government, 2013; Nugent and Casey, 2014).

The condition, community composition and arrangements of biomass and dead fuel mixes (vegetation and peat) can vary considerably across peatland sites according to past management, including grazing, drainage, and burning activities, as well as natural terrain, hydrological, and climatic constraints. Some vegetation like gorse species (genus: *Ulex*) can support high levels of dead vegetation on living shrubs, and this dead material can help initiate fire and induce live tissue to burn, even at high live moisture levels.

Figure 20.2.6. High upland fine fuel load over wet heath peatland, Killarney National Park, January 2011. The horizontal field of view in the background 1 km. Source: Forest Service, DAFM. Photo: Ciaran Nugent, Forest Service, Department of Agriculture, Food and the Marine (DAFM).

Forests

Due to historic forestry practice, there is a close spatial association between plantation forestry in upland and lowland peatland areas. As a consequence, fires on open peatland frequently impact forests in these nearby areas. Young establishment phase trees are especially vulnerable. Between 1930 and 2015, Irish forest cover increased from approximately 90,000 ha in 1930 to almost 719,374 ha in 2010.

The presence of fire in Irish upland and peatland areas is indicated by the area burned annually since 1930 (Figure 20.2.4), when public afforestation programs commenced. From the 1970s the area burned increased in line with rising afforestation rates, and reflecting the high level of exposure to fire risk of forests at the most vulnerable younger stages of development on fire prone marginal land types. This pattern is later repeated in private forest estate damage from the early 1990s when private afforestation commenced and planting levels increased.

Ignitions

Nowadays, most fires in Ireland appear to be caused by people. There are long-standing associations between people, fire, and peatlands. Ignitions by lightning and a "broken bottle" scenario are unlikely to occur in Ireland, based on meteorological conditions. Fire has traditionally been used in Irish peatlands as a means of managing heather moorland for sheep. It is locally known as "burning for grass." Since 1976 the Wildlife Act has restricted the dates during which this burning can occur in order to protect birds during the nesting season from March 1 to August 31. Despite this, burning has been carried out during other times of the year, illegally and in an uncontrolled fashion. This lack of control often causes fires that spread across ecologically sensitive heathland and into commercial forestry plantations. There is also a significant risk associated with turf cutting practices in some areas, particularly on raised bogs where removal of the scraw, or turf surface, over peat banks to facilitate turf fuel drying is carried out by burning (Malone and O'Connell, 2009).

Illegal dumping of household waste, malicious, illegal ignitions, or deliberate arson are all known factors in wildfire ignitions, particularly in areas where fires have occurred where agricultural grazing or turf cutting are not practiced; for example, national parks, military areas, and urban recreational areas. The proximity of many malicious fires to urban areas, population centers, and critical infrastructure can make these fires highly significant and often problematic. These types of fires often take place during summer months and are associated with increased public use of lands containing high levels of gorse accumulations.

Despite the known risks to peatland habitats arising from fire, fire management issues do not appear to feature in the current Irish Peatland Strategy and in general conservation plans for peatland reserves.

Future of Fire Regimes in Peatlands

Predicting future fire regimes and return intervals requires understanding a range of factors: local, regional, and global scales, from climate change and precipitation, to location of ignition, fuel load, and flammability. Moreover, a comparable study position in the past where environmental conditions were similar to those in which you are trying to predict would also be highly useful. With climate change predicted to increase temperatures, reduce precipitation, and lengthen fire seasons, it is important to study past fire patterns and gain an understanding of their interrelated factors in order to potentially guard against future detrimental effects. A combination of computer models, modern satellite data, climate data, and historical fire regime data is proving progressive in predicting future fire patterns; however, much work is still need, especially in Ireland.

As in other parts of Europe, there is increasing concern regarding increased accumulations of aboveground fuels as a result of major economic and social changes in rural areas and reductions in the intensity of farming practice and grazing regimes (Birot, 2009; Montiel and Kraus, 2010). As seen in other regions, climate change impacts on upland vegetation and increased growth rates are projected to give rise to higher and more dangerous fuel loads in upland vegetation that form the basis for wildfires. These factors can be further compounded by inconsistent policies in relation to fire use and grazing regimes, and the designation of naturally high fire risk land types for nature conservation.

A more extensive analysis of the evolution of fire regimes in Irish peatlands should lead to a set of appropriate measures to face a potential increase of wildfires (vegetation and peat fires) in Ireland during the next years. As an example, the low level of detected fire incidence during March, relative to the high level of activity in April and May, appears to support the argument for extending the open season for prescribed burning in Ireland to March under Irish legislation. This management measure would help control fuel loads in areas previously identified as strategic locations that can avoid more complex fires threatening society and nature.

Finally, the fact that nowadays a large number of ignitions are caused by humans highlights the importance of the social component of wildfires in Irish peatlands. Ireland's fire problem presents as a classic "wicked problem" issue—such problems are usually multicausal and multiscalar, and straddle organizational and disciplinary boundaries (Parrott, 2017). In addition to these features, wildfire in Ireland is seen as a transient, low human risk issue, with multiple, often conflicting stakeholder involvement and no central authority to lead or coordinate responses. In effect, wildfires are nobody's problem, until they are everybody's problem. Arising from these conditions, Irish firefighting services are poorly configured and equipped to deal with wildfire issues. This is largely due to the transient and episodic nature of the issue and a resulting poor evidence base that limits policy responses and preparedness and precludes meaningful strategic responses.

Recent public disquiet and discourse on the issue of wildfire in Ireland has clearly demonstrated the need for a reliable evidence base with which to guide and validate realistic wildfire management policies and programs.

Photo by Chris Uys, 2013.

20.3 Documented Peat Fires

Smoldering fire in Abbeyleix Bog on September 18, 2013. The fire was suppressed after consuming a 10 × 10 m area. The cause of the fire is unknown.

Introduction

We are going to document three examples of peatland fires in Ireland that happened between 2013 and 2015 (Figure 20.1.1). During these years many other peat fires have taken place in Ireland, but information was not always collected. These three examples have been chosen because scientists and technicians have been working on these case studies and therefore information about the evolution and the effects of these fires is available.

The common factor is that the three peat fires occurred in degraded peatlands. This means that each site had been impacted to varying degrees by drainage and peat exploitation over the last couple of decades. The fact that all three are degraded peatlands makes them more vulnerable to peat fires.

Wicklow Mountains National Park

The Ecosystem and Environmental Conditions

This fire took place on the slopes of Kippure Mountain, Co. Wicklow, Ireland (6° 21′ W, 53° 10′ N; Figure 20.1.1) on April 22, 2015. Kippure, a rounded, granite mountain, rises 757 m above sea level, and is covered in blanket peat of varying depths, up to 4–5 m (Cole and Mitchell, 2003). On the gentler eastern slopes this forms part of the Liffey Head Bog complex. It is the only actively growing blanket bog in the east of the country. The River Liffey rises here and flows through the Irish capital city, Dublin.

The steeper slopes of the mountain are covered in a mosaic of wet and dry heath. The characteristic ling heather *C. vulgaris* and bilberry *Vaccinium myrtyllis* mix gives way to cross-leaved heath *E. tetralix* and purple moorgrass *M. caerula* in wetter hollows. Granite boulders protrude here and there, providing footholds for bell heather *E. cinerea*. In places these peatland plant communities give way to acidic grasslands crisscrossed by flushes. The lower slopes of the mountain also bear witness to the last ice age in the form of small moraines and thin coverings of mineral soils, many of which have been planted with commercial conifers. The Coronation Plantation is an open woodland of oak and Scots pine on the southern edge of the study area. Planted in the 1830s it is of high conservation value and at extreme risk from fire. This land is all part of the Wicklow Mountains National Park. It is designated as a Special Area of Conservation for its habitats and as a Special Protection Area for Peregrine (*Falco peregrinus*) and Merlin falcons (*Falco columbarius*), which breed here (NPWS, 2017a). The commercial forestry consists of both state and privately owned plantations dominated by Sitka spruce *Picea sitchensis*.

On May 9, 2000, 673 ha were burned in the western section of this study area, including Coronation Woodland (Figure 20.3.1). A series of fires that originated on the northern slopes of Kippure in April 2003 spread over the mountain summit and burned 1250 ha (Figure 20.3.2). In spring 2015 a series of fires burned the southern slopes of

Figure 20.3.1. Aerial image of the Kippure area burned during the fire event. The horizontal field of view in the background is 2 km. Source: An Garda Síochána Air Support Unit, 2015.

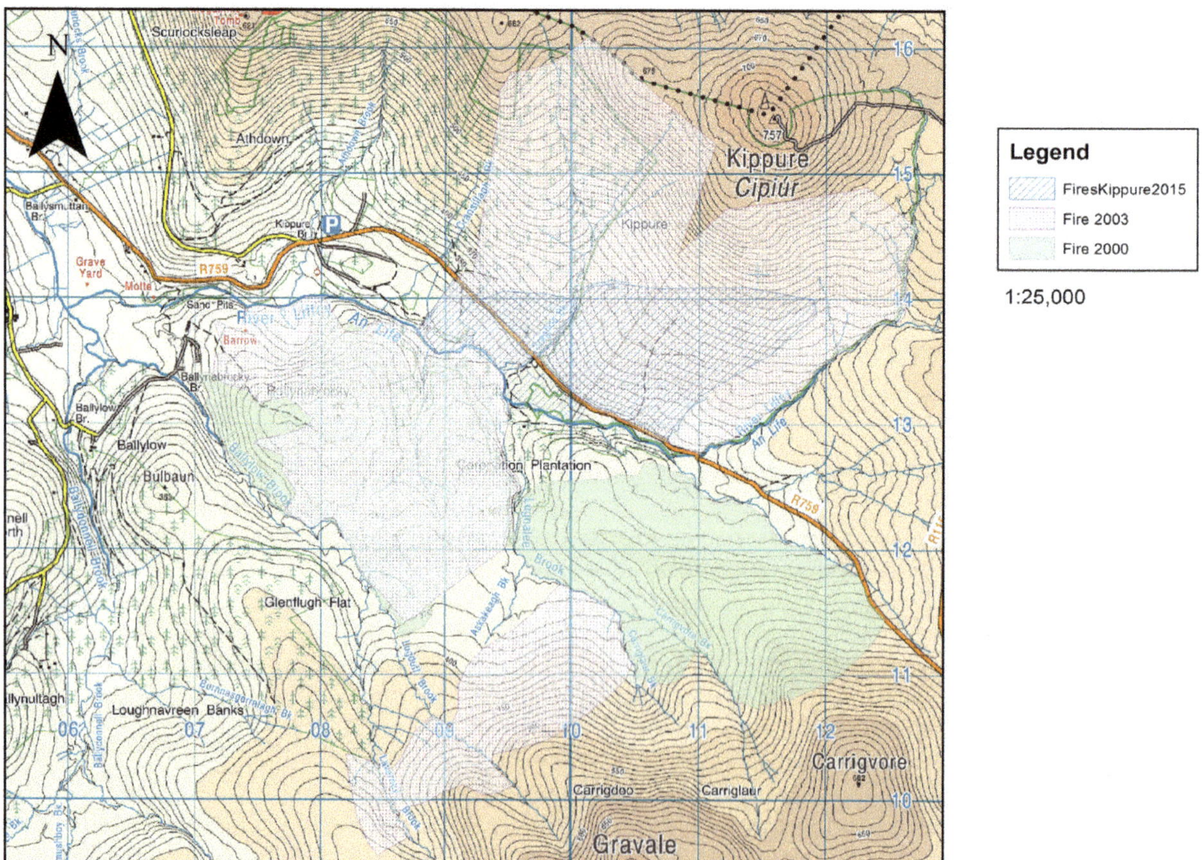

Figure 20.3.2. Map of the fire areas for three different events in 2000, 2003, and 2015, including where they overlap. This map shows the R759. Kippure Mountain is north of the road while Coronation Plantation is to the south (GR 095 130). Kippure House Estate lies between the River Liffey and the road at GR 090 140. The Sraghoe Brook can be seen at GR 100 140. Map Source: Enda Mullen, 2017. Ordinance Survey material by permission of the government (License no. EN 0059216).

Kippure. The first, in March, originated in the Kippure House Estate and burned 25 ha of young forestry. The most serious event was on April 22 and included both forestry and conservation heathlands. On that date a total of 282 ha burned (Figures 20.3.1 and 20.3.2). The weather station at Kippure East recorded only 55 mm rainfall in total during the 3 weeks prior to this last event. Most of this rain fell in the first 3 days of the month (Figure 20.3.3). The last rain was a week before the fire and only amounted to 1.6 mm (Source: Dublin City Council).

The Peat Fire Event

On April 22, 2015 a small fire near the Military Road – R115 (Figure 20.3.2), which borders the eastern and southern flanks of Kippure, was fanned by an easterly wind and quickly spread across the heath on the southern slopes of Kippure, in the Wicklow Mountains National Park. It jumped the adjacent R759 road and burned the land between it and the River Liffey. The river prevented it from entering the Coronation Plantation. On the other side of the road, the fire jumped the Sraghoe Brook and entered state-owned forestry. Progressing through the forestry it spread across the road near the Cransillagh Brook, where it burned the remains of the forestry in Kippure House

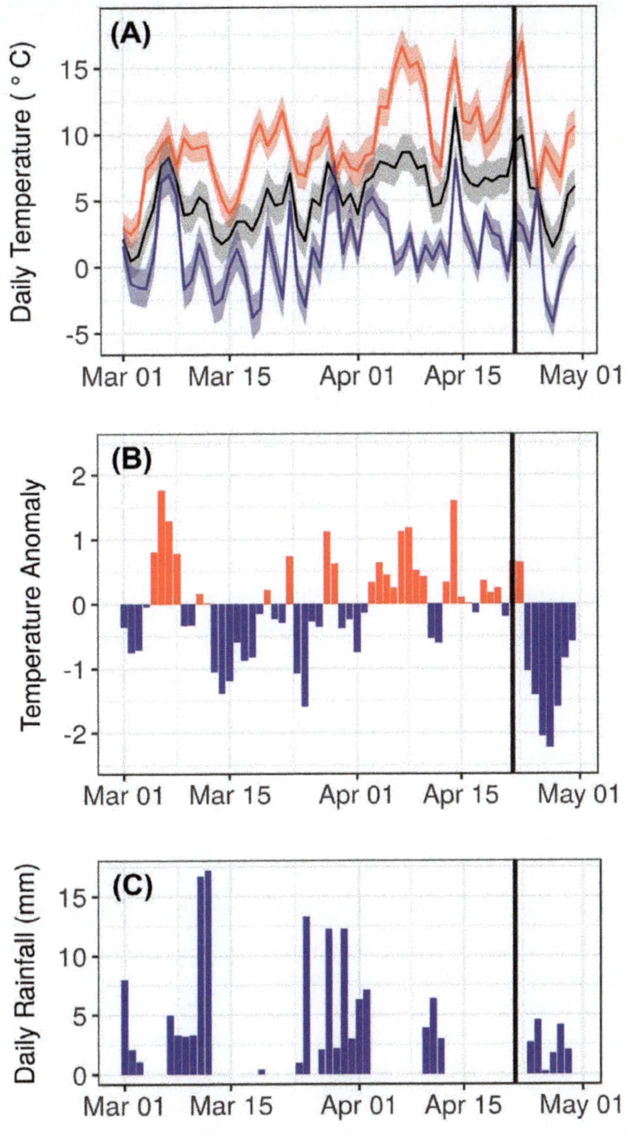

Figure 20.3.3. Daily temperature and rainfall data for March and April of 2015 in the region of Kippure Estate. The fire started on April 22, 2015 (vertical black line). Panel A, the mean (black line), maximum (red line) and minimum (blue line) daily temperatures (shaded regions show standard errors). Panel B, the daily mean temperature anomaly for 2015 relative to the 30-year average (1985–2015) in units of standard deviations across 30 years (red and blue bars correspond to positive and negative anomalies, respectively). Panel C, daily rainfall (mm). Data are from the E-OBS dataset (Haylock et al., 2008). Figure by Jonathan M. Yearsley, 2018.

Figure 20.3.4. Fire brigades extinguishing the smoldering combustion and small flames by swatting out the surface of the peat. Only smoke is released from smoldering combustion, making the smoldering front detection and fire extinction more difficult. Photo by Nuria Prat-Guitart, 2015.

Estate (which had previously ignited in March). The latter was an area prone to peat smoldering combustion due to the drainage undertaken for the tree plantation about 10 years earlier.

Two houses were endangered by the fire. Several units of the Co. Wicklow Fire Service fought the blaze close to these properties. In the forestry, workers used heavy machinery to cut a fire break in the trees, attempting to save the older timber. National Parks and Wildlife Service staff fought the fire in heather and *Molina* dominated habitats to prevent the blaze progressing into the Coronation Plantation. A helicopter dropped water on the fire in the trees close to the two houses. The fire was finally extinguished about 21:00 after a day of fighting it. However, the underlying peat in the forestry continued to smolder for another 6 days (Figure 20.3.4). Smoldering combustion occurred due to high drought stress in the ground layers. They were difficult to control and extinguish and were the source of new flaming spots causing the Fire Service more call-outs. Altogether approximately 282 ha of land was burned, including 220 ha of high conservation value heathland, 35 ha commercial forestry plantation, and many 200-year old Scots Pine trees.

Fire Behavior and Its Effects

Smoldering fire spread was observed 24 h after the fire event started (Figure 20.3.5). At that time, the charred or partially consumed vegetation was irregularly distributed on the landscape. The lack of flames and the release of smoke from the ground indicated that 24 h later there was still active consumption of peat layers due to smoldering fire fronts. Compared to flaming fires, smoldering is considered a type of low-intensity fire (Rein, 2016), meaning that fire slowly spreads and can be expected to last for several days, releasing small amounts of energy (Keeley, 2009). The Kippure Estate section of this fire continued smoldering for 6 days.

The spread of a smoldering front can happen in all directions, either at ground surface or in deep peat layers (Rein, 2016). The spread of the fire, remaining active days after the start of the fire, supports previous studies indicating a slow speed of the smoldering fires moving at rates of less than 9 cm h^{-1} (Prat-Guitart et al., 2016). A close-up photo of a peat surface shows the position of the smoldering front moving at the surface of the peat (Figure 20.3.6, left). The spread to deeper peat layers could also be happening simultaneously if the moisture content of the peat was low enough to sustain downward smoldering self-propagation (Huang and Rein, 2015); however, this was not directly observed during the fire event due to the difficulties of detecting deep smoldering from the surface.

Figure 20.3.5. Peat smoldering in a young pine plantation (between 2 and 4 m tall) 24 h after the fire event started. Since smoldering combustion is flameless, the presence of smoke is the main indicator that peat fire is still active. Surface consumption (charred areas) is irregularly distributed. The horizontal field of view in the background is 300 m). Photo by Nuria Prat-Guitart, 2015.

The irregular consumption of organic soil during the Wicklow Mountains Fire suggests that the distribution of the moisture content at the surface ground layers was also heterogeneously distributed (Figure 20.3.5). In drained peatlands however, the moisture distribution of the surface soil layers tends to be more homogeneous, thus the irregular consumption of the peat would indicate the moisture content distribution at the surface is strongly associated with the plant and root distribution.

Overlapping effects of the 2003 and the 2015 fires were also observed in the area: trees that were damaged in the 2003 fire were affected again by the 2015 events (Figure 20.3.6, right). The organic soil underneath the dead roots was consumed to ashes.

Infrared Imagery as a Tool for Analyzing Smoldering Peat

The infrared camera captures the energy that is released by a body across the wavelength between 7 and 14 μm (Minkina and Dudzik, 2009), becoming a useful tool to identify smoldering peat in dynamic environments. Infrared imagery reveals precise positioning of the smoldering fire front in flameless situations in the surface layer (Figure 20.3.7). The information provided by infrared imagery is often used for technical and research purposes to observe and analyze fire propagation and dynamics of smoldering fires during fire events and in laboratory environments (Belcher et al., 2010; Hadden et al., 2013; Prat-Guitart et al., 2015).

Even though the ground was partially covered with char and ashes, 24 h after the fire it was observed with the infrared camera that the soil was still under combustion (Figures 20.3.7 and 20.3.8), which could not be appreciated with the bare eye. In many cases smoldering fire fronts were still active in the base of small trees. This pattern is common in peat fires (Davies et al., 2013), where roots are potentially shallow and dry (Section 20.3). Even though temperatures were not measured in the field, it is known from past studies that smoldering combustion would range between 200°C and 600°C, well below usual temperatures from flaming fires, around 1500°C (Prat-Guitart, 2016; Rein, 2016).

Figure 20.3.6. Left: A smoldering front spreads downward through the peat layers on the side of a dry drainage ditch. Smoke is released from the active fire front. Black and gray colors indicate charred peat and ashes. Right: The smoldering fire consumes peat surrounding the roots of a dead tree. The tree in the foreground was dead from a previous fire event. Photos by Nuria Prat-Guitart, 2015.

Figure 20.3.7. Peat consumption at the base of a tree. Left, a photo shows the presence of char (black) and ash (gray and white colors). Right, an infrared image superimposed on the photo showing the smoldering fire that is actively consuming peat. Colors of the infrared image represent temperature, with dark blue corresponding to ambient background temperature. The photos were taken 24 h after the fire event started. Photo by Nuria Prat-Guitart, 2015.

Figure 20.3.8. An infrared image superimposed on a visual image shows hot-spots of peat smoldering in the surface of the Wicklow Mountains peatland (Kippure Estate). These hot-spots are difficult to detect without the infrared heat detection. The photos were taken 24 h after the fire event started. The trees are between 2 and 4 m tall, and the horizontal field of view in the background is 300 m. Photo by Nuria Prat-Guitart, 2015.

All Saints Bog

The Ecosystem and Environmental Conditions

All Saints Bog is an approximately 223 ha raised bog in County Offaly that is located 8 km north-northwest of Birr (7° 59′ W, 53° 9′ N; Figures 20.1.1 and 20.3.9; Valverde et al., 2014). The bog is a Special Area of Conservation (SAC 566, EC Habitats Directive (92/43/EEC) as amended). The high bog has 39.78 ha (17.8%) that is currently classified as active raised bog and 183.17 ha (82.2%) of degraded raised bog (Valverde et al., 2014). All Saints Bog also has 14.34 ha of bog woodland (predominantly birch and pine; Figures 20.3.10 and 20.3.11) and is the only site in Ireland with extensive wet bog woodland (Cross and Lynn, 2013). Pollen core analysis indicates that prior to 1800 All Saints Bog was in a landscape dominated by mixed woodland, but woodland cover has since declined (Cole and Mitchell, 2003). The wet bog woodland on All Saints Bog is likely a remnant of a much more extensive woodland from the past (Cole and Mitchell, 2003).

Peat cutting occurred on the All Saints Bog (Valverde et al., 2014). Since the 1840s, over 52% of the original high bog has been cut (Valverde et al., 2005). The most extensive peat cutting is along the northern edge where the cut peat is for domestic use. Commercial peat extraction (Erin peats) occurred to the east of the current high bog, leaving an area of bare peat with no vegetation (Figures 20.3.9 and 20.3.11), although commercial extraction has now ceased. Peat cutting has left over 27 km of drainage ditches in the high bog, which is modifying the bog's hydrology (Valverde et al., 2014). These drains are considered to have a high importance on the habitats of the high bog, but no drains have been blocked. Surrounding land use can also impact the hydrology of the bog, notably a quarry to the southwest of the bog.

Over the last 25 years several fire events have occurred on the high bog: 1991/1992 (Kelly et al., 1995), April 2003 (Valverde et al., 2005), and July 2013 (Hudspith et al., 2014). There was also a fire on the bare peat and adjacent land to the east of the high bog in March 2010. The fires in 2003 and 2013 were severe fires that covered 94 ha and

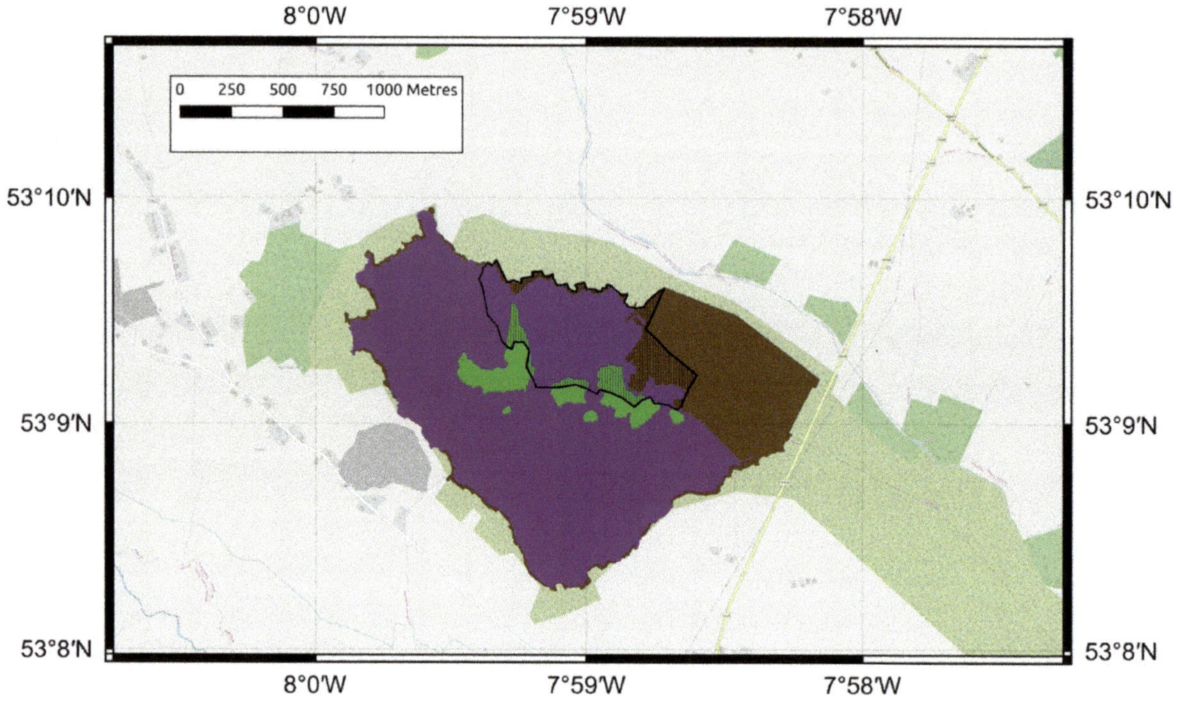

Figure 20.3.9. Current extent of the high bog (purple region) with the bog woodland in the center of the bog (dark green region) and the bare peat and face bank (brown region). The extent of the 2013 fire is shown as the hatched region with a black boundary. The background map is from OpenStreetMap contributors, 2017. Figure by Jonathan M. Yearsley, 2018.

Figure 20.3.10. The bog woodland on All Saints Bog, showing the extent of the 2013 fire into the woodland. Left is woodland affected by the fire (light and scorched burn severity) and right is woodland unaffected by the fire. The horizontal field of view in the foreground is 5 m. Photo by Jonathan M. Yearsley, 2013.

Figure 20.3.11. The bare peat left from commercial peat extraction in the eastern section of All Saints Bog. The bog woodland can be seen in the distance along the horizon. Fire damage from the 2013 fire can be seen on the right but did not spread onto the bare peat. The horizontal field of view in the foreground is 200 m. Photo by Jonathan M. Yearsley, 2013.

42 ha of the high bog, respectively. Both the 2003 and the 2013 fires spread across the high bog and into the bog woodland. Although fire has a long-term association with peatland ecosystems (Davies et al., 2016), the frequency of these recent fire events on All Saints Bog is high relative to historical fire frequency estimates for Ireland (ranging from 0.0002 to 0.01 fires yr^{-1}; Hawthorne and Mitchell, 2016). Increasing fire frequency may be related to human impacts on the bog (Davies et al., 2016).

The area of active raised bog on All Saints Bog has been in decline over the last 20 years, likely due to a combination of pressures from peat cutting, drainage, fire events, and local quarrying (Valverde et al., 2014). The quality of the bog woodland is also in decline, due primarily to drying of the bog and recent fire events. In 2005 All Saints Bog had an overall conservation status of unfavorable-declining (Valverde et al., 2005). Prior to the 2013 fire, the overall conservation status of All Saints Bog was classified as unfavorable bad-declining (Valverde et al., 2014).

The Peat Fire Event

The most recent fire started on July 14, 2013 and affected approximately 42 ha (19%) of the high bog (Figures 20.3.9 and 20.3.12). Weather conditions in July 2013 were exceptionally warm and dry (Figure 20.3.13), providing conditions that favored the spread of the fire. The average mean and maximum daily temperatures for the first 13 days of July 2013 were 17.3°C and 22.9°C, respectively (a temperature anomaly of over two standard deviations above the average for these days). The fire started toward the northern edge of the bog and spread southeastward. The fire was actively fought by the Fire Service and extinguished within 4 days. Protecting the bog woodland was the focus of the fire fighting. Some of the suppression methods used were trenches dug into the peat to act as fire breaks and water drops from helicopters.

Fire Behavior and Its Effects

Visual estimates of burn severity (the loss or change in organic matter aboveground and belowground, Keeley (2009); Table 20.3.1) found predominantly scorched, light, and moderate burns (Figure 20.3.12).

Interestingly all the deeply burned sites (Figures 20.3.12 and 20.3.14) were in the most degraded areas of the bog, which were also some of the driest areas of the bog. The fuel type varied across the burned site and both the fuel type and ecotope (ecotopes are regions of relatively homogeneous biotic and abiotic conditions; Schouten, 2002) were found to be associated with aspects of burn severity (Hudspith et al., 2014). Shrub and moss layers were highly impacted in the both moderately and deeply burned areas, which also contained the highest percentages of charcoal across the site (Hudspith et al., 2014).

A range of the charcoals were studied by Hudspith et al. (2014) using an approach that measures the light reflected back from the charcoals (charcoal reflectance). This approach has recently been studied by Belcher and Hudspith (2016) and utilized to understand tundra fires (Hudspith et al., 2017); both suggest that this approach implies that higher charcoal reflectance values relate to longer durations of heating of different areas. Hudspith et al. (2014) took 2766 measurements of byrophyte charcoal reflectance from All Saints Bog that covered areas of light, moderate, and deep burn severities. The highest burn severity class (deeply burned) was found to contain the most highly reflecting bryophyte char, suggesting that these areas were burned, or heated, for a longer duration. The most deeply burned areas and those containing the highest bryophyte reflectance values all occurred in the facebank ecotope of the bog; the area directly next to the main area of peat cutting and the area of the bog that has been the most disturbed (Figure 20.3.9). The most deeply burned areas of the bog also included charred peat clasts, providing evidence that pyrolysis of the peat surface itself occurred in these areas (Hudspith et al., 2014).

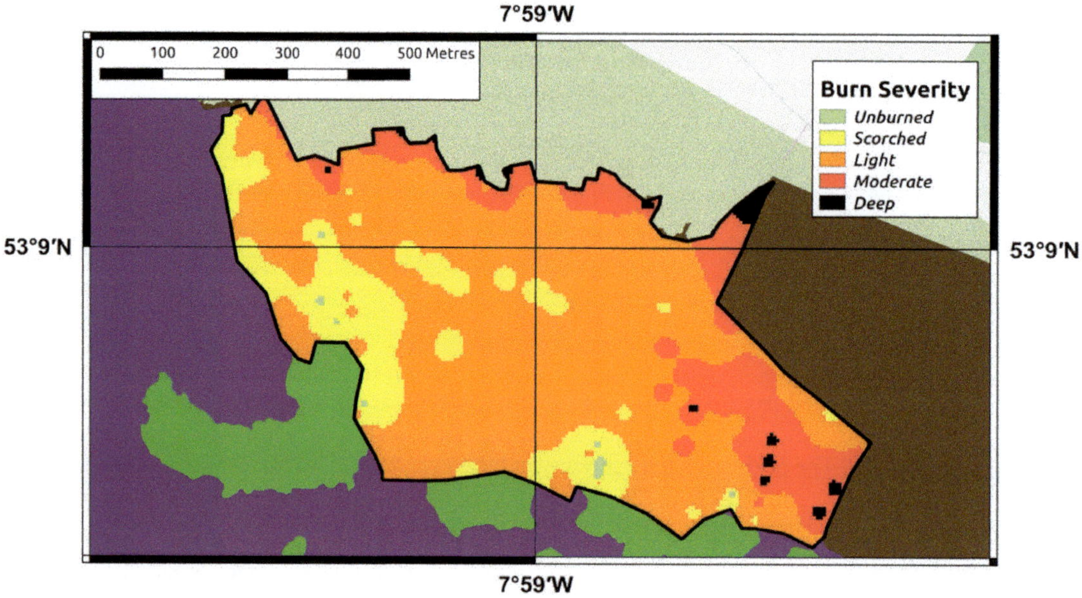

Figure 20.3.12. A close-up map of the 2013 fire on All Saints Bog (Figure 20.3.10) with visual estimates of the burn severity on a five-point scale (Table 20.3.1). Purple, green, and brown regions show the high bog, bog woodland, and bare peat, respectively. Figure by Jonathan M. Yearsley, 2018.

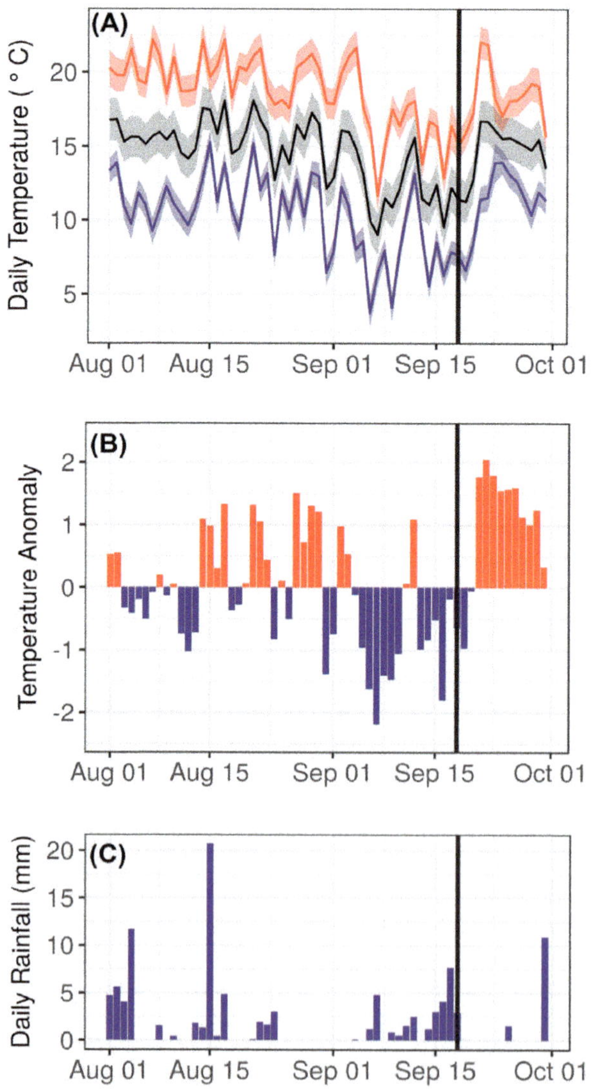

Figure 20.3.13. Daily temperature and rainfall data for June and July of 2013 in the region of All Saints Bog. The fire started on July 14, 2013 (vertical black line). Panel A, the mean (black line), maximum (red line), and minimum (blue line) daily temperatures (shaded regions show standard errors). Panel B, the daily mean temperature anomaly for 2013 relative to the 30-year average (1985–2015) in units of standard deviations across 30 years (red and blue bars correspond to positive and negative anomalies, respectively). Panel C, daily rainfall (mm). Data are from the E-OBS dataset (Haylock et al., 2008). Figure by Jonathan M. Yearsley, 2018.

Table 20.3.1
The burn severity classification for All Saints Bog (Hudspith et al., 2014); modified from Keeley (2009); the burn severity is obtained by a visual assessment of the surface.*

Burn Severity	Classification for All Saints Bog	Example Images
Unburned	Plant parts green and unaltered, no direct effect from heat.	 
Scorched	Vegetation on hummocks intact and grasses unaffected, but consumption of fine fuels in the shrub layer.	 
Light	Hummocks composed of sedge, sphagnum, and lichens killed by radiated heat, but uncharred. Fine fuel from shrub layer consumed (foliage and twigs) some larger stems scorched/partially charred.	 
Moderate	Understory shrubs (primarily *Calluna*) consumed. Charred bryophyte ground layer and surface peat.	 
Deep	Exposed tree roots and charred peat surfaces. Charred and/or consumed shrub layer and bryophyte ground layer.	 

*Photos by Jonathan M. Yearsley, 2013.

Figure 20.3.14. A site of deep burn severity on All Saints Bog. The roots of the tree in the center of the photo have been exposed by the fire. The horizontal field of view in the foreground is 70 m. Photo by Jonathan M. Yearsley, 2013.

Interestingly charcoals (charred Calluna and a range of woody material) can be found throughout the top 2 m depth of the peat at All Saints Bog (New et al., 2016), providing evidence that surface fires occurred relatively frequently. Charred peat agglomerates have also been described throughout the peat depth, implying that the peat itself also burned or at least was pyrolyzed by the radiant heat of passing surface fires (New et al., 2016). Variations in the abundance of plant and peat char throughout the 2-m depth of the peat reveals that the bog has burned periodically over a period of the last few thousand years, indicating that All Saints Bog appears to have had a relatively long history of fires (New et al., 2016).

Abbeyleix Bog

The Ecosystem and Environmental Conditions

The Killamuck Bog, also known as Abbeyleix Bog, is situated in County Laois, Ireland (52°53′57.6″N 7°21′07.8″W; Figure 20.1.1). The bog is currently classified as *Degraded Raised Bog*, extending to almost 110 ha of raised bog protected under the Habitats Directive (Council Directive 92/43/EEC) (Ecologic Environmental, 2009; NPWS, 2017b). The peatland is currently under the Abbeyleix Bog Project for conservation. Peat harvesting by locals was a common activity since the 1950s. Drain channels were built in subsequent years for industrial peat exploitation; however, this industrial harvesting activity never took place (Renou-Wilson et al., 2011). The old peat-cutting zone surrounds the entire high bog, and much of it is now planted with Scots Pine (*Pinus sylvestris*) and Downy Birch (*Betula pubescens*). The blocking of the dense drainage network associated with the peat-cutting was implemented in 2009 as a measure to restore the drained area of the bog.

Face-bank, marginal, and submarginal ecotopes characterize most of the Degraded Raised Bog, including Ling Heather (*C. vulgaris*), Bog Myrtle (*Myrica gale*), Purple Moor-grass (*M. caerulea*) poor *Sphagnum* cover, and flat microtopography (Ecologic Environmental, 2009). Woodland is developing on the old cutover bog dominated by Scots Pine (*P. sylvestris*) and Downy Birch (*B. pubescens*).

Abbeyleix Bog has no records of fire history (Ecologic Environmental, 2009). The fire event reported in Section 20.3 was the first ever recorded in the area. Weather conditions on the days previous to the fire were relatively dry and mild (Figure 20.3.15). The closest weather stations registered an average rainfall of 16 mm (previous 2 weeks), an average maximum temperature of 16.5°C, and average minimum temperatures of 7.7°C.

The Peat Fire Event

A smoldering fire took place in Abbeyleix Bog on September 18, 2013. Smoke from the ground was spotted by two ecologists surveying the bog. The total land affected by the fire was a relatively small area of 10 × 10 m. By the time the fire was spotted and suppressed it had already consumed the peat surface down to 1 m in places. The location of the ignition point is unknown, but it is thought unlikely that the fire was started deliberately by humans, as the area is not near any walkways. The fire brigades reached the area while the fire was spreading underground before the transition to flaming fire had taken place. Being aware of the potential of any smoldering fire to develop into a flaming fire (Rein, 2016), and also the environmental value of the bog, the local fire brigades immediately suppressed the fire. This was done by first swatting out the soil and later flooding the area to make sure no smoldering peat remained (Figure 20.3.16).

Fire Behavior and Its Effects

It can be seen (Figure 20.3.17, top) that the active smoldering front is underneath the surface of unconsumed peat. However, this consumption was irregular within the burned perimeter. Since smoldering fire propagation is largely influenced by the conditions of the peat and the surrounding environment, fire spread can happen simultaneously in any direction (Rein, 2016).

The root of plants within the consumed perimeter were damaged or burned by the smoldering fire, but the aboveground parts of these plants were only partially affected by the heat radiated from the smoldering (Figure 20.3.17, bottom). This is typical of smoldering peat fire events, where plants are commonly killed by the damage to their roots (Davies et al., 2013; Section 20.3).

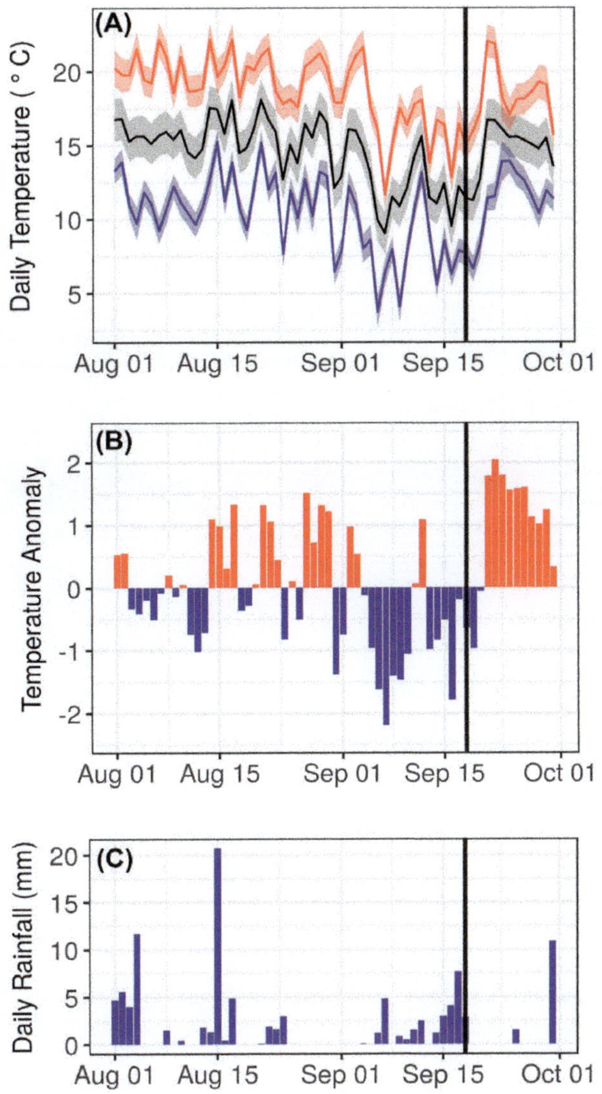

Figure 20.3.15. Daily temperature and rainfall data for August and September of 2013 in the region of Abbeyleix Bog. The fire started on September 18, 2013 (vertical black line). Panel A, the mean (black line), maximum (red line), and minimum (blue line) daily temperatures (shaded regions show standard errors). Panel B, the daily mean temperature anomaly for 2013 relative to the 30-year average (1985–2015) in units of standard deviations across 30 years (red and blue bars correspond to positive and negative anomalies, respectively). Panel C, daily rainfall (mm). Data are from the E-OBS dataset (Haylock et al., 2008). Figure by Jonathan M. Yearsley, 2018.

Figure 20.3.16. Left: fire brigades suppressing the smoldering peat fire. Right: 1 week after the fire was suppressed, the soil was stirred to assure an effective smoldering fire suppressing. Photos by Anthony Tynan (left), 2013; Nuria Prat-Guitart (right), 2013.

Figure 20.3.17. Examples of ongoing smoldering consumption of peat hummocks during the Abbeyleix Bog fire event in 2013. Top: Consumption affected the upper peat soil layers and rooting systems. Yellow rectangle: red coal undergoing oxidation. Bottom: Ash and char remaining after the consumption of a hummock. Leaves were affected by heat radiation. Photos by Chris Uys, 2013.

Photo by Nuria Prat-Guitart, 2013.

20.4 Peat Fire Effects on Irish Peatlands

Peat consumption in 2013 at the All Saints Bog Fire. The fire started on July 14th and was extinguished by the Offaly Fire and Rescue Service after 4 days of digging trenches into the peat and dropping water from helicopters. The fire affected an area of 42 ha. The horizontal field of view is 1 m.

Fire Effects

Due to the Atlantic weather conditions, the ground surface of peatlands in Ireland remains moist most of the year and the spread of the fire from flaming to smoldering peat fires in Ireland is limited to drought. During periods of drought and extreme drought the drying of the peat surface layers increase and, with it, the risk of peat combustion and sustained smoldering propagation during vegetation fires (Section 20.2). The status and condition of the peatland is a fundamental variable to evaluate the potential effects of peat fires.

In functioning bog systems, dry peat is limited to the uppermost layer, distributed in patchy mosaics between sphagnum, the peat-forming and moisture-preserving moss. Therefore, fires in bog systems are expected to lead to little peat consumption due to the limitation of the dry peat layer to patches of a few centimeters from the surface (Davies et al., 2015). However, degraded peatlands, principally those that have been drained, reduce the area occupied by peat-forming mosses while increasing the presence of other mosses and vascular plants. Thus degraded peatlands are vulnerable to more severe and frequent vegetation and smoldering fires, as has been reported for the All Saints Bog Fire (Figure 20.4.1; Hudspith et al., 2014), in several peatland studies worldwide, in tropical peatlands (Page et al., 2002; Van der Werf et al., 2008), and in boreal peatlands from North America (Turetsky et al., 2011; Wein, 1983).

Low-severity peat fires usually consume a few centimeters of the uppermost peat layer, having an overall small impact on the peatland surface (Sherwood et al., 2013) and to the atmospheric carbon (Turetsky et al., 2015). The carbon released from peat fires has not yet been studied in Ireland, however other studies have quantified that the carbon release from low-severity surface combustion is commonly between 0.1 and $9\,kg\,C\,m^{-2}$ (Benscoter and Wieder, 2003; Davies et al., 2013; Mack et al., 2011). Moreover, low-severity peat fires typically burn a patchy mosaic. This was observed in the Wicklow Mountains Fire (Figure 20.3.9). The patchy consumption of the peat may benefit the colonization of moss species from unburnt areas (Benscoter et al., 2015, 2005). Following these surface peat fires there is rapid resequestration of the carbon released by the recovering vegetation (Turetsky et al., 2015). After the uppermost dry layers have been consumed, the new surface is closer to the water-table position and protected from deep-burning peat fires.

A high frequency of peat fires, in spite of their severity, may have impacts at a local scale such as habitat loss. If the ecosystem lacks resilience to repeated fires, the response can be a shift in the vegetation composition. Short fire-return intervals in degraded bogs imply less time for sphagnum moss recovery after a fire event and other plants (moss and vascular plants) can colonize the burnt spaces (Figure 20.4.2). Under the circumstances of repeated fires, the peat being consumed in the surface layer is older, and has been stored for longer periods of time, having negative impacts on carbon sequestration and on the release of greenhouse gases (Turetsky et al., 2015; Wieder et al., 2009).

High-severity peat fires can remove several vertical meters of peat (Turetsky et al., 2015). These fires are often associated with poor ecosystem conditions before the fire, in addition to severe drought conditions. Deep peat consumption has not yet been documented in Ireland. From studies in peatlands worldwide, for example in Indonesia,

Figure 20.4.1. Top: The central part of All Saints Bog looking from the south toward the bog woodland on the horizon. This part of the bog was not affected by the fire in 2013 (burn severity of unburned; Table 20.3.1). The horizontal field of view in the background is 1 km. Bottom: The drained area burnt during the 2013 fire. The fire consumed surface layers of peat and ground vegetation exposing deeper layers of peat and damaged root systems. The horizontal field of views in the left and right photos are 1 m and 2 m, respectively. Photos by Jonathan M. Yearsley (top), 2013 and Nuria Prat-Guitart (bottom), 2013.

it is known that the consumption of peat stored in deep layers over many centuries through to millennia (Page et al., 2002) implies a substantial release of carbon (up to 315 kg C m^{-2}) compared to low-severity peat fires.

Habitat loss is also a short-term ecosystem response of high-severity fires. One example of fire-induced habitat loss in temperate bogs is the heating of rooting systems that kills vascular plants and reduces seedling establishment (Johnstone et al., 2010; Miyanishi and Johnson, 2002; O'Brien et al., 2010). This habitat loss can trigger long-term ecosystem responses such as changes in the vegetation succession (Benscoter and Vitt, 2008). These changes can ultimately lead to variation in the properties of the peat being accumulated and modification of the carbon sequestration regimes (Bond-Lamberty et al., 2004; Turetsky et al., 2002).

Figure 20.4.2. Area burnt during the Wicklow Mountains National Park Fire (Kippure Estate). The *M. caerula* grass densely colonized the peatland after the 2015 and 2003 fires. This grass then out-competes other heathland species and can lead to a loss of biodiversity. The horizontal field of view in the foreground is 500 m. Photo by Nuria Prat-Guitart, June 2015.

Acknowledgments

We thank C. Flynn, N. Bugler, and J. Ryan at the National Parks and Wildlife Services at the Department of Arts, Heritage and the Gaeltacht (DAHG), Ireland, for providing data about All Saints Bog and F. Fernandez for use of the ecotope map for the bog; and Chris Uys from the Abbeyleix Bog project for providing information about the fire. The authors acknowledge the use of data and imagery from LANCE FIRMS operated by the NASA/GSFC/Earth Science Data and Information System (ESDIS) with funding provided by NASA; the E-OBS dataset from the EU-FP6 project ENSEMBLES (http://ensembles-eu.metoffice.com); the data providers in the ECA&D project (http://www.ecad.eu); and map data copyrighted OpenStreetMap contributors (https://www.openstreetmap.org). CMB acknowledges a European Research Council Starter Grant ERC-2013-StG-335891-ECOFLAM. Jon M. Yearsley, Claire M. Belcher, and Nuria Prat-Guitart are grateful for funding from the Higher Education Authority (HEA) through the Programme for Research at Third Level Institutions, Cycle 5 (PRTLI-5) and cofunding by the European Regional Development Fund (ERDF).

References

Abbeyleix Bog Project, 2018. Community-Based Natural Resource Management Approach. http://www.abbeyleixbog.ie (accessed January 2018).

Allen, K.A., Denelle, P., Ruiz, F.M.S., Santana, V.M., Marrs, R.H., 2016. Prescribed moorland burning meets good practice guidelines: a monitoring case study using aerial photography in the Peak District, UK. Ecol. Indicat. 62, 76–85.

Belcher, C.M., Yearsley, J.M., Hadden, R.M., McElwain, J.C., Rein, G., 2010. Baseline intrinsic flammability of Earth's ecosystems estimated from paleoatmospheric oxygen over the past 350 million years. Proc. Natl. Acad. Sci. Unit. States Am. 107, 22448–22453.

Belcher, C.M., Hudspith, V.A., 2016. The formation of charcoal reflectance and its potential use in post-fire assessments. Int. J. Wildland Fire 25, 775.

Benscoter, B.W., Greenacre, D., Turetsky, M.R., 2015. Wildfire as a key determinant of peatland microtopography. Can. J. For. Res. 1137, 1133–1137.
Benscoter, B.W., Kelman-Wieder, R., Vitt, D.H., 2005. Linking microtopography with post-fire succession in bogs. J. Veg. Sci. 16, 453–460.
Benscoter, B.W., Vitt, D.H., 2008. Spatial patterns and temporal trajectories of the bog ground layer along a post-fire chronosequence. Ecosystems 11, 1054–1064.
Benscoter, B.W., Wieder, R.K., 2003. Variability in organic matter lost by combustion in a boreal bog during the 2001 Chisholm fire. Can. J. For. Res. 33, 2509–2513.
Living with wildfires – what science can tell us. In: Birot, Y. (Ed.), 2009. European Forest Institute Discussion Paper 15. Joensuu, Finland 82 pp.
Bistinas, I., Harrison, S.P., Prentice, I.C., Pereira, J.M.C., 2014. Causal relationships vs. emergent patterns in the global controls of fire frequency. Biogeosciences 11, 5087–5101.
Bond-Lamberty, B., Wang, C., Gower, S.T., 2004. Net primary production and net ecosystem production of a boreal black spruce wildfire chronosequence. Global Change Biol. 10, 473–487.
Cole, E.E., Mitchell, F.J.G., 2003. Human impact on the Irish landscape during the late Holocene inferred from palynological studies at three peatland sites. Holocene 13, 507–515.
Cross, J., Lynn, D., 2013. Results of a Monitoring Survey of Bog Woodland. Dublin, Ireland https://www.npws.ie/sites/default/files/publications/pdf/IWM%2069%20Bog%20woodland%20Monitoring.pdf (accessed February 2018).
Davies, G.M., Domènech, R., Gray, A., Johnson, P.C.D., 2015. Vegetation structure and fire weather influence variation in burn severity and fuel consumption during peatland wildfires. Biogeosci. Discuss. 12, 15737–15762.
Davies, G.M., Gray, A., Rein, G., Legg, C.J., 2013. Peat consumption and carbon loss due to smouldering wildfire in a temperate peatland. For. Ecol. Manag. 308, 169–177.
Davies, G.M., Kettridge, N., Stoof, C.R., Gray, A., Ascoli, D., Fernandes, P.M., Marrs, R., Allen, K.A., Doerr, S.H., Clay, G.D., McMorrow, J., Vandvik, V., 2016. The role of fire in UK peatland and moorland management: the need for informed, unbiased debate. Philos. Trans. R. Soc. Lond. B Biol. Sci. 371. https://doi.org/10.1098/rstb.2015.0342 (accessed February 2018).
Daniau, A.L., Bartlein, P.J., Harrison, S.P., Prentice, I.C., Brewer, S., Friedlingstein, P., Harrison-Prentice, T.I., Inoue, J., Izumi, K., Marlon, J.R., Mooney, S., 2012. Predictability of biomass burning in response to climate changes. Global Biogeochem. Cycles 26 (4), 1–12. https://doi.org/10.1029/2011GB004249 (accessed February 2018).
Dodson, J.R., 1990a. The Holocene vegetation of a prehistorically inhabited valley, Dingle peninsula, Co. Kerry. Proc. R. Ir. Acad. B 90, 151–174.
Dodson, J.R., 1990b. Fine resolution pollen analysis of vegettaion history in the Lough Adoon Valley, Co. Kerry, western Ireland. Rev. Palaeobot. Palynol. 64, 235–245.
Ecologic Environmental, Ecological Consultants Ltd, 2009. Killamuck Bog (Abbeyleix, County Laois) High Bog Ecological Survey.
Environmental Protetion Agency, 2017. National Soils Map. https://data.gov.ie/dataset/national-soils-map (accessed January, 2018).
Fogarty, P., 2017. Whittled Away – Ireland's Vanishing Nature. Collins Press. 300 pp.
Hadden, R.M., Rein, G., Belcher, C.M., 2013. Study of the competing chemical reactions in the initiation and spread of smouldering combustion in peat. Proc. Combust. Inst. 34, 2547–2553.
Hawthorne, D., Mitchell, F.J.G., 2016. Identifying past fire regimes throughout the Holocene in Ireland using new and established methods of charcoal analysis. Quat. Sci. Rev. 137, 45–53.
Hawthorne, D., Mitchell, F.J.G., 2018. Investigating patterns of wildfire in Ireland and their correlation with regional and global trends in fire history. Quat. Int. 488, 58–66. https://doi.org/10.1016/j.quaint.2017.06.067.
Haylock, M.R., Hofstra, N., Klein Tank, A.M.G., Klok, E.J., Jones, P.D., New, M., 2008. A European daily high-resolution gridded data set of surface temperature and precipitation for 1950–2006. J. Geophys. Res. 113, D20119.
Huang, X., Rein, G., 2015. Computational study of critical moisture content and depth of burn in peat fires. Int. J. Wildland Fire. 2015 (24), 798–808. https://doi.org/10.1071/WF14178.
Hudspith, V.A., Belcher, C.M., Yearsley, J.M., 2014. Charring temperatures are driven by the fuel types burned in a peatland wildfire. Front. Plant Sci. 5, 714.
Hudspith, V.A., Belcher, C.M., Barnes, J., Dash, C.B., Kelly, R., Hu, F.S., 2017. Charcoal reflectance suggests heating duration and fuel moisture affected burn severity in four Alaskan tundra wildfires. Int. J. Wildland Fire 26, 306.

Iglesias, V., Yospin, G.I., Whitlock, C., 2014. Reconstruction of fire regimes through integrated paleoecological proxy data and ecological modeling. Front. Plant Sci. 5 (785). http://doi.org/10.3389/fpls.2014.00785 (accessed February 2018).

Johnstone, J.F., Chapin, F.S., Hollingsworth, T.N., Mack, M.C., Romanovsky, V., Turetsky, M.R., 2010. Fire, climate change, and forest resilience in interior Alaska. Can. J. For. Res. 40, 1302–1312.

Keeley, J.E., 2009. Fire intensity, fire severity and burn severity: a brief review and suggested usage. Int. J. Wildland Fire 18, 116–126.

Kelly, L., Doak, M., Dromey, M., 1995. Raised Bog Restoration Project: An Investigation into the Conservation and Restoration of Selected Raised Bog Sites in Ireland (Dublin, Ireland).

Kloster, S., Brücher, T., Brovkin, V., Wilkenskjeld, S., 2015. Controls on fire activity over the Holocene. Clim. Past 11, 781–788.

Mack, M.C., Bret-Harte, M.S., Hollingsworth, T.N., Jandt, R.R., Schuur, E.A.G., Shaver, G.R., Verbyla, D.L., 2011. Carbon loss from an unprecedented Arctic tundra wildfire. Nature 475, 489–492.

Malone, S., O'Connell, C., 2009. Ireland's Peatland Conservation Action Plan 2020. Irish Peatland Conservation Council, Lullymore, Rathangan, Co., Kildare, Ireland. http://www.ipcc.ie/a-to-z-peatlands/peatland-action-plan (accessed February 2018).

Marlon, J.R., Bartlein, P.J., Daniau, A.L., Harrison, S.P., Maezumi, S.Y., Power, M.J., Tinner, W., Vanniére, B., 2013. Global biomass burning: a synthesis and review of Holocene paleofire records and their controls. Quat. Sci. Rev. 65, 5–25.

Marlon, J.R., Bartlein, P.J., Gavin, D.G., Long, C.J., Anderson, R.S., Briles, C.E., Brown, K.J., Colombaroli, D., Hallett, D.J., Power, M.J., Scharf, E.A., 2012. Long-term perspective on wildfires in the western USA. Proc. Natl. Acad. Sci. Unit. States Am. 109 (9), E535–E543.

Millin-Chalabi, G., McMorrow, J., Agnew, C., 2011. Using SAR Intensity and Coherence to Detect a Moorland Wildfire Scar in the Peak District. In: RSPSoc Annual Conference 2011. Bournemouth University, UK. https://www.research.manchester.ac.uk/portal/files/56661055/FULL_TEXT.PDF (accessed February 2018).

Minkina, W., Dudzik, S., 2009. I. Nfrared Thermograph Errors and Uncertainties. John Wiley & Sons, Ltd. 192 pp.

Miyanishi, K., Johnson, E.A., 2002. Process and patterns of duff consumption in the mixedwood boreal forest. Can. J. For. Res. 32, 1285–1295.

Molloy, K., O'Connell, M., 1995. Palaeoecological investigations towards the reconstruction of environment and land-use changes during prehistory at Céide Fields, western Ireland. Probl. Kustenforsch. Sudlichen Nord. 23, 187–225.

Montiel, C., Kraus, D., 2010. Best practices in fire use – prescribed burning and suppression fire programmes in selected case-study regions in Europe. In: European Forest Institute Research Report 24. Report Prepared under EU Sixth Framework Research Project FIRE PARADOX FP6-018505.

National Wildfire Coordinating Group, United States Department of the Interior, 346 pages, https://www.colorado firecamp.com/suppression-tactics/suppression-tactics-guide.pdf (accessed April 2018).

New, S.L., Belcher, C.M., Hudspith, V.A., Gallego-Sala, V., 2016. Holocene fire history: can evidence of peat burning be found in the palaeo-archive? Mires Peat 18, 1–11.

NPWS, 2017a. Wicklow Mountains SAC (Site Code 002122). Conservation Objectives Supporting Document - Blanket Bogs and Associated Habitats, Ireland. https://www.npws.ie/protected-sites/sac/002122 (accessed February 2018).

NPWS, 2017b. National Raised Bog Special Areas of Conservation Management Plan, National Parks and Wildlife Service, Ireland, https://www.npws.ie/peatlands-and-turf-cutting/management-plans (accessed April 2018).

Nugent, C., Casey, J., 2014. Prescribed fire as a land management tool in the Irish uplands. In: Proceedings of 2014 Teagasc Hill Sheep Conference, Bantry, County Cork. https://www.teagasc.ie/media/website/publications/201 4/Teagasc-Hill-Sheep-Conference–proceedings.pdf (accessed February 2018).

O'Brien, J.J., Hiers, J.K., Mitchell, R.J., Varner III, J.M., Mordecai, K., 2010. Acute physiological stress and mortality following fire in a long-unburned longleaf pine ecosystem. Fire Ecol. 6, 1–12.

OCarroll, E., Mitchell, F.J.G., 2016. Quantifying woodland resource usage and selection from Neolithic to post mediaeval times in the Irish midlands. Environ. Archaeol. 1–14.

O'Donnell, L., 2016. The power of the pyre - a holistic study of cremation focusing on charcoal remains. J. Archaeol. Sci. 65, 161–171.

O'Connell, M., 1990. Origins of Irish Lowland Blanket Bog. In: Doyle, G.J. (Ed.), Ecology and Conservation of Irish Peatlands in Ecology and Conservation of Irish Peatlands. Royal Irish Academy, Dublin, pp. 49–71.

O'Connell, M., Molloy, K., Bowler, M., 1988. Post-glacial landscape evolution in Connemara, western Ireland with particular reference to woodland history. In: Birks, H.H., Birks, H.J.B., Kaland, P.E., Moe, D. (Eds.), The Cultural Landscape: Past Present and Future. Cambridge University Press, pp. 267–288.

OpenStreetMap contributors, 2017. Planet Dump Illustration. retrieved from https://planet.osm.org (accessed February 2018).
Page, S., Siegert, F., Rieley, J.O., Boehm, H.V., Jayak, A., Limink, S., 2002. The amount of arbon released from peat and forest fires in Indonesia during 1997. Nature 420, 61–65.
Parrott, L., 2017. The modelling spiral for solving "wicked" environmental problems: guidance for stakeholder involvement and collaborative model development. Methods Ecol. Evol. 8, 1005–1011.
Power, M.J., Marlon, J., Ortiz, N., Bartlein, P.J., Harrison, S.P., Mayle, F.E., Ballouche, A., Bradshaw, R.H., Carcaillet, C., Cordova, C., Mooney, S., 2008. Changes in fire regimes since the Last Glacial Maximum: an assessment based on a global synthesis and analysis of charcoal data. Clim. Dynam. 30 (7–8), 887–907.
Prat-Guitart, N., Hadden, R.M., Belcher, C.M., Rein, G., Yearsley, J.M., 2015. Infrared image analysis as a tool for studying the horizontal smoldering propagation of laboratory peat fires. In: Stracher, G.B., Prakash, A., Rein, G. (Eds.), Coal and Peat Fires, a Global Perspective. Peat - Geology, Combustion and Case Studies. Elsevier, Amsterdam, pp. 121–139.
Prat-Guitart, N., Rein, G., Hadden, R.M., Belcher, C.M., Yearsley, J.M., 2016. Propagation probability and spread rates of self-sustained smouldering fires under controlled moisture content and bulk density conditions. Int. J. Wildland Fire 25, 456–465.
Rein, G., 2016. Smoldering combustion. In: Hurley, M.J. (Ed.), SFPE Handbook of Fire Protection Engineering, pp. 581–603. https://doi.org/10.1007/978-1-4939-2565-0.
Renou-Wilson, F., Bolger, T., Bullock, C., Convery, F.J., Curry, J., Ward, S., Wilson, D., Muller, C., 2011. BOGLAND : Sustainable Management of Peatlands in Ireland. Dublin.
Riley, K.L., Loehman, R.A., 2016. Mid-21st-century climate changes increase predicted fire occurrence and fire season length, Northern Rocky Mountains, United States. Ecosyst. Ecosph. 7 (11), e01543. https://doi.org/10.1002/ecs2.1543. (accessed February 2018).
Schmuck, G., San-Miguel-Ayanz, J., Camia, A., Durrant, T., Boca, R., Libertá, G., Schulte, E., 2013. Forest Fires in Europe, Middle East and North Africa 2012. Publications Office of the European Union, Luxembourg. https://publications.europa.eu/en/publication-detail/-/publication/079c0ec0-af32-499d-8864-67434f155b39/language-en (accessed February 2018).
Schouten, M.G.C. (Ed.), 2002. Conservation and Restoration of Raised Bogs: Geological, Hydrological and Ecological Studies. Dúchas - the Heritage Service of the Department of the Environment and Local Government, Ireland, Staatsbosbeheer, the Netherlands, Geological Survey of Ireland, Dublin. 220 pp.
Scottish Government, 2013. Wildfire Operational Guidance. Scotland, Edinburgh. ISBN: 978-1-78256-498-0. 357 pp. http://www.gov.scot/Publications/2013/10/6118/0 (accessed February 2018).
Sherwood, J.H., Kettridge, N., Thompson, D.K., Morris, P.J., Silins, U., Waddington, J.M., 2013. Effect of drainage and wildfire on peat hydrophysical properties. Hydrol. Process. 27, 1866–1874.
Turetsky, M.R., Benscoter, B., Page, S., Rein, G., Van der Werf, G.R., Watts, A., 2015. Global vulnerability of peatlands to fire and carbon loss. Nat. Geosci. 8, 11–14.
Turetsky, M.R., Donahue, W.F., Benscoter, B.W., 2011. Experimental drying intensifies burning and carbon losses in a northern peatland. Nat. Commun. 2, 514.
Turetsky, M.R., Wieder, K., Halsey, L., Vitt, D., 2002. Current disturbance and the diminishing peatland carbon sink. Geophys. Res. Lett. 29, 21-1–21-4. https://doi.org/10.1029/2001GL014000.
Valverde, F.F., Connolly, K., Crowley, W., Denyer, J., Duff, C., Smith, G., 2014. Raised BogMonitoring and Assessment Survey 2013. Dublin, Ireland https://www.npws.ie/sites/default/files/publications/pdf/IWM81_0.pdf (accessed February 2018).
Valverde, F.F., Fanning, M., Mccorry, M., Crowley, W., 2005. Raised Bog Monitoring Project 2004-2005. Dublin, Ireland https://www.npws.ie/sites/default/files/publications/pdf/Valverde_et_al_2005_Raised_Bog_Monitoring_-_P1_Summary_Report.pdf (accessed February 2018).
Van der Werf, G.R., Dempewolf, J., Trigg, S.N., Randerson, J.T., Kasibhatla, P.S., Giglio, L., Murdiyarso, D., Peters, W., Morton, D.C., Collatz, G.J., Dolman, a J., DeFries, R.S., 2008. Climate regulation of fire emissions and deforestation in equatorial Asia. Proc. Natl. Acad. Sci. U.S.A. 105, 20350–20355.
Van Wagner, C.E., 1987. Developement and Structure of the Canadian Forest FIre Weather Index System. Canadian Forestry Service, Ottawa. http://cfs.nrcan.gc.ca/pubwarehouse/pdfs/19927.pdf (accessed February 2018).
Wein, R.W., 1983. Fire behaviour and ecological effects in organic terrain. In: Wein, R.W., MacLean, D.A. (Eds.), The Role of Fire in Northern Circumpolar Ecosystems. John Wiley & Sons Ltd., pp. 81–95 (Chapter 5).
Wieder, R.K., Scott, K.D., Kamminga, K., Vile, M. a., Vitt, D.H., Bone, T., Xu, B., Benscoter, B.W., Bhatti, J.S., 2009. Postfire carbon balance in boreal bogs of Alberta, Canada. Global Change Biol. 15, 63–81.

Additional Case Studies

(1) Stracher, Glenn B. (editor), 2004, Coal Fires Burning around the World, International Journal of Coal Geology, v. 59, issues 1-2, 151 p.

(2) Stracher, Glenn B. (editor), 2007, Geology of Coal Fires: Case Studies from Around the World, Reviews in Engineering Geology XVIII, 283 p.

Author Index

Note: Page numbers followed by "f" indicate figures and "t" indicate tables.

H

Subject Index

Note: Page numbers followed by "f" indicate figures and "t" indicate tables.

CPI Antony Rowe
Chippenham, UK
2018-12-12 17:55

From chennai